MANUAL ON UNIFORM TRAFFIC CONTROL DEVICES
for Streets and Highways

2009 Edition

Including Revision 1 dated May 2012
and Revision 2 dated May 2012

The Manual on Uniform Traffic Control Devices (MUTCD) is approved by the Federal Highway Administrator as the National Standard in accordance with Title 23 U.S. Code, Sections 109(d), 114(a), 217, 315, and 402(a), 23 CFR 655, and 49 CFR 1.48(b)(8), 1.48(b)(33), and 1.48(c)(2).

Addresses for Publications Referenced in the MUTCD

American Automobile Association (AAA)
1000 AAA Drive
Heathrow, FL 32746
www.aaa.com
800-222-4357

American Association of State Highway and Transportation Officials (AASHTO)
444 North Capitol Street, NW, Suite 249
Washington, DC 20001
www.transportation.org
202-624-5800

American National Standards Institute (ANSI)
1819 L Street, NW, 6th Floor
Washington, DC 20036
www.ansi.org
202-293-8020

American Railway Engineering and Maintenance-of-Way Association (AREMA)
10003 Derekwood Lane, Suite 210
Lanham, MD 20706
www.arema.org
301-459-3200

Federal Highway Administration Report Center
Facsimile number: 814-239-2156
report.center@fhwa.dot.gov

Illuminating Engineering Society (IES)
120 Wall Street, Floor 17
New York, NY 10005
www.iesna.org
212-248-5000

Institute of Makers of Explosives
1120 19th Street, NW, Suite 310
Washington, DC 20036-3605
www.ime.org
202-429-9280

Institute of Transportation Engineers (ITE)
1099 14th Street, NW, Suite 300 West
Washington, DC 20005-3438
www.ite.org
202-289-0222

International Organization for Standardization
1, ch. de la Voie-Creuse
Case Postale 56
CH-1211
Geneva 20, Switzerland
www.iso.ch
011-41-22-749-0111

International Safety Equipment Association (ISEA)
1901 North Moore Street, Suite 808
Arlington, VA 22209
www.safetyequipment.org
703-525-1695

National Committee on Uniform Traffic Laws and Ordinances (NCUTLO)
107 South West Street, Suite 110
Alexandria, VA 22314
www.ncutlo.org
800-807-5290

National Electrical Manufacturers Association (NEMA)
1300 North 17th Street, Suite 1752
Rosslyn, VA 22209
www.nema.org
703-841-3200

Occupational Safety and Health Administration (OSHA)
U.S. Department of Labor
200 Constitution Avenue, NW
Washington, DC 20210
www.osha.gov
800-321-6742

Transportation Research Board (TRB)
The National Academies
500 Fifth Street, NW
Washington, DC 20001
www.nas.edu/trb
202-334-3072

U.S. Architectural and Transportation Barriers Compliance Board (The U.S. Access Board)
1331 F Street, NW, Suite 1000
Washington, DC 20004-1111
www.access-board.gov
202-272-0080

Acknowledgments

The Federal Highway Administration gratefully acknowledges the valuable assistance that it received from the National Committee on Uniform Traffic Control Devices and its more than 250 voluntary members in the development of this Manual.

MANUAL ON UNIFORM TRAFFIC CONTROL DEVICES
TABLE OF CONTENTS

Page

INTRODUCTION I-1

PART 1. GENERAL

CHAPTER 1A. GENERAL

Section 1A.01	Purpose of Traffic Control Devices	1
Section 1A.02	Principles of Traffic Control Devices	1
Section 1A.03	Design of Traffic Control Devices	1
Section 1A.04	Placement and Operation of Traffic Control Devices	2
Section 1A.05	Maintenance of Traffic Control Devices	2
Section 1A.06	Uniformity of Traffic Control Devices	2
Section 1A.07	Responsibility for Traffic Control Devices	2
Section 1A.08	Authority for Placement of Traffic Control Devices	3
Section 1A.09	Engineering Study and Engineering Judgment	4
Section 1A.10	Interpretations, Experimentations, Changes, and Interim Approvals	4
Section 1A.11	Relation to Other Publications	7
Section 1A.12	Color Code	10
Section 1A.13	Definitions of Headings, Words, and Phrases in this Manual	10
Section 1A.14	Meanings of Acronyms and Abbreviations in this Manual	23
Section 1A.15	Abbreviations Used on Traffic Control Devices	24

PART 2. SIGNS

CHAPTER 2A. GENERAL

Section 2A.01	Function and Purpose of Signs	27
Section 2A.02	Definitions	27
Section 2A.03	Standardization of Application	27
Section 2A.04	Excessive Use of Signs	27
Section 2A.05	Classification of Signs	28
Section 2A.06	Design of Signs	28
Section 2A.07	Retroreflectivity and Illumination	29
Section 2A.08	Maintaining Minimum Retroreflectivity	30
Section 2A.09	Shapes	32
Section 2A.10	Sign Colors	32
Section 2A.11	Dimensions	32
Section 2A.12	Symbols	34
Section 2A.13	Word Messages	35
Section 2A.14	Sign Borders	36
Section 2A.15	Enhanced Conspicuity for Standard Signs	36
Section 2A.16	Standardization of Location	37
Section 2A.17	Overhead Sign Installations	41
Section 2A.18	Mounting Height	42
Section 2A.19	Lateral Offset	43
Section 2A.20	Orientation	43
Section 2A.21	Posts and Mountings	44
Section 2A.22	Maintenance	44
Section 2A.23	Median Opening Treatments for Divided Highways with Wide Medians	44

CHAPTER 2B. REGULATORY SIGNS, BARRICADES, AND GATES

Section 2B.01	Application of Regulatory Signs	45
Section 2B.02	Design of Regulatory Signs	45
Section 2B.03	Size of Regulatory Signs	45
Section 2B.04	Right-of-Way at Intersections	49
Section 2B.05	STOP Sign (R1-1) and ALL WAY Plaque (R1-3P)	51
Section 2B.06	STOP Sign Applications	52
Section 2B.07	Multi-Way Stop Applications	52
Section 2B.08	YIELD Sign (R1-2)	53
Section 2B.09	YIELD Sign Applications	53
Section 2B.10	STOP Sign or YIELD Sign Placement	53
Section 2B.11	Yield Here To Pedestrians Signs and Stop Here For Pedestrians Signs (R1-5 Series)	54
Section 2B.12	In-Street and Overhead Pedestrian Crossing Signs (R1-6, R1-6a, R1-9, and R1-9a)	55
Section 2B.13	Speed Limit Sign (R2-1)	56
Section 2B.14	Truck Speed Limit Plaque (R2-2P)	58
Section 2B.15	Night Speed Limit Plaque (R2-3P)	58
Section 2B.16	Minimum Speed Limit Plaque (R2-4P)	59
Section 2B.17	Higher Fines Signs and Plaque (R2-6P, R2-10, and R2-11)	59
Section 2B.18	Movement Prohibition Signs (R3-1 through R3-4, R3-18, and R3-27)	60
Section 2B.19	Intersection Lane Control Signs (R3-5 through R3-8)	61
Section 2B.20	Mandatory Movement Lane Control Signs (R3-5, R3-5a, R3-7, and R3-20)	62
Section 2B.21	Optional Movement Lane Control Sign (R3-6)	63
Section 2B.22	Advance Intersection Lane Control Signs (R3-8 Series)	64
Section 2B.23	RIGHT (LEFT) LANE MUST EXIT Sign (R3-33)	64
Section 2B.24	Two-Way Left Turn Only Signs (R3-9a, R3-9b)	64
Section 2B.25	BEGIN and END Plaques (R3-9cP, R3-9dP)	64
Section 2B.26	Reversible Lane Control Signs (R3-9e through R3-9i)	65
Section 2B.27	Jughandle Signs (R3-23, R3-24, R3-25, and R3-26 Series)	67
Section 2B.28	DO NOT PASS Sign (R4-1)	72
Section 2B.29	PASS WITH CARE Sign (R4-2)	73
Section 2B.30	KEEP RIGHT EXCEPT TO PASS Sign (R4-16) and SLOWER TRAFFIC KEEP RIGHT Sign (R4-3)	73
Section 2B.31	TRUCKS USE RIGHT LANE Sign (R4-5)	73
Section 2B.32	Keep Right and Keep Left Signs (R4-7, R4-8)	73
Section 2B.33	STAY IN LANE Sign (R4-9)	74
Section 2B.34	RUNAWAY VEHICLES ONLY Sign (R4-10)	74
Section 2B.35	Slow Vehicle Turn-Out Signs (R4-12, R4-13, and R4-14)	74
Section 2B.36	DO NOT DRIVE ON SHOULDER Sign (R4-17) and DO NOT PASS ON SHOULDER Sign (R4-18)	75
Section 2B.37	DO NOT ENTER Sign (R5-1)	75
Section 2B.38	WRONG WAY Sign (R5-1a)	76
Section 2B.39	Selective Exclusion Signs	76
Section 2B.40	ONE WAY Signs (R6-1, R6-2)	77
Section 2B.41	Wrong-Way Traffic Control at Interchange Ramps	79
Section 2B.42	Divided Highway Crossing Signs (R6-3, R6-3a)	82
Section 2B.43	Roundabout Directional Arrow Signs (R6-4, R6-4a, and R6-4b)	84
Section 2B.44	Roundabout Circulation Plaque (R6-5P)	84
Section 2B.45	Examples of Roundabout Signing	84
Section 2B.46	Parking, Standing, and Stopping Signs (R7 and R8 Series)	88
Section 2B.47	Design of Parking, Standing, and Stopping Signs	89
Section 2B.48	Placement of Parking, Stopping, and Standing Signs	92
Section 2B.49	Emergency Restriction Signs (R8-4, R8-7, R8-8)	92
Section 2B.50	WALK ON LEFT FACING TRAFFIC and No Hitchhiking Signs (R9-1, R9-4, R9-4a)	92

Section 2B.51	Pedestrian Crossing Signs (R9-2, R9-3)	92
Section 2B.52	Traffic Signal Pedestrian and Bicycle Actuation Signs (R10-1 through R10-4, and R10-24 through R10-26)	94
Section 2B.53	Traffic Signal Signs (R10-5 through R10-30)	95
Section 2B.54	No Turn on Red Signs (R10-11 Series, R10-17a, and R10-30)	95
Section 2B.55	Photo Enforced Signs and Plaques (R10-18, R10-19P, R10-19aP)	97
Section 2B.56	Ramp Metering Signs (R10-28 and R10-29)	97
Section 2B.57	KEEP OFF MEDIAN Sign (R11-1)	97
Section 2B.58	ROAD CLOSED Sign (R11-2) and LOCAL TRAFFIC ONLY Signs (R11-3 Series, R11-4)	98
Section 2B.59	Weight Limit Signs (R12-1 through R12-5)	98
Section 2B.60	Weigh Station Signs (R13 Series)	99
Section 2B.61	TRUCK ROUTE Sign (R14-1)	99
Section 2B.62	Hazardous Material Signs (R14-2, R14-3)	99
Section 2B.63	National Network Signs (R14-4, R14-5)	100
Section 2B.64	Headlight Use Signs (R16-5 through R16-11)	100
Section 2B.65	FENDER BENDER Sign (R16-4)	101
Section 2B.66	Seat Belt Symbol	101
Section 2B.67	Barricades	101
Section 2B.68	Gates	101

CHAPTER 2C. WARNING SIGNS AND OBJECT MARKERS

Section 2C.01	Function of Warning Signs	103
Section 2C.02	Application of Warning Signs	103
Section 2C.03	Design of Warning Signs	103
Section 2C.04	Size of Warning Signs	103
Section 2C.05	Placement of Warning Signs	108
Section 2C.06	Horizontal Alignment Warning Signs	109
Section 2C.07	Horizontal Alignment Signs (W1-1 through W1-5, W1-11, W1-15)	110
Section 2C.08	Advisory Speed Plaque (W13-1P)	112
Section 2C.09	Chevron Alignment Sign (W1-8)	112
Section 2C.10	Combination Horizontal Alignment/Advisory Speed Signs (W1-1a, W1-2a)	113
Section 2C.11	Combination Horizontal Alignment/Intersection Signs (W1-10 Series)	113
Section 2C.12	One-Direction Large Arrow Sign (W1-6)	113
Section 2C.13	Truck Rollover Warning Sign (W1-13)	114
Section 2C.14	Advisory Exit and Ramp Speed Signs (W13-2 and W13-3)	114
Section 2C.15	Combination Horizontal Alignment/Advisory Exit and Ramp Speed Signs (W13-6 and W13-7)	115
Section 2C.16	Hill Signs (W7-1, W7-1a)	115
Section 2C.17	Truck Escape Ramp Signs (W7-4 Series)	115
Section 2C.18	HILL BLOCKS VIEW Sign (W7-6)	117
Section 2C.19	ROAD NARROWS Sign (W5-1)	117
Section 2C.20	NARROW BRIDGE Sign (W5-2)	118
Section 2C.21	ONE LANE BRIDGE Sign (W5-3)	118
Section 2C.22	Divided Highway Sign (W6-1)	119
Section 2C.23	Divided Highway Ends Sign (W6-2)	119
Section 2C.24	Freeway or Expressway Ends Signs (W19 Series)	119
Section 2C.25	Double Arrow Sign (W12-1)	119
Section 2C.26	DEAD END/NO OUTLET Signs (W14-1, W14-1a, W14-2, W14-2a)	119
Section 2C.27	Low Clearance Signs (W12-2 and W12-2a)	120
Section 2C.28	BUMP and DIP Signs (W8-1, W8-2)	120
Section 2C.29	SPEED HUMP Sign (W17-1)	120
Section 2C.30	PAVEMENT ENDS Sign (W8-3)	122
Section 2C.31	Shoulder Signs (W8-4, W8-9, W8-17, W8-23, and W8-25)	122
Section 2C.32	Surface Condition Signs (W8-5, W8-7, W8-8, W8-11, W8-13, and W8-14)	122

Section 2C.33	Warning Signs and Plaques for Motorcyclists (W8-15, W8-15P, and W8-16)	123
Section 2C.34	NO CENTER LINE Sign (W8-12)	123
Section 2C.35	Weather Condition Signs (W8-18, W8-19, W8-21, and W8-22)	123
Section 2C.36	Advance Traffic Control Signs (W3-1, W3-2, W3-3, W3-4)	123
Section 2C.37	Advance Ramp Control Signal Signs (W3-7 and W3-8)	124
Section 2C.38	Reduced Speed Limit Ahead Signs (W3-5, W3-5a)	124
Section 2C.39	DRAW BRIDGE Sign (W3-6)	125
Section 2C.40	Merge Signs (W4-1, W4-5)	125
Section 2C.41	Added Lane Signs (W4-3, W4-6)	126
Section 2C.42	Lane Ends Signs (W4-2, W9-1, W9-2)	126
Section 2C.43	RIGHT (LEFT) LANE EXIT ONLY AHEAD Sign (W9-7)	126
Section 2C.44	Two-Way Traffic Sign (W6-3)	127
Section 2C.45	NO PASSING ZONE Sign (W14-3)	127
Section 2C.46	Intersection Warning Signs (W2-1 through W2-8)	127
Section 2C.47	Two-Direction Large Arrow Sign (W1-7)	128
Section 2C.48	Traffic Signal Signs (W25-1, W25-2)	128
Section 2C.49	Vehicular Traffic Warning Signs (W8-6, W11-1, W11-5, W11-5a, W11-8, W11-10, W11-11, W11-12P, W11-14, W11-15, and W11-15a)	128
Section 2C.50	Non-Vehicular Warning Signs (W11-2, W11-3, W11-4, W11-6, W11-7, W11-9, and W11-16 through W11-22)	130
Section 2C.51	Playground Sign (W15-1)	131
Section 2C.52	NEW TRAFFIC PATTERN AHEAD Sign (W23-2)	131
Section 2C.53	Use of Supplemental Warning Plaques	131
Section 2C.54	Design of Supplemental Warning Plaques	132
Section 2C.55	Distance Plaques (W16-2 Series, W16-3 Series, W16-4P, W7-3aP)	132
Section 2C.56	Supplemental Arrow Plaques (W16-5P, W16-6P)	132
Section 2C.57	Hill-Related Plaques (W7-2 Series, W7-3 Series)	133
Section 2C.58	Advance Street Name Plaque (W16-8P, W16-8aP)	133
Section 2C.59	CROSS TRAFFIC DOES NOT STOP Plaque (W4-4P)	133
Section 2C.60	SHARE THE ROAD Plaque (W16-1P)	133
Section 2C.61	Photo Enforced Plaque (W16-10P)	134
Section 2C.62	NEW Plaque (W16-15P)	134
Section 2C.63	Object Marker Design and Placement Height	134
Section 2C.64	Object Markers for Obstructions Within the Roadway	135
Section 2C.65	Object Markers for Obstructions Adjacent to the Roadway	135
Section 2C.66	Object Markers for Ends of Roadways	136

CHAPTER 2D. GUIDE SIGNS—CONVENTIONAL ROADS

Section 2D.01	Scope of Conventional Road Guide Sign Standards	137
Section 2D.02	Application	137
Section 2D.03	Color, Retroreflection, and Illumination	137
Section 2D.04	Size of Signs	137
Section 2D.05	Lettering Style	138
Section 2D.06	Size of Lettering	138
Section 2D.07	Amount of Legend	140
Section 2D.08	Arrows	140
Section 2D.09	Numbered Highway Systems	142
Section 2D.10	Route Signs and Auxiliary Signs	142
Section 2D.11	Design of Route Signs	143
Section 2D.12	Design of Route Sign Auxiliaries	144
Section 2D.13	Junction Auxiliary Sign (M2-1)	144
Section 2D.14	Combination Junction Sign (M2-2)	145
Section 2D.15	Cardinal Direction Auxiliary Signs (M3-1 through M3-4)	145
Section 2D.16	Auxiliary Signs for Alternative Routes (M4 Series)	145

Section 2D.17	ALTERNATE Auxiliary Signs (M4-1, M4-1a)	145
Section 2D.18	BY-PASS Auxiliary Sign (M4-2)	146
Section 2D.19	BUSINESS Auxiliary Sign (M4-3)	146
Section 2D.20	TRUCK Auxiliary Sign (M4-4)	146
Section 2D.21	TO Auxiliary Sign (M4-5)	146
Section 2D.22	END Auxiliary Sign (M4-6)	146
Section 2D.23	BEGIN Auxiliary Sign (M4-14)	146
Section 2D.24	TEMPORARY Auxiliary Signs (M4-7, M4-7a)	147
Section 2D.25	Temporary Detour and Auxiliary Signs	147
Section 2D.26	Advance Turn Arrow Auxiliary Signs (M5-1, M5-2, and M5-3)	147
Section 2D.27	Lane Designation Auxiliary Signs (M5-4, M5-5, and M5-6)	148
Section 2D.28	Directional Arrow Auxiliary Signs (M6 Series)	148
Section 2D.29	Route Sign Assemblies	148
Section 2D.30	Junction Assembly	153
Section 2D.31	Advance Route Turn Assembly	153
Section 2D.32	Directional Assembly	153
Section 2D.33	Combination Lane-Use/Destination Overhead Guide Sign (D15-1)	154
Section 2D.34	Confirming or Reassurance Assemblies	155
Section 2D.35	Trailblazer Assembly	155
Section 2D.36	Destination and Distance Signs	156
Section 2D.37	Destination Signs (D1 Series)	156
Section 2D.38	Destination Signs at Circular Intersections	157
Section 2D.39	Destination Signs at Jughandles	158
Section 2D.40	Location of Destination Signs	158
Section 2D.41	Distance Signs (D2 Series)	161
Section 2D.42	Location of Distance Signs	161
Section 2D.43	Street Name Signs (D3-1 or D3-1a)	161
Section 2D.44	Advance Street Name Signs (D3-2)	163
Section 2D.45	Signing on Conventional Roads on Approaches to Interchanges	164
Section 2D.46	Freeway Entrance Signs (D13-3 and D13-3a)	170
Section 2D.47	Parking Area Guide Sign (D4-1)	171
Section 2D.48	PARK - RIDE Sign (D4-2)	171
Section 2D.49	Weigh Station Signing (D8 Series)	172
Section 2D.50	Community Wayfinding Signs	172
Section 2D.51	Truck, Passing, or Climbing Lane Signs (D17-1 and D17-2)	178
Section 2D.52	Slow Vehicle Turn-Out Sign (D17-7)	178
Section 2D.53	Signing of Named Highways	179
Section 2D.54	Crossover Signs (D13-1 and D13-2)	179
Section 2D.55	National Scenic Byways Signs (D6-4, D6-4a)	179

CHAPTER 2E. GUIDE SIGNS—FREEWAYS AND EXPRESSWAYS

Section 2E.01	Scope of Freeway and Expressway Guide Sign Standards	181
Section 2E.02	Freeway and Expressway Signing Principles	181
Section 2E.03	Guide Sign Classification	181
Section 2E.04	General	182
Section 2E.05	Color of Guide Signs	182
Section 2E.06	Retroreflection or Illumination	182
Section 2E.07	Characteristics of Urban Signing	182
Section 2E.08	Characteristics of Rural Signing	183
Section 2E.09	Signing of Named Highways	183
Section 2E.10	Amount of Legend on Guide Signs	183
Section 2E.11	Number of Signs at an Overhead Installation and Sign Spreading	183
Section 2E.12	Pull-Through Signs (E6-2, E6-2a)	184
Section 2E.13	Designation of Destinations	184

Section 2E.14	Size and Style of Letters and Signs	185
Section 2E.15	Interline and Edge Spacing	185
Section 2E.16	Sign Borders	192
Section 2E.17	Abbreviations	192
Section 2E.18	Symbols	192
Section 2E.19	Arrows for Interchange Guide Signs	192
Section 2E.20	Signing for Option Lanes at Splits and Multi-Lane Exits	193
Section 2E.21	Design of Overhead Arrow-per-Lane Guide Signs for Option Lanes	193
Section 2E.22	Design of Freeway and Expressway Diagrammatic Guide Signs for Option Lanes	198
Section 2E.23	Signing for Intermediate and Minor Interchange Multi-Lane Exits with an Option Lane	203
Section 2E.24	Signing for Interchange Lane Drops	203
Section 2E.25	Overhead Sign Installations	206
Section 2E.26	Lateral Offset	210
Section 2E.27	Route Signs and Trailblazer Assemblies	210
Section 2E.28	Eisenhower Interstate System Signs (M1-10, M1-10a)	211
Section 2E.29	Signs for Intersections at Grade	211
Section 2E.30	Interchange Guide Signs	211
Section 2E.31	Interchange Exit Numbering	212
Section 2E.32	Interchange Classification	216
Section 2E.33	Advance Guide Signs	216
Section 2E.34	Next Exit Plaques	218
Section 2E.35	Other Supplemental Guide Signs	218
Section 2E.36	Exit Direction Signs	220
Section 2E.37	Exit Gore Signs (E5-1 Series)	222
Section 2E.38	Post-Interchange Signs	222
Section 2E.39	Post-Interchange Distance Signs	223
Section 2E.40	Interchange Sequence Signs	223
Section 2E.41	Community Interchanges Identification Signs	225
Section 2E.42	NEXT XX EXITS Sign	225
Section 2E.43	Signing by Type of Interchange	226
Section 2E.44	Freeway-to-Freeway Interchange	226
Section 2E.45	Cloverleaf Interchange	226
Section 2E.46	Cloverleaf Interchange with Collector-Distributor Roadways	230
Section 2E.47	Partial Cloverleaf Interchange	230
Section 2E.48	Diamond Interchange	230
Section 2E.49	Diamond Interchange in Urban Area	234
Section 2E.50	Closely-Spaced Interchanges	234
Section 2E.51	Minor Interchange	234
Section 2E.52	Signing on Conventional Road Approaches and Connecting Roadways	235
Section 2E.53	Wrong-Way Traffic Control at Interchange Ramps	235
Section 2E.54	Weigh Station Signing	236

CHAPTER 2F. TOLL ROAD SIGNS

Section 2F.01	Scope	237
Section 2F.02	Sizes of Toll Road Signs	237
Section 2F.03	Use of Purple Backgrounds and Underlay Panels with ETC Account Pictographs	238
Section 2F.04	Size of ETC Pictographs	238
Section 2F.05	Regulatory Signs for Toll Plazas	238
Section 2F.06	Pay Toll Advance Warning Sign (W9-6)	240
Section 2F.07	Pay Toll Advance Warning Plaque (W9-6P)	241
Section 2F.08	Stop Ahead Pay Toll Warning Sign (W9-6a)	242
Section 2F.09	Stop Ahead Pay Toll Warning Plaque (W9-6aP)	242
Section 2F.10	LAST EXIT BEFORE TOLL Warning Plaque (W16-16P)	242
Section 2F.11	TOLL Auxiliary Sign (M4-15)	242

Section 2F.12	Electronic Toll Collection (ETC) Account-Only Auxiliary Signs (M4-16 and M4-20)......... 243
Section 2F.13	Toll Facility and Toll Plaza Guide Signs – General 243
Section 2F.14	Advance Signs for Conventional Toll Plazas.. 248
Section 2F.15	Advance Signs for Toll Plazas on Diverging Alignments from Open-Road ETC Account-Only Lanes 249
Section 2F.16	Toll Plaza Canopy Signs... 252
Section 2F.17	Guide Signs for Entrances to ETC Account-Only Facilities.. 252
Section 2F.18	ETC Program Information Signs 252

CHAPTER 2G. PREFERENTIAL AND MANAGED LANE SIGNS

Section 2G.01	Scope... 253
Section 2G.02	Sizes of Preferential and Managed Lane Signs 253
Section 2G.03	Regulatory Signs for Preferential Lanes – General 253
Section 2G.04	Preferential Lane Vehicle Occupancy Definition Regulatory Signs (R3-10 Series and R3-13 Series).. 258
Section 2G.05	Preferential Lane Periods of Operation Regulatory Signs (R3-11 Series and R3-14 Series)... 259
Section 2G.06	Preferential Lane Advance Regulatory Signs (R3-12, R3-12e, R3-12f, R3-15, R3-15a, and R3-15d) 263
Section 2G.07	Preferential Lane Ends Regulatory Signs (R3-12a, R3-12b, R3-12c, R3-12d, R3-12g, R3-12h, R3-15b, R3-15c, and R3-15e)... 263
Section 2G.08	Warning Signs on Median Barriers for Preferential Lanes 263
Section 2G.09	High-Occupancy Vehicle (HOV) Plaque (W16-11P) 264
Section 2G.10	Preferential Lane Guide Signs – General 265
Section 2G.11	Guide Signs for Initial Entry Points to Preferential Lanes 267
Section 2G.12	Guide Signs for Intermediate Entry Points to Preferential Lanes 268
Section 2G.13	Guide Signs for Egress from Preferential Lanes to General-Purpose Lanes 270
Section 2G.14	Guide Signs for Direct Entrances to Preferential Lanes from Another Highway 273
Section 2G.15	Guide Signs for Direct Exits from Preferential Lanes to Another Highway 273
Section 2G.16	Signs for Priced Managed Lanes – General 276
Section 2G.17	Regulatory Signs for Priced Managed Lanes 279
Section 2G.18	Guide Signs for Priced Managed Lanes 279

CHAPTER 2H. GENERAL INFORMATION SIGNS

Section 2H.01	Sizes of General Information Signs 292
Section 2H.02	General Information Signs (I Series).. 292
Section 2H.03	Traffic Signal Speed Sign (I1-1) 294
Section 2H.04	Miscellaneous Information Signs.. 294
Section 2H.05	Reference Location Signs (D10-1 through D10-3) and Intermediate Reference Location Signs (D10-1a through D10-3a).. 294
Section 2H.06	Enhanced Reference Location Signs (D10-4, D10-5)... 296
Section 2H.07	Auto Tour Route Signs 297
Section 2H.08	Acknowledgment Signs 297

CHAPTER 2I. GENERAL SERVICE SIGNS

Section 2I.01	Sizes of General Service Signs... 299
Section 2I.02	General Service Signs for Conventional Roads.. 300
Section 2I.03	General Service Signs for Freeways and Expressways 303
Section 2I.04	Interstate Oasis Signing.. 306
Section 2I.05	Rest Area and Other Roadside Area Signs... 307
Section 2I.06	Brake Check Area Signs (D5-13 and D5-14) 308
Section 2I.07	Chain-Up Area Signs (D5-15 and D5-16) 308
Section 2I.08	Tourist Information and Welcome Center Signs 308
Section 2I.09	Radio Information Signing... 310
Section 2I.10	TRAVEL INFO CALL 511 Signs (D12-5 and D12-5a).. 311
Section 2I.11	Carpool and Ridesharing Signing... 311

CHAPTER 2J. SPECIFIC SERVICE SIGNS

Section 2J.01	Eligibility	312
Section 2J.02	Application	313
Section 2J.03	Logos and Logo Sign Panels	313
Section 2J.04	Number and Size of Signs and Logo Sign Panels	317
Section 2J.05	Size of Lettering	317
Section 2J.06	Signs at Interchanges	317
Section 2J.07	Single-Exit Interchanges	317
Section 2J.08	Double-Exit Interchanges	318
Section 2J.09	Specific Service Trailblazer Signs	318
Section 2J.10	Signs at Intersections	319
Section 2J.11	Signing Policy	319

CHAPTER 2K. TOURIST-ORIENTED DIRECTIONAL SIGNS

Section 2K.01	Purpose and Application	320
Section 2K.02	Design	320
Section 2K.03	Style and Size of Lettering	323
Section 2K.04	Arrangement and Size of Signs	323
Section 2K.05	Advance Signs	323
Section 2K.06	Sign Locations	324
Section 2K.07	State Policy	324

CHAPTER 2L. CHANGEABLE MESSAGE SIGNS

Section 2L.01	Description of Changeable Message Signs	325
Section 2L.02	Applications of Changeable Message Signs	325
Section 2L.03	Legibility and Visibility of Changeable Message Signs	326
Section 2L.04	Design Characteristics of Changeable Message Signs	326
Section 2L.05	Message Length and Units of Information	328
Section 2L.06	Installation of Permanent Changeable Message Signs	329

CHAPTER 2M. RECREATIONAL AND CULTURAL INTEREST AREA SIGNS

Section 2M.01	Scope	330
Section 2M.02	Application of Recreational and Cultural Interest Area Signs	330
Section 2M.03	Regulatory and Warning Signs	330
Section 2M.04	General Design Requirements for Recreational and Cultural Interest Area Symbol Guide Signs	330
Section 2M.05	Symbol Sign Sizes	332
Section 2M.06	Use of Educational Plaques	332
Section 2M.07	Use of Prohibitive Circle and Diagonal Slash for Non-Road Applications	332
Section 2M.08	Placement of Recreational and Cultural Interest Area Symbol Signs	332
Section 2M.09	Destination Guide Signs	333
Section 2M.10	Memorial or Dedication Signing	339

CHAPTER 2N. EMERGENCY MANAGEMENT SIGNING

Section 2N.01	Emergency Management	342
Section 2N.02	Design of Emergency Management Signs	342
Section 2N.03	Evacuation Route Signs (EM-1 and EM-1a)	342
Section 2N.04	AREA CLOSED Sign (EM-2)	344
Section 2N.05	TRAFFIC CONTROL POINT Sign (EM-3)	344
Section 2N.06	MAINTAIN TOP SAFE SPEED Sign (EM-4)	344
Section 2N.07	ROAD (AREA) USE PERMIT REQUIRED FOR THRU TRAFFIC Sign (EM-5)	345
Section 2N.08	Emergency Aid Center Signs (EM-6 Series)	345
Section 2N.09	Shelter Directional Signs (EM-7 Series)	346

PART 3. MARKINGS

CHAPTER 3A. GENERAL

Section 3A.01	Functions and Limitations	347
Section 3A.02	Standardization of Application	347
Section 3A.03	Maintaining Minimum Pavement Marking Retroreflectivity	347
Section 3A.04	Materials	347
Section 3A.05	Colors	348
Section 3A.06	Functions, Widths, and Patterns of Longitudinal Pavement Markings	348

CHAPTER 3B. PAVEMENT AND CURB MARKINGS

Section 3B.01	Yellow Center Line Pavement Markings and Warrants	349
Section 3B.02	No-Passing Zone Pavement Markings and Warrants	352
Section 3B.03	Other Yellow Longitudinal Pavement Markings	354
Section 3B.04	White Lane Line Pavement Markings and Warrants	356
Section 3B.05	Other White Longitudinal Pavement Markings	370
Section 3B.06	Edge Line Pavement Markings	371
Section 3B.07	Warrants for Use of Edge Lines	371
Section 3B.08	Extensions Through Intersections or Interchanges	371
Section 3B.09	Lane-Reduction Transition Markings	374
Section 3B.10	Approach Markings for Obstructions	376
Section 3B.11	Raised Pavement Markers – General	376
Section 3B.12	Raised Pavement Markers as Vehicle Positioning Guides with Other Longitudinal Markings	379
Section 3B.13	Raised Pavement Markers Supplementing Other Markings	379
Section 3B.14	Raised Pavement Markers Substituting for Pavement Markings	380
Section 3B.15	Transverse Markings	381
Section 3B.16	Stop and Yield Lines	381
Section 3B.17	Do Not Block Intersection Markings	382
Section 3B.18	Crosswalk Markings	383
Section 3B.19	Parking Space Markings	385
Section 3B.20	Pavement Word, Symbol, and Arrow Markings	387
Section 3B.21	Speed Measurement Markings	393
Section 3B.22	Speed Reduction Markings	393
Section 3B.23	Curb Markings	394
Section 3B.24	Chevron and Diagonal Crosshatch Markings	395
Section 3B.25	Speed Hump Markings	395
Section 3B.26	Advance Speed Hump Markings	395

CHAPTER 3C. ROUNDABOUT MARKINGS

Section 3C.01	General	399
Section 3C.02	White Lane Line Pavement Markings for Roundabouts	413
Section 3C.03	Edge Line Pavement Markings for Roundabout Circulatory Roadways	413
Section 3C.04	Yield Lines for Roundabouts	413
Section 3C.05	Crosswalk Markings at Roundabouts	413
Section 3C.06	Word, Symbol, and Arrow Pavement Markings for Roundabouts	413
Section 3C.07	Markings for Other Circular Intersections	414

CHAPTER 3D MARKINGS FOR PREFERENTIAL LANES

Section 3D.01	Preferential Lane Word and Symbol Markings	415
Section 3D.02	Preferential Lane Longitudinal Markings for Motor Vehicles	416

CHAPTER 3E. MARKINGS FOR TOLL PLAZAS

Section 3E.01	Markings for Toll Plazas	423

CHAPTER 3F DELINEATORS
Section 3F.01 Delineators 424
Section 3F.02 Delineator Design 424
Section 3F.03 Delineator Application 424
Section 3F.04 Delineator Placement and Spacing 426

CHAPTER 3G COLORED PAVEMENTS
Section 3G.01 General 428

CHAPTER 3H CHANNELIZING DEVICES USED FOR EMPHASIS OF PAVEMENT MARKING PATTERNS
Section 3H.01 Channelizing Devices 429

CHAPTER 3I ISLANDS
Section 3I.01 General 430
Section 3I.02 Approach-End Treatment 430
Section 3I.03 Island Marking Application 430
Section 3I.04 Island Marking Colors 430
Section 3I.05 Island Delineation 431
Section 3I.06 Pedestrian Islands and Medians 431

CHAPTER 3J RUMBLE STRIP MARKINGS
Section 3J.01 Longitudinal Rumble Strip Markings 432
Section 3J.02 Transverse Rumble Strip Markings 432

PART 4 HIGHWAY TRAFFIC SIGNALS

CHAPTER 4A GENERAL
Section 4A.01 Types 433
Section 4A.02 Definitions Relating to Highway Traffic Signals 433

CHAPTER 4B TRAFFIC CONTROL SIGNALS—GENERAL
Section 4B.01 General 434
Section 4B.02 Basis of Installation or Removal of Traffic Control Signals 434
Section 4B.03 Advantages and Disadvantages of Traffic Control Signals 434
Section 4B.04 Alternatives to Traffic Control Signals 435
Section 4B.05 Adequate Roadway Capacity 435

CHAPTER 4C TRAFFIC CONTROL SIGNAL NEEDS STUDIES
Section 4C.01 Studies and Factors for Justifying Traffic Control Signals 436
Section 4C.02 Warrant 1, Eight-Hour Vehicular Volume 437
Section 4C.03 Warrant 2, Four-Hour Vehicular Volume 439
Section 4C.04 Warrant 3, Peak Hour 439
Section 4C.05 Warrant 4, Pedestrian Volume 442
Section 4C.06 Warrant 5, School Crossing 442
Section 4C.07 Warrant 6, Coordinated Signal System 445
Section 4C.08 Warrant 7, Crash Experience 445
Section 4C.09 Warrant 8, Roadway Network 446
Section 4C.10 Warrant 9, Intersection Near a Grade Crossing 446

CHAPTER 4D TRAFFIC CONTROL SIGNAL FEATURES
Section 4D.01 General 449
Section 4D.02 Responsibility for Operation and Maintenance 449
Section 4D.03 Provisions for Pedestrians 450
Section 4D.04 Meaning of Vehicular Signal Indications 450

Section 4D.05	Application of Steady Signal Indications	453
Section 4D.06	Signal Indications – Design, Illumination, Color, and Shape	456
Section 4D.07	Size of Vehicular Signal Indications	456
Section 4D.08	Positions of Signal Indications Within a Signal Face – General	457
Section 4D.09	Positions of Signal Indications Within a Vertical Signal Face	457
Section 4D.10	Positions of Signal Indications Within a Horizontal Signal Face	459
Section 4D.11	Number of Signal Faces on an Approach	459
Section 4D.12	Visibility, Aiming, and Shielding of Signal Faces	461
Section 4D.13	Lateral Positioning of Signal Faces	463
Section 4D.14	Longitudinal Positioning of Signal Faces	464
Section 4D.15	Mounting Height of Signal Faces	465
Section 4D.16	Lateral Offset (Clearance) of Signal Faces	465
Section 4D.17	Signal Indications for Left-Turn Movements – General	465
Section 4D.18	Signal Indications for Permissive Only Mode Left-Turn Movements	467
Section 4D.19	Signal Indications for Protected Only Mode Left-Turn Movements	469
Section 4D.20	Signal Indications for Protected/Permissive Mode Left-Turn Movements	471
Section 4D.21	Signal Indications for Right-Turn Movements – General	474
Section 4D.22	Signal Indications for Permissive Only Mode Right-Turn Movements	475
Section 4D.23	Signal Indications for Protected Only Mode Right-Turn Movements	478
Section 4D.24	Signal Indications for Protected/Permissive Mode Right-Turn Movements	480
Section 4D.25	Signal Indications for Approaches With Shared Left-Turn/Right-Turn Lanes and No Through Movement	484
Section 4D.26	Yellow Change and Red Clearance Intervals	485
Section 4D.27	Preemption and Priority Control of Traffic Control Signals	489
Section 4D.28	Flashing Operation of Traffic Control Signals – General	491
Section 4D.29	Flashing Operation – Transition Into Flashing Mode	491
Section 4D.30	Flashing Operation – Signal Indications During Flashing Mode	491
Section 4D.31	Flashing Operation – Transition Out of Flashing Mode	492
Section 4D.32	Temporary and Portable Traffic Control Signals	492
Section 4D.33	Lateral Offset of Signal Supports and Cabinets	493
Section 4D.34	Use of Signs at Signalized Locations	493
Section 4D.35	Use of Pavement Markings at Signalized Locations	494

CHAPTER 4E   PEDESTRIAN CONTROL FEATURES

Section 4E.01	Pedestrian Signal Heads	495
Section 4E.02	Meaning of Pedestrian Signal Head Indications	495
Section 4E.03	Application of Pedestrian Signal Heads	495
Section 4E.04	Size, Design, and Illumination of Pedestrian Signal Head Indications	496
Section 4E.05	Location and Height of Pedestrian Signal Heads	497
Section 4E.06	Pedestrian Intervals and Signal Phases	497
Section 4E.07	Countdown Pedestrian Signals	499
Section 4E.08	Pedestrian Detectors	500
Section 4E.09	Accessible Pedestrian Signals and Detectors – General	504
Section 4E.10	Accessible Pedestrian Signals and Detectors – Location	505
Section 4E.11	Accessible Pedestrian Signals and Detectors – Walk Indications	505
Section 4E.12	Accessible Pedestrian Signals and Detectors – Tactile Arrows and Locator Tones	507
Section 4E.13	Accessible Pedestrian Signals and Detectors – Extended Pushbutton Press Features	507

CHAPTER 4F   PEDESTRIAN HYBRID BEACONS

Section 4F.01	Application of Pedestrian Hybrid Beacons	509
Section 4F.02	Design of Pedestrian Hybrid Beacons	509
Section 4F.03	Operation of Pedestrian Hybrid Beacons	511

CHAPTER 4G TRAFFIC CONTROL SIGNALS AND HYBRID BEACONS FOR EMERGENCY-VEHICLE ACCESS

Section 4G.01	Application of Emergency-Vehicle Traffic Control Signals and Hybrid Beacons	513
Section 4G.02	Design of Emergency-Vehicle Traffic Control Signals	513
Section 4G.03	Operation of Emergency-Vehicle Traffic Control Signals	513
Section 4G.04	Emergency-Vehicle Hybrid Beacons	514

CHAPTER 4H TRAFFIC CONTROL SIGNALS FOR ONE-LANE, TWO-WAY FACILITIES

Section 4H.01	Application of Traffic Control Signals for One-Lane, Two-Way Facilities	516
Section 4H.02	Design of Traffic Control Signals for One-Lane, Two-Way Facilities	516
Section 4H.03	Operation of Traffic Control Signals for One-Lane, Two-Way Facilities	516

CHAPTER 4I TRAFFIC CONTROL SIGNALS FOR FREEWAY ENTRANCE RAMPS

Section 4I.01	Application of Freeway Entrance Ramp Control Signals	517
Section 4I.02	Design of Freeway Entrance Ramp Control Signals	517
Section 4I.03	Operation of Freeway Entrance Ramp Control Signals	518

CHAPTER 4J TRAFFIC CONTROL FOR MOVABLE BRIDGES

Section 4J.01	Application of Traffic Control for Movable Bridges	519
Section 4J.02	Design and Location of Movable Bridge Signals and Gates	519
Section 4J.03	Operation of Movable Bridge Signals and Gates	521

CHAPTER 4K HIGHWAY TRAFFIC SIGNALS AT TOLL PLAZAS

Section 4K.01	Traffic Signals at Toll Plazas	522
Section 4K.02	Lane-Use Control Signals at or Near Toll Plazas	522
Section 4K.03	Warning Beacons at Toll Plazas	522

CHAPTER 4L FLASHING BEACONS

Section 4L.01	General Design and Operation of Flashing Beacons	523
Section 4L.02	Intersection Control Beacon	523
Section 4L.03	Warning Beacon	523
Section 4L.04	Speed Limit Sign Beacon	524
Section 4L.05	Stop Beacon	524

CHAPTER 4M LANE-USE CONTROL SIGNALS

Section 4M.01	Application of Lane-Use Control Signals	525
Section 4M.02	Meaning of Lane-Use Control Signal Indications	525
Section 4M.03	Design of Lane-Use Control Signals	526
Section 4M.04	Operation of Lane-Use Control Signals	527

CHAPTER 4N IN-ROADWAY LIGHTS

Section 4N.01	Application of In-Roadway Lights	528
Section 4N.02	In-Roadway Warning Lights at Crosswalks	528

PART 5 TRAFFIC CONTROL DEVICES FOR LOW-VOLUME ROADS

CHAPTER 5A GENERAL

Section 5A.01	Function	531
Section 5A.02	Application	531
Section 5A.03	Design	531
Section 5A.04	Placement	533

CHAPTER 5B REGULATORY SIGNS

Section 5B.01	Introduction	534
Section 5B.02	STOP and YIELD Signs (R1-1 and R1-2)	534

Section 5B.03	Speed Limit Signs (R2 Series)	534
Section 5B.04	Traffic Movement and Prohibition Signs (R3, R4, R5, R6, R9, R10, R11, R12, R13, and R14 Series)	535
Section 5B.05	Parking Signs (R8 Series)	535
Section 5B.06	Other Regulatory Signs	535

CHAPTER 5C WARNING SIGNS

Section 5C.01	Introduction	536
Section 5C.02	Horizontal Alignment Signs (W1-1 through W1-8)	536
Section 5C.03	Intersection Warning Signs (W2-1 through W2-6)	537
Section 5C.04	Stop Ahead and Yield Ahead Signs (W3-1, W3-2)	537
Section 5C.05	NARROW BRIDGE Sign (W5-2)	537
Section 5C.06	ONE LANE BRIDGE Sign (W5-3)	537
Section 5C.07	Hill Sign (W7-1)	537
Section 5C.08	PAVEMENT ENDS Sign (W8-3)	537
Section 5C.09	Vehicular Traffic Warning and Non-Vehicular Warning Signs (W11 Series and W8-6)	537
Section 5C.10	Advisory Speed Plaque (W13-1P)	539
Section 5C.11	DEAD END or NO OUTLET Signs (W14-1, W14-1a, W14-2, W14-2a)	539
Section 5C.12	NO TRAFFIC SIGNS Sign (W18-1)	539
Section 5C.13	Other Warning Signs	539
Section 5C.14	Object Markers and Barricades	539

CHAPTER 5D GUIDE SIGNS

| Section 5D.01 | Introduction | 540 |

CHAPTER 5E MARKINGS

Section 5E.01	Introduction	541
Section 5E.02	Center Line Markings	541
Section 5E.03	Edge Line Markings	541
Section 5E.04	Delineators	541
Section 5E.05	Other Markings	541

CHAPTER 5F TRAFFIC CONTROL FOR HIGHWAY-RAIL GRADE CROSSINGS

Section 5F.01	Introduction	542
Section 5F.02	Grade Crossing (Crossbuck) Sign and Number of Tracks Plaque (R15-1, R15-2P)	542
Section 5F.03	Grade Crossing Advance Warning Signs (W10 Series)	542
Section 5F.04	STOP and YIELD Signs (R1-1, R1-2)	543
Section 5F.05	Pavement Markings	543
Section 5F.06	Other Traffic Control Devices	543

CHAPTER 5G TEMPORARY TRAFFIC CONTROL ZONES

Section 5G.01	Introduction	544
Section 5G.02	Applications	544
Section 5G.03	Channelization Devices	544
Section 5G.04	Markings	545
Section 5G.05	Other Traffic Control Devices	545

CHAPTER 5H TRAFFIC CONTROL FOR SCHOOL AREAS

| Section 5H.01 | Introduction | 546 |

PART 6   TEMPORARY TRAFFIC CONTROL

CHAPTER 6A   GENERAL
Section 6A.01   General 547

CHAPTER 6B   FUNDAMENTAL PRINCIPLES
Section 6B.01   Fundamental Principles of Temporary Traffic Control 549

CHAPTER 6C   TEMPORARY TRAFFIC CONTROL ELEMENTS
Section 6C.01   Temporary Traffic Control Plans 551
Section 6C.02   Temporary Traffic Control Zones 552
Section 6C.03   Components of Temporary Traffic Control Zones 552
Section 6C.04   Advance Warning Area 552
Section 6C.05   Transition Area 554
Section 6C.06   Activity Area 554
Section 6C.07   Termination Area 555
Section 6C.08   Tapers 555
Section 6C.09   Detours and Diversions 558
Section 6C.10   One-Lane, Two-Way Traffic Control 558
Section 6C.11   Flagger Method of One-Lane, Two-Way Traffic Control 558
Section 6C.12   Flag Transfer Method of One-Lane, Two-Way Traffic Control 558
Section 6C.13   Pilot Car Method of One-Lane, Two-Way Traffic Control 560
Section 6C.14   Temporary Traffic Control Signal Method of One-Lane, Two-Way Traffic Control 560
Section 6C.15   Stop or Yield Control Method of One-Lane, Two-Way Traffic Control 560

CHAPTER 6D   PEDESTRIAN AND WORKER SAFETY
Section 6D.01   Pedestrian Considerations 561
Section 6D.02   Accessibility Considerations 563
Section 6D.03   Worker Safety Considerations 564

CHAPTER 6E   FLAGGER CONTROL
Section 6E.01   Qualifications for Flaggers 566
Section 6E.02   High-Visibility Safety Apparel 566
Section 6E.03   Hand-Signaling Devices 566
Section 6E.04   Automated Flagger Assistance Devices 567
Section 6E.05   STOP/SLOW Automated Flagger Assistance Devices 569
Section 6E.06   Red/Yellow Lens Automated Flagger Assistance Devices 571
Section 6E.07   Flagger Procedures 573
Section 6E.08   Flagger Stations 575

CHAPTER 6F   TEMPORARY TRAFFIC CONTROL ZONE DEVICES
Section 6F.01   Types of Devices 576
Section 6F.02   General Characteristics of Signs 576
Section 6F.03   Sign Placement 577
Section 6F.04   Sign Maintenance 583
Section 6F.05   Regulatory Sign Authority 583
Section 6F.06   Regulatory Sign Design 583
Section 6F.07   Regulatory Sign Applications 583
Section 6F.08   ROAD (STREET) CLOSED Sign (R11-2) 583
Section 6F.09   Local Traffic Only Signs (R11-3a, R11-4) 585
Section 6F.10   Weight Limit Signs (R12-1, R12-2, R12-5) 585
Section 6F.11   STAY IN LANE Sign (R4-9) 586
Section 6F.12   Work Zone and Higher Fines Signs and Plaques 586
Section 6F.13   PEDESTRIAN CROSSWALK Sign (R9-8) 586
Section 6F.14   SIDEWALK CLOSED Signs (R9-9, R9-10, R9-11, R9-11a) 586

Section 6F.15	Special Regulatory Signs	587
Section 6F.16	Warning Sign Function, Design, and Application	587
Section 6F.17	Position of Advance Warning Signs	587
Section 6F.18	ROAD (STREET) WORK Sign (W20-1)	591
Section 6F.19	DETOUR Sign (W20-2)	591
Section 6F.20	ROAD (STREET) CLOSED Sign (W20-3)	591
Section 6F.21	ONE LANE ROAD Sign (W20-4)	591
Section 6F.22	Lane(s) Closed Signs (W20-5, W20-5a)	591
Section 6F.23	CENTER LANE CLOSED AHEAD Sign (W9-3)	592
Section 6F.24	Lane Ends Sign (W4-2)	592
Section 6F.25	ON RAMP Plaque (W13-4P)	592
Section 6F.26	RAMP NARROWS Sign (W5-4)	592
Section 6F.27	SLOW TRAFFIC AHEAD Sign (W23-1)	592
Section 6F.28	EXIT OPEN and EXIT CLOSED Signs (E5-2, E5-2a)	592
Section 6F.29	EXIT ONLY Sign (E5-3)	593
Section 6F.30	NEW TRAFFIC PATTERN AHEAD Sign (W23-2)	593
Section 6F.31	Flagger Signs (W20-7, W20-7a)	593
Section 6F.32	Two-Way Traffic Sign (W6-3)	593
Section 6F.33	Workers Signs (W21-1, W21-1a)	593
Section 6F.34	FRESH OIL (TAR) Sign (W21-2)	593
Section 6F.35	ROAD MACHINERY AHEAD Sign (W21-3)	593
Section 6F.36	Motorized Traffic Signs (W8-6, W11-10)	594
Section 6F.37	Shoulder Work Signs (W21-5, W21-5a, W21-5b)	594
Section 6F.38	SURVEY CREW Sign (W21-6)	594
Section 6F.39	UTILITY WORK Sign (W21-7)	594
Section 6F.40	Signs for Blasting Areas	594
Section 6F.41	BLASTING ZONE AHEAD Sign (W22-1)	595
Section 6F.42	TURN OFF 2-WAY RADIO AND CELL PHONE Sign (W22-2)	595
Section 6F.43	END BLASTING ZONE Sign (W22-3)	595
Section 6F.44	Shoulder Signs and Plaque (W8-4, W8-9, W8-17, and W8-17P)	595
Section 6F.45	UNEVEN LANES Sign (W8-11)	595
Section 6F.46	STEEL PLATE AHEAD Sign (W8-24)	595
Section 6F.47	NO CENTER LINE Sign (W8-12)	595
Section 6F.48	Reverse Curve Signs (W1-4 Series)	596
Section 6F.49	Double Reverse Curve Signs (W24-1 Series)	596
Section 6F.50	Other Warning Signs	596
Section 6F.51	Special Warning Signs	596
Section 6F.52	Advisory Speed Plaque (W13-1P)	596
Section 6F.53	Supplementary Distance Plaque (W7-3aP)	597
Section 6F.54	Motorcycle Plaque (W8-15P)	597
Section 6F.55	Guide Signs	597
Section 6F.56	ROAD WORK NEXT XX MILES Sign (G20-1)	597
Section 6F.57	END ROAD WORK Sign (G20-2)	598
Section 6F.58	PILOT CAR FOLLOW ME Sign (G20-4)	598
Section 6F.59	Detour Signs (M4-8, M4-8a, M4-8b, M4-9, M4-9a, M4-9b, M4-9c, and M4-10)	598
Section 6F.60	Portable Changeable Message Signs	598
Section 6F.61	Arrow Boards	601
Section 6F.62	High-Level Warning Devices (Flag Trees)	603
Section 6F.63	Channelizing Devices	604
Section 6F.64	Cones	606
Section 6F.65	Tubular Markers	606
Section 6F.66	Vertical Panels	607
Section 6F.67	Drums	607

Section 6F.68	Type 1, 2, or 3 Barricades	607
Section 6F.69	Direction Indicator Barricades	609
Section 6F.70	Temporary Traffic Barriers as Channelizing Devices	609
Section 6F.71	Longitudinal Channelizing Devices	609
Section 6F.72	Temporary Lane Separators	610
Section 6F.73	Other Channelizing Devices	610
Section 6F.74	Detectable Edging for Pedestrians	610
Section 6F.75	Temporary Raised Islands	611
Section 6F.76	Opposing Traffic Lane Divider and Sign (W6-4)	611
Section 6F.77	Pavement Markings	612
Section 6F.78	Temporary Markings	612
Section 6F.79	Temporary Raised Pavement Markers	613
Section 6F.80	Delineators	613
Section 6F.81	Lighting Devices	614
Section 6F.82	Floodlights	614
Section 6F.83	Warning Lights	614
Section 6F.84	Temporary Traffic Control Signals	615
Section 6F.85	Temporary Traffic Barriers	616
Section 6F.86	Crash Cushions	617
Section 6F.87	Rumble Strips	618
Section 6F.88	Screens	618

CHAPTER 6G TYPE OF TEMPORARY TRAFFIC CONTROL ZONE ACTIVITIES

Section 6G.01	Typical Applications	619
Section 6G.02	Work Duration	619
Section 6G.03	Location of Work	621
Section 6G.04	Modifications To Fulfill Special Needs	621
Section 6G.05	Work Affecting Pedestrian and Bicycle Facilities	622
Section 6G.06	Work Outside of the Shoulder	622
Section 6G.07	Work on the Shoulder with No Encroachment	623
Section 6G.08	Work on the Shoulder with Minor Encroachment	624
Section 6G.09	Work Within the Median	624
Section 6G.10	Work Within the Traveled Way of a Two-Lane Highway	624
Section 6G.11	Work Within the Traveled Way of an Urban Street	625
Section 6G.12	Work Within the Traveled Way of a Multi-Lane, Non-Access Controlled Highway	625
Section 6G.13	Work Within the Traveled Way at an Intersection	626
Section 6G.14	Work Within the Traveled Way of a Freeway or Expressway	627
Section 6G.15	Two-Lane, Two-Way Traffic on One Roadway of a Normally Divided Highway	628
Section 6G.16	Crossovers	628
Section 6G.17	Interchanges	628
Section 6G.18	Work in the Vicinity of a Grade Crossing	629
Section 6G.19	Temporary Traffic Control During Nighttime Hours	629

CHAPTER 6H TYPICAL APPLICATIONS

| Section 6H.01 | Typical Applications | 631 |

CHAPTER 6I CONTROL OF TRAFFIC THROUGH TRAFFIC INCIDENT MANAGEMENT AREAS

Section 6I.01	General	726
Section 6I.02	Major Traffic Incidents	727
Section 6I.03	Intermediate Traffic Incidents	728
Section 6I.04	Minor Traffic Incidents	728
Section 6I.05	Use of Emergency-Vehicle Lighting	729

PART 7   TRAFFIC CONTROL FOR SCHOOL AREAS

CHAPTER 7A   GENERAL

Section 7A.01	Need for Standards	731
Section 7A.02	School Routes and Established School Crossings	731
Section 7A.03	School Crossing Control Criteria	731
Section 7A.04	Scope	732

CHAPTER 7B   SIGNS

Section 7B.01	Size of School Signs	733
Section 7B.02	Illumination and Reflectorization	734
Section 7B.03	Position of Signs	734
Section 7B.04	Height of Signs	734
Section 7B.05	Installation of Signs	734
Section 7B.06	Lettering	734
Section 7B.07	Sign Color for School Warning Signs	734
Section 7B.08	School Sign (S1-1) and Plaques	734
Section 7B.09	School Zone Sign (S1-1) and Plaques (S4-3P, S4-7P) and END SCHOOL ZONE Sign (S5-2)	736
Section 7B.10	Higher Fines Zone Signs (R2-10, R2-11) and Plaques	736
Section 7B.11	School Advance Crossing Assembly	736
Section 7B.12	School Crossing Assembly	741
Section 7B.13	School Bus Stop Ahead Sign (S3-1)	742
Section 7B.14	SCHOOL BUS TURN AHEAD Sign (S3-2)	742
Section 7B.15	School Speed Limit Assembly (S4-1P, S4-2P, S4-3P, S4-4P, S4-6P, S5-1) and END SCHOOL SPEED LIMIT Sign (S5-3)	742
Section 7B.16	Reduced School Speed Limit Ahead Sign (S4-5, S4-5a)	743
Section 7B.17	Parking and Stopping Signs (R7 and R8 Series)	743

CHAPTER 7C   MARKINGS

Section 7C.01	Functions and Limitations	744
Section 7C.02	Crosswalk Markings	744
Section 7C.03	Pavement Word, Symbol, and Arrow Markings	744

CHAPTER 7D   CROSSING SUPERVISION

Section 7D.01	Types of Crossing Supervision	745
Section 7D.02	Adult Crossing Guards	745
Section 7D.03	Qualifications of Adult Crossing Guards	745
Section 7D.04	Uniform of Adult Crossing Guards	745
Section 7D.05	Operating Procedures for Adult Crossing Guards	745

PART 8   TRAFFIC CONTROL FOR RAILROAD AND LIGHT RAIL TRANSIT GRADE CROSSINGS

CHAPTER 8A   GENERAL

Section 8A.01	Introduction	747
Section 8A.02	Use of Standard Devices, Systems, and Practices at Highway-Rail Grade Crossings	747
Section 8A.03	Use of Standard Devices, Systems, and Practices at Highway-LRT Grade Crossings	748
Section 8A.04	Uniform Provisions	749
Section 8A.05	Grade Crossing Elimination	749
Section 8A.06	Illumination at Grade Crossings	750
Section 8A.07	Quiet Zone Treatments at Highway-Rail Grade Crossings	750
Section 8A.08	Temporary Traffic Control Zones	750

CHAPTER 8B   SIGNS AND MARKINGS

Section 8B.01	Purpose	751
Section 8B.02	Sizes of Grade Crossing Signs	751
Section 8B.03	Grade Crossing (Crossbuck) Sign (R15-1) and Number of Tracks Plaque (R15-2P) at Active and Passive Grade Crossings	751
Section 8B.04	Crossbuck Assemblies with YIELD or STOP Signs at Passive Grade Crossings	754
Section 8B.05	Use of STOP (R1-1) or YIELD (R1-2) Signs without Crossbuck Signs at Highway-LRT Grade Crossings	758
Section 8B.06	Grade Crossing Advance Warning Signs (W10 Series)	758
Section 8B.07	EXEMPT Grade Crossing Plaques (R15-3P, W10-1aP)	759
Section 8B.08	Turn Restrictions During Preemption	760
Section 8B.09	DO NOT STOP ON TRACKS Sign (R8-8)	760
Section 8B.10	TRACKS OUT OF SERVICE Sign (R8-9)	760
Section 8B.11	STOP HERE WHEN FLASHING Signs (R8-10, R8-10a)	761
Section 8B.12	STOP HERE ON RED Signs (R10-6, R10-6a)	761
Section 8B.13	Light Rail Transit Only Lane Signs (R15-4 Series)	761
Section 8B.14	Do Not Pass Light Rail Transit Signs (R15-5, R15-5a)	761
Section 8B.15	No Motor Vehicles On Tracks Signs (R15-6, R15-6a)	762
Section 8B.16	Divided Highway with Light Rail Transit Crossing Signs (R15-7 Series)	762
Section 8B.17	LOOK Sign (R15-8)	762
Section 8B.18	Emergency Notification Sign (I-13)	762
Section 8B.19	Light Rail Transit Approaching-Activated Blank-Out Warning Sign (W10-7)	763
Section 8B.20	TRAINS MAY EXCEED 80 MPH Sign (W10-8)	763
Section 8B.21	NO TRAIN HORN Sign or Plaque (W10-9, W10-9P)	763
Section 8B.22	NO GATES OR LIGHTS Plaque (W10-13P)	763
Section 8B.23	Low Ground Clearance Grade Crossing Sign (W10-5)	763
Section 8B.24	Storage Space Signs (W10-11, W10-11a, W10-11b)	764
Section 8B.25	Skewed Crossing Sign (W10-12)	764
Section 8B.26	Light Rail Transit Station Sign (I-12)	764
Section 8B.27	Pavement Markings	764
Section 8B.28	Stop and Yield Lines	766
Section 8B.29	Dynamic Envelope Markings	767

CHAPTER 8C   FLASHING-LIGHT SIGNALS, GATES, AND TRAFFIC CONTROL SIGNALS

Section 8C.01	Introduction	769
Section 8C.02	Flashing-Light Signals	769
Section 8C.03	Flashing-Light Signals at Highway-LRT Grade Crossings	772
Section 8C.04	Automatic Gates	772
Section 8C.05	Use of Automatic Gates at LRT Grade Crossings	773
Section 8C.06	Four-Quadrant Gate Systems	773
Section 8C.07	Wayside Horn Systems	775
Section 8C.08	Rail Traffic Detection	775
Section 8C.09	Traffic Control Signals at or Near Highway-Rail Grade Crossings	776
Section 8C.10	Traffic Control Signals at or Near Highway-LRT Grade Crossings	777
Section 8C.11	Use of Traffic Control Signals for Control of LRT Vehicles at Grade Crossings	778
Section 8C.12	Grade Crossings Within or In Close Proximity to Circular Intersections	780
Section 8C.13	Pedestrian and Bicycle Signals and Crossings at LRT Grade Crossings	780

CHAPTER 8D   PATHWAY GRADE CROSSINGS

Section 8D.01	Purpose	786
Section 8D.02	Use of Standard Devices, Systems, and Practices	786
Section 8D.03	Pathway Grade Crossing Signs and Markings	786
Section 8D.04	Stop Lines, Edge Lines, and Detectable Warnings	786
Section 8D.05	Passive Devices for Pathway Grade Crossings	787
Section 8D.06	Active Traffic Control Systems for Pathway Grade Crossings	788

PART 9 TRAFFIC CONTROL FOR BICYCLE FACILITIES

CHAPTER 9A GENERAL

Section 9A.01	Requirements for Bicyclist Traffic Control Devices	789
Section 9A.02	Scope	789
Section 9A.03	Definitions Relating to Bicycles	789
Section 9A.04	Maintenance	789
Section 9A.05	Relation to Other Documents	789
Section 9A.06	Placement Authority	789
Section 9A.07	Meaning of Standard, Guidance, Option, and Support	789
Section 9A.08	Colors	789

CHAPTER 9B SIGNS

Section 9B.01	Application and Placement of Signs	790
Section 9B.02	Design of Bicycle Signs	790
Section 9B.03	STOP and YIELD Signs (R1-1, R1-2)	792
Section 9B.04	Bike Lane Signs and Plaques (R3-17, R3-17aP, R3-17bP)	794
Section 9B.05	BEGIN RIGHT TURN LANE YIELD TO BIKES Sign (R4-4)	794
Section 9B.06	Bicycles May Use Full Lane Sign (R4-11)	794
Section 9B.07	Bicycle WRONG WAY Sign and RIDE WITH TRAFFIC Plaque (R5-1b, R9-3cP)	794
Section 9B.08	NO MOTOR VEHICLES Sign (R5-3)	795
Section 9B.09	Selective Exclusion Signs	795
Section 9B.10	No Parking Bike Lane Signs (R7-9, R7-9a)	795
Section 9B.11	Bicycle Regulatory Signs (R9-5, R9-6, R10-4, R10-24, R10-25, and R10-26)	795
Section 9B.12	Shared-Use Path Restriction Sign (R9-7)	795
Section 9B.13	Bicycle Signal Actuation Sign (R10-22)	796
Section 9B.14	Other Regulatory Signs	796
Section 9B.15	Turn or Curve Warning Signs (W1 Series)	796
Section 9B.16	Intersection Warning Signs (W2 Series)	796
Section 9B.17	Bicycle Surface Condition Warning Sign (W8-10)	796
Section 9B.18	Bicycle Warning and Combined Bicycle/Pedestrian Signs (W11-1 and W11-15)	796
Section 9B.19	Other Bicycle Warning Signs	798
Section 9B.20	Bicycle Guide Signs (D1-1b, D1-1c, D1-2b, D1-2c, D1-3b, D1-3c, D11-1, D11-1c)	798
Section 9B.21	Bicycle Route Signs (M1-8, M1-8a, M1-9)	800
Section 9B.22	Bicycle Route Sign Auxiliary Plaques	802
Section 9B.23	Bicycle Parking Area Sign (D4-3)	804
Section 9B.24	Reference Location Signs (D10-1 through D10-3) and Intermediate Reference Location Signs (D10-1a through D10-3a)	804
Section 9B.25	Mode-Specific Guide Signs for Shared-Use Paths (D11-1a, D11-2, D11-3, D11-4)	805
Section 9B.26	Object Markers	805

CHAPTER 9C MARKINGS

Section 9C.01	Functions of Markings	806
Section 9C.02	General Principles	806
Section 9C.03	Marking Patterns and Colors on Shared-Use Paths	806
Section 9C.04	Markings For Bicycle Lanes	806
Section 9C.05	Bicycle Detector Symbol	810
Section 9C.06	Pavement Markings for Obstructions	810
Section 9C.07	Shared Lane Marking	810

CHAPTER 9D SIGNALS

Section 9D.01	Application	816
Section 9D.02	Signal Operations for Bicycles	816

APPENDIX A1. CONGRESSIONAL LEGISLATION A1-1
APPENDIX A2. METRIC CONVERSIONS A2-1

FIGURES   Page

Figure 1A-1	Process for Requesting and Conducting Experimentations for New Traffic Control Devices	5
Figure 1A-2	Process for Incorporating New Traffic Control Devices into the MUTCD	8
Figure 2A-1	Examples of Enhanced Conspicuity for Signs	37
Figure 2A-2	Examples of Heights and Lateral Locations of Sign Installations	38
Figure 2A-3	Examples of Locations for Some Typical Signs at Intersections	39
Figure 2A-4	Relative Locations of Regulatory, Warning, and Guide Signs on an Intersection Approach	40
Figure 2B-1	STOP and YIELD Signs and Plaques	51
Figure 2B-2	Unsignalized Pedestrian Crosswalk Signs	55
Figure 2B-3	Speed Limit and Photo Enforcement Signs and Plaques	57
Figure 2B-4	Movement Prohibition and Lane Control Signs and Plaques	60
Figure 2B-5	Intersection Lane Control Sign Arrow Options for Roundabouts	62
Figure 2B-6	Center and Reversible Lane Control Signs and Plaques	65
Figure 2B-7	Location of Reversible Two-Way Left-Turn Signs	66
Figure 2B-8	Jughandle Regulatory Signs	68
Figure 2B-9	Examples of Applications of Jughandle Regulatory and Guide Signing	69
Figure 2B-10	Passing, Keep Right, and Slow Traffic Signs	72
Figure 2B-11	Selective Exclusion Signs	75
Figure 2B-12	Locations of Wrong-Way Signing for Divided Highways with Median Widths of 30 Feet or Wider	76
Figure 2B-13	ONE WAY and Divided Highway Crossing Signs	78
Figure 2B-14	Locations of ONE WAY Signs	79
Figure 2B-15	ONE WAY Signing for Divided Highways with Median Widths of 30 Feet or Wider	80
Figure 2B-16	ONE WAY Signing for Divided Highways with Median Widths Narrower Than 30 Feet	81
Figure 2B-17	ONE WAY Signing for Divided Highways with Median Widths Narrower Than 30 Feet and Separated Left-Turn Lanes	82
Figure 2B-18	Example of Application of Regulatory Signing and Pavement Markings at an Exit Ramp Termination to Deter Wrong-Way Entry	83
Figure 2B-19	Example of Application of Regulatory Signing and Pavement Markings at an Entrance Ramp Terminal Where the Design Does Not Clearly Indicate the Direction of Flow	83
Figure 2B-20	Roundabout Signs and Plaques	84
Figure 2B-21	Example of Regulatory and Warning Signs for a Mini-Roundabout	85
Figure 2B-22	Example of Regulatory and Warning Signs for a One-Lane Roundabout	86
Figure 2B-23	Example of Regulatory and Warning Signs for a Two-Lane Roundabout with Consecutive Double Lefts	87
Figure 2B-24	Parking and Standing Signs and Plaques (R7 Series)	88
Figure 2B-25	Parking and Stopping Signs and Plaques (R8 Series)	90
Figure 2B-26	Pedestrian Signs and Plaques	93
Figure 2B-27	Traffic Signal Signs and Plaques	96
Figure 2B-28	Ramp Metering Signs	97
Figure 2B-29	Road Closed and Weight Limit Signs	98
Figure 2B-30	Truck Signs	99
Figure 2B-31	Headlight Use Signs	100
Figure 2B-32	Other Regulatory Signs and Symbols	101
Figure 2C-1	Horizontal Alignment Signs and Plaques	109
Figure 2C-2	Example of Warning Signs for a Turn	111
Figure 2C-3	Example of Advisory Speed Signing for an Exit Ramp	116
Figure 2C-4	Vertical Grade Signs and Plaques	117
Figure 2C-5	Miscellaneous Warning Signs	118

Figure 2C-6	Roadway and Weather Condition and Advance Traffic Control Signs and Plaques	121
Figure 2C-7	Reduced Speed Limit Ahead Signs	124
Figure 2C-8	Merging and Passing Signs and Plaques	125
Figure 2C-9	Intersection Warning Signs and Plaques	127
Figure 2C-10	Vehicular Traffic Warning Signs and Plaques	129
Figure 2C-11	Non-Vehicular Warning Signs	130
Figure 2C-12	Supplemental Warning Plaques	132
Figure 2C-13	Object Markers	135
Figure 2D-1	Examples of Color-Coded Destination Guide Signs	138
Figure 2D-2	Arrows for Use on Guide Signs	141
Figure 2D-3	Route Signs	143
Figure 2D-4	Route Sign Auxiliaries	145
Figure 2D-5	Advance Turn and Directional Arrow Auxiliary Signs	147
Figure 2D-6	Illustration of Directional Assemblies and Other Route Signs (for One Direction of Travel Only)	149
Figure 2D-7	Destination and Distance Signs	155
Figure 2D-8	Destination Signs for Roundabouts	158
Figure 2D-9	Examples of Guide Signs for Roundabouts	159
Figure 2D-10	Street Name and Parking Signs	162
Figure 2D-11	Example of Interchange Crossroad Signing for a One-Lane Approach	165
Figure 2D-12	Example of Minor Interchange Crossroad Signing	166
Figure 2D-13	Examples of Multi-Lane Crossroad Signing for a Diamond Interchange	167
Figure 2D-14	Examples of Multi-Lane Crossroad Signing for a Partial Cloverleaf Interchange	168
Figure 2D-15	Examples of Multi-Lane Crossroad Signing for a Cloverleaf Interchange	169
Figure 2D-16	Example of Crossroad Signing for an Entrance Ramp with a Nearby Frontage Road	170
Figure 2D-17	Example of Weigh Station Signing	173
Figure 2D-18	Examples of Community Wayfinding Guide Signs	174
Figure 2D-19	Example of a Community Wayfinding Guide Sign System Showing Direction from a Freeway or Expressway	175
Figure 2D-20	Example of a Color-Coded Community Wayfinding Guide Sign System	176
Figure 2D-21	Crossover, Truck Lane, and Slow Vehicle Signs	178
Figure 2D-22	Examples of Use of the National Scenic Byways Sign	180
Figure 2E-1	Example of Guide Sign Spreading	184
Figure 2E-2	Pull-Through Signs	184
Figure 2E-3	Overhead Arrow-per-Lane Guide Sign for a Multi-Lane Exit with an Option Lane	194
Figure 2E-4	Overhead Arrow-per-Lane Guide Signs for a Two-Lane Exit to the Right with an Option Lane	195
Figure 2E-5	Overhead Arrow-per-Lane Guide Signs for a Two-Lane Exit to the Right with an Option Lane (Through Lanes Curve to the Left)	196
Figure 2E-6	Overhead Arrow-per-Lane Guide Signs for a Split with an Option Lane	197
Figure 2E-7	Diagrammatic Guide Sign for a Multi-Lane Exit with an Option Lane	199
Figure 2E-8	Diagrammatic Guide Signs for a Two-Lane Exit to the Right with an Option Lane	200
Figure 2E-9	Diagrammatic Guide Signs for a Two-Lane Exit to the Right with an Option Lane (Through Lanes Curve to the Left)	201
Figure 2E-10	Diagrammatic Guide Signs for a Split with an Option Lane	202
Figure 2E-11	Example of Signing for a Two-Lane Intermediate or Minor Interchange Exit with an Option Lane and a Dropped Lane	204
Figure 2E-12	Example of Signing for a Two-Lane Intermediate or Minor Interchange Exit with Option and Auxiliary Lanes	205
Figure 2E-13	EXIT ONLY and LEFT Sign Panels	206
Figure 2E-14	Guide Signs for a Split with Dedicated Lanes	207
Figure 2E-15	Guide Signs for a Single-Lane Exit to the Left with a Dropped Lane	208
Figure 2E-16	Guide Signs for a Single-Lane Exit to the Right with a Dropped Lane	209
Figure 2E-17	Interstate, Off-Interstate, and U.S. Route Signs	210

Figure 2E-18	Eisenhower Interstate System Signs	211
Figure 2E-19	Example of Interchange Numbering for Mainline and Circumferential Routes	213
Figure 2E-20	Example of Interchange Numbering for Mainline, Loop, and Spur Routes	214
Figure 2E-21	Example of Interchange Numbering for Overlapping Routes	215
Figure 2E-22	Examples of Interchange Advance Guide Signs, Exit Number Plaques, and LEFT Plaque	217
Figure 2E-23	Next Exit Plaques	218
Figure 2E-24	Supplemental Guide Sign for a Multi-Exit Interchange	219
Figure 2E-25	Supplemental Guide Sign for a Park – Ride Facility	219
Figure 2E-26	Examples of Interchange Exit Direction Signs	220
Figure 2E-27	Interchange Exit Direction Sign with an Advisory Speed Panel	221
Figure 2E-28	Exit Gore Signs	222
Figure 2E-29	Post-Interchange Distance Sign	223
Figure 2E-30	Example of Using an Interchange Sequence Sign for Closely-Spaced Interchanges	224
Figure 2E-31	Interchange Sequence Sign	225
Figure 2E-32	Community Interchanges Identification Sign	225
Figure 2E-33	NEXT EXITS Sign	225
Figure 2E-34	Examples of Guide Signs for a Freeway-to-Freeway Interchange	227
Figure 2E-35	Examples of Guide Signs for a Full Cloverleaf Interchange	229
Figure 2E-36	Examples of Guide Signs for a Full Cloverleaf Interchange with Collector-Distributor Roadways	231
Figure 2E-37	Examples of Guide Signs for a Partial Cloverleaf Interchange	232
Figure 2E-38	Examples of Guide Signs for a Diamond Interchange	233
Figure 2E-39	Examples of Guide Signs for a Diamond Interchange in an Urban Area	235
Figure 2E-40	Examples of Guide Signs for a Minor Interchange	236
Figure 2F-1	Examples of ETC Account Pictographs and Use of Purple Backgrounds and Underlay Panels	239
Figure 2F-2	Toll Plaza Regulatory Signs and Plaques	240
Figure 2F-3	Toll Plaza Warning Signs and Plaques	241
Figure 2F-4	ETC Account-Only Auxiliary Signs for Use in Route Sign Assemblies	243
Figure 2F-5	Examples of Guide Signs for Entrances to Toll Highways or Ramps	245
Figure 2F-6	Examples of Guide Signs for the Entrance to a Toll Highway on which Tolls are Collected Electronically Only	246
Figure 2F-7	Examples of Guide Signs for Alternative Toll and Non-Toll Ramp Connections to a Non-Toll Highway	247
Figure 2F-8	Examples of Conventional Toll Plaza Advance Signs	248
Figure 2F-9	Examples of Toll Plaza Canopy Signs	248
Figure 2F-10	Examples of Mainline Toll Plaza Approach and Canopy Signing	250
Figure 2F-11	Examples of Guide Signs for a Mainline Toll Plaza on a Diverging Alignment from Open-Road ETC Lanes	251
Figure 2G-1	Preferential Lane Regulatory Signs and Plaques	255
Figure 2G-2	Example of Signing for an Added Continuous-Access Contiguous or Buffer-Separated HOV Lane	261
Figure 2G-3	Example of Signing for a General-Purpose Lane that Becomes a Continuous-Access Contiguous or Buffer-Separated HOV Lane	262
Figure 2G-4	Examples of Warning Signs and Plaques Applicable Only to Preferential Lanes	264
Figure 2G-5	Example of an Overhead Advance Guide Sign for a Preferential Lane Entrance	267
Figure 2G-6	Examples of Overhead or Post-Mounted Preferential Lane Entrance Direction Signs	267
Figure 2G-7	Entrance Gore Signs for Barrier-Separated Preferential Lanes	268
Figure 2G-8	Example of Signing for an Entrance to Access-Restricted HOV Lanes	269
Figure 2G-9	Example of Signing for an Intermediate Entry to a Barrier- or Buffer-Separated HOV Lane	271
Figure 2G-10	Example of Signing for the Intermediate Entry to, Egress from, and End of Access-Restricted HOV Lanes	272

Figure 2G-11	Examples of Barrier-Mounted Guide Signs for an Intermediate Egress from Preferential Lanes	273
Figure 2G-12	Examples of Guide Signs for an Intermediate Egress from a Barrier- or Buffer-Separated HOV Lane	274
Figure 2G-13	Example of Signing for a Direct Entrance Ramp to an HOV Lane from a Park-and-Ride Facility and a Local Street	275
Figure 2G-14	Exit Gore Sign for a Direct Exit from a Preferential Lane	276
Figure 2G-15	Examples of Guide Signs for Direct HOV Lane Entrance and Exit Ramps	277
Figure 2G-16	Examples of Guide Signs for a Direct Access Ramp between HOV Lanes on Separate Freeways	278
Figure 2G-17	Regulatory Signs for Managed Lanes	280
Figure 2G-18	Examples of Guide Signs for Entrances to Priced Managed Lanes	281
Figure 2G-19	Example of an Exit Destinations Sign for a Managed Lane	282
Figure 2G-20	Example of a Comparative Travel Time Information Sign for Preferential or Managed Lanes	282
Figure 2G-21	Example of Signing for the Entrance to an Access-Restricted Priced Managed Lane	283
Figure 2G-22	Example of Signing for the Entrance to an Access-Restricted Priced Managed Lane Where a General-Purpose Lane Becomes the Managed Lane	284
Figure 2G-23	Example of Signing for an Intermediate Entry to a Barrier- or Buffer-Separated Priced Managed Lane	285
Figure 2G-24	Example of Signing for the Intermediate Entry to, Egress from, and End of Access-Restricted Priced Managed Lanes	286
Figure 2G-25	Examples of Guide Signs for an Intermediate Egress from a Barrier- or Buffer-Separated HOV Lane	287
Figure 2G-26	Examples of Guide Signs for Direct Managed Lane Entrance and Exit Ramps	288
Figure 2G-27	Examples of Guide Signs for a Direct Access Ramp between Managed Lanes on Separate Freeways	289
Figure 2G-28	Examples of Guide Signs for a Direct Entrance Ramp to a Priced Managed Lane and Trailblazing to a Nearby Entrance to the General-Purpose Lanes	290
Figure 2G-29	Examples of Guide Signs for Separate Entrance Ramps to General-Purpose and Priced Managed Lanes from the Same Crossroad	291
Figure 2H-1	General Information and Miscellaneous Information Signs	293
Figure 2H-2	Reference Location Signs	295
Figure 2H-3	Intermediate Reference Location Signs	295
Figure 2H-4	Enhanced Reference Location Signs	296
Figure 2H-5	Examples of Acknowledgment Sign Designs	298
Figure 2I-1	General Service Signs and Plaques	301
Figure 2I-2	Example of Next Services Plaque	302
Figure 2I-3	Examples of General Service Signs with and without Exit Numbering	304
Figure 2I-4	Examples of Interstate Oasis Signs and Plaques	306
Figure 2I-5	Rest Area and Other Roadside Area Signs	307
Figure 2I-6	Brake Check Area and Chain-Up Area Signs	308
Figure 2I-7	Examples of Tourist Information and Welcome Center Signs	309
Figure 2I-8	Radio, Telephone, and Carpool Information Signs	310
Figure 2J-1	Examples of Specific Service Signs	314
Figure 2J-2	Examples of Specific Service Sign Locations	315
Figure 2J-3	Examples of Supplemental Messages on Logo Sign Panels	316
Figure 2J-4	Examples of RV Access Supplemental Messages on Logo Sign Panels	316
Figure 2J-5	Examples of Specific Service Trailblazer Signs	319
Figure 2K-1	Examples of Tourist-Oriented Directional Signs	321
Figure 2K-2	Examples of Intersection Approach Signs and Advance Signs for Tourist-Oriented Directional Signs	322
Figure 2M-1	Examples of Use of Arrows, Educational Plaques, and Prohibitory Slashes	333

Figure 2M-2	Examples of Recreational and Cultural Interest Area Guide Signs	334
Figure 2M-3	Arrangement, Height, and Lateral Position of Signs Located Within Recreational and Cultural Interest Areas	335
Figure 2M-4	Examples of Symbol and Destination Guide Signing Layout	336
Figure 2M-5	Recreational and Cultural Interest Area Symbol Signs for General Applications	337
Figure 2M-6	Recreational and Cultural Interest Area Symbol Signs for Accommodations	338
Figure 2M-7	Recreational and Cultural Interest Area Symbol Signs for Services	338
Figure 2M-8	Recreational and Cultural Interest Area Symbol Signs for Land Recreation	339
Figure 2M-9	Recreational and Cultural Interest Area Symbol Signs for Water Recreation	340
Figure 2M-10	Recreational and Cultural Interest Area Symbol Signs for Winter Recreation	341
Figure 2N-1	Emergency Management Signs	343
Figure 3B-1	Examples of Two-Lane, Two-Way Marking Applications	350
Figure 3B-2	Examples of Four-or-More Lane, Two-Way Marking Applications	351
Figure 3B-3	Examples of Three-Lane, Two-Way Marking Applications	352
Figure 3B-4	Method of Locating and Determining the Limits of No-Passing Zones at Curves	353
Figure 3B-5	Example of Application of Three-Lane, Two-Way Marking for Changing Direction of the Center Lane	355
Figure 3B-6	Example of Reversible Lane Marking Application	356
Figure 3B-7	Example of Two-Way Left-Turn Lane Marking Applications	357
Figure 3B-8	Examples of Dotted Line and Channelizing Line Applications for Exit Ramp Markings	358
Figure 3B-9	Examples of Dotted Line and Channelizing Line Applications for Entrance Ramp Markings	360
Figure 3B-10	Examples of Applications of Freeway and Expressway Lane-Drop Markings	363
Figure 3B-11	Examples of Applications of Conventional Road Lane-Drop Markings	368
Figure 3B-12	Example of Solid Double White Lines Used to Prohibit Lane Changing	370
Figure 3B-13	Examples of Line Extensions through Intersections	372
Figure 3B-14	Examples of Applications of Lane-Reduction Transition Markings	375
Figure 3B-15	Examples of Applications of Markings for Obstructions in the Roadway	377
Figure 3B-16	Recommended Yield Line Layouts	382
Figure 3B-17	Examples of Yield Lines at Unsignalized Midblock Crosswalks	383
Figure 3B-18	Do Not Block Intersection Markings	384
Figure 3B-19	Examples of Crosswalk Markings	384
Figure 3B-20	Example of Crosswalk Markings for an Exclusive Pedestrian Phase that Permits Diagonal Crossing	385
Figure 3B-21	Examples of Parking Space Markings	386
Figure 3B-22	International Symbol of Accessibility Parking Space Marking	387
Figure 3B-23	Example of Elongated Letters for Word Pavement Markings	387
Figure 3B-24	Examples of Standard Arrows for Pavement Markings	388
Figure 3B-25	Examples of Elongated Route Shields for Pavement Markings	390
Figure 3B-26	Yield Ahead Triangle Symbols	391
Figure 3B-27	Examples of Lane-Use Control Word and Arrow Pavement Markings	392
Figure 3B-28	Example of the Application of Speed Reduction Markings	394
Figure 3B-29	Pavement Markings for Speed Humps without Crosswalks	396
Figure 3B-30	Pavement Markings for Speed Tables or Speed Humps with Crosswalks	397
Figure 3B-31	Advance Warning Markings for Speed Humps	398
Figure 3C-1	Example of Markings for Approach and Circulatory Roadways at a Roundabout	399
Figure 3C-2	Lane-Use Arrow Pavement Marking Options for Roundabout Approaches	400
Figure 3C-3	Example of Markings for a One-Lane Roundabout	400
Figure 3C-4	Example of Markings for a Two-Lane Roundabout with One- and Two-Lane Approaches	401
Figure 3C-5	Example of Markings for a Two-Lane Roundabout with One-Lane Exits	403
Figure 3C-6	Example of Markings for a Two-Lane Roundabout with Two-Lane Exits	404
Figure 3C-7	Example of Markings for a Two-Lane Roundabout with a Double Left Turn	405
Figure 3C-8	Example of Markings for a Two-Lane Roundabout with a Double Right Turn	406
Figure 3C-9	Example of Markings for a Two-Lane Roundabout with Consecutive Double Lefts	407

Figure 3C-10	Example of Markings for a Three-Lane Roundabout with Two- and Three-Lane Approaches	408
Figure 3C-11	Example of Markings for a Three-Lane Roundabout with Three-Lane Approaches	409
Figure 3C-12	Example of Markings for a Three-Lane Roundabout with Two-Lane Exits	410
Figure 3C-13	Example of Markings for Two Linked Roundabouts	411
Figure 3C-14	Example of Markings for a Diamond Interchange with Two Circular-Shaped Roundabout Ramp Terminals	412
Figure 3D-1	Markings for Barrier-Separated Preferential Lanes	418
Figure 3D-2	Markings for Buffer-Separated Preferential Lanes	418
Figure 3D-3	Markings for Contiguous Preferential Lanes	420
Figure 3D-4	Markings for Counter-Flow Preferential Lanes on Divided Highways	422
Figure 3F-1	Examples of Delineator Placement	425
Figure 3J-1	Examples of Longitudinal Rumble Strip Markings	432
Figure 4C-1	Warrant 2, Four-Hour Vehicular Volume	440
Figure 4C-2	Warrant 2, Four-Hour Vehicular Volume (70% Factor)	440
Figure 4C-3	Warrant 3, Peak Hour	441
Figure 4C-4	Warrant 3, Peak Hour (70% Factor)	441
Figure 4C-5	Warrant 4, Pedestrian Four-Hour Volume	443
Figure 4C-6	Warrant 4, Pedestrian Four-Hour Volume (70% Factor)	443
Figure 4C-7	Warrant 4, Pedestrian Peak Hour	444
Figure 4C-8	Warrant 4, Pedestrian Peak Hour (70% Factor)	444
Figure 4C-9	Warrant 9, Intersection Near a Grade Crossing (One Approach Lane at the Track Crossing)	447
Figure 4C-10	Warrant 9, Intersection Near a Grade Crossing (Two or More Approach Lanes at the Track Crossing)	447
Figure 4D-1	Example of U-Turn Signal Face	456
Figure 4D-2	Typical Arrangements of Signal Sections in Signal Faces That Do Not Control Turning Movements	458
Figure 4D-3	Recommended Vehicular Signal Faces for Approaches with Posted, Statutory, or 85th-Percentile Speed of 45 mph or Higher	460
Figure 4D-4	Lateral and Longitudinal Location of Primary Signal Faces	463
Figure 4D-5	Maximum Mounting Height of Signal Faces Located Between 40 Feet and 53 Feet from Stop Line	465
Figure 4D-6	Typical Position and Arrangements of Shared Signal Faces for Permissive Only Mode Left Turns	467
Figure 4D-7	Typical Position and Arrangements of Separate Signal Faces with Flashing Yellow Arrow for Permissive Only Mode Left Turns	468
Figure 4D-8	Typical Position and Arrangements of Separate Signal Faces with Flashing Red Arrow for Permissive Only Mode and Protected/Permissive Mode Left Turns	469
Figure 4D-9	Typical Positions and Arrangements of Shared Signal Faces for Protected Only Mode Left Turns	470
Figure 4D-10	Typical Position and Arrangements of Separate Signal Faces for Protected Only Mode Left Turns	471
Figure 4D-11	Typical Position and Arrangements of Shared Signal Faces for Protected/Permissive Mode Left Turns	472
Figure 4D-12	Typical Position and Arrangements of Separate Signal Faces with Flashing Yellow Arrow for Protected/Permissive Mode and Protected Only Mode Left Turns	473
Figure 4D-13	Typical Positions and Arrangements of Shared Signal Faces for Permissive Only Mode Right Turns	476
Figure 4D-14	Typical Position and Arrangements of Separate Signal Faces with Flashing Yellow Arrow for Permissive Only Mode Right Turns	477
Figure 4D-15	Typical Position and Arrangements of Separate Signal Faces with Flashing Red Arrow for Permissive Only Mode and Protected/Permissive Mode Right Turns	478

Figure 4D-16	Typical Positions and Arrangements of Shared Signal Faces for Protected Only Mode Right Turns	479
Figure 4D-17	Typical Position and Arrangements of Separate Signal Faces for Protected Only Mode Right Turns	480
Figure 4D-18	Typical Positions and Arrangements of Shared Signal Faces for Protected/Permissive Mode Right Turns	481
Figure 4D-19	Typical Position and Arrangements of Separate Signal Faces with Flashing Yellow Arrow for Protected/Permissive Mode and Protected Only Mode Right Turns	482
Figure 4D-20	Signal Indications for Approaches with a Shared Left-Turn/Right-Turn Lane and No Through Movement	486
Figure 4E-1	Typical Pedestrian Signal Indications	496
Figure 4E-2	Pedestrian Intervals	498
Figure 4E-3	Pushbutton Location Area	501
Figure 4E-4	Typical Pushbutton Locations	502
Figure 4F-1	Guidelines for the Installation of Pedestrian Hybrid Beacons on Low-Speed Roadways	510
Figure 4F-2	Guidelines for the Installation of Pedestrian Hybrid Beacons on High-Speed Roadways	510
Figure 4F-3	Sequence for a Pedestrian Hybrid Beacon	511
Figure 4G-1	Sequence for an Emergency-Vehicle Hybrid Beacon	515
Figure 4M-1	Left-Turn Lane-Use Control Signals	526
Figure 5B-1	Regulatory Signs on Low-Volume Roads	534
Figure 5B-2	Parking Signs and Plaques on Low-Volume Roads	535
Figure 5C-1	Horizontal Alignment and Intersection Warning Signs and Plaques and Object Markers on Low-Volume Roads	536
Figure 5C-2	Other Warning Signs and Plaques on Low-Volume Roads	538
Figure 5F-1	Highway-Rail Grade Crossing Signs and Plaques for Low-Volume Roads	542
Figure 5G-1	Temporary Traffic Control Signs and Plaques on Low-Volume Roads	545
Figure 6C-1	Component Parts of a Temporary Traffic Control Zone	553
Figure 6C-2	Types of Tapers and Buffer Spaces	556
Figure 6C-3	Example of a One-Lane, Two-Way Traffic Taper	559
Figure 6E-1	Example of the Use of a STOP/SLOW Automated Flagger Assistance Device (AFAD)	570
Figure 6E-2	Example of the Use of a Red/Yellow Lens Automated Flagger Assistance Device (AFAD)	572
Figure 6E-3	Use of Hand-Signaling Devices by Flaggers	574
Figure 6F-1	Height and Lateral Location of Signs—Typical Installations	581
Figure 6F-2	Methods of Mounting Signs Other Than on Posts	582
Figure 6F-3	Regulatory Signs and Plaques in Temporary Traffic Control Zones	584
Figure 6F-4	Warning Signs and Plaques in Temporary Traffic Control Zones	588
Figure 6F-5	Exit Open and Closed and Detour Signs	592
Figure 6F-6	Advance Warning Arrow Board Display Specifications	602
Figure 6F-7	Channelizing Devices	605
Figure 6H-1	Work Beyond the Shoulder (TA-1)	635
Figure 6H-2	Blasting Zone (TA-2)	637
Figure 6H-3	Work on the Shoulders (TA-3)	639
Figure 6H-4	Short-Duration or Mobile Operation on a Shoulder (TA-4)	641
Figure 6H-5	Shoulder Closure on a Freeway (TA-5)	643
Figure 6H-6	Shoulder Work with Minor Encroachment (TA-6)	645
Figure 6H-7	Road Closure with a Diversion (TA-7)	647
Figure 6H-8	Road Closure with an Off-Site Detour (TA-8)	649
Figure 6H-9	Overlapping Routes with a Detour (TA-9)	651
Figure 6H-10	Lane Closure on a Two-Lane Road Using Flaggers (TA-10)	653
Figure 6H-11	Lane Closure on a Two-Lane Road with Low Traffic Volumes (TA-11)	655
Figure 6H-12	Lane Closure on a Two-Lane Road Using Traffic Control Signals (TA-12)	657
Figure 6H-13	Temporary Road Closure (TA-13)	659
Figure 6H-14	Haul Road Crossing (TA-14)	661
Figure 6H-15	Work in the Center of a Road with Low Traffic Volumes (TA-15)	663

Figure 6H-16	Surveying Along the Center Line of a Road with Low Traffic Volumes (TA-16)	665
Figure 6H-17	Mobile Operations on a Two-Lane Road (TA-17)	667
Figure 6H-18	Lane Closure on a Minor Street (TA-18)	669
Figure 6H-19	Detour for One Travel Direction (TA-19)	671
Figure 6H-20	Detour for a Closed Street (TA-20)	673
Figure 6H-21	Lane Closure on the Near Side of an Intersection (TA-21)	675
Figure 6H-22	Right-Hand Lane Closure on the Far Side of an Intersection (TA-22)	677
Figure 6H-23	Left-Hand Lane Closure on the Far Side of an Intersection (TA-23)	679
Figure 6H-24	Half Road Closure on the Far Side of an Intersection (TA-24)	681
Figure 6H-25	Multiple Lane Closures at an Intersection (TA-25)	683
Figure 6H-26	Closure in the Center of an Intersection (TA-26)	685
Figure 6H-27	Closure at the Side of an Intersection (TA-27)	687
Figure 6H-28	Sidewalk Detour or Diversion (TA-28)	689
Figure 6H-29	Crosswalk Closures and Pedestrian Detours (TA-29)	691
Figure 6H-30	Interior Lane Closure on a Multi-Lane Street (TA-30)	693
Figure 6H-31	Lane Closures on a Street with Uneven Directional Volumes (TA-31)	695
Figure 6H-32	Half Road Closure on a Multi-Lane, High-Speed Highway (TA-32)	697
Figure 6H-33	Stationary Lane Closure on a Divided Highway (TA-33)	699
Figure 6H-34	Lane Closure with a Temporary Traffic Barrier (TA-34)	701
Figure 6H-35	Mobile Operation on a Multi-Lane Road (TA-35)	703
Figure 6H-36	Lane Shift on a Freeway (TA-36)	705
Figure 6H-37	Double Lane Closure on a Freeway (TA-37)	707
Figure 6H-38	Interior Lane Closure on a Freeway (TA-38)	709
Figure 6H-39	Median Crossover on a Freeway (TA-39)	711
Figure 6H-40	Median Crossover for an Entrance Ramp (TA-40)	713
Figure 6H-41	Median Crossover for an Exit Ramp (TA-41)	715
Figure 6H-42	Work in the Vicinity of an Exit Ramp (TA-42)	717
Figure 6H-43	Partial Exit Ramp Closure (TA-43)	719
Figure 6H-44	Work in the Vicinity of an Entrance Ramp (TA-44)	721
Figure 6H-45	Temporary Reversible Lane Using Movable Barriers (TA-45)	723
Figure 6H-46	Work in the Vicinity of a Grade Crossing (TA-46)	725
Figure 6I-1	Examples of Traffic Incident Management Area Signs	727
Figure 7A-1	Example of School Route Plan Map	732
Figure 7B-1	School Area Signs	735
Figure 7B-2	Example of Signing for a Higher Fines School Zone without a School Crossing	737
Figure 7B-3	Example of Signing for a Higher Fines School Zone with a School Speed Limit	738
Figure 7B-4	Example of Signing for a School Crossing Outside of a School Zone	739
Figure 7B-5	Example of Signing for a School Zone with a School Speed Limit and a School Crossing	740
Figure 7B-6	In-Street Signs in School Areas	741
Figure 7C-1	Two-Lane Pavement Marking of "SCHOOL"	744
Figure 8B-1	Regulatory Signs and Plaques for Grade Crossings	753
Figure 8B-2	Crossbuck Assembly with a YIELD or STOP Sign on the Crossbuck Sign Support	754
Figure 8B-3	Crossbuck Assembly with a YIELD or STOP Sign on a Separate Sign Support	755
Figure 8B-4	Warning Signs and Plaques for Grade Crossings	759
Figure 8B-5	Example of an Emergency Notification Sign	762
Figure 8B-6	Example of Placement of Warning Signs and Pavement Markings at Grade Crossings	765
Figure 8B-7	Grade Crossing Pavement Markings	766
Figure 8B-8	Example of Dynamic Envelope Pavement Markings at Grade Crossings	767
Figure 8B-9	Examples of Light Rail Transit Vehicle Dynamic Envelope Markings for Mixed-Use Alignments	768
Figure 8C-1	Composite Drawing of Active Traffic Control Devices for Grade Crossings Showing Clearances	770
Figure 8C-2	Example of Location Plan for Flashing-Light Signals and Four-Quadrant Gates	774
Figure 8C-3	Light Rail Transit Signals	779

Figure 8C-4	Example of Flashing-Light Signal Assembly for Pedestrian Crossings	781
Figure 8C-5	Example of a Shared Pedestrian/Roadway Gate	782
Figure 8C-6	Example of a Separate Pedestrian Gate	782
Figure 8C-7	Examples of Placement of Pedestrian Gates	783
Figure 8C-8	Example of Swing Gates	784
Figure 8C-9	Example of Pedestrian Barriers at an Offset Grade Crossing	784
Figure 8C-10	Examples of Pedestrian Barrier Installation at an Offset Non-Intersection Grade Crossing	785
Figure 8D-1	Example of Signing and Markings for a Pathway Grade Crossing	787
Figure 9B-1	Sign Placement on Shared-Use Paths	790
Figure 9B-2	Regulatory Signs and Plaques for Bicycle Facilities	793
Figure 9B-3	Warning Signs and Plaques and Object Markers for Bicycle Facilities	797
Figure 9B-4	Guide Signs and Plaques for Bicycle Facilities	799
Figure 9B-5	Example of Signing for the Beginning and End of a Designated Bicycle Route on a Shared-Use Path	801
Figure 9B-6	Example of Bicycle Guide Signing	802
Figure 9B-7	Examples of Signing and Markings for a Shared-Use Path Crossing	803
Figure 9B-8	Example of Mode-Specific Guide Signing on a Shared-Use Path	805
Figure 9C-1	Example of Intersection Pavement Markings—Designated Bicycle Lane with Left-Turn Area, Heavy Turn Volumes, Parking, One-Way Traffic, or Divided Highway	807
Figure 9C-2	Examples of Center Line Markings for Shared-Use Paths	808
Figure 9C-3	Word, Symbol, and Arrow Pavement Markings for Bicycle Lanes	809
Figure 9C-4	Example of Bicycle Lane Treatment at a Right Turn Only Lane	811
Figure 9C-5	Example of Bicycle Lane Treatment at Parking Lane into a Right Turn Only Lane	812
Figure 9C-6	Example of Pavement Markings for Bicycle Lanes on a Two-Way Street	813
Figure 9C-7	Bicycle Detector Pavement Marking	814
Figure 9C-8	Examples of Obstruction Pavement Markings	815
Figure 9C-9	Shared Lane Marking	815

TABLES

		Page
Table I-1	Evolution of the MUTCD	I-2
Table I-2	Target Compliance Dates Established by the FHWA	I-4
Table 1A-1	Acceptable Abbreviations	24
Table 1A-2	Abbreviations that Shall be Used Only on Portable Changeable Message Signs	25
Table 1A-3	Unacceptable Abbreviations	26
Table 2A-1	Illumination of Sign Elements	29
Table 2A-2	Retroreflection of Sign Elements	29
Table 2A-3	Minimum Maintained Retroreflectivity Levels	31
Table 2A-4	Use of Sign Shapes	32
Table 2A-5	Common Uses of Sign Colors	33
Table 2B-1	Regulatory Sign and Plaque Sizes	46
Table 2B-2	Meanings of Symbols and Legends on Reversible Lane Control Signs	65
Table 2C-1	Categories of Warning Signs and Plaques	104
Table 2C-2	Warning Sign and Plaque Sizes	105
Table 2C-3	Minimum Size of Supplemental Warning Plaques	107
Table 2C-4	Guidelines for Advance Placement of Warning Signs	108
Table 2C-5	Horizontal Alignment Sign Selection	110
Table 2C-6	Approximate Spacing of Chevron Alignment Signs on Horizontal Curves	113
Table 2D-1	Conventional Road Guide Sign Sizes	139
Table 2D-2	Recommended Minimum Letter Heights on Street Name Signs	163
Table 2E-1	Freeway or Expressway Guide Sign and Plaque Sizes	186

Table 2E-2	Minimum Letter and Numeral Sizes for Expressway Guide Signs According to Interchange Classification	188
Table 2E-3	Minimum Letter and Numeral Sizes for Expressway Guide Signs According to Sign Type	189
Table 2E-4	Minimum Letter and Numeral Sizes for Freeway Guide Signs According to Interchange Classification	190
Table 2E-5	Minimum Letter and Numeral Sizes for Freeway Guide Signs According to Sign Type	191
Table 2F-1	Toll Facility Sign and Plaque Minimum Sizes	237
Table 2G-1	Managed and Preferential Lanes Sign and Plaque Minimum Sizes	254
Table 2H-1	General Information Sign Sizes	292
Table 2I-1	General Service Sign and Plaque Sizes	299
Table 2J-1	Minimum Letter and Numeral Sizes for Specific Service Signs According to Sign Type	316
Table 2L-1	Example of Units of Information	328
Table 2M-1	Category Chart for Recreational and Cultural Interest Area Symbols	331
Table 2N-1	Emergency Management Sign Sizes	343
Table 3B-1	Minimum Passing Sight Distances for No-Passing Zone Markings	352
Table 3D-1	Standard Edge Line and Lane Line Markings for Preferential Lanes	417
Table 3F-1	Approximate Spacing for Delineators on Horizontal Curves	427
Table 4C-1	Warrant 1, Eight-Hour Vehicular Volume	438
Table 4C-2	Warrant 9, Adjustment Factor for Daily Frequency of Rail Traffic	448
Table 4C-3	Warrant 9, Adjustment Factor for Percentage of High-Occupancy Buses	448
Table 4C-4	Warrant 9, Adjustment Factor for Percentage of Tractor-Trailer Trucks	448
Table 4D-1	Recommended Minimum Number of Primary Signal Faces for Through Traffic on Approaches with Posted, Statutory, or 85th-Percentile Speed of 45 mph or Higher	461
Table 4D-2	Minimum Sight Distance for Signal Visibility	461
Table 5A-1	Sign and Plaque Sizes on Low-Volume Roads	532
Table 6C-1	Recommended Advance Warning Sign Minimum Spacing	554
Table 6C-2	Stopping Sight Distance as a Function of Speed	555
Table 6C-3	Taper Length Criteria for Temporary Traffic Control Zones	557
Table 6C-4	Formulas for Determining Taper Length	557
Table 6E-1	Stopping Sight Distance as a Function of Speed	575
Table 6F-1	Temporary Traffic Control Zone Sign and Plaque Sizes	578
Table 6H-1	Index to Typical Applications	632
Table 6H-2	Meaning of Symbols on Typical Application Diagrams	633
Table 6H-3	Meaning of Letter Codes on Typical Application Diagrams	633
Table 6H-4	Formulas for Determining Taper Length	633
Table 7B-1	School Area Sign and Plaque Sizes	733
Table 8B-1	Grade Crossing Sign and Plaque Minimum Sizes	752
Table 9B-1	Bicycle Facility Sign and Plaque Minimum Sizes	791
Table A2-1	Conversion of Inches to Millimeters	A2-1
Table A2-2	Conversion of Feet to Meters	A2-1
Table A2-3	Conversion of Miles to Kilometers	A2-1
Table A2-4	Conversion of Miles per Hour to Kilometers/Hour	A2-1

MANUAL ON UNIFORM TRAFFIC CONTROL DEVICES
INTRODUCTION

Standard:

01 Traffic control devices shall be defined as all signs, signals, markings, and other devices used to regulate, warn, or guide traffic, placed on, over, or adjacent to a street, highway, pedestrian facility, bikeway, or private road open to public travel (see definition in Section 1A.13) by authority of a public agency or official having jurisdiction, or, in the case of a private road, by authority of the private owner or private official having jurisdiction.

02 The Manual on Uniform Traffic Control Devices (MUTCD) is incorporated by reference in 23 Code of Federal Regulations (CFR), Part 655, Subpart F and shall be recognized as the national standard for all traffic control devices installed on any street, highway, bikeway, or private road open to public travel (see definition in Section 1A.13) in accordance with 23 U.S.C. 109(d) and 402(a). The policies and procedures of the Federal Highway Administration (FHWA) to obtain basic uniformity of traffic control devices shall be as described in 23 CFR 655, Subpart F.

03 In accordance with 23 CFR 655.603(a), for the purposes of applicability of the MUTCD:
   A. Toll roads under the jurisdiction of public agencies or authorities or public-private partnerships shall be considered to be public highways;
   B. Private roads open to public travel shall be as defined in Section 1A.13; and
   C. Parking areas, including the driving aisles within those parking areas, that are either publicly or privately owned shall not be considered to be "open to public travel" for purposes of MUTCD applicability.

04 Any traffic control device design or application provision contained in this Manual shall be considered to be in the public domain. Traffic control devices contained in this Manual shall not be protected by a patent, trademark, or copyright, except for the Interstate Shield and any items owned by FHWA.

Support:

05 Pictographs, as defined in Section 1A.13, are embedded in traffic control devices but the pictographs themselves are not considered traffic control devices for the purposes of Paragraph 4.

06 The need for uniform standards was recognized long ago. The American Association of State Highway Officials (AASHO), now known as the American Association of State Highway and Transportation Officials (AASHTO), published a manual for rural highways in 1927, and the National Conference on Street and Highway Safety (NCSHS) published a manual for urban streets in 1930. In the early years, the necessity for unification of the standards applicable to the different classes of road and street systems was obvious. To meet this need, a joint committee of AASHO and NCSHS developed and published the original edition of this Manual on Uniform Traffic Control Devices (MUTCD) in 1935. That committee, now called the National Committee on Uniform Traffic Control Devices (NCUTCD), though changed from time to time in name, organization, and personnel, has been in continuous existence and has contributed to periodic revisions of this Manual. The FHWA has administered the MUTCD since the 1971 edition. The FHWA and its predecessor organizations have participated in the development and publishing of the previous editions. There were nine previous editions of the MUTCD, and several of those editions were revised one or more times. Table I-1 traces the evolution of the MUTCD, including the two manuals developed by AASHO and NCSHS.

Standard:

07 The U.S. Secretary of Transportation, under authority granted by the Highway Safety Act of 1966, decreed that traffic control devices on all streets and highways open to public travel in accordance with 23 U.S.C. 109(d) and 402(a) in each State shall be in substantial conformance with the Standards issued or endorsed by the FHWA.

Support:

08 The "Uniform Vehicle Code (UVC)" is one of the publications referenced in the MUTCD. The UVC contains a model set of motor vehicle codes and traffic laws for use throughout the United States.

Guidance:

09 *The States should adopt Section 15-116 of the UVC, which states that, "No person shall install or maintain in any area of private property used by the public any sign, signal, marking, or other device intended to regulate, warn, or guide traffic unless it conforms with the State manual and specifications adopted under Section 15-104."*

Table I-1. Evolution of the MUTCD

Year	Name	Month / Year Revised
1927	Manual and Specifications for the Manufacture, Display, and Erection of U.S. Standard Road Markers and Signs (for rural roads)	4/29, 12/31
1930	Manual on Street Traffic Signs, Signals, and Markings (for urban streets)	No revisions
1935	Manual on Uniform Traffic Control Devices for Streets and Highways (MUTCD)	2/39
1942	Manual on Uniform Traffic Control Devices for Streets and Highways — War Emergency Edition	No revisions
1948	Manual on Uniform Traffic Control Devices for Streets and Highways	9/54
1961	Manual on Uniform Traffic Control Devices for Streets and Highways	No revisions
1971	Manual on Uniform Traffic Control Devices for Streets and Highways	11/71, 4/72, 3/73, 10/73, 6/74, 6/75, 9/76, 12/77
1978	Manual on Uniform Traffic Control Devices for Streets and Highways	12/79, 12/83, 9/84, 3/86
1988	Manual on Uniform Traffic Control Devices for Streets and Highways	1/90, 3/92, 9/93, 11/94, 12/96, 6/98, 1/00
2000	Manual on Uniform Traffic Control Devices for Streets and Highways — Millennium Edition	7/02
2003	Manual on Uniform Traffic Control Devices for Streets and Highways	11/04, 12/07
2009	Manual on Uniform Traffic Control Devices for Streets and Highways	

Support:

10   The Standard, Guidance, Option, and Support material described in this edition of the MUTCD provide the transportation professional with the information needed to make appropriate decisions regarding the use of traffic control devices on streets, highways, bikeways, and private roads open to public travel (see definition in Section 1A.13).

11   Throughout this Manual the headings Standard, Guidance, Option, and Support are used to classify the nature of the text that follows. Figures and tables, including the notes contained therein, supplement the text and might constitute a Standard, Guidance, Option, or Support. The user needs to refer to the appropriate text to classify the nature of the figure, table, or note contained therein.

Standard:

12   **When used in this Manual, the text headings of Standard, Guidance, Option, and Support shall be as defined in Paragraph 1 of Section 1A.13.**

Support:

13   Throughout this Manual all dimensions and distances are provided in English units. Appendix A2 contains tables for converting each of the English unit numerical values that are used in this Manual to the equivalent Metric (International System of Units) values.

Guidance:

14   *If Metric units are to be used in laying out distances or determining sizes of devices, such units should be specified on plan drawings and made known to those responsible for designing, installing, or maintaining traffic control devices.*

15   *Except when a specific numeral is required or recommended by the text of a Section of this Manual, numerals displayed on the images of devices in the figures that specify quantities such as times, distances, speed limits, and weights should be regarded as examples only. When installing any of these devices, the numerals should be appropriately altered to fit the specific situation.*

Support:

16   The following information will be useful when reference is being made to a specific portion of text in this Manual.

17   There are nine Parts in this Manual and each Part is comprised of one or more Chapters. Each Chapter is comprised of one or more Sections. Parts are given a numerical identification, such as Part 2 – Signs. Chapters are identified by the Part number and a letter, such as Chapter 2B – Regulatory Signs, Barricades, and Gates. Sections are identified by the Chapter number and letter followed by a decimal point and a number, such as Section 2B.03 – Size of Regulatory Signs.

18. Each Section is comprised of one or more paragraphs. The paragraphs are indented and are identified by a number. Paragraphs are counted from the beginning of each Section without regard to the intervening text headings (Standard, Guidance, Option, or Support). Some paragraphs have lettered or numbered items. As an example of how to cite this Manual, the phrase "Not less than 40 feet beyond the stop line" that appears in Section 4D.14 of this Manual would be referenced in writing as "Section 4D.14, P1, A.1," and would be verbally referenced as "Item A.1 of Paragraph 1 of Section 4D.14."

Standard:

19. **In accordance with 23 CFR 655.603(b)(3), States or other Federal agencies that have their own MUTCDs or Supplements shall revise these MUTCDs or Supplements to be in substantial conformance with changes to the National MUTCD within 2 years of the effective date of the Final Rule for the changes. Substantial conformance of such State or other Federal agency MUTCDs or Supplements shall be as defined in 23 CFR 655.603(b)(1).**

20. **After the effective date of a new edition of the MUTCD or a revision thereto, or after the adoption thereof by the State, whichever occurs later, new or reconstructed devices installed shall be in compliance with the new edition or revision.**

21. **In cases involving Federal-aid projects for new highway or bikeway construction or reconstruction, the traffic control devices installed (temporary or permanent) shall be in conformance with the most recent edition of the National MUTCD before that highway is opened or re-opened to the public for unrestricted travel [23 CFR 655.603(d)(2) and (d)(3)].**

22. **Unless a particular device is no longer serviceable, non-compliant devices on existing highways and bikeways shall be brought into compliance with the current edition of the National MUTCD as part of the systematic upgrading of substandard traffic control devices (and installation of new required traffic control devices) required pursuant to the Highway Safety Program, 23 U.S.C. §402(a). The FHWA has the authority to establish other target compliance dates for implementation of particular changes to the MUTCD [23 CFR 655.603(d)(1)]. These target compliance dates established by the FHWA shall be as shown in Table I-2.**

23. **Except as provided in Paragraph 24, when a non-compliant traffic control device is being replaced or refurbished because it is damaged, missing, or no longer serviceable for any reason, it shall be replaced with a compliant device.**

Option:

24. A damaged, missing, or otherwise non-serviceable device that is non-compliant may be replaced in kind if engineering judgment indicates that:
    A. One compliant device in the midst of a series of adjacent non-compliant devices would be confusing to road users; and/or
    B. The schedule for replacement of the whole series of non-compliant devices will result in achieving timely compliance with the MUTCD.

Table I-2. Target Compliance Dates Established by the FHWA

2009 MUTCD Section Number(s)	2009 MUTCD Section Title	Specific Provision	Compliance Date
2A.08	Maintaining Minimum Retroreflectivity	Implementation and continued use of an assessment or management method that is designed to maintain regulatory and warning sign retroreflectivity at or above the established minimum levels (see Paragraph 2)	2 years from the effective date of this revision of the 2009 MUTCD*
2A.19	Lateral Offset	Crashworthiness of sign supports on roads with posted speed limit of 50 mph or higher (see Paragraph 2)	January 17, 2013 (date established in the 2000 MUTCD)
2B.40	ONE WAY Signs (R6-1, R6-2)	New requirements in the 2009 MUTCD for the number and locations of ONE WAY signs (see Paragraphs 4, 9, and 10)	December 31, 2019
2C.06 through 2C.14	Horizontal Alignment Warning Signs	Revised requirements in the 2009 MUTCD regarding the use of various horizontal alignment signs (see Table 2C-5)	December 31, 2019
2E.31, 2E.33, and 2E.36	Plaques for Left-Hand Exits	New requirement in the 2009 MUTCD to use E1-5aP and E1-5bP plaques for left-hand exits	December 31, 2014
4D.26	Yellow Change and Red Clearance Intervals	New requirement in the 2009 MUTCD that durations of yellow change and red clearance intervals shall be determined using engineering practices (see Paragraphs 3 and 6)	5 years from the effective date of this revision of the 2009 MUTCD, or when timing adjustments are made to the individual intersection and/or corridor, whichever occurs first
4E.06	Pedestrian Intervals and Signal Phases	New requirement in the 2009 MUTCD that the pedestrian change interval shall not extend into the red clearance interval and shall be followed by a buffer interval of at least 3 seconds (see Paragraph 4)	5 years from the effective date of this revision of the 2009 MUTCD, or when timing adjustments are made to the individual intersection and/or corridor, whichever occurs first
6D.03**	Worker Safety Considerations	New requirement in the 2009 MUTCD that all workers within the right-of-way shall wear high-visibility apparel (see Paragraphs 4, 6, and 7)	December 31, 2011
6E.02**	High-Visibility Safety Apparel	New requirement in the 2009 MUTCD that all flaggers within the right-of-way shall wear high-visibility apparel	December 31, 2011
7D.04**	Uniform of Adult Crossing Guards	New requirement in the 2009 MUTCD for high-visibility apparel for adult crossing guards	December 31, 2011
8B.03, 8B.04	Grade Crossing (Crossbuck) Signs and Supports	Retroreflective strip on Crossbuck sign and support (see Paragraph 7 in Section 8B.03 and Paragraphs 15 and 18 in Section 8B.04)	December 31, 2019
8B.04	Crossbuck Assemblies with YIELD or STOP Signs at Passive Grade Crossings	New requirement in the 2009 MUTCD for the use of STOP or YIELD signs with Crossbuck signs at passive grade crossings	December 31, 2019

* Types of signs other than regulatory or warning are to be added to an agency's management or assessment method as resources allow.

** MUTCD requirement is a result of a legislative mandate.

Note: All compliance dates that were previously published in Table I-2 of the 2009 MUTCD and that do not appear in this revised table have been eliminated.

PART 1
GENERAL

CHAPTER 1A. GENERAL

Section 1A.01 Purpose of Traffic Control Devices

Support:

01 The purpose of traffic control devices, as well as the principles for their use, is to promote highway safety and efficiency by providing for the orderly movement of all road users on streets, highways, bikeways, and private roads open to public travel throughout the Nation.

02 Traffic control devices notify road users of regulations and provide warning and guidance needed for the uniform and efficient operation of all elements of the traffic stream in a manner intended to minimize the occurrences of crashes.

Standard:

03 **Traffic control devices or their supports shall not bear any advertising message or any other message that is not related to traffic control.**

Support:

04 Tourist-oriented directional signs and Specific Service signs are not considered advertising; rather, they are classified as motorist service signs.

Section 1A.02 Principles of Traffic Control Devices

Support:

01 This Manual contains the basic principles that govern the design and use of traffic control devices for all streets, highways, bikeways, and private roads open to public travel (see definition in Section 1A.13) regardless of type or class or the public agency, official, or owner having jurisdiction. This Manual's text specifies the restriction on the use of a device if it is intended for limited application or for a specific system. It is important that these principles be given primary consideration in the selection and application of each device.

Guidance:

02 *To be effective, a traffic control device should meet five basic requirements:*
   A. *Fulfill a need;*
   B. *Command attention;*
   C. *Convey a clear, simple meaning;*
   D. *Command respect from road users; and*
   E. *Give adequate time for proper response.*

03 *Design, placement, operation, maintenance, and uniformity are aspects that should be carefully considered in order to maximize the ability of a traffic control device to meet the five requirements listed in the previous paragraph. Vehicle speed should be carefully considered as an element that governs the design, operation, placement, and location of various traffic control devices.*

Support:

04 The definition of the word "speed" varies depending on its use. The definitions of specific speed terms are contained in Section 1A.13.

Guidance:

05 *The actions required of road users to obey regulatory devices should be specified by State statute, or in cases not covered by State statute, by local ordinance or resolution. Such statutes, ordinances, and resolutions should be consistent with the "Uniform Vehicle Code" (see Section 1A.11).*

06 *The proper use of traffic control devices should provide the reasonable and prudent road user with the information necessary to efficiently and lawfully use the streets, highways, pedestrian facilities, and bikeways.*

Support:

07 Uniformity of the meaning of traffic control devices is vital to their effectiveness. The meanings ascribed to devices in this Manual are in general accord with the publications mentioned in Section 1A.11.

Section 1A.03 Design of Traffic Control Devices

Guidance:

01 *Devices should be designed so that features such as size, shape, color, composition, lighting or retroreflection, and contrast are combined to draw attention to the devices; that size, shape, color, and simplicity of message combine to produce a clear meaning; that legibility and size combine with placement to permit adequate time for response; and that uniformity, size, legibility, and reasonableness of the message combine to command respect.*

02 *Aspects of a device's standard design should be modified only if there is a demonstrated need.*

Support:

03 An example of modifying a device's design would be to modify the Combination Horizontal Alignment/Intersection (W1-10) sign to show intersecting side roads on both sides rather than on just one side of the major road within the curve.

Option:

04 With the exception of symbols and colors, minor modifications in the specific design elements of a device may be made provided the essential appearance characteristics are preserved.

Section 1A.04 Placement and Operation of Traffic Control Devices

Guidance:

01 *Placement of a traffic control device should be within the road user's view so that adequate visibility is provided. To aid in conveying the proper meaning, the traffic control device should be appropriately positioned with respect to the location, object, or situation to which it applies. The location and legibility of the traffic control device should be such that a road user has adequate time to make the proper response in both day and night conditions.*

02 *Traffic control devices should be placed and operated in a uniform and consistent manner.*

03 *Unnecessary traffic control devices should be removed. The fact that a device is in good physical condition should not be a basis for deferring needed removal or change.*

Section 1A.05 Maintenance of Traffic Control Devices

Guidance:

01 *Functional maintenance of traffic control devices should be used to determine if certain devices need to be changed to meet current traffic conditions.*

02 *Physical maintenance of traffic control devices should be performed to retain the legibility and visibility of the device, and to retain the proper functioning of the device.*

Support:

03 Clean, legible, properly mounted devices in good working condition command the respect of road users.

Section 1A.06 Uniformity of Traffic Control Devices

Support:

01 Uniformity of devices simplifies the task of the road user because it aids in recognition and understanding, thereby reducing perception/reaction time. Uniformity assists road users, law enforcement officers, and traffic courts by giving everyone the same interpretation. Uniformity assists public highway officials through efficiency in manufacture, installation, maintenance, and administration. Uniformity means treating similar situations in a similar way. The use of uniform traffic control devices does not, in itself, constitute uniformity. A standard device used where it is not appropriate is as objectionable as a non-standard device; in fact, this might be worse, because such misuse might result in disrespect at those locations where the device is needed and appropriate.

Section 1A.07 Responsibility for Traffic Control Devices

Standard:

01 **The responsibility for the design, placement, operation, maintenance, and uniformity of traffic control devices shall rest with the public agency or the official having jurisdiction, or, in the case of private roads open to public travel, with the private owner or private official having jurisdiction. 23 CFR 655.603 adopts the MUTCD as the national standard for all traffic control devices installed on any street, highway, bikeway, or private road open to public travel (see definition in Section 1A.13). When a State or other Federal agency manual or supplement is required, that manual or supplement shall be in substantial conformance with the National MUTCD.**

02 **23 CFR 655.603 also states that traffic control devices on all streets, highways, bikeways, and private roads open to public travel in each State shall be in substantial conformance with standards issued or endorsed by the Federal Highway Administrator.**

Support:

03 The Introduction of this Manual contains information regarding the meaning of substantial conformance and the applicability of the MUTCD to private roads open to public travel.

04 The "Uniform Vehicle Code" (see Section 1A.11) has the following provision in Section 15-104 for the adoption of a uniform manual:

"(a) The [State Highway Agency] shall adopt a manual and specification for a uniform system of traffic control devices consistent with the provisions of this code for use upon highways within this State. Such uniform system shall correlate with and so far as possible conform to the system set forth in the most recent edition of the Manual on Uniform Traffic Control Devices for Streets and Highways, and other standards issued or endorsed by the Federal Highway Administrator."

"(b) The Manual adopted pursuant to subsection (a) shall have the force and effect of law."

05 All States have officially adopted the National MUTCD either in its entirety, with supplemental provisions, or as a separate published document.

Guidance:

06 *These individual State manuals or supplements should be reviewed for specific provisions relating to that State.*

Support:

07 The National MUTCD has also been adopted by the National Park Service, the U.S. Forest Service, the U.S. Military Command, the Bureau of Indian Affairs, the Bureau of Land Management, and the U.S. Fish and Wildlife Service.

Guidance:

08 *States should adopt Section 15-116 of the "Uniform Vehicle Code," which states that, "No person shall install or maintain in any area of private property used by the public any sign, signal, marking, or other device intended to regulate, warn, or guide traffic unless it conforms with the State manual and specifications adopted under Section 15-104."*

Section 1A.08 Authority for Placement of Traffic Control Devices

Standard:

01 **Traffic control devices, advertisements, announcements, and other signs or messages within the highway right-of-way shall be placed only as authorized by a public authority or the official having jurisdiction, or, in the case of private roads open to public travel, by the private owner or private official having jurisdiction, for the purpose of regulating, warning, or guiding traffic.**

02 **When the public agency or the official having jurisdiction over a street or highway or, in the case of private roads open to public travel, the private owner or private official having jurisdiction, has granted proper authority, others such as contractors and public utility companies shall be permitted to install temporary traffic control devices in temporary traffic control zones. Such traffic control devices shall conform with the Standards of this Manual.**

03 **All regulatory traffic control devices shall be supported by laws, ordinances, or regulations.**

Support:

04 Provisions of this Manual are based upon the concept that effective traffic control depends upon both appropriate application of the devices and reasonable enforcement of the regulations.

05 Although some highway design features, such as curbs, median barriers, guardrails, speed humps or tables, and textured pavement, have a significant impact on traffic operations and safety, they are not considered to be traffic control devices and provisions regarding their design and use are generally not included in this Manual.

06 Certain types of signs and other devices that do not have any traffic control purpose are sometimes placed within the highway right-of-way by or with the permission of the public agency or the official having jurisdiction over the street or highway. Most of these signs and other devices are not intended for use by road users in general, and their message is only important to individuals who have been instructed in their meanings. These signs and other devices are not considered to be traffic control devices and provisions regarding their design and use are not included in this Manual. Among these signs and other devices are the following:

   A. Devices whose purpose is to assist highway maintenance personnel. Examples include markers to guide snowplow operators, devices that identify culvert and drop inlet locations, and devices that precisely identify highway locations for maintenance or mowing purposes.
   B. Devices whose purpose is to assist fire or law enforcement personnel. Examples include markers that identify fire hydrant locations, signs that identify fire or water district boundaries, speed measurement pavement markings, small indicator lights to assist in enforcement of red light violations, and photo enforcement systems.
   C. Devices whose purpose is to assist utility company personnel and highway contractors, such as markers that identify underground utility locations.
   D. Signs posting local non-traffic ordinances.
   E. Signs giving civic organization meeting information.

Standard:

07 **Signs and other devices that do not have any traffic control purpose that are placed within the highway right-of-way shall not be located where they will interfere with, or detract from, traffic control devices.**

Guidance:

08 *Any unauthorized traffic control device or other sign or message placed on the highway right-of-way by a private organization or individual constitutes a public nuisance and should be removed. All unofficial or non-essential traffic control devices, signs, or messages should be removed.*

Section 1A.09 Engineering Study and Engineering Judgment

Support:

01 Definitions of an engineering study and engineering judgment are contained in Section 1A.13.

Standard:

02 **This Manual describes the application of traffic control devices, but shall not be a legal requirement for their installation.**

Guidance:

03 *The decision to use a particular device at a particular location should be made on the basis of either an engineering study or the application of engineering judgment. Thus, while this Manual provides Standards, Guidance, and Options for design and applications of traffic control devices, this Manual should not be considered a substitute for engineering judgment. Engineering judgment should be exercised in the selection and application of traffic control devices, as well as in the location and design of roads and streets that the devices complement.*

04 *Early in the processes of location and design of roads and streets, engineers should coordinate such location and design with the design and placement of the traffic control devices to be used with such roads and streets.*

05 *Jurisdictions, or owners of private roads open to public travel, with responsibility for traffic control that do not have engineers on their staffs who are trained and/or experienced in traffic control devices should seek engineering assistance from others, such as the State transportation agency, their county, a nearby large city, or a traffic engineering consultant.*

Support:

06 As part of the Federal-aid Program, each State is required to have a Local Technology Assistance Program (LTAP) and to provide technical assistance to local highway agencies. Requisite technical training in the application of the principles of the MUTCD is available from the State's Local Technology Assistance Program for needed engineering guidance and assistance.

Section 1A.10 Interpretations, Experimentations, Changes, and Interim Approvals

Standard:

01 **Design, application, and placement of traffic control devices other than those adopted in this Manual shall be prohibited unless the provisions of this Section are followed.**

Support:

02 Continuing advances in technology will produce changes in the highway, vehicle, and road user proficiency; therefore, portions of the system of traffic control devices in this Manual will require updating. In addition, unique situations often arise for device applications that might require interpretation or clarification of this Manual. It is important to have a procedure for recognizing these developments and for introducing new ideas and modifications into the system.

Standard:

03 **Except as provided in Paragraph 4, requests for any interpretation, permission to experiment, interim approval, or change shall be submitted electronically to the Federal Highway Administration (FHWA), Office of Transportation Operations, MUTCD team, at the following e-mail address: MUTCDofficialrequest@dot.gov.**

Option:

04 If electronic submittal is not possible, requests for interpretations, permission to experiment, interim approvals, or changes may instead be mailed to the Office of Transportation Operations, HOTO-1, Federal Highway Administration, 1200 New Jersey Avenue, SE, Washington, DC 20590.

Support:

05 Communications regarding other MUTCD matters that are not related to official requests will receive quicker attention if they are submitted electronically to the MUTCD Team Leader or to the appropriate individual MUTCD team member. Their e-mail addresses are available through the links contained on the "Who's Who" page on the MUTCD website at http://mutcd.fhwa.dot.gov/team.htm.

06 An interpretation includes a consideration of the application and operation of standard traffic control devices, official meanings of standard traffic control devices, or the variations from standard device designs.

Guidance:

07 *Requests for an interpretation of this Manual should contain the following information:*
   A. *A concise statement of the interpretation being sought;*
   B. *A description of the condition that provoked the need for an interpretation;*
   C. *Any illustration that would be helpful to understand the request; and*
   D. *Any supporting research data that is pertinent to the item to be interpreted.*

Support:

08 Requests to experiment include consideration of field deployment for the purpose of testing or evaluating a new traffic control device, its application or manner of use, or a provision not specifically described in this Manual.

09 A request for permission to experiment will be considered only when submitted by the public agency or toll facility operator responsible for the operation of the road or street on which the experiment is to take place. For a private road open to public travel, the request will be considered only if it is submitted by the private owner or private official having jurisdiction.

10 A diagram indicating the process for experimenting with traffic control devices is shown in Figure 1A-1.

Figure 1A-1. Process for Requesting and Conducting Experimentations for New Traffic Control Devices

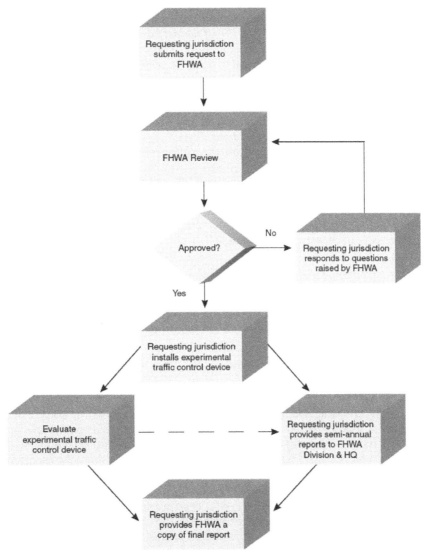

Guidance:

11 The request for permission to experiment should contain the following:
   A. A statement indicating the nature of the problem.
   B. A description of the proposed change to the traffic control device or application of the traffic control device, how it was developed, the manner in which it deviates from the standard, and how it is expected to be an improvement over existing standards.
   C. Any illustration that would be helpful to understand the traffic control device or use of the traffic control device.
   D. Any supporting data explaining how the traffic control device was developed, if it has been tried, in what ways it was found to be adequate or inadequate, and how this choice of device or application was derived.
   E. A legally binding statement certifying that the concept of the traffic control device is not protected by a patent or copyright. (An example of a traffic control device concept would be countdown pedestrian signals in general. Ordinarily an entire general concept would not be patented or copyrighted, but if it were it would not be acceptable for experimentation unless the patent or copyright owner signs a waiver of rights acceptable to the FHWA. An example of a patented or copyrighted specific device within the general concept of countdown pedestrian signals would be a manufacturer's design for its specific brand of countdown signal, including the design details of the housing or electronics that are unique to that manufacturer's product. As long as the general concept is not patented or copyrighted, it is acceptable for experimentation to incorporate the use of one or more patented devices of one or several manufacturers.)
   F. The time period and location(s) of the experiment.
   G. A detailed research or evaluation plan that must provide for close monitoring of the experimentation, especially in the early stages of its field implementation. The evaluation plan should include before and after studies as well as quantitative data describing the performance of the experimental device.
   H. An agreement to restore the site of the experiment to a condition that complies with the provisions of this Manual within 3 months following the end of the time period of the experiment. This agreement must also provide that the agency sponsoring the experimentation will terminate the experimentation at any time that it determines significant safety concerns are directly or indirectly attributable to the experimentation. The FHWA's Office of Transportation Operations has the right to terminate approval of the experimentation at any time if there is an indication of safety concerns. If, as a result of the experimentation, a request is made that this Manual be changed to include the device or application being experimented with, the device or application will be permitted to remain in place until an official rulemaking action has occurred.
   I. An agreement to provide semi-annual progress reports for the duration of the experimentation, and an agreement to provide a copy of the final results of the experimentation to the FHWA's Office of Transportation Operations within 3 months following completion of the experimentation. The FHWA's Office of Transportation Operations has the right to terminate approval of the experimentation if reports are not provided in accordance with this schedule.

Support:

12 A change includes consideration of a new device to replace a present standard device, an additional device to be added to the list of standard devices, or a revision to a traffic control device application or placement criteria.

Guidance:

13 *Requests for a change to this Manual should contain the following information:*
   A. *A statement indicating what change is proposed;*
   B. *Any illustration that would be helpful to understand the request; and*
   C. *Any supporting research data that is pertinent to the item to be reviewed.*

Support:

14 Interim approval allows interim use, pending official rulemaking, of a new traffic control device, a revision to the application or manner of use of an existing traffic control device, or a provision not specifically described in this Manual. The FHWA issues an Interim Approval by official memorandum signed by the Associate Administrator for Operations and posts this memorandum on the MUTCD website. the issuance by FHWA of an interim approval will typically result in the traffic control device or application being placed into the next scheduled rulemaking process for revisions to this Manual.

15 Interim approval is considered based on the results of successful experimentation, results of analytical or laboratory studies, and/or review of non-U.S. experience with a traffic control device or application. Interim approval considerations include an assessment of relative risks, benefits, costs, impacts, and other factors.

16 Interim approval allows for optional use of a traffic control device or application and does not create a new mandate or recommendation for use. Interim approval includes conditions that jurisdictions agree to comply with in order to use the traffic control device or application until an official rulemaking action has occurred.

Standard:

17 **A jurisdiction, toll facility operator, or owner of a private road open to public travel that desires to use a traffic control device for which FHWA has issued an interim approval shall request permission from FHWA.**

Guidance:

18 *The request for permission to place a traffic control device under an interim approval should contain the following:*
   A. *A description of where the device will be used, such as a list of specific locations or highway segments or types of situations, or a statement of the intent to use the device jurisdiction-wide;*
   B. *An agreement to abide by the specific conditions for use of the device as contained in the FHWA's interim approval document;*
   C. *An agreement to maintain and continually update a list of locations where the device has been installed; and*
   D. *An agreement to:*
      1. *Restore the site(s) of the interim approval to a condition that complies with the provisions in this Manual within 3 months following the issuance of a Final Rule on this traffic control device; and*
      2. *Terminate use of the device or application installed under the interim approval at any time that it determines significant safety concerns are directly or indirectly attributable to the device or application. The FHWA's Office of Transportation Operations has the right to terminate the interim approval at any time if there is an indication of safety concerns.*

Option:

19 A State may submit a request for the use of a device under interim approval for all jurisdictions in that State, as long as the request contains the information listed in Paragraph 18.

Guidance:

20 *A local jurisdiction, toll facility operator, or owner of a private road open to public travel using a traffic control device or application under an interim approval that was granted by FHWA either directly or on a statewide basis based on the State's request should inform the State of the locations of such use.*

21 *A local jurisdiction, toll facility operator, or owner of a private road open to public travel that is requesting permission to experiment or permission to use a device or application under an interim approval should first check for any State laws and/or directives covering the application of the MUTCD provisions that might exist in their State.*

Option:

22 A device or application installed under an interim approval may remain in place, under the conditions established in the interim approval, until an official rulemaking action has occurred.

Support:

23 A diagram indicating the process for incorporating new traffic control devices into this Manual is shown in Figure 1A-2.

24 For additional information concerning interpretations, experimentation, changes, or interim approvals, visit the MUTCD website at http://mutcd.fhwa.dot.gov.

Section 1A.11 Relation to Other Publications

Standard:

01 **To the extent that they are incorporated by specific reference, the latest editions of the following publications, or those editions specifically noted, shall be a part of this Manual: "Standard Highway Signs and Markings" book (FHWA); and "Color Specifications for Retroreflective Sign and Pavement Marking Materials" (appendix to subpart F of Part 655 of Title 23 of the Code of Federal Regulations).**

Support:

02 The "Standard Highway Signs and Markings" book includes standard alphabets and symbols and arrows for signs and pavement markings.

03 For information about the publications mentioned in Paragraph 1, visit the Federal Highway Administration's MUTCD website at http://mutcd.fhwa.dot.gov, or write to the FHWA, 1200 New Jersey Avenue, SE, HOTO, Washington, DC 20590.

Figure 1A-2. Process for Incorporating New Traffic Control Devices into the MUTCD

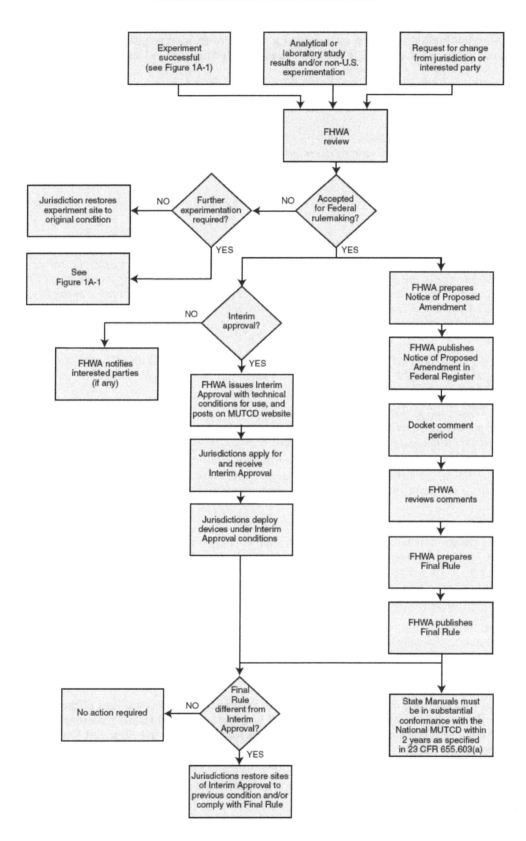

₀₄ Other publications that are useful sources of information with respect to the use of this Manual are listed in this paragraph. See Page i of this Manual for ordering information for the following publications (later editions might also be available as useful sources of information):

1. "AAA School Safety Patrol Operations Manual," 2006 Edition (American Automobile Association—AAA)
2. "A Policy on Geometric Design of Highways and Streets," 2004 Edition (American Association of State Highway and Transportation Officials—AASHTO)
3. "Guide for the Development of Bicycle Facilities," 1999 Edition (AASHTO)
4. "Guide for the Planning, Design, and Operation of Pedestrian Facilities," 2004 Edition (AASHTO)
5. "Guide to Metric Conversion," 1993 Edition (AASHTO)
6. "Guidelines for the Selection of Supplemental Guide Signs for Traffic Generators Adjacent to Freeways," 4th Edition/Guide Signs, Part II: Guidelines for Airport Guide Signing/Guide Signs, Part III: List of Control Cities for Use in Guide Signs on Interstate Highways," Item Code: GSGLC-4, 2001 Edition (AASHTO)
7. "Roadside Design Guide," 2006 Edition (AASHTO)
8. "Standard Specifications for Movable Highway Bridges," 1988 Edition (AASHTO)
9. "Traffic Engineering Metric Conversion Folders—Addendum to the Guide to Metric Conversion," 1993 Edition (AASHTO)
10. "2009 AREMA Communications & Signals Manual," (American Railway Engineering & Maintenance-of-Way Association—AREMA)
11. "Changeable Message Sign Operation and Messaging Handbook (FHWA-OP-03-070)," 2004 Edition (Federal Highway Administration—FHWA)
12. "Designing Sidewalks and Trails for Access—Part 2—Best Practices Design Guide (FHWA-EP-01-027)," 2001 Edition (FHWA)
13. "Federal-Aid Highway Program Guidance on High Occupancy Vehicle (HOV) Lanes," 2001 (FHWA)
14. "Maintaining Traffic Sign Retroreflectivity," 2007 Edition (FHWA)
15. "Railroad-Highway Grade Crossing Handbook—Revised Second Edition (FHWA-SA-07-010)," 2007 Edition (FHWA)
16. "Ramp Management and Control Handbook (FHWA-HOP-06-001)," 2006 Edition (FHWA)
17. "Roundabouts-An Informational Guide (FHWA-RD-00-067)," 2000 Edition (FHWA)
18. "Signal Timing Manual (FHWA-HOP-08-024)," 2008 Edition (FHWA)
19. "Signalized Intersections: an Informational Guide (FHWA-HRT-04-091)," 2004 Edition (FHWA)
20. "Travel Better, Travel Longer: A Pocket Guide to Improving Traffic Control and Mobility for Our Older Population (FHWA-OP-03-098)," 2003 Edition (FHWA)
21. "Practice for Roadway Lighting," RP-8, 2001 (Illuminating Engineering Society—IES)
22. "Safety Guide for the Prevention of Radio Frequency Radiation Hazards in the Use of Commercial Electric Detonators (Blasting Caps)," Safety Library Publication No. 20, July 2001 Edition (Institute of Makers of Explosives)
23. "American National Standard for High-Visibility Public Safety Vests," (ANSI/ISEA 207-2006), 2006 Edition (International Safety Equipment Association—ISEA)
24. "American National Standard for High-Visibility Safety Apparel and Headwear," (ANSI/ISEA 107-2004), 2004 Edition (ISEA)
25. "Manual of Traffic Signal Design," 1998 Edition (Institute of Transportation Engineers—ITE)
26. "Manual of Transportation Engineering Studies," 1994 Edition (ITE)
27. "Pedestrian Traffic Control Signal Indications," Part 1—1985 Edition; Part 2 (LED Pedestrian Traffic Signal Modules)—2004 Edition (ITE)
28. "Preemption of Traffic Signals Near Railroad Crossings," 2006 Edition (ITE)
29. "Purchase Specification for Flashing and Steady Burn Warning Lights," 1981 Edition (ITE)
30. "Traffic Control Devices Handbook," 2001 Edition (ITE)
31. "Traffic Detector Handbook," 1991 Edition (ITE)
32. "Traffic Engineering Handbook," 2009 Edition (ITE)
33. "Traffic Signal Lamps," 1980 Edition (ITE)
34. "Vehicle Traffic Control Signal Heads," Part 1—1985 Edition; Part 2 (LED Circular Signal Supplement)—2005 Edition; Part 3 (LED Vehicular Arrow Traffic Signal Supplement)—2004 Edition (ITE)
35. "Uniform Vehicle Code (UVC) and Model Traffic Ordinance," 2000 Edition (National Committee on Uniform Traffic Laws and Ordinances—NCUTLO)
36. "NEMA Standards Publication TS 4-2005 Hardware Standards for Dynamic Message Signs (DMS) With NTCIP Requirements," 2005 Edition (National Electrical Manufacturers Association—NEMA)
37. "Occupational Safety and Health Administration Regulations (Standards - 29 CFR), General Safety and Health Provisions - 1926.20," amended June 30, 1993 (Occupational Safety and Health Administration—OSHA)
38. "Accessible Pedestrian Signals—A Guide to Best Practices (NCHRP Web-Only Document 117A)," 2008 Edition (Transportation Research Board—TRB)

39. "Guidelines for Accessible Pedestrian Signals (NCHRP Web-Only Document 117B)," 2008 Edition (TRB)
40. "Highway Capacity Manual," 2000 Edition (TRB)
41. "Recommended Procedures for the Safety Performance Evaluation of Highway Features," (NCHRP Report 350), 1993 Edition (TRB)
42. "The Americans with Disabilities Act Accessibility Guidelines for Buildings and Facilities (ADAAG)," July 1998 Edition (The U.S. Access Board)

Section 1A.12 Color Code

Support:

01 The following color code establishes general meanings for 11 colors of a total of 13 colors that have been identified as being appropriate for use in conveying traffic control information. tolerance limits for each color are contained in 23 CFR Part 655, Appendix to Subpart F and are available at the Federal Highway Administration's MUTCD website at http://mutcd.fhwa.dot.gov or by writing to the FHWA, Office of Safety Research and Development (HRD-T-301), 6300 Georgetown Pike, McLean, VA 22101.

02 The two colors for which general meanings have not yet been assigned are being reserved for future applications that will be determined only by FHWA after consultation with the States, the engineering community, and the general public. The meanings described in this Section are of a general nature. More specific assignments of colors are given in the individual Parts of this Manual relating to each class of devices.

Standard:

03 **The general meaning of the 13 colors shall be as follows:**
   A. **Black—regulation**
   B. **Blue—road user services guidance, tourist information, and evacuation route**
   C. **Brown—recreational and cultural interest area guidance**
   D. **Coral—unassigned**
   E. **Fluorescent Pink—incident management**
   F. **Fluorescent Yellow-Green—pedestrian warning, bicycle warning, playground warning, school bus and school warning**
   G. **Green—indicated movements permitted, direction guidance**
   H. **Light Blue—unassigned**
   I. **Orange—temporary traffic control**
   J. **Purple—lanes restricted to use only by vehicles with registered electronic toll collection (ETC) accounts**
   K. **Red—stop or prohibition**
   L. **White—regulation**
   M. **Yellow—warning**

Section 1A.13 Definitions of Headings, Words, and Phrases in this Manual

Standard:

01 **When used in this Manual, the text headings of Standard, Guidance, Option, and Support shall be defined as follows:**

   A. **Standard—a statement of required, mandatory, or specifically prohibitive practice regarding a traffic control device. All Standard statements are labeled, and the text appears in bold type. The verb "shall" is typically used. The verbs "should" and "may" are not used in Standard statements. Standard statements are sometimes modified by Options.**

   B. **Guidance—a statement of recommended, but not mandatory, practice in typical situations, with deviations allowed if engineering judgment or engineering study indicates the deviation to be appropriate. All Guidance statements are labeled, and the text appears in unbold type. The verb "should" is typically used. The verbs "shall" and "may" are not used in Guidance statements. Guidance statements are sometimes modified by Options.**

   C. **Option—a statement of practice that is a permissive condition and carries no requirement or recommendation. Option statements sometime contain allowable modifications to a Standard or Guidance statement. All Option statements are labeled, and the text appears in unbold type. The verb "may" is typically used. The verbs "shall" and "should" are not used in Option statements.**

   D. **Support—an informational statement that does not convey any degree of mandate, recommendation, authorization, prohibition, or enforceable condition. Support statements are labeled, and the text appears in unbold type. The verbs "shall," "should," and "may" are not used in Support statements.**

⁰² Unless otherwise defined in this Section, or in other Parts of this Manual, words or phrases shall have the meaning(s) as defined in the most recent editions of the "Uniform Vehicle Code," "AASHTO Transportation Glossary (Highway Definitions)," and other publications mentioned in Section 1A.11.

⁰³ The following words and phrases, when used in this Manual, shall have the following meanings:

1. **Accessible Pedestrian Signal**—a device that communicates information about pedestrian signal timing in non-visual format such as audible tones, speech messages, and/or vibrating surfaces.
2. **Accessible Pedestrian Signal Detector**—a device designated to assist the pedestrian who has visual or physical disabilities in activating the pedestrian phase.
3. **Active Grade Crossing Warning System**—the flashing-light signals, with or without warning gates, together with the necessary control equipment used to inform road users of the approach or presence of rail traffic at grade crossings.
4. **Actuated Operation**—a type of traffic control signal operation in which some or all signal phases are operated on the basis of actuation.
5. **Actuation**—initiation of a change in or extension of a traffic signal phase through the operation of any type of detector.
6. **Advance Preemption**—the notification of approaching rail traffic that is forwarded to the highway traffic signal controller unit or assembly by the railroad or light rail transit equipment in advance of the activation of the railroad or light rail transit warning devices.
7. **Advance Preemption Time**—the period of time that is the difference between the required maximum highway traffic signal preemption time and the activation of the railroad or light rail transit warning devices.
8. **Advisory Speed**—a recommended speed for all vehicles operating on a section of highway and based on the highway design, operating characteristics, and conditions.
9. **Alley**—a street or highway intended to provide access to the rear or side of lots or buildings in urban areas and not intended for the purpose of through vehicular traffic.
10. **Altered Speed Zone**—a speed limit, other than a statutory speed limit, that is based upon an engineering study.
11. **Approach**—all lanes of traffic moving toward an intersection or a midblock location from one direction, including any adjacent parking lane(s).
12. **Arterial Highway (Street)**—a general term denoting a highway primarily used by through traffic, usually on a continuous route or a highway designated as part of an arterial system.
13. **Attended Lane (Manual Lane)**—a toll lane adjacent to a toll booth occupied by a human toll collector who makes change, issues receipts, and perform other toll-related functions. Attended lanes at toll plazas typically require vehicles to stop to pay the toll.
14. **Automatic Lane**—see Exact Change Lane.
15. **Average Annual Daily Traffic (AADT)**—the total volume of traffic passing a point or segment of a highway facility in both directions for one year divided by the number of days in the year. Normally, periodic daily traffic volumes are adjusted for hours of the day counted, days of the week, and seasons of the year to arrive at average annual daily traffic.
16. **Average Daily Traffic (ADT)**—the average 24 hour volume, being the total volume during a stated period divided by the number of days in that period. Normally, this would be periodic daily traffic volumes over several days, not adjusted for days of the week or seasons of the year.
17. **Average Day**—a day representing traffic volumes normally and repeatedly found at a location, typically a weekday when volumes are influenced by employment or a weekend day when volumes are influenced by entertainment or recreation.
18. **Backplate**—see Signal Backplate.
19. **Barrier-Separated Lane**—a preferential lane or other special purpose lane that is separated from the adjacent general-purpose lane(s) by a physical barrier.
20. **Beacon**—a highway traffic signal with one or more signal sections that operates in a flashing mode.
21. **Bicycle**—a pedal-powered vehicle upon which the human operator sits.
22. **Bicycle Facilities**—a general term denoting improvements and provisions that accommodate or encourage bicycling, including parking and storage facilities, and shared roadways not specifically defined for bicycle use.
23. **Bicycle Lane**—a portion of a roadway that has been designated for preferential or exclusive use by bicyclists by pavement markings and, if used, signs.
24. **Bikeway**—a generic term for any road, street, path, or way that in some manner is specifically designated for bicycle travel, regardless of whether such facilities are designated for the exclusive use of bicycles or are to be shared with other transportation modes.

25. **Buffer-Separated Lane**—a preferential lane or other special purpose lane that is separated from the adjacent general-purpose lane(s) by a pattern of standard longitudinal pavement markings that is wider than a normal or wide lane line marking. The buffer area might include rumble strips, textured pavement, or channelizing devices such as tubular markers or traversable curbs, but does not include a physical barrier.
26. **Cantilevered Signal Structure**—a structure, also referred to as a mast arm, that is rigidly attached to a vertical pole and is used to provide overhead support of highway traffic signal faces or grade crossing signal units.
27. **Center Line Markings**—the yellow pavement marking line(s) that delineates the separation of traffic lanes that have opposite directions of travel on a roadway. These markings need not be at the geometrical center of the pavement.
28. **Changeable Message Sign**—a sign that is capable of displaying more than one message (one of which might be a "blank" display), changeable manually, by remote control, or by automatic control. Electronic-display changeable message signs are referred to as Dynamic Message Signs in the National Intelligent Transportation Systems (ITS) Architecture and are referred to as Variable Message Signs in the National Electrical Manufacturers Association (NEMA) standards publication.
29. **Channelizing Line Markings**—a wide or double solid white line used to form islands where traffic in the same direction of travel is permitted on both sides of the island.
30. **Circular Intersection**—an intersection that has an island, generally circular in design, located in the center of the intersection where traffic passes to the right of the island. Circular intersections include roundabouts, rotaries, and traffic circles.
31. **Circulatory Roadway**—the roadway within a circular intersection on which traffic travels in a counterclockwise direction around an island in the center of the circular intersection.
32. **Clear Storage Distance**—when used in Part 8, the distance available for vehicle storage measured between 6 feet from the rail nearest the intersection to the intersection stop line or the normal stopping point on the highway. At skewed grade crossings and intersections, the 6-foot distance shall be measured perpendicular to the nearest rail either along the center line or edge line of the highway, as appropriate, to obtain the shorter distance. Where exit gates are used, the distance available for vehicle storage is measured from the point where the rear of the vehicle would be clear of the exit gate arm. In cases where the exit gate arm is parallel to the track(s) and is not perpendicular to the highway, the distance is measured either along the center line or edge line of the highway, as appropriate, to obtain the shorter distance.
33. **Clear Zone**—the total roadside border area, starting at the edge of the traveled way, that is available for an errant driver to stop or regain control of a vehicle. This area might consist of a shoulder, a recoverable slope, and/or a non-recoverable, traversable slope with a clear run-out area at its toe.
34. **Collector Highway**—a term denoting a highway that in rural areas connects small towns and local highways to arterial highways, and in urban areas provides land access and traffic circulation within residential, commercial, and business areas and connects local highways to the arterial highways.
35. **Concurrent Flow Preferential Lane**—a preferential lane that is operated in the same direction as the adjacent mixed flow lanes, separated from the adjacent general-purpose freeway lanes by a standard lane stripe, painted buffer, or barrier.
36. **Conflict Monitor**—a device used to detect and respond to improper or conflicting signal indications and improper operating voltages in a traffic controller assembly.
37. **Constant Warning Time Detection**—a means of detecting rail traffic that provides relatively uniform warning time for the approach of trains or light rail transit traffic that are not accelerating or decelerating after being detected.
38. **Contiguous Lane**—a lane, preferential or otherwise, that is separated from the adjacent lane(s) only by a normal or wide lane line marking.
39. **Controller Assembly**—a complete electrical device mounted in a cabinet for controlling the operation of a highway traffic signal.
40. **Controller Unit**—that part of a controller assembly that is devoted to the selection and timing of the display of signal indications.
41. **Conventional Road**—a street or highway other than a low-volume road (as defined in Section 5A.01), expressway, or freeway.
42. **Counter-Flow Lane**—a lane operating in a direction opposite to the normal flow of traffic designated for peak direction of travel during at least a portion of the day. Counter-flow lanes are usually separated from the off-peak direction lanes by tubular markers or other flexible channelizing devices, temporary lane separators, or movable or permanent barrier.

43. Crashworthy—a characteristic of a roadside appurtenance that has been successfully crash tested in accordance with a national standard such as the National Cooperative Highway Research Program Report 350, "Recommended Procedures for the Safety Performance Evaluation of Highway Features."
44. Crosswalk—(a) that part of a roadway at an intersection included within the connections of the lateral lines of the sidewalks on opposite sides of the highway measured from the curbs or in the absence of curbs, from the edges of the traversable roadway, and in the absence of a sidewalk on one side of the roadway, the part of a roadway included within the extension of the lateral lines of the sidewalk at right angles to the center line; (b) any portion of a roadway at an intersection or elsewhere distinctly indicated as a pedestrian crossing by pavement marking lines on the surface, which might be supplemented by contrasting pavement texture, style, or color.
45. Crosswalk Lines—white pavement marking lines that identify a crosswalk.
46. Cycle Length—the time required for one complete sequence of signal indications.
47. Dark Mode—the lack of all signal indications at a signalized location. (The dark mode is most commonly associated with power failures, ramp meters, hybrid beacons, beacons, and some movable bridge signals.)
48. Delineator—a retroreflective device mounted on the roadway surface or at the side of the roadway in a series to indicate the alignment of the roadway, especially at night or in adverse weather.
49. Design Vehicle—the longest vehicle permitted by statute of the road authority (State or other) on that roadway.
50. Designated Bicycle Route—a system of bikeways designated by the jurisdiction having authority with appropriate directional and informational route signs, with or without specific bicycle route numbers.
51. Detectable—having a continuous edge within 6 inches of the surface so that pedestrians who have visual disabilities can sense its presence and receive usable guidance information.
52. Detector—a device used for determining the presence or passage of vehicles or pedestrians.
53. Downstream—a term that refers to a location that is encountered by traffic subsequent to an upstream location as it flows in an "upstream to downstream" direction. For example, "the downstream end of a lane line separating the turn lane from a through lane on the approach to an intersection" is the end of the lane line that is closest to the intersection.
54. Dropped Lane—a through lane that becomes a mandatory turn lane on a conventional roadway, or a through lane that becomes a mandatory exit lane on a freeway or expressway. The end of an acceleration lane and reductions in the number of through lanes that do not involve a mandatory turn or exit are not considered dropped lanes.
55. Dual-Arrow Signal Section—a type of signal section designed to include both a yellow arrow and a green arrow.
56. Dynamic Envelope—the clearance required for light rail transit traffic or a train and its cargo overhang due to any combination of loading, lateral motion, or suspension failure (see Figure 8B-8).
57. Dynamic Exit Gate Operating Mode—a mode of operation where the exit gate operation is based on the presence of vehicles within the minimum track clearance distance.
58. Edge Line Markings—white or yellow pavement marking lines that delineate the right or left edge(s) of a traveled way.
59. Electronic Toll Collection (ETC)—a system for automated collection of tolls from moving or stopped vehicles through wireless technologies such as radio-frequency communication or optical scanning. ETC systems are classified as one of the following: (1) systems that require users to have registered toll accounts, with the use of equipment inside or on the exterior of vehicles, such as a transponder or barcode decal, that communicates with or is detected by roadside or overhead receiving equipment, or with the use of license plate optical scanning, to automatically deduct the toll from the registered user account, or (2) systems that do not require users to have registered toll accounts because vehicle license plates are optically scanned and invoices for the toll amount are sent through postal mail to the address of the vehicle owner.
60. Electronic Toll Collection (ETC) Account-Only Lane—a non-attended toll lane that is restricted to use only by vehicles with a registered toll payment account.
61. Emergency-Vehicle Hybrid Beacon—a special type of hybrid beacon used to warn and control traffic at an unsignalized location to assist authorized emergency vehicles in entering or crossing a street or highway.
62. Emergency-Vehicle Traffic Control Signal—a special traffic control signal that assigns the right-of-way to an authorized emergency vehicle.
63. End-of-Roadway Marker—a device used to warn and alert road users of the end of a roadway in other than temporary traffic control zones.

64. Engineering Judgment—the evaluation of available pertinent information, and the application of appropriate principles, provisions, and practices as contained in this Manual and other sources, for the purpose of deciding upon the applicability, design, operation, or installation of a traffic control device. Engineering judgment shall be exercised by an engineer, or by an individual working under the supervision of an engineer, through the application of procedures and criteria established by the engineer. Documentation of engineering judgment is not required.
65. Engineering Study—the comprehensive analysis and evaluation of available pertinent information, and the application of appropriate principles, provisions, and practices as contained in this Manual and other sources, for the purpose of deciding upon the applicability, design, operation, or installation of a traffic control device. An engineering study shall be performed by an engineer, or by an individual working under the supervision of an engineer, through the application of procedures and criteria established by the engineer. An engineering study shall be documented.
66. Entrance Gate—an automatic gate that can be lowered across the lanes approaching a grade crossing to block road users from entering the grade crossing.
67. Exact Change Lane (Automatic Lane)—a non-attended toll lane that has a receptacle into which road users deposit coins totaling the exact amount of the toll. Exact Change lanes at toll plazas typically require vehicles to stop to pay the toll.
68. Exit Gate—an automatic gate that can be lowered across the lanes departing a grade crossing to block road users from entering the grade crossing by driving in the opposing traffic lanes.
69. Exit Gate Clearance Time—for Four-Quadrant Gate systems at grade crossings, the amount of time provided to delay the descent of the exit gate arm(s) after entrance gate arm(s) begin to descend.
70. Exit Gate Operating Mode—for Four-Quadrant Gate systems at grade crossings, the mode of control used to govern the operation of the exit gate arms.
71. Expressway—a divided highway with partial control of access.
72. Flagger—a person who actively controls the flow of vehicular traffic into and/or through a temporary traffic control zone using hand-signaling devices or an Automated Flagger Assistance Device (AFAD).
73. Flasher—a device used to turn highway traffic signal indications on and off at a repetitive rate of approximately once per second.
74. Flashing—an operation in which a light source, such as a traffic signal indication, is turned on and off repetitively.
75. Flashing-Light Signals—a warning device consisting of two red signal indications arranged horizontally that are activated to flash alternately when rail traffic is approaching or present at a grade crossing.
76. Flashing Mode—a mode of operation in which at least one traffic signal indication in each vehicular signal face of a highway traffic signal is turned on and off repetitively.
77. Freeway—a divided highway with full control of access.
78. Full-Actuated Operation—a type of traffic control signal operation in which all signal phases function on the basis of actuation.
79. Gate—an automatically-operated or manually-operated traffic control device that is used to physically obstruct road users such that they are discouraged from proceeding past a particular point on a roadway or pathway, or such that they are discouraged from entering a particular grade crossing, ramp, lane, roadway, or facility.
80. Grade Crossing—the general area where a highway and a railroad and/or light rail transit route cross at the same level, within which are included the tracks, highway, and traffic control devices for traffic traversing that area.
81. Guide Sign—a sign that shows route designations, destinations, directions, distances, services, points of interest, or other geographical, recreational, or cultural information.
82. High-Occupancy Vehicle (HOV)—a motor vehicle carrying at least two or more persons, including carpools, vanpools, and buses.
83. Highway—a general term for denoting a public way for purposes of vehicular travel, including the entire area within the right-of-way.
84. Highway-Light Rail Transit Grade Crossing—the general area where a highway and a light rail transit route cross at the same level, within which are included the light rail transit tracks, highway, and traffic control devices for traffic traversing that area.
85. Highway-Rail Grade Crossing—the general area where a highway and a railroad cross at the same level, within which are included the railroad tracks, highway, and traffic control devices for highway traffic traversing that area.

86. Highway Traffic Signal—a power-operated traffic control device by which traffic is warned or directed to take some specific action. These devices do not include power-operated signs, steadily-illuminated pavement markers, warning lights (see Section 6F.83), or steady burning electric lamps.
87. HOV Lane—any preferential lane designated for exclusive use by high-occupancy vehicles for all or part of a day—including a designated lane on a freeway, other highway, street, or independent roadway on a separate right-of-way.
88. Hybrid Beacon—a special type of beacon that is intentionally placed in a dark mode (no indications displayed) between periods of operation and, when operated, displays both steady and flashing traffic control signal indications.
89. Inherently Low Emission Vehicle (ILEV)—any kind of vehicle that, because of inherent properties of the fuel system design, will not have significant evaporative emissions, even if its evaporative emission control system has failed.
90. In-Roadway Lights—a special type of highway traffic signal installed in the roadway surface to warn road users that they are approaching a condition on or adjacent to the roadway that might not be readily apparent and might require the road users to slow down and/or come to a stop.
91. Interchange—a system of interconnecting roadways providing for traffic movement between two or more highways that do not intersect at grade.
92. Interconnection—when used in Part 8, the electrical connection between the railroad or light rail transit active warning system and the highway traffic signal controller assembly for the purpose of preemption.
93. Intermediate Interchange—an interchange with an urban or rural route that is not a major or minor interchange as defined in this Section.
94. Intersection—intersection is defined as follows:
    (a) The area embraced within the prolongation or connection of the lateral curb lines, or if none, the lateral boundary lines of the roadways of two highways that join one another at, or approximately at, right angles, or the area within which vehicles traveling on different highways that join at any other angle might come into conflict.
    (b) The junction of an alley or driveway with a roadway or highway shall not constitute an intersection, unless the roadway or highway at said junction is controlled by a traffic control device.
    (c) If a highway includes two roadways that are 30 feet or more apart (see definition of Median), then every crossing of each roadway of such divided highway by an intersecting highway shall be a separate intersection.
    (d) If both intersecting highways include two roadways that are 30 feet or more apart, then every crossing of any two roadways of such highways shall be a separate intersection.
    (e) At a location controlled by a traffic control signal, regardless of the distance between the separate intersections as defined in (c) and (d) above:
        (1) If a stop line, yield line, or crosswalk has not been designated on the roadway (within the median) between the separate intersections, the two intersections and the roadway (median) between them shall be considered as one intersection;
        (2) Where a stop line, yield line, or crosswalk is designated on the roadway on the intersection approach, the area within the crosswalk and/or beyond the designated stop line or yield line shall be part of the intersection; and
        (3) Where a crosswalk is designated on a roadway on the departure from the intersection, the intersection shall include the area extending to the far side of such crosswalk.
95. Intersection Control Beacon—a beacon used only at an intersection to control two or more directions of travel.
96. Interval—the part of a signal cycle during which signal indications do not change.
97. Interval Sequence—the order of appearance of signal indications during successive intervals of a signal cycle.
98. Island—a defined area between traffic lanes for control of vehicular movements, for toll collection, or for pedestrian refuge. It includes all end protection and approach treatments. Within an intersection area, a median or an outer separation is considered to be an island.
99. Lane Drop—see Dropped Lane.
100. Lane Line Markings—white pavement marking lines that delineate the separation of traffic lanes that have the same direction of travel on a roadway.
101. Lane-Use Control Signal—a signal face displaying indications to permit or prohibit the use of specific lanes of a roadway or to indicate the impending prohibition of such use.

102. Legend—see Sign Legend.
103. Lens—see Signal Lens.
104. Light Rail Transit Traffic (Light Rail Transit Equipment)—every device in, upon, or by which any person or property can be transported on light rail transit tracks, including single-unit light rail transit cars (such at streetcars and trolleys) and assemblies of multiple light rail transit cars coupled together.
105. Locomotive Horn—an air horn, steam whistle, or similar audible warning device (see 49 CFR Part 229.129) mounted on a locomotive or control cab car. The terms "locomotive horn," "train whistle," "locomotive whistle," and "train horn" are used interchangeably in the railroad industry.
106. Logo—a distinctive emblem or trademark that identifies a commercial business and/or the product or service offered by the business.
107. Longitudinal Markings—pavement markings that are generally placed parallel and adjacent to the flow of traffic such as lane lines, center lines, edge lines, channelizing lines, and others.
108. Louver—see Signal Louver.
109. Major Interchange—an interchange with another freeway or expressway, or an interchange with a high-volume multi-lane highway, principal urban arterial, or major rural route where the interchanging traffic is heavy or includes many road users unfamiliar with the area.
110. Major Street—the street normally carrying the higher volume of vehicular traffic.
111. Malfunction Management Unit—same as Conflict Monitor.
112. Managed Lane—a highway lane or set of lanes, or a highway facility, for which variable operational strategies such as direction of travel, tolling, pricing, and/or vehicle type or occupancy requirements are implemented and managed in real-time in response to changing conditions. Managed lanes are typically buffer- or barrier-separated lanes parallel to the general-purpose lanes of a highway in which access is restricted to designated locations. There are also some highways on which all lanes are managed.
113. Manual Lane—see Attended Lane.
114. Maximum Highway Traffic Signal Preemption Time—the maximum amount of time needed following initiation of the preemption sequence for the highway traffic signals to complete the timing of the right-of-way transfer time, queue clearance time, and separation time.
115. Median—the area between two roadways of a divided highway measured from edge of traveled way to edge of traveled way. The median excludes turn lanes. The median width might be different between intersections, interchanges, and at opposite approaches of the same intersection.
116. Minimum Track Clearance Distance—for standard two-quadrant warning devices, the minimum track clearance distance is the length along a highway at one or more railroad or light rail transit tracks, measured from the highway stop line, warning device, or 12 feet perpendicular to the track center line, to 6 feet beyond the track(s) measured perpendicular to the far rail, along the center line or edge line of the highway, as appropriate, to obtain the longer distance. For Four-Quadrant Gate systems, the minimum track clearance distance is the length along a highway at one or more railroad or light rail transit tracks, measured either from the highway stop line or entrance warning device, to the point where the rear of the vehicle would be clear of the exit gate arm. In cases where the exit gate arm is parallel to the track(s) and is not perpendicular to the highway, the distance is measured either along the center line or edge line of the highway, as appropriate, to obtain the longer distance.
117. Minimum Warning Time—when used in Part 8, the least amount of time active warning devices shall operate prior to the arrival of rail traffic at a grade crossing.
118. Minor Interchange—an interchange where traffic is local and very light, such as interchanges with land service access roads. Where the sum of the exit volumes is estimated to be lower than 100 vehicles per day in the design year, the interchange is classified as local.
119. Minor Street—the street normally carrying the lower volume of vehicular traffic.
120. Movable Bridge Resistance Gate—a type of traffic gate, which is located downstream of the movable bridge warning gate, that provides a physical deterrent to vehicle and/or pedestrian traffic when placed in the appropriate position.
121. Movable Bridge Signal—a highway traffic signal installed at a movable bridge to notify traffic to stop during periods when the roadway is closed to allow the bridge to open.
122. Movable Bridge Warning Gate—a type of traffic gate designed to warn, but not primarily to block, vehicle and/or pedestrian traffic when placed in the appropriate position.
123. Multi-Lane—more than one lane moving in the same direction. A multi-lane street, highway, or roadway has a basic cross-section comprised of two or more through lanes in one or both directions. A multi-lane approach has two or more lanes moving toward the intersection, including turning lanes.

124. Neutral Area—the paved area between the channelizing lines separating an entrance or exit ramp or a channelized turn lane or channelized entering lane from the adjacent through lane(s).
125. Object Marker—a device used to mark obstructions within or adjacent to the roadway.
126. Occupancy Requirement—any restriction that regulates the use of a facility or one or more lanes of a facility for any period of the day based on a specified number of persons in a vehicle.
127. Occupant—a person driving or riding in a car, truck, bus, or other vehicle.
128. Open-Road ETC Lane—a non-attended lane that is designed to allow toll payments to be electronically collected from vehicles traveling at normal highway speeds. Open-Road ETC lanes are typically physically separated from the toll plaza, often following the alignment of the mainline lanes, with toll plaza lanes for cash toll payments being on a different alignment after diverging from the mainline lanes or a subset thereof.
129. Open-Road Tolling—a system designed to allow electronic toll collection (ETC) from vehicles traveling at normal highway speeds. Open-Road Tolling might be used on toll roads or toll facilities in conjunction with toll plazas. Open-Road Tolling is also typically used on managed lanes and on toll facilities that only accept payment by ETC.
130. Open-Road Tolling Point—the location along an Open-Road ETC lane at which roadside or overhead detection and receiving equipment are placed and vehicles are electronically assessed a toll.
131. Opposing Traffic—vehicles that are traveling in the opposite direction. At an intersection, vehicles entering from an approach that is approximately straight ahead would be considered to be opposing traffic, but vehicles entering from approaches on the left or right would not be considered to be opposing traffic.
132. Overhead Sign—a sign that is placed such that a portion or the entirety of the sign or its support is directly above the roadway or shoulder such that vehicles travel below it. Typical installations include signs placed on cantilever arms that extend over the roadway or shoulder, on sign support structures that span the entire width of the pavement, on mast arms or span wires that also support traffic control signals, and on highway bridges that cross over the roadway.
133. Parking Area—a parking lot or parking garage that is separated from a roadway. Parallel or angle parking spaces along a roadway are not considered a parking area.
134. Passive Grade Crossing—a grade crossing where none of the automatic traffic control devices associated with an Active Grade Crossing Warning System are present and at which the traffic control devices consist entirely of signs and/or markings.
135. Pathway—a general term denoting a public way for purposes of travel by authorized users outside the traveled way and physically separated from the roadway by an open space or barrier and either within the highway right-of-way or within an independent alignment. Pathways include shared-use paths, but do not include sidewalks.
136. Pathway Grade Crossing—the general area where a pathway and railroad or light rail transit tracks cross at the same level, within which are included the tracks, pathway, and traffic control devices for pathway traffic traversing that area.
137. Paved—a bituminous surface treatment, mixed bituminous concrete, or Portland cement concrete roadway surface that has both a structural (weight bearing) and a sealing purpose for the roadway.
138. Pedestrian—a person on foot, in a wheelchair, on skates, or on a skateboard.
139. Pedestrian Change Interval—an interval during which the flashing UPRAISED HAND (symbolizing DONT WALK) signal indication is displayed.
140. Pedestrian Clearance Time—the time provided for a pedestrian crossing in a crosswalk, after leaving the curb or shoulder, to travel to the far side of the traveled way or to a median.
141. Pedestrian Facilities—a general term denoting improvements and provisions made to accommodate or encourage walking.
142. Pedestrian Hybrid Beacon— a special type of hybrid beacon used to warn and control traffic at an unsignalized location to assist pedestrians in crossing a street or highway at a marked crosswalk.
143. Pedestrian Signal Head—a signal head, which contains the symbols WALKING PERSON (symbolizing WALK) and UPRAISED HAND (symbolizing DONT WALK), that is installed to direct pedestrian traffic at a traffic control signal.
144. Permissive Mode—a mode of traffic control signal operation in which left or right turns are permitted to be made after yielding to pedestrians, if any, and/or opposing traffic, if any. When a CIRCULAR GREEN signal indication is displayed, both left and right turns are permitted unless otherwise prohibited by another traffic control device. When a flashing YELLOW ARROW or flashing RED ARROW signal indication is displayed, the turn indicated by the arrow is permitted.

145. **Physical Gore**—a longitudinal point where a physical barrier or the lack of a paved surface inhibits road users from crossing from a ramp or channelized turn lane or channelized entering lane to the adjacent through lane(s) or vice versa.
146. **Pictograph**—a pictorial representation used to identify a governmental jurisdiction, an area of jurisdiction, a governmental agency, a military base or branch of service, a governmental-approved university or college, a toll payment system, or a government-approved institution.
147. **Plaque**—a traffic control device intended to communicate specific information to road users through a word, symbol, or arrow legend that is placed immediately adjacent to a sign to supplement the message on the sign. The difference between a plaque and a sign is that a plaque cannot be used alone. The designation for a plaque includes a "P" suffix.
148. **Platoon**—a group of vehicles or pedestrians traveling together as a group, either voluntarily or involuntarily, because of traffic signal controls, geometrics, or other factors.
149. **Portable Traffic Control Signal**—a temporary traffic control signal that is designed so that it can be easily transported and reused at different locations.
150. **Post-Mounted Sign**—a sign that is placed to the side of the roadway such that no portion of the sign or its support is directly above the roadway or shoulder.
151. **Posted Speed Limit**—a speed limit determined by law or regulation and displayed on Speed Limit signs.
152. **Preemption**—the transfer of normal operation of a traffic control signal to a special control mode of operation.
153. **Preferential Lane**—a highway lane reserved for the exclusive use of one or more specific types of vehicles or vehicles with at least a specific number of occupants.
154. **Pre-Signal**—traffic control signal faces that control traffic approaching a grade crossing in conjunction with the traffic control signal faces that control traffic approaching a highway-highway intersection beyond the tracks. Supplemental near-side traffic control signal faces for the highway-highway intersection are not considered pre-signals. Pre-signals are typically used where the clear storage distance is insufficient to store one or more design vehicles.
155. **Pretimed Operation**—a type of traffic control signal operation in which none of the signal phases function on the basis of actuation.
156. **Primary Signal Face**—one of the required or recommended minimum number of signal faces for a given approach or separate turning movement, but not including near-side signal faces required as a result of the far-side signal faces exceeding the maximum distance from the stop line.
157. **Principal Legend**—place names, street names, and route numbers placed on guide signs.
158. **Priority Control**—a means by which the assignment of right-of-way is obtained or modified.
159. **Private Road Open to Public Travel**—private toll roads and roads (including any adjacent sidewalks that generally run parallel to the road) within shopping centers, airports, sports arenas, and other similar business and/or recreation facilities that are privately owned, but where the public is allowed to travel without access restrictions. Roads within private gated properties (except for gated toll roads) where access is restricted at all times, parking areas, driving aisles within parking areas, and private grade crossings shall not be included in this definition.
160. **Protected Mode**—a mode of traffic control signal operation in which left or right turns are permitted to be made when a left or right GREEN ARROW signal indication is displayed.
161. **Public Road**—any road, street, or similar facility under the jurisdiction of and maintained by a public agency and open to public travel.
162. **Pushbutton**—a button to activate a device or signal timing for pedestrians, bicyclists, or other road users.
163. **Pushbutton Information Message**—a recorded message that can be actuated by pressing a pushbutton when the walk interval is not timing and that provides the name of the street that the crosswalk associated with that particular pushbutton crosses and can also provide other information about the intersection signalization or geometry.
164. **Pushbutton Locator Tone**—a repeating sound that informs approaching pedestrians that a pushbutton exists to actuate pedestrian timing or receive additional information and that enables pedestrians who have visual disabilities to locate the pushbutton.
165. **Queue Clearance Time**—when used in Part 8, the time required for the design vehicle of maximum length stopped just inside the minimum track clearance distance to start up and move through and clear the entire minimum track clearance distance. If pre-signals are present, this time shall be long enough to allow the vehicle to move through the intersection, or to clear the tracks if there is sufficient clear storage distance. If a Four-Quadrant Gate system is present, this time shall be long enough to permit the exit gate arm to lower after the design vehicle is clear of the minimum track clearance distance.

166. Quiet Zone—a segment of a rail line, with one or a number of consecutive public highway-rail grade crossings at which locomotive horns are not routinely sounded per 49 CFR Part 222.
167. Rail Traffic—every device in, upon, or by which any person or property can be transported on rails or tracks and to which all other traffic must yield the right-of-way by law at grade crossings, including trains, one or more locomotives coupled (with or without cars), other railroad equipment, and light rail transit operating in exclusive or semi-exclusive alignments. Light rail transit operating in a mixed-use alignment, to which other traffic is not required to yield the right-of-way by law, is a vehicle and is not considered to be rail traffic.
168. Raised Pavement Marker—a device mounted on or in a road surface that has a height generally not exceeding approximately 1 inch above the road surface for a permanent marker, or not exceeding approximately 2 inches above the road surface for a temporary flexible marker, and that is intended to be used as a positioning guide and/or to supplement or substitute for pavement markings.
169. Ramp Control Signal—a highway traffic signal installed to control the flow of traffic onto a freeway at an entrance ramp or at a freeway-to-freeway ramp connection.
170. Ramp Meter—see Ramp Control Signal.
171. Red Clearance Interval—an interval that follows a yellow change interval and precedes the next conflicting green interval.
172. Regulatory Sign—a sign that gives notice to road users of traffic laws or regulations.
173. Retroreflectivity—a property of a surface that allows a large portion of the light coming from a point source to be returned directly back to a point near its origin.
174. Right-of-Way [Assignment]—the permitting of vehicles and/or pedestrians to proceed in a lawful manner in preference to other vehicles or pedestrians by the display of a sign or signal indications.
175. Right-of-Way Transfer Time—when used in Part 8, the maximum amount of time needed for the worst case condition, prior to display of the track clearance green interval. This includes any railroad or light rail transit or highway traffic signal control equipment time to react to a preemption call, and any traffic control signal green, pedestrian walk and clearance, yellow change, and red clearance intervals for conflicting traffic.
176. Road—see Roadway.
177. Road User—a vehicle operator, bicyclist, or pedestrian, including persons with disabilities, within the highway or on a private road open to public travel.
178. Roadway—that portion of a highway improved, designed, or ordinarily used for vehicular travel and parking lanes, but exclusive of the sidewalk, berm, or shoulder even though such sidewalk, berm, or shoulder is used by persons riding bicycles or other human-powered vehicles. In the event a highway includes two or more separate roadways, the term roadway as used in this Manual shall refer to any such roadway separately, but not to all such roadways collectively.
179. Roadway Network—a geographical arrangement of intersecting roadways.
180. Roundabout—a circular intersection with yield control at entry, which permits a vehicle on the circulatory roadway to proceed, and with deflection of the approaching vehicle counter-clockwise around a central island.
181. Rumble Strip—a series of intermittent, narrow, transverse areas of rough-textured, slightly raised, or depressed road surface that extend across the travel lane to alert road users to unusual traffic conditions or are located along the shoulder, along the roadway center line, or within islands formed by pavement markings to alert road users that they are leaving the travel lanes.
182. Rural Highway—a type of roadway normally characterized by lower volumes, higher speeds, fewer turning conflicts, and less conflict with pedestrians.
183. Safe-Positioned—the positioning of emergency vehicles at an incident in a manner that attempts to protect both the responders performing their duties and road users traveling through the incident scene, while minimizing, to the extent practical, disruption of the adjacent traffic flow.
184. School—a public or private educational institution recognized by the State education authority for one or more grades K through 12 or as otherwise defined by the State.
185. School Zone—a designated roadway segment approaching, adjacent to, and beyond school buildings or grounds, or along which school related activities occur.
186. Semi-Actuated Operation—a type of traffic control signal operation in which at least one, but not all, signal phases function on the basis of actuation.
187. Separate Turn Signal Face—a signal face that exclusively controls a turn movement and that displays signal indications that are applicable only to the turn movement.
188. Separation Time—the component of maximum highway traffic signal preemption time during which the minimum track clearance distance is clear of vehicular traffic prior to the arrival of rail traffic.

189. **Shared Roadway**—a roadway that is officially designated and marked as a bicycle route, but which is open to motor vehicle travel and upon which no bicycle lane is designated.
190. **Shared Turn Signal Face**—a signal face, for controlling both a turn movement and the adjacent through movement, that always displays the same color of circular signal indication that the adjacent through signal face or faces display.
191. **Shared-Use Path**—a bikeway outside the traveled way and physically separated from motorized vehicular traffic by an open space or barrier and either within the highway right-of-way or within an independent alignment. Shared-use paths are also used by pedestrians (including skaters, users of manual and motorized wheelchairs, and joggers) and other authorized motorized and non-motorized users.
192. **Sidewalk**—that portion of a street between the curb line, or the lateral line of a roadway, and the adjacent property line or on easements of private property that is paved or improved and intended for use by pedestrians.
193. **Sign**—any traffic control device that is intended to communicate specific information to road users through a word, symbol, and/or arrow legend. Signs do not include highway traffic signals, pavement markings, delineators, or channelization devices.
194. **Sign Assembly**—a group of signs, located on the same support(s), that supplement one another in conveying information to road users.
195. **Sign Illumination**—either internal or external lighting that shows similar color by day or night. Street or highway lighting shall not be considered as meeting this definition.
196. **Sign Legend**—all word messages, logos, pictographs, and symbol and arrow designs that are intended to convey specific meanings. The border, if any, on a sign is not considered to be a part of the legend.
197. **Sign Panel**—a separate panel or piece of material containing a word, symbol, and/or arrow legend that is affixed to the face of a sign.
198. **Signal Backplate**—a thin strip of material that extends outward from and parallel to a signal face on all sides of a signal housing to provide a background for improved visibility of the signal indications.
199. **Signal Coordination**—the establishment of timed relationships between adjacent traffic control signals.
200. **Signal Face**—an assembly of one or more signal sections that is provided for controlling one or more traffic movements on a single approach.
201. **Signal Head**—an assembly of one or more signal faces that is provided for controlling traffic movements on one or more approaches.
202. **Signal Housing**—that part of a signal section that protects the light source and other required components.
203. **Signal Indication**—the illumination of a signal lens or equivalent device.
204. **Signal Lens**—that part of the signal section that redirects the light coming directly from the light source and its reflector, if any.
205. **Signal Louver**—a device that can be mounted inside a signal visor to restrict visibility of a signal indication from the side or to limit the visibility of the signal indication to a certain lane or lanes, or to a certain distance from the stop line.
206. **Signal Phase**—the right-of-way, yellow change, and red clearance intervals in a cycle that are assigned to an independent traffic movement or combination of movements.
207. **Signal Section**—the assembly of a signal housing, signal lens, if any, and light source with necessary components to be used for displaying one signal indication.
208. **Signal System**—two or more traffic control signals operating in signal coordination.
209. **Signal Timing**—the amount of time allocated for the display of a signal indication.
210. **Signal Visor**—that part of a signal section that directs the signal indication specifically to approaching traffic and reduces the effect of direct external light entering the signal lens.
211. **Signing**—individual signs or a group of signs, not necessarily on the same support(s), that supplement one another in conveying information to road users.
212. **Simultaneous Preemption**—notification of approaching rail traffic is forwarded to the highway traffic signal controller unit or assembly and railroad or light rail transit active warning devices at the same time.
213. **Special Purpose Road**—a low-volume, low-speed road that serves recreational areas or resource development activities.

214. Speed—speed is defined based on the following classifications:
    (a) Average Speed—the summation of the instantaneous or spot-measured speeds at a specific location of vehicles divided by the number of vehicles observed.
    (b) Design Speed—a selected speed used to determine the various geometric design features of a roadway.
    (c) 85th-Percentile Speed—the speed at or below which 85 percent of the motor vehicles travel.
    (d) Operating Speed—a speed at which a typical vehicle or the overall traffic operates. Operating speed might be defined with speed values such as the average, pace, or 85th-percentile speeds.
    (e) Pace—the 10 mph speed range representing the speeds of the largest percentage of vehicles in the traffic stream.
215. Speed Limit—the maximum (or minimum) speed applicable to a section of highway as established by law or regulation.
216. Speed Limit Sign Beacon—a beacon used to supplement a SPEED LIMIT sign.
217. Speed Measurement Markings—a white transverse pavement marking placed on the roadway to assist the enforcement of speed regulations.
218. Speed Zone—a section of highway with a speed limit that is established by law or regulation, but which might be different from a legislatively specified statutory speed limit.
219. Splitter Island—a median island used to separate opposing directions of traffic entering and exiting a roundabout.
220. Station Crossing—a pathway grade crossing that is associated with a station platform.
221. Statutory Speed Limit—a speed limit established by legislative action that typically is applicable for a particular class of highways with specified design, functional, jurisdictional and/or location characteristics and that is not necessarily displayed on Speed Limit signs.
222. Steady (Steady Mode)—the continuous display of a signal indication for the duration of an interval, signal phase, or consecutive signal phases.
223. Stop Beacon—a beacon used to supplement a STOP sign, a DO NOT ENTER sign, or a WRONG WAY sign.
224. Stop Line—a solid white pavement marking line extending across approach lanes to indicate the point at which a stop is intended or required to be made.
225. Street—see Highway.
226. Supplemental Signal Face—a signal face that is not a primary signal face but which is provided for a given approach or separate turning movement to enhance visibility or conspicuity.
227. Symbol—the approved design of a pictorial representation of a specific traffic control message for signs, pavement markings, traffic control signals, or other traffic control devices, as shown in the MUTCD.
228. Temporary Traffic Control Signal—a traffic control signal that is installed for a limited time period.
229. Temporary Traffic Control Zone—an area of a highway where road user conditions are changed because of a work zone or incident by the use of temporary traffic control devices, flaggers, uniformed law enforcement officers, or other authorized personnel.
230. Theoretical Gore—a longitudinal point at the upstream end of a neutral area at an exit ramp or channelized turn lane where the channelizing lines that separate the ramp or channelized turn lane from the adjacent through lane(s) begin to diverge, or a longitudinal point at the downstream end of a neutral area at an entrance ramp or channelized entering lane where the channelizing lines that separate the ramp or channelized entering lane from the adjacent through lane(s) intersect each other.
231. Timed Exit Gate Operating Mode—a mode of operation where the exit gate descent at a grade crossing is based on a predetermined time interval.
232. Toll Booth—a shelter where a toll attendant is stationed to collect tolls or issue toll tickets. A toll booth is located adjacent to a toll lane and is typically set on a toll island.
233. Toll Island—a raised island on which a toll booth or other toll collection and related equipment are located.
234. Toll Lane—an individual lane located within a toll plaza in which a toll payment is collected or, for toll-ticket systems, a toll ticket is issued.
235. Toll Plaza—the location at which tolls are collected consisting of a grouping of toll booths, toll islands, toll lanes, and, typically, a canopy. Toll plazas might be located on highway mainlines or on interchange ramps. A mainline toll plaza is sometimes referred to as a barrier toll plaza because it interrupts the traffic flow.

236. Toll-Ticket System—a system in which the user of a toll road receives a ticket from a machine or toll booth attendant upon entering a toll system. The ticket denotes the user's point of entry and, upon exiting the toll system, the user surrenders the ticket and is charged a toll based on the distance traveled between the points of entry and exit.
237. Traffic—pedestrians, bicyclists, ridden or herded animals, vehicles, streetcars, and other conveyances either singularly or together while using for purposes of travel any highway or private road open to public travel.
238. Traffic Control Device—a sign, signal, marking, or other device used to regulate, warn, or guide traffic, placed on, over, or adjacent to a street, highway, private road open to public travel, pedestrian facility, or shared-use path by authority of a public agency or official having jurisdiction, or, in the case of a private road open to public travel, by authority of the private owner or private official having jurisdiction.
239. Traffic Control Signal (Traffic Signal)—any highway traffic signal by which traffic is alternately directed to stop and permitted to proceed.
240. Train—one or more locomotives coupled, with or without cars, that operates on rails or tracks and to which all other traffic must yield the right-of-way by law at highway-rail grade crossings.
241. Transverse Markings—pavement markings that are generally placed perpendicular and across the flow of traffic such as shoulder markings; word, symbol, and arrow markings; stop lines; crosswalk lines; speed measurement markings; parking space markings; and others.
242. Traveled Way—the portion of the roadway for the movement of vehicles, exclusive of the shoulders, berms, sidewalks, and parking lanes.
243. Turn Bay—a lane for the exclusive use of turning vehicles that is formed on the approach to the location where the turn is to be made. In most cases where turn bays are provided, drivers who desire to turn must move out of a through lane into the newly formed turn bay in order to turn. A through lane that becomes a turn lane is considered to be a dropped lane rather than a turn bay.
244. Upstream—a term that refers to a location that is encountered by traffic prior to a downstream location as it flows in an "upstream to downstream" direction. For example, "the upstream end of a lane line separating the turn lane from a through lane on the approach to an intersection" is the end of the line that is furthest from the intersection.
245. Urban Street—a type of street normally characterized by relatively low speeds, wide ranges of traffic volumes, narrower lanes, frequent intersections and driveways, significant pedestrian traffic, and more businesses and houses.
246. Vehicle—every device in, upon, or by which any person or property can be transported or drawn upon a highway, except trains and light rail transit operating in exclusive or semi-exclusive alignments. Light rail transit equipment operating in a mixed-use alignment, to which other traffic is not required to yield the right-of-way by law, is a vehicle.
247. Vibrotactile Pedestrian Device—an accessible pedestrian signal feature that communicates, by touch, information about pedestrian timing using a vibrating surface.
248. Visibility-Limited Signal Face or Visibility-Limited Signal Section—a type of signal face or signal section designed (or shielded, hooded, or louvered) to restrict the visibility of a signal indication from the side, to a certain lane or lanes, or to a certain distance from the stop line.
249. Walk Interval—an interval during which the WALKING PERSON (symbolizing WALK) signal indication is displayed.
250. Warning Beacon—a beacon used only to supplement an appropriate warning or regulatory sign or marker.
251. Warning Light—a portable, powered, yellow, lens-directed, enclosed light that is used in a temporary traffic control zone in either a steady burn or a flashing mode.
252. Warning Sign—a sign that gives notice to road users of a situation that might not be readily apparent.
253. Warrant—a warrant describes a threshold condition based upon average or normal conditions that, if found to be satisfied as part of an engineering study, shall result in analysis of other traffic conditions or factors to determine whether a traffic control device or other improvement is justified. Warrants are not a substitute for engineering judgment. The fact that a warrant for a particular traffic control device is met is not conclusive justification for the installation of the device.
254. Wayside Equipment—the signals, switches, and/or control devices for railroad or light rail transit operations housed within one or more enclosures located along the railroad or light rail transit right-of-way and/or on railroad or light rail transit property.
255. Wayside Horn System—a stationary horn (or series of horns) located at a grade crossing that is used in conjunction with train-activated or light rail transit-activated warning systems to provide audible warning of approaching rail traffic to road users on the highway or pathway approaches to a grade crossing, either as a supplement or alternative to the sounding of a locomotive horn.

256. Worker—a person on foot whose duties place him or her within the right-of-way of a street, highway, or pathway, such as street, highway, or pathway construction and maintenance forces, survey crews, utility crews, responders to incidents within the street, highway, or pathway right-of-way, and law enforcement personnel when directing traffic, investigating crashes, and handling lane closures, obstructed roadways, and disasters within the right-of-way of a street, highway, or pathway.
257. Wrong-Way Arrow—a slender, elongated, white pavement marking arrow placed upstream from the ramp terminus to indicate the correct direction of traffic flow. Wrong-way arrows are intended primarily to warn wrong-way road users that they are going in the wrong direction.
258. Yellow Change Interval—the first interval following the green or flashing arrow interval during which the steady yellow signal indication is displayed.
259. Yield Line—a row of solid white isosceles triangles pointing toward approaching vehicles extending across approach lanes to indicate the point at which the yield is intended or required to be made.

Section 1A.14 Meanings of Acronyms and Abbreviations in this Manual

Standard:

01   The following acronyms and abbreviations, when used in this Manual, shall have the following meanings:
1. AADT—annual average daily traffic
2. AASHTO—American Association of State Highway and Transportation Officials
3. ADA—Americans with Disabilities Act
4. ADAAG—Americans with Disabilities Accessibility Guidelines
5. ADT—average daily traffic
6. AFAD—Automated Flagger Assistance Device
7. ANSI—American National Standards Institute
8. CFR—Code of Federal Regulations
9. CMS—changeable message sign
10. dBA—A-weighted decibels
11. EPA—Environmental Protection Agency
12. ETC—electronic toll collection
13. EV—electric vehicle
14. FHWA—Federal Highway Administration
15. FRA—Federal Railroad Administration
16. FTA—Federal Transit Administration
17. HOT—high occupancy tolls
18. HOTM—FHWA's Office of Transportation Management
19. HOTO—FHWA's Office of Transportation Operations
20. HOV—high-occupancy vehicle
21. ILEV—inherently low emission vehicle
22. ISEA—International Safety Equipment Association
23. ITE—Institute of Transportation Engineers
24. ITS—intelligent transportation systems
25. LED—light emitting diode
26. LP—liquid petroleum
27. MPH or mph—miles per hour
28. MUTCD—Manual on Uniform Traffic Control Devices
29. NCHRP—National Cooperative Highway Research Program
30. ORT—open-road tolling
31. PCMS—portable changeable message sign
32. PRT—perception-response time
33. RPM—raised pavement marker
34. RRPM—raised retroreflective pavement marker
35. RV—recreational vehicle
36. TDD—telecommunication devices for the deaf
37. TRB—Transportation Research Board
38. TTC—temporary traffic control
39. U.S.—United States
40. U.S.C.—United States Code
41. USDOT—United States Department of Transportation
42. UVC—Uniform Vehicle Code
43. VPH or vph—vehicles per hour

Section 1A.15 Abbreviations Used on Traffic Control Devices

Standard:

01 When the word messages shown in Table 1A-1 need to be abbreviated in connection with traffic control devices, the abbreviations shown in Table 1A-1 shall be used.

02 When the word messages shown in Table 1A-2 need to be abbreviated on a portable changeable message sign, the abbreviations shown in Table 1A-2 shall be used. Unless indicated by an asterisk, these abbreviations shall only be used on portable changeable message signs.

Guidance:

03 *The abbreviations for the words listed in Table 1A-2 that also show a prompt word should not be used on a portable changeable message sign unless the prompt word shown in Table 1A-2 either precedes or follows the abbreviation, as applicable.*

Standard:

04 The abbreviations shown in Table 1A-3 shall not be used in connection with traffic control devices because of their potential to be misinterpreted by road users.

Guidance:

05 *If multiple abbreviations are permitted in Table 1A-1 or 1A-2, the same abbreviation should be used throughout a single jurisdiction.*

06 *Except as otherwise provided in Table 1A-1 or 1A-2 or unless necessary to avoid confusion, periods, commas, apostrophes, question marks, ampersands, and other punctuation marks or characters that are not letters or numerals should not be used in any abbreviation.*

Table 1A-1. Acceptable Abbreviations

Word Message	Standard Abbreviation
Afternoon / Evening	PM
Alternate	ALT
AM Radio	AM
Avenue	AVE, AV
Bicycle	BIKE
Boulevard	BLVD*
Bridge	(See Table 1A-2)
CB Radio	CB
Center (as part of a place name)	CTR
Circle	CIR*
Civil Defense	CD
Compressed Natural Gas	CNG
Court	CT*
Crossing (other than highway-rail)	X-ING
Drive	DR*
East	E
Electric Vehicle	EV
Expressway	EXPWY*
Feet	FT
FM Radio	FM
Freeway	FRWY, FWY*
Friday	FRI
Hazardous Material	HAZMAT
High Occupancy Vehicle	HOV

Word Message	Standard Abbreviation
Highway	HWY*
Hospital	HOSP
Hour(s)	HR, HRS
Information	INFO
Inherently Low Emission Vehicle	ILEV
International	INTL
Interstate	(See Table 1A-2)
Junction / Intersection	JCT
Lane	(See Table 1A-2)
Liquid Propane Gas	LP-GAS
Maximum	MAX
Mile(s)	MI
Miles Per Hour	MPH
Minimum	MIN
Minute(s)	MIN
Monday	MON
Morning / Late Night	AM
Mount	MT
Mountain	MTN
National	NATL
North	N
Parkway	PKWY*
Pedestrian	PED
Place	PL*

Word Message	Standard Abbreviation
Pounds	LBS
Road	RD*
Saint	ST
Saturday	SAT
South	S
State, county, or other non-US or non-Interstate numbered route	(See Table 1A-2)
Street	ST*
Sunday	SUN
Telephone	PHONE
Temporary	TEMP
Terrace	TER*
Thursday	THURS
Thruway	THWY*
Tons of Weight	T
Trail	TR*
Tuesday	TUES
Turnpike	TPK*
Two-Way Intersection	2-WAY
US Numbered Route	US
Wednesday	WED
West	W

*This abbreviation shall not be used for any application other than the name of a roadway.

Table 1A-2. Abbreviations That Shall be Used Only on Portable Changeable Message Signs

Word Message	Standard Abbreviation	Prompt Word That Should Precede the Abbreviation	Prompt Word That Should Follow the Abbreviation
Access	ACCS	—	Road
Ahead	AHD	Fog	—
Blocked	BLKD	Lane	—
Bridge	BR*	[Name]	—
Cannot	CANT	—	—
Center	CNTR	—	Lane
Chemical	CHEM	—	Spill
Condition	COND	Traffic	—
Congested	CONG	Traffic	—
Construction	CONST	—	Ahead
Crossing	XING	—	—
Do Not	DONT	—	—
Downtown	DWNTN	—	Traffic
Eastbound	E-BND	—	—
Emergency	EMER	—	—
Entrance, Enter	ENT	—	—
Exit	EX	Next	—
Express	EXP	—	Lane
Frontage	FRNTG	—	Road
Hazardous	HAZ	—	Driving
Highway-Rail Grade Crossing	RR XING	—	—
Interstate	I-*	—	[Number]
It Is	ITS	—	—
Lane	LN	[Roadway Name]*, Right, Left, Center	—
Left	LFT	—	—
Local	LOC	—	Traffic
Lower	LWR	—	Level
Maintenance	MAINT	—	—
Major	MAJ	—	Accident
Minor	MNR	—	Accident
Normal	NORM	—	—
Northbound	N-BND	—	—
Oversized	OVRSZ	—	Load
Parking	PKING	—	—
Pavement	PVMT	Wet	—
Prepare	PREP	—	To Stop
Quality	QLTY	Air	—
Right	RT	Keep, Next	—
Right	RT	—	Lane
Roadwork	RDWK	—	Ahead, [Distance]
Route	RT, RTE	Best	—
Service	SERV	—	—
Shoulder	SHLDR	—	—
Slippery	SLIP	—	—
Southbound	S-BND	—	—
Speed	SPD	—	—
State, county, or other non-US or non-Interstate numbered route	[Route Abbreviation determined by highway agency]**	—	[Number]
Tires With Lugs	LUGS	—	—
Traffic	TRAF	—	—
Travelers	TRVLRS	—	—
Two-Wheeled Vehicles	CYCLES	—	—
Upper	UPR	—	Level
Vehicle(s)	VEH, VEHS	—	—
Warning	WARN	—	—
Westbound	W-BND	—	—
Will Not	WONT	—	—

* This abbreviation, when accompanied by the prompt word, may be used on traffic control devices other than portable changeable message signs.
** A space and no dash shall be placed between the abbreviation and the number of the route.

Table 1A-3. Unacceptable Abbreviations

Abbreviation	Intended Word	Common Misinterpretation
ACC	Accident	Access (Road)
CLRS	Clears	Colors
DLY	Delay	Daily
FDR	Feeder	Federal
L	Left	Lane (Merge)
LT	Light (Traffic)	Left
PARK	Parking	Park
POLL	Pollution (Index)	Poll
RED	Reduce	Red
STAD	Stadium	Standard
WRNG	Warning	Wrong

PART 2
SIGNS

CHAPTER 2A. GENERAL

Section 2A.01 Function and Purpose of Signs

Support:

01 This Manual contains Standards, Guidance, and Options for the signing of all types of highways, and private roads open to public travel. The functions of signs are to provide regulations, warnings, and guidance information for road users. Words, symbols, and arrows are used to convey the messages. Signs are not typically used to confirm rules of the road.

02 Detailed sign requirements are located in the following Chapters of Part 2:
- Chapter 2B — Regulatory Signs, Barricades, and Gates
- Chapter 2C — Warning Signs and Object Markers
- Chapter 2D — Guide Signs for Conventional Roads
- Chapter 2E — Guide Signs for Freeways and Expressways
- Chapter 2F — Toll Road Signs
- Chapter 2G — Preferential and Managed Lane Signs
- Chapter 2H — General Information Signs
- Chapter 2I — General Service Signs
- Chapter 2J — Specific Service (Logo) Signs
- Chapter 2K — Tourist-Oriented Directional Signs
- Chapter 2L — Changeable Message Signs
- Chapter 2M — Recreational and Cultural Interest Area Signs
- Chapter 2N — Emergency Management Signs

Standard:

03 **Because the requirements and standards for signs depend on the particular type of highway upon which they are to be used, the definitions for freeway, expressway, conventional road, and special purpose road given in Section 1A.13 shall apply in Part 2.**

Section 2A.02 Definitions

Support:

01 Definitions and acronyms that are applicable to signs are given in Sections 1A.13 and 1A.14.

Section 2A.03 Standardization of Application

Support:

01 It is recognized that urban traffic conditions differ from those in rural environments, and in many instances signs are applied and located differently. Where pertinent and practical, this Manual sets forth separate recommendations for urban and rural conditions.

Guidance:

02 *Signs should be used only where justified by engineering judgment or studies, as provided in Section 1A.09.*

03 *Results from traffic engineering studies of physical and traffic factors should indicate the locations where signs are deemed necessary or desirable.*

04 *Roadway geometric design and sign application should be coordinated so that signing can be effectively placed to give the road user any necessary regulatory, warning, guidance, and other information.*

Standard:

05 **Each standard sign shall be displayed only for the specific purpose as prescribed in this Manual. Determination of the particular signs to be applied to a specific condition shall be made in accordance with the provisions set forth in Part 2. Before any new highway, private road open to public travel (see definition in Section 1A.13), detour, or temporary route is opened to public travel, all necessary signs shall be in place. Signs required by road conditions or restrictions shall be removed when those conditions cease to exist or the restrictions are withdrawn.**

Section 2A.04 Excessive Use of Signs

Guidance:

01 *Regulatory and warning signs should be used conservatively because these signs, if used to excess, tend to lose their effectiveness. If used, route signs and directional guide signs should be used frequently because their use promotes efficient operations by keeping road users informed of their location.*

Section 2A.05 Classification of Signs
Standard:

01 **Signs shall be defined by their function as follows:**
   A. Regulatory signs give notice of traffic laws or regulations.
   B. Warning signs give notice of a situation that might not be readily apparent.
   C. Guide signs show route designations, destinations, directions, distances, services, points of interest, and other geographical, recreational, or cultural information.

Support:

02 Object markers are defined in Section 2C.63.

Section 2A.06 Design of Signs
Support:

01 This Manual shows many typical standard signs and object markers approved for use on streets, highways, bikeways, and pedestrian crossings.

02 In the specifications for individual signs and object markers, the general appearance of the legend, color, and size are shown in the accompanying tables and illustrations, and are not always detailed in the text.

03 Detailed drawings of standard signs, object markers, alphabets, symbols, and arrows (see Figure 2D-2) are shown in the "Standard Highway Signs and Markings" book. Section 1A.11 contains information regarding how to obtain this publication.

04 The basic requirements of a sign are that it be legible to those for whom it is intended and that it be understandable in time to permit a proper response. Desirable attributes include:
   A. High visibility by day and night; and
   B. High legibility (adequately sized letters, symbols, or arrows, and a short legend for quick comprehension by a road user approaching a sign).

05 Standardized colors and shapes are specified so that the several classes of traffic signs can be promptly recognized. Simplicity and uniformity in design, position, and application are important.

Standard:

06 **The term legend shall include all word messages and symbol and arrow designs that are intended to convey specific meanings.**

07 **Uniformity in design shall include shape, color, dimensions, legends, borders, and illumination or retroreflectivity.**

08 **Standardization of these designs does not preclude further improvement by minor changes in the proportion or orientation of symbols, width of borders, or layout of word messages, but all shapes and colors shall be as indicated.**

09 **All symbols shall be unmistakably similar to, or mirror images of, the adopted symbol signs, all of which are shown in the "Standard Highway Signs and Markings" book (see Section 1A.11). Symbols and colors shall not be modified unless otherwise provided in this Manual. All symbols and colors for signs not shown in the "Standard Highway Signs and Markings" book shall follow the procedures for experimentation and change described in Section 1A.10.**

Option:

10 Although the standard design of symbol signs cannot be modified, the orientation of the symbol may be changed to better reflect the direction of travel, if appropriate.

Standard:

11 **Where a standard word message is applicable, the wording shall be as provided in this Manual.**

12 **In situations where word messages are required other than those provided in this Manual, the signs shall be of the same shape and color as standard signs of the same functional type.**

Option:

13 State and local highway agencies may develop special word message signs in situations where roadway conditions make it necessary to provide road users with additional regulatory, warning, or guidance information, such as when road users need to be notified of special regulations or warned about a situation that might not be readily apparent. Unlike colors that have not been assigned or symbols that have not been approved for signs, new word message signs may be used without the need for experimentation.

Standard:

14 **Except as provided in Paragraph 16 and except for the Carpool Information (D12-2) sign (see Section 2I.11), Internet addresses and e-mail addresses, including domain names and uniform resource locators (URL), shall not be displayed on any sign, supplemental plaque, sign panel (including logo sign panels on Specific Service signs), or changeable message sign.**

Guidance:

15 *Unless otherwise provided in this Manual for a specific sign, and except as provided in Paragraph 16, telephone numbers of more than four characters should not be displayed on any sign, supplemental plaque, sign panel (including logo sign panels on Specific Service signs), or changeable message sign.*

Option:

16 Internet addresses, e-mail addresses, or telephone numbers with more than four characters may be displayed on signs, supplemental plaques, sign panels, and changeable message signs that are intended for viewing only by pedestrians, bicyclists, occupants of parked vehicles, or drivers of vehicles on low-speed roadways where engineering judgment indicates that an area is available for drivers to stop out of the traffic flow to read the message.

Standard:

17 **Pictographs (see definition in Section 1A.13) shall not be displayed on signs except as specifically provided in this Manual. Pictographs shall be simple, dignified, and devoid of any advertising. When used to represent a political jurisdiction (such as a State, county, or municipal corporation) the pictograph shall be the official designation adopted by the jurisdiction. When used to represent a college or university, the pictograph shall be the official seal adopted by the institution. Pictorial representations of university or college programs shall not be permitted to be displayed on a sign.**

Section 2A.07 Retroreflectivity and Illumination

Support:

01 There are many materials currently available for retroreflection and various methods currently available for the illumination of signs and object markers. New materials and methods continue to emerge. New materials and methods can be used as long as the signs and object markers meet the standard requirements for color, both by day and by night.

Standard:

02 **Regulatory, warning, and guide signs and object markers shall be retroreflective (see Section 2A.08) or illuminated to show the same shape and similar color by both day and night, unless otherwise provided in the text discussion in this Manual for a particular sign or group of signs.**

03 **The requirements for sign illumination shall not be considered to be satisfied by street or highway lighting.**

Option:

04 Sign elements may be illuminated by the means shown in Table 2A-1.

05 Retroreflection of sign elements may be accomplished by the means shown in Table 2A-2.

06 Light Emitting Diode (LED) units may be used individually within the legend or symbol of a sign and in the border of a sign, except for changeable message signs, to improve the conspicuity, increase the legibility of sign legends and borders, or provide a changeable message.

Table 2A-1. Illumination of Sign Elements

Means of Illumination	Sign Element to be Illuminated
Light behind the sign face	• Symbol or word message • Background • Symbol, word message, and background (through a translucent material)
Attached or independently mounted light source designed to direct essentially uniform illumination onto the sign face	• Entire sign face
Light emitting diodes (LEDs)	• Symbol or word message • Portions of the sign border

Other devices, or treatments that highlight the sign shape, color, or message:
  Luminous tubing
  Fiber optics
  Incandescent light bulbs
  Luminescent panels
	• Symbol or word message
	• Entire sign face

Table 2A-2. Retroreflection of Sign Elements

Means of Retroreflection	Sign Element
Reflector "buttons" or similar units	Symbol Word message Border
A material that has a smooth, sealed outer surface over a microstructure that reflects light	Symbol Word message Border Background

Standard:

07 **Except as provided in Paragraphs 11 and 12, neither individual LEDs nor groups of LEDs shall be placed within the background area of a sign.**

08 **If used, the LEDs shall have a maximum diameter of 1/4 inch and shall be the following colors based on the type of sign:**
   A. **White or red, if used with STOP or YIELD signs.**
   B. **White, if used with regulatory signs other than STOP or YIELD signs.**
   C. **White or yellow, if used with warning signs.**
   D. **White, if used with guide signs.**
   E. **White, yellow, or orange, if used with temporary traffic control signs.**
   F. **White or yellow, if used with school area signs.**

09 **If flashed, all LED units shall flash simultaneously at a rate of more than 50 and less than 60 times per minute.**

10 **The uniformity of the sign design shall be maintained without any decrease in visibility, legibility, or driver comprehension during either daytime or nighttime conditions.**

Option:

11 For STOP and YIELD signs, LEDs may be placed within the border or within one border width within the background of the sign.

12 For STOP/SLOW paddles (see Section 6E.03) used by flaggers and the STOP paddles (see Section 7D.05) used by adult crossing guards, individual LEDs or groups of LEDs may be used.

Support:

13 Other methods of enhancing the conspicuity of standard signs are described in Section 2A.15.

14 Information regarding the use of retroreflective material on the sign support is contained in Section 2A.21.

Section 2A.08 Maintaining Minimum Retroreflectivity

Support:

01 Retroreflectivity is one of several factors associated with maintaining nighttime sign visibility (see Section 2A.22).

Standard:

02 **Public agencies or officials having jurisdiction shall use an assessment or management method that is designed to maintain sign retroreflectivity at or above the minimum levels in Table 2A-3.**

Support:

03 Compliance with the Standard in Paragraph 2 is achieved by having a method in place and using the method to maintain the minimum levels established in Table 2A-3. Provided that an assessment or management method is being used, an agency or official having jurisdiction would be in compliance with the Standard in Paragraph 2 even if there are some individual signs that do not meet the minimum retroreflectivity levels at a particular point in time.

Guidance:

04 *Except for those signs specifically identified in Paragraph 6, one or more of the following assessment or management methods should be used to maintain sign retroreflectivity:*
   A. *Visual Nighttime Inspection—The retroreflectivity of an existing sign is assessed by a trained sign inspector conducting a visual inspection from a moving vehicle during nighttime conditions. Signs that are visually identified by the inspector to have retroreflectivity below the minimum levels should be replaced.*
   B. *Measured Sign Retroreflectivity—Sign retroreflectivity is measured using a retroreflectometer. Signs with retroreflectivity below the minimum levels should be replaced.*
   C. *Expected Sign Life—When signs are installed, the installation date is labeled or recorded so that the age of a sign is known. The age of the sign is compared to the expected sign life. The expected sign life is based on the experience of sign retroreflectivity degradation in a geographic area compared to the minimum levels. Signs older than the expected life should be replaced.*
   D. *Blanket Replacement—All signs in an area/corridor, or of a given type, should be replaced at specified intervals. This eliminates the need to assess retroreflectivity or track the life of individual signs. The replacement interval is based on the expected sign life, compared to the minimum levels, for the shortest-life material used on the affected signs.*

Table 2A-3. Minimum Maintained Retroreflectivity Levels[1]

Sign Color	Sheeting Type (ASTM D4956-04)				Additional Criteria
	Beaded Sheeting			Prismatic Sheeting	
	I	II	III	III, IV, VI, VII, VIII, IX, X	
White on Green	W*; G ≥ 7	W*; G ≥ 15	W*; G ≥ 25	W ≥ 250; G ≥ 25	Overhead
	W*; G ≥ 7	W ≥ 120; G ≥ 15			Post-mounted
Black on Yellow or Black on Orange	Y*; O*	Y ≥ 50; O ≥ 50			2
	Y*; O*	Y ≥ 75; O ≥ 75			3
White on Red	W ≥ 35; R ≥ 7				4
Black on White	W ≥ 50				—

[1] The minimum maintained retroreflectivity levels shown in this table are in units of cd/lx/m² measured at an observation angle of 0.2° and an entrance angle of -4.0°.
[2] For text and fine symbol signs measuring at least 48 inches and for all sizes of bold symbol signs
[3] For text and fine symbol signs measuring less than 48 inches
[4] Minimum sign contrast ratio ≥ 3:1 (white retroreflectivity ÷ red retroreflectivity)
* This sheeting type shall not be used for this color for this application.

Bold Symbol Signs

- W1-1,2 – Turn and Curve
- W1-3,4 – Reverse Turn and Curve
- W1-5 – Winding Road
- W1-6,7 – Large Arrow
- W1-8 – Chevron
- W1-10 – Intersection in Curve
- W1-11 – Hairpin Curve
- W1-15 – 270 Degree Loop
- W2-1 – Cross Road
- W2-2,3 – Side Road
- W2-4,5 – T and Y Intersection
- W2-6 – Circular Intersection
- W2-7,8 – Double Side Roads

- W3-1 – Stop Ahead
- W3-2 – Yield Ahead
- W3-3 – Signal Ahead
- W4-1 – Merge
- W4-2 – Lane Ends
- W4-3 – Added Lane
- W4-5 – Entering Roadway Merge
- W4-6 – Entering Roadway Added Lane
- W6-1,2 – Divided Highway Begins and Ends
- W6-3 – Two-Way Traffic
- W10-1,2,3,4,11,12 – Grade Crossing Advance Warning

- W11-2 – Pedestrian Crossing
- W11-3,4,16-22 – Large Animals
- W11-5 – Farm Equipment
- W11-6 – Snowmobile Crossing
- W11-7 – Equestrian Crossing
- W11-8 – Fire Station
- W11-10 – Truck Crossing
- W12-1 – Double Arrow
- W16-5P,6P,7P – Pointing Arrow Plaques
- W20-7 – Flagger
- W21-1 – Worker

Fine Symbol Signs (symbol signs not listed as bold symbol signs)

Special Cases

- W3-1 – Stop Ahead: Red retroreflectivity ≥ 7
- W3-2 – Yield Ahead: Red retroreflectivity ≥ 7; White retroreflectivity ≥ 35
- W3-3 – Signal Ahead: Red retroreflectivity ≥ 7; Green retroreflectivity ≥ 7
- W3-5 – Speed Reduction: White retroreflectivity ≥ 50
- For non-diamond shaped signs, such as W14-3 (No Passing Zone), W4-4P (Cross Traffic Does Not Stop), or W13-1P,2,3,6,7 (Speed Advisory Plaques), use the largest sign dimension to determine the proper minimum retroreflectivity level.

E. *Control Signs—Replacement of signs in the field is based on the performance of a sample of control signs. The control signs might be a small sample located in a maintenance yard or a sample of signs in the field. The control signs are monitored to determine the end of retroreflective life for the associated signs. All field signs represented by the control sample should be replaced before the retroreflectivity levels of the control sample reach the minimum levels.*

F. *Other Methods—Other methods developed based on engineering studies can be used.*

Support:

Additional information about these methods is contained in the 2007 Edition of FHWA's "Maintaining Traffic Sign Retroreflectivity" (see Section 1A.11).

Option:

Highway agencies may exclude the following signs from the retroreflectivity maintenance guidelines described in this Section:

A. Parking, Standing, and Stopping signs (R7 and R8 series)
B. Walking/Hitchhiking/Crossing signs (R9 series, R10-1 through R10-4b)
C. Acknowledgment signs
D. All signs with blue or brown backgrounds
E. Bikeway signs that are intended for exclusive use by bicyclists or pedestrians

Section 2A.09 Shapes
Standard:

01　Particular shapes, as shown in Table 2A-4, shall be used exclusively for specific signs or series of signs, unless otherwise provided in the text discussion in this Manual for a particular sign or class of signs.

Section 2A.10 Sign Colors
Standard:

01　The colors to be used on standard signs and their specific use on these signs shall be as provided in the applicable Sections of this Manual. The color coordinates and values shall be as described in 23 CFR, Part 655, Subpart F, Appendix.

Table 2A-4. Use of Sign Shapes

Shape	Signs
Octagon	Stop*
Equilateral Triangle (1 point down)	Yield*
Circle	Grade Crossing Advance Warning*
Pennant Shape/Isosceles Triangle (longer axis horizontal)	No Passing*
Pentagon (pointed up)	School Advance Warning Sign (squared bottom corners)* County Route Sign (tapered bottom corners)*
Crossbuck (two rectangles in an "X" configuration)	Grade Crossing*
Diamond	Warning Series
Rectangle (including square)	Regulatory Series Guide Series** Warning Series
Trapezoid	Recreational and Cultural Interest Area Series National Forest Route Sign

* This sign shall be exclusively the shape shown.

** Guide series includes general service, specific service, tourist-oriented directional, general information, recreational and cultural interest area, and emergency management signs.

Support:

02　As a quick reference, common uses of sign colors are shown in Table 2A-5. Color schemes on specific signs are shown in the illustrations located in each appropriate Chapter.

03　Whenever white is specified in this Manual or in the "Standard Highway Signs and Markings" book (see Section 1A.11) as a color, it is understood to include silver-colored retroreflective coatings or elements that reflect white light.

04　The colors coral and light blue are being reserved for uses that will be determined in the future by the Federal Highway Administration.

05　Information regarding color coding of destinations on guide signs, including community wayfinding signs, is contained in Chapter 2D.

Option:

06　The approved fluorescent version of the standard red, yellow, green, or orange color may be used as an alternative to the corresponding standard color.

Section 2A.11 Dimensions
Support:

01　The "Standard Highway Signs and Markings" book (see Section 1A.11) prescribes design details for up to five different sizes depending on the type of traffic facility, including bikeways. Smaller sizes are designed to be used on bikeways and some other off-road applications. Larger sizes are designed for use on freeways and expressways, and can also be used to enhance road user safety and convenience on other facilities, especially on multi-lane divided highways and on undivided highways having five or more lanes of traffic and/or high speeds. The intermediate sizes are designed to be used on other highway types.

Standard:

02　**The sign dimensions prescribed in the sign size tables in the various Parts and Chapters in this Manual and in the "Standard Highway Signs and Markings" book (see Section 1A.11) shall be used unless engineering judgment determines that other sizes are appropriate. Except as provided in Paragraph 3, where engineering judgment determines that sizes smaller than the prescribed dimensions are appropriate for use, the sign dimensions shall not be less than the minimum dimensions specified in this Manual. The sizes shown in the Minimum columns that are smaller than the sizes shown in the Conventional Road columns in the various sign size tables in this Manual shall only be used on low-speed roadways, alleys, and private roads open to public travel where the reduced legend size would be adequate for the regulation or warning or where physical conditions preclude the use of larger sizes.**

Table 2A-5. Common Uses of Sign Colors

Type of Sign	Legend								Background										
	Black	Green	Red	White	Yellow	Orange	Fluorescent Yellow-Green	Fluorescent Pink	Black	Blue	Brown	Green	Orange*	Red*	White	Yellow*	Purple	Fluorescent Yellow-Green	Fluorescent Pink
Regulatory	X		X	X					X					X	X				
Prohibitive			X	X										X	X				
Permissive		X													X				
Warning	X															X			
Pedestrian	X															X		X	
Bicycle	X															X		X	
Guide				X								X							
Interstate Route				X						X					X				
State Route	X														X				
U.S. Route	X														X				
County Route					X					X									
Forest Route				X							X								
Street Name				X								X							
Destination				X								X							
Reference Location				X								X							
Information				X						X		X							
Evacuation Route				X						X									
Road User Service				X						X									
Recreational				X							X	X							
Temporary Traffic Control	X												X						
Incident Management	X													X					X
School	X																	X	
ETC-Account Only	X																X****		
Changeable Message Signs																			
Regulatory			X***	X					X										
Warning					X				X										
Temporary Traffic Control					X	X			X										
Guide				X					X			X**							
Motorist Services				X					X	X**									
Incident Management					X			X	X										
School, Pedestrian, Bicycle					X		X		X										

* Fluorescent versions of these background colors may also be used.

** These alternative background colors would be provided by blue or green lighted pixels such that the entire CMS would be lighted, not just the legend.

*** Red is used only for the circle and slash or other red elements of a similar static regulatory sign.

**** The use of the color purple on signs is restricted per the provisions of Paragraph 1 of Section 2F.03.

Option:

03 For alleys with restrictive physical conditions and vehicle usage that limits installation of the Minimum size sign (or the Conventional Road size sign if no Minimum size is shown), both the sign height and the sign width may be decreased by up to 6 inches.

Guidance:

04 *The sizes shown in the Freeway and Expressway columns in the various sign size tables in this Manual should be used on freeways and expressways, and for other higher-speed applications based upon engineering judgment, to provide larger signs for increased visibility and recognition.*

05 *The sizes shown in the Oversized columns in the various sign size tables in this Manual size should be used for those special applications where speed, volume, or other factors result in conditions where increased emphasis, improved recognition, or increased legibility is needed, as determined by engineering judgment or study.*

06 *Increases above the prescribed sizes should be used where greater legibility or emphasis is needed. If signs larger than the prescribed sizes are used, the overall sign dimensions should be increased in 6-inch increments.*

Standard:

07 **Where engineering judgment determines that sizes that are different than the prescribed dimensions are appropriate for use, standard shapes and colors shall be used and standard proportions shall be retained as much as practical.**

Guidance:

08 *When supplemental plaques are installed with larger sized signs, a corresponding increase in the size of the plaque and its legend should also be made. The resulting plaque size should be approximately in the same relative proportion to the larger sized sign as the conventional sized plaque is to the conventional sized sign.*

Section 2A.12 Symbols

Standard:

01 **Symbol designs shall in all cases be unmistakably similar to those shown in this Manual and in the "Standard Highway Signs and Markings" book (see Section 1A.11).**

Support:

02 New symbol designs are adopted by the Federal Highway Administration based on research evaluations to determine road user comprehension, sign conspicuity, and sign legibility.

03 Sometimes a change from word messages to symbols requires significant time for public education and transition. Therefore, this Manual sometimes includes the practice of using educational plaques to accompany new symbol signs.

Guidance:

04 *New warning or regulatory symbol signs not readily recognizable by the public should be accompanied by an educational plaque.*

Option:

05 Educational plaques may be left in place as long as they are in serviceable condition.

06 State and/or local highway agencies may conduct research studies to determine road user comprehension, sign conspicuity, and sign legibility.

Guidance:

07 *Although most standard symbols are oriented facing left, mirror images of these symbols should be used where the reverse orientation might better convey to road users a direction of movement.*

Standard:

08 **A symbol used for a given category of signs (regulatory, warning, or guide) shall not be used for a different category of signs, except as specifically authorized in this Manual.**

09 **Except as provided in Paragraph 11, a recreational and cultural interest area symbol (see Chapter 2M) shall not be used on streets or highways outside of recreational and cultural interest areas.**

10 **A recreational and cultural interest area guide sign symbol (see Chapter 2M) shall not be used on any regulatory or warning sign on any street, road, or highway.**

Option:

11 A recreational and cultural interest area guide sign symbol (see Section 2M.04) may be used on a highway guide sign outside of a recreational and cultural interest area to supplement a comparable word message for which there is no approved symbol for that message in Chapters 2B through 2I or 2N.

Section 2A.13 Word Messages

Standard:

01 **Except as provided in Section 2A.06, all word messages shall use standard wording and letters as shown in this Manual and in the "Standard Highway Signs and Markings" book (see Section 1A.11).**

Guidance:

02 *Word messages should be as brief as possible and the lettering should be large enough to provide the necessary legibility distance. A minimum specific ratio of 1 inch of letter height per 30 feet of legibility distance should be used.*

03 *Abbreviations (see Section 1A.15) should be kept to a minimum.*

04 *Word messages should not contain periods, apostrophes, question marks, ampersands, or other punctuation or characters that are not letters, numerals, or hyphens unless necessary to avoid confusion.*

05 *The solidus (slanted line or forward slash) is intended to be used for fractions only and should not be used to separate words on the same line of legend. Instead, a hyphen should be used for this purpose, such as "TRUCKS - BUSES."*

Standard:

06 **Fractions shall be displayed with the numerator and denominator diagonally arranged about the solidus (slanted line or forward slash). The overall height of the fraction is measured from the top of the numerator to the bottom of the denominator, each of which is vertically aligned with the upper and lower ends of the solidus. The overall height of the fraction shall be determined by the height of the numerals within the fraction, and shall be 1.5 times the height of an individual numeral within the fraction.**

Support:

07 The "Standard Highway Signs and Markings" book (see Section 1A.11) contains details regarding the layouts of fractions on signs.

Guidance:

08 *When initials are used to represent an abbreviation for separate words (such as "U S" for a United States route), the initials should be separated by a space of between 1/2 and 3/4 of the letter height of the initials.*

09 *When an Interstate route is displayed in text form instead of using the route shield, a hyphen should be used for clarity, such as "I-50."*

Standard:

10 **All sign lettering shall be in upper-case letters as provided in the "Standard Highway Signs and Markings" book (see Section 1A.11), unless otherwise provided in this Manual for a particular sign or type of message.**

11 **The sign lettering for names of places, streets, and highways shall be composed of a combination of lower-case letters with initial upper-case letters.**

Support:

12 Letter height is expressed in terms of the height of an upper-case letter. For mixed-case legends (those composed of an initial upper-case letter followed by lower-case letters), the height of the lower-case letters is derived from the specified height of the initial upper-case letter based on a prescribed ratio. Letter heights for mixed-case legends might be expressed in terms of both the upper- and lower-case letters, or in terms of the initial upper-case letter alone. When the height of a lower-case letter is specified or determined from the prescribed ratio, the reference is to the nominal loop height of the letter. The term loop height refers to the portion of a lower-case letter that excludes any ascending or descending stems or tails of the letter, such as with the letters "d" or "q." The nominal loop height is equal to the actual height of a non-rounded lower-case letter whose form does not include ascending or descending stems or tails, such as the letter "x." The rounded portions of a lower-case letter extend slightly above and below the baselines projected from the top and bottom of such a non-rounded letter so that the appearance of a uniform letter height within a word is achieved. The actual loop height of a rounded lower-case letter is slightly greater than the nominal loop height and this additional height is excluded from the expression of the lower-case letter height.

Standard:

13 **When a mixed-case legend is used, the height of the lower-case letters shall be 3/4 of the height of the initial upper-case letter.**

14 **The unique letter forms for each of the Standard Alphabet series shall not be stretched, compressed, warped, or otherwise manipulated.**

Support:

15 Section 2D.04 contains information regarding the acceptable methods of modifying the length of a word for a given letter height and series.

Section 2A.14 Sign Borders
Standard:

01 **Unless otherwise provided, each sign illustrated in this Manual shall have a border of the same color as the legend, at or just inside the edge.**

02 **The corners of all sign borders shall be rounded, except for STOP signs.**

Guidance:

03 *A dark border on a light background should be set in from the edge, while a light border on a dark background should extend to the edge of the sign. A border for 30-inch signs with a light background should be from 1/2 to 3/4 inch in width, 1/2 inch from the edge. For similar signs with a light border, a width of 1 inch should be used. For other sizes, the border width should be of similar proportions, but should not exceed the stroke-width of the major lettering of the sign. On signs exceeding 72 x 120 inches in size, the border should be 2 inches wide, or on larger signs, 3 inches wide. Except for STOP signs and as otherwise provided in Section 2E.16, the corners of the sign should be rounded to a radius that is concentric with that of the border.*

Section 2A.15 Enhanced Conspicuity for Standard Signs
Option:

01 Based upon engineering judgment, where the improvement of the conspicuity of a standard regulatory, warning, or guide sign is desired, any of the following methods may be used, as appropriate, to enhance the sign's conspicuity (see Figure 2A-1):

    A. Increasing the size of a standard regulatory, warning, or guide sign.
    B. Doubling-up of a standard regulatory, warning, or guide sign by adding a second identical sign on the left-hand side of the roadway.
    C. Adding a solid yellow or fluorescent yellow rectangular "header panel" above a standard regulatory sign, with the width of the panel corresponding to the width of the standard regulatory sign. A legend of "NOTICE," "STATE LAW," or other appropriate text may be added in black letters within the header panel for a period of time determined by engineering judgment.
    D. Adding a NEW plaque (see Section 2C.62) above a new standard regulatory or warning sign, for a period of time determined by engineering judgment, to call attention to the new sign.
    E. Adding one or more red or orange flags (cloth or retroreflective sheeting) above a standard regulatory or warning sign, with the flags oriented so as to be at 45 degrees to the vertical.
    F. Adding a solid yellow, a solid fluorescent yellow, or a diagonally striped black and yellow (or black and fluorescent yellow) strip of retroreflective sheeting at least 3 inches wide around the perimeter of a standard warning sign. This may be accomplished by affixing the standard warning sign on a background that is 6 inches larger than the size of the standard warning sign.
    G. Adding a warning beacon (see Section 4L.03) to a standard regulatory (other than a STOP or a Speed Limit sign), warning, or guide sign.
    H. Adding a speed limit sign beacon (see Section 4L.04) to a standard Speed Limit sign.
    I. Adding a stop beacon (see Section 4L.05) to a STOP sign.
    J. Adding light emitting diode (LED) units within the symbol or legend of a sign or border of a standard regulatory, warning, or guide sign, as provided in Section 2A.07.
    K. Adding a strip of retroreflective material to the sign support in compliance with the provisions of Section 2A.21.
    L. Using other methods that are specifically allowed for certain signs as described elsewhere in this Manual.

Support:

02 Sign conspicuity improvements can also be achieved by removing non-essential and illegal signs from the right-of-way (see Section 1A.08), and by relocating signs to provide better spacing.

Standard:

03 **The NEW plaque (see Section 2C.62) shall not be used alone.**

04 **Strobe lights shall not be used to enhance the conspicuity of highway signs.**

Figure 2A-1. Examples of Enhanced Conspicuity for Signs

A – W16-15P plaque above a regulatory or warning sign if the regulation or condition is new

B – Red or orange flags above a regulatory, warning, or guide sign

C – W16-18P plaque above a regulatory sign

D – Solid yellow, solid fluorescent yellow, or diagonally striped black and yellow (or black and fluorescent yellow) strip of retroreflective sheeting around a warning sign

E – Vertical retroreflective strip on sign support

F – Supplemental beacon

Section 2A.16 Standardization of Location

Support:

01 Standardization of position cannot always be attained in practice. Examples of heights and lateral locations of signs for typical installations are illustrated in Figure 2A-2, and examples of locations for some typical signs at intersections are illustrated in Figures 2A-3 and 2A-4.

02 Examples of advance signing on an intersection approach are illustrated in Figure 2A-4. Chapters 2B, 2C, and 2D contain provisions regarding the application of regulatory, warning, and guide signs, respectively.

Standard:

03 **Signs requiring separate decisions by the road user shall be spaced sufficiently far apart for the appropriate decisions to be made. One of the factors considered when determining the appropriate spacing shall be the posted or 85th-percentile speed.**

Guidance:

04 *Signs should be located on the right-hand side of the roadway where they are easily recognized and understood by road users. Signs in other locations should be considered only as supplementary to signs in the normal locations, except as otherwise provided in this Manual.*

05 *Signs should be individually installed on separate posts or mountings except where:*

    A. *One sign supplements another;*
    B. *Route or directional signs are grouped to clarify information to motorists;*

Figure 2A-2. Examples of Heights and Lateral Locations of Sign Installations

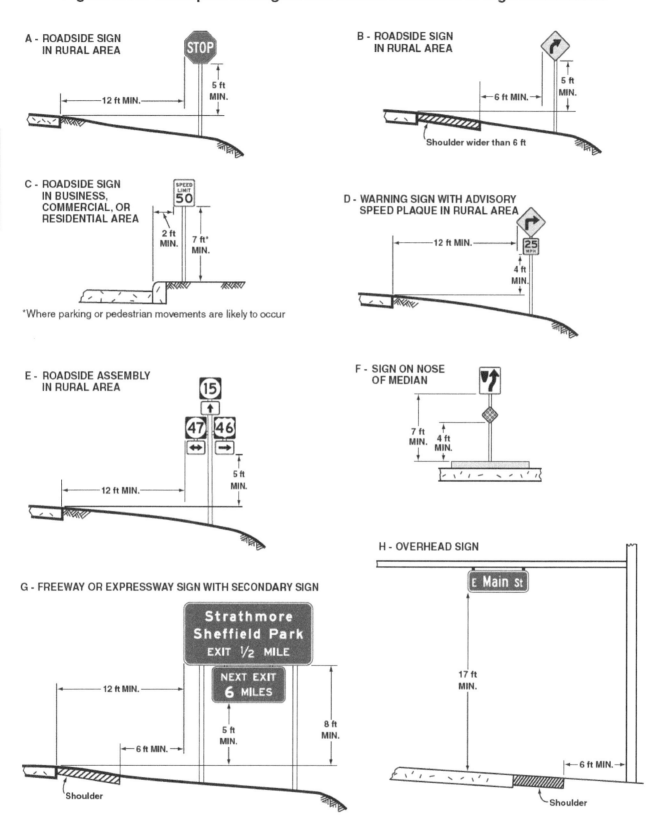

Note:
See Section 2A.19 for reduced lateral offset distances that may be used in areas where lateral offsets are limited, and in business, commercial, or residential areas where sidewalk width is limited or where existing poles are close to the curb.

Figure 2A-3. Examples of Locations for Some Typical Signs at Intersections

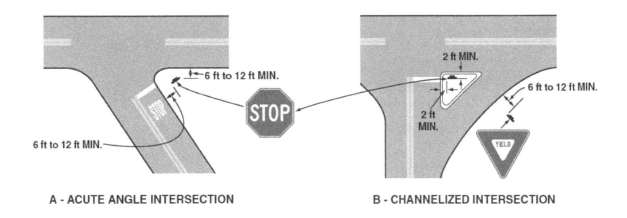

A - ACUTE ANGLE INTERSECTION

B - CHANNELIZED INTERSECTION

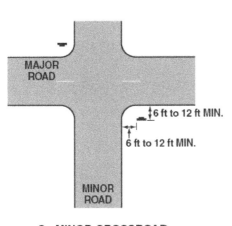

C - MINOR CROSSROAD

D - URBAN INTERSECTION

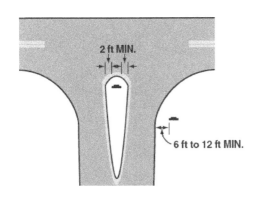

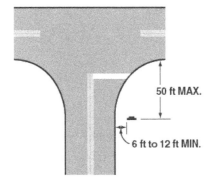

E - DIVISIONAL ISLAND

F - WIDE THROAT INTERSECTION

Note: Lateral offset is a minimum of 6 feet measured from the edge of the shoulder, or 12 feet measured from the edge of the traveled way. See Section 2A.19 for lower minimums that may be used in urban areas, or where lateral offset space is limited.

Figure 2A-4. Relative Locations of Regulatory, Warning, and Guide Signs on an Intersection Approach

A – Single-lane approach

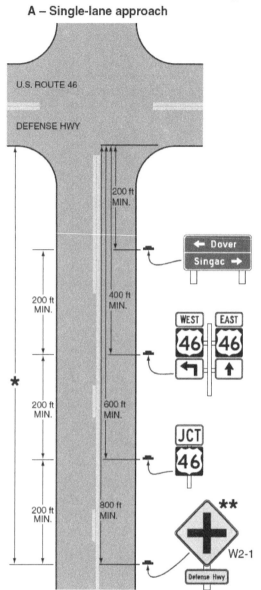

B – Multi-lane approach

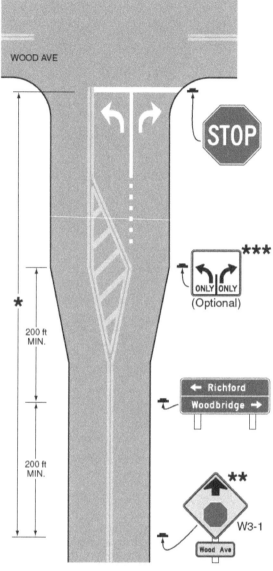

★ See Table 2C-4 for the recommended minimum distance
★★ See Section 2C.46 for the application of the W2-1 sign and Section 2C.36 for the application of the W3-1 sign
★★★ See Section 2B.22 for the application of Intersection Lane Control signs

Note: See Chapter 2D for information on guide signs and Part 3 for information on pavement markings

---

C. Regulatory signs that do not conflict with each other are grouped, such as Turn Prohibition signs posted with ONE WAY signs or a parking regulation sign posted with a Speed Limit sign; or
D. Street Name signs are posted with a STOP or YIELD sign.

06 Signs should be located so that they:

A. Are outside the clear zone unless placed on a breakaway or yielding support (see Section 2A.19),
B. Optimize nighttime visibility,
C. Minimize the effects of mud splatter and debris,
D. Do not obscure each other,
E. Do not obscure the sight distance to approaching vehicles on the major street for drivers who are stopped on minor-street approaches, and
F. Are not hidden from view.

Support:

07 The clear zone is the total roadside border area, starting at the edge of the traveled way, available for use by errant vehicles. The width of the clear zone is dependent upon traffic volumes, speeds, and roadside geometry. Additional information can be found in AASHTO's "Roadside Design Guide" (see Section 1A.11).

Guidance:

08 *With the increase in traffic volumes and the desire to provide road users regulatory, warning, and guidance information, an order of priority for sign installation should be established.*

Support:

09 An order of priority is especially critical where space is limited for sign installation and there is a demand for several different types of signs. Overloading road users with too much information is not desirable.

Guidance:

10 *Because regulatory and warning information is more critical to the road user than guidance information, regulatory and warning signing whose location is critical should be displayed rather than guide signing in cases where conflicts occur. Community wayfinding and acknowledgment guide signs should have a lower priority as to placement than other guide signs. Information of a less critical nature should be moved to less critical locations or omitted.*

Option:

11 Under some circumstances, such as on curves to the right, signs may be placed on median islands or on the left-hand side of the road. A supplementary sign located on the left-hand side of the roadway may be used on a multi-lane road where traffic in a lane to the right might obstruct the view to the right.

Guidance:

12 *In urban areas where crosswalks exist, signs should not be placed within 4 feet in advance of the crosswalk (see Drawing D in Figure 2A-3).*

Section 2A.17 Overhead Sign Installations

Guidance:

01 *Overhead signs should be used on freeways and expressways, at locations where some degree of lane-use control is desirable, and at locations where space is not available at the roadside.*

Support:

02 The operational requirements of the present highway system are such that overhead signs have value at many locations. The factors to be considered for the installation of overhead sign displays are not definable in specific numerical terms.

Option:

03 The following conditions (not in priority order) may be considered in an engineering study to determine if overhead signs would be beneficial:
- A. Traffic volume at or near capacity,
- B. Complex interchange design,
- C. Three or more lanes in each direction,
- D. Restricted sight distance,
- E. Closely-spaced interchanges,
- F. Multi-lane exits,
- G. Large percentage of trucks,
- H. Street lighting background,
- I. High-speed traffic,
- J. Consistency of sign message location through a series of interchanges,
- K. Insufficient space for post-mounted signs,
- L. Junction of two freeways, and
- M. Left exit ramps.

04 Over-crossing structures may be used to support overhead signs.

Support:

05 Under some circumstances, the use of over-crossing structures as sign supports might be the only practical solution that will provide adequate viewing distance. The use of such structures as sign supports might eliminate the need for the foundations and sign supports along the roadside.

Section 2A.18 Mounting Height

Standard:

01 **The provisions of this Section shall apply unless specifically stated otherwise for a particular sign or object marker elsewhere in this Manual.**

Support:

02 The mounting height requirements for object markers are provided in Chapter 2C.

03 In addition to the provisions of this Section, information affecting the minimum mounting height of signs as a function of crash performance can be found in AASHTO's "Roadside Design Guide" (see Section 1A.11).

Standard:

04 **The minimum height, measured vertically from the bottom of the sign to the elevation of the near edge of the pavement, of signs installed at the side of the road in rural areas shall be 5 feet (see Figure 2A-2).**

05 **The minimum height, measured vertically from the bottom of the sign to the top of the curb, or in the absence of curb, measured vertically from the bottom of the sign to the elevation of the near edge of the traveled way, of signs installed at the side of the road in business, commercial, or residential areas where parking or pedestrian movements are likely to occur, or where the view of the sign might be obstructed, shall be 7 feet (see Figure 2A-2).**

Option:

06 The height to the bottom of a secondary sign mounted below another sign may be 1 foot less than the height specified in Paragraphs 4 and 5.

Standard:

07 **The minimum height, measured vertically from the bottom of the sign to the sidewalk, of signs installed above sidewalks shall be 7 feet.**

08 **If the bottom of a secondary sign that is mounted below another sign is mounted lower than 7 feet above a pedestrian sidewalk or pathway (see Section 6D.02), the secondary sign shall not project more than 4 inches into the pedestrian facility.**

Option:

09 Signs that are placed 30 feet or more from the edge of the traveled way may be installed with a minimum height of 5 feet, measured vertically from the bottom of the sign to the elevation of the near edge of the pavement.

Standard:

10 **Directional signs on freeways and expressways shall be installed with a minimum height of 7 feet, measured vertically from the bottom of the sign to the elevation of the near edge of the pavement. All route signs, warning signs, and regulatory signs on freeways and expressways shall be installed with a minimum height of 7 feet, measured vertically from the bottom of the sign to the elevation of the near edge of the pavement. If a secondary sign is mounted below another sign on a freeway or expressway, the major sign shall be installed with a minimum height of 8 feet and the secondary sign shall be installed with a minimum height of 5 feet, measured vertically from the bottom of the sign to the elevation of the near edge of the pavement.**

11 **Where large signs having an area exceeding 50 square feet are installed on multiple breakaway posts, the clearance from the ground to the bottom of the sign shall be at least 7 feet.**

Option:

12 A route sign assembly consisting of a route sign and auxiliary signs (see Section 2D.31) may be treated as a single sign for the purposes of this Section.

13 The mounting height may be adjusted when supports are located near the edge of the right-of-way on a steep backslope in order to avoid the sometimes less desirable alternative of placing the sign closer to the roadway.

Standard:

14 **Overhead signs shall provide a vertical clearance of not less than 17 feet to the sign, light fixture, or sign bridge over the entire width of the pavement and shoulders except where the structure on which the overhead signs are to be mounted or other structures along the roadway near the sign structure have a lesser vertical clearance.**

Option:

15 If the vertical clearance of other structures along the roadway near the sign structure is less than 16 feet, the vertical clearance to an overhead sign structure or support may be as low as 1 foot higher than the vertical clearance of the other structures in order to improve the visibility of the overhead signs.

16 In special cases it may be necessary to reduce the clearance to overhead signs because of substandard dimensions in tunnels and other major structures such as double-deck bridges.

Support:

17 Figure 2A-2 illustrates some examples of the mounting height requirements contained in this Section.

Section 2A.19 Lateral Offset

Standard:

01 **For overhead sign supports, the minimum lateral offset from the edge of the shoulder (or if no shoulder exists, from the edge of the pavement) to the near edge of overhead sign supports (cantilever or sign bridges) shall be 6 feet. Overhead sign supports shall have a barrier or crash cushion to shield them if they are within the clear zone.**

02 **Post-mounted sign and object marker supports shall be crashworthy (breakaway, yielding, or shielded with a longitudinal barrier or crash cushion) if within the clear zone.**

Guidance:

03 *For post-mounted signs, the minimum lateral offset should be 12 feet from the edge of the traveled way. If a shoulder wider than 6 feet exists, the minimum lateral offset for post-mounted signs should be 6 feet from the edge of the shoulder.*

Support:

04 The minimum lateral offset requirements for object markers are provided in Chapter 2C.

05 The minimum lateral offset is intended to keep trucks and cars that use the shoulders from striking the signs or supports.

Guidance:

06 *All supports should be located as far as practical from the edge of the shoulder. Advantage should be taken to place signs behind existing roadside barriers, on over-crossing structures, or other locations that minimize the exposure of the traffic to sign supports.*

Option:

07 Where permitted, signs may be placed on existing supports used for other purposes, such as highway traffic signal supports, highway lighting supports, and utility poles.

Standard:

08 **If signs are placed on existing supports, they shall meet other placement criteria contained in this Manual.**

Option:

09 Lesser lateral offsets may be used on connecting roadways or ramps at interchanges, but not less than 6 feet from the edge of the traveled way.

10 On conventional roads in areas where it is impractical to locate a sign with the lateral offset prescribed by this Section, a lateral offset of at least 2 feet may be used.

11 A lateral offset of at least 1 foot from the face of the curb may be used in business, commercial or residential areas where sidewalk width is limited or where existing poles are close to the curb.

Guidance:

12 *Overhead sign supports and post-mounted sign and object marker supports should not intrude into the usable width of a sidewalk or other pedestrian facility.*

Support:

13 Figures 2A-2 and 2A-3 illustrate some examples of the lateral offset requirements contained in this Section.

Section 2A.20 Orientation

Guidance:

01 *Unless otherwise provided in this Manual, signs should be vertically mounted at right angles to the direction of, and facing, the traffic that they are intended to serve.*

02 *Where mirror reflection from the sign face is encountered to such a degree as to reduce legibility, the sign should be turned slightly away from the road. Signs that are placed 30 feet or more from the pavement edge should be turned toward the road. On curved alignments, the angle of placement should be determined by the direction of approaching traffic rather than by the roadway edge at the point where the sign is located.*

Option:

03 On grades, sign faces may be tilted forward or back from the vertical position to improve the viewing angle.

Section 2A.21 Posts and Mountings

Standard:

01 **Sign posts, foundations, and mountings shall be so constructed as to hold signs in a proper and permanent position, and to resist swaying in the wind or displacement by vandalism.**

Support:

02 The latest edition of AASHTO's "Specifications for Structural Supports for Highway Signs, Luminaires, and Traffic Signals" contains additional information regarding posts and mounting (see Page i for AASHTO's address).

Option:

03 Where engineering judgment indicates a need to draw attention to the sign during nighttime conditions, a strip of retroreflective material may be used on regulatory and warning sign supports.

Standard:

04 **If a strip of retroreflective material is used on the sign support, it shall be at least 2 inches in width, it shall be placed for the full length of the support from the sign to within 2 feet above the edge of the roadway, and its color shall match the background color of the sign, except that the color of the strip for the YIELD and DO NOT ENTER signs shall be red.**

Section 2A.22 Maintenance

Guidance:

01 *Maintenance activities should consider proper position, cleanliness, legibility, and daytime and nighttime visibility (see Section 2A.09). Damaged or deteriorated signs, gates, or object markers should be replaced.*

02 *To assure adequate maintenance, a schedule for inspecting (both day and night), cleaning, and replacing signs, gates, and object markers should be established. Employees of highway, law enforcement, and other public agencies whose duties require that they travel on the roadways should be encouraged to report any damaged, deteriorated, or obscured signs, gates, or object markers at the first opportunity.*

03 *Steps should be taken to see that weeds, trees, shrubbery, and construction, maintenance, and utility materials and equipment do not obscure the face of any sign or object marker.*

04 *A regular schedule of replacement of lighting elements for illuminated signs should be maintained.*

Section 2A.23 Median Opening Treatments for Divided Highways with Wide Medians

Guidance:

01 *Where divided highways are separated by median widths at the median opening itself of 30 feet or more, median openings should be signed as two separate intersections.*

CHAPTER 2B. REGULATORY SIGNS, BARRICADES, AND GATES

Section 2B.01 Application of Regulatory Signs
Standard:

01 **Regulatory signs shall be used to inform road users of selected traffic laws or regulations and indicate the applicability of the legal requirements.**

02 **Regulatory signs shall be installed at or near where the regulations apply. The signs shall clearly indicate the requirements imposed by the regulations and shall be designed and installed to provide adequate visibility and legibility in order to obtain compliance.**

03 **Regulatory signs shall be retroreflective or illuminated (see Section 2A.07) to show the same shape and similar color by both day and night, unless specifically stated otherwise in the text discussion in this Manual for a particular sign or group of signs.**

04 **The requirements for sign illumination shall not be considered to be satisfied by street or highway lighting.**

Support:

05 Section 1A.09 contains information regarding the assistance that is available to jurisdictions that do not have engineers on their staffs who are trained and/or experienced in traffic control devices.

Section 2B.02 Design of Regulatory Signs
Standard:

01 **Regulatory signs shall be rectangular unless specifically designated otherwise. Regulatory signs shall be designed in accordance with the sizes, shapes, colors, and legends contained in the "Standard Highway Signs and Markings" book (see Section 1A.11).**

Option:

02 Regulatory word message signs other than those classified and specified in this Manual and the "Standard Highways Signs and Markings" book (see Section 1A.11) may be developed to aid the enforcement of other laws or regulations.

03 Except for symbols on regulatory signs, minor modifications may be made to the design provided that the essential appearance characteristics are met.

Support:

04 The use of educational plaques to supplement symbol signs is described in Section 2A.12.

Guidance:

05 *Changeable message signs displaying a regulatory message incorporating a prohibitory message that includes a red circle and slash on a static sign should display a red symbol that approximates the same red circle and slash as closely as possible.*

Section 2B.03 Size of Regulatory Signs
Standard:

01 **Except as provided in Section 2A.11, the sizes for regulatory signs shall be as shown in Table 2B-1.**

Support:

02 Section 2A.11 contains information regarding the applicability of the various columns in Table 2B-1.

Standard:

03 **Except as provided in Paragraphs 4 and 5, the minimum sizes for regulatory signs facing traffic on multi-lane conventional roads shall be as shown in the Multi-lane column of Table 2B-1.**

Option:

04 Where the posted speed limit is 35 mph or less on a multi-lane highway or street, other than for a STOP sign, the minimum size shown in the Single Lane column in Table 2B-1 may be used.

05 Where a regulatory sign, other than a STOP sign, is placed on the left-hand side of a multi-lane roadway in addition to the installation of the same regulatory sign on the right-hand side or the roadway, the size shown in the Single Lane column in Table 2B-1 may be used for both the sign on the right-hand side and the sign on the left-hand side of the roadway.

Standard:

06 **A minimum size of 36 x 36 inches shall be used for STOP signs that face multi-lane approaches.**

Table 2B-1. Regulatory Sign and Plaque Sizes (Sheet 1 of 4)

Sign or Plaque	Sign Designation	Section	Conventional Road Single Lane	Conventional Road Multi-Lane	Expressway	Freeway	Minimum	Oversized
Stop	R1-1	2B.05	30 x 30*	36 x 36	36 x 36	—	30 x 30*	48 x 48
Yield	R1-2	2B.08	36 x 36 x 36*	48 x 48 x 48	48 x 48 x 48	60 x 60 x 60	30 x 30 x 30*	—
To Oncoming Traffic (plaque)	R1-2aP	2B.10	24 x 18	24 x 18	36 x 30	48 x 36	24 x 18	—
All Way (plaque)	R1-3P	2B.05	18 x 6	18 x 6	—	—	—	30 x 12
Yield Here to Peds	R1-5	2B.11	—	36 x 36	—	—	—	36 x 36
Yield Here to Pedestrians	R1-5a	2B.11	—	36 x 48	—	—	—	36 x 48
Stop Here for Peds	R1-5b	2B.11	—	36 x 36	—	—	—	36 x 36
Stop Here for Pedestrians	R1-5c	2B.11	—	36 x 48	—	—	—	36 x 48
In-Street Ped Crossing	R1-6,6a	2B.12	12 x 36	12 x 36	—	—	—	—
Overhead Ped Crossing	R1-9,9a	2B.12	90 x 24	90 x 24	—	—	—	—
Except Right Turn (plaque)	R1-10P	2B.05	24 x 18	24 x 18	—	—	—	—
Speed Limit	R2-1	2B.13	24 x 30*	30 x 36	36 x 48	48 x 60	18 x 24*	30 x 36
Truck Speed Limit (plaque)	R2-2P	2B.14	24 x 24	24 x 24	36 x 36	48 x 48	—	36 x 36
Night Speed Limit (plaque)	R2-3P	2B.15	24 x 24	24 x 24	36 x 36	48 x 48	—	36 x 36
Minimum Speed Limit (plaque)	R2-4P	2B.16	24 x 30	24 x 30	36 x 48	48 x 60	—	36 x 48
Combined Speed Limit	R2-4a	2B.16	24 x 48	24 x 48	36 x 72	48 x 96	—	36 x 72
Unless Otherwise Posted (plaque)	R2-5P	2B.13	24 x 18	24 x 18	—	—	—	—
Citywide (plaque)	R2-5aP	2B.13	24 x 6	24 x 6	—	—	—	—
Neighborhood (plaque)	R2-5bP	2B.13	24 x 6	24 x 6	—	—	—	—
Residential (plaque)	R2-5cP	2B.13	24 x 6	24 x 6	—	—	—	—
Fines Higher (plaque)	R2-6P	2B.17	24 x 18	24 x 18	36 x 24	48 x 36	—	36 x 24
Fines Double (plaque)	R2-6aP	2B.17	24 x 18	24 x 18	36 x 24	48 x 36	—	36 x 24
$XX Fine (plaque)	R2-6bP	2B.17	24 x 18	24 x 18	36 x 24	48 x 36	—	36 x 24
Begin Higher Fines Zone	R2-10	2B.17	24 x 30	24 x 30	36 x 48	48 x 60	—	36 x 48
End Higher Fines Zone	R2-11	2B.17	24 x 30	24 x 30	36 x 48	48 x 60	—	36 x 48
Movement Prohibition	R3-1,2,3,4,18,27	2B.18	24 x 24*	36 x 36	36 x 36	—	—	48 x 48
Mandatory Movement Lane Control	R3-5,5a	2B.20	30 x 36	30 x 36	—	—	—	—
Left Lane (plaque)	R3-5bP	2B.20	30 x 12	30 x 12	—	—	—	—
HOV 2+ (plaque)	R3-5cP	2B.20	24 x 12	24 x 12	—	—	—	—
Taxi Lane (plaque)	R3-5dP	2B.20	30 x 12	30 x 12	—	—	—	—
Center Lane (plaque)	R3-5eP	2B.20	30 x 12	30 x 12	—	—	—	—
Right Lane (plaque)	R3-5fP	2B.20	30 x 12	30 x 12	—	—	—	—
Bus Lane (plaque)	R3-5gP	2B.20	30 x 12	30 x 12	—	—	—	—
Optional Movement Lane Control	R3-6	2B.21	30 x 36	30 x 36	—	—	—	—
Right (Left) Lane Must Turn Right (Left)	R3-7	2B.20	30 x 30*	36 x 36	—	—	—	—
Advance Intersection Lane Control	R3-8,8a,8b	2B.22	Varies x 30	Varies x 30	—	—	—	Varies x 36
Two-Way Left Turn Only (overhead)	R3-9a	2B.24	30 x 36	30 x 36	—	—	—	—
Two-Way Left Turn Only (post-mounted)	R3-9b	2B.24	24 x 36	24 x 36	—	—	—	36 x 48
BEGIN	R3-9cP	2B.25	30 x 12	30 x 12	—	—	—	—
END	R3-9dP	2B.25	30 x 12	30 x 12	—	—	—	—
Reversible Lane Control (symbol)	R3-9e	2B.26	108 x 48	108 x 48	—	—	—	—
Reversible Lane Control (post-mounted)	R3-9f	2B.26	30 x 42*	36 x 54	—	—	—	—
Advance Reversible Lane Control Transition Signing	R3-9g,9h	2B.26	108 x 36	108 x 36	—	—	—	—
End Reverse Lane	R3-9i	2B.26	108 x 48	108 x 48	—	—	—	—
Begin Right (Left) Turn Lane	R3-20	2B.20	24 x 36	24 x 36	—	—	—	—
All Turns (U Turn) from Right Lane	R3-23,23a	2B.27	60 x 36	60 x 36	—	—	—	—
All Turns (U Turn) with arrow	R3-24,24b, 25,25b,26a	2B.27	72 x 18	72 x 18	—	—	—	—
U and Left Turns with arrow	R3-24a,25a,26	2B.27	60 x 24	60 x 24	—	—	—	—
Right Lane Must Exit	R3-33	2B.23	—	—	78 x 36	78 x 36	—	—

Table 2B-1. Regulatory Sign and Plaque Sizes (Sheet 2 of 4)

Sign or Plaque	Sign Designation	Section	Conventional Road Single Lane	Conventional Road Multi-Lane	Expressway	Freeway	Minimum	Oversized
Do Not Pass	R4-1	2B.28	24 x 30	24 x 30	36 x 48	48 x 60	18 x 24	36 x 48
Pass With Care	R4-2	2B.29	24 x 30	24 x 30	36 x 48	48 x 60	18 x 24	36 x 48
Slower Traffic Keep Right	R4-3	2B.30	24 x 30	24 x 30	36 x 48	48 x 60	18 x 24	36 x 48
Trucks Use Right Lane	R4-5	2B.31	24 x 30	24 x 30	36 x 48	48 x 60	—	36 x 48
Keep Right	R4-7,7a,7b	2B.32	24 x 30	24 x 30	36 x 48	48 x 60	18 x 24	36 x 48
Narrow Keep Right	R4-7c	2B.32	18 x 30	18 x 30	—	—	—	—
Keep Left	R4-8,8a,8b	2B.32	24 x 30	24 x 30	36 x 48	48 x 60	18 x 24	36 x 48
Narrow Keep Left	R4-8c	2B.32	18 x 30	18 x 30	—	—	—	—
Stay in Lane	R4-9	2B.33	24 x 30	24 x 30	36 x 48	48 x 60	18 x 24	36 x 48
Runaway Vehicles Only	R4-10	2B.34	48 x 48	48 x 48	—	—	—	—
Slow Vehicles with XX or More Following Vehicles Must Use Turn-Out	R4-12	2B.35	42 x 24	42 x 24	—	—	—	—
Slow Vehicles Must Use Turn-Out Ahead	R4-13	2B.35	42 x 24	42 x 24	—	—	—	—
Slow Vehicles Must Turn Out	R4-14	2B.35	30 x 42	30 x 42	—	—	—	—
Keep Right Except to Pass	R4-16	2B.30	24 x 30	24 x 30	36 x 48	48 x 60	18 x 24	36 x 48
Do Not Drive on Shoulder	R4-17	2B.36	24 x 30	24 x 30	36 x 48	48 x 60	18 x 24	36 x 48
Do Not Pass on Shoulder	R4-18	2B.36	24 x 30	24 x 30	36 x 48	48 x 60	18 x 24	36 x 48
Do Not Enter	R5-1	2B.37	30 x 30*	36 x 36	36 x 36	48 x 48	—	36 x 36
Wrong Way	R5-1a	2B.38	36 x 24*	42 x 30	36 x 24*	42 x 30	30 x 18*	42 x 30
No Trucks	R5-2,2a	2B.39	24 x 24	24 x 24	30 x 30	36 x 36	—	36 x 36
No Motor Vehicles	R5-3	2B.39	24 x 24	24 x 24	—	—	24 x 24	—
No Commercial Vehicles	R5-4	2B.39	24 x 30	24 x 30	36 x 48	36 x 48	—	—
No Vehicles with Lugs	R5-5	2B.39	24 x 30	24 x 30	36 x 48	48 x 60	—	—
No Bicycles	R5-6	2B.39	24 x 24	24 x 24	30 x 30	36 x 36	24 x 24	48 x 48
No Non-Motorized Traffic	R5-7	2B.39	30 x 24	30 x 24	42 x 24	48 x 30	—	42 x 24
No Motor-Driven Cycles	R5-8	2B.39	30 x 24	30 x 24	42 x 24	48 x 30	—	42 x 24
No Pedestrians, Bicycles, Motor-Driven Cycles	R5-10a	2B.39	30 x 36	30 x 36	—	—	—	—
No Pedestrians or Bicycles	R5-10b	2B.39	30 x 18	30 x 18	—	—	—	—
No Pedestrians	R5-10c	2B.39	24 x 12	24 x 12	—	—	—	—
Authorized Vehicles Only	R5-11	2B.39	30 x 24	30 x 24	—	—	—	—
One Way	R6-1	2B.40	36 x 12*	54 x 18	54 x 18	54 x 18	—	54 x 18
One Way	R6-2	2B.40	24 x 30*	30 x 36	36 x 48	48 x 60	18 x 24*	36 x 48
Divided Highway Crossing	R6-3,3a	2B.42	30 x 24	30 x 24	36 x 30	—	—	36 x 30
Roundabout Directional (2 chevrons)	R6-4	2B.43	30 x 24	30 x 24	—	—	—	—
Roundabout Directional (3 chevrons)	R6-4a	2B.43	48 x 24	48 x 24	—	—	—	—
Roundabout Directional (4 chevrons)	R6-4b	2B.43	60 x 24	60 x 24	—	—	—	—
Roundabout Circulation (plaque)	R6-5P	2B.44	30 x 30	30 x 30	—	—	—	—
BEGIN ONE WAY	R6-6	2B.40	24 x 30	30 x 36	—	—	—	—
END ONE WAY	R6-7	2B.40	24 x 30	30 x 36	—	—	—	—
Parking Restrictions	R7-1, 2,2a,3,4,5,6,7,8, 21,21a,22,23, 23a,107,108	2B.46	12 x 18	12 x 18	—	—	—	—
Van Accessible (plaque)	R7-8P	2B.46	18 x 9	18 x 9	—	—	—	—
Fee Station	R7-20	2B.46	24 x 18	24 x 18	—	—	—	—
No Parking (with transit logo)	R7-107a	2B.46	12 x 30	12 x 30	—	—	—	—
No Parking/Restricted Parking (combined sign)	R7-200	2B.46	24 x 18	24 x 18	—	—	—	—
No Parking/Restricted Parking (combined sign)	R7-200a	2B.46	12 x 30	12 x 30	—	—	—	—
Tow Away Zone (plaque)	R7-201P,201aP	2B.46	12 x 6	12 x 6	—	—	—	—
This Side of Sign (plaque)	R7-202P	2B.46	12 x 6	12 x 6	—	—	—	—

Table 2B-1. Regulatory Sign and Plaque Sizes (Sheet 3 of 4)

Sign or Plaque	Sign Designation	Section	Conventional Road Single Lane	Conventional Road Multi-Lane	Expressway	Freeway	Minimum	Oversized
Emergency Snow Route	R7-203	2B.46	18 x 24	18 x 24	—	—	—	24 x 30
No Parking on Pavement	R8-1	2B.46	24 x 30	24 x 30	36 x 48	48 x 60	—	36 x 48
No Parking Except on Shoulder	R8-2	2B.46	24 x 30	24 x 30	36 x 48	48 x 60	—	36 x 48
No Parking (symbol)	R8-3	2B.46	24 x 24*	30 x 30	36 x 36	48 x 48	12 x 12*	36 x 36
No Parking	R8-3a	2B.46	24 x 30	24 x 30	36 x 36	48 x 48	18 x 24	36 x 36
Except Sundays and Holidays (plaque)	R8-3bP	2B.46	24 x 18	24 x 18	—	—	12 x 9	30 x 24
On Pavement (plaque)	R8-3cP	2B.46	24 x 18	24 x 18	—	—	12 x 9	30 x 24
On Bridge (plaque)	R8-3dP	2B.46	24 x 18	24 x 18	—	—	12 x 9	30 x 24
On Tracks (plaque)	R8-3eP	2B.46	12 x 9	12 x 9	—	—	—	30 x 24
Except on Shoulder (plaque)	R8-3fP	2B.46	24 x 18	24 x 18	—	—	12 x 9	30 x 24
Loading Zone (plaque)	R8-3gP	2B.46	24 x 18	24 x 18	—	—	12 x 9	30 x 24
Times of Day (plaque)	R8-3hP	2B.46	24 x 18	24 x 18	—	—	12 x 9	30 x 24
Emergency Parking Only	R8-4	2B.49	30 x 24	30 x 24	30 x 24	48 x 36	—	48 x 36
No Stopping on Pavement	R8-5	2B.46	24 x 30	24 x 30	36 x 48	48 x 60	—	36 x 48
No Stopping Except on Shoulder	R8-6	2B.46	24 x 30	24 x 30	36 x 48	48 x 60	—	36 x 48
Emergency Stopping Only	R8-7	2B.49	30 x 24	30 x 24	48 x 36	48 x 36	—	48 x 36
Walk on Left Facing Traffic	R9-1	2B.50	18 x 24	18 x 24	—	—	—	—
Cross Only at Crosswalks	R9-2	2B.51	12 x 18	12 x 18	—	—	—	—
No Pedestrian Crossing (symbol)	R9-3	2B.51	18 x 18	18 x 18	24 x 24	30 x 30	—	30 x 30
No Pedestrian Crossing	R9-3a	2B.51	12 x 18	12 x 18	—	—	—	—
Use Crosswalk (plaque)	R9-3bP	2B.51	18 x 12	18 x 12	—	—	—	—
No Hitchhiking (symbol)	R9-4	2B.50	18 x 18	18 x 18	—	—	—	24 x 24
No Hitchhiking	R9-4a	2B.50	18 x 24	18 x 24	—	—	12 x 18	—
No Skaters	R9-13	2B.39	18 x 18	18 x 18	24 x 24	30 x 30	—	30 x 30
No Equestrians	R9-14	2B.39	18 x 18	18 x 18	24 x 24	30 x 30	—	30 x 30
Cross Only On Green	R10-1	2B.52	12 x 18	12 x 18	—	—	—	—
Pedestrian Signs and Plaques	R10-2,3,3b,3c,3d,4	2B.52	9 x 12	9 x 12	—	—	—	—
Pedestrian Signs	R10-3a,3e,3f,3g,3h,3i,4a	2B.52	9 x 15	9 x 15	—	—	—	—
Left on Green Arrow Only	R10-5	2B.53	30 x 36	30 x 36	48 x 60	—	24 x 30	48 x 60
Stop Here on Red	R10-6	2B.53	24 x 36	24 x 36	—	—	—	36 x 48
Stop Here on Red	R10-6a	2B.53	24 x 30	24 x 30	—	—	—	36 x 42
Do Not Block Intersection	R10-7	2B.53	24 x 30	24 x 30	—	—	—	—
Use Lane with Green Arrow	R10-8	2B.53	36 x 42	36 x 42	36 x 42	—	—	60 x 72
Left (Right) Turn Signal	R10-10	2B.53	30 x 36	30 x 36	—	—	—	—
No Turn on Red	R10-11	2B.54	24 x 30*	36 x 48	—	—	—	36 x 48
No Turn on Red	R10-11a	2B.54	30 x 36*	36 x 48	—	—	—	—
No Turn on Red	R10-11b	2B.54	36 x 36	36 x 36	—	—	—	—
No Turn on Red Except From Right Lane	R10-11c	2B.54	30 x 42	30 x 42	—	—	—	—
No Turn on Red From This Lane	R10-11d	2B.54	30 x 42	30 x 42	—	—	—	—
Left Turn Yield on Green	R10-12	2B.53	30 x 36	30 x 36	—	—	—	—
Emergency Signal	R10-13	2B.53	42 x 30	42 x 30	—	—	—	—
Emergency Signal - Stop on Flashing Red	R10-14	2B.53	36 x 42	36 x 42	—	—	—	—
Emergency Signal - Stop on Flashing Red (overhead)	R10-14a	2B.53	60 x 24	60 x 24	—	—	—	—
Turning Vehicles Yield to Peds	R10-15	2B.53	30 x 30	30 x 30	—	—	—	—
U-Turn Yield to Right Turn	R10-16	2B.53	30 x 36	30 x 36	—	—	—	—
Right on Red Arrow After Stop	R10-17a	2B.54	36 x 48	36 x 48	—	—	—	—
Traffic Laws Photo Enforced	R10-18	2B.55	36 x 24	36 x 24	48 x 30	54 x 36	—	54 x 36
Photo Enforced (symbol plaque)	R10-19P	2B.55	24 x 12	24 x 12	36 x 18	48 x 24	—	48 x 24
Photo Enforced (plaque)	R10-19aP	2B.55	24 x 18	24 x 18	36 x 30	48 x 36	—	48 x 36
MON—FRI (and times) (3 lines) (plaque)	R10-20aP	2B.53	24 x 24	24 x 24	—	—	—	—

Table 2B-1. Regulatory Sign and Plaque Sizes (Sheet 4 of 4)

Sign or Plaque	Sign Designation	Section	Conventional Road		Expressway	Freeway	Minimum	Oversized
			Single Lane	Multi-Lane				
SUNDAY (and times) (2 lines) (plaque)	R10-20aP	2B.53	24 x 18	24 x 18	—	—	—	—
Crosswalk, Stop on Red	R10-23	2B.53	24 x 30	24 x 30	—	—	—	—
Push Button To Turn On Warning Lights	R10-25	2B.52	9 x 12	9 x 12	—	—	—	—
Left Turn Yield on Flashing Red Arrow After Stop	R10-27	2B.53	30 x 36	30 x 36	—	—	—	—
XX Vehicles Per Green	R10-28	2B.56	24 x 30	24 x 30	—	—	—	—
XX Vehicles Per Green Each Lane	R10-29	2B.56	36 x 24	36 x 24	—	—	—	—
Right Turn on Red Must Yield to U-Turn	R10-30	2B.54	30 x 36	30 x 36	—	—	—	—
At Signal (plaque)	R10-31P	2B.53	24 x 9	24 x 9	—	—	—	—
Push Button for 2 Seconds for Extra Crossing Time	R10-32P	2B.52	9 x 12	9 x 12	—	—	—	—
Keep Off Median	R11-1	2B.57	24 x 30	24 x 30	—	—	—	—
Road Closed	R11-2	2B.58	48 x 30	48 x 30	—	—	—	—
Road Closed - Local Traffic Only	R11-3a,3b,4	2B.58	60 x 30	60 x 30	—	—	—	—
Weight Limit	R12-1,2	2B.59	24 x 30	24 x 30	36 x 48	—	—	36 x 48
Weight Limit	R12-3	2B.59	24 x 36	24 x 36	—	—	—	—
Weight Limit	R12-4	2B.59	36 x 24	36 x 24	—	—	—	—
Weight Limit	R12-5	2B.59	24 x 36	24 x 36	36 x 48	48 x 60	—	—
Weigh Station	R13-1	2B.60	72 x 54	72 x 54	96 x 72	120 x 90	—	—
Truck Route	R14-1	2B.61	24 x 18	24 x 18	—	—	—	—
Hazardous Material	R14-2,3	2B.62	24 x 24	24 x 24	30 x 30	36 x 36	—	42 x 42
National Network	R14-4,5	2B.63	30 x 30	30 x 30	36 x 36	36 x 36	—	42 x 42
Fender Bender Move Vehicles	R16-4	2B.65	36 x 24	36 x 24	48 x 36	60 x 48	—	48 x 36
Lights On When Using Wipers or Raining	R16-5,6	2B.64	24 x 30	24 x 30	36 x 48	48 x 60	—	36 x 48
Turn On Headlights Next XX Miles	R16-7	2B.64	48 x 15	48 x 15	72 x 24	96 x 30	—	72 x 24
Turn On, Check Headlights	R16-8,9	2B.64	30 x 15	30 x 15	48 x 24	60 x 30	—	48 x 24
Begin, End Daytime Headlight Section	R16-10,11	2B.64	48 x 15	48 x 15	72 x 24	96 x 30	—	72 x 24

* See Table 9B-1 for minimum size required for signs on bicycle facilities

Notes: 1. Larger signs may be used when appropriate
2. Dimensions in inches are shown as width x height

---

07 **Where side roads intersect a multi-lane street or highway that has a speed limit of 45 mph or higher, the minimum size of the STOP signs facing the side road approaches, even if the side road only has one approach lane, shall be 36 x 36 inches.**

08 **Where side roads intersect a multi-lane street or highway that has a speed limit of 40 MPH or lower, the minimum size of the STOP signs facing the side road approaches shall be as shown in the Single Lane or Multi-lane columns of Table 2B-1 based on the number of approach lanes on the side street approach.**

Guidance:

09 *The minimum sizes for regulatory signs facing traffic on exit and entrance ramps should be as shown in the column of Table 2B-1 that corresponds to the mainline roadway classification (Expressway or Freeway). If a minimum size is not provided in the Freeway column, the minimum size in the Expressway column should be used. If a minimum size is not provided in the Freeway or Expressway Column, the size in the Oversized column should be used.*

Section 2B.04 Right-of-Way at Intersections

Support:

01 State or local laws written in accordance with the "Uniform Vehicle Code" (see Section 1A.11) establish the right-of-way rule at intersections having no regulatory traffic control signs such that the driver of a vehicle approaching an intersection must yield the right-of-way to any vehicle or pedestrian already in the intersection.

When two vehicles approach an intersection from different streets or highways at approximately the same time, the right-of-way rule requires the driver of the vehicle on the left to yield the right-of-way to the vehicle on the right. The right-of-way can be modified at through streets or highways by placing YIELD (R1-2) signs (see Sections 2B.08 and 2B.09) or STOP (R1-1) signs (see Sections 2B.05 through 2B.07) on one or more approaches.

Guidance:

02 *Engineering judgment should be used to establish intersection control. The following factors should be considered:*

   A. *Vehicular, bicycle, and pedestrian traffic volumes on all approaches;*
   B. *Number and angle of approaches;*
   C. *Approach speeds;*
   D. *Sight distance available on each approach; and*
   E. *Reported crash experience.*

03 *YIELD or STOP signs should be used at an intersection if one or more of the following conditions exist:*

   A. *An intersection of a less important road with a main road where application of the normal right-of-way rule would not be expected to provide reasonable compliance with the law;*
   B. *A street entering a designated through highway or street; and/or*
   C. *An unsignalized intersection in a signalized area.*

04 *In addition, the use of YIELD or STOP signs should be considered at the intersection of two minor streets or local roads where the intersection has more than three approaches and where one or more of the following conditions exist:*

   A. *The combined vehicular, bicycle, and pedestrian volume entering the intersection from all approaches averages more than 2,000 units per day;*
   B. *The ability to see conflicting traffic on an approach is not sufficient to allow a road user to stop or yield in compliance with the normal right-of-way rule if such stopping or yielding is necessary; and/or*
   C. *Crash records indicate that five or more crashes that involve the failure to yield the right-of-way at the intersection under the normal right-of-way rule have been reported within a 3-year period, or that three or more such crashes have been reported within a 2-year period.*

05 *YIELD or STOP signs should not be used for speed control.*

Support:

06 Section 2B.07 contains provisions regarding the application of multi-way STOP control at an intersection.

Guidance:

07 *Once the decision has been made to control an intersection, the decision regarding the appropriate roadway to control should be based on engineering judgment. In most cases, the roadway carrying the lowest volume of traffic should be controlled.*

08 *A YIELD or STOP sign should not be installed on the higher volume roadway unless justified by an engineering study.*

Support:

09 The following are considerations that might influence the decision regarding the appropriate roadway upon which to install a YIELD or STOP sign where two roadways with relatively equal volumes and/or characteristics intersect:

   A. Controlling the direction that conflicts the most with established pedestrian crossing activity or school walking routes;
   B. Controlling the direction that has obscured vision, dips, or bumps that already require drivers to use lower operating speeds; and
   C. Controlling the direction that has the best sight distance from a controlled position to observe conflicting traffic.

Standard:

10 **Because the potential for conflicting commands could create driver confusion, YIELD or STOP signs shall not be used in conjunction with any traffic control signal operation, except in the following cases:**

   A. **If the signal indication for an approach is a flashing red at all times;**
   B. **If a minor street or driveway is located within or adjacent to the area controlled by the traffic control signal, but does not require separate traffic signal control because an extremely low potential for conflict exists; or**
   C. **If a channelized turn lane is separated from the adjacent travel lanes by an island and the channelized turn lane is not controlled by a traffic control signal.**

2009 Edition   Page 51

11 Except as provided in Section 2B.09, STOP signs and YIELD signs shall not be installed on different approaches to the same unsignalized intersection if those approaches conflict with or oppose each other.

12 Portable or part-time STOP or YIELD signs shall not be used except for emergency and temporary traffic control zone purposes.

13 A portable or part-time (folding) STOP sign that is manually placed into view and manually removed from view shall not be used during a power outage to control a signalized approach unless the maintaining agency establishes that the signal indication that will first be displayed to that approach upon restoration of power is a flashing red signal indication and that the portable STOP sign will be manually removed from view prior to stop-and-go operation of the traffic control signal.

Option:

14 A portable or part-time (folding) STOP sign that is electrically or mechanically operated such that it only displays the STOP message during a power outage and ceases to display the STOP message upon restoration of power may be used during a power outage to control a signalized approach.

Support:

15 Section 9B.03 contains provisions regarding the assignment of priority at a shared-use path/roadway intersection.

Section 2B.05 STOP Sign (R1-1) and ALL WAY Plaque (R1-3P)

Standard:

01 When it is determined that a full stop is always required on an approach to an intersection, a STOP (R1-1) sign (see Figure 2B-1) shall be used.

02 The STOP sign shall be an octagon with a white legend and border on a red background.

03 Secondary legends shall not be used on STOP sign faces.

04 At intersections where all approaches are controlled by STOP signs (see Section 2B.07), an ALL WAY supplemental plaque (R1-3P) shall be mounted below each STOP sign. The ALL WAY plaque (see Figure 2B-1) shall have a white legend and border on a red background.

05 The ALL WAY plaque shall only be used if all intersection approaches are controlled by STOP signs.

06 Supplemental plaques with legends such as 2-WAY, 3-WAY, 4-WAY, or other numbers of ways shall not be used with STOP signs.

Support:

07 The use of the CROSS TRAFFIC DOES NOT STOP (W4-4P) plaque (and other plaques with variations of this word message) is described in Section 2C.59.

Guidance:

08 *Plaques with the appropriate alternative messages of TRAFFIC FROM LEFT (RIGHT) DOES NOT STOP (W4-4aP) or ONCOMING TRAFFIC DOES NOT STOP (W4-4bP) should be used at intersections where STOP signs control all but one approach to the intersection, unless the only non-stopped approach is from a one-way street.*

Option:

09 An EXCEPT RIGHT TURN (R1-10P) plaque (see Figure 2B-1) may be mounted below the STOP sign if an engineering study determines that a special combination of geometry and traffic volumes is present that makes it possible for right-turning traffic on the approach to be permitted to enter the intersection without stopping.

Support:

10 The design and application of Stop Beacons are described in Section 4L.05.

Figure 2B-1. STOP and YIELD Signs and Plaques

R1-1

R1-3P

R1-2   R1-2aP   R1-10P

Section 2B.06 STOP Sign Applications

Guidance:

01 At intersections where a full stop is not necessary at all times, consideration should first be given to using less restrictive measures such as YIELD signs (see Sections 2B.08 and 2B.09).

02 The use of STOP signs on the minor-street approaches should be considered if engineering judgment indicates that a stop is always required because of one or more of the following conditions:
   A. The vehicular traffic volumes on the through street or highway exceed 6,000 vehicles per day;
   B. A restricted view exists that requires road users to stop in order to adequately observe conflicting traffic on the through street or highway; and/or
   C. Crash records indicate that three or more crashes that are susceptible to correction by the installation of a STOP sign have been reported within a 12-month period, or that five or more such crashes have been reported within a 2-year period. Such crashes include right-angle collisions involving road users on the minor-street approach failing to yield the right-of-way to traffic on the through street or highway.

Support:

03 The use of STOP signs at grade crossings is described in Sections 8B.04 and 8B.05.

Section 2B.07 Multi-Way Stop Applications

Support:

01 Multi-way stop control can be useful as a safety measure at intersections if certain traffic conditions exist. Safety concerns associated with multi-way stops include pedestrians, bicyclists, and all road users expecting other road users to stop. Multi-way stop control is used where the volume of traffic on the intersecting roads is approximately equal.

02 The restrictions on the use of STOP signs described in Section 2B.04 also apply to multi-way stop applications.

Guidance:

03 The decision to install multi-way stop control should be based on an engineering study.

04 The following criteria should be considered in the engineering study for a multi-way STOP sign installation:
   A. Where traffic control signals are justified, the multi-way stop is an interim measure that can be installed quickly to control traffic while arrangements are being made for the installation of the traffic control signal.
   B. Five or more reported crashes in a 12-month period that are susceptible to correction by a multi-way stop installation. Such crashes include right-turn and left-turn collisions as well as right-angle collisions.
   C. Minimum volumes:
      1. The vehicular volume entering the intersection from the major street approaches (total of both approaches) averages at least 300 vehicles per hour for any 8 hours of an average day; and
      2. The combined vehicular, pedestrian, and bicycle volume entering the intersection from the minor street approaches (total of both approaches) averages at least 200 units per hour for the same 8 hours, with an average delay to minor-street vehicular traffic of at least 30 seconds per vehicle during the highest hour; but
      3. If the 85th-percentile approach speed of the major-street traffic exceeds 40 mph, the minimum vehicular volume warrants are 70 percent of the values provided in Items 1 and 2.
   D. Where no single criterion is satisfied, but where Criteria B, C.1, and C.2 are all satisfied to 80 percent of the minimum values. Criterion C.3 is excluded from this condition.

Option:

05 Other criteria that may be considered in an engineering study include:
   A. The need to control left-turn conflicts;
   B. The need to control vehicle/pedestrian conflicts near locations that generate high pedestrian volumes;
   C. Locations where a road user, after stopping, cannot see conflicting traffic and is not able to negotiate the intersection unless conflicting cross traffic is also required to stop; and
   D. An intersection of two residential neighborhood collector (through) streets of similar design and operating characteristics where multi-way stop control would improve traffic operational characteristics of the intersection.

Section 2B.08 YIELD Sign (R1-2)
Standard:

01 **The YIELD (R1-2) sign (see Figure 2B-1) shall be a downward-pointing equilateral triangle with a wide red border and the legend YIELD in red on a white background.**

Support:

02 The YIELD sign assigns right-of-way to traffic on certain approaches to an intersection. Vehicles controlled by a YIELD sign need to slow down to a speed that is reasonable for the existing conditions or stop when necessary to avoid interfering with conflicting traffic.

Section 2B.09 YIELD Sign Applications
Option:

01 YIELD signs may be installed:
   A. On the approaches to a through street or highway where conditions are such that a full stop is not always required.
   B. At the second crossroad of a divided highway, where the median width at the intersection is 30 feet or greater. In this case, a STOP or YIELD sign may be installed at the entrance to the first roadway of a divided highway, and a YIELD sign may be installed at the entrance to the second roadway.
   C. For a channelized turn lane that is separated from the adjacent travel lanes by an island, even if the adjacent lanes at the intersection are controlled by a highway traffic control signal or by a STOP sign.
   D. At an intersection where a special problem exists and where engineering judgment indicates the problem to be susceptible to correction by the use of the YIELD sign.
   E. Facing the entering roadway for a merge-type movement if engineering judgment indicates that control is needed because acceleration geometry and/or sight distance is not adequate for merging traffic operation.

Standard:

02 **A YIELD (R1-2) sign shall be used to assign right-of-way at the entrance to a roundabout. YIELD signs at roundabouts shall be used to control the approach roadways and shall not be used to control the circulatory roadway.**

03 **Other than for all of the approaches to a roundabout, YIELD signs shall not be placed on all of the approaches to an intersection.**

Section 2B.10 STOP Sign or YIELD Sign Placement
Standard:

01 **The STOP or YIELD sign shall be installed on the near side of the intersection on the right-hand side of the approach to which it applies. When the STOP or YIELD sign is installed at this required location and the sign visibility is restricted, a Stop Ahead sign (see Section 2C.36) shall be installed in advance of the STOP sign or a Yield Ahead sign (see Section 2C.36) shall be installed in advance of the YIELD sign.**

02 **The STOP or YIELD sign shall be located as close as practical to the intersection it regulates, while optimizing its visibility to the road user it is intended to regulate.**

03 **STOP signs and YIELD signs shall not be mounted on the same post.**

04 **No items other than inventory stickers, sign installation dates, and bar codes shall be affixed to the fronts of STOP or YIELD signs, and the placement of these items shall be in the border of the sign.**

05 **No items other than official traffic control signs, inventory stickers, sign installation dates, anti-vandalism stickers, and bar codes shall be mounted on the backs of STOP or YIELD signs.**

06 **No items other than retroreflective strips (see Section 2A.21) or official traffic control signs shall be mounted on the fronts or backs of STOP or YIELD signs supports.**

Guidance:

07 *STOP or YIELD signs should not be placed farther than 50 feet from the edge of the pavement of the intersected roadway (see Drawing F in Figure 2A-3).*

08 *A sign that is mounted back-to-back with a STOP or YIELD sign should stay within the edges of the STOP or YIELD sign. If necessary, the size of the STOP or YIELD sign should be increased so that any other sign installed back-to-back with a STOP or YIELD sign remains within the edges of the STOP or YIELD sign.*

Option:

09 Where drivers proceeding straight ahead must yield to traffic approaching from the opposite direction, such as at a one-lane bridge, a TO ONCOMING TRAFFIC (R1-2aP) plaque may be mounted below the YIELD sign.

Support:

10 Figure 2A-3 shows examples of some typical placements of STOP signs and YIELD signs.

11 Section 2A.16 contains additional information about separate and combined mounting of other signs with STOP or YIELD signs.

Guidance:

12 *Stop lines that are used to supplement a STOP sign should be located as described in Section 3B.16. Yield lines that are used to supplement a YIELD sign should be located as described in Section 3B.16.*

13 *Where there is a marked crosswalk at the intersection, the STOP sign should be installed in advance of the crosswalk line nearest to the approaching traffic.*

14 *Except at roundabouts, where there is a marked crosswalk at the intersection, the YIELD sign should be installed in advance of the crosswalk line nearest to the approaching traffic.*

15 *Where two roads intersect at an acute angle, the STOP or YIELD sign should be positioned at an angle, or shielded, so that the legend is out of view of traffic to which it does not apply.*

16 *If a raised splitter island is available on the left-hand side of a multi-lane roundabout approach, an additional YIELD sign should be placed on the left-hand side of the approach.*

Option:

17 If a raised splitter island is available on the left-hand side of a single lane roundabout approach, an additional YIELD sign may be placed on the left-hand side of the approach.

18 At wide-throat intersections or where two or more approach lanes of traffic exist on the signed approach, observance of the right-of-way control may be improved by the installation of an additional STOP or YIELD sign on the left-hand side of the road and/or the use of a stop or yield line. At channelized intersections or at divided roadways separated by a median, the additional STOP or YIELD sign may be placed on a channelizing island or in the median. An additional STOP or YIELD sign may also be placed overhead facing the approach at the intersection to improve observance of the right-of-way control.

Standard:

19 **More than one STOP sign or more than one YIELD sign shall not be placed on the same support facing in the same direction.**

Option:

20 For a yield-controlled channelized right-turn movement onto a roadway without an acceleration lane and for an entrance ramp onto a freeway or expressway without an acceleration lane, a NO MERGE AREA (W4-5P) supplemental plaque (see Section 2C.40) may be mounted below a Yield Ahead (W3-2) sign and/or below a YIELD (R1-2) sign when engineering judgment indicates that road users would expect an acceleration lane to be present.

Section 2B.11 Yield Here To Pedestrians Signs and Stop Here For Pedestrians Signs (R1-5 Series)

Standard:

01 **Yield Here To (Stop Here For) Pedestrians (R1-5, R1-5a, R1-5b, or R1-5c) signs (see Figure 2B-2) shall be used if yield (stop) lines are used in advance of a marked crosswalk that crosses an uncontrolled multi-lane approach. The Stop Here for Pedestrians signs shall only be used where the law specifically requires that a driver must stop for a pedestrian in a crosswalk. The legend STATE LAW may be displayed at the top of the R1-5, R1-5a, R1-5b, and R1-5c signs, if applicable.**

Guidance:

02 *If yield (stop) lines and Yield Here To (Stop Here For) Pedestrians signs are used in advance of a crosswalk that crosses an uncontrolled multi-lane approach, they should be placed 20 to 50 feet in advance of the nearest crosswalk line (see Section 3B.16 and Figure 3B-17), and parking should be prohibited in the area between the yield (stop) line and the crosswalk.*

03 *Yield (stop) lines and Yield Here To (Stop Here For) Pedestrians signs should not be used in advance of crosswalks that cross an approach to or departure from a roundabout.*

Option:

04 Yield Here To (Stop Here For) Pedestrians signs may be used in advance of a crosswalk that crosses an uncontrolled multi-lane approach to indicate to road users where to yield (stop) even if yield (stop) lines are not used.

Figure 2B-2. Unsignalized Pedestrian Crosswalk Signs

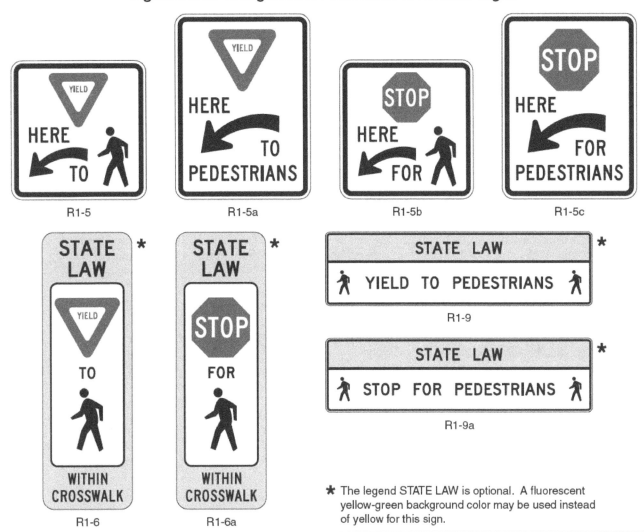

05 A Pedestrian Crossing (W11-2) warning sign may be placed overhead or may be post-mounted with a diagonal downward pointing arrow (W16-7P) plaque at the crosswalk location where Yield Here To (Stop Here For) Pedestrians signs have been installed in advance of the crosswalk.

Standard:

06 **If a W11-2 sign has been post-mounted at the crosswalk location where a Yield Here To (Stop Here For) Pedestrians sign is used on the approach, the Yield Here To (Stop Here For) Pedestrians sign shall not be placed on the same post as or block the road user's view of the W11-2 sign.**

Option:

07 An advance Pedestrian Crossing (W11-2) warning sign with an AHEAD or a distance supplemental plaque may be used in conjunction with a Yield Here To (Stop Here For) Pedestrians sign on the approach to the same crosswalk.

08 In-Street Pedestrian Crossing signs and Yield Here To (Stop Here For) Pedestrians signs may be used together at the same crosswalk.

Section 2B.12 In-Street and Overhead Pedestrian Crossing Signs (R1-6, R1-6a, R1-9, and R1-9a)

Option:

01 The In-Street Pedestrian Crossing (R1-6 or R1-6a) sign (see Figure 2B-2) or the Overhead Pedestrian Crossing (R1-9 or R1-9a) sign (see Figure 2B-2) may be used to remind road users of laws regarding right-of-way at an unsignalized pedestrian crosswalk. The legend STATE LAW may be displayed at the top of the R1-6, R1-6a, R1-9, and R1-9a signs, if applicable. On the R1-6 and R1-6a signs, the legends STOP or YIELD may be used instead of the appropriate STOP sign or YIELD sign symbol.

02 Highway agencies may develop and apply criteria for determining the applicability of In-Street Pedestrian Crossing signs.

Standard:

03 **If used, the In-Street Pedestrian Crossing sign shall be placed in the roadway at the crosswalk location on the center line, on a lane line, or on a median island. The In-Street Pedestrian Crossing sign shall not be post-mounted on the left-hand or right-hand side of the roadway.**

04 **If used, the Overhead Pedestrian Crossing sign shall be placed over the roadway at the crosswalk location.**

05 **An In-Street or Overhead Pedestrian Crossing sign shall not be placed in advance of the crosswalk to educate road users about the State law prior to reaching the crosswalk, nor shall it be installed as an educational display that is not near any crosswalk.**

Guidance:

06 *If an island (see Chapter 3I) is available, the In-Street Pedestrian Crossing sign, if used, should be placed on the island.*

Option:

07 If a Pedestrian Crossing (W11-2) warning sign is used in combination with an In-Street or an Overhead Pedestrian Crossing sign, the W11-2 sign with a diagonal downward pointing arrow (W16-7P) plaque may be post-mounted on the right-hand side of the roadway at the crosswalk location.

Standard:

08 **The In-Street Pedestrian Crossing sign and the Overhead Pedestrian Crossing sign shall not be used at signalized locations.**

09 **The STOP FOR legend shall only be used in States where the State law specifically requires that a driver must stop for a pedestrian in a crosswalk.**

10 **The In-Street Pedestrian Crossing sign shall have a black legend (except for the red STOP or YIELD sign symbols) and border on a white background, surrounded by an outer yellow or fluorescent yellow-green background area (see Figure 2B-2). The Overhead Pedestrian Crossing sign shall have a black legend and border on a yellow or fluorescent yellow-green background at the top of the sign and a black legend and border on a white background at the bottom of the sign (see Figure 2B-2).**

11 **Unless the In-Street Pedestrian Crossing sign is placed on a physical island, the sign support shall be designed to bend over and then bounce back to its normal vertical position when struck by a vehicle.**

Support:

12 The Provisions of Section 2A.18 concerning mounting height are not applicable for the In-Street Pedestrian Crossing sign.

Standard:

13 **The top of an In-Street Pedestrian Crossing sign shall be a maximum of 4 feet above the pavement surface. The top of an In-Street Pedestrian Crossing sign placed in an island shall be a maximum of 4 feet above the island surface.**

Option:

14 The In-Street Pedestrian Crossing sign may be used seasonally to prevent damage in winter because of plowing operations, and may be removed at night if the pedestrian activity at night is minimal.

15 In-Street Pedestrian Crossing signs, Overhead Pedestrian Crossing signs, and Yield Here To (Stop Here For) Pedestrians signs may be used together at the same crosswalk.

Section 2B.13 Speed Limit Sign (R2-1)

Standard:

01 **Speed zones (other than statutory speed limits) shall only be established on the basis of an engineering study that has been performed in accordance with traffic engineering practices. The engineering study shall include an analysis of the current speed distribution of free-flowing vehicles.**

02 **The Speed Limit (R2-1) sign (see Figure 2B-3) shall display the limit established by law, ordinance, regulation, or as adopted by the authorized agency based on the engineering study. The speed limits displayed shall be in multiples of 5 mph.**

03 **Speed Limit (R2-1) signs, indicating speed limits for which posting is required by law, shall be located at the points of change from one speed limit to another.**

Figure 2B-3. Speed Limit and Photo Enforcement Signs and Plaques

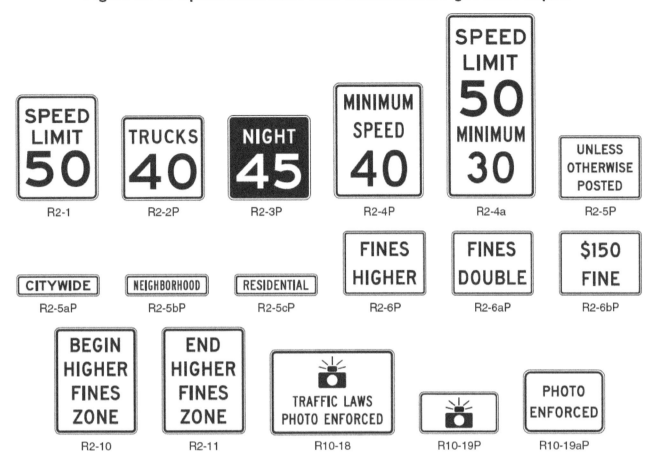

⁰⁴ At the downstream end of the section to which a speed limit applies, a Speed Limit sign showing the next speed limit shall be installed. Additional Speed Limit signs shall be installed beyond major intersections and at other locations where it is necessary to remind road users of the speed limit that is applicable.

⁰⁵ Speed Limit signs indicating the statutory speed limits shall be installed at entrances to the State and, where appropriate, at jurisdictional boundaries in urban areas.

Support:

⁰⁶ In general, the maximum speed limits applicable to rural and urban roads are established:
   A. Statutorily – a maximum speed limit applicable to a particular class of road, such as freeways or city streets, that is established by State law; or
   B. As altered speed zones – based on engineering studies.

⁰⁷ State statutory limits might restrict the maximum speed limit that can be established on a particular road, notwithstanding what an engineering study might indicate.

Option:

⁰⁸ If a jurisdiction has a policy of installing Speed Limit signs in accordance with statutory requirements only on the streets that enter a city, neighborhood, or residential area to indicate the speed limit that is applicable to the entire city, neighborhood, or residential area unless otherwise posted, a CITYWIDE (R2-5aP), NEIGHBORHOOD (R2-5bP), or RESIDENTIAL (R2-5cP) plaque may be mounted above the Speed Limit sign and an UNLESS OTHERWISE POSTED (R2-5P) plaque may be mounted below the Speed Limit sign (see Figure 2B-3).

Guidance:

09 *A Reduced Speed Limit Ahead (W3-5 or W3-5a) sign (see Section 2C.38) should be used to inform road users of a reduced speed zone where the speed limit is being reduced by more than 10 mph, or where engineering judgment indicates the need for advance notice to comply with the posted speed limit ahead.*

10 *States and local agencies should conduct engineering studies to reevaluate non-statutory speed limits on segments of their roadways that have undergone significant changes since the last review, such as the addition or elimination of parking or driveways, changes in the number of travel lanes, changes in the configuration of bicycle lanes, changes in traffic control signal coordination, or significant changes in traffic volumes.*

11 *No more than three speed limits should be displayed on any one Speed Limit sign or assembly.*

12 *When a speed limit within a speed zone is posted, it should be within 5 mph of the 85th-percentile speed of free-flowing traffic.*

13 *Speed studies for signalized intersection approaches should be taken outside the influence area of the traffic control signal, which is generally considered to be approximately 1/2 mile, to avoid obtaining skewed results for the 85th-percentile speed.*

Support:

14 Advance warning signs and other traffic control devices to attract the motorist's attention to a signalized intersection are usually more effective than a reduced speed limit zone.

Guidance:

15 *An advisory speed plaque (see Section 2C.08) mounted below a warning sign should be used to warn road users of an advisory speed for a roadway condition. A Speed Limit sign should not be used for this situation.*

Option:

16 Other factors that may be considered when establishing or reevaluating speed limits are the following:
   A. Road characteristics, shoulder condition, grade, alignment, and sight distance;
   B. The pace;
   C. Roadside development and environment;
   D. Parking practices and pedestrian activity; and
   E. Reported crash experience for at least a 12-month period.

17 Two types of Speed Limit signs may be used: one to designate passenger car speeds, including any nighttime information or minimum speed limit that might apply; and the other to show any special speed limits for trucks and other vehicles.

18 A changeable message sign that changes the speed limit for traffic and ambient conditions may be installed provided that the appropriate speed limit is displayed at the proper times.

19 A changeable message sign that displays to approaching drivers the speed at which they are traveling may be installed in conjunction with a Speed Limit sign.

Guidance:

20 *If a changeable message sign displaying approach speeds is installed, the legend YOUR SPEED XX MPH or such similar legend should be displayed. The color of the changeable message legend should be a yellow legend on a black background or the reverse of these colors.*

Support:

21 Advisory Speed signs and plaques are discussed in Sections 2C.08 and 2C.14. Temporary Traffic Control Zone Speed signs are discussed in Part 6. The WORK ZONE (G20-5aP) plaque intended for installation above a Speed Limit sign is discussed in Section 6F.12. School Speed Limit signs are discussed in Section 7B.15.

Section 2B.14 Truck Speed Limit Plaque (R2-2P)
Standard:

01 **Where a special speed limit applies to trucks or other vehicles, the legend TRUCKS XX or such similar legend shall be displayed below the legend Speed Limit XX on the same sign or on a separate R2-2P plaque (see Figure 2B-3) below the standard legend.**

Section 2B.15 Night Speed Limit Plaque (R2-3P)
Standard:

01 **Where different speed limits are prescribed for day and night, both limits shall be posted.**

Guidance:

02 *A Night Speed Limit (R2-3P) plaque (see Figure 2B-3) should be reversed using a white retroreflectorized legend and border on a black background.*

Option:

03 A Night Speed Limit plaque may be combined with or installed below the standard Speed Limit (R2-1) sign.

Section 2B.16 Minimum Speed Limit Plaque (R2-4P)

Standard:

01 **A Minimum Speed Limit (R2-4P) plaque (see Figure 2B-3) shall be displayed only in combination with a Speed Limit sign.**

Option:

02 Where engineering judgment determines that slow speeds on a highway might impede the normal and reasonable movement of traffic, the Minimum Speed Limit plaque may be installed below a Speed Limit (R2-1) sign to indicate the minimum legal speed. If desired, the Speed Limit sign and the Minimum Speed Limit plaque may be combined on the R2-4a sign (see Figure 2B-3).

Section 2B.17 Higher Fines Signs and Plaque (R2-6P, R2-10, and R2-11)

Standard:

01 **If increased fines are imposed for traffic violations within a designated zone of a roadway, a BEGIN HIGHER FINES ZONE (R2-10) sign (see Figure 2B-3) or a FINES HIGHER (R2-6P) plaque (see Figure 2B-3) shall be used to provide notice to road users. If used, the FINES HIGHER plaque shall be mounted below an applicable regulatory or warning sign in a temporary traffic control zone, a school zone, or other applicable designated zone.**

02 **If an R2-10 sign or an R2-6P plaque is posted to provide notice of increased fines for traffic violations, an END HIGHER FINES ZONE (R2-11) sign (see Figure 2B-3) shall be installed at the downstream end of the zone to provide notice to road users of the termination of the increased fines zone.**

Guidance:

03 *If used, the BEGIN HIGHER FINES ZONE sign or FINES HIGHER plaque should be located at the beginning of the temporary traffic control zone, school zone, or other applicable designated zone and just beyond any interchanges, major intersections, or other major traffic generators.*

Standard:

04 **The Higher Fines signs and plaque shall have a black legend and border on a white rectangular background. All supplemental plaques mounted below the Higher Fines signs and plaque shall have a black legend and border on a white rectangular background.**

Guidance:

05 *Agencies should limit the use of the Higher Fines signs and plaque to locations where work is actually underway, or to locations where the roadway, shoulder, or other conditions, including the presence of a school zone and/or a reduced school speed limit zone, require a speed reduction or extra caution on the part of the road user.*

Option:

06 Alternate legends such as BEGIN (or END) DOUBLE FINES ZONE may also be used for the R2-10 and R2-11 signs.

07 The legend FINES HIGHER on the R2-6P plaque may be replaced by FINES DOUBLE (R2-6aP), $XX FINE (R2-6bP), or another legend appropriate to the specific regulation (see Figure 2B-3).

08 The following may be mounted below an R2-10 sign or R2-6P plaque:
  A. A supplemental plaque specifying the times that the higher fines are in effect (similar to the S4-1P plaque shown in Figure 7B-1), or
  B. A supplemental plaque WHEN CHILDREN (WORKERS) ARE PRESENT, or
  C. A supplemental plaque WHEN FLASHING (similar to the S4-4P plaque shown in Figure 7B-1) if used in conjunction with a yellow flashing beacon.

Support:

09 Section 6F.12 contains information regarding other signs and plaques associated with increased fines for traffic violations in temporary traffic control zones. Section 7B.10 contains information regarding other signs and plaques associated with increased fines for traffic violations in designated school zones.

Section 2B.18 Movement Prohibition Signs (R3-1 through R3-4, R3-18, and R3-27)

Standard:

01 Except as provided in Paragraphs 11 and 13, where specific movements are prohibited, Movement Prohibition signs shall be installed.

Guidance:

02 *Movement Prohibition signs should be placed where they will be most easily seen by road users who might be intending to make the movement.*

03 *If No Right Turn (R3-1) signs (see Figure 2B-4) are used, at least one should be placed either over the roadway or at a right-hand corner of the intersection.*

04 *If No Left Turn (R3-2) signs (see Figure 2B-4) are used, at least one should be placed over the roadway, at the far left-hand corner of the intersection, on a median, or in conjunction with the STOP sign or YIELD sign located on the near right-hand corner.*

Figure 2B-4. Movement Prohibition and Lane Control Signs and Plaques

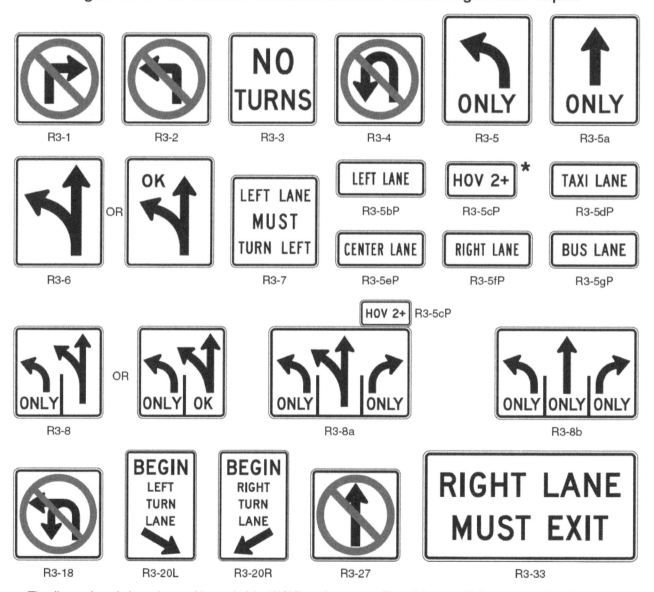

* The diamond symbol may be used instead of the "HOV" word message. The minimum vehicle occupancy level may vary, such as 2+, 3+, 4+. The words "LANE" or "ONLY" may be used with this sign when appropriate.

05 *Except as provided in Item C of Paragraph 9 for signalized locations, if NO TURNS (R3-3) signs (see Figure 2B-4) are used, two signs should be used, one at a location specified for a No Right Turn sign and one at a location specified for a No Left Turn sign.*

06 *If No U-Turn (R3-4) signs (see Figure 2B-4) or combination No U-Turn/No Left Turn (R3-18) signs (see Figure 2B-4) are used, at least one should be used at a location specified for No Left Turn signs.*

Option:

07 If both left turns and U-turns are prohibited, the combination No U-Turn/No Left Turn (R3-18) sign (see Figure 2B-4) may be used instead of separate R3-2 and R3-4 signs.

Guidance:

08 *If No Straight Through (R3-27) signs (see Figure 2B-4) are used, at least one should be placed either over the roadway or at a location where it can be seen by road users who might be intending to travel straight through the intersection.*

09 *If turn prohibition signs are installed in conjunction with traffic control signals:*
   A. *The No Right Turn sign should be installed adjacent to a signal face viewed by road users in the right-hand lane.*
   B. *The No Left Turn (or No U-Turn or combination No U-Turn/No Left Turn) sign should be installed adjacent to a signal face viewed by road users in the left-hand lane.*
   C. *A NO TURNS sign should be placed adjacent to a signal face viewed by all road users on that approach, or two signs should be used.*

Option:

10 If turn prohibition signs are installed in conjunction with traffic control signals, an additional Movement Prohibition sign may be post-mounted to supplement the sign mounted overhead.

11 Where ONE WAY signs are used (see Section 2B.40), No Left Turn and No Right Turn signs may be omitted.

12 When the movement restriction applies during certain time periods only, the following Movement Prohibition signing alternatives may be used and are listed in order of preference:
   A. Changeable message signs, especially at signalized intersections.
   B. Permanently mounted signs incorporating a supplementary legend showing the hours and days during which the prohibition is applicable.
   C. Portable signs, installed by proper authority, located off the roadway at each corner of the intersection. The portable signs are only to be used during the time that the movement prohibition is applicable.

13 Movement Prohibition signs may be omitted at a ramp entrance to an expressway or a channelized intersection where the design is such as to indicate clearly the one-way traffic movement on the ramp or turning lane.

Standard:

14 **The No Left Turn (R3-2) sign, the No U-Turn (R3-4) sign, and the combination No U-Turn/No Left Turn (R3-18) sign shall not be used at approaches to roundabouts to prohibit drivers from turning left onto the circulatory roadway of a roundabout.**

Support:

15 At roundabouts, the use of R3-2, R3-4, or R3-18 signs to prohibit left turns onto the circulatory roadway might confuse drivers about the possible legal turning movements around the roundabout. Roundabout Directional Arrow (R6-4 series) signs (see Section 2B.43) and/or ONE WAY (R6-1R or R6-2R) signs are the appropriate signs to indicate the travel direction within a roundabout.

Section 2B.19 Intersection Lane Control Signs (R3-5 through R3-8)

Standard:

01 **Intersection Lane Control signs, if used, shall require road users in certain lanes to turn, shall permit turns from a lane where such turns would otherwise not be permitted, shall require a road user to stay in the same lane and proceed straight through an intersection, or shall indicate permitted movements from a lane.**

02 **Intersection Lane Control signs (see Figure 2B-4) shall have three applications:**
   A. **Mandatory Movement Lane Control (R3-5, R3-5a, and R3-7) signs,**
   B. **Optional Movement Lane Control (R3-6) sign, and**
   C. **Advance Intersection Lane Control (R3-8 series) signs.**

Guidance:

03 *When Intersection Lane Control signs are mounted overhead, each sign should be placed over the lane or a projection of the lane to which it applies.*

04 *On signalized approaches where through lanes that become mandatory turn lanes, multiple-lane turns that include shared lanes for through and turning movements, or other lane-use regulations are present that would be unexpected by unfamiliar road users, overhead lane control signs should be installed at the signalized location over the appropriate lanes or projections thereof and in advance of the intersection over the appropriate lanes.*

05 *Where overhead mounting on the approach is impractical for the advance and/or intersection lane-use signs, one of the following alternatives should be employed:*

   A. *At locations where through lanes become mandatory turn lanes, a mandatory movement lane control (R3-7) sign should be post-mounted on the left-hand side of the roadway where a through lane is becoming a mandatory left-turn lane on a one-way street or where a median of sufficient width for the signs is available, or on the right-hand side of the roadway where a through lane is becoming a mandatory right-turn lane.*

   B. *At locations where a through lane is becoming a mandatory left-turn lane on a two-way street where a median of sufficient width for the signs is not available, and at locations where multiple-lane turns that include shared lanes for through and turning movements are present, an Advance Intersection Lane Control (R3-8 series) sign should be post-mounted in a prominent location in advance of the intersection, and consideration should be given to the use of an oversized version in accordance with Table 2B-1.*

Standard:

06 **Use of an overhead sign for one approach lane shall not require installation of overhead signs for the other lanes of that approach.**

Option:

07 Where the number of through lanes on an approach is two or less, the Intersection Lane Control signs (R3-5, R3-6, or R3-8) may be overhead or post-mounted.

08 Intersection Lane Control signs may be omitted where:

   A. A turn bay has been provided by physical construction or pavement markings, and
   B. Only the road users using such turn bays are permitted to make a turn in that direction.

09 At roundabouts, Intersection Lane Control (R3-5, R3-6, and R3-8 series) signs may display any of the arrow symbol options shown in Figure 2B-5.

Section 2B.20 Mandatory Movement Lane Control Signs (R3-5, R3-5a, R3-7, and R3-20)

Standard:

01 **If used, the Mandatory Movement Lane Control (R3-5, R3-5a, and R3-7) sign (see Figure 2B-4) shall indicate only the single vehicle movement that is required from the lane. If used, the Mandatory Movement Lane Control sign shall be located in advance of the intersection, such as near the upstream end of the mandatory movement lane, and/or at the intersection where the regulation applies. When the mandatory movement applies to lanes exclusively designated for HOV traffic, the R3-5cP supplemental plaque shall be used. When the mandatory movement applies to lanes that are not HOV facilities, but are lanes exclusively designated for buses and/or taxis, the word message R3-5dP and/or R3-5gP supplemental plaques shall be used.**

Figure 2B-5. Intersection Lane Control Sign Arrow Options for Roundabouts

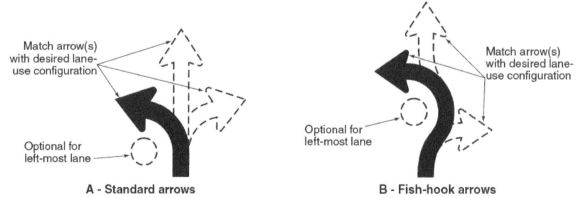

A - Standard arrows

B - Fish-hook arrows

02 The Mandatory Movement Lane Control (R3-7) sign shall include the legend RIGHT (LEFT) LANE MUST TURN RIGHT (LEFT). The Mandatory Movement Lane Control (R3-5 and R3-5a) symbol signs shall include the legend ONLY.

03 The R3-7 word message sign shall be for post-mounting only.

04 Where the number of lanes available to through traffic on an approach is three or more, Mandatory Movement Lane Control (R3-5 and R3-5a) symbol signs, if used, shall be mounted overhead over the specific lanes to which they apply (see Section 2B.19).

Guidance:

05 *If the R3-5 or R3-5a sign is post-mounted on an approach with two or fewer through lanes, a supplemental plaque (see Figure 2B-4), such as LEFT LANE (R3-5bP), HOV 2+ (R3-5cP), TAXI LANE (R3-5dP), CENTER LANE (R3-5eP), RIGHT LANE (R3-5fP), BUS LANE (R3-5gP), or BOTH LANES, should be added above the sign to indicate the specific lane to which the mandatory movement applies. If Mandatory Lane Movement Control (R3-5) symbol signs with supplemental R3-5bP or R3-5fP plaques are used, they should be mounted adjacent to and along only the full width portion of the turn lane.*

06 *The use of the Mandatory Movement Lane Control (R3-7) word message sign should be limited to only locations that are adjacent to the full-width portion of a mandatory turn lane. The R3-7 sign should not be installed adjacent to a through lane in advance of a turn bay taper or adjacent to a turn bay taper.*

07 *Mandatory Movement Lane Control signs should be accompanied by lane-use arrow markings, especially where traffic volumes are high, where there is a high percentage of commercial vehicles, or where other distractions exist.*

Option:

08 The Straight Through Only (R3-5a) sign may be used to require a road user in a particular lane to proceed straight through an intersection.

09 When the Mandatory Movement Lane Control sign for a left-turn lane is installed back-to-back with a Keep Right (R4-7) sign, the dimensions of the Mandatory Movement Lane Control (R3-5) sign may be the same as the Keep Right sign.

10 The diamond symbol may be used instead of the word message HOV on the R3-5cP supplemental plaque.

11 The BEGIN RIGHT TURN LANE (R3-20R) sign (see Figure 2B-4) may be post-mounted on the right-hand side of the roadway at the upstream end of the turn lane taper of a mandatory right-turn lane. The BEGIN LEFT TURN LANE (R3-20L) sign (see Figure 2B-4) may be post-mounted on a median (or on the left-hand side of the roadway for a one-way street) at the upstream end of the turn lane taper of a mandatory left-turn lane.

Section 2B.21 Optional Movement Lane Control Sign (R3-6)

Standard:

01 **If used, the Optional Movement Lane Control (R3-6) sign (see Figure 2B-4) shall be used for two or more movements from a specific lane or to emphasize permitted movements. If used, the Optional Movement Lane Control sign shall be located in advance of the intersection, such as near the upstream end of an adjacent mandatory movement lane, and/or at the intersection where the regulation applies.**

02 **If used, the Optional Movement Lane Control sign shall indicate all permissible movements from specific lanes.**

03 **Optional Movement Lane Control signs shall be used for two or more movements from a specific lane where a movement, not normally allowed, is permitted.**

04 **The Optional Movement Lane Control sign shall not be used alone to effect a turn prohibition.**

05 **Where the number of lanes available to through traffic on an approach is three or more, an Optional Movement Lane Control (R3-6) sign, if used, shall be mounted overhead over the specific lane to which it applies (see Section 2B.19).**

Guidance:

06 *If the Optional Movement Lane Control sign is post-mounted on an approach with two or fewer through lanes, a supplemental plaque (see Figure 2B-4), such as LEFT LANE (R3-5bP), HOV 2+ (R3-5cP), TAXI LANE (R3-5dP), CENTER LANE (R3-5eP), RIGHT LANE (R3-5fP), or BUS LANE (R3-5gP), should be added above the R3-6 sign to indicate the specific lane from which the optional movements can be made.*

Option:

07 The word message OK may be used within the border in combination with the arrow symbols of the R3-6 sign.

Standard:

08 **Because more than one movement is permitted from the lane, the word message ONLY shall not be used on an Optional Movement Lane Control sign.**

Section 2B.22 Advance Intersection Lane Control Signs (R3-8 Series)

Option:

01 Advance Intersection Lane Control (R3-8, R3-8a, and R3-8b) signs (see Figure 2B-4) may be used to indicate the configuration of all lanes ahead.

02 The word messages ONLY, OK, THRU, ALL, or HOV 2+ may be used within the border in combination with the arrow symbols of the R3-8 sign series. The HOV 2+ (R3-5cP) supplemental plaque may be installed at the top outside border of the R3-8 sign over the applicable lane designation on the sign. The diamond symbol may be used instead of the word message HOV. The minimum allowable vehicle occupancy requirement may vary based on the level established for a particular facility.

Guidance:

03 *If used, an Advance Intersection Lane Control sign should be placed at an adequate distance in advance of the intersection so that road users can select the appropriate lane (see Figure 2A-4). If used, the Advance Intersection Lane Control sign should be installed either in advance of the tapers or at the beginning of the turn lane.*

Option:

04 An Advance Intersection Lane Control sign may be repeated closer to the intersection for additional emphasis.

Standard:

05 **Where three or more approach lanes are available to traffic, Advance Intersection Lane Control (R3-8 series) signs, if used, shall be post-mounted in advance of the intersection and shall not be mounted overhead (see Section 2B.19).**

Section 2B.23 RIGHT (LEFT) LANE MUST EXIT Sign (R3-33)

Option:

01 A RIGHT (LEFT) LANE MUST EXIT (R3-33) sign (see Figure 2B-4) may be used to supplement an overhead EXIT ONLY guide sign to inform road users that traffic in the right-hand (left-hand) lane of a roadway that is approaching a grade-separated interchange is required to depart the roadway on the exit ramp at the next interchange.

Support:

02 Section 2C.43 contains information regarding a warning sign that can be used in advance of lane drops at grade-separated interchanges.

Section 2B.24 Two-Way Left Turn Only Signs (R3-9a, R3-9b)

Guidance:

01 *Two-Way Left Turn Only (R3-9a or R3-9b) signs (see Figure 2B-6) should be used in conjunction with the required pavement markings where a non-reversible lane is reserved for the exclusive use of left-turning vehicles in either direction and is not used for passing, overtaking, or through travel.*

Option:

02 The post-mounted R3-9b sign may be used as an alternate to or a supplement to the overhead R3-9a sign. The legend BEGIN or END may be used within the border of the main sign itself, or on an R3-9cP or R3-9dP plaque (see Figure 2B-6) mounted immediately above it.

Support:

03 Signing is especially helpful to drivers in areas where the two-way left turn only maneuver is new, in areas subject to environmental conditions that frequently obscure the pavement markings, and on peripheral streets with two-way left turn only lanes leading to an extensive system of routes with two-way left turn only lanes.

Section 2B.25 BEGIN and END Plaques (R3-9cP, R3-9dP)

Option:

01 The BEGIN (R3-9cP) or END (R3-9dP) plaque (see Figure 2B-6) may be used to supplement a regulatory sign to inform road users of the location where a regulatory condition begins or ends.

Standard:

02 **If used, the BEGIN or END plaque shall be mounted directly above a regulatory sign.**

Figure 2B-6. Center and Reversible Lane Control Signs and Plaques

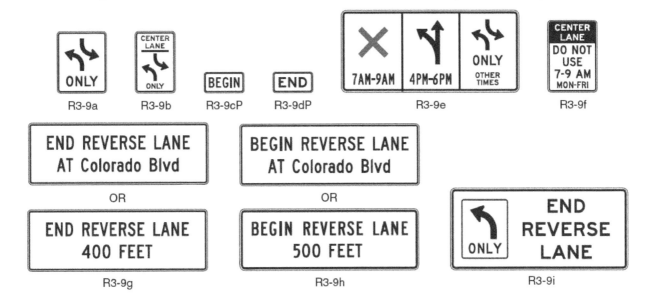

Section 2B.26 Reversible Lane Control Signs (R3-9e through R3-9i)

Option:

01 A reversible lane may be used for through traffic (with left turns either permitted or prohibited) in alternating directions during different periods of the day, and the lane may be used for exclusive left turns in one or both directions during other periods of the day as well. Reversible Lane Control (R3-9e through R3-9i) signs (see Figure 2B-6) may be either static type or changeable message type. These signs may be either post-mounted or overhead.

Standard:

02 **Post-mounted Reversible Lane Control signs shall be used only as a supplement to overhead signs or signals. post-mounted signs shall be identical in design to the overhead signs and an additional legend such as CENTER LANE shall be added to the sign (R3-9f) to indicate which lane is controlled. For both word messages and symbols, this legend shall be at the top of the sign.**

03 **Where it is determined by an engineering study that lane-use control signals or physical barriers are not necessary, the lane shall be controlled by overhead Reversible Lane Control signs (see Figure 2B-7).**

Option:

04 Reversing traffic flow may be controlled with pavement markings and Reversible Lane Control signs (without the use of lane control signals), when all of the following conditions are met:
   A. Only one lane is being reversed,
   B. An engineering study indicates that the use of Reversible Lane Control signs alone would result in an acceptable level of safety and efficiency, and
   C. There are no unusual or complex operations in the reversible lane pattern.

Standard:

05 **Reversible Lane Control signs shall contain the legend or symbols designating the allowable uses of the lane and the time periods such uses are allowed. Where symbols and legends are used, their meanings shall be as shown in Table 2B-2.**

Table 2B-2. Meanings of Symbols and Legends on Reversible Lane Control Signs

Symbol / Word Message	Meaning
Red X on white background	Lane closed
Upward pointing black arrow on white background (if left turns are permitted, the arrow shall be modified to show left / through arrow)	Lane open for through travel and any turns not otherwise prohibited
Black two-way left-turn arrows on white background and legend ONLY	Lane may be used only for left turns in either direction (i.e., as a two-way left-turn lane)
Black single left-turn arrow on white background and legend ONLY	Lane may be used only for left turns in one direction (without opposing left turns in the same lane)

Figure 2B-7. Location of Reversible Two-Way Left-Turn Signs

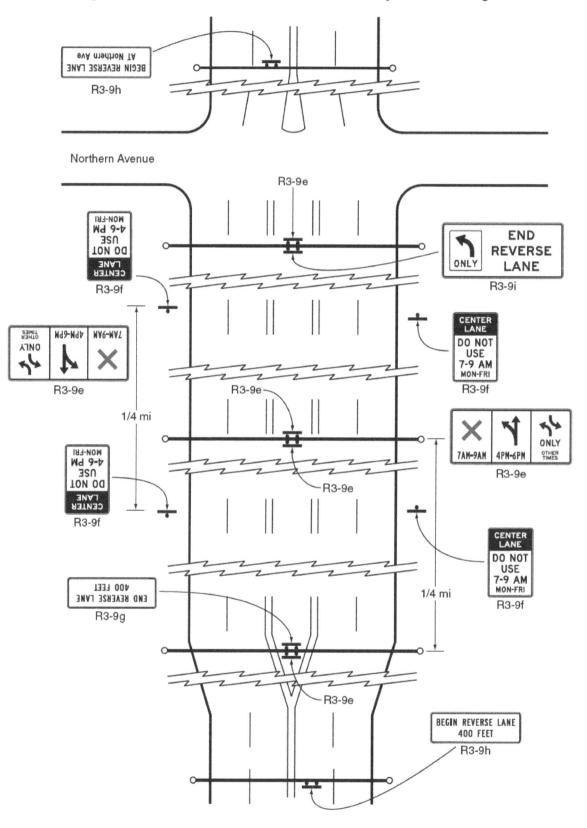

06 **Reversible Lane Control signs shall consist of a white background with a black legend and border, except for the R3-9d sign, where the color red is used.**

07 **Symbol signs, such as the R3-9d sign, shall consist of the appropriate symbol in the upper portion of the sign with the appropriate times of the day and days of the week below it. All times of the day and days of the week shall be accounted for on the sign to eliminate confusion to the road user.**

08 **In situations where more than one message is conveyed to the road user, such as on the R3-9d sign, the sign legend shall be arranged as follows:**
   A. **The prohibition or restriction message is the primary legend and shall be on the top for word message signs and to the far left for symbol signs,**
   B. **The permissive use message shall be displayed as the second legend, and**
   C. **The OTHER TIMES message shall be displayed at the bottom for word message signs and to the far right for symbol signs.**

Option:

09 The symbol signs may also include a downward pointing arrow with the legend THIS LANE. The term OTHER TIMES may be used for either the symbol or word message sign.

Standard:

10 **A Reversible Lane Control sign shall be mounted over the center of the lane that is being reversed and shall be perpendicular to the roadway alignment.**

11 **If the vertical or horizontal alignment is curved to the degree that a driver would be unable to see at least one sign, and preferably two signs, then additional overhead signs shall be installed. The placement of the signs shall be such that the driver will have a definite indication of the lanes specifically reserved for use at any given time. Special consideration shall be given to major generators introducing traffic between the normal sign placement.**

12 **Transitions at the entry to and exit from a section of roadway with reversible lanes shall be carefully reviewed, and advance signs shall be installed to notify or warn drivers of the boundaries of the reversible lane controls. The R3-9g or R3-9h signs shall be used for this purpose.**

Option:

13 More than one sign may be used at the termination of the reversible lane to emphasize the importance of the message (R3-9i).

Standard:

14 **Flashing beacons, if used to accentuate the overhead Reversible Lane Control signs, shall comply with the applicable requirements for flashing beacons in Chapter 4L.**

15 **When used in conjunction with Reversible Lane Control signs, the Turn Prohibition signs (R3-1 to R3-4, R3-18) shall be mounted overhead and separate from the Reversible Lane Control signs. The Turn Prohibition signs shall be designed and installed in accordance with Section 2B.18.**

Guidance:

16 *For additional emphasis, a supplemental plaque stating the distance of the prohibition, such as NEXT 1 MILE, should be added to the Turn Prohibition signs that are used in conjunction with Reversible Lane Control signs.*

17 *If used, overhead signs should be located at intervals not greater than 1/4 mile. The bottom of the overhead Reversible Lane Control signs should not be more than 19 feet above the pavement grade.*

18 *Where more than one sign is used at the termination of a reversible lane, they should be at least 250 feet apart. Longer distances between signs are appropriate for streets with speeds over 35 mph, but the separation should not exceed 1,000 feet.*

19 *Because left-turning vehicles have a significant impact on the safety and efficiency of a reversible lane operation, if an exclusive left-turn lane or two-way left-turn lane cannot be incorporated into the lane-use pattern for a particular peak or off-peak period, consideration should be given to prohibiting left turns and U-turns during that time period.*

Section 2B.27 Jughandle Signs (R3-23, R3-24, R3-25, and R3-26 Series)

Support:

01 A jughandle turn is a left-turn or U-turn that because of special geometry is made by initially making a right turn. This type of turn can increase the operational efficiency of a roadway by eliminating the need for exclusive left-turn lanes and can increase the operational efficiency of a traffic control signal by eliminating the need for protected left-turn phases. A jughandle turn can also provide an opportunity for trucks and commercial vehicles to make a U-turn where the median and roadway are not of sufficient width to accommodate a traditional U-turn by these vehicles.

02 Figure 2B-8 shows the various signs that can be used for signing jughandle turns. Figure 2B-9 shows examples of regulatory and destination guide signing for various types of jughandle turns.

Standard:

03 **On multi-lane roadways, since road users generally anticipate that they need to be in the left-hand lane when approaching a location where they desire to turn left or make a U-turn, an ALL TURNS FROM RIGHT LANE (R3-23) or a U TURN FROM RIGHT LANE (R3-23a) sign (see Figure 2B-9) shall be installed in advance of the location to inform drivers that left turns and/or U-turns will be made from the right-hand lane.**

Option:

04 Where a median of sufficient width is available, supplemental regulatory or guide signs may also be placed on the left-hand side of the roadway.

Standard

05 **An R3-24 series sign with an upward diagonal arrow pointing to the right if the jughandle entrance is designed as an exit ramp (see Drawings A and B of Figure 2B-9) or an R3-25 series sign with a horizontal arrow pointing to the right if the jughandle entrance is designed as an intersection shall be installed on the right-hand side of the roadway at the entrance to the jughandle. The legend on the sign shall be ALL TURNS, U TURN, or U AND LEFT TURNS, as appropriate.**

06 **If the jughandle is designed such that the jughandle entrance is downstream of the location where the turn would normally have been made (see Drawing C of Figure 2B-9), an R3-26 series sign with an arrow pointing straight upward shall be installed on the right-hand side of the roadway at the intersection to inform road users that they need to proceed straight through the intersection in order to make a left turn or U-turn. The legend on the sign shall be U TURN or U AND LEFT TURNS, as appropriate.**

Support:

07 The R3-24, R3-25, and R3-26 series of signs are designed to be mounted below conventional guide signs.

08 Section 2C.14 contains information regarding the use of advisory exit and ramp speed signs for exit ramps.

09 Section 2D.39 contains information regarding the use of guide signs for jughandles.

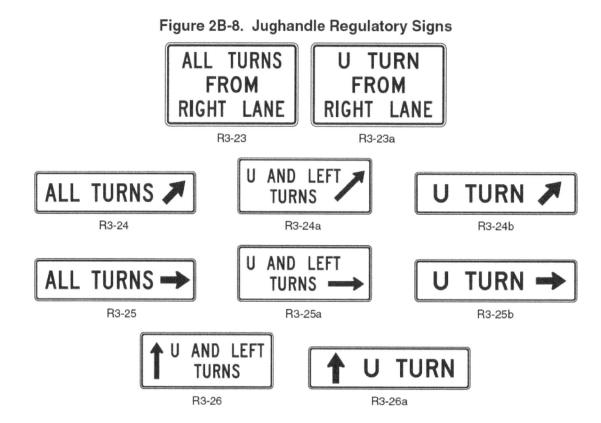

Figure 2B-8. Jughandle Regulatory Signs

Figure 2B-9. Examples of Applications of Jughandle Regulatory and Guide Signing
(Sheet 1 of 3)

A – Turns made prior to the intersection

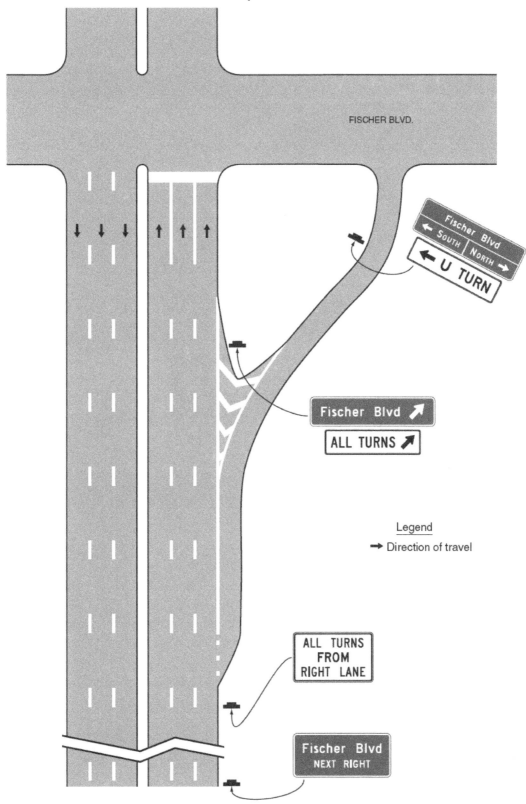

Figure 2B-9. Examples of Applications of Jughandle Regulatory and Guide Signing
(Sheet 2 of 3)

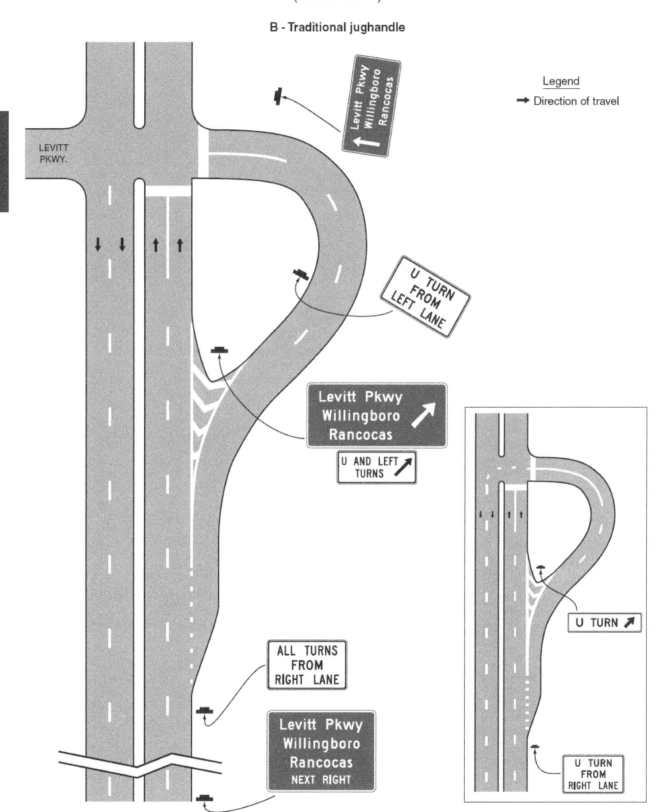

Figure 2B-9. Examples of Applications of Jughandle Regulatory and Guide Signing
(Sheet 3 of 3)

C - Turns made beyond the intersection

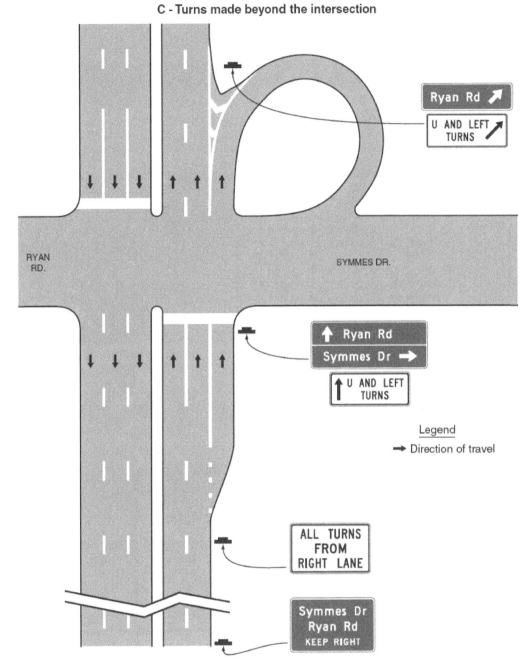

Page 72
2009 Edition

Section 2B.28 DO NOT PASS Sign (R4-1)

Option:

01 The Do Not Pass (R4-1) sign (see Figure 2B-10) may be used in addition to pavement markings (see Section 3B.02) to emphasize the restriction on passing. The Do Not Pass sign may be used at the beginning of, and at intervals within, a zone through which sight distance is restricted or where other conditions make overtaking and passing inappropriate.

02 If signing is needed on the left-hand side of the roadway for additional emphasis, NO PASSING ZONE (W14-3) signs may be used (see Section 2C.45).

Support:

03 Standards for determining the location and extent of no-passing zone pavement markings are set forth in Section 3B.02.

Figure 2B-10. Passing, Keep Right, and Slow Traffic Signs

R4-1

R4-2

R4-3

R4-5

R4-7

R4-7a

R4-7b

R4-7c

R4-8

R4-8a

R4-8b

R4-8c

R4-9

R4-10

R4-12

R4-13

R4-14

R4-16

R4-17

R4-18

Sect. 2B.28
December 2009

Section 2B.29 PASS WITH CARE Sign (R4-2)

Guidance:

01 *The PASS WITH CARE (R4-2) sign (see Figure 2B-10) should be installed at the downstream end of a no-passing zone if a DO NOT PASS sign has been installed at the upstream end of the zone.*

Section 2B.30 KEEP RIGHT EXCEPT TO PASS Sign (R4-16) and SLOWER TRAFFIC KEEP RIGHT Sign (R4-3)

Option:

01 The KEEP RIGHT EXCEPT TO PASS (R4-16) sign (see Figure 2B-10) may be used on multi-lane roadways to direct drivers to stay in the right-hand lane except when they are passing another vehicle.

Guidance:

02 *If used, the KEEP RIGHT EXCEPT TO PASS sign should be installed just beyond the beginning of a multi-lane roadway and at selected locations along multi-lane roadways for additional emphasis.*

Option:

03 The SLOWER TRAFFIC KEEP RIGHT (R4-3) sign (see Figure 2B-10) may be used on multi-lane roadways to reduce unnecessary lane changing.

Guidance:

04 *If used, the SLOWER TRAFFIC KEEP RIGHT sign should be installed just beyond the beginning of a multi-lane pavement, and at selected locations where there is a tendency on the part of some road users to drive in the left-hand lane (or lanes) below the normal speed of traffic. This sign should not be used on the approach to an interchange or through an interchange area.*

Section 2B.31 TRUCKS USE RIGHT LANE Sign (R4-5)

Guidance:

01 *If an extra lane has been provided for trucks and other slow-moving traffic, a SLOWER TRAFFIC KEEP RIGHT (R4-3) sign (see Figure 2B-10), TRUCKS USE RIGHT LANE (R4-5) sign (see Figure 2B-10), or other appropriate sign should be installed at the beginning of the lane.*

Option:

02 The SLOWER TRAFFIC KEEP RIGHT sign may be used as a supplement or as an alternative to the TRUCKS USE RIGHT LANE sign. Both signs may be used on multi-lane roadways to improve capacity and reduce lane changing.

03 The TRUCKS USE RIGHT LANE (R4-5) sign may be used on multi-lane roadways to reduce unnecessary lane changing.

Guidance:

04 *If an extra lane has been provided for trucks and other slow-moving traffic, a Lane Ends sign (see Section 2C.42) should be installed in advance of the point where the extra lane ends. Appropriate pavement markings should be installed at both the upstream and downstream ends of the extra lane (see Section 3B.09 and Figure 3B-13).*

Support:

05 Section 2D.51 contains information regarding advance information signs for extra lanes that have been provided for trucks and other slow-moving traffic.

Section 2B.32 Keep Right and Keep Left Signs (R4-7, R4-8)

Option:

01 The Keep Right (R4-7) sign (see Figure 2B-10) may be used at locations where it is necessary for traffic to pass only to the right-hand side of a roadway feature or obstruction. The Keep Left (R4-8) sign (see Figure 2B-10) may be used at locations where it is necessary for traffic to pass only to the left-hand side of a roadway feature or obstruction.

Guidance:

02 *At locations where it is not readily apparent that traffic is required to keep to the right, a Keep Right sign should be used.*

03 *If used, the Keep Right sign should be installed as close as practical to approach ends of raised medians, parkways, islands, and underpass piers. The sign should be mounted on the face of or just in front of a pier or other obstruction separating opposite directions of traffic in the center of the highway such that traffic will have to pass to the right-hand side of the sign.*

Standard:

04 **The Keep Right sign shall not be installed on the right-hand side of the roadway in a position where traffic must pass to the left-hand side of the sign.**

Option:

05 The Keep Right sign may be omitted at intermediate ends of divisional islands and medians.

06 Word message KEEP RIGHT (LEFT) with an arrow (R4-7a or R4-7b) signs (see Figure 2B-10) may be used instead of the R4-7 or R4-8 symbol signs.

07 Where the obstruction obscures the Keep Right sign, the minimum placement height may be increased for better sign visibility.

08 A narrow Keep Right (R4-7c) sign (see Figure 2B-10) may be installed on the approach end of a median island that is less than 4 feet wide at the point where the sign is to be located.

Standard:

09 **A narrow Keep Right (R4-7c) sign shall not be installed on a median island that has a width of 4 feet or more at the point where the sign is to be located.**

Section 2B.33 STAY IN LANE Sign (R4-9)

Option:

01 A STAY IN LANE (R4-9) sign (see Figure 2B-10) may be used on multi-lane highways to direct road users to stay in their lane until conditions permit shifting to another lane.

Guidance:

02 *If a STAY IN LANE sign is used, it should be accompanied by a double solid white lane line(s) to prohibit lane changing.*

Section 2B.34 RUNAWAY VEHICLES ONLY Sign (R4-10)

Guidance:

01 *A RUNAWAY VEHICLES ONLY (R4-10) sign (see Figure 2B-10) should be installed near a truck escape (or runaway truck) ramp entrance to discourage other road users from entering the ramp.*

Section 2B.35 Slow Vehicle Turn-Out Signs (R4-12, R4-13, and R4-14)

Support:

01 On two-lane highways in areas where traffic volumes and/or vertical or horizontal curvature make passing difficult, turn-out areas are sometimes provided for the purpose of giving a group of faster vehicles an opportunity to pass a slow-moving vehicle.

Option:

02 A SLOW VEHICLES WITH XX OR MORE FOLLOWING VEHICLES MUST USE TURN-OUT (R4-12) sign (see Figure 2B-10) may be installed in advance of a turn-out area to inform drivers who are driving so slow that they have accumulated a specific number of vehicles behind them that they are required by the traffic laws of that State to use the turn-out to allow the vehicles following them to pass.

Support:

03 The specific number of vehicles displayed on the R4-12 sign provides law enforcement personnel with the information they need to enforce this regulation.

Option:

04 If an R4-12 sign has been installed in advance of a turn-out area, a SLOW VEHICLES MUST USE TURN-OUT AHEAD (R4-13) sign (see Figure 2B-10) may also be installed downstream from the R4-12 sign, but upstream from the turn-out area, to remind slow drivers that they are required to use a turn-out that is a short distance ahead.

Standard:

05 **If an R4-12 sign has been installed in advance of a turn-out area, a SLOW VEHICLES MUST TURN OUT (with arrow) (R4-14) sign (see Figure 2B-10) shall be installed at the entry point of the turn-out area.**

Support:

06 Section 2D.52 contains information regarding advance information signs for slow vehicle turn-out areas.

Section 2B.36 DO NOT DRIVE ON SHOULDER Sign (R4-17) and DO NOT PASS ON SHOULDER Sign (R4-18)

Option:
01 The DO NOT DRIVE ON SHOULDER (R4-17) sign (see Figure 2B-10) may be installed to inform road users that using the shoulder of a roadway as a travel lane is prohibited.
02 The DO NOT PASS ON SHOULDER (R4-18) sign (see Figure 2B-10) may be installed to inform road users that using the shoulder of a roadway to pass other vehicles is prohibited.

Section 2B.37 DO NOT ENTER Sign (R5-1)

Standard:
01 **The DO NOT ENTER (R5-1) sign (see Figure 2B-11) shall be used where traffic is prohibited from entering a restricted roadway.**
Guidance:
02 *The DO NOT ENTER sign, if used, should be placed directly in view of a road user at the point where a road user could wrongly enter a divided highway, one-way roadway, or ramp (see Figure 2B-12). The sign should be mounted on the right-hand side of the roadway, facing traffic that might enter the roadway or ramp in the wrong direction.*
03 *If the DO NOT ENTER sign would be visible to traffic to which it does not apply, the sign should be turned away from, or shielded from, the view of that traffic.*
Option:
04 The DO NOT ENTER sign may be installed where it is necessary to emphasize the one-way traffic movement on a ramp or turning lane.
05 A second DO NOT ENTER sign on the left-hand side of the roadway may be used, particularly where traffic approaches from an intersecting roadway (see Figure 2B-12).

Figure 2B-11. Selective Exclusion Signs

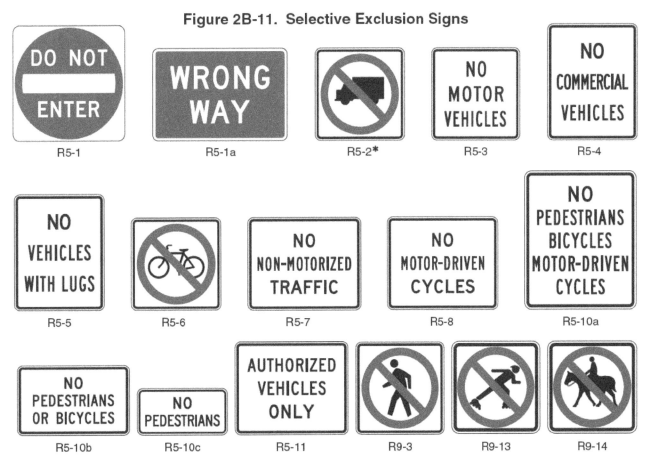

* An optional word message sign is shown in the "Standard Highway Signs and Markings" book

Figure 2B-12. Locations of Wrong-Way Signing for Divided Highways with Median Widths of 30 Feet or Wider

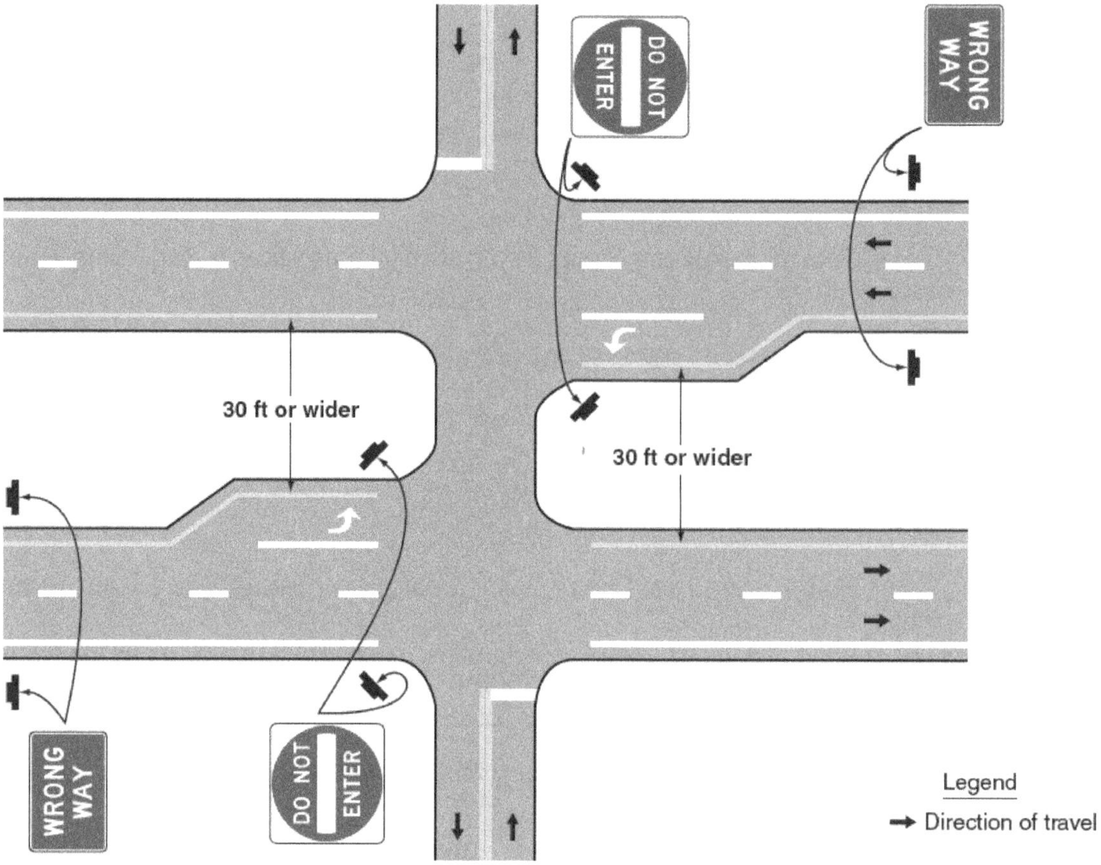

Legend
→ Direction of travel

Support:

06 Section 2B.41 contains information regarding an optional lower mounting height for DO NOT ENTER signs that are located along an exit ramp facing a road user who is traveling in the wrong direction.

Section 2B.38 WRONG WAY Sign (R5-1a)

Option:

01 The WRONG WAY (R5-1a) sign (see Figure 2B-11) may be used as a supplement to the DO NOT ENTER sign where an exit ramp intersects a crossroad or a crossroad intersects a one-way roadway in a manner that does not physically discourage or prevent wrong-way entry (see Figure 2B-12).

Guidance:

02 *If used, the WRONG WAY sign should be placed at a location along the exit ramp or the one-way roadway farther from the crossroad than the DO NOT ENTER sign (see Section 2B.41).*

Support:

03 Section 2B.41 contains information regarding an optional lower mounting height for WRONG WAY signs that are located along an exit ramp facing a road user who is traveling in the wrong direction.

Section 2B.39 Selective Exclusion Signs

Support:

01 Selective Exclusion signs (see Figure 2B-11) give notice to road users that State or local statutes or ordinances exclude designated types of traffic from using particular roadways or facilities.

Standard:

02 **If used, Selective Exclusion signs shall clearly indicate the type of traffic that is excluded.**

Support:

03   Typical exclusion messages include:
- A. No Trucks (R5-2),
- B. NO MOTOR VEHICLES (R5-3),
- C. NO COMMERCIAL VEHICLES (R5-4),
- D. NO TRUCKS (VEHICLES) WITH LUGS (R5-5),
- E. No Bicycles (R5-6),
- F. NO NON-MOTORIZED TRAFFIC (R5-7),
- G. NO MOTOR-DRIVEN CYCLES (R5-8),
- H. No Pedestrians (R9-3),
- I. No Skaters (R9-13),
- J. No Equestrians (R9-14), and
- K. No Hazardous Material (R14-3) (see Section 2B.62).

Option:

04   Appropriate combinations or groupings of these legends into a single sign, such as NO PEDESTRIANS BICYCLES MOTOR-DRIVEN CYCLES (R5-10a), or NO PEDESTRIANS OR BICYCLES (R5-10b) may be used.

Guidance:

05   *If an exclusion is governed by vehicle weight, a Weight Limit sign (see Section 2B.59) should be used instead of a Selective Exclusion sign.*

06   *If used on a freeway or expressway ramp, the NO PEDESTRIANS OR BICYCLES (R5-10b) sign should be installed in a location where it is clearly visible to any pedestrian or bicyclist attempting to enter the limited access facility from a street intersecting the exit ramp.*

07   *The Selective Exclusion sign should be placed on the right-hand side of the roadway at an appropriate distance from the intersection so as to be clearly visible to all road users turning into the roadway that has the exclusion. The NO PEDESTRIANS (R5-10c) or No Pedestrian Crossing (R9-3) sign (see Section 2B.51) should be installed so as to be clearly visible to pedestrians who are at a location where an alternative route is available.*

Option:

08   The NO PEDESTRIANS (R5-10c) or No Pedestrian Crossing (R9-3) sign may also be used at underpasses or elsewhere where pedestrian facilities are not provided.

09   The NO TRUCKS (R5-2a) word message sign may be used as an alternate to the No Trucks (R5-2) symbol sign.

10   The AUTHORIZED VEHICLES ONLY (R5-11) sign may be used at median openings and other locations to prohibit vehicles from using the median opening or facility unless they have special permission (such as law enforcement vehicles or emergency vehicles) or are performing official business (such as highway agency vehicles).

Section 2B.40 ONE WAY Signs (R6-1, R6-2)

Standard:

01   **Except as provided in Paragraph 6, the ONE WAY (R6-1 or R6-2) sign (see Figure 2B-13) shall be used to indicate streets or roadways upon which vehicular traffic is allowed to travel in one direction only.**

02   **ONE WAY signs shall be placed parallel to the one-way street at all alleys and roadways that intersect one-way roadways as shown in Figure 2B-14.**

03   **At an intersection with a divided highway that has a median width at the intersection itself of 30 feet or more, ONE WAY signs shall be placed, visible to each crossroad approach, on the near right and far left corners of each intersection with the directional roadways (see Figure 2B-15).**

04   **At an intersection with a divided highway that has a median width at the intersection itself of less than 30 feet, Keep Right (R4-7) signs and/or ONE WAY signs shall be installed (see Figures 2B-16 and 2B-17). If Keep Right signs are installed, they shall be placed as close as practical to the approach ends of the medians and shall be visible to traffic on the divided highway and each crossroad approach. If ONE WAY signs are installed, they shall be placed on the near right and far left corners of the intersection and shall be visible to each crossroad approach.**

Option:

05   At an intersection with a divided highway that has a median width at the intersection itself of less than 30 feet, ONE WAY signs may also be placed on the far right corner of the intersection as shown in Figures 2B-16 and 2B-17.

06   ONE WAY signs may be omitted on the one-way roadways of divided highways, where the design of interchanges indicates the direction of traffic on the separate roadways.

Figure 2B-13. ONE WAY and Divided Highway Crossing Signs

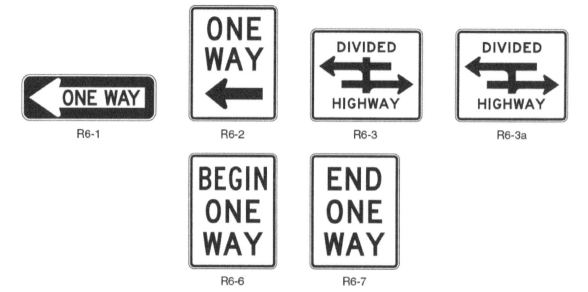

Standard:

07 If used at unsignalized intersections with one-way streets, ONE WAY signs shall be placed on the near right and the far left corners of the intersection facing traffic entering or crossing the one-way street (see Figure 2B-14).

08 If used at signalized intersections with one-way streets, ONE WAY signs shall be placed near the appropriate signal faces, on the poles holding the traffic signals, on the mast arm or span wire holding the signals, or at the locations specified for unsignalized intersections.

09 At unsignalized T-intersections where the roadway at the top of the T-intersection is a one-way roadway, ONE WAY signs shall be placed on the near right and the far side of the intersection facing traffic on the stem approach (see Figure 2B-14).

10 At signalized T-intersections where the roadway at the top of the T-intersection is a one-way roadway, ONE WAY signs shall be placed near the appropriate signal faces, on the poles holding the traffic signals, on the mast arm or span wire holding the signals, or at the locations specified for unsignalized intersections.

Option:

11 Where the central island of a roundabout allows for the installation of signs, ONE WAY signs may be used instead of or in addition to Roundabout Directional Arrow (R6-4 series) signs (see Section 2B.43) to direct traffic counter-clockwise around the central island.

Guidance:

12 *Where used on the central island of a roundabout, the mounting height of a ONE WAY sign should be at least 4 feet, measured vertically from the bottom of the sign to the elevation of the near edge of the traveled way.*

Support:

13 Using ONE WAY signs on the central island of a roundabout might result in some drivers incorrectly concluding that the cross street is a one-way street. Using Roundabout Directional Arrow signs might reduce this confusion. However, using ONE WAY signs might be necessary in States that have defined a roundabout as a series of T-intersections.

Option:

14 The BEGIN ONE WAY (R6-6) sign (see Figure 2B-13) may be used notify road users of the beginning point of a one direction of travel restriction on the street or roadway. The END ONE WAY (R6-7) sign (see Figure 2B-13) may be used notify road users of the ending point of a one direction of travel restriction on the street or roadway.

Figure 2B-14. Locations of ONE WAY Signs

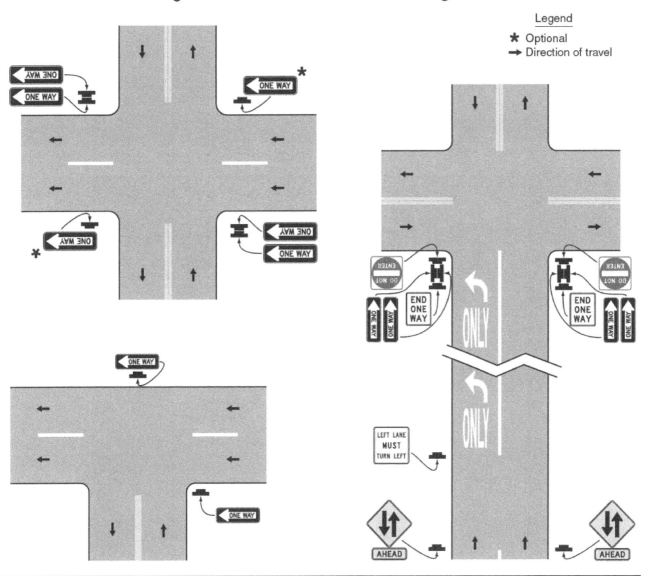

Section 2B.41 Wrong-Way Traffic Control at Interchange Ramps

Standard:

01 At interchange exit ramp terminals where the ramp intersects a crossroad in such a manner that wrong-way entry could inadvertently be made, the following signs shall be used (see Figure 2B-18):

   A. At least one ONE WAY sign for each direction of travel on the crossroad shall be placed where the exit ramp intersects the crossroad.
   B. At least one DO NOT ENTER sign shall be conspicuously placed near the downstream end of the exit ramp in positions appropriate for full view of a road user starting to enter wrongly from the crossroad.
   C. At least one WRONG WAY sign shall be placed on the exit ramp facing a road user traveling in the wrong direction.

Guidance:

02 *In addition, the following pavement markings should be used (see Figure 2B-18):*

   A. *On two-lane paved crossroads at interchanges, double solid yellow lines should be used as a center line for an adequate distance on both sides approaching the ramp intersections.*
   B. *Where crossroad channelization or ramp geometrics do not make wrong-way movements difficult, a lane-use arrow should be placed in each lane of an exit ramp near the crossroad terminal where it will be clearly visible to a potential wrong-way road user.*

Figure 2B-15. ONE WAY Signing for Divided Highways with Median Widths of 30 Feet or Wider

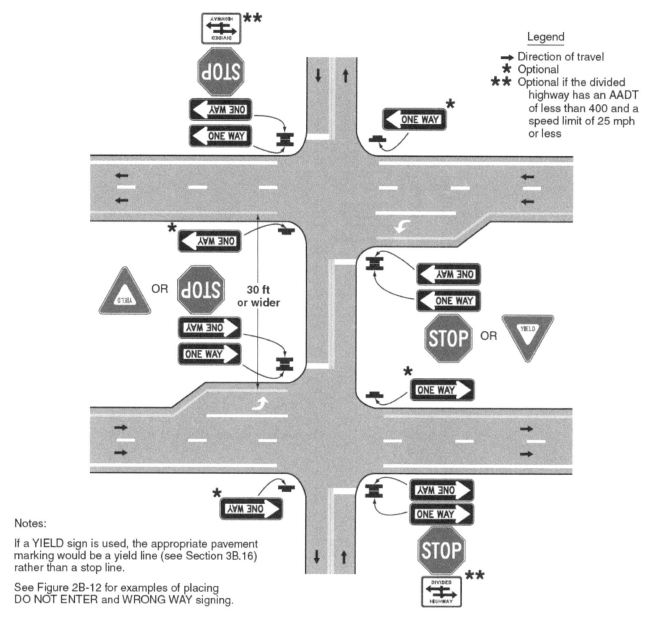

Notes:
If a YIELD sign is used, the appropriate pavement marking would be a yield line (see Section 3B.16) rather than a stop line.

See Figure 2B-12 for examples of placing DO NOT ENTER and WRONG WAY signing.

Option:

03   The following traffic control devices may be used to supplement the signs and pavement markings described in Paragraphs 1 and 2:
   A. Additional ONE WAY signs may be placed, especially on two-lane rural crossroads, appropriately in advance of the ramp intersection to supplement the required ONE WAY sign(s).
   B. Additional WRONG WAY signs may be used.
   C. Slender, elongated wrong-way arrow pavement markings (see Figure 3B-24) intended primarily to warn wrong-way road users that they are traveling in the wrong direction may be placed upstream from the ramp terminus (see Figure 2B-18) to indicate the correct direction of traffic flow. Wrong-way arrow pavement markings may also be placed on the exit ramp at appropriate locations near the crossroad junction to indicate wrong-way movement. The wrong-way arrow markings may consist of pavement markings or bidirectional red-and-white raised pavement markers or other units that show red to wrong-way road users and white to other road users (see Figure 3B-24).

Sect. 2B.41

Figure 2B-16. ONE WAY Signing for Divided Highways with Median Widths Narrower Than 30 Feet

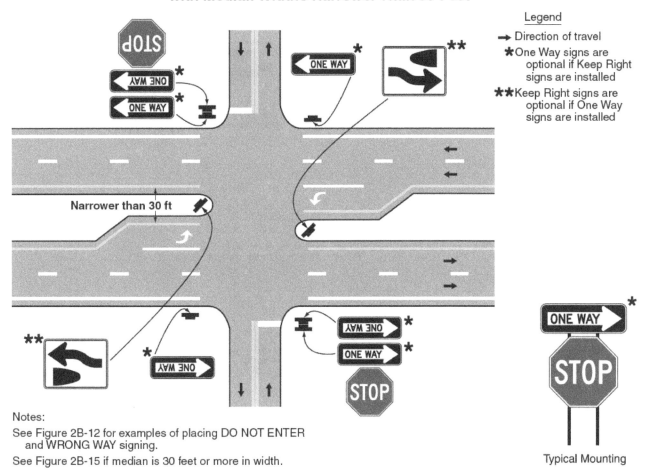

Notes:
See Figure 2B-12 for examples of placing DO NOT ENTER and WRONG WAY signing.
See Figure 2B-15 if median is 30 feet or more in width.

D. Lane-use arrow pavement markings may be placed on the exit ramp and crossroad near their intersection to indicate the permissive direction of flow.

E. Freeway entrance signs (see Section 2D.46) may be used.

Guidance:

04 *On interchange entrance ramps where the ramp merges with the through roadway and the design of the interchange does not clearly make evident the direction of traffic on the separate roadways or ramps, a ONE WAY sign visible to traffic on the entrance ramp and through roadway should be placed on each side of the through roadway near the entrance ramp merging point as illustrated in Figure 2B-19.*

Option:

05 At locations where engineering judgment determines that a special need exists, other standard warning or prohibitive methods and devices may be used as a deterrent to the wrong-way movement.

06 Where there are no parked cars, pedestrian activity or other obstructions such as snow or vegetation, and if an engineering study indicates that a lower mounting height would address wrong-way movements on freeway or expressway exit ramps, a DO NOT ENTER sign(s) and/or a WRONG WAY sign(s) that is located along the exit ramp facing a road user who is traveling in the wrong direction may be installed at a minimum mounting height of 3 feet, measured vertically from the bottom of the sign to the elevation of the near edge of the pavement.

Support:

07 Section 2B.41 contains further information on signing to avoid wrong-way movements at at-grade intersections on expressways.

Figure 2B-17. ONE WAY Signing for Divided Highways with Median Widths Narrower Than 30 Feet and Separated Left-Turn Lanes

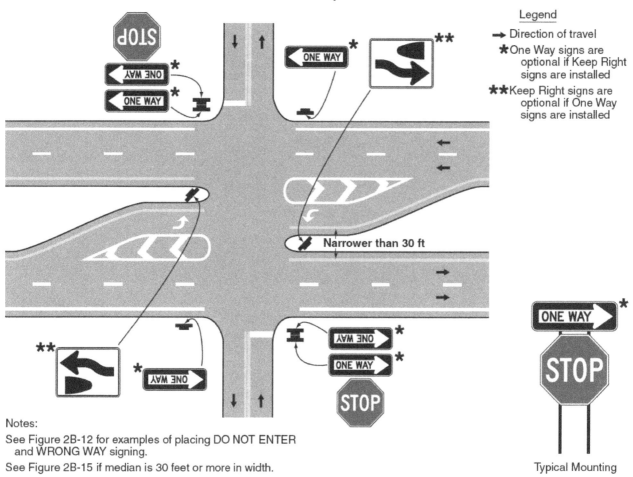

Notes:
See Figure 2B-12 for examples of placing DO NOT ENTER and WRONG WAY signing.
See Figure 2B-15 if median is 30 feet or more in width.

Section 2B.42 Divided Highway Crossing Signs (R6-3, R6-3a)

Standard:

01　On unsignalized minor-street approaches from which both left turns and right turns are permitted onto a divided highway that has a median width at the intersection itself of 30 feet or more, except as provided in Paragraph 2, a Divided Highway Crossing (R6-3 or R6-3a) sign (see Figure 2B-13) shall be used to advise road users that they are approaching an intersection with a divided highway (see Figure 2B-15).

Option:

02　If the divided highway that has a median width at the intersection itself of 30 feet or more has a traffic volume of less than 400 AADT and a speed limit of 25 mph or less, the Divided Highway Crossing signs facing the unsignalized minor-street approaches may be omitted.

03　A Divided Highway Crossing sign may be used on signalized minor-street approaches from which both left turns and right turns are permitted onto a divided highway to advise road users that they are approaching an intersection with a divided highway.

Standard:

04　**If a Divided Highway Crossing sign is used at a four-legged intersection, the R6-3 sign shall be used. If used at a T-intersection, the R6-3a sign shall be used.**

05　**The Divided Highway Crossing sign shall be located on the near right corner of the intersection, mounted beneath a STOP or YIELD sign or on a separate support.**

Option:

06　An additional Divided Highway Crossing sign may be installed on the left-hand side of the approach to supplement the Divided Highway Crossing sign on the near right corner of the intersection.

Figure 2B-18. Example of Application of Regulatory Signing and Pavement Markings at an Exit Ramp Termination to Deter Wrong-Way Entry

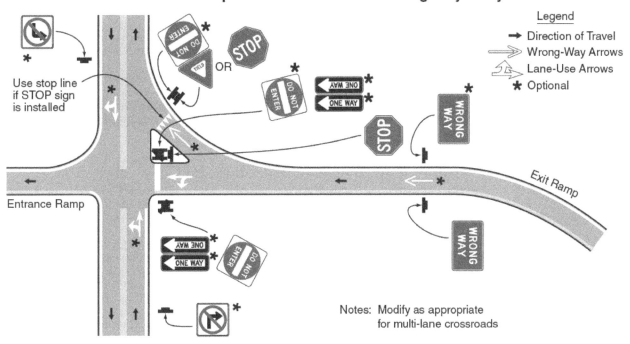

Figure 2B-19. Example of Application of Regulatory Signing and Pavement Markings at an Entrance Ramp Terminal Where the Design Does Not Clearly Indicate the Direction of Flow

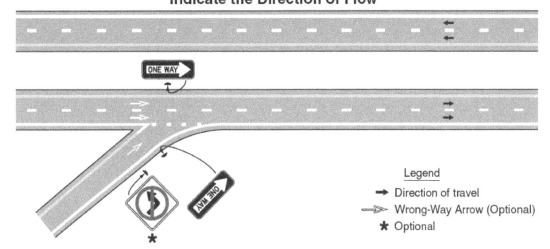

Section 2B.43 Roundabout Directional Arrow Signs (R6-4, R6-4a, and R6-4b)

Guidance:

01 *Where the central island of a roundabout allows for the installation of signs, Roundabout Directional Arrow (R6-4 series) signs (see Figure 2B-20) should be used in the central island to direct traffic counter-clockwise around the central island, except as provided in Paragraph 11 in Section 2B.40.*

Standard:

02 **The R6-4 sign shall be a horizontal rectangle with two black chevron symbols pointing to the right on a white background. The R6-4a sign shall be a horizontal rectangle with three black chevron symbols pointing to the right on a white background. The R6-4b sign shall be a horizontal rectangle with four black chevron symbols pointing to the right on a white background. No border shall be used on the Roundabout Directional Arrow signs.**

03 **Roundabout Directional Arrow signs shall be used only at roundabouts and other circular intersections.**

Guidance:

04 *When used on the central island of a roundabout, the mounting height of a Roundabout Directional Arrow sign should be at least 4 feet, measured vertically from the bottom of the sign to the elevation of the near edge of the traveled way.*

Option:

05 More than one Roundabout Directional Arrow sign and/or R6-4a or R6-4b signs may be used facing high-speed approaches, facing approaches with limited visibility, or in other circumstances as determined by engineering judgment where increased sign visibility would be appropriate.

Section 2B.44 Roundabout Circulation Plaque (R6-5P)

Guidance:

01 *Where the central island of a roundabout does not provide a reasonable place to install a sign, Roundabout Circulation (R6-5P) plaques (see Figure 2B-20) should be placed below the YIELD signs on each approach.*

Option:

02 At roundabouts where Roundabout Directional Arrow signs and/or ONE WAY signs have been installed in the central island, Roundabout Circulation plaques may be placed below the YIELD signs on approaches to roundabouts to supplement the central island signs.

03 The Roundabout Circulation plaque may be used at any type of circular intersection.

Section 2B.45 Examples of Roundabout Signing

Support:

01 Figures 2B-21 through 2B-23 illustrate examples of regulatory and warning signing for roundabouts of various configurations.

02 Section 2D.38 contains information regarding guide signing at roundabouts and Chapter 3C contains information regarding pavement markings at roundabouts.

Figure 2B-20. Roundabout Signs and Plaques

Figure 2B-21. Example of Regulatory and Warning Signs for a Mini-Roundabout

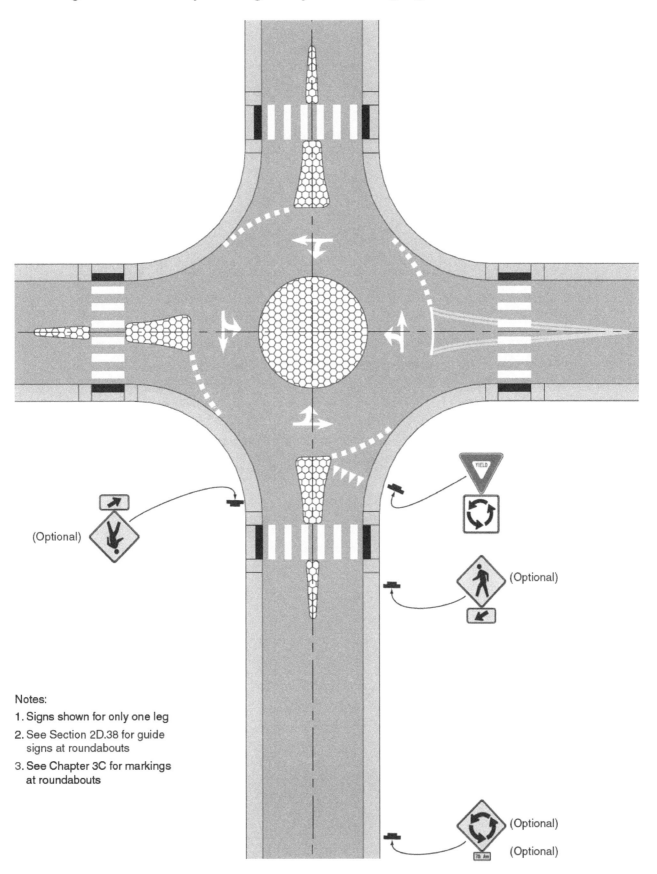

Notes:
1. Signs shown for only one leg
2. See Section 2D.38 for guide signs at roundabouts
3. See Chapter 3C for markings at roundabouts

Figure 2B-22. Example of Regulatory and Warning Signs for a One-Lane Roundabout

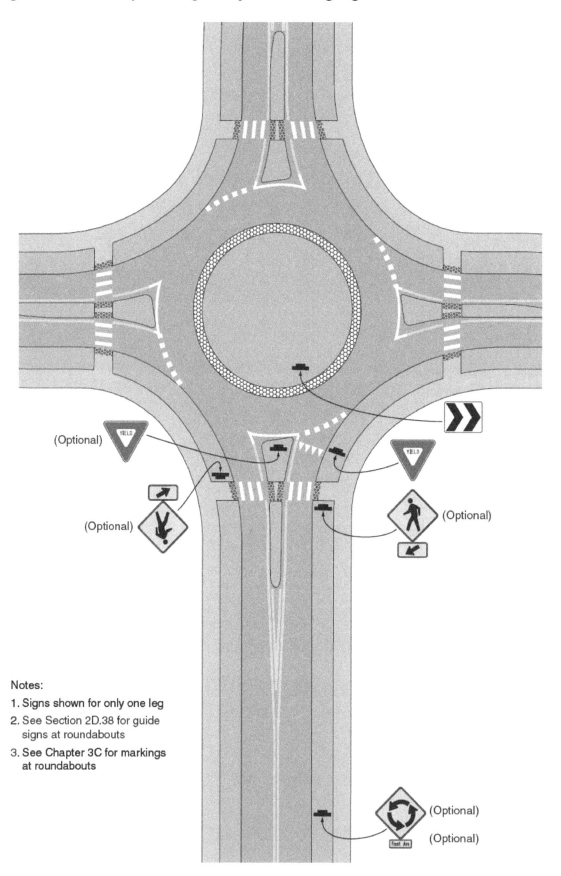

Notes:
1. Signs shown for only one leg
2. See Section 2D.38 for guide signs at roundabouts
3. See Chapter 3C for markings at roundabouts

Figure 2B-23. Example of Regulatory and Warning Signs for a Two-Lane Roundabout with Consecutive Double Lefts

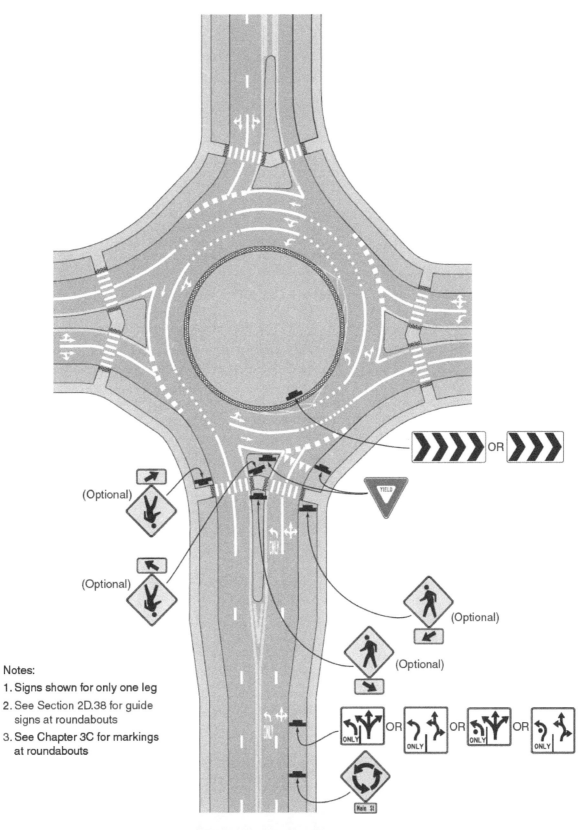

Notes:
1. Signs shown for only one leg
2. See Section 2D.38 for guide signs at roundabouts
3. See Chapter 3C for markings at roundabouts

Page 88   2009 Edition

Section 2B.46 Parking, Standing, and Stopping Signs (R7 and R8 Series)

Support:
01   Signs governing the parking, stopping, and standing of vehicles cover a wide variety of regulations, and only general guidance can be provided here. The word "standing" when used on the R7 and R8 series of signs refers to the practice of a driver keeping the vehicle in a stationary position while continuing to occupy the vehicle. Typical examples of parking, stopping, and standing signs and plaques (see Figures 2B-24 and 2B-25) are as follows:

Figure 2B-24. Parking and Standing Signs and Plaques (R7 Series) (Sheet 1 of 2)

R7-23

R7-23a

R7-107

R7-107a

R7-108

1. NO PARKING ANY TIME (R7-1);
2. NO PARKING X:XX AM TO X:XX PM (R7-2, R7-2a);
3. NO PARKING EXCEPT SUNDAYS AND HOLIDAYS (R7-3);
4. NO STANDING ANY TIME (R7-4);
5. XX HOUR PARKING X:XX AM – X:XX PM (R7-5);
6. NO PARKING LOADING ZONE (R7-6);
7. NO PARKING BUS STOP (R7-7, R7-107, R7-107a);
8. RESERVED PARKING for persons with disabilities (R7-8);
9. VAN ACCESSIBLE (R7-8P);
10. Pay Station (R7-20);
11. Pay Parking (R7-21, R7-21a, R7-22);
12. Parking Permitted X:XX AM TO X:XX PM (R7-23);
13. Parking Permitted XX HOUR(S) XX AM – XX PM (R7-23a);
14. XX HR PARKING X:XX AM TO X:XX PM (R7-108);
15. NO PARKING ANYTIME/XX HOUR PARKING X:XX AM – X:XX PM (R7-200, R7-200a);
16. TOW-AWAY ZONE (R7-201P, R7-201aP);
17. THIS SIDE OF SIGN (R7-202P);
18. EMERGENCY SNOW ROUTE NO PARKING IF OVER XX INCHES (R7-203);
19. NO PARKING ON PAVEMENT (R8-1);
20. NO PARKING EXCEPT ON SHOULDER (R8-2);
21. No Parking (R8-3, R8-3a);
22. EXCEPT SUNDAYS AND HOLIDAYS (R8-3bP);
23. ON PAVEMENT (R8-3cP);
24. ON BRIDGE (R8-3dP);
25. ON TRACKS (R8-3eP);
26. EXCEPT ON SHOULDER (R8-3fP);
27. LOADING ZONE (R8-3gP);
28. X:XX AM TO X:XX PM (R8-3hP);
29. EMERGENCY PARKING ONLY (R8-4);
30. NO STOPPING ON PAVEMENT (R8-5);
31. NO STOPPING EXCEPT ON SHOULDER (R8-6); and
32. EMERGENCY STOPPING ONLY (R8-7).

Figure 2B-24. Parking and Standing Signs and Plaques (R7 Series) (Sheet 2 of 2)

Section 2B.47 Design of Parking, Standing, and Stopping Signs

Support:
01   Discussions of parking signs and parking regulations in this Section apply not only to parking, but also to standing and stopping.

Standard:
02   **The legend on parking signs shall state applicable regulations. Parking signs (see Figures 2B-24 and 2B-25) shall comply with the standards of shape, color, and location.**

Figure 2B-25. Parking and Stopping Signs and Plaques (R8 Series)

03 Where parking is prohibited at all times or at specific times, the basic design for parking signs shall have a red legend and border on a white background (Parking Prohibition signs), except that the R8-4 and R8-7 signs and the alternate design for the R7-201aP plaque shall have a black legend and border on a white background, and the R8-3 sign shall have a black legend and border and a red circle and slash on a white background.

04 Where only limited-time parking or parking in a particular manner are permitted, the signs shall have a green legend and border on a white background (Permissive Parking signs).

Guidance:

05 *Parking signs should display the following information from top to bottom of the sign, in the order listed:*
   A. *The restriction or prohibition;*
   B. *The times of the day that it is applicable, if not at all hours; and*
   C. *The days of the week that it is applicable, if not every day.*

06 *If the parking restriction applies to a limited area or zone, the limits of the restriction should be shown by arrows or supplemental plaques. If arrows are used and if the sign is at the end of a parking zone, there should be a single-headed arrow pointing in the direction that the regulation is in effect. If the sign is at an intermediate point in a zone, there should be a double-headed arrow pointing both ways. When a single sign is used at the transition point between two parking zones, it should display a right and left arrow pointing in the direction that the respective restrictions apply.*

07 *Where special parking restrictions are imposed during heavy snowfall, Emergency Snow Route (R7-203) signs (see Figure 2B-24) should be installed. The legend will vary according to the regulations, but the signs should be vertical rectangles, having a white background with the upper part of the plate a red background.*

Standard:

08 Where parking spaces that are reserved for persons with disabilities are designated to accommodate wheelchair vans, a VAN ACCESSIBLE (R7-8P) plaque shall be mounted below the R7-8 sign. The R7-8 sign (see Figure 2B-24) shall have a green legend and border and a white wheelchair symbol on a blue square, all on a white background. The R7-8P plaque (see Figure 2B-24) shall have a green legend and border on a white background.

Option:

09 To minimize the number of parking signs, blanket regulations that apply to a given district may, if legal, be posted at district boundary lines.

10 As an alternate to the use of arrows to show designated restriction zones, word messages such as BEGIN, END, HERE TO CORNER, HERE TO ALLEY, THIS SIDE OF SIGN, or BETWEEN SIGNS may be used.

11 Where parking is prohibited during certain hours and time-limited parking or parking in a particular manner is permitted during certain other time periods, the red Parking Prohibition and green Permissive Parking signs may be designed as follows:
   A. Two 12 x 18-inch parking signs may be used with the red Parking Prohibition sign installed above or to the left of the green Permissive Parking sign; or
   B. The red Parking Prohibition sign and the green Permissive Parking sign may be combined (see Figure 2B-24) to form an R7-200 sign on a single 24 x 18-inch sign, or an R7-200a sign on a single 12 x 30-inch sign.

12 At the transition point between two parking zones, a single sign or two signs mounted side by side may be used.

13 The words NO PARKING may be used as an alternative to the No Parking symbol. The supplemental educational plaque, NO PARKING, with a red legend and border on a white background, may be used above signs incorporating the No Parking symbol.

14 Alternate designs for the R7-107 sign may be developed such as the R7-107a sign (see Figure 2B-24). Alternate designs may include, on a single sign, a transit logo, an approved bus symbol, a parking prohibition, the words BUS STOP, and an arrow.
The preferred bus symbol color is black, but other dark colors may be used. Additionally, the transit logo may be displayed on the bus face in the appropriate colors instead of placing the logo separately. The reverse side of the sign may contain bus routing information.

15 To make the parking regulations more effective and to improve public relations by giving a definite warning, a TOW-AWAY ZONE (R7-201P) plaque (see Figure 2B-24) may be appended to, or incorporated in, any parking prohibition sign. The Tow-Away Zone (R7-201aP) symbol plaque may be used instead of the R7-201P word message plaque. The R7-201aP plaque may have either a black or red legend and border on a white background.

Guidance:

16 *If a fee is charged for parking and a midblock pay station is used instead of individual parking meters for each parking space, pay parking signs should be used. Pay Parking (R7-22) signs (see Figure 2B-24) should be used to define the area where the pay station parking applies. Pay Station (R7-20) signs (see Figure 2B-24) should be used at the pay station or to direct road users to the pay station.*

Standard:

17 **If the pay parking is subject to a maximum time limit, the appropriate time limit (number of hours or minutes) shall be displayed on the Pay Parking (R7-21 or R7-21a) and Pay Station (R7-20) signs.**

Option:

18 In rural areas (see Figure 2B-25), the legends NO PARKING ON PAVEMENT (R8-1) or NO STOPPING ON PAVEMENT (R8-5) are generally suitable and may be used. If a roadway has paved shoulders, the NO PARKING EXCEPT ON SHOULDER sign (R8-2) or the NO STOPPING EXCEPT ON SHOULDER sign (R8-6) may be used as these signs would be less likely to cause confusion. The R8-3 symbol sign or the word message NO PARKING (R8-3a) sign may be used to prohibit any parking along a given highway. Word message supplemental plaques may be mounted below the R8-3 or R8-3a sign. These word message supplemental plaques may include legends such as EXCEPT SUNDAYS AND HOLIDAYS (R8-3bP), ON PAVEMENT (R8-3cP), ON BRIDGE (R8-3dP), ON TRACKS (R8-3eP), EXCEPT ON SHOULDERS (R8-3fP), LOADING ZONE (with arrow) (R8-3gP), and X:XX AM TO X:XX PM (with arrow) (R8-3hP).

19 Colors that are in compliance with the provisions of Section 2A.10 may be used for color coding of parking time limits.

Guidance:

20 *If colors are used for color coding of parking time limits, the colors green, red, and black should be the only colors that are used.*

Section 2B.48 Placement of Parking, Stopping, and Standing Signs

Guidance:

01 *When signs with arrows are used to indicate the extent of the restricted zones, the signs should be set at an angle of not less than 30 degrees or more than 45 degrees with the line of traffic flow in order to be visible to approaching traffic.*

02 *Spacing of signs should be based on legibility and sign orientation.*

03 *If the zone is unusually long, signs showing a double arrow should be used at intermediate points within the zone.*

Standard:

04 **If the signs are mounted at an angle of 90 degrees to the curb line, two signs shall be mounted back to back at the transition point between two parking zones, each with an appended THIS SIDE OF SIGN (R7-202P) supplemental plaque.**

Guidance:

05 *If the signs are mounted at an angle of 90 degrees to the curb line, signs without any arrows or appended plaques should be used at intermediate points within a parking zone, facing in the direction of approaching traffic. Otherwise the standards of placement should be the same as for signs using directional arrows.*

Section 2B.49 Emergency Restriction Signs (R8-4, R8-7, R8-8)

Option:

01 The EMERGENCY PARKING ONLY (R8-4) sign (see Figure 2B-25) or the EMERGENCY STOPPING ONLY (R8-7) sign (see Figure 2B-25) may be used to discourage or prohibit shoulder parking, particularly where scenic or other attractions create a tendency for road users to stop temporarily.

02 The DO NOT STOP ON TRACKS (R8-8) sign (see Figure 8B-1) may be used to discourage or prohibit parking or stopping on railroad or light rail transit tracks (see Section 8B.09).

Standard:

03 **Emergency Restriction signs shall be rectangular and shall have a red or black legend and border on a white background.**

Section 2B.50 WALK ON LEFT FACING TRAFFIC and No Hitchhiking Signs (R9-1, R9-4, R9-4a)

Option:

01 The WALK ON LEFT FACING TRAFFIC (R9-1) sign (see Figure 2B-26) may be used on highways where no sidewalks are provided.

Standard:

02 **If used, the WALK ON LEFT FACING TRAFFIC sign shall be installed on the right-hand side of the road where pedestrians walk on the pavement or shoulder in the absence of pedestrian pathways or sidewalks.**

Option:

03 The No Hitchhiking (R9-4) sign (see Figure 2B-26) may be used to prohibit standing in or adjacent to the roadway for the purpose of soliciting a ride. The R9-4a word message sign (see Figure 2B-26) may be used as an alternate to the R9-4 symbol sign.

Section 2B.51 Pedestrian Crossing Signs (R9-2, R9-3)

Option:

01 Pedestrian Crossing signs (see Figure 2B-26) may be used to limit pedestrian crossing to specific locations.

Standard:

02 **If used, Pedestrian Crossing signs shall be installed to face pedestrian approaches.**

Option:

03 Where crosswalks are clearly defined, the CROSS ONLY AT CROSSWALKS (R9-2) sign may be used to prohibit pedestrians from crossing at locations away from crosswalks.

04 The No Pedestrian Crossing (R9-3) sign may be used to prohibit pedestrians from crossing a roadway at an undesirable location or in front of a school or other public building where a crossing is not designated.

05 The NO PEDESTRIAN CROSSING (R9-3a) word message sign may be used as an alternate to the R9-3 symbol sign. The USE CROSSWALK (R9-3bP) supplemental plaque, along with an arrow, may be installed below either sign to designate the direction of the crossing.

Figure 2B-26. Pedestrian Signs and Plaques (Sheet 1 of 2)

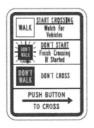

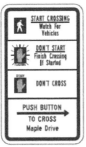
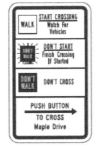
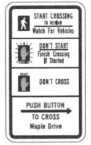

Figure 2B-26. Pedestrian Signs and Plaques (Sheet 2 of 2)

R10-4

R10-4a

R10-25

R10-32P

Support:

06 One of the most frequent uses of the Pedestrian Crossing signs is at signalized intersections that have three crossings that can be used and one leg that cannot be crossed.

Guidance:

07 *The R9-3bP plaque should not be installed in combination with educational plaques.*

Section 2B.52 Traffic Signal Pedestrian and Bicycle Actuation Signs (R10-1 through R10-4, and R10-24 through R10-26)

Standard:

01 **Traffic Signal signs applicable to pedestrian actuation (see Figure 2B-26) or bicyclist actuation (see Figure 9B-2) shall be mounted immediately above or incorporated into the pushbutton detector units (see Section 4E.08).**

Support:

02 Traffic Signal signs applicable to pedestrians include:
   A. CROSS ONLY ON GREEN (symbolic circular green) (R10-1);
   B. CROSS ONLY ON (symbolic walk indication) SIGNAL (R10-2);
   C. Push Button for Walk Signal (R10-3 series); and
   D. Push Button for Green Signal (R10-4 series).

Option:

03 The following signs may be used as an alternate for the R10-3 and R10-4 signs:
   A. Push Button to Cross Street Wait for Walk Signal (R10-3a); or
   B. Push Button to Cross Street Wait for Green Signal (R10-4a).

04 The name of the street to be crossed may be substituted for the word STREET in the legends on the R10-3a and R10-4a signs.

Guidance:

05 *The finger in the pushbutton symbol on the R10-3, R10-3a, R10-4, and R10-4a signs should point in the same direction as the arrow on the sign.*

Option:

06 Where symbol-type pedestrian signal indications are used, an educational sign (R10-3b) may be used instead of the R10-3 sign to improve pedestrian understanding of pedestrian indications at signalized intersections. Where word-type pedestrian signal indications are being retained for the remainder of their useful service life, the legends WALK/DONT WALK may be substituted for the symbols on the educational sign R10-3b, thus creating educational sign R10-3c. The R10-3d educational sign may be used to inform pedestrians that the pedestrian clearance time is sufficient only for the pedestrian to cross to the median at locations where pedestrians cross in two stages using a median refuge island. The R10-3e educational sign may be used where countdown pedestrian signals have been provided. In order to assist the pedestrian in understanding which pushbutton to push, the R10-3f to R10-3i educational signs that provide the name of the street to be crossed may be used instead of the R10-3b to R10-3e educational signs.

07 The R10-24 or R10-26 sign (see Section 9B.11) may be used where a pushbutton detector has been installed exclusively to actuate a green phase for bicyclists.

08 The R10-25 sign (see Figure 2B-26) may be used where a pushbutton detector has been installed for pedestrians to activate In-Roadway Warning Lights (see Chapter 4N) or flashing beacons that have been added to the pedestrian warning signs.

Support:

09 Section 4E.08 contains information regarding the application of the R10-32P plaque.

Section 2B.53 Traffic Signal Signs (R10-5 through R10-30)

Option:

01   To supplement traffic signal control, Traffic Signal signs R10-5 through R10-30 may be used to regulate road users.

02   Traffic Signal signs (see Figure 2B-27) may be installed at certain locations to clarify signal control. Among the legends that may be used for this purpose are LEFT ON GREEN ARROW ONLY (R10-5), STOP HERE ON RED (R10-6 or R10-6a) for observance of stop lines, DO NOT BLOCK INTERSECTION (R10-7) for avoidance of traffic obstructions, USE LANE(S) WITH GREEN ARROW (R10-8) for obedience to lane-use control signals (see Chapter 4M), LEFT TURN YIELD ON GREEN (symbolic circular green) (R10-12), and LEFT TURN YIELD ON FLASHING RED ARROW AFTER STOP (R10-27).

Guidance:

03   *If used, the LEFT ON GREEN ARROW ONLY (R10-5) sign, the LEFT TURN YIELD ON GREEN (symbolic circular green) (R10-12) sign, or the LEFT TURN YIELD ON FLASHING RED ARROW AFTER STOP (R10-27) sign should be located adjacent to the left-turn signal face.*

Option:

04   If needed for additional emphasis, an additional LEFT TURN YIELD ON GREEN (symbolic circular green) (R10-12) sign with an AT SIGNAL (R10-31P) supplemental plaque (see Figure 2B-27) may be installed in advance of the intersection.

05   In situations where traffic control signals are coordinated for progressive timing, the Traffic Signal Speed (I1-1) sign may be used (see Section 2H.03).

Standard:

06   **The CROSSWALK STOP ON RED (symbolic circular red) (R10-23) sign (see Figure 2B-27) shall only be used in conjunction with pedestrian hybrid beacons (see Section 4F.02).**

07   **The EMERGENCY SIGNAL (R10-13) sign (see Figure 2B-27) shall be used in conjunction with emergency-vehicle traffic control signals (see Section 4G.02).**

08   **The EMERGENCY SIGNAL—STOP ON FLASHING RED (R10-14 or R10-14a) sign (see Figure 2B-27) shall be used in conjunction with emergency-vehicle hybrid beacons (see Section 4G.04).**

Option:

09   In order to remind drivers who are making turns to yield to pedestrians, a Turning Vehicles Yield to Pedestrians (R10-15) sign (see Figure 2B-27) may be used.

10   A U-TURN YIELD TO RIGHT TURN (R10-16) sign (see Figure 2B-27) may be installed near the left-turn signal face if U-turns are allowed on a protected left-turn movement on an approach from which a right-turn GREEN ARROW signal indication is simultaneously being displayed to drivers making a right turn from the conflicting approach to their left.

Section 2B.54 No Turn on Red Signs (R10-11 Series, R10-17a, and R10-30)

Standard:

01   **Where a right turn on red (or a left turn on red from a one-way street to a one-way street) is to be prohibited, a symbolic NO TURN ON RED (symbolic circular red) (R10-11) sign (see Figure 2B-27) or a NO TURN ON RED (R10-11a, R10-11b) word message sign (see Figure 2B-27) shall be used.**

Guidance:

02   *If used, the No Turn on Red sign should be installed near the appropriate signal head.*

03   *A No Turn on Red sign should be considered when an engineering study finds that one or more of the following conditions exists:*
   A. *Inadequate sight distance to vehicles approaching from the left (or right, if applicable);*
   B. *Geometrics or operational characteristics of the intersection that might result in unexpected conflicts;*
   C. *An exclusive pedestrian phase;*
   D. *An unacceptable number of pedestrian conflicts with right-turn-on-red maneuvers, especially involving children, older pedestrians, or persons with disabilities;*
   E. *More than three right-turn-on-red accidents reported in a 12-month period for the particular approach; or*
   F. *The skew angle of the intersecting roadways creates difficulty for drivers to see traffic approaching from their left.*

Figure 2B-27. Traffic Signal Signs and Plaques

* A fluorescent yellow-green background color may be used instead of yellow for this sign.

Option:

04 A supplemental R10-20aP plaque (see Figure 2B-27) showing times of day (similar to the S4-1P plaque shown in Figure 7B-1) with a black legend and border on a white background may be mounted below a No Turn on Red sign to indicate that the restriction is in place only during certain times.

05 Alternatively, a blank-out sign may be used instead of a static NO TURN ON RED sign, to display either the NO TURN ON RED legend or the No Right Turn symbol or word message, as appropriate, only at certain times during the day or during one or more portion(s) of a particular cycle of the traffic signal.

06 On signalized approaches with more than one right-turn lane, a NO TURN ON RED EXCEPT FROM RIGHT LANE (R10-11c) sign (see Figure 2B-27) may be post-mounted at the intersection or a NO TURN ON RED FROM THIS LANE (with down arrow) (R10-11d) sign (see Figure 2B-27) may be mounted directly over the center of the lane from which turns on red are prohibited.

Guidance:

07 *Where turns on red are permitted and the signal indication is a steady RED ARROW, the RIGHT (LEFT) ON RED ARROW AFTER STOP (R10-17a) sign (see Figure 2B-27) should be installed adjacent to the RED ARROW signal indication.*

Option:

08 A RIGHT TURN ON RED MUST YIELD TO U-TURN (R10-30) sign (see Figure 2B-27) may be installed to remind road users that they must yield to conflicting u-turn traffic on the street or highway onto which they are turning right on a red signal after stopping.

Section 2B.55 Photo Enforced Signs and Plaques (R10-18, R10-19P, R10-19aP)

Option:

01 A TRAFFIC LAWS PHOTO ENFORCED (R10-18) sign (see Figure 2B-3) may be installed at a jurisdictional boundary to advise road users that some of the traffic regulations within that jurisdiction are being enforced by photographic equipment.

02 A Photo Enforced (R10-19P) plaque or a PHOTO ENFORCED (R10-19aP) word message plaque (see Figure 2B-3) may be mounted below a regulatory sign to advise road users that the regulation is being enforced by photographic equipment.

Standard:

03 **If used below a regulatory sign, the Photo Enforced (R10-19P or R10-19aP) plaque shall be a rectangle with a black legend and border on a white background.**

Section 2B.56 Ramp Metering Signs (R10-28 and R10-29)

Option:

01 When ramp control signals (see Chapter 4I) are used to meter traffic on a freeway or expressway entrance ramp, regulatory signs with legends appropriate to the control may be installed adjacent to the ramp control signal faces.

02 For entrance ramps with only one controlled lane, an XX VEHICLE(S) PER GREEN (R10-28) sign (see Figure 2B-28) may be used to inform road users of the number of vehicles that are permitted to proceed during each short display of the green signal indication. For entrance ramps with more than one controlled lane, an XX VEHICLE(S) PER GREEN Each Lane (R10-29) (see Figure 2B-28) sign may be used to inform road users of the number of vehicles that are permitted to proceed from each lane during each short display of the green signal indication.

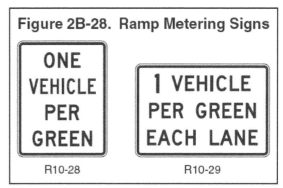

Figure 2B-28. Ramp Metering Signs

Section 2B.57 KEEP OFF MEDIAN Sign (R11-1)

Option:

01 The KEEP OFF MEDIAN (R11-1) sign (see Figure 2B-29) may be used to prohibit driving into or parking on the median.

Guidance:

02 *The KEEP OFF MEDIAN sign should be installed on the left of the roadway within the median at random intervals as needed wherever there is a tendency for encroachment.*

Figure 2B-29. Road Closed and Weight Limit Signs

Section 2B.58 ROAD CLOSED Sign (R11-2) and LOCAL TRAFFIC ONLY Signs (R11-3 Series, R11-4)

Guidance:

01 *The ROAD CLOSED (R11-2) sign should be installed where roads have been closed to all traffic (except authorized vehicles).*

02 *ROAD CLOSED – LOCAL TRAFFIC ONLY (R11-3) or ROAD CLOSED TO THRU TRAFFIC (R11-4) signs should be used where through traffic is not permitted, or for a closure some distance beyond the sign, but where the highway is open for local traffic up to the point of closure.*

Standard:

03 **The Road Closed (R11-2, R11-3 series, and R11-4) signs (see Figure 2B-29) shall be designed as horizontal rectangles. These signs shall be preceded by the applicable Advance Road Closed warning sign with the secondary legend AHEAD and, if applicable, an Advance Detour warning sign (see Section 6F.19).**

Option:

04 An intersecting street name or a well-known destination may be substituted for the XX MILES AHEAD legend in urban areas.

05 The word message BRIDGE OUT may be substituted for the ROAD CLOSED legend where applicable.

Section 2B.59 Weight Limit Signs (R12-1 through R12-5)

Option:

01 The Weight Limit (R12-1) sign carrying the legend WEIGHT LIMIT XX TONS may be used to indicate vehicle weight restrictions including load.

02 Where the restriction applies to axle weight rather than gross load, the legend may be AXLE WEIGHT LIMIT XX TONS or AXLE WEIGHT LIMIT XX LBS (R12-2).

03 To restrict trucks of certain sizes by reference to empty weight in residential areas, the legend may be NO TRUCKS OVER XX TONS EMPTY WT or NO TRUCKS OVER XX LBS EMPTY WT (R12-3).

04 In areas where multiple regulations of the type described in Paragraphs 1 through 3 are applicable, a sign combining the necessary messages on a single sign may be used, such as WEIGHT LIMIT XX TONS PER AXLE, XX TONS GROSS (R12-4).

05 Posting of specific load limits may be accomplished by use of the Weight Limit symbol sign (R12-5). A sign containing the legend WEIGHT LIMIT on the top two lines, and showing three different truck symbols and their respective weight limits for which restrictions apply may be used, with the weight limits displayed to the right of each symbol as XX T. A bottom line of legend stating GROSS WT may be included if needed for enforcement purposes.

Standard:

06 **If used, the Weight Limit sign (see Figure 2B-29) shall be located in advance of the applicable section of highway or structure.**

Guidance:

07 *If used, the Weight Limit sign with an advisory distance ahead legend should be placed at approach road intersections or other points where prohibited vehicles can detour or turn around.*

Section 2B.60 Weigh Station Signs (R13 Series)

Guidance:

01 *An R13-1 sign with the legend TRUCKS OVER XX TONS MUST ENTER WEIGH STATION NEXT RIGHT (see Figure 2B-30) should be used to direct appropriate traffic into a weigh station.*

02 *The R13-1 sign should be supplemented by the D8 series of guide signs (see Section 2D.49).*

Option:

03 The reverse color combination, a white legend and border on a black background, may be used for the R13-1 sign.

Section 2B.61 TRUCK ROUTE Sign (R14-1)

Guidance:

01 *The TRUCK ROUTE (R14-1) sign (see Figure 2B-30) should be used to mark a route that has been designated to allow truck traffic.*

Option:

02 On a numbered highway, the TRUCK (M4-4) auxiliary sign may be used (see Section 2D.20).

Section 2B.62 Hazardous Material Signs (R14-2, R14-3)

Option:

01 The Hazardous Material Route (R14-2) sign (see Figure 2B-30) may be used to identify routes that have been designated by proper authority for vehicles transporting hazardous material.

02 On routes where the transporting of hazardous material is prohibited, the Hazardous Material Prohibition (R14-3) sign (see Figure 2B-30) may be used.

Figure 2B-30. Truck Signs

R14-1

R14-2

R14-3

R14-4

R14-5

R13-1*

*The R13-1 sign may be black-on-white or white-on-black

Guidance:

03 *If used, the Hazardous Material Prohibition sign should be installed on a street or roadway at a point where vehicles transporting hazardous material have the opportunity to take an alternate route.*

Section 2B.63 National Network Signs (R14-4, R14-5)

Support:

01 The signing of the National Network routes for trucking is optional.

Standard:

02 **When a National Network route is signed, the National Network (R14-4) sign (see Figure 2B-30) shall be used.**

Option:

03 The National Network Prohibition (R14-5) sign (see Figure 2B-30) may be used to identify routes, portions of routes, and ramps where trucks are prohibited. The R14-5 sign may also be used to mark the ends of designated routes.

Section 2B.64 Headlight Use Signs (R16-5 through R16-11)

Support:

01 Some States require road users to turn on their vehicle headlights under certain weather conditions, as a safety improvement measure on roadways experiencing high crash rates, or in special situations such as when driving through a tunnel.

02 Figure 2B-31 shows the various signs that can be used for informing motorists of these requirements.

Option:

03 A LIGHTS ON WHEN USING WIPERS (R16-5) sign or a LIGHTS ON WHEN RAINING (R16-6) sign may be installed to inform road users of State laws regarding headlight use. Although these signs are typically installed facing traffic entering the State just inside the State border, they also may be installed at other locations within the State.

Guidance:

04 *If a particular section of roadway has been designated as a safety improvement zone within which headlight use is required, a TURN ON HEADLIGHTS NEXT XX MILES (R16-7) sign or a BEGIN DAYTIME HEADLIGHT SECTION (R16-10) sign should be installed at the upstream end of the section, and a END DAYTIME HEADLIGHT SECTION (R16-11) sign should be installed at the downstream end of the section.*

Option:

05 A TURN ON HEADLIGHTS (R16-8) sign may be installed to require road users to turn on their headlights in special situations such as when driving through a tunnel. A CHECK HEADLIGHTS (R16-9) sign may be installed downstream from the special situation to inform drivers that the using their headlights is no longer required.

Figure 2B-31. Headlight Use Signs

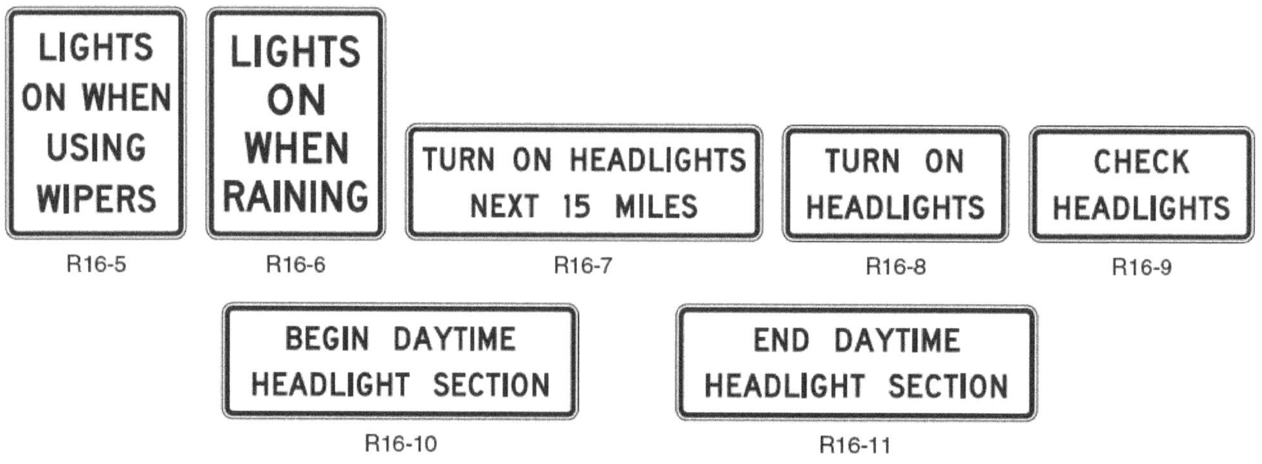

Section 2B.65 FENDER BENDER Sign (R16-4)
Option:
01 A FENDER BENDER MOVE VEHICLES FROM TRAVEL LANES (R16-4) sign (see Figure 2B-32) may be installed to require motorists to move their vehicle out of the travel lanes if they have been involved in a crash.

Section 2B.66 Seat Belt Symbol
Standard:
01 **When a seat belt symbol is used, the symbol shown in Figure 2B-32 shall be used.**
Guidance:
02 *The seat belt symbol should not be used alone. If used, the seat belt symbol should be incorporated into regulatory sign messages for mandatory seat belt use.*

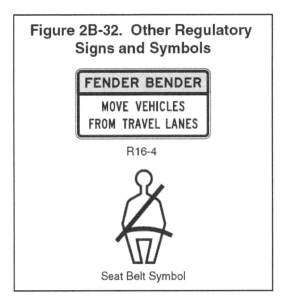

Figure 2B-32. Other Regulatory Signs and Symbols

Section 2B.67 Barricades
Option:
01 Barricades may be used to mark any of the following conditions:
   A. A roadway ends,
   B. A ramp or lane closed for operational purposes, or
   C. The permanent or semi-permanent closure or termination of a roadway.

Standard:
02 **When used to warn and alert road users of the terminus of a roadway in other than temporary traffic control zones, barricades shall meet the design criteria of Section 6F.68 for a Type 3 Barricade, except that the colors of the stripes shall be retroreflective white and retroreflective red.**
Option:
03 An end-of-roadway marker or markers may be used as described in Section 2C.66.
Guidance:
04 *Appropriate advance warning signs (see Chapter 2C) should be used.*

Section 2B.68 Gates
Support:
01 Gates described in this section used for weather or other emergency conditions are typically permanently installed to enable the gate to be immediately deployed as needed to prohibit the entry of traffic to the highway segment(s).

02 A gate typically features a gate arm that is moved from a vertical to a horizontal position or is rotated in a horizontal plane from parallel to traffic to perpendicular to traffic. Traffic is obstructed and required to stop when the gate arm is placed in a horizontal position perpendicular to traffic. Another type of gate consists of a segment of fence (usually on rollers) that swings open and closed, or that is retracted to open and then extended to close.

03 Gates are sometimes used to enforce a required stop. Some examples of such uses are the following:
   A. Parking facility entrances and exits,
   B. Private community entrances and exits,
   C. Military base entrances and exits,
   D. Toll plaza lanes,
   E. Movable bridges (see Chapter 4J),
   F. Automated Flagger Assistance Devices (see Chapter 6E), and
   G. Grade crossings (see Part 8).

04 Gates are sometimes used to periodically close a roadway or a ramp. Some examples of such uses are the following:
   A. Closing ramps to implement counter-flow operations for evacuations,
   B. Closing ramps that lead to reversible lanes, and
   C. Closing roadways for weather events such as snow, ice, or flooding, or for other emergencies.

Standard:

05 **Except as provided in Paragraph 6, gate arms, if used, shall be fully retroreflectorized on both sides, have vertical stripes alternately red and white at 16-inch intervals measured horizontally as shown in Figure 8C-1.**

Option:

06 If used on a one-way roadway or ramp, the retroreflectorization may be omitted on the side of the gate facing away from approaching traffic.

07 Where gate arms are used to block off ramps into reversible lanes or to redirect approaching traffic, the red and white striping may be angled such that the stripes slope downward at an angle of 45 degrees toward the side of the gate arm on which traffic is to pass.

Standard:

08 **The gate arm shall extend across the approaching lane or lanes of traffic to effectively block motor vehicle and/or pedestrian travel as appropriate.**

09 **When gate arms are in the vertical position or rotated to an open position, the closest part of the gate arm and support shall have a lateral offset of at least 2 feet from the face of the curb or the edge of the traveled way.**

10 **When gate arms that are located in the median or on an island are in the horizontal position or rotated to a closed position, the closest part of the counterweight or its supports shall have a lateral offset of at least 2 feet from the face of the curb or the edge of the traveled way of the open roadway on the opposite side of the median or island.**

Guidance:

11 *When a gate that is rotated in a horizontal plane is in the position where it is parallel to traffic (indicating that the roadway is open), the outer end of the gate arm should be rotated to the downstream direction (from the perspective of traffic in the lane adjacent to the gate support) to prevent spearing if the gate is struck by an errant vehicle.*

12 *If a pedestrian route is present and if it is not intended that pedestrian traffic be controlled by the gate, a minimum of 2 feet of lateral offset from supports, posts, counterweights, and gate mechanisms should be provided when the gate arm is in the open position and when the gate arm is in the closed position such that pedestrian travel is not impeded.*

Option:

13 Red lights may be attached to traffic gates.

Standard:

14 **If red lights are attached to a traffic gate, the red lights shall be steadily illuminated or flashed only during the period when the gate is in the horizontal or closed position and when the gate is in the process of being opened or closed.**

15 **Except as provided in Paragraph 16, rolling sections of fence, if used, shall include either a horizontal strip of retroreflectorized sheeting on both sides of the fence with vertical stripes alternately red and white at 16-inch intervals measured horizontally to simulate the appearance of a gate arm in the horizontal position, or one or more Type 4 object markers (see Section 2C.66), or both. If a horizontal strip of retroreflectorized sheeting is used, the bottom of the sheeting shall be located 3.5 to 4.5 feet above the roadway surface.**

Option:

16 If used on a one-way roadway or ramp, the retroreflectorization may be omitted on the side of the fence facing away from approaching traffic.

CHAPTER 2C. WARNING SIGNS AND OBJECT MARKERS

Section 2C.01 Function of Warning Signs

Support:

01　Warning signs call attention to unexpected conditions on or adjacent to a highway, street, or private roads open to public travel and to situations that might not be readily apparent to road users. Warning signs alert road users to conditions that might call for a reduction of speed or an action in the interest of safety and efficient traffic operations.

Section 2C.02 Application of Warning Signs

Standard:

01　**The use of warning signs shall be based on an engineering study or on engineering judgment.**

Guidance:

02　*The use of warning signs should be kept to a minimum as the unnecessary use of warning signs tends to breed disrespect for all signs. In situations where the condition or activity is seasonal or temporary, the warning sign should be removed or covered when the condition or activity does not exist.*

Option:

03　Consistent with the provisions of Chapter 2L, changeable message signs may be used to display a warning message.

04　Consistent with the provisions of Chapter 4L, a Warning Beacon may be used in combination with a standard warning sign.

Support:

05　The categories of warning signs are shown in Table 2C-1.

06　Warning signs provided in this Manual cover most of the conditions that are likely to be encountered. Additional warning signs for low-volume roads (as defined in Section 5A.01), temporary traffic control zones, school areas, grade crossings, and bicycle facilities are discussed in Parts 5 through 10, respectively.

07　Section 1A.09 contains information regarding the assistance that is available to jurisdictions that do not have engineers on their staffs who are trained and/or experienced in traffic control devices.

Section 2C.03 Design of Warning Signs

Standard:

01　**Except as provided in Paragraph 2 or unless specifically designated otherwise, all warning signs shall be diamond-shaped (square with one diagonal vertical) with a black legend and border on a yellow background. Warning signs shall be designed in accordance with the sizes, shapes, colors, and legends contained in the "Standard Highway Signs and Markings" book (see Section 1A.11).**

Option:

02　A warning sign that is larger than the size shown in the Oversized column in Table 2C-2 for that particular sign may be diamond-shaped or may be rectangular or square in shape.

03　Except for symbols on warning signs, minor modifications may be made to the design provided that the essential appearance characteristics are met. Modifications may be made to the symbols shown on combined horizontal alignment/intersection signs (see Section 2C.11) and intersection warning signs (see Section 2C.46) in order to approximate the geometric configuration of the intersecting roadway(s).

04　Word message warning signs other than those provided in this Manual may be developed and installed by State and local highway agencies.

05　Warning signs regarding conditions associated with pedestrians, bicyclists, and playgrounds may have a black legend and border on a yellow or fluorescent yellow-green background.

Standard:

06　**Warning signs regarding conditions associated with school buses and schools and their related supplemental plaques shall have a black legend and border on a fluorescent yellow-green background (see Section 7B.07).**

Section 2C.04 Size of Warning Signs

Standard:

01　**Except as provided in Section 2A.11, the sizes for warning signs shall be as shown in Table 2C-2.**

Table 2C-1. Categories of Warning Signs and Plaques

Category	Group	Section	Signs or Plaques	Sign Designations
Roadway Related	Changes in Horizontal Alignment	2C.07	Turn, Curve, Reverse Turn, Reverse Curve, Winding Road, Hairpin Curve, 270-Degree Curve	W1-1,2,3,4,5,11,15
		2C.08	Advisory Speed	W13-1P
		2C.09	Chevron Alignment	W1-8
		2C.10	Combination Horizontal Alignment/Advisory Speed	W1-1a,2a
		2C.11	Combination Horizontal Alignment/Intersection	W1-10,10a,10b,10c,10d
		2C.12	Large Arrow (one direction)	W1-6
		2C.13	Truck Rollover	W1-13
		2C.14	Advisory Exit or Ramp Speed	W13-2,3
		2C.15	Combination Horizontal Alignment/Advisory Exit or Ramp Speed	W13-6,7
	Vertical Alignment	2C.16	Hill	W7-1,1a,2P,2bP,3P,3aP,3bP
		2C.17	Truck Escape Ramp	W7-4,4b,4c,4dP,4eP,4fP
		2C.18	Hill Blocks View	W7-6
	Cross Section	2C.19	Road Narrows	W5-1
		2C.20,21	Narrow Bridge, One Lane Bridge	W5-2,3
		2C.22,23,25	Divided Highway, Divided Highway Ends, Double Arrow	W6-1,2; W12-1
		2C.24	Freeway or Expressway Ends, All Traffic Must Exit	W19-1,2,3,4,5
		2C.26	Dead End, No Outlet	W14-1,1a,2,2a
		2C.27	Low Clearance	W12-2,2a
	Roadway Surface Condition	2C.28,29	Bump, Dip, Speed Hump	W8-1,2; W17-1
		2C.30	Pavement Ends	W8-3
		2C.31	Shoulder, Uneven Lanes	W8-4,9,11,17,17P,23,25
		2C.32	Slippery When Wet, Loose Gravel, Rough Road, Bridge Ices Before Road, Fallen Rocks	W8-5,7,8,13,14
		2C.33	Grooved Pavement, Metal Bridge Deck	W8-15,15P,16
		2C.34	No Center Line	W8-12
	Weather	2C.35	Road May Flood, Flood Gauge, Gusty Winds Area, Fog Area	W8-18,19,21,22
Traffic Related	Advance Traffic Control	2C.36-39	Stop Ahead, Yield Ahead, Signal Ahead, Be Prepared To Stop, Speed Reduction, Drawbridge Ahead, Ramp Meter Ahead	W3-1,2,3,4,5,5a,6,7,8
	Traffic Flow	2C.40-45	Merge, No Merge Area, Lane Ends, Added Lane, Two-Way Traffic, Right Lane Exit Only Ahead, No Passing Zone	W4-1,2,3,5,5P,6; W6-3; W9-1,2,7; W14-3
	Intersections	2C.46	Cross Road, Side Road, T, Y, Circular Intersection, Side Roads	W2-1,2,3,4,5,6,7,8; W16-12P,17P
		2C.47	Large Arrow (two directions)	W1-7
		2C.48	Oncoming Extended Green	W25-1,2
	Vehicular Traffic	2C.49	Truck Crossing, Truck (symbol), Emergency Vehicle, Tractor, Bicycle, Golf Cart, Horse-Drawn Vehicle, Trail Crossing	W8-6; W11-1,5,5a,8,10, 11,12P,14,15,15P,15a; W16-13P
	Non-Vehicular	2C.50,51	Pedestrian, Deer, Cattle, Snowmobile, Equestrian, Wheelchair, Large Animals, Playground	W11-2,3,4,6,7,9,16,17,18,19, 20,21,22; W15-1; W16-13P
	New	2C.52	New Traffic Pattern Ahead	W23-2
Other Supplemental Plaques	Location	2C.53	Downward Diagonal Arrow, Ahead	W16-7P,9P
	HOV	2C.53	High-Occupancy Vehicle	W16-11P
	Distance	2C.55	XX Feet, XX Miles, Next XX Feet, Next XX Miles	W7-3aP; W16-2P,2aP,3P,3aP,4P
	Arrow	2C.56	Advance Arrow, Directional Arrow	W16-5P,6P
	Street Name Plaque	2C.58	Advance Street Name	W16-8P,8aP
	Intersection	2C.59	Cross Traffic Does Not Stop	W4-4P,4aP,4bP
	Share The Road	2C.60	Share The Road	W16-1P
	Photo Enforced	2C.61	Photo Enforced	W16-10P,10aP
	New	2C.62	New	W16-15P

Table 2C-2. Warning Sign and Plaque Sizes (Sheet 1 of 3)

Sign or Plaque	Sign Designation	Section	Conventional Road Single Lane	Conventional Road Multi-Lane	Expressway	Freeway	Minimum	Oversized
Horizontal Alignment	W1-1,2,3,4,5	2C.07	30 x 30*	36 x 36	36 x 36	36 x 36	—	48 x 48
Combination Horizontal Alignment/Advisory Speed	W1-1a,2a	2C.10	36 x 36	36 x 36	48 x 48	48 x 48	—	48 x 48
One-Direction Large Arrow	W1-6	2C.12	48 x 24	48 x 24	60 x 30	60 x 30	—	60 x 30
Two-Direction Large Arrow	W1-7	2C.47	48 x 24	48 x 24	—	—	—	60 x 30
Chevron Alignment	W1-8	2C.09	18 x 24	18 x 24	30 x 36	36 x 48	—	24 x 30
Combination Horizontal Alignment/Intersection	W1-10,10a, 10b,10c,10d, 10e	2C.11	36 x 36	36 x 36	36 x 36	48 x 48	—	—
Hairpin Curve	W1-11	2C.07	30 x 30	30 x 30	36 x 36	48 x 48	—	48 x 48
Truck Rollover	W1-13	2C.13	36 x 36	36 x 36	36 x 36	48 x 48	—	36 x 36
270-degree Loop	W1-15	2C.07	30 x 30	30 x 30	36 x 36	48 x 48	—	48 x 48
Intersection Warning	W2-1, 2,3,4,5,6,7,8	2C.46	30 x 30	30 x 30	36 x 36	—	24 x 24	48 x 48
Advanced Traffic Control	W3-1,2,3	2C.36	30 x 30	30 x 30	48 x 48	48 x 48	30 x 30	—
Be Prepared to Stop	W3-4	2C.36	36 x 36	36 x 36	48 x 48	48 x 48	30 x 30	—
Reduced Speed Limit Ahead	W3-5	2C.38	36 x 36	36 x 36	48 x 48	48 x 48	—	—
XX MPH Speed Zone Ahead	W3-5a	2C.38	36 x 36	36 x 36	48 x 48	48 x 48	—	—
Draw Bridge	W3-6	2C.39	36 x 36	36 x 36	48 x 48	—	—	60 x 60
Ramp Meter Ahead	W3-7	2C.37	36 x 36	36 x 36	—	—	—	—
Ramp Metered When Flashing	W3-8	2C.37	36 x 36	36 x 36	—	—	—	—
Merge	W4-1	2C.40	36 x 36	36 x 36	48 x 48	48 x 48	30 x 30*	—
Lane Ends	W4-2	2C.42	36 x 36	36 x 36	48 x 48	48 x 48	30 x 30*	—
Added Lane	W4-3	2C.41	36 x 36	36 x 36	48 x 48	48 x 48	30 x 30*	—
Cross Traffic Does Not Stop (plaque)	W4-4P	2C.59	24 x 12	24 x 12	36 x 18	—	—	48 x 24
Traffic From Left (Right) Does Not Stop (plaque)	W4-4aP	2C.59	24 x 12	24 x 12	36 x 18	—	—	48 x 24
Oncoming Traffic Does Not Stop (plaque)	W4-4bP	2C.59	24 x 12	24 x 12	36 x 18	—	—	48 x 24
Entering Roadway Merge	W4-5	2C.40	36 x 36	36 x 36	48 x 48	—	—	—
No Merge Area (plaque)	W4-5P	2C.40	18 x 24	18 x 24	24 x 30	—	—	—
Entering Roadway Added Lane	W4-6	2C.41	36 x 36	36 x 36	48 x 48	—	—	—
Road Narrows	W5-1	2C.19	36 x 36	36 x 36	48 x 48	48 x 48	30 x 30*	—
Narrow Bridge	W5-2	2C.20	36 x 36	36 x 36	48 x 48	48 x 48	30 x 30*	—
One Lane Bridge	W5-3	2C.21	36 x 36	36 x 36	48 x 48	48 x 48	30 x 30*	—
Divided Highway	W6-1	2C.22	36 x 36	36 x 36	48 x 48	48 x 48	—	—
Divided Highway Ends	W6-2	2C.23	36 x 36	36 x 36	48 x 48	48 x 48	—	—
Two-Way Traffic	W6-3	2C.44	36 x 36	36 x 36	48 x 48	48 x 48	—	—
Hill	W7-1	2C.16	30 x 30*	36 x 36	36 x 36	36 x 36	24 x 24*	48 x 48
Hill with Grade	W7-1a	2C.16	30 x 30*	36 x 36	36 x 36	36 x 36	24 x 24*	48 x 48
Use Low Gear (plaque)	W7-2P	2C.57	24 x 18	24 x 18	—	—	—	—
Trucks Use Lower Gear (plaque)	W7-2bP	2C.57	24 x 18	24 x 18	—	—	—	—
XX% Grade (plaque)	W7-3P	2C.57	24 x 18	24 x 18	—	—	—	—
Next XX Miles (plaque)	W7-3aP	2C.55	24 x 18	24 x 18	—	—	—	—
XX% Grade, XX Miles (plaque)	W7-3bP	2C.57	24 x 18	24 x 18	—	—	—	—
Runaway Truck Ramp XX Miles	W7-4	2C.17	78 x 48	78 x 48	78 x 48	78 x 48	—	—
Runaway Truck Ramp (with arrow)	W7-4b	2C.17	78 x 60	78 x 60	78 x 60	78 x 60	—	—
Truck Escape Ramp	W7-4c	2C.17	78 x 60	78 x 60	78 x 60	78 x 60	—	—
Sand, Gravel, Paved (plaques)	W7-4dP, 4eP,4fP	2C.17	24 x 12	24 x 12	24 x 12	24 x 12	—	—
Hill Blocks View	W7-6	2C.18	30 x 30*	36 x 36	36 x 36	—	—	48 x 48
Bump or Dip	W8-1,2	2C.28	30 x 30*	36 x 36	36 x 36	48 x 48	24 x 24*	48 x 48

Table 2C-2. Warning Sign and Plaque Sizes (Sheet 2 of 3)

Sign or Plaque	Sign Designation	Section	Conventional Road Single Lane	Conventional Road Multi-Lane	Expressway	Freeway	Minimum	Oversized
Pavement Ends	W8-3	2C.30	36 x 36	36 x 36	48 x 48	—	30 x 30*	—
Soft Shoulder	W8-4	2C.31	36 x 36	36 x 36	48 x 48	48 x 48	24 x 24*	48 x 48
Slippery When Wet	W8-5	2C.32	30 x 30*	36 x 36	36 x 36	48 x 48	24 x 24*	48 x 48
Road Condition (plaques)	W8-5P,5bP,5cP	2C.32	24 x 18	24 x 18	30 x 24	36 x 30	—	36 x 30
Ice	W8-5aP	2C.32	24 x 12	24 x 12	30 x 18	30 x 18	—	—
Truck Crossing	W8-6	2C.49	36 x 36	36 x 36	36 x 36	48 x 48	24 x 24*	48 x 48
Loose Gravel	W8-7	2C.32	36 x 36	36 x 36	36 x 36	—	24 x 24*	48 x 48
Rough Road	W8-8	2C.32	36 x 36	36 x 36	36 x 36	48 x 48	24 x 24*	48 x 48
Low Shoulder	W8-9	2C.31	36 x 36	36 x 36	36 x 36	48 x 48	24 x 24*	48 x 48
Uneven Lanes	W8-11	2C.32	36 x 36	36 x 36	36 x 36	48 x 48	—	48 x 48
No Center Line	W8-12	2C.34	36 x 36	36 x 36	36 x 36	48 x 48	—	—
Bridge Ices Before Road	W8-13	2C.32	36 x 36	36 x 36	36 x 36	48 x 48	24 x 24*	48 x 48
Fallen Rocks	W8-14	2C.32	30 x 30*	36 x 36	36 x 36	48 x 48	24 x 24*	48 x 48
Grooved Pavement	W8-15	2C.33	30 x 30*	36 x 36	36 x 36	48 x 48	24 x 24*	48 x 48
Motorcycle (plaque)	W8-15P	2C.33	24 x 18	24 x 18	30 x 24	36 x 30	—	36 x 30
Metal Bridge Deck	W8-16	2C.33	30 x 30*	36 x 36	36 x 36	48 x 48	24 x 24*	48 x 48
Shoulder Drop Off (symbol)	W8-17	2C.31	30 x 30*	36 x 36	36 x 36	48 x 48	24 x 24*	48 x 48
Shoulder Drop-Off (plaque)	W8-17P	2C.31	24 x 18	24 x 18	30 x 24	36 x 30	—	36 x 30
Road May Flood	W8-18	2C.35	36 x 36	36 x 36	36 x 36	48 x 48	24 x 24*	48 x 48
Flood Gauge	W8-19	2C.35	12 x 72	12 x 72	—	—	—	—
Gusty Winds Area	W8-21	2C.35	36 x 36	36 x 36	36 x 36	48 x 48	24 x 24*	48 x 48
Fog Area	W8-22	2C.35	36 x 36	36 x 36	36 x 36	48 x 48	24 x 24*	48 x 48
No Shoulder	W8-23	2C.31	36 x 36	36 x 36	36 x 36	48 x 48	24 x 24*	48 x 48
Shoulder Ends	W8-25	2C.31	30 x 30*	36 x 36	36 x 36	48 x 48	24 x 24*	48 x 48
Left (Right) Lane Ends	W9-1	2C.42	36 x 36	36 x 36	36 x 36	48 x 48	30 x 30*	48 x 48
Lane Ends Merge Left (Right)	W9-2	2C.42	36 x 36	36 x 36	36 x 36	48 x 48	30 x 30*	48 x 48
Right (Left) Lane Exit Only Ahead	W9-7	2C.43	132 x 72	132 x 72	132 x 72	132 x 72	—	—
Bicycle	W11-1	2C.49	30 x 30	30 x 30	36 x 36	—	24 x 24*	48 x 48
Pedestrian	W11-2	2C.50	30 x 30*	36 x 36	36 x 36	—	24 x 24*	48 x 48
Large Animals	W11-3,4,16,17,18,19,20,21,22	2C.50	30 x 30*	36 x 36	36 x 36	—	24 x 24*	48 x 48
Farm Vehicle	W11-5,5a	2C.49	30 x 30*	36 x 36	36 x 36	—	24 x 24*	48 x 48
Snowmobile	W11-6	2C.50	30 x 30*	36 x 36	36 x 36	—	24 x 24*	48 x 48
Equestrian	W11-7	2C.50	30 x 30*	36 x 36	36 x 36	—	24 x 24*	48 x 48
Emergency Vehicle	W11-8	2C.49	30 x 30*	36 x 36	36 x 36	—	24 x 24*	48 x 48
Handicapped	W11-9	2C.50	30 x 30*	36 x 36	36 x 36	—	—	48 x 48
Truck	W11-10	2C.49	30 x 30*	36 x 36	36 x 36	—	24 x 24*	48 x 48
Golf Cart	W11-11	2C.49	30 x 30*	36 x 36	36 x 36	—	24 x 24*	48 x 48
Emergency Signal Ahead (plaque)	W11-12P	2C.49	36 x 30	36 x 30	36 x 30	—	—	—
Horse-Drawn Vehicle	W11-14	2C.49	30 x 30*	36 x 36	36 x 36	—	24 x 24*	48 x 48
Bicycle / Pedestrian	W11-15	2C.49	30 x 30*	36 x 36	36 x 36	—	24 x 24*	48 x 48
Trail Crossing	W11-15a	2C.49	30 x 30*	36 x 36	36 x 36	—	24 x 24*	48 x 48
Trail X-ing (plaque)	W11-15P	2C.49	24 x 18	24 x 18	30 x 24	—	—	36 x 30
Double Arrow	W12-1	2C.25	30 x 30*	36 x 36	36 x 36	—	—	—
Low Clearance (with arrows)	W12-2	2C.27	36 x 36	36 x 36	48 x 48	48 x 48	30 x 30*	—
Low Clearance	W12-2a	2C.27	78 x 24	78 x 24	—	—	—	—
Advisory Speed (plaque)	W13-1P	2C.08	18 x 18	18 x 18	24 x 24	30 x 30	—	30 x 30
Advisory Exit or Ramp Speed	W13-2,3	2C.14	24 x 30	24 x 30	36 x 48	36 x 48	—	48 x 60
Combination Horizontal Alignment/Advisory Exit or Ramp Speed	W13-6,7	2C.15	24 x 42	24 x 42	36 x 60	36 x 60	—	48 x 84
Dead End, No Outlet	W14-1,2	2C.26	30 x 30*	36 x 36	36 x 36	—	24 x 24*	48 x 48

Table 2C-2. Warning Sign and Plaque Sizes (Sheet 3 of 3)

Sign or Plaque	Sign Designation	Section	Conventional Road Single Lane	Conventional Road Multi-Lane	Expressway	Freeway	Minimum	Oversized
Dead End, No Outlet (with arrow)	W14-1a,2a	2C.26	36 x 8	36 x 8	—	—	—	—
No Passing Zone (pennant)	W14-3	2C.45	48 x 48 x 36	48 x 48 x 36	—	—	40 x 40 x 30	64 x 64 x 48
Playground	W15-1	2C.51	30 x 30*	36 x 36	36 x 36	—	24 x 24*	48 x 48
Share the Road (plaque)	W16-1P	2C.60	18 x 24	18 x 24	24 x 30	—	—	24 x 30
XX Feet	W16-2P	2C.55	24 x 18	24 x 18	—	—	—	30 x 24
XX Ft	W16-2aP	2C.55	24 x 12	24 x 12	—	—	—	30 x 18
XX Miles (2-line plaque)	W16-3P	2C.55	30 x 24	30 x 24	—	—	—	—
XX Miles (1-line plaque)	W16-3aP	2C.55	30 x 12	30 x 12	—	—	—	—
Next XX Feet (plaque)	W16-4P	2C.55	30 x 24	30 x 24	—	—	—	—
Supplemental Arrow (plaque)	W16-5P,6P	2C.56	24 x 18	24 x 18	—	—	—	—
Downward Diagonal Arrow (plaque)	W16-7P	2C.50	24 x 12	24 x 12	—	—	—	30 x 18
Advance Street Name (1-line plaque)	W16-8P	2C.58	Varies x 8	Varies x 8	—	—	—	—
Advance Street Name (2-line plaque)	W16-8aP	2C.58	Varies x 15	Varies x 15	—	—	—	—
Ahead (plaque)	W16-9P	2C.50	24 x 12	24 x 12	30 x 18	—	—	—
Photo Enforced (symbol plaque)	W16-10P	2C.61	24 x 12	24 x 12	36 x 18	—	—	48 x 24
Photo Enforced (plaque)	W16-10aP	2C.61	24 x 18	24 x 18	36 x 30	—	—	48 x 36
HOV (plaque)	W16-11P	2G.09	24 x 12	24 x 12	30 x 18	—	—	30 x 18
Traffic Circle (plaque)	W16-12P	2C.46	24 x 18	24 x 18	—	—	—	—
When Flashing (plaque)	W16-13P	2C.50	24 x 18	24 x 18	—	—	—	—
New (plaque)	W16-15P	2C.62	24 x 12	24 x 12	—	—	—	—
Roundabout (plaque)	W16-17P	2C.46	24 x 12	24 x 12	—	—	—	—
NOTICE	W16-18P	2A.15	24 x 12	24 x 12	—	—	—	—
Speed Hump	W17-1	2C.29	30 x 30*	36 x 36	—	—	24 x 24*	48 x 48
Freeway Ends XX Miles	W19-1	2C.24	—	—	—	144 x 48	—	—
Expressway Ends XX Miles	W19-2	2C.24	—	—	144 x 48	—	—	—
Freeway Ends	W19-3	2C.24	—	—	—	48 x 48	—	—
Expressway Ends	W19-4	2C.24	—	—	48 x 48	—	—	—
All Traffic Must Exit	W19-5	2C.24	—	—	90 x 48	90 x 48	—	—
New Traffic Pattern Ahead	W23-2	2C.52	36 x 36	36 x 36	—	—	—	—
Traffic Signal Extended Green	W25-1,2	2C.48	24 x 30	24 x 30	—	—	—	—

* The minimum size required for diamond-shaped warning signs facing traffic on multi-lane conventional roads shall be 36 x 36 per Section 2C.04

Notes: 1. Larger signs may be used when appropriate
2. Dimensions in inches are shown as width x height

Support:

01 Section 2A.11 contains information regarding the applicability of the various columns in Table 2C-2.

Standard:

03 **Except as provided in Paragraph 5, the minimum size for all diamond-shaped warning signs facing traffic on a multi-lane conventional road where the posted speed limit is higher than 35 mph shall be 36 x 36 inches.**

04 **The minimum size for supplemental warning plaques that are not included in Table 2C-2 shall be as shown in Table 2C-3.**

Option:

05 If a diamond-shaped warning sign is placed on the left-hand side of a multi-lane roadway to supplement the installation of the same warning sign on the right-hand side of the roadway, the minimum size identified in the Single Lane column in Table 2C-2 may be used.

Table 2C-3. Minimum Size of Supplemental Warning Plaques

Size of Warning Sign	Size of Supplemental Plaque Rectangular			Size of Supplemental Plaque Square
	1 Line	2 Lines	Arrow	
24 x 24	24 x 12	24 x 18	24 x 12	18 x 18
30 x 30				
36 x 36	30 x 18	30 x 24	30 x 18	24 x 24
48 x 48				

Notes: 1. Larger supplemental plaques may be used when appropriate
2. Dimensions in inches are shown as width x height

06 Signs and plaques larger than those shown in Tables 2C-2 and 2C-3 may be used (see Section 2A.11).

Guidance:

07 *The minimum size for all diamond-shaped warning signs facing traffic on exit and entrance ramps should be the size identified in Table 2C-2 for the mainline roadway classification (Expressway or Freeway). If a minimum size is not provided in the Freeway Column, the Expressway size should be used. If a minimum size is not provided in the Freeway or the Expressway Column, the Oversized size should be used.*

Section 2C.05 Placement of Warning Signs

Support:

01 For information on placement of warning signs, see Sections 2A.16 to 2A.21.

02 The time needed for detection, recognition, decision, and reaction is called the Perception-Response Time (PRT). Table 2C-4 is provided as an aid for determining warning sign location. The distances shown in Table 2C-4 can be adjusted for roadway features, other signing, and to improve visibility.

Guidance:

03 *Warning signs should be placed so that they provide an adequate PRT. The distances contained in Table 2C-4 are for guidance purposes and should be applied with engineering judgment. Warning signs should not be placed too far in advance of the condition, such that drivers might tend to forget the warning because of other driving distractions, especially in urban areas.*

Table 2C-4. Guidelines for Advance Placement of Warning Signs

Posted or 85th-Percentile Speed	Advance Placement Distance[1]								
	Condition A: Speed reduction and lane changing in heavy traffic[2]	Condition B: Deceleration to the listed advisory speed (mph) for the condition							
		0^3	10^4	20^4	30^4	40^4	50^4	60^4	70^4
20 mph	225 ft	100 ft[6]	N/A[5]	—	—	—	—	—	—
25 mph	325 ft	100 ft[6]	N/A[5]	N/A[5]	—	—	—	—	—
30 mph	460 ft	100 ft[6]	N/A[5]	N/A[5]	—	—	—	—	—
35 mph	565 ft	100 ft[6]	N/A[5]	N/A[5]	N/A[5]	—	—	—	—
40 mph	670 ft	125 ft	100 ft[6]	100 ft[6]	N/A[5]	—	—	—	—
45 mph	775 ft	175 ft	125 ft	100 ft[6]	100 ft[6]	N/A[5]	—	—	—
50 mph	885 ft	250 ft	200 ft	175 ft	125 ft	100 ft[6]	—	—	—
55 mph	990 ft	325 ft	275 ft	225 ft	200 ft	125 ft	N/A[5]	—	—
60 mph	1,100 ft	400 ft	350 ft	325 ft	275 ft	200 ft	100 ft[6]	—	—
65 mph	1,200 ft	475 ft	450 ft	400 ft	350 ft	275 ft	200 ft	100 ft[6]	—
70 mph	1,250 ft	550 ft	525 ft	500 ft	450 ft	375 ft	275 ft	150 ft	—
75 mph	1,350 ft	650 ft	625 ft	600 ft	550 ft	475 ft	375 ft	250 ft	100 ft[6]

[1] The distances are adjusted for a sign legibility distance of 180 feet for Condition A. The distances for Condition B have been adjusted for a sign legibility distance of 250 feet, which is appropriate for an alignment warning symbol sign. For Conditions A and B, warning signs with less than 6-inch legend or more than four words, a minimum of 100 feet should be added to the advance placement distance to provide adequate legibility of the warning sign.

[2] Typical conditions are locations where the road user must use extra time to adjust speed and change lanes in heavy traffic because of a complex driving situation. Typical signs are Merge and Right Lane Ends. The distances are determined by providing the driver a PRT of 14.0 to 14.5 seconds for vehicle maneuvers (2005 AASHTO Policy, Exhibit 3-3, Decision Sight Distance, Avoidance Maneuver E) minus the legibility distance of 180 feet for the appropriate sign.

[3] Typical condition is the warning of a potential stop situation. Typical signs are Stop Ahead, Yield Ahead, Signal Ahead, and Intersection Warning signs. The distances are based on the 2005 AASHTO Policy, Exhibit 3-1, Stopping Sight Distance, providing a PRT of 2.5 seconds, a deceleration rate of 11.2 feet/second^2, minus the sign legibility distance of 180 feet.

[4] Typical conditions are locations where the road user must decrease speed to maneuver through the warned condition. Typical signs are Turn, Curve, Reverse Turn, or Reverse Curve. The distance is determined by providing a 2.5 second PRT, a vehicle deceleration rate of 10 feet/second^2, minus the sign legibility distance of 250 feet.

[5] No suggested distances are provided for these speeds, as the placement location is dependent on site conditions and other signing. An alignment warning sign may be placed anywhere from the point of curvature up to 100 feet in advance of the curve. However, the alignment warning sign should be installed in advance of the curve and at least 100 feet from any other signs.

[6] The minimum advance placement distance is listed as 100 feet to provide adequate spacing between signs.

2009 Edition   Page 109

04 *Minimum spacing between warning signs with different messages should be based on the estimated PRT for driver comprehension of and reaction to the second sign.*

05 *The effectiveness of the placement of warning signs should be periodically evaluated under both day and night conditions.*

Option:

06 Warning signs that advise road users about conditions that are not related to a specific location, such as Deer Crossing or SOFT SHOULDER, may be installed in an appropriate location, based on engineering judgment, since they are not covered in Table 2C-4.

Section 2C.06 Horizontal Alignment Warning Signs

Support:

01 A variety of horizontal alignment warning signs (see Figure 2C-1), pavement markings (see Chapter 3B), and delineation (see Chapter 3F) can be used to advise motorists of a change in the roadway alignment. Uniform application of these traffic control devices with respect to the amount of change in the roadway alignment conveys a consistent message establishing driver expectancy and promoting effective roadway operations. The design and application of horizontal alignment warning signs to meet those requirements are addressed in Sections 2C.06 through 2C.15.

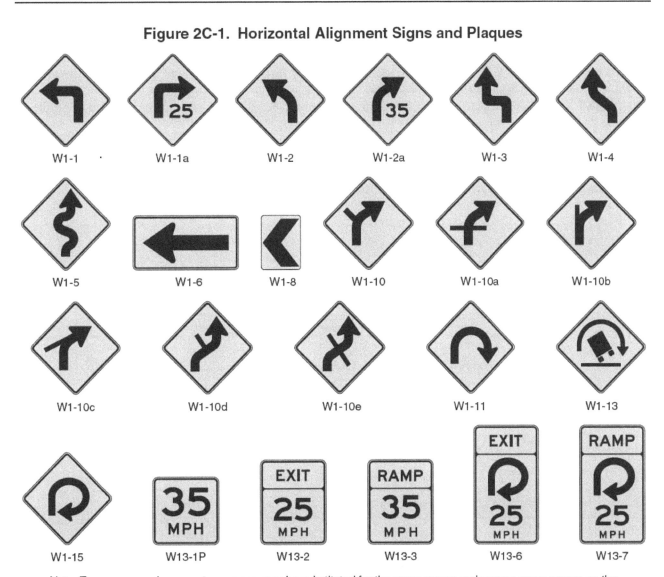

Figure 2C-1. Horizontal Alignment Signs and Plaques

Note: Turn arrows and reverse turn arrows may be substituted for the curve arrows and reverse curve arrows on the W1-10 series signs where appropriate.

Standard:

02 **In advance of horizontal curves on freeways, on expressways, and on roadways with more than 1,000 AADT that are functionally classified as arterials or collectors, horizontal alignment warning signs shall be used in accordance with Table 2C-5 based on the speed differential between the roadway's posted or statutory speed limit or 85th-percentile speed, whichever is higher, or the prevailing speed on the approach to the curve, and the horizontal curve's advisory speed.**

Option:

03 Horizontal Alignment Warning signs may also be used on other roadways or on arterial and collector roadways with less than 1,000 AADT based on engineering judgment.

Section 2C.07 Horizontal Alignment Signs (W1-1 through W1-5, W1-11, W1-15)

Standard:

01 **If Table 2C-5 indicates that a horizontal alignment sign (see Figure 2C-1) is required, recommended, or allowed, the sign installed in advance of the curve shall be a Curve (W1-2) sign unless a different sign is recommended or allowed by the provisions of this Section.**

02 **A Turn (W1-1) sign shall be used instead of a Curve sign in advance of curves that have advisory speeds of 30 mph or less (see Figure 2C-2).**

Guidance:

03 *Where there are two changes in roadway alignment in opposite directions that are separated by a tangent distance of less than 600 feet, the Reverse Turn (W1-3) sign should be used instead of multiple Turn (W1-1) signs and the Reverse Curve (W1-4) sign should be used instead of multiple Curve (W1-2) signs.*

Option:

04 A Winding Road (W1-5) sign may be used instead of multiple Turn (W1-1) or Curve (W1-2) signs where there are three or more changes in roadway alignment each separated by a tangent distance of less than 600 feet.

05 A NEXT XX MILES (W7-3aP) supplemental distance plaque (see Section 2C.55) may be installed below the Winding Road sign where continuous roadway curves exist for a specific distance.

06 If the curve has a change in horizontal alignment of 135 degrees or more, the Hairpin Curve (W1-11) sign may be used instead of a Curve or Turn sign.

07 If the curve has a change of direction of approximately 270 degrees, such as on a cloverleaf interchange ramp, the 270-degree Loop (W1-15) sign may be used instead of a Curve or Turn sign.

Guidance:

08 *When the Hairpin Curve sign or the 270-degree Loop sign is installed, either a One-Direction Large Arrow (W1-6) sign or Chevron Alignment (W1-8) signs should be installed on the outside of the turn or curve.*

Table 2C-5. Horizontal Alignment Sign Selection

Type of Horizontal Alignment Sign	Difference Between Speed Limit and Advisory Speed				
	5 mph	10 mph	15 mph	20 mph	25 mph or more
Turn (W1-1), Curve (W1-2), Reverse Turn (W1-3), Reverse Curve (W1-4), Winding Road (W1-5), and Combination Horizontal Alignment/Intersection (W10-1) (see Section 2C.07 to determine which sign to use)	Recommended	Required	Required	Required	Required
Advisory Speed Plaque (W13-1P)	Recommended	Required	Required	Required	Required
Chevrons (W1-8) and/or One Direction Large Arrow (W1-6)	Optional	Recommended	Required	Required	Required
Exit Speed (W13-2) and Ramp Speed (W13-3) on exit ramp	Optional	Optional	Recommended	Required	Required

Note: Required means that the sign and/or plaque shall be used, recommended means that the sign and/or plaque should be used, and optional means that the sign and/or plaque may be used.

See Section 2C.06 for roadways with less than 1,000 ADT.

Figure 2C-2. Example of Warning Signs for a Turn

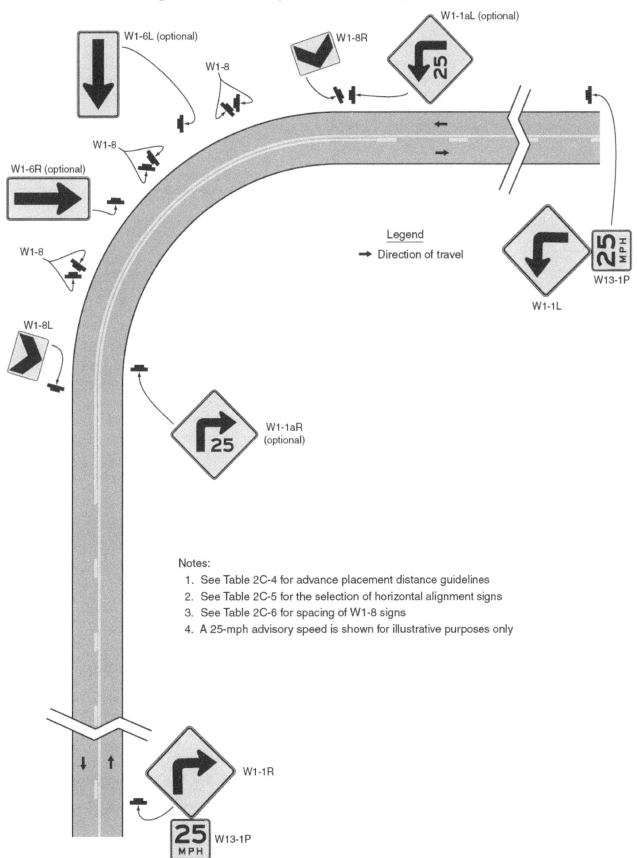

Notes:
1. See Table 2C-4 for advance placement distance guidelines
2. See Table 2C-5 for the selection of horizontal alignment signs
3. See Table 2C-6 for spacing of W1-8 signs
4. A 25-mph advisory speed is shown for illustrative purposes only

Section 2C.08 Advisory Speed Plaque (W13-1P)

Option:

01 The Advisory Speed (W13-1P) plaque (see Figure 2C-1) may be used to supplement any warning sign to indicate the advisory speed for a condition.

Standard:

02 **The use of the Advisory Speed plaque for horizontal curves shall be in accordance with the information shown in Table 2C-5. The Advisory Speed plaque shall also be used where an engineering study indicates a need to advise road users of the advisory speed for other roadway conditions.**

03 **If used, the Advisory Speed plaque shall carry the message XX MPH. The speed displayed shall be a multiple of 5 mph.**

04 **Except in emergencies or when the condition is temporary, an Advisory Speed plaque shall not be installed until the advisory speed has been determined by an engineering study.**

05 **The Advisory Speed plaque shall only be used to supplement a warning sign and shall not be installed as a separate sign installation.**

06 **The advisory speed shall be determined by an engineering study that follows established engineering practices.**

Support:

07 Among the established engineering practices that are appropriate for the determination of the recommended advisory speed for a horizontal curve are the following:

    A. An accelerometer that provides a direct determination of side friction factors
    B. A design speed equation
    C. A traditional ball-bank indicator using the following criteria:
        1. 16 degrees of ball-bank for speeds of 20 mph or less
        2. 14 degrees of ball-bank for speeds of 25 to 30 mph
        3. 12 degrees of ball-bank for speeds of 35 mph and higher

08 The 16, 14, and 12 degrees of ball-bank criteria are comparable to the current AASHTO horizontal curve design guidance. Research has shown that drivers often exceed existing posted advisory curve speeds by 7 to 10 mph.

Guidance:

09 *The advisory speed should be determined based on free-flowing traffic conditions.*

10 *Because changes in conditions, such as roadway geometrics, surface characteristics, or sight distance, might affect the advisory speed, each location should be evaluated periodically or when conditions change.*

Section 2C.09 Chevron Alignment Sign (W1-8)

Standard:

01 **The use of the Chevron Alignment (W1-8) sign (see Figures 2C-1 and 2C-2) to provide additional emphasis and guidance for a change in horizontal alignment shall be in accordance with the information shown in Table 2C-5.**

Option:

02 When used, Chevron Alignment signs may be used instead of or in addition to standard delineators.

Standard:

03 **The Chevron Alignment sign shall be a vertical rectangle. No border shall be used on the Chevron Alignment sign.**

04 **If used, Chevron Alignment signs shall be installed on the outside of a turn or curve, in line with and at approximately a right angle to approaching traffic. Chevron Alignment signs shall be installed at a minimum height of 4 feet, measured vertically from the bottom of the sign to the elevation of the near edge of the traveled way.**

Guidance:

05 *The approximate spacing of Chevron Alignment signs on the turn or curve measured from the point of curvature (PC) should be as shown in Table 2C-6.*

06 *If used, Chevron Alignment signs should be visible for a sufficient distance to provide the road user with adequate time to react to the change in alignment.*

Standard:

07 **Chevron Alignment signs shall not be placed on the far side of a T-intersection facing traffic on the stem approach to warn drivers that a through movement is not physically possible, as this is the function of a Two-Direction (or One-Direction) Large Arrow sign.**

08 **Chevron Alignment signs shall not be used to mark obstructions within or adjacent to the roadway, including the beginning of guardrails or barriers, as this is the function of an object marker (see Section 2C.63).**

Table 2C-6. Typical Spacing of Chevron Alignment Signs on Horizontal Curves

Advisory Speed	Curve Radius	Sign Spacing
15 mph or less	Less than 200 feet	40 feet
20 to 30 mph	200 to 400 feet	80 feet
35 to 45 mph	401 to 700 feet	120 feet
50 to 60 mph	701 to 1,250 feet	160 feet
More than 60 mph	More than 1,250 feet	200 feet

Note: The relationship between the curve radius and the advisory speed shown in this table should not be used to determine the advisory speed.

Section 2C.10 Combination Horizontal Alignment/Advisory Speed Signs (W1-1a, W1-2a)

Option:

01 The Turn (W1-1) sign or the Curve (W1-2) sign may be combined with the Advisory Speed (W13-1P) plaque (see Section 2C.08) to create a combination Turn/Advisory Speed (W1-1a) sign or combination Curve/Advisory Speed (W1-2a) sign (see Figure 2C-1).

02 The combination Horizontal Alignment/Advisory Speed sign may be used to supplement the advance Horizontal Alignment warning sign and Advisory Speed plaque based upon an engineering study.

Standard:

03 **If used, the combination Horizontal Alignment/Advisory Speed sign shall not be used alone and shall not be used as a substitute for a Horizontal Alignment warning sign and Advisory Speed plaque at the advance warning location. The combination Horizontal Alignment/Advisory Speed sign shall only be used as a supplement to the advance Horizontal Alignment warning sign. If used, the combination Horizontal Alignment/Advisory Speed sign shall be installed at the beginning of the turn or curve.**

Guidance:

04 *The advisory speed displayed on the combination Horizontal Alignment/Advisory Speed sign should be based on the advisory speed for the horizontal curve using recommended engineering practices (see Section 2C.08).*

Section 2C.11 Combination Horizontal Alignment/Intersection Signs (W1-10 Series)

Option:

01 The Turn (W1-1) sign or the Curve (W1-2) sign may be combined with the Cross Road (W2-1) sign or the Side Road (W2-2 or W2-3) sign to create a combination Horizontal Alignment/Intersection (W1-10 series) sign (see Figure 2C-1) that depicts the condition where an intersection occurs within or immediately adjacent to a turn or curve.

Guidance:

02 *Elements of the combination Horizontal Alignment/Intersection sign related to horizontal alignment should comply with the provisions of Section 2C.07, and elements related to intersection configuration should comply with the provisions of Section 2C.46. The symbol design should approximate the configuration of the intersecting roadway(s). No more than one Cross Road or two Side Road symbols should be displayed on any one combination Horizontal Alignment/Intersection sign.*

Standard:

03 **The use of the combination Horizontal Alignment/Intersection sign shall be in accordance with the appropriate Turn or Curve sign information shown in Table 2C-5.**

Section 2C.12 One-Direction Large Arrow Sign (W1-6)

Option:

01 A One-Direction Large Arrow (W1-6) sign (see Figure 2C-1) may be used either as a supplement or alternative to Chevron Alignment signs in order to delineate a change in horizontal alignment (see Figure 2C-2).

02 A One-Direction Large Arrow (W1-6) sign may be used to supplement a Turn or Reverse Turn sign (see Figure 2C-2) to emphasize the abrupt curvature.

Standard:

03 **The One-Direction Large Arrow sign shall be a horizontal rectangle with an arrow pointing to the left or right.**

04 **The use of the One-Direction Large Arrow sign shall be in accordance with the information shown in Table 2C-5.**

05 **If used, the One-Direction Large Arrow sign shall be installed on the outside of a turn or curve in line with and at approximately a right angle to approaching traffic.**

06 **The One-Direction Large Arrow sign shall not be used where there is no alignment change in the direction of travel, such as at the beginnings and ends of medians or at center piers.**

07 **The One-Direction Large Arrow sign directing traffic to the right shall not be used in the central island of a roundabout.**

Guidance:

08 *If used, the One-Direction Large Arrow sign should be visible for a sufficient distance to provide the road user with adequate time to react to the change in alignment.*

Section 2C.13 Truck Rollover Warning Sign (W1-13)

Option:

01 A Truck Rollover Warning (W1-13) sign (see Figure 2C-1) may be used to warn drivers of vehicles with a high center of gravity, such as trucks, tankers, and recreational vehicles, of a curve or turn where geometric conditions might contribute to a loss of control and a rollover as determined by an engineering study.

Support:

02 Among the established engineering practices that are appropriate for the determination of the truck rollover potential of a horizontal curve are the following:
- A. An accelerometer that provides a direct determination of side friction factors
- B. A design speed equation
- C. A traditional ball-bank indicator using 10 degrees of ball-bank

Standard:

03 **If a Truck Rollover Warning (W1-13) sign is used, it shall be accompanied by an Advisory Speed (W13-1P) plaque indicating the recommended speed for vehicles with a higher center of gravity.**

Option:

04 The Truck Rollover Warning sign may be displayed as a static sign, as a static sign supplemented by a flashing warning beacon, or as a changeable message sign activated by the detection of an approaching vehicle with a high center of gravity that is traveling in excess of the recommended speed for the condition.

Support:

05 The curved arrow on the Truck Rollover Warning sign shows the direction of roadway curvature. The truck tips in the opposite direction.

Section 2C.14 Advisory Exit and Ramp Speed Signs (W13-2 and W13-3)

Standard:

01 **Advisory Exit Speed (W13-2) and Advisory Ramp Speed (W13-3) signs (see Figure 2C-1) shall be vertical rectangles. The use of Advisory Exit Speed and Advisory Ramp Speed signs on freeway and expressway ramps shall be in accordance with the information shown in Table 2C-5.**

Guidance:

02 *If used, the Advisory Exit Speed sign should be installed along the deceleration lane and the advisory speed displayed should be based on an engineering study. When a Truck Rollover (W1-13) sign (see Section 2C.13) is also installed for the ramp, the advisory exit speed should be based on the truck advisory speed for the horizontal alignment using recommended engineering practices.*

03 *If used, the Advisory Exit Speed sign should be visible in time for the road user to decelerate and make an exiting maneuver.*

Support:

04 table 2C-4 lists recommended advance sign placement distances for deceleration to various advisory speeds.

Guidance:

05 *If used, the Advisory Ramp Speed sign should be installed on the ramp to confirm the ramp advisory speed.*

06 *If used, Chevron Alignment (W1-8) signs and/or One-Direction Large Arrow (W1-6) signs should be installed on the outside of the exit curve as described in Sections 2C.09 and 2C.12.*

Option:

07 Where there is a need to remind road users of the recommended advisory speed, a horizontal alignment warning sign with an advisory speed plaque may be installed at or beyond the beginning of the exit curve or on the outside of the curve, provided that it is apparent that the sign applies only to exiting traffic. These signs may also be used at intermediate points along the ramp, especially if the ramp curvature changes and the subsequent curves on the ramp have a different advisory speed than the initial ramp curve.

Support:

08 Figure 2C-3 shows an example of advisory speed signing for an exit ramp.

Section 2C.15 Combination Horizontal Alignment/Advisory Exit and Ramp Speed Signs (W13-6 and W13-7)

Option:

01 A horizontal alignment sign (see Section 2C.07) may be combined with an Advisory Exit Speed or Advisory Ramp Speed sign to create a combination Horizontal Alignment/Advisory Exit Speed (W13-6) sign or a combination Horizontal Alignment/Advisory Ramp Speed (W13-7) sign (see Figure 2C-1). These combination signs may be used where the severity of the exit ramp curvature might not be apparent to road users in the deceleration lane or where the curvature needs to be specifically identified as being on the exit ramp rather than on the mainline.

Section 2C.16 Hill Signs (W7-1, W7-1a)

Guidance:

01 *The Hill (W7-1) sign (see Figure 2C-4) should be used in advance of a downgrade where the length, percent of grade, horizontal curvature, and/or other physical features require special precautions on the part of road users.*

02 *The Hill sign and supplemental grade (W7-3P) plaque (see Section 2C.57) used in combination, or the W7-1a sign used alone, should be installed in advance of downgrades for the following conditions:*

   A. *5% grade that is more than 3,000 feet in length,*
   B. *6% grade that is more than 2,000 feet in length,*
   C. *7% grade that is more than 1,000 feet in length,*
   D. *8% grade that is more than 750 feet in length, or*
   E. *9% grade that is more than 500 feet in length.*

03 *These signs should also be installed for steeper grades or where crash experience and field observations indicate a need.*

04 *Supplemental plaques (see Section 2C.57) and larger signs should be used for emphasis or where special hill characteristics exist. On longer grades, the use of the Hill sign with a distance (W7-3aP) plaque or the combination distance/grade (W7-3bP) plaque at periodic intervals of approximately 1-mile spacing should be considered.*

Standard:

05 **If the percent grade is displayed on a supplemental plaque, the plaque shall be placed below the Hill (W7-1) sign.**

Option:

06 A USE LOW GEAR (W7-2P) or TRUCKS USE LOWER GEAR (W7-2bP) supplemental plaque (see Figure 2C-4) may be used to indicate a situation where downshifting as well as braking might be advisable.

Section 2C.17 Truck Escape Ramp Signs (W7-4 Series)

Guidance:

01 *Where applicable, truck escape (or runaway truck) ramp advance warning signs (see Figure 2C-4) should be located approximately 1 mile, and 1/2 mile in advance of the grade, and of the ramp. A sign also should be placed at the gore. A RUNAWAY VEHICLES ONLY (R4-10) sign (see Section 2B.35) should be installed near the ramp entrance to discourage other road users from entering the ramp. No Parking (R8-3) signs should be placed near the ramp entrance.*

Standard:

02 **When truck escape ramps are installed, at least one of the W7-4 series signs shall be used.**

Figure 2C-3. Example of Advisory Speed Signing for an Exit Ramp

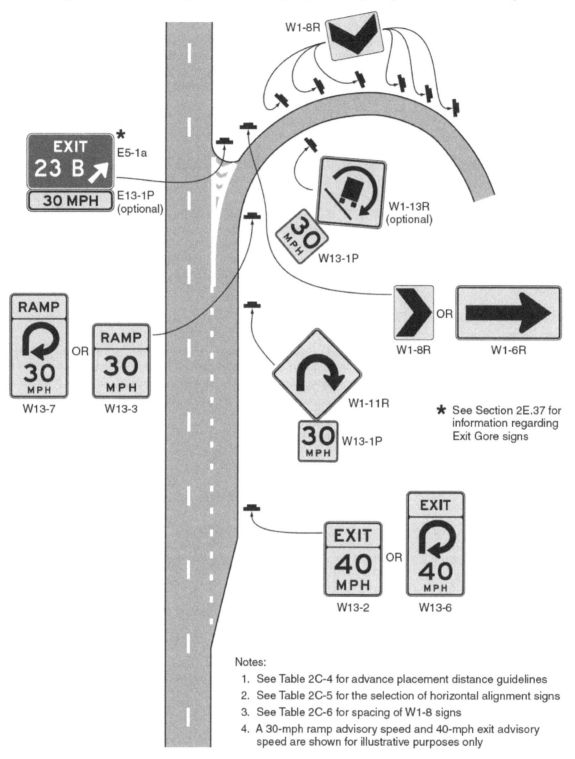

Notes:
1. See Table 2C-4 for advance placement distance guidelines
2. See Table 2C-5 for the selection of horizontal alignment signs
3. See Table 2C-6 for spacing of W1-8 signs
4. A 30-mph ramp advisory speed and 40-mph exit advisory speed are shown for illustrative purposes only

Figure 2C-4. Vertical Grade Signs and Plaques

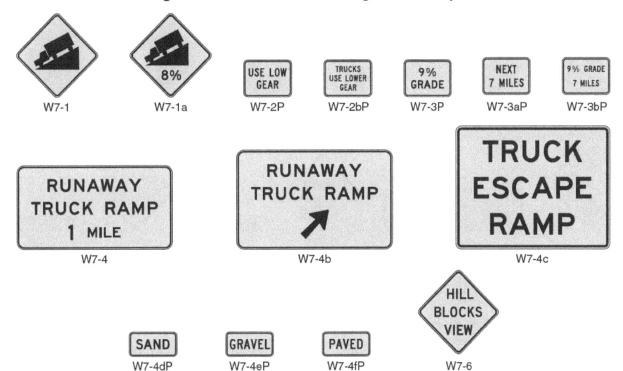

Option:

03 A SAND (W7-4dP), GRAVEL (W7-4eP), or PAVED (W7-4fP) supplemental plaque (see Figure 2C-4) may be used to describe the ramp surface. State and local highway agencies may develop appropriate word message signs for the specific situation.

Section 2C.18 HILL BLOCKS VIEW Sign (W7-6)

Option:

01 A HILL BLOCKS VIEW (W7-6) sign (see Figure 2C-4) may be used in advance of a crest vertical curve to advise road users to reduce speed as they approach and traverse the hill as only limited stopping sight distance is available.

Guidance:

02 *When a HILL BLOCKS VIEW sign is used, it should be supplemented by an Advisory Speed (W13-1P) plaque indicating the recommended speed for traveling over the hillcrest based on available stopping sight distance.*

Section 2C.19 ROAD NARROWS Sign (W5-1)

Guidance:

01 *Except as provided in Paragraph 2, a ROAD NARROWS (W5-1) sign (see Figure 2C-5) should be used in advance of a transition on two-lane roads where the pavement width is reduced abruptly to a width such that vehicles traveling in opposite directions cannot simultaneously travel through the narrow portion of the roadway without reducing speed.*

Option:

02 The ROAD NARROWS (W5-1) sign may be omitted on low-volume local streets that have speed limits of 30 mph or less.

03 Additional emphasis may be provided by the use of object markers and delineators (see Sections 2B.63 through 2B.65 and Chapter 3F). The Advisory Speed (W13-1P) plaque (see Section 2C.08) may be used to indicate the recommended speed.

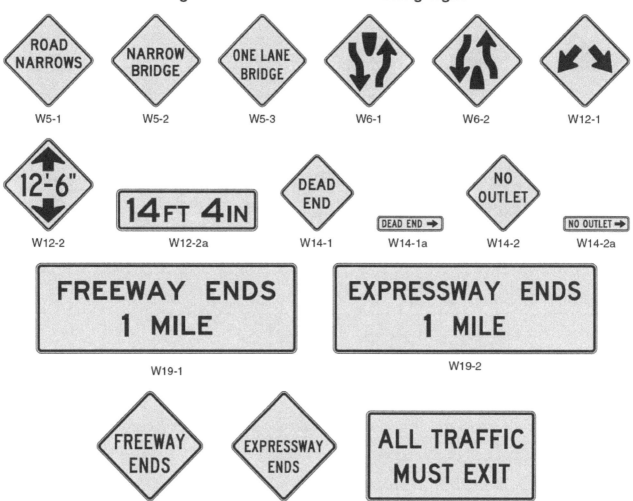

Figure 2C-5. Miscellaneous Warning Signs

Section 2C.20 NARROW BRIDGE Sign (W5-2)

Guidance:

01　A NARROW BRIDGE (W5-2) sign (see Figure 2C-5) should be used in advance of any bridge or culvert having a two-way roadway clearance width of 16 to 18 feet, or any bridge or culvert having a roadway clearance less than the width of the approach travel lanes.

02　Additional emphasis should be provided by the use of object markers, delineators, and/or pavement markings.

Option:

03　A NARROW BRIDGE sign may be used in advance of a bridge or culvert on which the approach shoulders are narrowed or eliminated.

Section 2C.21 ONE LANE BRIDGE Sign (W5-3)

Guidance:

01　A ONE LANE BRIDGE (W5-3) sign (see Figure 2C-5) should be used on two-way roadways in advance of any bridge or culvert:

　　A. Having a clear roadway width of less than 16 feet, or
　　B. Having a clear roadway width of less than 18 feet when commercial vehicles constitute a high proportion of the traffic, or
　　C. Having a clear roadway width of 18 feet or less where the sight distance is limited on the approach to the structure.

02　Additional emphasis should be provided by the use of object markers, delineators, and/or pavement markings.

Section 2C.22 Divided Highway Sign (W6-1)

Guidance:

01   A Divided Highway (W6-1) sign (see Figure 2C-5) should be used on the approaches to a section of highway (not an intersection or junction) where the opposing flows of traffic are separated by a median or other physical barrier.

Standard:

02   **The Divided Highway (W6-1) sign shall not be used instead of a Keep Right (R4-7 series) sign on the approach end of a median island.**

Section 2C.23 Divided Highway Ends Sign (W6-2)

Guidance:

01   A Divided Highway Ends (W6-2) sign (see Figure 2C-5) should be used in advance of the end of a section of physically divided highway (not an intersection or junction) as a warning of two-way traffic ahead.

02   The Two-Way Traffic (W6-3) sign (see Section 2C.44) should be used to give warning and notice of the transition to a two-lane, two-way section.

Section 2C.24 Freeway or Expressway Ends Signs (W19 Series)

Option:

01   A FREEWAY ENDS XX MILES (W19-1) sign or a FREEWAY ENDS (W19-3) sign (see Figure 2C-5) may be used in advance of the end of a freeway.

02   An EXPRESSWAY ENDS XX MILES (W19-2) sign or an EXPRESSWAY ENDS (W19-4) sign (see Figure 2C-5) may be used in advance of the end of an expressway.

03   The rectangular W19-1 and W19-2 signs may be post-mounted or may be mounted overhead for increased emphasis.

Guidance:

04   If the reason that the freeway is ending is that the next portion of the freeway is not yet constructed and as a result all traffic must use an exit ramp to leave the freeway, an ALL TRAFFIC MUST EXIT (W19-5) sign (see Figure 2C-5) should be used in addition to the Freeway Ends signs in advance of the downstream end of the freeway.

Section 2C.25 Double Arrow Sign (W12-1)

Option:

01   The Double Arrow (W12-1) sign (see Figure 2C-5) may be used to advise road users that traffic is permitted to pass on either side of an island, obstruction, or gore in the roadway. Traffic separated by this sign may either rejoin or change directions.

Guidance:

02   If used on an island, the Double Arrow sign should be mounted near the approach end.

03   If used in front of a pier or obstruction, the Double Arrow sign should be mounted on the face of, or just in front of, the obstruction. Where stripe markings are used on the obstruction, they should be discontinued to leave a 3-inch space around the outside of the sign.

Section 2C.26 DEAD END/NO OUTLET Signs (W14-1, W14-1a, W14-2, W14-2a)

Option:

01   The DEAD END (W14-1) sign (see Figure 2C-5) may be used at the entrance of a single road or street that terminates in a dead end or cul-de-sac. The NO OUTLET (W14-2) sign (see Figure 2C-5) may be used at the entrance to a road or road network from which there is no other exit.

02   DEAD END (W14-1a) or NO OUTLET (W14-2a) signs (see Figure 2C-5) may be used in combination with Street Name (D3-1) signs (see Section 2D.43) to warn turning traffic that the cross street ends in the direction indicated by the arrow.

03   At locations where the cross street does not have a name, the W14-1a or W14-2a signs may be used alone in place of a street name sign.

Standard:

04   **The DEAD END (W14-1a) and NO OUTLET (W14-2a) signs shall be horizontal rectangles with an arrow pointing to the left or right.**

05 When the W14-1 or W14-2 sign is used, the sign shall be posted as near as practical to the entry point or at a sufficient advance distance to permit the road user to avoid the dead end or no outlet condition by turning at the nearest intersecting street.

06 The DEAD END (W14-1a) or NO OUTLET (W14-2a) signs shall not be used instead of the W14-1 or W14-2 signs where traffic can proceed straight through the intersection into the dead end street or no outlet area.

Section 2C.27 Low Clearance Signs (W12-2 and W12-2a)

Standard:

01 **The Low Clearance (W12-2) sign (see Figure 2C-5) shall be used to warn road users of clearances less than 12 inches above the statutory maximum vehicle height.**

Guidance:

02 *The actual clearance should be displayed on the Low Clearance sign to the nearest 1 inch not exceeding the actual clearance. However, in areas that experience changes in temperature causing frost action, a reduction, not exceeding 3 inches, should be used for this condition.*

03 *Where the clearance is less than the legal maximum vehicle height, the W12-2 sign with a supplemental distance plaque should be placed at the nearest intersecting road or wide point in the road at which a vehicle can detour or turn around.*

04 *In the case of an arch or other structure under which the clearance varies greatly, two or more signs should be used as necessary on the structure itself to give information as to the clearances over the entire roadway.*

05 *Clearances should be evaluated periodically, particularly when resurfacing operations have occurred.*

Option:

06 The Low Clearance sign may be installed on or in advance of the structure. If a sign is placed on the structure, it may be a rectangular shape (W12-2a) with the appropriate legend (see Figure 2C-5).

Section 2C.28 BUMP and DIP Signs (W8-1, W8-2)

Guidance:

01 *BUMP (W8-1) and DIP (W8-2) signs (see Figure 2C-6) should be used to give warning of a sharp rise or depression in the profile of the road.*

Option:

02 These signs may be supplemented with an Advisory Speed plaque (see Section 2C.08).

Standard:

03 **The DIP sign shall not be used at a short stretch of depressed alignment that might momentarily hide a vehicle.**

Guidance:

04 *A short stretch of depressed alignment that might momentarily hide a vehicle should be treated as a no-passing zone when center line striping is provided on a two-lane or three-lane road (see Section 3B.02).*

Section 2C.29 SPEED HUMP Sign (W17-1)

Guidance:

01 *The SPEED HUMP (W17-1) sign (see Figure 2C-6) should be used to give warning of a vertical deflection in the roadway that is designed to limit the speed of traffic.*

02 *If used, the SPEED HUMP sign should be supplemented by an Advisory Speed plaque (see Section 2C.08).*

Option:

03 If a series of speed humps exists in close proximity, an Advisory Speed plaque may be eliminated on all but the first SPEED HUMP sign in the series.

04 The legend SPEED BUMP may be used instead of the legend SPEED HUMP on the W17-1 sign.

Support:

05 Speed humps generally provide more gradual vertical deflection than speed bumps. Speed bumps limit the speed of traffic more severely than speed humps. Other forms of speed humps include speed tables and raised intersections. However, these differences in engineering terminology are not well known by the public, so for signing purposes these terms are interchangeable.

Figure 2C-6. Roadway and Weather Condition and Advance Traffic Control Signs and Plaques

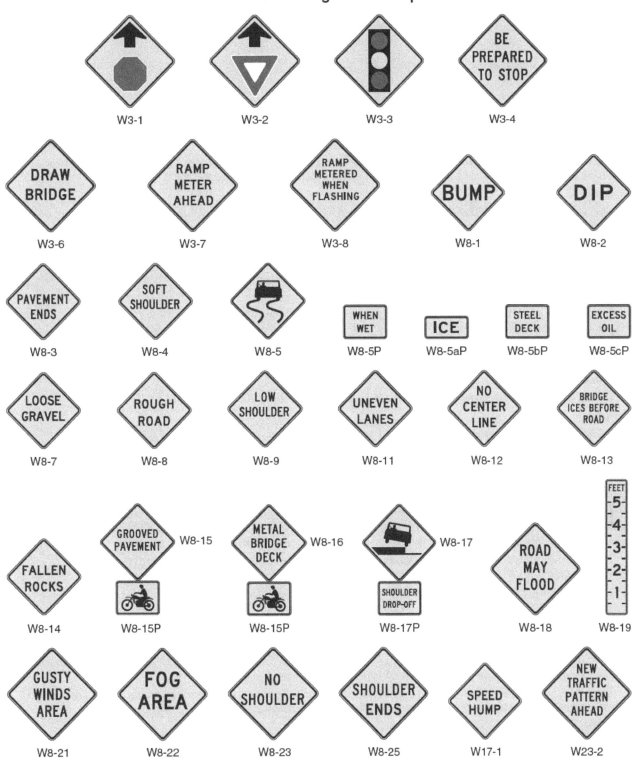

Section 2C.30 PAVEMENT ENDS Sign (W8-3)

Guidance:

01 A PAVEMENT ENDS (W8-3) word message sign (see Figure 2C-6) should be used where a paved surface changes to either a gravel treated surface or an earth road surface.

Option:

02 An Advisory Speed plaque (see Section 2C.08) may be used when the change in roadway condition requires a reduced speed.

Section 2C.31 Shoulder Signs (W8-4, W8-9, W8-17, W8-23, and W8-25)

Option:

01 The SOFT SHOULDER (W8-4) sign (see Figure 2C-6) may be used to warn of a soft shoulder condition.

02 The LOW SHOULDER (W8-9) sign (see Figure 2C-6) may be used to warn of a shoulder condition where there is an elevation difference of less than 3 inches between the shoulder and the travel lane.

Guidance:

03 The Shoulder Drop Off (W8-17) sign (see Figure 2C-6) should be used where an unprotected shoulder drop-off, adjacent to the travel lane, exceeds 3 inches in depth for a significant continuous length along the roadway, based on engineering judgment.

Option:

04 A SHOULDER DROP-OFF (W8-17P) supplemental plaque (see Figure 2C-6) may be mounted below the W8-17 sign.

05 The NO SHOULDER (W8-23) sign (see Figure 2C-6) may be used to warn road users that a shoulder does not exist along a portion of the roadway.

06 The SHOULDER ENDS (W8-25) sign (see Figure 2C-6) may be used to warn road users that a shoulder is ending.

Standard:

07 **When used, shoulder signs shall be placed in advance of the condition (see Table 2C-4).**

Guidance:

08 Additional shoulder signs should be placed at appropriate intervals along the road where the condition continually exists.

Section 2C.32 Surface Condition Signs (W8-5, W8-7, W8-8, W8-11, W8-13, and W8-14)

Option:

01 The Slippery When Wet (W8-5) sign (see Figure 2C-6) may be used to warn of unexpected slippery conditions. Supplemental plaques with legends such as ICE, WHEN WET, STEEL DECK, or EXCESS OIL may be used with the W8-5 sign to indicate the reason that the slippery conditions might be present.

02 The LOOSE GRAVEL (W8-7) sign (see Figure 2C-6) may be used to warn of loose gravel on the roadway surface.

03 The ROUGH ROAD (W8-8) sign (see Figure 2C-6) may be used to warn of a rough roadway surface.

04 An UNEVEN LANES (W8-11) sign (see Figure 2C-6) may be used to warn of a difference in elevation between travel lanes.

05 The BRIDGE ICES BEFORE ROAD (W8-13) sign (see Figure 2C-6) may be used in advance of bridges to advise bridge users of winter weather conditions. The BRIDGE ICES BEFORE ROAD sign may be removed or covered during seasons of the year when its message is not relevant.

06 The FALLEN ROCKS (W8-14) sign (see Figure 2C-6) may be used in advance of an area that is adjacent to a hillside, mountain, or cliff where rocks frequently fall onto the roadway.

Guidance:

07 When used, Surface Condition signs should be placed in advance of the beginning of the affected section (see Table 2C-4), and additional signs should be placed at appropriate intervals along the road where the condition exists.

2009 Edition   Page 123

Section 2C.33 Warning Signs and Plaques for Motorcyclists (W8-15, W8-15P, and W8-16)

Support:

01   The signs and plaques described in this Section are intended to give motorcyclists advance notice of surface conditions that might adversely affect their ability to maintain control of their motorcycle under wet or dry conditions. The use of some of the advance surface condition warning signs described in Section 2C.32, such as Slippery When Wet, LOOSE GRAVEL, or ROUGH ROAD, can also be helpful to motorcyclists if those conditions exist.

Option:

02   If a portion of a street or highway features a roadway pavement surface that is grooved or textured instead of smooth, such as a grooved skid resistance treatment for a horizontal curve or a brick pavement surface, a GROOVED PAVEMENT (W8-15) sign (see Figure 2C-6) may be used to provide advance warning of this condition to motorcyclists, bicyclists, and other road users. Alternate legends such as TEXTURED PAVEMENT or BRICK PAVEMENT may also be used on the W8-15 sign.

03   If a bridge or a portion of a bridge includes a metal or grated surface, a METAL BRIDGE DECK (W8-16) sign (see Figure 2C-6) may be used to provide advance warning of this condition to motorcyclists, bicyclists, and other road users.

04   A Motorcycle (W8-15P) plaque (see Figure 2C-6) may be mounted below or above a W8-15 or W8-16 sign if the warning is intended to be directed primarily to motorcyclists.

Section 2C.34 NO CENTER LINE Sign (W8-12)

Option:

01   The NO CENTER LINE (W8-12) sign (see Figure 2C-6) may be used to warn of a roadway without center line pavement markings.

Section 2C.35 Weather Condition Signs (W8-18, W8-19, W8-21, and W8-22)

Option:

01   The ROAD MAY FLOOD (W8-18) sign (see Figure 2C-6) may be used to warn road users that a section of roadway is subject to frequent flooding. A Depth Gauge (W8-19) sign (see Figure 2C-6) may also be installed within a roadway section that frequently floods.

Standard:

02   **If used, the Depth Gauge sign shall be in addition to the ROAD MAY FLOOD sign and shall indicate the depth of the water at the deepest point on the roadway.**

Option:

03   The GUSTY WINDS AREA (W8-21) sign (see Figure 2C-6) may be used to warn road users that wind gusts frequently occur along a section of highway that are strong enough to impact the stability of trucks, recreational vehicles, and other vehicles with high centers of gravity. A NEXT XX MILES (W7-3a) supplemental plaque may be mounted below the W8-21 sign to inform road users of the length of roadway that frequently experiences strong wind gusts.

04   The FOG AREA (W8-22) sign (see Figure 2C-6) may be used to warn road users that foggy conditions frequently reduce visibility along a section of highway. A NEXT XX MILES (W7-3a) supplemental plaque may be mounted below the W8-22 sign to inform road users of the length of roadway that frequently experiences foggy conditions.

Section 2C.36 Advance Traffic Control Signs (W3-1, W3-2, W3-3, W3-4)

Standard:

01   **The Advance Traffic Control symbol signs (see Figure 2C-6) include the Stop Ahead (W3-1), Yield Ahead (W3-2), and Signal Ahead (W3-3) signs. These signs shall be installed on an approach to a primary traffic control device that is not visible for a sufficient distance to permit the road user to respond to the device (see Table 2C-4). The visibility criteria for a traffic control signal shall be based on having a continuous view of at least two signal faces for the distance specified in Table 4D-2.**

Support:

02   Figure 2A-4 shows the typical placement of an Advance Traffic Control sign.

03   Permanent obstructions causing the limited visibility might include roadway alignment or structures. Intermittent obstructions might include foliage or parked vehicles.

Guidance:

04 *Where intermittent obstructions occur, engineering judgment should determine the treatment to be implemented.*

Option:

05 An Advance Traffic Control sign may be used for additional emphasis of the primary traffic control device, even when the visibility distance to the device is satisfactory.

06 An advance street name plaque (see Section 2C.58) may be installed above or below an Advance Traffic Control sign.

07 A warning beacon may be used with an Advance Traffic Control sign.

08 A BE PREPARED TO STOP (W3-4) sign (see Figure 2C-6) may be used to warn of stopped traffic caused by a traffic control signal or in advance of a section of roadway that regularly experiences traffic congestion.

Standard:

09 **When a BE PREPARED TO STOP sign is used in advance of a traffic control signal, it shall be used in addition to a Signal Ahead sign and shall be placed downstream from the Signal Ahead (W3-3) sign.**

Option:

10 The BE PREPARED TO STOP sign may be supplemented with a warning beacon (see Section 4L.03).

Guidance:

11 *When the warning beacon is interconnected with a traffic control signal or queue detection system, the BE PREPARED TO STOP sign should be supplemented with a WHEN FLASHING (W16-13P) plaque (see Figure 2C-12).*

Support:

12 Section 2C.40 contains information regarding the use of a NO MERGE AREA (W4-5P) supplemental plaque in conjunction with a Yield Ahead sign.

Section 2C.37 Advance Ramp Control Signal Signs (W3-7 and W3-8)

Option:

01 A RAMP METER AHEAD (W3-7) sign (see Figure 2C-6) may be used to warn road users that a freeway entrance ramp is metered and that they will encounter a ramp control signal (see Chapter 4I).

Guidance:

02 *When the ramp control signals are operated only during certain periods of the day, a RAMP METERED WHEN FLASHING (W3-8) sign (see Figure 2C-6) should be installed in advance of the ramp control signal near the entrance to the ramp, or on the arterial on the approach to the ramp, to alert road users to the presence and operation of ramp meters.*

Standard:

03 **The RAMP METERED WHEN FLASHING sign shall be supplemented with a warning beacon (see Section 4L.03) that flashes when the ramp control signal is in operation.**

Section 2C.38 Reduced Speed Limit Ahead Signs (W3-5, W3-5a)

Guidance:

01 *A Reduced Speed Limit Ahead (W3-5 or W3-5a) sign (see Figure 2C-7) should be used to inform road users of a reduced speed zone where the speed limit is being reduced by more than 10 mph, or where engineering judgment indicates the need for advance notice to comply with the posted speed limit ahead.*

Standard:

02 **If used, Reduced Speed Limit Ahead signs shall be followed by a Speed Limit (R2-1) sign installed at the beginning of the zone where the speed limit applies.**

03 **The speed limit displayed on the Reduced Speed Limit Ahead sign shall be identical to the speed limit displayed on the subsequent Speed Limit sign.**

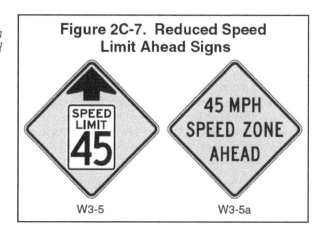

Figure 2C-7. Reduced Speed Limit Ahead Signs

Section 2C.39 DRAW BRIDGE Sign (W3-6)
Standard:

01 A DRAW BRIDGE (W3-6) sign (see Figure 2C-6) shall be used in advance of movable bridge signals and gates (see Section 4J.02) to give warning to road users, except in urban conditions where such signing would not be practical.

Section 2C.40 Merge Signs (W4-1, W4-5)
Option:

01 A Merge (W4-1) sign (see Figure 2C-8) may be used to warn road users on the major roadway that merging movements might be encountered in advance of a point where lanes from two separate roadways converge as a single traffic lane and no turning conflict occurs.

02 A Merge sign may also be installed on the side of the entering roadway to warn road users on the entering roadway of the merge condition.

Guidance:

03 *The Merge sign should be installed on the side of the major roadway where merging traffic will be encountered and in such a position as to not obstruct the road user's view of entering traffic.*

04 *Where two roadways of approximately equal importance converge, a Merge sign should be placed on each roadway.*

05 *When a Merge sign is to be installed on an entering roadway that curves before merging with the major roadway, such as a ramp with a curving horizontal alignment as it approaches the major roadway, the Entering Roadway Merge (W4-5) sign (see Figure 2C-8) should be used to better portray the actual geometric conditions to road users on the entering roadway.*

06 *The Merge sign should not be used where two roadways converge and merging movements are not required.*

07 *The Merge sign should not be used in place of a Lane Ends sign (see Section 2C.42) where lanes of traffic moving on a single roadway must merge because of a reduction in the actual or usable pavement width.*

Option:

08 An Entering Roadway Merge (W4-5) sign with a NO MERGE AREA (W4-5P) supplemental plaque (see Figure 2C-8) mounted below it may be used to warn road users on an entering roadway that they will encounter an abrupt merging situation without an acceleration lane at the downstream end of the ramp.

09 A Merge (W4-1) sign with a NO MERGE AREA (W4-5P) supplemental plaque mounted below it may be used to warn road users on the major roadway that traffic on an entering roadway will encounter an abrupt merging situation without an acceleration lane at the downstream end of the ramp.

Figure 2C-8. Merging and Passing Signs and Plaques

10   For a yield-controlled channelized right-turn movement onto a roadway without an acceleration lane, a NO MERGE AREA (W4-5P) supplemental plaque may be mounted below a Yield Ahead (W3-2) sign and/or below a YIELD (R1-2) sign when engineering judgment indicates that road users would expect an acceleration lane to be present.

Section 2C.41 Added Lane Signs (W4-3, W4-6)

Guidance:

01   *The Added Lane (W4-3) sign (see Figure 2C-8) should be installed in advance of a point where two roadways converge and merging movements are not required. When possible, the Added Lane sign should be placed such that it is visible from both roadways; if this is not possible, an Added Lane sign should be placed on the side of each roadway.*

02   *When an Added Lane sign is to be installed on a roadway that curves before converging with another roadway that has a tangent alignment at the point of convergence, the Entering Roadway Added Lane (W4-6) sign (see Figure 2C-8) should be used to better portray the actual geometric conditions to road users on the curving roadway.*

Section 2C.42 Lane Ends Signs (W4-2, W9-1, W9-2)

Guidance:

01   *The LANE ENDS MERGE LEFT (RIGHT) (W9-2) sign or the Lane Ends (W4-2) sign should be used to warn of the reduction in the number of traffic lanes in the direction of travel on a multi-lane highway (see Figure 2C-8).*

Option:

02   The RIGHT (LEFT) LANE ENDS (W9-1) sign (see Figure 2C-8) may be used in advance of the Lane Ends (W4-2) sign or the LANE ENDS MERGE LEFT (RIGHT) (W9-2) sign as additional warning or to emphasize that the traffic lane is ending and that a merging maneuver will be required.

Guidance:

03   *If used, the RIGHT (LEFT) LANE ENDS (W9-1) sign should be installed adjacent to the Lane-Reduction Arrow pavement markings.*

Option:

04   On one-way streets or on divided highways where the width of the median will permit, two Lane Ends signs may be placed facing approaching traffic, one on the right-hand side and the other on the left-hand side or median.

Support:

05   Section 3B.09 contains information regarding the use of pavement markings in conjunction with a lane reduction.

Guidance:

06   *Where an extra lane has been provided for slower moving traffic (see Section 2B.31), a Lane Ends word sign or a Lane Ends (W4-2) symbol sign should be installed in advance of the downstream end of the extra lane.*

07   *Lane Ends signs should not be installed in advance of the downstream end of an acceleration lane.*

Standard:

08   **In dropped lane situations, regulatory signs (see Section 2B.20) shall be used to inform road users that a through lane is becoming a mandatory turn lane. The W4-2, W9-1, and W9-2 signs shall not be used in dropped lane situations.**

Section 2C.43 RIGHT (LEFT) LANE EXIT ONLY AHEAD Sign (W9-7)

Option:

01   The RIGHT (LEFT) LANE EXIT ONLY AHEAD (W9-7) sign (see Figure 2C-8) may be used to provide advance warning to road users that traffic in the right-hand (left-hand) lane of a roadway that is approaching a grade-separated interchange will be required to depart the roadway on an exit ramp at the next interchange.

Standard:

02   **The W9-7 sign shall be a horizontal rectangle with a black legend and border on a yellow background.**

Guidance:

03   *If used, the W9-7 sign should be installed upstream from the first overhead guide sign that contains an EXIT ONLY sign panel or upstream from the first RIGHT (LEFT) LANE MUST EXIT (R3-33) regulatory sign, whichever is farther upstream from the exit.*

2009 Edition   Page 127

Support:

04   Section 2B.23 contains information regarding a regulatory sign that can also be used for lane drops at grade-separated interchanges.

Section 2C.44 Two-Way Traffic Sign (W6-3)

Guidance:

01   *A Two-Way Traffic (W6-3) sign (see Figure 2C-8) should be used to warn road users of a transition from a multi-lane divided section of roadway to a two-lane, two-way section of roadway.*

02   *A Two-Way Traffic (W6-3) sign with an AHEAD (W16-9P) plaque (see Figure 2C-12) should be used to warn road users of a transition from a one-way street to a two-lane, two-way section of roadway (see Figure 2B-14).*

Option:

03   The Two-Way Traffic sign may be used at intervals along a two-lane, two-way roadway and may be used to supplement the Divided Highway (Road) Ends (W6-2) sign discussed in Section 2C.23.

Section 2C.45 NO PASSING ZONE Sign (W14-3)

Standard:

01   **The NO PASSING ZONE (W14-3) sign (see Figure 2C-8) shall be a pennant-shaped isosceles triangle with its longer axis horizontal and pointing to the right. When used, the NO PASSING ZONE sign shall be installed on the left side of the roadway at the beginning of no-passing zones identified by pavement markings or DO NOT PASS signs or both (see Sections 2B.28 and 3B.02).**

Section 2C.46 Intersection Warning Signs (W2-1 through W2-8)

Option:

01   A Cross Road (W2-1) symbol, Side Road (W2-2 or W2-3) symbol, T-Symbol (W2-4), or Y-Symbol (W2-5) sign (see Figure 2C-9) may be used in advance of an intersection to indicate the presence of an intersection and the possibility of turning or entering traffic.

Figure 2C-9. Intersection Warning Signs and Plaques

W1-7   W2-1   W2-2   W2-3   W2-4

W2-5   W2-6 W16-17P (optional) / W16-12P (optional)   W2-7L   W2-7R

W2-8   W4-4P   W4-4aP   W4-4bP   W25-1   W25-2

02 The Circular Intersection (W2-6) symbol sign (see Figure 2C-9) may be installed in advance of a circular intersection (see Figures 2B-21 through 2B-23).

Guidance:

03 *If an approach to a roundabout has a statutory or posted speed limit of 40 mph or higher, the Circular Intersection (W2-6) symbol sign should be installed in advance of the circular intersection.*

Option:

04 An educational plaque (see Figure 2C-9) with a legend such as ROUNDABOUT (W16-17P) or TRAFFIC CIRCLE (W16-12P) may be mounted below a Circular Intersection symbol sign.

05 The relative importance of the intersecting roadways may be shown by different widths of lines in the symbol.

06 An advance street name plaque (see Section 2C.58) may be installed above or below an Intersection Warning sign.

Guidance:

07 *The Intersection Warning sign should illustrate and depict the general configuration of the intersecting roadway, such as cross road, side road, T-intersection, or Y-intersection.*

08 *Intersection Warning signs, other than the Circular Intersection (W2-6) symbol sign and the T-intersection (W2-4) symbol sign should not be used on approaches controlled by STOP signs, YIELD signs, or signals.*

09 *If an Intersection Warning sign is used where the side roads are not opposite of each other, the Offset Side Roads (W2-7) symbol sign (see Figure 2C-9) should be used instead of the Cross Road symbol sign.*

10 *If an Intersection Warning sign is used where two closely-spaced side roads are on the same side of the highway, the Double Side Roads (W2-8) symbol sign (see Figure 2C-9) should be used instead of the Side Road symbol sign.*

11 *No more than two side road symbols should be displayed on the same side of the highway on a W2-7 or W2-8 symbol sign, and no more than three side road symbols should be displayed on a W2-7 or W2-8 symbol sign.*

Support:

12 Figure 2A-4 shows the typical placement of an Intersection Warning sign.

Section 2C.47 Two-Direction Large Arrow Sign (W1-7)

Standard:

01 **The Two-Direction Large Arrow (W1-7) sign (see Figure 2C-9) shall be a horizontal rectangle.**

02 **If used, it shall be installed on the far side of a T-intersection in line with, and at approximately a right angle to, traffic approaching from the stem of the T-intersection.**

03 **The Two-Direction Large Arrow sign shall not be used where there is no change in the direction of travel such as at the beginnings and ends of medians or at center piers.**

04 **The Two-Direction Large Arrow sign directing traffic to the left and right shall not be used in the central island of a roundabout.**

Guidance:

05 *The Two-Direction Large Arrow sign should be visible for a sufficient distance to provide the road user with adequate time to react to the intersection configuration.*

Section 2C.48 Traffic Signal Signs (W25-1, W25-2)

Standard:

01 **At locations where either a W25-1 or a W25-2 sign is required based on the provisions in Section 4D.05, the W25-1 or W25-2 sign (see Figure 2C-9) shall be installed near the left-most signal head. The W25-1 and W25-2 signs shall be vertical rectangles.**

Section 2C.49 Vehicular Traffic Warning Signs (W8-6, W11-1, W11-5, W11-5a, W11-8, W11-10, W11-11, W11-12P, W11-14, W11-15, and W11-15a)

Option:

01 Vehicular Traffic Warning (W8-6, W11-1, W11-5, W11-5a, W11-8, W11-10, W11-11, W11-12P, W11-14, W11-15, and W11-15a) signs (see Figure 2C-10) may be used to alert road users to locations where unexpected entries into the roadway by trucks, bicyclists, farm vehicles, emergency vehicles, golf carts, horse-drawn vehicles, or other vehicles might occur. The TRUCK CROSSING (W8-6) word message sign may be used as an alternate to the Truck Crossing (W11-10) symbol sign.

Figure 2C-10. Vehicular Traffic Warning Signs and Plaques

* A fluorescent yellow-green background color may be used for this sign or plaque.

Support:
01 These locations might be relatively confined or might occur randomly over a segment of roadway.

Guidance:
02 *Vehicular Traffic Warning signs should be used only at locations where the road user's sight distance is restricted, or the condition, activity, or entering traffic would be unexpected.*

03 *If the condition or activity is seasonal or temporary, the Vehicular Traffic Warning sign should be removed or covered when the condition or activity does not exist.*

Option:
05 The combined Bicycle/Pedestrian (W11-15) sign may be used where both bicyclists and pedestrians might be crossing the roadway, such as at an intersection with a shared-use path. A TRAIL X-ING (W11-15P) supplemental plaque (see Figure 2C-10) may be mounted below the W11-15 sign. The TRAIL CROSSING (W11-15a) sign may be used to warn of shared-use path crossings where pedestrians, bicyclists, and other user groups might be crossing the roadway.

06 The W11-1, W11-15, and W11-15a signs and their related supplemental plaques may have a fluorescent yellow-green background with a black legend and border.

07 Supplemental plaques (see Section 2C.53) with legends such as AHEAD, XX FEET, NEXT XX MILES, or SHARE THE ROAD may be mounted below Vehicular Traffic Warning signs to provide advance notice to road users of unexpected entries.

Guidance:
08 *If used in advance of a pedestrian and bicycle crossing, a W11-15 or W11-15a sign should be supplemented with an AHEAD or XX FEET plaque to inform road users that they are approaching a point where crossing activity might occur.*

Standard:
09 **If a post-mounted W11-1, W11-11, W11-15, or W11-15a sign is placed at the location of the crossing point where golf carts, pedestrians, bicyclists, or other shared-use path users might be crossing the roadway, a diagonal downward pointing arrow (W16-7P) plaque (see Figure 2C-12) shall be mounted below the sign. If the W11-1, W11-11, W11-15, or W11-15a sign is mounted overhead, the W16-7P supplemental plaque shall not be used.**

Option:
10 The crossing location identified by a W11-1, W11-11, W11-15, or W11-15a sign may be defined with crosswalk markings (see Section 3B.18).

Standard:

11 **The Emergency Vehicle (W11-8) sign (see Figure 2C-10) with the EMERGENCY SIGNAL AHEAD (W11-12P) supplemental plaque (see Figure 2C-10) shall be placed in advance of all emergency-vehicle traffic control signals (see Chapter 4G).**

Option:

12 The Emergency Vehicle (W11-8) sign, or a word message sign indicating the type of emergency vehicle (such as rescue squad), may be used in advance of the emergency-vehicle station when no emergency-vehicle traffic control signal is present.

13 A Warning Beacon (see Section 4L.03) may be used with any Vehicular Traffic Warning sign to indicate specific periods when the condition or activity is present or is likely to be present, or to provide enhanced sign conspicuity.

14 A supplemental WHEN FLASHING (W16-13P) plaque (see Figure 2C-12) may be used with any Vehicular Traffic Warning sign that is supplemented with a Warning Beacon to indicate specific periods when the condition or activity is present or is likely to be present.

Section 2C.50 Non-Vehicular Warning Signs (W11-2, W11-3, W11-4, W11-6, W11-7, W11-9, and W11-16 through W11-22)

Option:

01 Non-Vehicular Warning (W11-2, W11-3, W11-4, W11-6, W11-7, W11-9, and W11-16 through W11-22) signs (see Figure 2C-11) may be used to alert road users in advance of locations where unexpected entries into the roadway might occur or where shared use of the roadway by pedestrians, animals, or equestrians might occur.

Support:

02 These conflicts might be relatively confined, or might occur randomly over a segment of roadway.

Guidance:

03 *If used in advance of a pedestrian, snowmobile, or equestrian crossing, the W11-2, W11-6, W11-7, and W11-9 signs should be supplemented with plaques (see Section 2C.55) with the legend AHEAD or XX FEET to inform road users that they are approaching a point where crossing activity might occur.*

Figure 2C-11. Non-Vehicular Warning Signs

* A fluorescent yellow-green background color may be used for this sign or plaque.

Standard:

04 **If a post-mounted W11-2, W11-6, W11-7, or W11-9 sign is placed at the location of the crossing point where pedestrians, snowmobilers, or equestrians might be crossing the roadway, a diagonal downward pointing arrow (W16-7P) plaque (see Figure 2C-12) shall be mounted below the sign. If the W11-2, W11-6, W11-7, or W11-9 sign is mounted overhead, the W16-7P plaque shall not be used.**

Option:

05 A Pedestrian Crossing (W11-2) sign may be placed overhead or may be post-mounted with a diagonal downward pointing arrow (W16-7P) plaque at the crosswalk location where Yield Here To (Stop Here For) Pedestrians signs (see Section 2B.11) have been installed in advance of the crosswalk.

Standard:

06 **If a W11-2 sign has been post-mounted at the crosswalk location where a Yield Here To (Stop Here For) Pedestrians sign is used on the approach, the Yield Here To (Stop Here For) Pedestrians sign shall not be placed on the same post as or block the road user's view of the W11-2 sign.**

Option:

07 An advance Pedestrian Crossing (W11-2) sign with an AHEAD or a distance supplemental plaque may be used in conjunction with a Yield Here To (Stop Here For) Pedestrians sign on the approach to the same crosswalk.

08 The crossing location identified by a W11-2, W11-6, W11-7, or W11-9 sign may be defined with crosswalk markings (see Section 3B.18).

09 The W11-2 and W11-9 signs and their related supplemental plaques may have a fluorescent yellow-green background with a black legend and border.

Guidance:

10 *When a fluorescent yellow-green background is used, a systematic approach featuring one background color within a zone or area should be used. The mixing of standard yellow and fluorescent yellow-green backgrounds within a selected site area should be avoided.*

Option:

11 A Warning Beacon (see Section 4L.03) may be used with any Non-Vehicular Warning sign to indicate specific periods when the condition or activity is present or is likely to be present, or to provide enhanced sign conspicuity.

12 A supplemental WHEN FLASHING (W16-13P) plaque (see Figure 2C-12) may be used with any Non-Vehicular Warning sign that is supplemented with a Warning Beacon to indicate specific periods when the condition or activity is present or is likely to be present.

Section 2C.51 Playground Sign (W15-1)

Option:

01 The Playground (W15-1) sign (see Figure 2C-11) may be used to give advance warning of a designated children's playground that is located adjacent to the road.

02 The Playground sign may have a fluorescent yellow-green background with a black legend and border.

Guidance:

03 *If the access to the playground area requires a roadway crossing, the application of crosswalk pavement markings (see Section 3B.18) and Non-Vehicular Warning signs (see Section 2C.50) should be considered.*

Section 2C.52 NEW TRAFFIC PATTERN AHEAD Sign (W23-2)

Option:

01 A NEW TRAFFIC PATTERN AHEAD (W23-2) sign (see Figure 2C-6) may be used on the approach to an intersection or along a section of roadway to provide advance warning of a change in traffic patterns, such as revised lane usage, roadway geometry, or signal phasing.

Guidance:

02 *The NEW TRAFFIC PATTERN AHEAD sign should be removed when the traffic pattern returns to normal, when the changed pattern is no longer considered to be new, or within six months.*

Section 2C.53 Use of Supplemental Warning Plaques

Option:

01 A supplemental warning plaque (see Figure 2C-12) may be displayed with a warning or regulatory sign when engineering judgment indicates that road users require additional warning information beyond that contained in the main message of the warning or regulatory sign.

Figure 2C-12. Supplemental Warning Plaques

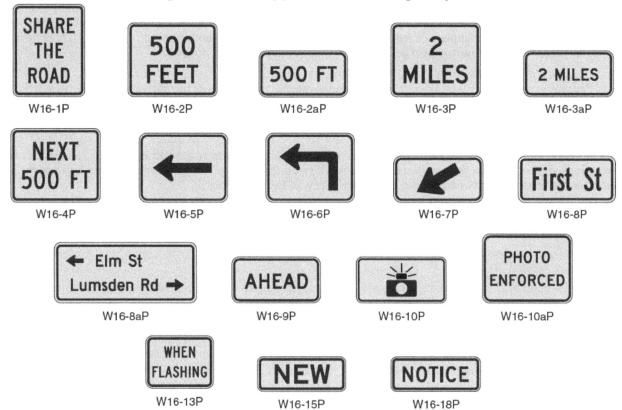

Note: The background color (yellow or fluorescent yellow-green) shall match the color of the warning sign that it supplements.

Standard:

02 Supplemental warning plaques shall be used only in combination with warning or regulatory signs. They shall not be mounted alone or displayed alone. If used, a supplemental warning plaque shall be installed on the same post(s) as the warning or regulatory sign that it supplements.

03 Unless otherwise provided in this Manual for a particular plaque, supplemental warning plaques shall be mounted below the sign they supplement.

Section 2C.54 Design of Supplemental Warning Plaques

Standard:

01 A supplemental warning plaque used with a warning sign shall have the same legend, border, and background color as the warning sign with which it is displayed. A supplemental warning plaque used with a regulatory sign shall have a black legend and border on a yellow background.

02 Supplemental warning plaques shall be square or rectangular.

Section 2C.55 Distance Plaques (W16-2 Series, W16-3 Series, W16-4P, W7-3aP)

Option:

01 The Distance Ahead (W16-2 series and W16-3 series) plaques (see Figure 2C-12) may be used to inform the road user of the distance to the condition indicated by the warning sign.

02 The Next Distance (W7-3aP and W16-4P) plaques (see Figures 2C-4 and 2C-12) may be used to inform road users of the length of roadway over which the condition indicated by the warning sign exists.

Section 2C.56 Supplemental Arrow Plaques (W16-5P, W16-6P)

Guidance:

01 *If the condition indicated by a warning sign is located on an intersecting road and the distance between the intersection and condition is not sufficient to provide adequate advance placement of the warning sign, a Supplemental Arrow (W16-5P or W16-6P) plaque (see Figure 2C-12) should be used below the warning sign.*

Standard:

02 **Supplemental Arrow plaques shall have the same legend design as the Advance Turn Arrow and Directional Arrow auxiliary signs (see Sections 2D.26 and 2D.28) except that they shall have a black legend and border on a yellow or fluorescent yellow-green background, as appropriate.**

Section 2C.57 Hill-Related Plaques (W7-2 Series, W7-3 Series)

Guidance:

01 *Hill-Related (W7-2 series, W7-3 series) plaques (see Figure 2C-4) or other appropriate legends and larger signs should be used for emphasis or where special hill characteristics exist.*

02 *On longer grades, the use of the distance plaque (W7-3aP or W7-3bP) at periodic intervals of approximately 1-mile spacing should be considered.*

Section 2C.58 Advance Street Name Plaque (W16-8P, W16-8aP)

Option:

01 An Advance Street Name (W16-8P or W16-8aP) plaque (see Figure 2C-12) may be used with any Intersection sign (W2 series, W10-2, W10-3, or W10-4) or Advance Traffic Control (W3 series) sign to identify the name of the intersecting street.

Standard:

02 **The lettering on Advance Street Name plaques shall be composed of a combination of lower-case letters with initial upper-case letters.**

03 **If two street names are used on the Advance Street Name plaque, a directional arrow pointing in the direction of the street shall be placed next to each street name. Arrows pointing to the left shall be placed to the left of the street name, and arrows pointing to the right shall be placed to the right of the street name.**

Guidance:

04 *If two street names are used on the Advance Street Name plaque, the street names and associated arrows should be displayed in the following order:*
   A. For a single intersection, the name of the street to the left should be displayed above the name of the street to the right; or
   B. For two sequential intersections, such as where the plaque is used with an Offset Side Roads (W2-7) or a Double Side Road (W2-8) symbol sign, the name of the first street encountered should be displayed above the name of the second street encountered, and the arrow associated with the second street encountered should be an advance arrow, such as the arrow shown on the W16-6P arrow plaque (see Figure 2C-12).

Section 2C.59 CROSS TRAFFIC DOES NOT STOP Plaque (W4-4P)

Option:

01 The CROSS TRAFFIC DOES NOT STOP (W4-4P) plaque (see Figure 2C-9) may be used in combination with a STOP sign when engineering judgment indicates that conditions are present that are causing or could cause drivers to misinterpret the intersection as an all-way stop.

02 Alternative messages (see Figure 2C-9) such as TRAFFIC FROM LEFT (RIGHT) DOES NOT STOP (W4-4aP) or ONCOMING TRAFFIC DOES NOT STOP (W4-4bP) may be used when such messages more accurately describe the traffic controls established at the intersection.

Guidance:

03 *Plaques with the appropriate alternative messages of TRAFFIC FROM LEFT (RIGHT) DOES NOT STOP or ONCOMING TRAFFIC DOES NOT STOP should be used at intersections where STOP signs control all but one approach to the intersection, unless the only non-stopped approach is from a one-way street.*

Standard:

04 **If a W4-4P plaque or a plaque with an alternative message is used, it shall be mounted below the STOP sign.**

Section 2C.60 SHARE THE ROAD Plaque (W16-1P)

Option:

01 In situations where there is a need to warn drivers to watch for other slower forms of transportation traveling along the highway, such as bicycles, golf carts, horse-drawn vehicles, or farm machinery, a SHARE THE ROAD (W16-1P) plaque (see Figure 2C-12) may be used.

Standard:

02 A W16-1P plaque shall not be used alone. If a W16-1P plaque is used, it shall be mounted below either a Vehicular Traffic Warning sign (see Section 2C.49) or a Non-Vehicular Warning sign (see Section 2C.50). The background color of the W16-1P plaque shall match the background color of the warning sign with which it is displayed.

Section 2C.61 Photo Enforced Plaque (W16-10P)

Option:

01 A Photo Enforced (W16-10P) plaque or a PHOTO ENFORCED (W16-10aP) word message plaque (see Figure 2C-12) may be mounted below a warning sign to advise road users that the regulations associated with the condition being warned about (such as a traffic control signal or a toll plaza) are being enforced by photographic equipment.

Standard:

02 If used below a warning sign, the Photo Enforced (W16-10P or W16-10aP) plaque shall be a rectangle with a black legend and border on a yellow background.

Section 2C.62 NEW Plaque (W16-15P)

Option:

01 A NEW (W16-15P) plaque (see Figure 2C-12) may be mounted above a regulatory sign when a new regulation takes effect in order to alert road users to the new traffic regulation. A NEW plaque may also be mounted above an advance warning sign (such as a Signal Ahead sign for a newly-installed traffic control signal) for a new traffic regulation.

Standard:

02 The NEW plaque shall not be used alone.

03 The NEW plaque shall be removed no later than 6 months after the regulation has been in effect.

Section 2C.63 Object Marker Design and Placement Height

Support:

01 Type 1, 2, and 3 object markers are used to mark obstructions within or adjacent to the roadway. Type 4 object markers are used to mark the end of a roadway.

Standard:

02 When used, object markers (see Figure 2C-13) shall not have a border and shall consist of an arrangement of one or more of the following types:

Type 1—a diamond-shaped sign, at least 18 inches on a side, consisting of either a yellow (OM1-1) or black (OM1-2) sign with nine yellow retroreflective devices, each with a minimum diameter of 3 inches, mounted symmetrically on the sign, or an all-yellow retroreflective sign (OM1-3).

Type 2—either a marker (OM2-1V or OM2-1H) consisting of three yellow retroreflective devices, each with a minimum diameter of 3 inches, arranged either horizontally or vertically on a white sign measuring at least 6 x 12 inches; or an all-yellow horizontal or vertical retroreflective sign (OM2-2V or OM2-2H), measuring at least 6 x 12 inches.

Type 3—a striped marker, 12 x 36 inches, consisting of a vertical rectangle with alternating black and retroreflective yellow stripes sloping downward at an angle of 45 degrees toward the side of the obstruction on which traffic is to pass. The minimum width of the yellow and black stripes shall be 3 inches.

Type 4—a diamond-shaped sign, at least 18 inches on a side, consisting of either a red (OM4-1) or black (OM4-2) sign with nine red retroreflective devices, each with a minimum diameter of 3 inches, mounted symmetrically on the sign, or an all-red retroreflective sign (OM4-3).

Support:

03 A better appearance can be achieved if the black stripes are wider than the yellow stripes.

04 Type 3 object markers with stripes that begin at the upper right side and slope downward to the lower left side are designated as right object markers (OM3-R). Object markers with stripes that begin at the upper left side and slope downward to the lower right side are designated as left object markers (OM3-L).

Guidance:

05 *When used for marking obstructions within the roadway or obstructions that are 8 feet or less from the shoulder or curb, the minimum mounting height, measured from the bottom of the object marker to the elevation*

of the near edge of the traveled way, should be 4 feet.

06 *When used to mark obstructions more than 8 feet from the shoulder or curb, the clearance from the ground to the bottom of the object marker should be at least 4 feet.*

07 *Object markers should not present a vertical or horizontal clearance obstacle for pedestrians.*

Option:

08 When object markers or markings are applied to an obstruction that by its nature requires a lower or higher mounting, the vertical mounting height may vary according to need.

Support:

09 Section 9B.26 contains information regarding the use of object markers on shared-use paths.

Section 2C.64 Object Markers for Obstructions Within the Roadway

Standard:

01 **Obstructions within the roadway shall be marked with a Type 1 or Type 3 object marker. In addition to markers on the face of the obstruction, warning of approach to the obstruction shall be given by appropriate pavement markings (see Section 3B.10).**

Option:

02 To provide additional emphasis, a Type 1 or Type 3 object marker may be installed at or near the approach end of a median island.

03 To provide additional emphasis, large surfaces such as bridge piers may be painted with diagonal stripes, 12 inches or greater in width, similar in design to the Type 3 object marker.

Standard:

04 **The alternating black and retroreflective yellow stripes (OM3-L, OM3-R) shall be sloped down at an angle of 45 degrees toward the side on which traffic is to pass the obstruction. If traffic can pass to either side of the obstruction, the alternating black and retroreflective yellow stripes (OM3-C) shall form chevrons that point upwards.**

Option:

05 Appropriate signs (see Sections 2B.32 and 2C.25) directing traffic to one or both sides of the obstruction may be used instead of the object marker.

Figure 2C-13. Object Markers

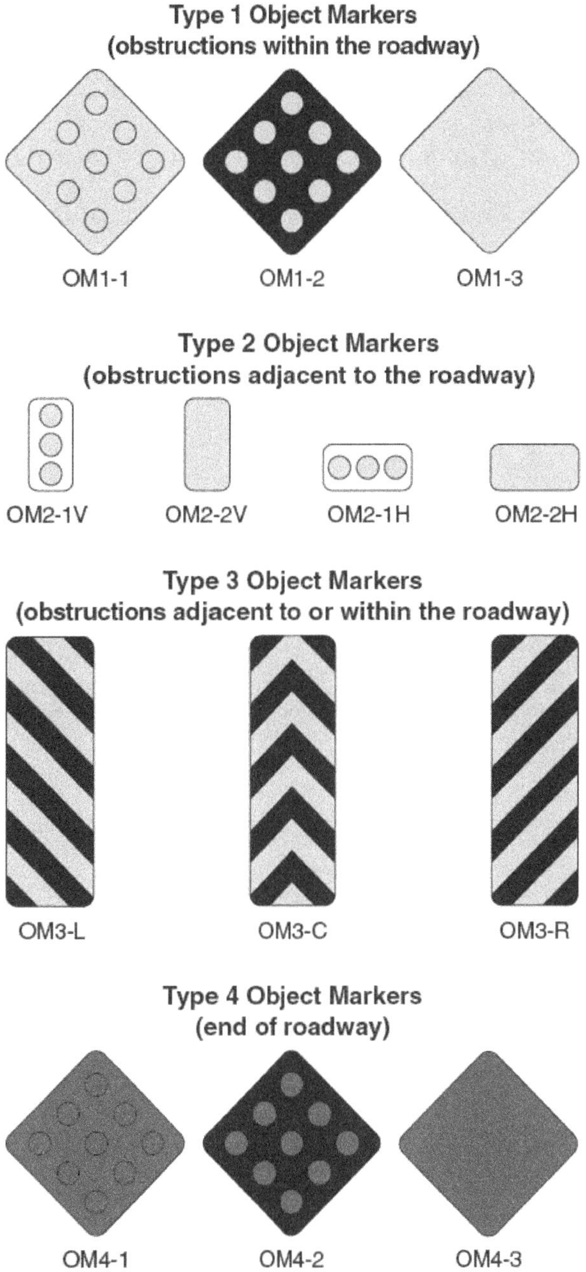

Type 1 Object Markers
(obstructions within the roadway)

OM1-1   OM1-2   OM1-3

Type 2 Object Markers
(obstructions adjacent to the roadway)

OM2-1V   OM2-2V   OM2-1H   OM2-2H

Type 3 Object Markers
(obstructions adjacent to or within the roadway)

OM3-L   OM3-C   OM3-R

Type 4 Object Markers
(end of roadway)

OM4-1   OM4-2   OM4-3

Section 2C.65 Object Markers for Obstructions Adjacent to the Roadway

Support:

01 Obstructions not actually within the roadway are sometimes so close to the edge of the road that they need a marker. These include underpass piers, bridge abutments, handrails, ends of traffic barriers, utility poles, and culvert headwalls. In other cases there might not be a physical object involved, but other roadside conditions exist, such as narrow shoulders, drop-offs, gores, small islands, and abrupt changes in the roadway alignment, that might make it undesirable for a road user to leave the roadway, and therefore would create a need for a marker.

Standard:

02 **If a Type 2 or Type 3 object marker is used to mark an obstruction adjacent to the roadway, the edge of the object marker that is closest to the road user shall be installed in line with the closest edge of the obstruction.**

03 **Where Type 3 object markers are applied to the approach ends of guardrail and other roadside appurtances, sheeting without a substrate shall be directly affixed to the approach end of the guardrail in a rectangular shape conforming to the size of the approach end of the guardrail with alternating black and retroreflective yellow stripes sloping downward at a angle of 45 degrees toward the side of the obstruction on which traffic is to pass.**

04 **Type 1 and Type 4 object markers shall not be used to mark obstructions adjacent to the roadway.**

Guidance:

05 *Standard warning signs in this Chapter should also be used where applicable.*

Section 2C.66 Object Markers for Ends of Roadways

Support:

01 The Type 4 object marker is used to warn and alert road users of the end of a roadway in other than construction or maintenance areas.

Standard:

02 **If an object marker is used to mark the end of a roadway, a Type 4 object marker shall be used.**

Option:

03 The Type 4 object marker may be used in instances where there are no alternate vehicular paths.

04 Where conditions warrant, more than one marker, or a larger marker with or without a Type 3 Barricade (see Section 2B.67), may be used at the end of the roadway.

Standard:

05 **The minimum mounting height, measured vertically from the bottom of a Type 4 object marker to the elevation of the near edge of the traveled way, shall be 4 feet.**

Guidance:

06 *Appropriate advance warning signs in this Chapter should be used.*

CHAPTER 2D. GUIDE SIGNS—CONVENTIONAL ROADS

Section 2D.01 Scope of Conventional Road Guide Sign Standards

Standard:

01 The provisions of this Chapter shall apply to any road or street other than low-volume roads (as defined in Section 5A.01), expressways, and freeways.

Section 2D.02 Application

Support:

01 Guide signs are essential to direct road users along streets and highways, to inform them of intersecting routes, to direct them to cities, towns, villages, or other important destinations, to identify nearby rivers and streams, parks, forests, and historical sites, and generally to give such information as will help them along their way in the most simple, direct manner possible.

02 Chapter 2A addresses placement, location, and other general criteria for signs.

Section 2D.03 Color, Retroreflection, and Illumination

Support:

01 Requirements for illumination, retroreflection, and color are stated under the specific headings for individual guide signs or groups of signs. General provisions are given in Sections 2A.07, 2A.08, and 2A.10.

Standard:

02 **Except where otherwise provided in this Manual for individual signs or groups of signs, guide signs on streets and highways shall have a white message and border on a green background. All messages, borders, and legends shall be retroreflective and all backgrounds shall be retroreflective or illuminated.**

Support:

03 Color coding is sometimes used to help road users distinguish between multiple potentially confusing destinations. Examples of valuable uses of color coding include guide signs for roadways approaching or inside an airport property with multiple terminals serving multiple airlines, and community wayfinding guide signs for various traffic generator destinations within a community or area.

Standard:

04 **Except where otherwise provided in this Manual, different color sign backgrounds shall not be used to provide color coding of destinations. The color coding shall be accomplished by the use of different colored square or rectangular sign panels on the face of the guide signs.**

Option:

05 The different colored sign panels may include a black or white (whichever provides the better contrast with the panel color) letter, numeral, or other appropriate designation to identify an airport terminal or other destination.

Support:

06 Two examples of color-coded sign assemblies are shown in Figure 2D-1. Section 2D.50 contains specific provisions regarding Community Wayfinding guide signs.

Section 2D.04 Size of Signs

Standard:

01 **Except as provided in Section 2A.11, the sizes of conventional road guide signs that have standardized designs shall be as shown in Table 2D-1.**

Support:

02 Section 2A.11 contains information regarding the applicability of the various columns in Table 2D-1.

Option:

03 Signs larger than those shown in Table 2D-1 may be used (see Section 2A.11).

Support:

04 For other guide signs, the legends are so variable that a standardized design or size is not appropriate. The sign size is determined primarily by the length of the message, and the size of lettering and spacing necessary for proper legibility.

Option:

05 Reduced letter height, reduced interline spacing, and reduced edge spacing may be used on guide signs if sign size must be limited by factors such as lane width or vertical or lateral clearance.

Figure 2D-1. Examples of Color-Coded Destination Guide Signs

A - Freeway or Expressway – Airport Terminals

B - Conventional Road or Street – Urban Areas

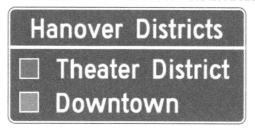

Guidance:

06 *Reduced spacing between the letters or words on a line of legend should not be used as a means of reducing the overall size of a guide sign, except where determined necessary by engineering judgment to meet unusual lateral space constraints. In such cases, the legibility distance of the sign legend should be the primary consideration in determining whether to reduce the spacing between the letters or the words or between the words and the sign border, or to reduce the letter height.*

07 *When a reduction in the prescribed size is necessary, the design used should be as similar as possible to the design for the standard size.*

Section 2D.05 Lettering Style
Standard:

01 The design of upper-case letters, lower-case letters, numerals, route shields, and spacing shall be as provided in the "Standard Highway Signs and Markings" book (see Section 1A.11).

02 The lettering for names of places, streets, and highways on conventional road guide signs shall be a combination of lower-case letters with initial upper-case letters (see Section 2A.13). The nominal loop height of the lower-case letters shall be 3/4 the height of the initial upper-case letter. When a mixed-case legend letter height is specified referring only to the initial upper-case letter, the height of the lower-case letters that follow shall be determined by this proportion. When the height of a lower-case letter is referenced, the reference is made to the nominal loop height and the height of the initial upper-case letter shall also be determined by this proportion.

03 All other word legends on conventional road guide signs shall be in upper-case letters.

04 The unique letter forms for each of the Standard Alphabet series shall not be stretched, compressed, warped, or otherwise manipulated. Modifications to the length of a word for a given letter height and series shall be accomplished only by the methods described in Section 2D.04.

Section 2D.06 Size of Lettering
Support:

01 Sign legibility is a direct function of letter size and spacing. Legibility distance has to be sufficient to give road users enough time to read and comprehend the sign. Under optimum conditions, a guide sign message can be read and understood in a brief glance. The legibility distance takes into account factors such as inattention, blocking of view by other vehicles, unfavorable weather, inferior eyesight, or other causes for delayed or slow reading. Where conditions permit, repetition of guide information on successive signs gives the road user more than one opportunity to obtain the information needed.

Table 2D-1. Conventional Road Guide Sign Sizes

Sign	Sign Designation	Section	Conventional Road	Minimum	Oversized
Interstate Route Sign (1 or 2 digits)	M1-1	2D.11	24 x 24	24 x 24	36 x 36
Interstate Route Sign (3 digits)	M1-1	2D.11	30 x 24	30 x 24	45 x 36
Off-Interstate Route Sign (1 or 2 digits)	M1-2,3	2D.11	24 x 24	24 x 24	36 x 36
Off-Interstate Route Sign (3 digits)	M1-2,3	2D.11	30 x 24	30 x 24	45 x 36
U.S. Route Sign (1 or 2 digits)	M1-4	2D.11	24 x 24	24 x 24	36 x 36
U.S. Route Sign (3 digits)	M1-4	2D.11	30 x 24	30 x 24	45 x 36
State Route Sign (1 or 2 digits)	M1-5	2D.11	24 x 24	24 x 24	36 x 36
State Route Sign (3 digits)	M1-5	2D.11	30 x 24	30 x 24	45 x 36
County Route Sign (1, 2, or 3 digits)	M1-6	2D.11	24 x 24	24 x 24	36 x 36
Forest Route (1, 2, or 3 digits)	M1-7	2D.11	24 x 24	18 x 18	36 x 36
Junction	M2-1	2D.13	21 x 15	21 x 15	30 x 21
Combination Junction (2 route signs)	M2-2	2D.14	60 x 48*	—	—
Cardinal Direction	M3-1,2,3,4	2D.15	24 x 12	24 x 12	36 x 18
Alternate	M4-1,1a	2D.17	24 x 12	24 x 12	36 x 18
By-Pass	M4-2	2D.18	24 x 12	24 x 12	36 x 18
Business	M4-3	2D.19	24 x 12	24 x 12	36 x 18
Truck	M4-4	2D.20	24 x 12	24 x 12	36 x 18
To	M4-5	2D.21	24 x 12	24 x 12	36 x 18
End	M4-6	2D.22	24 x 12	24 x 12	36 x 18
Temporary	M4-7,7a	2D.24	24 x 12	24 x 12	36 x 18
Begin	M4-14	2D.23	24 x 12	24 x 12	36 x 18
Advance Turn Arrow	M5-1,2,3	2D.28	21 x 15	21 x 15	—
Lane Designation	M5-4,5,6	2D.33	24 x 18	24 x 18	36 x 24
Directional Arrow	M6-1,2,2a,3,4,5,6,7	2D.29	21 x 15	21 x 15	30 x 21
Destination (1 line)	D1-1	2D.39	Varies x 18	Varies x 18	—
Destination and Distance (1 line)	D1-1a	2D.39	Varies x 18	Varies x 18	—
Circluar Intersection Destination (1 line)	D1-1d	2D.40	Varies x 18	Varies x 18	—
Circluar Intersection Departure Guide	D1-1e	2D.40	Varies x 42*	—	—
Destination (2 lines)	D1-2	2D.39	Varies x 30	Varies x 30	—
Destination and Distance (2 lines)	D1-2a	2D.39	Varies x 30	Varies x 30	—
Circluar Intersection Destination (2 lines)	D1-2d	2D.40	Varies x 30	Varies x 30	—
Destination (3 lines)	D1-3	2D.39	Varies x 42	Varies x 42	—
Destination and Distance (3 lines)	D1-3a	2D.39	Varies x 42	Varies x 42	—
Circluar Intersection Destination (3 lines)	D1-3d	2D.40	Varies x 42	Varies x 42	—
Distance (1 line)	D2-1	2D.43	Varies x 18	Varies x 18	—
Distance (2 lines)	D2-2	2D.43	Varies x 30	Varies x 30	—
Distance (3 lines)	D2-3	2D.43	Varies x 42	Varies x 42	—
Street Name (1 line)	D3-1,1a	2D.45	Varies x 12	Varies x 8	Varies x 18
Advance Street Name (2 lines)	D3-2	2D.46	Varies x 30*	—	—
Advance Street Name (3 lines)	D3-2	2D.46	Varies x 42*	—	—
Advance Street Name (4 lines)	D3-2	2D.46	Varies x 60*	—	—
Parking Area	D4-1	2D.49	30 x 24	18 x 15	—
Park - Ride	D4-2	2D.50	30 x 36	24 x 30	36 x 48
National Scenic Byways	D6-4	2D.56	24 x 24	24 x 24	—
National Scenic Byways	D6-4a	2D.56	24 x 12	24 x 12	—
Weigh Station XX Miles	D8-1	2D.51	78 x 60	60 x 48	96 x 72
Weigh Station Next Right	D8-2	2D.51	84 x 72	66 x 54	108 x 90
Weigh Station (with arrow)	D8-3	2D.51	66 x 60	48 x 42	84 x 78
Crossover	D13-1,2	2D.55	60 x 30	60 x 30	78 x 42
Freeway Entrance	D13-3	2D.48	48 x 30	48 x 30	—
Freeway Entrance (with arrow)	D13-3a	2D.48	48 x 42	48 x 42	—
Combination Lane Use / Destination	D15-1	2D.35	Varies x 96	Varies x 96	—
Next Truck Lane XX Miles	D17-1	2D.53	42 x 48	42 x 48	60 x 66
Truck Lane XX Miles	D17-2	2D.53	42 x 42	42 x 42	60 x 54
Slow Vehicle Turn-Out XX Miles	D17-7	2D.54	72 x 42	72 x 42	96 x 54

*The size shown is for a typical sign. The size should be appropriately based on the amount of legend required for the sign.

Notes: 1. Larger signs may be used when appropriate
2. Dimensions in inches are shown as width x height

Standard:

02 **Design layouts for conventional road guide signs showing interline spacing, edge spacing, and other specification details shall be as shown in the "Standard Highway Signs and Markings" book (see Section 1A.11).**

03 **The principal legend on guide signs shall be in letters and numerals at least 6 inches in height for all upper-case letters, or a combination of 6 inches in height for upper-case letters and 4.5 inches in height for lower-case letters. On low-volume roads (as defined in Section 5A.01) with speeds of 25 mph or less, and on urban streets with speeds of 25 mph or less, the principal legend shall be in letters at least 4 inches in height for all upper-case letters, or a combination of 4 inches in height for upper-case letters and 3 inches in height for lower-case letters.**

Guidance:

04 *Lettering sizes should be consistent on any particular class of highway.*

05 *The minimum lettering sizes provided in this Manual should be exceeded where conditions indicate a need for greater legibility.*

Section 2D.07 Amount of Legend

Support:

01 The longer the legend on a guide sign, the longer it will take road users to comprehend it, regardless of letter size.

Guidance:

02 *Except where otherwise provided in this Manual, guide signs should be limited to no more than three lines of destinations, which include place names, route numbers, street names, and cardinal directions. Where two or more signs are included in the same overhead display, the amount of legend should be further minimized. Where appropriate, a distance message or action information, such as an exit number, NEXT RIGHT, or directional arrows, should be provided on guide signs in addition to the destinations.*

Section 2D.08 Arrows

Support:

01 Arrows are used for lane assignment and to indicate the direction toward designated routes or destinations. Figure 2D-2 shows the various standard arrow designs that have been approved for use on guide signs. Detailed drawings and standardized sizes based on ranges of letter heights are shown for these arrows in the "Standard Highway Signs and Markings" book (see Section 1A.11).

Standard:

02 **On overhead signs where it is desirable to indicate a lane to be followed, a down arrow shall be positioned approximately over the center of the lane and shall point vertically downward toward the approximate center of that lane. Down arrows shall be used only on overhead guide signs that restrict the use of specific lanes to traffic bound for the destination(s) and/or route(s) indicated by these arrows. Down arrows shall not be used unless an arrow can be located over and pointed to the approximate center of each lane that can be used to reach the destination displayed on the sign.**

03 **If down arrows are used, having more than one down arrow pointing to the same lane on a single overhead sign (or on multiple signs on the same overhead sign structure) shall not be permitted.**

04 **Where a roadway is leaving the through lanes, a directional arrow shall point upward at an angle that approximates the alignment of the exit roadway.**

Option:

05 Curved-stem arrows (see Figure 2D-8) that represent the intended driver paths to destinations involving left-turn movements may be used on guide signs on approaches to circular intersections.

Standard:

06 **Curved-stem arrows shall not be used on any sign that is not associated with a circular intersection.**

Guidance:

07 *If curved-stem arrows are used, the principles set forth in Sections 2D.26 through 2D.29 should be followed.*

08 *The Type A directional arrow should be used on guide signs on freeways, expressways, and conventional roads to indicate the direction to a specific destination or group of destinations, except as otherwise provided in this Section and in Section 2E.19.*

09 *When a directional arrow in a vertical, upward-pointing orientation is placed to the side of a group of destinations to indicate a through movement, the Type A directional arrow should be used. When a directional arrow in a vertical, upward-pointing orientation is placed to the side of a single destination or under a destination or group of destinations, the Type B directional arrow should be used.*

Figure 2D-2. Arrows for Use on Guide Signs

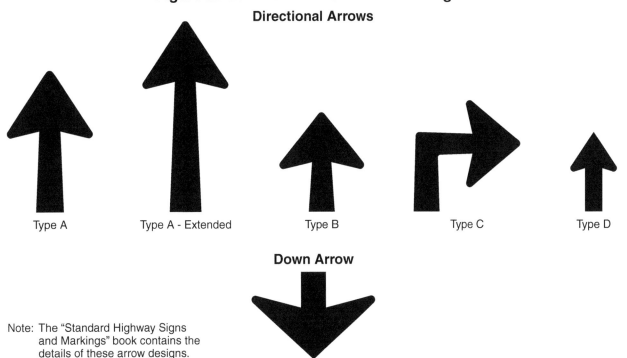

Note: The "Standard Highway Signs and Markings" book contains the details of these arrow designs.

---

10   *The Type B directional arrow should be used on guide signs on conventional roads when placed at any angle to the side of a single destination or when placed in a horizontal orientation to the side of a group of destinations.*

11   *The Type C advance turn directional arrow should be used on conventional road guide signs placed in advance of an intersection where a turn must be made to reach a posted destination or group of destinations.*

12   *The Type D directional arrow should be used primarily for sign applications other than guide signs, except as provided in Paragraph 15.*

Option:

13   The Type A-Extended directional arrow may be used on guide signs where additional emphasis regarding the direction is needed relative to the amount of legend on the sign.

14   The Type C directional arrow may be used to the side of the legend of an overhead guide sign to accentuate a sharp turn exit maneuver from a mainline roadway (see Section 2E.36 for additional information regarding Exit Direction signs for low advisory ramp speeds).

15   On conventional roads on the approach to an intersection where the Combination Lane-Use/Destination overhead guide sign (see Section 2D.33) is not used, the Type C advance turn directional arrow may be used beneath the legend of an overhead guide sign to indicate the fact that a turn must be made from a mandatory movement lane over which the sign is placed to reach the destination or destinations displayed on the sign.

16   The Type D directional arrow may be used on post-mounted guide signs on conventional roads with lower operating speeds if the height of the text on the sign is 8 inches or less.

17   The directional and down arrows shown in Figure 2D-2 may be used on signs other than guide signs for the purposes of providing directional guidance and lane assignment.

Guidance:

18   *Arrows used on guide signs to indicate the directions toward designated routes or destinations should be pointed at the appropriate angle to clearly convey the direction to be taken. A horizontally oriented directional arrow design should be used at right-angle intersections.*

19   *On a post-mounted guide sign, a directional arrow for a straight-through movement should point upward. Except as provided in Section 2D.46, for a turn, the arrow on a guide sign should point horizontally or at an upward angle that approximates the sharpness of the turn.*

20   *At an exit, an arrow should be placed at the side of the sign that will reinforce the movement of exiting traffic. The directional arrow design should be used.*

Option:
21 Arrows may be placed below the principal sign legend or on the appropriate side of the legend.
22 On a post-mounted sign at an exit where placement of the arrow to the side of the legend farthest from the roadway would create an unusually wide sign that limits the road user's view of the arrow, the directional arrow may be placed at the bottom portion of the sign, centered under the legend.

Guidance:
23 *The width across the arrowhead for the Types A, B, and C directional arrows should be between 1.5 and 1.75 times the height of the upper-case letters of the principal legend on the sign. The width across the arrowhead for the Type D directional arrow should be at least equal to the height of the upper-case letters of the principal legend on the sign. For down arrows used on overhead signs, the width across the arrowhead should be approximately two times the height of the upper-case letters of the principal legend on the sign.*
24 *Arrows used in Overhead Arrow-per-Lane and Diagrammatic guide signing, if used on conventional roads, except for signs on approaches to roundabouts, should follow the principles set forth in Section 2E.19. Arrows used in Diagrammatic guide signing on approaches to roundabouts should follow the principles set forth in Section 2D.38.*

Support:
25 The "Standard Highway Signs and Markings" book (see Section 1A.11) contains design details and standardized sizes of the various arrows based on ranges of letter heights of principal legends.

Section 2D.09 Numbered Highway Systems

Support:
01 The purpose of numbering and signing highway systems is to identify routes and facilitate travel.
02 The Interstate and United States (U.S.) highway systems are numbered by the American Association of State Highway and Transportation Officials (AASHTO) upon recommendations of the State highway organizations because the respective States own these systems. State and county road systems are numbered by the appropriate authorities.
03 The basic policy for numbering the Interstate and U.S. highway systems is contained in the following Purpose and Policy statements published by AASHTO (see Page i for AASHTO's address):
  A. "Establishment and Development of United States Numbered Highways," and
  B. "Establishment of a Marking System of the Routes Comprising the National System of Interstate and Defense Highways."

Guidance:
04 *The principles of these policies should be followed in establishing the highway systems described in Paragraph 2 and any other systems, with effective coordination between adjacent jurisdictions. Care should be taken to avoid the use of numbers or other designations that have been assigned to Interstate, U.S., or State routes in the same geographic area. Overlapping numbered routes should be kept to a minimum.*

Standard:
05 **Route systems shall be given preference in this order: Interstate, United States, State, and county. The preference shall be given by installing the highest-priority legend on the top or the left of the sign.**

Support:
06 Section 2D.53 contains information regarding the signing of unnumbered highways to enhance route guidance and facilitate travel.

Section 2D.10 Route Signs and Auxiliary Signs

Standard:
01 **All numbered highway routes shall be identified by route signs and auxiliary signs.**
02 **The signs for each system of numbered highways, which are distinctive in shape and color, shall be used only on that system and the approaches thereto.**

Option:
03 Route signs and auxiliary signs may be proportionally enlarged where greater legibility is needed.

Support:
04 Route signs are typically mounted in assemblies with auxiliary signs.
05 Section 2D.55 contains information regarding the signing for National Scenic Byways.
06 Section 2H.07 contains information regarding the signing for Auto Tour Routes.

Section 2D.11 Design of Route Signs

Standard:

01　The "Standard Highway Signs and Markings" book (see Section 1A.11) shall be used for designing route signs. Other route sign designs shall be established by the authority having jurisdiction.

02　**Interstate Route signs** (see Figure 2D-3) shall consist of a cutout shield, with the route number in white letters on a blue background, the word INTERSTATE in white upper-case letters on a red background, and a white border. This sign shall be used on all Interstate routes and in connection with route sign assemblies on intersecting highways.

03　A 24 x 24-inch minimum sign size shall be used for Interstate route numbers with one or two digits, and a 30 x 24-inch minimum sign size shall be used for Interstate route numbers having three digits.

Option:

04　Interstate Route signs may contain the State name in white upper-case letters on a blue background.

Standard:

05　**Off-Interstate Business Route signs** (see Figure 2D-3) shall consist of a cutout shield carrying the number of the connecting Interstate route and the words BUSINESS and either LOOP or SPUR in upper-case letters. The legend and border shall be white on a green background, and the shield shall be the same shape and dimensions as the Interstate Route sign. In no instance shall the word INTERSTATE appear on the Off-Interstate Business Route sign.

Option:

06　The Off-Interstate Business Route sign may be used on a major highway that is not a part of the Interstate system, but one that serves the business area of a city from an interchange on the system.

07　When used on a green guide sign, a white square or rectangle may be placed behind the shield to improve contrast.

Standard:

08　**U.S. Route signs** (see Figure 2D-3) shall consist of black numerals on a white shield surrounded by a rectangular black background without a border. This sign shall be used on all U.S. routes and in connection with route sign assemblies on intersecting highways.

09　A 24 x 24-inch minimum sign size shall be used for U.S. route numbers with one or two digits, and a 30 x 24-inch minimum sign size shall be used for U.S. route numbers having three digits.

10　**State Route signs** shall be designed by the individual State highway agencies.

Guidance:

11　*State Route signs (see Figure 2D-3) should be rectangular and should be approximately the same size as the U.S. Route sign. State Route signs should also be similar to the U.S. Route sign by containing approximately the same size black numerals on a white area surrounded by a rectangular black background without a border. The shape of the white area should be circular in the absence of any determination to the contrary by the individual State concerned.*

12　*Where U.S. or State Route signs are used as components of guide signs, only the distinctive shape of the shield itself and the route numerals within should be used. The rectangular background upon which the distinctive shape of the shield is mounted, such as the black area around the outside of the shields on the M1-4 and standard M1-5 signs, should not be included on the guide sign. Where U.S. or State Route signs are used as components of other signs of non-contrasting background colors, the rectangular background should be used to so that recognition of the distinctive shape of the shield can be maintained.*

Figure 2D-3. Route Signs

- Interstate Route Sign M1-1
- Off-Interstate Business Route Sign M1-2 (Loop), M1-3 (Spur)
- U.S. Route Sign M1-4
- State Route Sign M1-5
- County Route Sign M1-6
- Forest Route Sign M1-7

Standard:

13 If county road authorities elect to establish and identify a special system of important county roads, a statewide policy for such signing shall be established that includes a uniform numbering system to uniquely identify each route. The County Route (M1-6) sign (see Figure 2D-3) shall consist of a pentagon shape with a yellow county name and route number and border on a blue background. County Route signs displaying two digits or the equivalent (letter and numeral, or two letters) shall be a minimum size of 18 x 18 inches; those carrying three digits or the equivalent shall be a minimum size of 24 x 24 inches.

14 If a jurisdiction uses letters instead of numbers to identify routes, all references to numbered routes in this Chapter shall be interpreted to also include lettered routes.

Guidance:

15 *If used with other route signs in common assemblies, the County Route sign should be of a size compatible with that of the other route signs.*

Option:

16 When used on a green guide sign, a yellow square or rectangle may be placed behind the County Route sign to improve contrast.

Standard:

17 Route signs (see Figure 2D-3) for park and forest roads shall be designed with adequate distinctiveness and legibility and of a size compatible with other route signs used in common assemblies.

Section 2D.12 Design of Route Sign Auxiliaries

Standard:

01 Route sign auxiliaries carrying word legends, except the JCT sign, shall have a standard size of 24 x 12 inches. Those carrying arrow symbols, or the JCT sign, shall have a standard size of 21 x 15 inches. All route sign auxiliaries shall match the color combination of the route sign that they supplement.

Guidance:

02 *With route signs of larger heights, auxiliary signs should be suitably enlarged, but not such that they exceed the width of the route sign.*

03 *The background, legend, and border of a route sign auxiliary should have the same colors as those of the route sign with which the auxiliary is mounted in a route sign assembly (see Section 2D.29). For a route sign design that uses multiple background colors, such as the Interstate route sign, the background color of the corresponding auxiliary should be that of the background area on which the route number is placed on the route sign.*

Option:

04 A route sign and any auxiliary signs used with it may be combined on a single sign as a guide sign.

Guidance:

05 *If a route sign and its auxiliary signs are combined to form a single guide sign, the background color of the sign should be green and the design should comply with the basic principles for the design of guide signs.*

Standard:

06 If a route sign and its auxiliary signs are combined on a single sign with a green background, the auxiliary messages shall be white legends placed directly on the green background. Auxiliary signs shall not be mounted directly to a guide sign or other type of sign.

Support:

07 Chapter 2F contains information regarding auxiliary signs for toll highways.

Section 2D.13 Junction Auxiliary Sign (M2-1)

Standard:

01 The Junction (M2-1) auxiliary sign (see Figure 2D-4) shall carry the abbreviated legend JCT and shall be mounted at the top of an assembly (see Section 2D.30) directly above the route sign, the sign for an alternative route (see Section 2D.17) that is part of the route designation, or the Cardinal Direction auxiliary sign where access is available only to one direction of the intersected route. The minimum size of the Junction auxiliary sign shall be 21 x 15 inches for compatibility with auxiliary signs carrying arrow symbols.

Figure 2D-4. Route Sign Auxiliaries

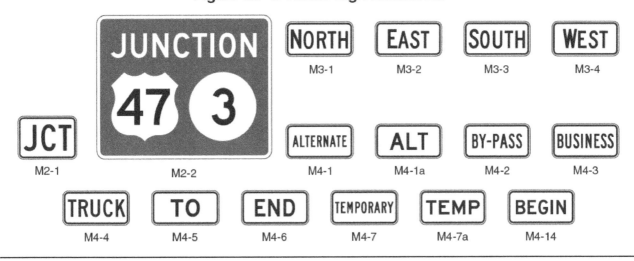

Section 2D.14 Combination Junction Sign (M2-2)

Option:

01 As an alternative to the standard Junction assembly where more than one route is to be intersected or joined, a rectangular guide sign may be used carrying the word JUNCTION above the route numbers.

Standard:

02 **The Combination Junction (M2-2) sign (see Figure 2D-4) shall have a green background with white border and lettering for the word JUNCTION.**

Guidance:

03 *The Combination Junction sign should comply with the specific provisions of Section 2D.11 regarding the incorporation of the route signs as components of guide signs.*

04 *Although the size of the Combination Junction sign will depend on the number of routes involved, the numerals should be large enough for clear legibility and should be of a size comparable with those in the individual route signs.*

Section 2D.15 Cardinal Direction Auxiliary Signs (M3-1 through M3-4)

Guidance:

01 *Cardinal Direction auxiliary signs (see Figure 2D-4) carrying the legend NORTH, EAST, SOUTH, or WEST should be used to indicate the general direction of the entire route.*

Standard:

02 **To improve the readability and recognition of the cardinal directions, the first letter of the cardinal direction words shall be ten percent larger, rounded up to the nearest whole number size.**

03 **If used, the Cardinal Direction auxiliary sign shall be mounted directly above a route sign or, if used, an auxiliary sign for an alternative route.**

Section 2D.16 Auxiliary Signs for Alternative Routes (M4 Series)

Option:

01 Auxiliary signs, carrying legends such as ALTERNATE, BY-PASS, BUSINESS, or TRUCK, may be used to indicate an alternate route of the same number between two points on that route.

Standard:

02 **If used, the auxiliary signs for alternative routes shall be mounted directly above a route sign.**

Section 2D.17 ALTERNATE Auxiliary Signs (M4-1, M4-1a)

Option:

01 The ALTERNATE (M4-1) or the ALT (M4-1a) auxiliary sign (see Figure 2D-4) may be used to indicate an officially designated alternate routing of a numbered route between two points on that route.

Standard:

02 **If used, the ALTERNATE or ALT auxiliary sign shall be mounted directly above a route sign.**

Guidance:

03 *The shorter (time or distance) or better-constructed route should retain the regular route number, and the longer or worse-constructed route should be designated as the alternate route.*

Section 2D.18 BY-PASS Auxiliary Sign (M4-2)

Option:

01 The BY-PASS (M4-2) auxiliary sign (see Figure 2D-4) may be used to designate a route that branches from the numbered route through a city, bypasses a part of the city or congested area, and rejoins the numbered route beyond the city.

Standard:

02 **If used, the BY-PASS auxiliary sign shall be mounted directly above a route sign.**

Section 2D.19 BUSINESS Auxiliary Sign (M4-3)

Option:

01 The BUSINESS (M4-3) auxiliary sign (see Figure 2D-4) may be used to designate an alternate route that branches from a numbered route, passes through the business portion of a city, and rejoins the numbered route beyond that area.

Standard:

02 **If used, the BUSINESS auxiliary sign shall be mounted directly above a route sign.**

Section 2D.20 TRUCK Auxiliary Sign (M4-4)

Option:

01 The TRUCK (M4-4) auxiliary sign (see Figure 2D-4) may be used to designate an alternate route that branches from a numbered route, when it is desirable to encourage or require commercial vehicles to use the alternate route.

Standard:

02 **If used, the TRUCK auxiliary sign shall be mounted directly above a route sign.**

Section 2D.21 TO Auxiliary Sign (M4-5)

Option:

01 The TO (M4-5) auxiliary sign (see Figure 2D-4) may be used to provide directional guidance to a particular road facility from other highways in the vicinity (see Section 2D.35).

Standard:

02 **If used, the TO auxiliary sign shall be mounted directly above a route sign or an auxiliary sign for an alternative route. If a Cardinal Direction auxiliary sign is also included in the assembly, the TO auxiliary sign shall be mounted directly above the Cardinal Direction auxiliary sign.**

Section 2D.22 END Auxiliary Sign (M4-6)

Guidance:

01 *The END (M4-6) auxiliary sign (see Figure 2D-4) should be used where the route being traveled ends, usually at a junction with another route.*

Standard:

02 **If used, the END auxiliary sign shall be mounted either directly above a route sign or above a sign for an alternative route that is part of the designation of the route being terminated.**

Section 2D.23 BEGIN Auxiliary Sign (M4-14)

Option:

01 The BEGIN (M4-14) auxiliary sign (see Figure 2D-4) may be used where a route begins, usually at a junction with another route.

Standard:

02 **If used, the BEGIN auxiliary sign shall be mounted at the top of the first Confirming assembly (see Section 2D.34) for the route that is beginning.**

Guidance:

03 *If a BEGIN auxiliary sign is included in the first Confirming assembly, a Cardinal Direction auxiliary sign should also be included in the assembly.*

Standard:

04 If a Cardinal Direction auxiliary sign is also included in the assembly, the BEGIN auxiliary sign shall be mounted directly above the Cardinal Direction auxiliary sign.

Section 2D.24 TEMPORARY Auxiliary Signs (M4-7, M4-7a)

Option:

01 The TEMPORARY (M4-7) or the TEMP (M4-7a) auxiliary sign (see Figure 2D-4) may be used for an interim period to designate a section of highway that is not planned as a permanent part of a numbered route, but that connects completed portions of that route.

Standard:

02 If used, the TEMPORARY or TEMP auxiliary sign shall be mounted directly above the route sign, above a Cardinal Direction sign, or above a sign for an alternate route that is a part of the route designation.

03 TEMPORARY or TEMP auxiliary signs shall be promptly removed when the temporary route is abandoned.

Section 2D.25 Temporary Detour and Auxiliary Signs

Support:

01 Chapter 6F contains information regarding Temporary Detour and Auxiliary signs.

Section 2D.26 Advance Turn Arrow Auxiliary Signs (M5-1, M5-2, and M5-3)

Standard:

01 If used, the Advance Turn Arrow auxiliary sign (see Figure 2D-5) shall be mounted directly below the route sign in Advance Route Turn assemblies, and displays a right or left arrow, the shaft of which is bent at a 90-degree angle (M5-1) or at a 45-degree angle (M5-2).

02 If used, the curved-stem Advance Turn Arrow auxiliary (M5-3) sign shall be used only on the approach to a circular intersection to depict a movement along the circulatory roadway around the central island and to the left, relative to the approach roadway and entry into the intersection.

Guidance:

03 *If the M5-3 sign is used, then this arrow type should also be used consistently on any regulatory lane-use signs (see Chapter 2B), Destination signs (see Section 2D.37), and pavement markings (see Part 3) for a particular destination or movement.*

Figure 2D-5. Advance Turn and Directional Arrow Auxiliary Signs

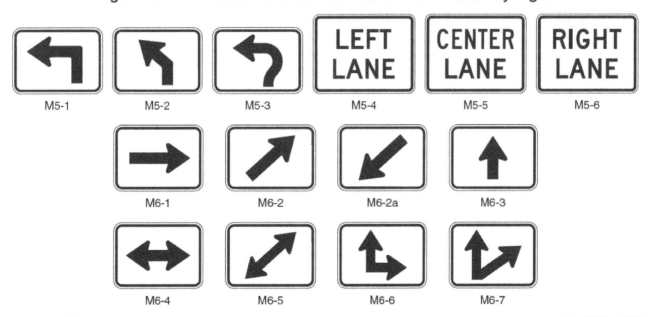

Section 2D.27 Lane Designation Auxiliary Signs (M5-4, M5-5, and M5-6)

Option:

01 A Lane Designation (M5-4, M5-5, or M5-6) auxiliary sign (see Figure 2D-5) may be mounted directly below the route sign in an Advance Route Turn assembly on multi-lane roadways to allow road users to move into the appropriate lane prior to reaching the intersection or interchange.

Standard:

02 **If used, the Lane Designation auxiliary signs shall be used only where the designated lane is a mandatory movement lane and shall be located adjacent to the full-width portion of the mandatory movement lane. The Lane Designation auxiliary signs shall not be installed adjacent to a through lane in advance of a lane that is being added or along the taper for a lane that is being added.**

Section 2D.28 Directional Arrow Auxiliary Signs (M6 Series)

Standard:

01 **If used, the Directional Arrow auxiliary sign (see Figure 2D-5) shall be mounted below the route sign and any other auxiliary signs in Directional assemblies (see Section 2D.32), and displays a single- or double-headed arrow pointing in the general direction that the route follows.**

02 **A Directional Arrow auxiliary sign that displays a double-headed arrow shall not be mounted in any Directional assembly in advance of or at a circular intersection.**

Option:

03 The downward pointing diagonal arrow auxiliary (M6-2a) sign may be used in a Directional assembly at the far corner of an intersection to indicate the immediate entry point to a freeway or expressway entrance ramp (see Section 2D.46).

Standard:

04 **The M6-2a sign shall not be used on the approach to or on the near side of an intersection, such as to designate an approach lane.**

Section 2D.29 Route Sign Assemblies

Standard:

01 **A Route Sign assembly shall consist of a route sign and auxiliary signs that further identify the route and indicate the direction. Route Sign assemblies shall be installed on all approaches to numbered routes that intersect with other numbered routes.**

02 **Where two or more routes follow the same section of highway, the route signs for Interstate, U.S., State, and county routes shall be mounted in that order from the left in horizontal arrangements and from the top in vertical arrangements. Subject to this order of precedence, route signs for lower-numbered routes shall be placed at the left or top.**

03 **Within groups of assemblies, information for routes intersecting from the left shall be mounted at the left in horizontal arrangements and at the top or center of vertical arrangements. Similarly, information for routes intersecting from the right shall be at the right or bottom, and for straight-through routes at the center in horizontal arrangements or top in vertical arrangements.**

04 **Route Sign assemblies shall be mounted in accordance with the general specifications for signs (Chapter 2A), with the lowest sign in the assembly at the height prescribed for single signs.**

Guidance:

05 *Assemblies for two or more routes, or for different directions on the same route, should be mounted in groups on a common support.*

Option:

06 Route Sign assemblies may be installed on the approaches to numbered routes on unnumbered roads and streets that carry an appreciable amount of traffic destined for the numbered route.

07 The diagrammatic route guide sign format, such as the D1-5 and D1-5a signs shown in Figure 2D-8, may be used on approaches to roundabouts.

08 If engineering judgment indicates that groups of assemblies that include overlapping routes or multiple turns might be confusing, route signs or auxiliary signs may be omitted or combined, provided that clear directions are given to road users.

Support:

09 Figure 2D-6 shows typical placements of route signs.

Figure 2D-6. Illustration of Directional Assemblies and Other Route Signs (for One Direction of Travel Only) (Sheet 1 of 4)

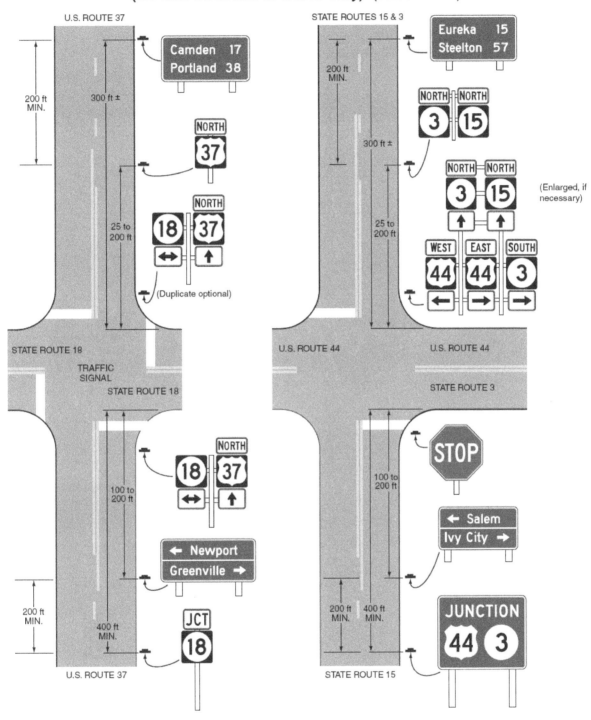

Note: The spacings shown on this figure are for rural intersections.
See Sections 2D.29, 2D.30, 2D.32, 2D.34, 2D.40, and 2D.42 for low-speed and/or urban conditions.

Figure 2D-6. Illustration of Directional Assemblies and Other Route Signs (for One Direction of Travel Only) (Sheet 2 of 4)

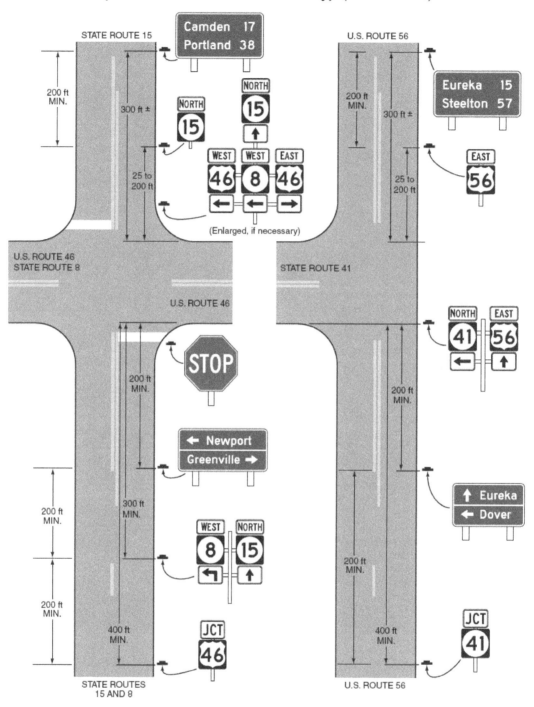

Note: The spacings shown on this figure are for rural intersections.
See Sections 2D.29, 2D.30, 2D.32, 2D.34, 2D.40, and 2D.42 for low-speed and/or urban conditions.

Figure 2D-6. Illustration of Directional Assemblies and Other Route Signs (for One Direction of Travel Only) (Sheet 3 of 4)

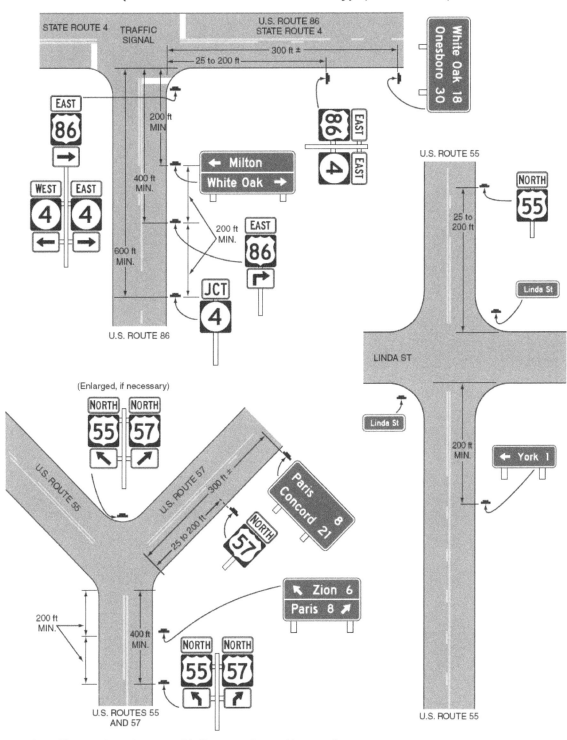

Note: The spacings shown on this figure are for rural intersections.
See Sections 2D.29, 2D.30, 2D.32, 2D.34, 2D.40, and 2D.42 for low-speed and/or urban conditions.

Figure 2D-6. Illustration of Directional Assemblies and Other Route Signs (for One Direction of Travel Only) (Sheet 4 of 4)

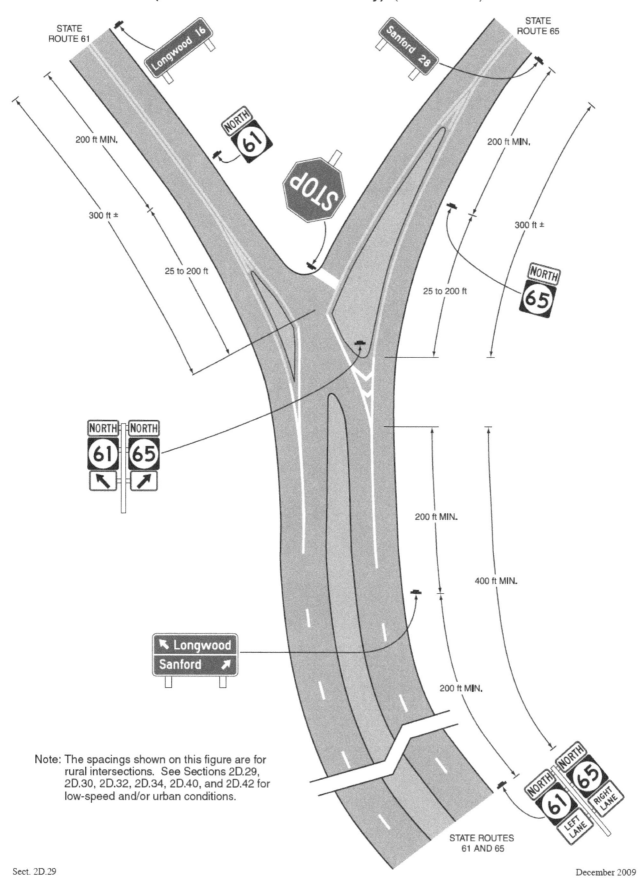

Section 2D.30 Junction Assembly

Standard:

01 A Junction assembly shall consist of a Junction auxiliary sign and a route sign. The route sign shall carry the number of the intersected or joined route.

02 The Junction assembly shall be installed in advance of every intersection where a numbered route is intersected or joined by another numbered route.

Guidance:

03 *In urban areas, the Junction assembly should be installed in the block preceding the intersection. In urban areas where speeds are low, the Junction assembly should not be installed more than 300 feet in advance of the intersection.*

04 *In rural areas, the Junction assembly should be installed at least 400 feet in advance of the intersection. In rural areas, the minimum distance between a Junction assembly and either a Destination sign or an Advance Route Turn assembly should be 200 feet.*

05 *Where speeds are high, greater spacings should be used.*

Option:

06 Where two or more routes are to be indicated, a single Junction auxiliary sign may be used for the assembly and all route signs grouped in a single mounting, or a Combination Junction (M2-2) sign (see Section 2D.14) may be used.

Section 2D.31 Advance Route Turn Assembly

Standard:

01 An Advance Route Turn assembly shall consist of a route sign, an Advance Turn Arrow or word message auxiliary sign, and a Cardinal Direction auxiliary sign, if needed. It shall be installed in advance of an intersection where a turn must be made to remain on the indicated route.

Option:

02 The Advance Route Turn assembly may be used to supplement the required Junction assembly in advance of intersecting routes.

Guidance:

03 *Where a multiple-lane highway approaches an interchange or intersection with a numbered route, the Advance Route Turn assembly should be used to pre-position turning vehicles in the correct lanes from which to make their turn.*

Option:

04 Lane Designation auxiliary signs (see Section 2D.27) may be used in Advance Route Turn Assemblies in place of the Advance Turn Arrow auxiliary signs where engineering judgment indicates that specific lane information associated with each route is needed and overhead signing is not practical and the designated lane is a mandatory movement lane. An assembly with the Lane Designation auxiliary signs may supplement or substitute for an assembly with Advance Turn Arrow auxiliary signs.

Guidance:

05 *In low-speed areas, the Advance Route Turn assembly should be installed not less than 200 feet in advance of the turn. In high-speed areas, the Advance Route Turn assembly should be installed not less than 300 feet in advance of the turn. In rural areas, the minimum distance between an Advance Route Turn assembly and either a Destination sign or a Junction assembly should be 200 feet.*

Standard:

06 An assembly that includes an Advance Turn Arrow auxiliary sign shall not be placed where there is an intersection between it and the designated turn.

Guidance:

07 *Sufficient distance should be allowed between the assembly and any preceding intersection that could be mistaken for the indicated turn.*

Section 2D.32 Directional Assembly

Standard:

01 A Directional assembly shall consist of a Cardinal Direction auxiliary sign, if needed; a route sign; and a Directional Arrow auxiliary sign. The various uses of Directional assemblies shall be as provided in Items A through D:

A. Turn movements (indicated in advance by an Advance Route Turn assembly) shall be marked by a Directional assembly with a route sign displaying the number of the turning route and a single-headed arrow pointing in the direction of the turn.
B. The beginning of a route (indicated in advance by a Junction assembly) shall be marked by a Directional assembly with a route sign displaying the number of that route and a single-headed arrow pointing in the direction of the route.
C. An intersected route (indicated in advance by a Junction assembly) on a crossroad where the route is designated on both legs shall be designated by:
   1. Two Directional assemblies, each with a route sign displaying the number of the intersected route, a Cardinal Direction auxiliary sign, and a single-headed arrow pointing in the direction of movement on that route; or
   2. A Directional assembly with a route sign displaying the number of the intersected route and a double-headed arrow, pointing at appropriate angles to the left, right, or ahead.
D. An intersected route (indicated in advance by a Junction assembly) on a side road or on a crossroad where the route is designated only on one of the legs shall be designated by a Directional assembly with a route sign displaying the number of the intersected route, a Cardinal Direction auxiliary sign, and a single-headed arrow pointing in the direction of movement on that route.

Guidance:

02 *Straight-through movements should be indicated by a Directional assembly with a route sign displaying the number of the continuing route and a vertical arrow. A Directional assembly should not be used for a straight-through movement in the absence of other assemblies indicating right or left turns, as the Confirming assembly sign beyond the intersection normally provides adequate guidance.*

03 *Directional assemblies should be located on the near right corner of the intersection. At major intersections and at Y or offset intersections, additional Directional assemblies should be installed on the far right or left corner to confirm the near-side assemblies. When the near-corner position is not practical for Directional assemblies, the far right corner should be the preferred alternative, with oversized signs, if necessary, for legibility. Where unusual conditions exist, the location of a Directional assembly should be determined by engineering judgment with the goal being to provide the best possible combination of view and safety.*

Support:

04 It is more important that guide signs be readable, and that the information and direction displayed thereon be readily understood, at the appropriate time and place than to be located with absolute uniformity.

05 Figure 2D-6 shows typical placements of Directional assemblies.

Section 2D.33 Combination Lane-Use/Destination Overhead Guide Sign (D15-1)

Option:

01 At complex intersection approaches involving multiple turn lanes and destinations, a Combination Lane-Use/Destination (D15-1) overhead guide sign that combines a lane-use regulatory sign with destination information such as a cardinal direction, a route number, a street name, and/or a place name may be used.

Support:

02 At such locations, the combined information on the D15-1 signs can be even more effective than separate lane-use and guide signs for conveying to unfamiliar drivers which lane or lanes to use for a particular destination.

03 Figure 2D-7 shows an example of a D15-1 sign that combines lane-use and route number information and an example of a D15-1 sign that combines lane-use and street name information.

Standard:

04 **The Combination Lane-Use/Destination (D15-1) overhead guide sign shall be used only where the designated lane is a mandatory movement lane. The D15-1 sign shall not be used for lanes with optional movements.**

05 **The D15-1 sign shall have a green background with a white border. As shown in Figure 2D-7, the lane-use sign (see Chapter 2B) shall be placed near the bottom of the sign and the destination information shall be placed near the top of the sign. The D15-1 sign shall be located approximately over the center of the lane to which it applies.**

Figure 2D-7. Destination and Distance Signs

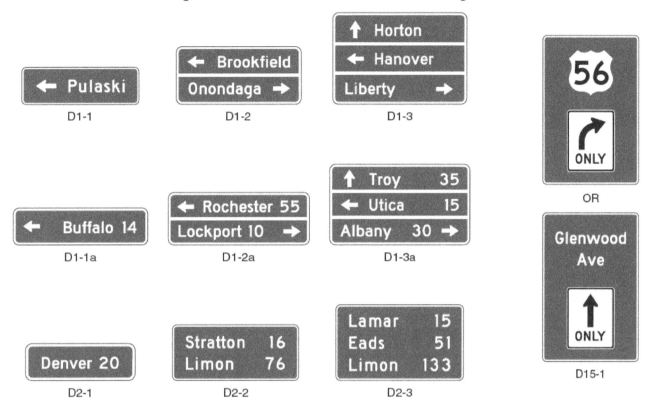

Section 2D.34 Confirming or Reassurance Assemblies
Standard:

01 **If used, Confirming or Reassurance assemblies shall consist of a Cardinal Direction auxiliary sign and a route sign. Where the Confirming or Reassurance assembly is for an alternative route, the appropriate auxiliary sign for an alternative route (see Section 2D.16) shall also be included in the assembly.**
Guidance:

02 *A Confirming assembly should be installed just beyond intersections of numbered routes. It should be placed 25 to 200 feet beyond the far shoulder or curb line of the intersected highway.*

03 *If used, Reassurance assemblies should be installed between intersections in urban areas as needed, and beyond the built-up area of any incorporated city or town.*

04 *Route signs for either confirming or reassurance purposes should be spaced at such intervals as necessary to keep road users informed of their routes.*

Section 2D.35 Trailblazer Assembly
Support:

01 Trailblazer assemblies provide directional guidance to a particular road facility from other highways in the vicinity. This guidance is accomplished by installing Trailblazer assemblies at strategic locations to indicate the direction to the nearest or most convenient point of access. The use of the word TO indicates that the road or street where the sign is posted is not a part of the indicated route, and that a road user is merely being directed progressively to the route.

Standard:

02 **A Trailblazer assembly shall consist of a TO auxiliary sign, a route sign for a numbered or named highway (see Section 2D.53) or an Auto Tour Route sign (see Section 2H.07), and a single-headed Directional Arrow auxiliary sign pointing in the direction leading to the route. Where the Trailblazer assembly is for an alternative route, the appropriate auxiliary sign for an alternative route (see Section 2D.16) shall also be included in the assembly.**

Option:

03 A Cardinal Direction auxiliary sign may be used with a Trailblazer assembly.

Guidance:

04 *The TO auxiliary sign, Cardinal Direction auxiliary sign, and Directional Arrow auxiliary sign should be of the standard size provided for auxiliary signs of their respective type. The route sign should be the size provided in Section 2D.11.*

Option:

05 Trailblazer assemblies may be installed with other Route Sign assemblies, or alone, in the immediate vicinity of the designated facilities.

Section 2D.36 Destination and Distance Signs

Support:

01 In addition to guidance by route numbers, it is desirable to supply the road user information concerning the destinations that can be reached by way of numbered or unnumbered routes. This is done by means of Destination signs and Distance signs.

Option:

02 Route shields and cardinal directions may be included on the Destination sign with the destinations and arrows.

Guidance:

03 *If Route shields and cardinal directions are included on a Destination sign, the height of the route shields should be at least two times the height of the upper-case letters of the principal legend and not less than 18 inches, and the cardinal directions should be in all upper-case letters that are at least the minimum height specified for these signs.*

Section 2D.37 Destination Signs (D1 Series)

Standard:

01 **Except on approaches to interchanges (see Section 2D.45), the Destination (D1-1 through D1-3) sign (see Figure 2D-7), if used, shall be a horizontal rectangle displaying the name of a city, town, village, or other traffic generator, and a directional arrow.**

Option:

02 The distance (see Section 2D.41) to the place named may also be displayed on the Destination (D1-1a through D1-3a) sign (see Figure 2D-7). If several destinations are to be displayed at a single point, the several names may be placed on a single sign with an arrow (and the distance, if desired) for each name. If more than one destination lies in the same direction, a single arrow may be used for such a group of destinations.

Guidance:

03 *Adequate separation should be made between any destinations or group of destinations in one direction and those in other directions by suitable design of the arrow, spacing of lines of legend, heavy lines entirely across the sign, or separate signs.*

Support:

04 Separation of destinations by direction by the use of a horizontal separator line can enhance the readability of a Destination sign by relating an arrow and its corresponding destination(s) and by eliminating the need for multiple arrows that point in the same direction and excessive space between lines of legend.

Standard:

05 **Except as otherwise provided in this Manual, an arrow pointing to the right shall be at the extreme right of the sign, and an arrow pointing left or up shall be at the extreme left. The distance numerals, if used, shall be placed to the right of the destination names.**

Option:

06 An arrow pointing up may be placed at the extreme right of the sign when the sign is mounted to the left of the traffic to which it applies.

Guidance:

07 *Unless a sloping arrow will convey a clearer indication of the direction to be followed, the directional arrows should be horizontal or vertical.*

08 *If several individual name signs are assembled into a group, all signs in the assembly should be of the same horizontal width.*

09 *Destination signs should be used:*
   A. *At the intersections of U.S. or State numbered routes with Interstate, U.S., or State numbered routes; and*
   B. *At points where they serve to direct traffic from U.S. or State numbered routes to the business section of towns, or to other destinations reached by unnumbered routes.*

Standard:

10 **Where a total of three or less destinations are provided on the Advance Guide (see Section 2E.33) and Supplemental Guide (see Section 2E.35) signs, no more than three destination names shall be used on a Destination sign. Where four destinations are provided by the Advance Guide and Supplemental Guide signs, no more than four destination names shall be used on a Destination sign.**

Guidance:

11 *If space permits, four destinations should be displayed as two separate signs at two separate locations.*

Option:

12 Where space does not permit, or where all four destinations are in one direction, a single sign may be used. Where a single sign is used and all destinations are in the same direction, the arrow may be placed below the destinations for the purpose of enhancing the conspicuity of the arrow.

Standard:

13 **Where a single four-name sign assembly is used, a heavy line entirely across the sign or separate signs shall be used to separate destinations by direction.**

Guidance:

14 *The closest destination lying straight ahead should be at the top of the sign or assembly, and below it the closest destinations to the left and to the right, in that order. The destination displayed for each direction should ordinarily be the next county seat or the next principal city, rather than a more distant destination. In the case of overlapping routes, only one destination should be displayed in each direction for each route.*

Standard:

15 **If more than one destination is displayed in the same direction, the name of a nearer destination shall be displayed above the name of a destination that is further away.**

Section 2D.38 Destination Signs at Circular Intersections

Standard:

01 **Destination signs that are used at circular intersections shall comply with the provisions of Section 2D.37, except as provided in this Section.**

Option:

02 Exit destination (D1-1d, D1-1e) signs (see Figure 2D-8) with diagonal upward-pointing arrows or Directional assemblies (see Section 2D.32) may be used to designate a particular exit from a circular intersection.

03 Exit destination (D1-2d, D1-3d) signs (see Figure 2D-8) with curved-stem arrows may be used on approaches to circular intersections to represent the left-turn movements.

04 Curved-stem arrows on circular intersection destination signs may point in diagonal directions to depict the location of an exit relative to the approach roadway and entry into the intersection.

05 Exit destination (D1-5 or D1-5a) signs (see Figure 2D-8) with a diagram of the circular intersection may be used on approaches to circular intersections.

Guidance:

06 *If curved-stem arrows are used on destination signs, then this arrow type should also be used consistently on any regulatory lane-use signs (see Chapter 2B), Directional assemblies (see Section 2D.32), and pavement markings (see Part 3) for a particular destination or movement.*

Support:

07 Figure 2D-9 illustrates two examples of guide signing for circular intersections.

08 Diagrammatic guide signs might be preferable where space is available and where the geometry of the circular intersection is non-typical, such as where more than four legs are present or where the legs are not at approximately 90-degree angles to each other.

Standard:

09 **If used, diagrammatic guide signs for circular intersections shall not depict the number of lanes within the intersection circulatory roadway, or on its approaches or exits, through the use of lane lines, multiple arrow shafts for the same movement, or other methods.**

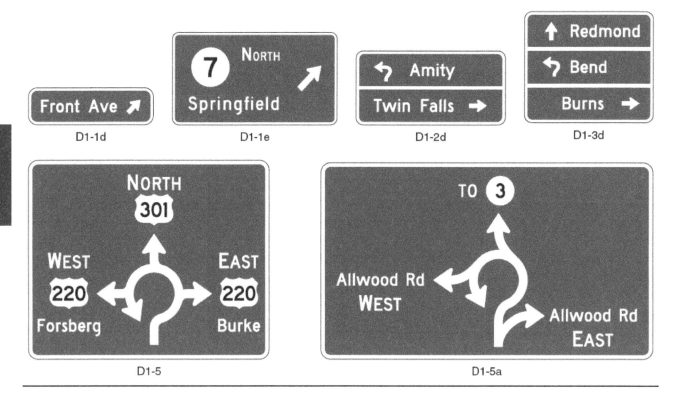

Figure 2D-8. Destination Signs for Roundabouts

Support:

10 Chapter 2B contains information regarding regulatory signs at circular intersections, Chapter 2C contains information regarding warning signs at circular intersections, and Chapter 3C contains information regarding pavement markings at circular intersections.

Section 2D.39 Destination Signs at Jughandles
Standard:

01 **Destination signs that are used at jughandles shall comply with the provisions of Section 2D.37, except as provided in this Section.**

Option:

02 If engineering judgment indicates that standard destination signs alone are insufficient to direct road users to their destinations at a jughandle, a diagrammatic guide sign depicting the appropriate geometry may be used to supplement the normal destination signs.

Support:

03 Section 2B.27 contains information regarding regulatory signs for jughandle turns. Figure 2B-9 shows examples of regulatory and destination guide signing for various types of jughandle turns.

Section 2D.40 Location of Destination Signs
Guidance:

01 *When used in high-speed areas, Destination signs should be located 200 feet or more in advance of the intersection, and following any Junction or Advance Route Turn assemblies that might be required. In rural areas, the minimum distance between a Destination sign and either an Advance Route Turn assembly or a Junction assembly should be 200 feet.*

Option:

02 In urban areas, shorter advance distances may be used.

03 Because the Destination sign is of lesser importance than the Junction, Advance Route Turn, or Directional assemblies, the Destination sign may be eliminated when sign spacing is critical.

Support:

04 Figure 2D-6 shows typical placements of Destination signs.

Figure 2D-9. Examples of Guide Signs for Roundabouts (Sheet 1 of 2)

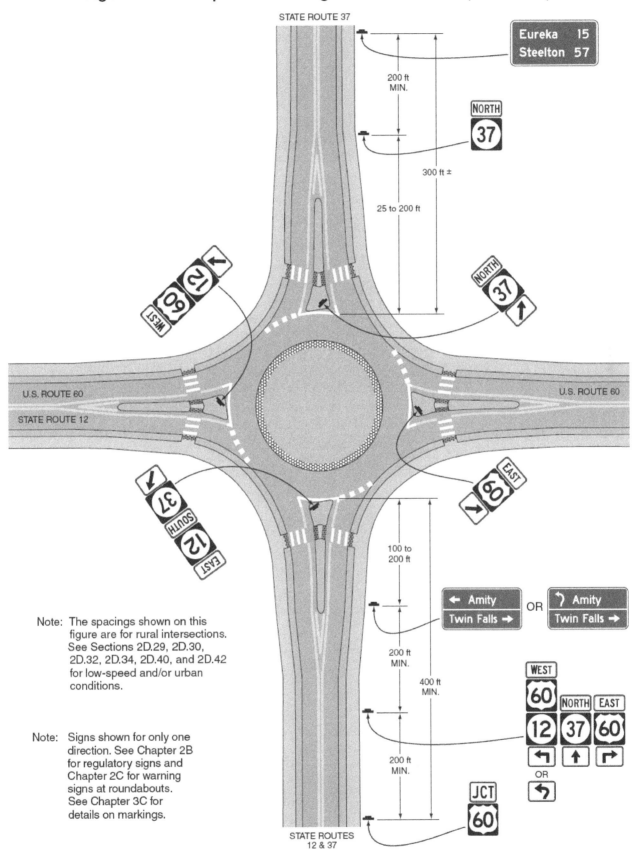

Note: The spacings shown on this figure are for rural intersections. See Sections 2D.29, 2D.30, 2D.32, 2D.34, 2D.40, and 2D.42 for low-speed and/or urban conditions.

Note: Signs shown for only one direction. See Chapter 2B for regulatory signs and Chapter 2C for warning signs at roundabouts. See Chapter 3C for details on markings.

Figure 2D-9. Examples of Guide Signs for Roundabouts (Sheet 2 of 2)

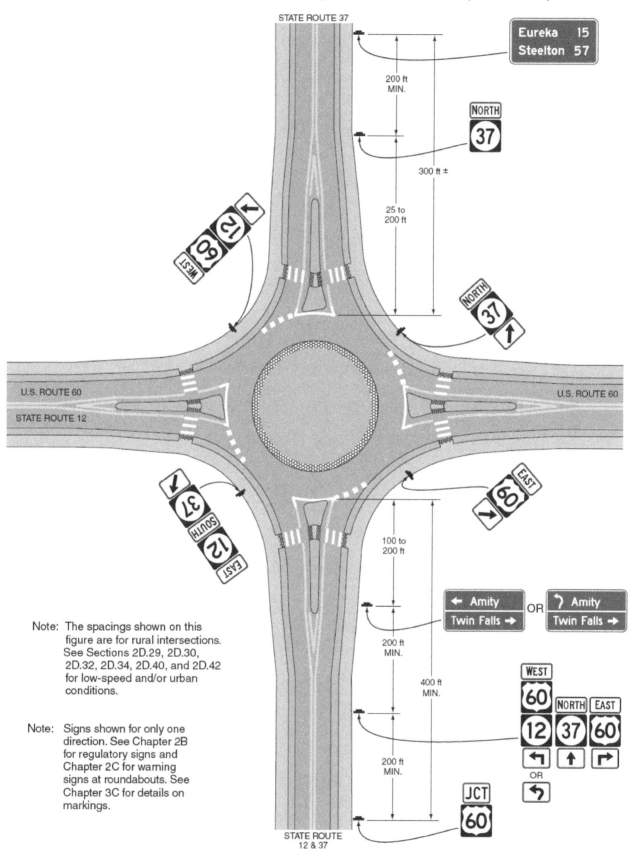

Section 2D.41 Distance Signs (D2 Series)
Standard:

01　If used, the Distance (D2-1 through D2-3) sign (see Figure 2D-7) shall be a horizontal rectangle of a size appropriate for the required legend, carrying the names of no more than three cities, towns, junctions, or other traffic generators, and the distance (to the nearest mile) to those places.

02　The distance numerals shall be placed to the right of the destination names as shown in Figure 2D-7.

Guidance:

03　*The distance displayed should be selected on a case-by-case basis by the jurisdiction that owns the road or by statewide policy. A well-defined central area or central business district should be used where one exists. In other cases, the layout of the community should be considered in relation to the highway being signed and the decision based on where it appears that most drivers would feel that they are in the center of the community in question.*

04　*The top name on the Distance sign should be that of the next place on the route having a post office or a railroad station, a route number or name of an intersected highway, or any other significant geographical identity. The bottom name on the sign should be that of the next major destination or control city. If three destinations are displayed, the middle line should be used to indicate communities of general interest along the route or important route junctions.*

Option:

05　The choice of names for the middle line may be varied on successive Distance signs to give road users additional information concerning communities served by the route.

Guidance:

06　*The control city should remain the same on all successive Distance signs throughout the length of the route until that city is reached.*

Option:

07　If more than one distant point may properly be designated, such as where the route divides at some distance ahead to serve two destinations of similar importance, and if these two destinations cannot appear on the same sign, the two names may be alternated on successive signs.

08　On a route continuing into another State, destinations in the adjacent State may be displayed.

Section 2D.42 Location of Distance Signs
Guidance:

01　*If used, Distance signs should be installed on important routes leaving municipalities and just beyond intersections of numbered routes in rural areas. If used, they should be placed just outside the municipal limits or at the edge of the built-up area if it extends beyond the limits.*

02　*Where overlapping routes separate a short distance from the municipal limits, the Distance sign at the municipal limits should be omitted. The Distance sign should be installed approximately 300 feet beyond the separation of the two routes.*

03　*Where, just outside of an incorporated municipality, two routes are concurrent and continue concurrently to the next incorporated municipality, the top name on the Distance sign should be that of the place where the routes separate; the bottom name should be that of the city to which the greater part of the through traffic is destined.*

Support:

04　Figure 2D-6 shows typical placements of Distance signs.

Section 2D.43 Street Name Signs (D3-1 or D3-1a)
Guidance:

01　*Street Name (D3-1 or D3-1a) signs (see Figure 2D-10) should be installed in urban areas at all street intersections regardless of other route signs that might be present and should be installed in rural areas to identify important roads that are not otherwise signed.*

Option:

02　For streets that are part of a U.S., State, or county numbered route, a D3-1a Street Name sign (see Figure 2D-10) that incorporates a route shield may be used to assist road users who might not otherwise be able to associate the name of the street with the route number.

Standard:

03　The lettering for names of streets and highways on Street Name signs shall be composed of a combination of lower-case letters with initial upper-case letters (see Section 2A.13).

Figure 2D-10. Street Name and Parking Signs

Guidance:

04 Lettering on post-mounted Street Name signs should be composed of initial upper-case letters at least 6 inches in height and lower-case letters at least 4.5 inches in height.

05 On multi-lane streets with speed limits greater than 40 mph, the lettering on post-mounted Street Name signs should be composed of initial upper-case letters at least 8 inches in height and lower-case letters at least 6 inches in height.

Option:

06 For local roads with speed limits of 25 mph or less, the lettering on post-mounted Street Name signs may be composed of initial upper-case letters at least 4 inches in height and lower-case letters at least 3 inches in height.

Guidance:

07 If overhead Street Name signs are used, the lettering should be composed of initial upper-case letters at least 12 inches in height and lower-case letters at least 9 inches in height.

Support:

08 The recommended minimum letter heights for Street Name signs are summarized in Table 2D-2.

Option:

09 Supplementary lettering to indicate the type of street (such as Street, Avenue, or Road) or the section of the city (such as NW) on the D3-1 and D3-1a signs may be in smaller lettering, composed of initial upper-case letters at least 3 inches in height and lower-case letters at least 2.25 inches in height. Conventional abbreviations (see Section 1A.15) may be used except for the street name itself.

10 A pictograph (see definition in Section 1A.13) may be used on a D3-1 sign.

Standard:

11 **Pictographs shall not be displayed on D3-1a or Advance Street Name (D3-2) signs (see Section 2D.44).**

12 **If a pictograph is used on a D3-1 sign, the height and width of the pictograph shall not exceed the upper-case letter height of the principal legend of the sign.**

Guidance:

13 The pictograph should be positioned to the left of the street name.

Standard:

14 **The Street Name sign shall be retroreflective or illuminated to show the same shape and similar color both day and night. The color of the legend (and border, if used) shall contrast with the background color of the sign.**

Option:

15 The border may be omitted from a Street Name sign.

Table 2D-2. Recommended Minimum Letter Heights on Street Name Signs

Type of Mounting	Type of Street or Highway	Speed Limit	Recommended Minimum Letter Height	
			Initial Upper-Case	Lower-Case
Overhead	All types	All speed limits	12 inches	9 inches
Post-mounted	Multi-lane	More than 40 mph	8 inches	6 inches
Post-mounted	Multi-lane	40 mph or less	6 inches	4.5 inches
Post-mounted	2-lane	All speed limits	6 inches*	4.5 inches*

* On local two-lane streets with speed limits of 25 mph or less, 4-inch initial upper-case letters with 3-inch lower-case letters may be used.

16 An alternative background color other than the normal guide sign color of green may be used for Street Name (D3-1 or D3-1a) signs where the highway agency determines this is necessary to assist road users in determining jurisdictional authority for roads.

Standard:

17 **Alternative background colors shall not be used for Advance Street Name (D3-2) signs (see Section 2D.44).**

18 **The only acceptable alternative background colors for Street Name (D3-1 or D3-1a) signs shall be blue, brown, or white. Regardless of whether green, blue, or brown is used as the background color for Street Name (D3-1 or D3-1a) signs, the legend (and border, if used) shall be white. For Street Name signs that use a white background, the legend (and border, if used) shall be black.**

Guidance:

19 *An alternative background color for Street Name signs, if used, should be applied to the Street Name (D3-1 or D3-1a) signs on all roadways under the jurisdiction of a particular highway agency.*

20 *In business or commercial areas and on principal arterials, Street Name signs should be placed at least on diagonally opposite corners. In residential areas, at least one Street Name sign should be mounted at each intersection. Signs naming both streets should be installed at each intersection. They should be mounted with their faces parallel to the streets they name.*

Option:

21 To optimize visibility, Street Name signs may be mounted overhead. Street Name signs may also be placed above a regulatory or STOP or YIELD sign with no required vertical separation.

Guidance:

22 *In urban or suburban areas, especially where Advance Street Name signs for signalized and other major intersections are not used, the use of overhead Street Name signs should be strongly considered.*

Option:

23 At intersection crossroads where the same road has two different street names for each direction of travel, both street names may be displayed on the same sign along with directional arrows.

24 On lower speed roadways, historic street name signs within locally identified historic districts that are consistent with the criteria contained in 36 CFR 60.4 for such structures and districts may be used without complying with the provisions of Paragraphs 3, 4, 6, 9, 12 through 14, and 18 through 20 of this section.

Support:

25 Information regarding the use of street names on supplemental plaques for use with intersection-related warning signs is contained in Section 2C.58.

Section 2D.44 Advance Street Name Signs (D3-2)

Support:

01 Advance Street Name (D3-2) signs (see Figure 2D-10) identify an upcoming intersection. Although this is often the next intersection, it could also be several intersections away in cases where the next signalized intersection is referenced.

Standard:

02 **Advance Street Name (D3-2) signs, if used, shall supplement rather than be used instead of the Street Name (D3-1) signs at the intersection.**

Option:

03 Advance Street Name (D3-2) signs may be installed in advance of signalized or unsignalized intersections to provide road users with advance information to identify the name(s) of the next intersecting street to prepare for crossing traffic and to facilitate timely deceleration and/or lane changing in preparation for a turn.

Guidance:

04 On arterial highways in rural areas, Advance Street Name signs should be used in advance of all signalized intersections and in advance of all intersections with exclusive turn lanes.

05 In urban areas, Advance Street Name signs should be used in advance of all signalized intersections on major arterial streets, except where signalized intersections are so closely spaced that advance placement of the signs is impractical.

06 The heights of the letters on Advance Street Name signs should be the same as those used for Street Name signs (see Section 2D.43).

Standard:

07 **If used, Advance Street Name signs shall have a white legend and border on a green background.**

08 **If used, Advance Street Name signs shall provide the name(s) of the intersecting street(s) on the top line(s) of the legend and the distance to the intersecting streets or messages such as NEXT SIGNAL, NEXT INTERSECTION, NEXT ROUNDABOUT, or directional arrow(s) on the bottom line of the legend.**

09 **Pictographs shall not be displayed on Advance Street Name signs.**

Option:

10 Directional arrow(s) may be placed to the right or left of the street name or message such as NEXT SIGNAL, as appropriate, rather than on the bottom line of the legend. Curved-stem arrows may be used on Advance Street Name signs on approaches to circular intersections.

11 For intersecting crossroads where the same road has a different street name for each direction of travel, the different street names may be displayed on the same Advance Street Name sign along with directional arrows.

12 In advance of two closely-spaced intersections where it is not practical to install separate Advance Street Name signs, the Advance Street Name sign may include the street names for both intersections along with appropriate supplemental legends for both street names, such as NEXT INTERSECTION, 2ND INTERSECTION, or NEXT LEFT and NEXT RIGHT, or directional arrows.

Guidance:

13 If two street names are used on the Advance Street Name sign, the street names should be displayed in the following order:
   A. For a single intersection where the same road has a different street name for each direction of travel, the name of the street to the left should be displayed above the name of the street to the right; or
   B. For two closely-spaced intersections, the name of the first street encountered should be displayed above the name of the second street encountered, and the arrow associated with the second street encountered should be an advance arrow, such as the arrow shown on the W16-6P arrow plaque (see Figure 2C-12).

Option:

14 An Advance Street Name (W16-8P or W16-8aP) plaque (see Section 2C.58) with black legend on a yellow background, installed supplemental to an Intersection (W2 series) or Advance Traffic Control (W3 series) warning sign may be used instead of an Advance Street Name guide sign.

Section 2D.45 Signing on Conventional Roads on Approaches to Interchanges

Support:

01 Because there are a number of different ramp configurations that are commonly used at interchanges with conventional roads, drivers on the conventional road cannot reliably predict whether they will be required to turn left or right in order to enter the correct ramp to access the freeway or expressway in the desired direction of travel. Consistently applied signing for conventional road approaches to freeway or expressway interchanges is highly desirable.

Standard:

02 **On multi-lane conventional roads approaching an interchange, guide signs shall be provided to identify which direction of turn is to be made and/or which specific lane to use for ramp access to each direction of the freeway or expressway.**

Guidance:

03 The signing of conventional roads with one lane of traffic approaching an interchange should consist of a sequence containing the following signs (see Figure 2D-11):
   A. Junction Assembly
   B. Destination sign
   C. Directional Assembly or Entrance Direction sign for the first ramp
   D. Advance Route Turn Assembly or Advance Entrance Direction sign with an advance turn arrow
   E. Directional Assembly or Entrance Direction sign for the second ramp

Figure 2D-11. Example of Interchange Crossroad Signing for a One-Lane Approach

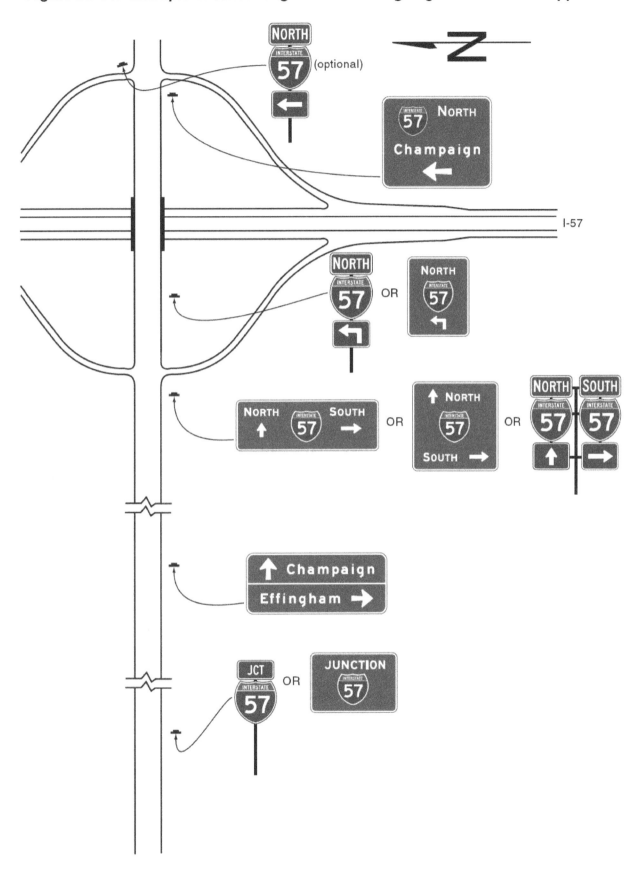

Standard:

04 **If used, the Entrance Direction sign shall consist of a white legend and border on a green background. It shall contain the freeway or expressway route shield(s), cardinal direction, and directional arrow(s).**

Option:

05 The Entrance Direction sign may contain a destination(s) and/or an action message such as NEXT RIGHT.

06 At minor interchanges, the following sequence of signs may be used (see Figure 2D-12):
   A. Junction Assembly
   B. Directional Assembly for the first ramp
   C. Directional Assembly for the second ramp

Guidance:

07 *On multi-lane conventional roads approaching an interchange, the sign sequence should contain the following signs (see Figures 2D-13 through 2D-15):*
   A. *Junction Assembly*
   B. *Advance Entrance Direction sign(s) for both directions (if applicable) of travel on the freeway or expressway*
   C. *Entrance Direction sign for first ramp*
   D. *Advance Turn Assembly*
   E. *Entrance Direction sign for the second ramp*

Support:

08 Advance Entrance Direction signs are used to direct road users to the appropriate lane(s).

Standard:

09 **The Advance Entrance Direction sign shall consist of a white legend and border on a green background. It shall contain the freeway or expressway route shield(s) and cardinal direction(s).**

Option:

10 The Advance Entrance Direction sign may have destinations, directional arrows, and/or an action message such as KEEP LEFT, NEXT LEFT, or SECOND RIGHT. Signs in this sequence may be mounted overhead to improve visibility as shown in Figures 2D-13 through 2D-15.

Support:

11 A post-mounted Advance Entrance Direction diagrammatic guide sign (see Figure 2D-16), within the sequence of approach guide signing described in Paragraphs 3, 6, and 7, might be helpful in depicting the location of a freeway or expressway entrance ramp that is in close proximity to an intervening intersection on the same side of the approach roadway and where signing for only the ramp might cause confusion to road users.

Figure 2D-12. Example of Minor Interchange Crossroad Signing

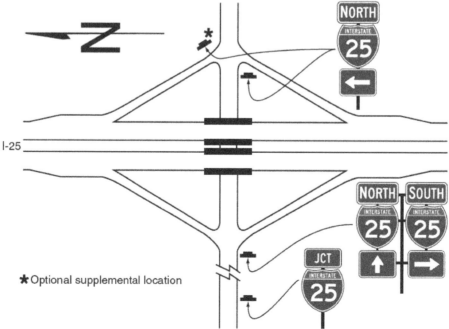

Figure 2D-13. Examples of Multi-Lane Crossroad Signing for a Diamond Interchange

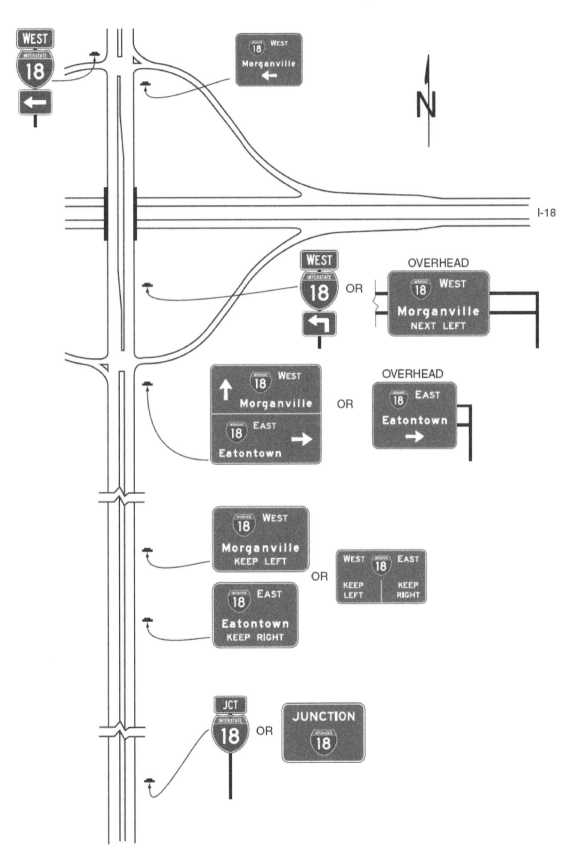

Figure 2D-14. Examples of Multi-Lane Crossroad Signing for a Partial Cloverleaf Interchange

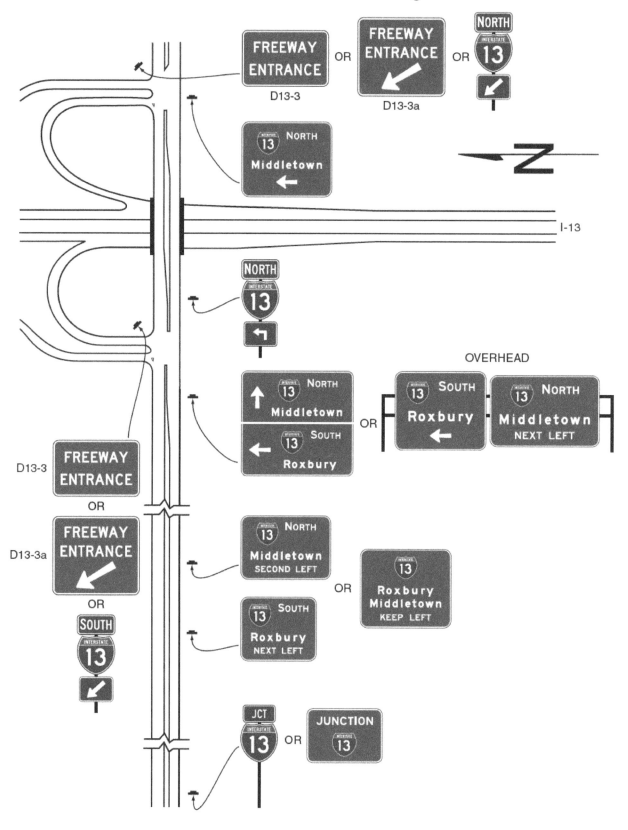

Figure 2D-15. Examples of Multi-Lane Crossroad Signing for a Cloverleaf Interchange

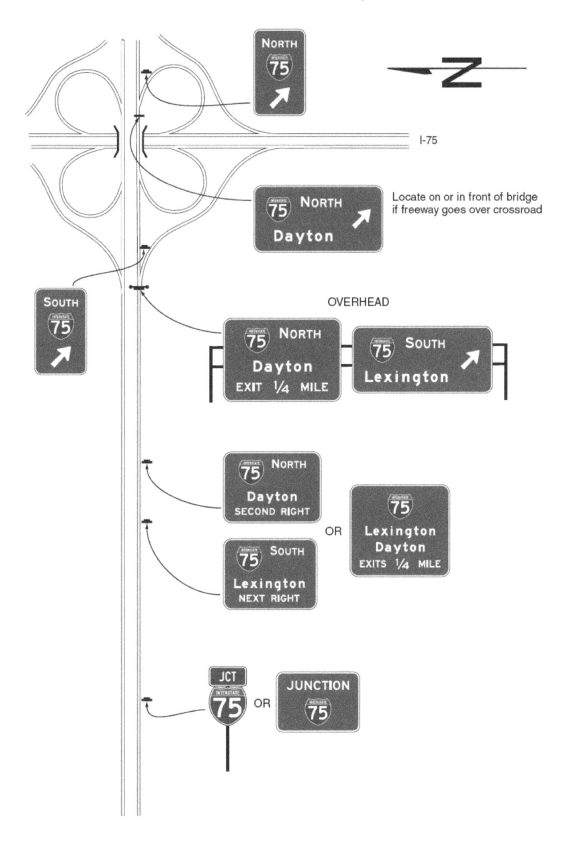

Figure 2D-16. Example of Crossroad Signing for an Entrance Ramp with a Nearby Frontage Road

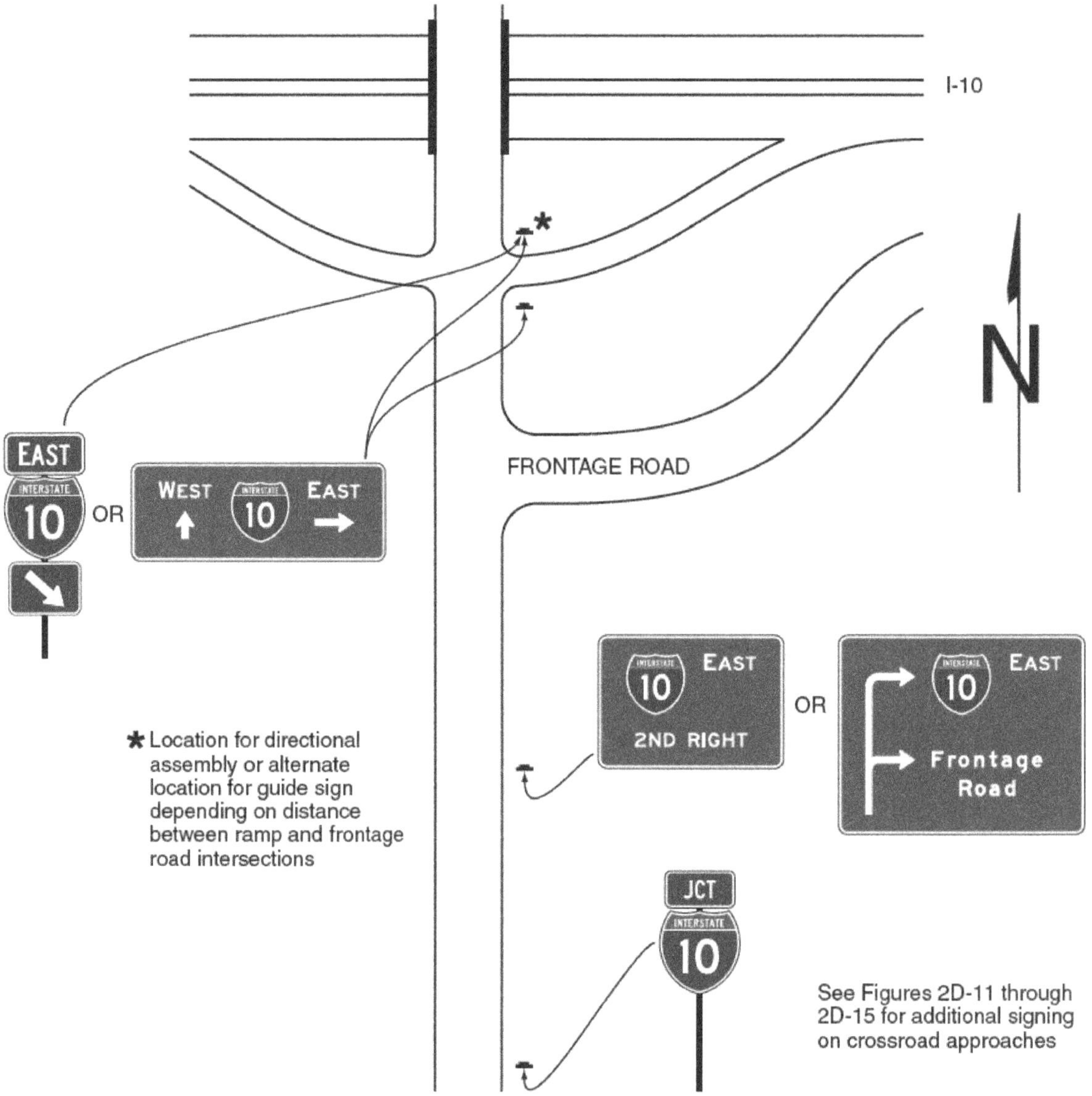

*Location for directional assembly or alternate location for guide sign depending on distance between ramp and frontage road intersections

See Figures 2D-11 through 2D-15 for additional signing on crossroad approaches

Standard:

12   If used, the post-mounted Advance Entrance Direction diagrammatic guide sign shall display only the two successive turns from the same side of the roadway, one of which shall be the entrance ramp. The post-mounted Advance Entrance Direction sign shall depict only the successive turns and shall not depict lane use with lane lines, multiple arrow shafts for the approach roadway, action messages, or other representations.

Support:

13   Section 2D.46 contains information regarding the use of a Directional assembly or a FREEWAY ENTRANCE sign to mark the entrance to a freeway or expressway at the far corner of an intersection.

Section 2D.46 Freeway Entrance Signs (D13-3 and D13-3a)

Option:

01   FREEWAY ENTRANCE (D13-3) signs or FREEWAY ENTRANCE with downward pointing diagonal arrow (D13-3a) signs (see Figure 2D-14) may be used on entrance ramps near the crossroad to inform road users of the freeway or expressway entrance, as appropriate.

₀₂ The D13-3 and D13-3a signs may display an alternate legend in place of FREEWAY, such as EXPRESSWAY or PARKWAY, as appropriate, or may display the name of an unnumbered highway.

₀₃ A Directional assembly (see Section 2D.32) with a downward pointing diagonal arrow auxiliary (M6-2a) sign (see Section 2D.28) may be used at the far left-hand corner of an intersection with a freeway or expressway entrance ramp as an alternative to the D13-3a sign, facing left-turning traffic on the conventional road approach to indicate the immediate point of entry to the freeway or expressway and distinguish the entrance ramp from an adjoining exit ramp terminal at the same intersection with the conventional road (see Figure 2D-14). A similar Directional assembly may be used at the far right-hand corner of an intersection with a freeway or expressway entrance ramp where the entrance ramp and a crossroad or side road follow one another in close succession on the conventional road approach and the point of entry to the freeway or expressway might be difficult for the road user to distinguish from the crossroad or side road on the conventional road approach (see Figure 2D-14).

Support:

₀₄ Section 2B.41 contains information regarding the use of regulatory signs to deter wrong-way movements at intersections of freeway or expressway ramps with conventional roads, and in the area where entrance ramps intersect with the mainline lanes.

Section 2D.47 Parking Area Guide Sign (D4-1)

Option:

₀₁ The Parking Area (D4-1) guide sign (see Figure 2D-10) may be used to show the direction to a nearby public parking area or parking facility.

Standard:

₀₂ **If used, the Parking Area (D4-1) guide sign shall be a horizontal rectangle with a standard size of 30 x 24 inches, or with a smaller size of 18 x 15 inches for minor, low-speed streets. It shall carry the word PARKING, with the letter P five times the height of the remaining letters, and a directional arrow. The legend and border shall be green on a retroreflectorized white background.**

Guidance:

₀₃ *If used, the Parking Area guide sign should be installed on major thoroughfares at the nearest point of access to the parking facility and where it can advise drivers of a place to park. The sign should not be used more than four blocks from the parking area.*

Section 2D.48 PARK - RIDE Sign (D4-2)

Option:

₀₁ PARK - RIDE (D4-2) signs (see Figure 2D-10) may be used to direct road users to park - ride facilities.

Standard:

₀₂ **The signs shall contain the word message PARK - RIDE and direction information (arrow or word message).**

Option:

₀₃ PARK - RIDE signs may contain the local transit pictograph and/or carpool symbol on the sign.

Standard:

₀₄ **If used, the local transit pictograph and/or carpool symbol shall be located in the top part of the sign above the message PARK - RIDE. In no case shall the vertical dimension of the local transit pictograph and/or carpool symbol exceed 18 inches.**

Guidance:

₀₅ *If the function of the parking facility is to provide parking for persons using public transportation, the local transit pictograph should be used on the guide sign. If the function of the parking facility is to serve carpool riders, the carpool symbol should be used on the guide sign. If the parking facility serves both functions, both the pictograph and carpool symbol should be used.*

Standard:

₀₆ **These signs shall have a retroreflective white legend and border on a rectangular green background. The carpool symbol shall be as shown for the D4-2 sign. The color of the local transit pictograph shall be selected by the local transit authority.**

Option:

₀₇ To increase the target value and contrast of the local transit pictograph, and to allow the local transit pictograph to retain its distinctive color and shape, the pictograph may be included within a white border or placed on a white background.

Section 2D.49 Weigh Station Signing (D8 Series)

Support:

01 The general concept for Weigh Station signing is similar to Rest Area signing (see Section 2I.05) because in both cases traffic using either area remains within the right-of-way.

Standard:

02 **The standard installation for Weigh Station signing shall include three basic signs:**
   A. Advance sign (D8-1),
   B. Exit Direction sign (D8-2), and
   C. Exit Gore sign (D8-3).

Support:

03 Example locations of these signs are shown in Figure 2D-17.

Option:

04 Where State law requires a regulatory sign (R13-1) in advance of the Weigh Station, a fourth sign (see Section 2B.60) may be located following the Advance sign.

Guidance:

05 *The Exit Direction sign (D8-2) or the Advance sign (D8-1) should display, either within the sign border or on a supplemental plaque or sign panel, the changeable message OPEN or CLOSED.*

Section 2D.50 Community Wayfinding Signs

Support:

01 Community wayfinding guide signs are part of a coordinated and continuous system of signs that direct tourists and other road users to key civic, cultural, visitor, and recreational attractions and other destinations within a city or a local urbanized or downtown area.

02 Community wayfinding guide signs are a type of destination guide sign for conventional roads with a common color and/or identification enhancement marker for destinations within an overall wayfinding guide sign plan for an area.

03 Figures 2D-18 through 2D-20 illustrate various examples of the design and application of community wayfinding guide signs.

Standard:

04 **The use of community wayfinding guide signs shall be limited to conventional roads. Community wayfinding guide signs shall not be installed on freeway or expressway mainlines or ramps. Direction to community wayfinding destinations from a freeway or expressway shall be limited to the use of a Supplemental Guide sign (see Section 2E.35) on the mainline and a Destination sign (see Section 2D.37) on the ramp to direct road users to the area or areas within which community wayfinding guide signs are used. The individual wayfinding destinations shall not be displayed on the Supplemental Guide and Destination signs except where the destinations are in accordance with the State or agency policy on Supplemental Guide signs.**

05 **Community wayfinding guide signs shall not be used to provide direction to primary destinations or highway routes or streets. Destination or other guide signs shall be used for this purpose as described elsewhere in this Chapter and shall have priority over any community wayfinding sign in placement, prominence, and conspicuity.**

06 **Because regulatory, warning, and other guide signs have a higher priority, community wayfinding guide signs shall not be installed where adequate spacing cannot be provided between the community wayfinding guide sign and other higher priority signs. Community wayfinding guide signs shall not be installed in a position where they would obscure the road users' view of other traffic control devices.**

07 **Community wayfinding guide signs shall not be mounted overhead.**

Guidance:

08 *If used, a community wayfinding guide sign system should be established on a local municipal or equivalent jurisdictional level or for an urbanized area of adjoining municipalities or equivalent that form an identifiable geographic entity that is conducive to a cohesive and continuous system of signs. Community wayfinding guide signs should not be used on a regional or statewide basis where infrequent or sparse placement does not contribute to a continuous or coordinated system of signing that is readily identifiable as such to the road user. In such cases, Destination or other guide signs detailed in this Chapter should be used to direct road users to an identifiable area in which the type of eligible destination described in Paragraph 1 is located.*

Figure 2D-17. Example of Weigh Station Signing

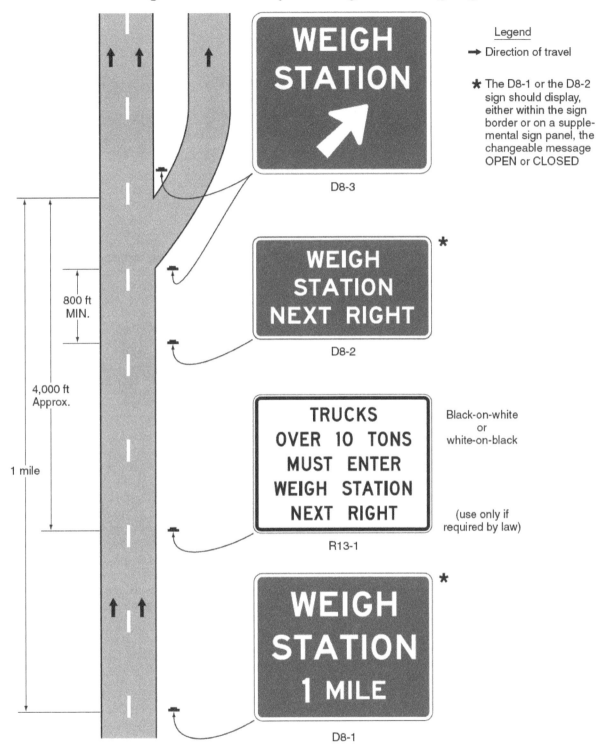

Support:
09 The specific provisions of this Section regarding the design of community wayfinding sign legends apply to vehicular community wayfinding signs and do not apply to those signs that are intended only to provide information or direction to pedestrians or other users of a sidewalk or roadside area.

Figure 2D-18. Examples of Community Wayfinding Guide Signs

A - Community Wayfinding Guide Signs with Enhancement Markers

B - Destination Guide Signs for Color-Coded Community Wayfinding System

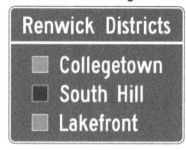 

Guidance:

10   Because pedestrian wayfinding signs typically use smaller legends that are inadequately sized for viewing by vehicular traffic and because they can provide direction to pedestrians that might conflict with that appropriate for vehicular traffic, wayfinding signs designed for and intended to provide direction to pedestrians or other users of a sidewalk or other roadside area should be located to minimize their conspicuity to vehicular traffic. Such signs should be located as far as practical from the street, such as at the far edge of the sidewalk. Where locating such signs farther from the roadway is not practical, the pedestrian wayfinding signs should have their conspicuity to vehicular traffic minimized by employing one or a combination of the following methods:

   A. Locating signs away from intersections where high-priority traffic control devices are present.
   B. Facing the pedestrian message toward the sidewalk and away from the street.
   C. Cantilevering the sign over the sidewalk if the pedestrian wayfinding sign is mounted at a height consistent with vehicular traffic signs, removing the pedestrian wayfinding signs from the line of sight in a sequence of vehicular signs.

11   To further minimize their conspicuity to vehicular traffic during nighttime conditions, pedestrian wayfinding signs should not be retroreflective.

Support:

12   Color coding is sometimes used on community wayfinding guide signs to help road users distinguish between multiple potentially confusing traffic generator destinations located in different neighborhoods or subareas within a community or area.

Option:

13   At the boundaries of the geographical area within which community wayfinding guide signing is used, an informational guide sign (see Figures 2D-18 and 2D-20) may be posted to inform road users about the presence of wayfinding signing and to identify the meanings of the various color codes or pictographs that are being used.

Standard:

14   **These informational guide signs shall have a white legend and border on a green background and shall have a design similar to that illustrated in Figures 2D-1 and 2D-18 and shall be consistent with the basic design principles for guide signs. These informational guide signs shall not be installed on freeway or expressway mainlines or ramps.**

Figure 2D-19. Example of a Community Wayfinding Guide Sign System Showing Direction from a Freeway or Expressway

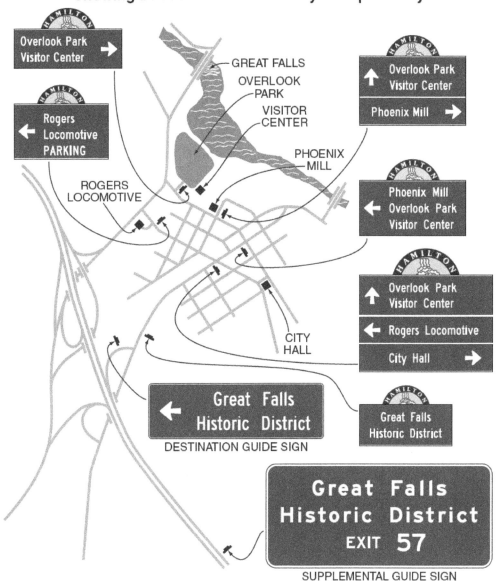

15 The color coding or a pictograph of the identification enhancement markers of the community wayfinding guide signing system shall be included on the informational guide sign posted at the boundary of the community wayfinding guide signing area. The color coding or pictographs shall apply to a specific, identifiable neighborhood or geographical subarea within the overall area covered by the community wayfinding guide signing. Color coding or pictographs shall not be used to distinguish between different types of destinations that are within the same designated neighborhood or subarea. The color coding shall be accomplished by the use of different colored square or rectangular panels on the face of the informational guide sign, each positioned to the left of the neighborhood or named geographic area to which the color-coding panel applies. The height of the colored square or rectangular panels shall not exceed two times the height of the upper-case letters of the principal legend on the sign.

Option:

16 The different colored square or rectangular panels may include either a black or a white (whichever provides the better contrast with the color of the panel) letter, numeral, or other appropriate designation to identify the destination.

Figure 2D-20. Example of a Color-Coded Community Wayfinding Guide Sign System

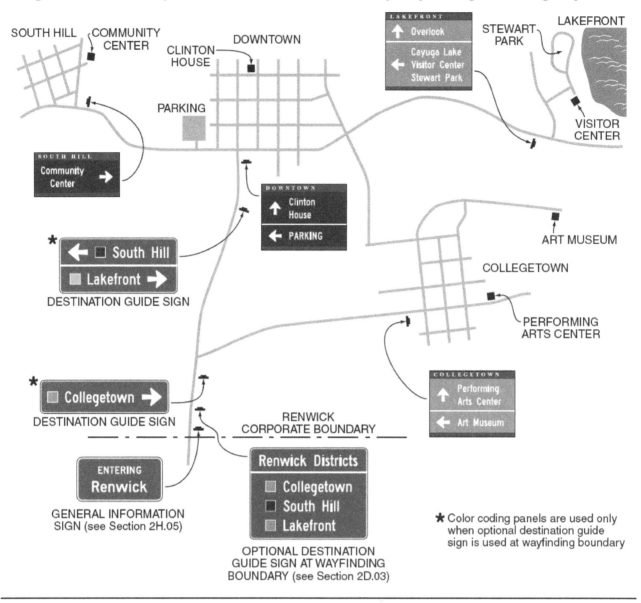

17   Except for the informational guide sign posted at the boundary of the wayfinding guide sign area, community wayfinding guide signs may use background colors other than green in order to provide a color identification for the wayfinding destinations by geographical area within the overall wayfinding guide signing system. Color-coded community wayfinding guide signs may be used with or without the boundary informational guide sign displaying corresponding color-coding panels described in Paragraphs 13 through 16. Except as provided in Paragraphs 18 and 19, in addition to the colors that are approved in this Manual for use on official traffic control signs (see Section 2A.10), other background colors may also be used for the color coding of community wayfinding guide signs.

Standard:

18   **The standard colors of red, orange, yellow, purple, or the fluorescent versions thereof, fluorescent yellow-green, and fluorescent pink shall not be used as background colors for community wayfinding guide signs, in order to minimize possible confusion with critical, higher-priority regulatory and warning sign color meanings readily understood by road users.**

19   **The minimum luminance ratio of legend to background for community wayfinding guide signs shall be 3:1.**

20   **All messages, borders, legends, and backgrounds of community wayfinding guide signs and any identification enhancement markers shall be retroreflective (see Sections 2A.07 and 2A.08).**

Guidance:

21 *Community wayfinding guide signs, exclusive of any identification enhancement marker used, should be rectangular in shape. Simplicity and uniformity in design, position, and application as described in Section 2A.06 are important and should be incorporated into the community wayfinding guide sign design and location plans for the area.*

22 *Community wayfinding guide signs should be limited to three destinations per sign (see Section 2D.07).*

23 *Abbreviations (see Section 1A.15) should be kept to a minimum, and should include only those that are commonly recognized and understood.*

24 *Horizontal lines of a color that contrasts with the sign background color should be used to separate groups of destinations by direction from each other.*

Support:

25 The basic requirement for all highway signs, including community wayfinding signs, is that they be legible to those for whom they are intended and that they be understandable in time to permit a proper response. Section 2A.06 contains additional information on the design of signs, including desirable attributes of effective designs.

Guidance:

26 *Word messages should be as brief as practical and the lettering should be large enough to provide the necessary legibility distance.*

Standard:

27 **The minimum specific ratio of letter height to legibility distance shall comply with the provisions of Section 2A.13. The size of lettering used for destination and directional legends on community wayfinding signs shall comply with the provisions of minimum letter heights as provided in Section 2D.06.**

28 **Interline and edge spacing shall comply with the provisions of Section 2D.06.**

29 **Except as provided in Paragraph 31, the lettering style used for destination and directional legends on community wayfinding guide signs shall comply with the provisions of Section 2D.05.**

30 **The lettering for destinations on community wayfinding guide signs shall be a combination of lower-case letters with initial upper-case letters (see Section 2D.05). All other word messages on community wayfinding guide signs shall be in all upper-case letters.**

Option:

31 A lettering style other than the Standard Alphabets provided in the "Standard Highway Signs and Markings" book may be used on community wayfinding guide signs if an engineering study determines that the legibility and recognition values for the chosen lettering style meet or exceed the values for the Standard Alphabets for the same legend height and stroke width.

Standard:

32 **Except for signs that are intended to be viewed only by pedestrians, bicyclists stopped out of the flow of traffic, or occupants of parked vehicles, Internet and e-mail addresses, including domain names and uniform resource locators (URL), shall not be displayed on any community wayfinding guide sign or sign assembly.**

33 **The arrow location and priority order of destinations shall follow the provisions described in Sections 2D.08 and 2D.37. Arrows shall be of the designs provided in Section 2D.08.**

Option:

34 Pictographs (see definition in Section 1A.13) may be used on community wayfinding guide signs.

Standard:

35 **If a pictograph is used, its height shall not exceed two times the height of the upper-case letters of the principal legend on the sign.**

36 **Except for pictographs, symbols that are not approved in this Manual for use on guide signs shall not be used on community wayfinding guide signs.**

37 **Business logos, commercial graphics, or other forms of advertising (see Section 1A.01) shall not be used on community wayfinding guide signs or sign assemblies.**

Option:

38 Other graphics that specifically identify the wayfinding system, including identification enhancement markers, may be used on the overall sign assembly and sign supports.

Support:

39 An enhancement marker consists of a shape, color, and/or pictograph that is used as a visual identifier for the community wayfinding guide signing system for an area. Figure 2D-18 shows examples of identification enhancement marker designs that can be used with community wayfinding guide signs.

Option:

40 An identification enhancement marker may be used in a community wayfinding guide sign assembly, or may be incorporated into the overall design of a community wayfinding guide sign, as a means of visually identifying the sign as part of an overall system of community wayfinding signs and destinations.

Standard:

41 **The sizes and shapes of identification enhancement markers shall be smaller than the community wayfinding guide signs themselves. Identification enhancement markers shall not be designed to have an appearance that could be mistaken by road users as being a traffic control device.**

Guidance:

42 *The area of the identification enhancement marker should not exceed 1/5 of the area of the community wayfinding guide sign with which it is mounted in the same sign assembly.*

Section 2D.51 Truck, Passing, or Climbing Lane Signs (D17-1 and D17-2)

Guidance:

01 *If an extra lane has been provided for trucks and other slow-moving traffic, a NEXT TRUCK LANE XX MILES (D17-1) sign and/or a TRUCK LANE XX MILES (D17-2) sign (see Figure 2D-21) should be installed in advance of the lane.*

Option:

02 Alternative legends such as PASSING LANE or CLIMBING LANE may be used instead of TRUCK LANE.

03 Section 2B.31 contains information regarding regulatory signs for these types of lanes.

Section 2D.52 Slow Vehicle Turn-Out Sign (D17-7)

Guidance:

01 *If a slow vehicle turn-out area has been provided for slow-moving traffic, a SLOW VEHICLE TURN-OUT XX MILES (D17-7) sign (see Figure 2D-21) should be installed in advance of the turn-out area.*

Option:

02 Section 2B.35 contains information regarding regulatory signs for slow vehicle turn-out areas.

Figure 2D-21. Crossover, Truck Lane, and Slow Vehicle Signs

* The words PASSING or CLIMBING may be substituted for the word TRUCK on the D17-1 and D17-2 signs.

Section 2D.53 Signing of Named Highways

Option:

01 Guide signs may contain street or highway names if the purpose is to enhance driver communication and guidance; however, they are to be considered as supplemental information to route numbers.

Standard:

02 **Highway names shall not replace official numeral designations.**

03 **Memorial names (see Section 2M.10) shall not appear on supplemental signs or on any other information sign on or along the highway or its intersecting routes.**

04 **The use of route signs shall be restricted to signs officially used for guidance of traffic in accordance with this Manual and the "Purpose and Policy" statement of the American Association of State Highway and Transportation Officials that applies to Interstate and U.S. numbered routes (see Page i for AASHTO's address).**

Option:

05 Unnumbered routes having major importance to proper guidance of traffic may be signed if carried out in accordance with the aforementioned policies. For unnumbered highways, a name to enhance route guidance may be used where the name is applied consistently throughout its length.

Guidance:

06 *Only one name should be used to identify any highway, whether numbered or unnumbered.*

Section 2D.54 Crossover Signs (D13-1 and D13-2)

Option:

01 Crossover signs may be installed on divided highways to identify median openings not otherwise identified by warning or other guide signs.

Standard:

02 **A CROSSOVER (D13-1) sign (see Figure 2D-21) shall not be used to identify a median opening that is permitted to be used only by official or authorized vehicles. If used, the sign shall be a horizontal rectangle of appropriate size to carry the word CROSSOVER and a horizontal directional arrow. The CROSSOVER sign shall have a white legend and border on a green background.**

Guidance:

03 *If used, the CROSSOVER sign should be installed immediately beyond the median opening, either on the right-hand side of the roadway or in the median.*

Option:

04 The Advance Crossover (D13-2) sign (see Figure 2D-21) may be installed in advance of the CROSSOVER sign to provide advance notice of the crossover.

Standard:

05 **If used, the Advance Crossover sign shall be a horizontal rectangle of appropriate size to carry the word CROSSOVER and the distance to the median opening. The sign shall have white legend and border on a green background.**

Guidance:

06 *The distance displayed on the Advance Crossover sign should be 1 MILE, 1/2 MILE, or 1/4 MILE, unless unusual conditions require some other distance. If used, the sign should be installed either on the right-hand side of the roadway or in the median at approximately the distance displayed on the sign.*

Section 2D.55 National Scenic Byways Signs (D6-4, D6-4a)

Support:

01 Certain roads have been designated by the U.S. Secretary of Transportation as National Scenic Byways or All-American Roads based on their archeological, cultural, historic, natural, recreational, or scenic qualities.

Option:

02 State and local highway agencies may install the National Scenic Byways (D6-4 or D6-4a) signs at entrance points to a route that has been recognized by the U.S. Secretary of Transportation as a National Scenic Byway or an All-American Road. The D6-4 or D6-4a sign may be installed on route sign assemblies (see Figure 2D-22) or as part of larger roadside structures. National Scenic Byways signs may also be installed at periodic intervals along the designated route and at intersections where the designated route turns or follows a different numbered highway. At locations where roadside features have been developed to enhance the traveler's experience such as rest areas, historic sites, interpretive facilities, or scenic overlooks, the National Scenic Byways sign may be placed on the associated sign assembly to inform travelers that the site contributes to the byway travel experience.

Figure 2D-22. Examples of Use of the National Scenic Byways Sign

Standard:

03 When a National Scenic Byways sign is installed on a National Scenic Byway or an All-American Road, the design shown for the D6-4 or D6-4a sign in Figure 2D-22 shall be used. Use of this design shall be limited to routes that have been designated as a National Scenic Byway or All-American Road by the U.S. Secretary of Transportation.

04 If used, the D6-4 or D6-4a sign shall be placed such that the roadway route signs have primary visibility for the road user.

CHAPTER 2E. GUIDE SIGNS—FREEWAYS AND EXPRESSWAYS

Section 2E.01 Scope of Freeway and Expressway Guide Sign Standards
Support:
01 The provisions of this Chapter provide a uniform and effective system of signing for high-volume, high-speed motor vehicle traffic on freeways and expressways. The requirements and specifications for expressway signing exceed those for conventional roads (see Chapter 2D), but are less than those for freeway signing. Since there are many geometric design variables to be found in existing roads, a signing concept commensurate with prevailing conditions is the primary consideration. Section 1A.13 includes definitions of freeway and expressway.

02 Guide signs for freeways and expressways are primarily identified by the name of the sign rather than by an assigned sign designation. Guidelines for the design of guide signs for freeways and expressways are provided in the "Standard Highway Signs and Markings" book (see Section 1A.11).

Standard:
03 **The provisions of this Chapter shall apply to any highway that meets the definition of freeway or expressway facilities.**

Section 2E.02 Freeway and Expressway Signing Principles
Support:
01 The development of a signing system for freeways and expressways is approached on the premise that the signing is primarily for the benefit and direction of road users who are not familiar with the route or area. The signing furnishes road users with clear instructions for orderly progress to their destinations. Sign installations are an integral part of the facility and, as such, are best planned concurrently with the development of highway location and geometric design. For optimal results, plans for signing are analyzed during the earliest stages of preliminary design, and details are correlated as final design is developed. The excessive signing found on many major highways usually is the result of using a multitude of signs that are too small and that are poorly designed and placed to accomplish the intended purpose.

02 Freeway and expressway signing is to be considered and developed as a planned system of installations. An engineering study is sometimes necessary for proper solution of the problems of many individual locations, but, in addition, consideration of an entire route is necessary.

Guidance
03 *Road users should be guided with consistent signing on the approaches to interchanges, when they drive from one State to another, and when driving through rural or urban areas. Because geographical, geometric, and operating factors regularly create significant differences between urban and rural conditions, the signing should take these conditions into account.*

04 *Guide signs on freeways and expressways should serve distinct functions as follows:*
   A. *Give directions to destinations, or to streets or highway routes, at intersections or interchanges;*
   B. *Furnish advance notice of the approach to intersections or interchanges;*
   C. *Direct road users into appropriate lanes in advance of diverging or merging movements;*
   D. *Identify routes and directions on those routes;*
   E. *Show distances to destinations;*
   F. *Indicate access to general motorist services, rest, scenic, and recreational areas; and*
   G. *Provide other information of value to the road user.*

Section 2E.03 Guide Sign Classification
Support:
01 Freeway and expressway guide signs are classified and treated in the following categories:
   A. Route signs and Trailblazer Assemblies (see Section 2E.27),
   B. At-Grade Intersection signs (see Section 2E.29),
   C. Interchange signs (see Sections 2E.30 through 2E.39),
   D. Interchange Sequence signs (see Section 2E.40),
   E. Community Interchanges Identification signs (see Section 2E.41),
   F. NEXT XX EXITS signs (see Section 2E.42),
   G. Weigh Station signing (see Section 2E.54),
   H. Miscellaneous Information signs (see Section 2H.04)
   I. Reference Location signs (see Section 2H.05),
   J. General Service signs (see Chapter 2I),
   K. Rest and Scenic Area signs (see Section 2I.05),
   L. Tourist Information and Welcome Center signs (see Section 2I.08),
   M. Radio Information signing (see Section 2I.09)
   N. Carpool and Ridesharing signing (see Section 2I.11),
   O. Specific Service signs (see Chapter 2J), and
   P. Recreational and Cultural Interest Area signs (see Chapter 2M).

Section 2E.04 General
Support:

01 Signs are designed so that they are legible to road users approaching them and readable in time to permit proper responses. Desired design characteristics include: (a) long visibility distances, (b) large lettering, symbols, and arrows, and (c) short legends for quick comprehension.

Standard:

02 **Standard shapes and colors shall be used so that traffic signs can be promptly recognized by road users.**

Section 2E.05 Color of Guide Signs
Standard:

01 **Guide signs on freeways and expressways, except as otherwise provided in this Manual, shall have white letters, symbols, arrows, and borders on a green background.**

Support:

02 Color requirements for route signs and trailblazers, signs with blank-out or changeable messages, signs for services, rest areas, park and recreational areas, and for certain miscellaneous signs are provided in the individual Sections dealing with the particular sign or sign group.

Section 2E.06 Retroreflection or Illumination
Standard:

01 **Letters, numerals, symbols, arrows, and borders of all guide signs shall be retroreflectorized. The background of all guide signs that are not independently illuminated shall be retroreflective.**

Support:

02 Where there is no serious interference from extraneous light sources, retroreflectorized post-mounted signs usually provide adequate nighttime visibility.

03 On freeways and expressways where much driving at night is done with low-beam headlights, the amount of headlight illumination incident to an overhead sign display is relatively small.

Guidance:

04 *Overhead sign installations should be illuminated unless an engineering study shows that retroreflectorization alone will perform effectively. The type of illumination chosen should provide effective and reasonably uniform illumination of the sign face and message.*

Section 2E.07 Characteristics of Urban Signing
Support:

01 Urban conditions are characterized not so much by city limits or other arbitrary boundaries, as by the following features:
- A. Mainline roadways with more than two lanes in each direction;
- B. High traffic volumes on the through roadways;
- C. High volumes of traffic entering and leaving interchanges;
- D. Interchanges closely spaced;
- E. Roadway and interchange lighting;
- F. Three or more interchanges serving the major city;
- G. A loop, circumferential, or spur serving a sizable portion of the urban population; and
- H. Visual clutter from roadside development.

02 Operating conditions and road geometrics on urban freeways and expressways usually make special sign treatments desirable, including:
- A. Use of Interchange Sequence signs (see Section 2E.40);
- B. Use of sign spreading to the maximum extent possible (see Section 2E.11);
- C. Elimination of General or Specific Service signing (see Chapters 2I and 2J);
- D. Reduction to a minimum of post-interchange signs (see Section 2E.38);
- E. Display of advance signs at distances closer to the interchange, with appropriate adjustments in the legend (see Section 2E.33);
- F. Use of overhead signs on roadway structures and independent sign supports (see Section 2E.25);
- G. Use of Overhead Arrow-per-Lane or Diagrammatic guide signs in advance of intersections and interchanges (see Sections 2E.21 and 2E.22); and
- H. Frequent use of street names as the principal message in guide signs.

03   Lower speeds which are often characteristic of urban operations do not justify lower signing standards. Typical traffic patterns are more complex for the road user to negotiate, and large, easy-to-read legends are, therefore, just as necessary as on rural highways.

Section 2E.08 Characteristics of Rural Signing

Support:

01   Rural areas ordinarily have greater distances between interchanges, which permits adequate spacing for the sequences of signs on the approach to and departure from each interchange. However, the absence of traffic in adjoining lanes and on entering or exiting ramps often adds monotony or inattention to rural driving. This increases the importance of signs that call for decisions or actions.

Guidance:

02   *Where there are long distances between interchanges and the alignment is relatively unchanging, signs should be positioned for their best effect on road users. The tendency to group all signing in the immediate vicinity of rural interchanges should be avoided by considering the entire route in the development of signing plans. Extra effort should be given to the placement of signs at natural target locations to command the attention of the road user, particularly when the message requires an action by the road user.*

Section 2E.09 Signing of Named Highways

Support:

01   Section 2D.53 contains information, which is also applicable to freeways and expressways, regarding the use of highway names on the signing for unnumbered highways to enhance route guidance and facilitate travel.

02   Section 2M.10 contains information regarding memorial signing of routes, bridges, or highway components.

Section 2E.10 Amount of Legend on Guide Signs

Guidance:

01   *No more than two destination names or street names should be displayed on any Advance Guide sign or Exit Direction sign. A city name and street name on the same sign should be avoided. Where two or three signs are placed on the same supports, destinations or names should be limited to one per sign, or to a total of three in the display. Sign legends should not exceed three lines of copy, exclusive of the exit number and action or distance information.*

Section 2E.11 Number of Signs at an Overhead Installation and Sign Spreading

Guidance:

01   *If overhead signs are warranted, as set forth in Section 2A.17, the number of signs at these locations should be limited to only those essential in communicating pertinent destination information to the road user. Exit Direction signs for a single exit and the Advance Guide signs should have only one sign with one or two destinations. Regulatory signs, such as speed limits, should not be used in conjunction with overhead guide sign installations. Because road users have limited time to read and comprehend sign messages, there should not be more than three guide signs displayed at any one location either on the overhead structure or its support.*

Option:

02   At overhead locations, more than one sign may be installed to advise of a multiple exit condition at an interchange. If the roadway ramp or crossing roadway has complex or unusual geometrics, additional signs with confirming messages may be provided to properly guide the road user.

Support:

03   Sign spreading is a concept where major overhead signs are spaced so that road users are not overloaded with a group of signs at a single location. Figure 2E-1 illustrates an example of sign spreading.

Guidance:

04   *Where overhead signing is used, sign spreading should be used at all single exit interchanges and to the extent possible at multi-exit interchanges. Sign spreading should be accomplished by use of the following:*

   A. The Exit Direction sign should be the only sign used in the vicinity of the gore (other than the Exit Gore sign). It should be located overhead near the theoretical gore and generally on an overhead sign support structure.

   B. The Advance Guide sign to indicate the next interchange exit should be placed near the crossroad location. If the crossroad goes over the mainline, the Advance Guide sign should be placed on the overcrossing structure or on a separate structure immediately in front of the overcrossing structure.

Figure 2E-1. Example of Guide Sign Spreading

Section 2E.12 Pull-Through Signs (E6-2, E6-2a)

Support:

01   Pull-Through (E6-2, E6-2a) signs (see Figure 2E-2) are overhead guide signs intended for through traffic.

Guidance:

02   *Pull-Through signs should be used where the geometrics of a given interchange are such that it is not clear to the road user as to which is the through roadway, or where additional route guidance is desired. Pull-Through signs with down arrows should be used where the alignment of the through lanes is curved and the exit direction is straight ahead, where the number of through lanes is not readily evident, and at multi-lane exits where there is a reduction in the number of through lanes.*

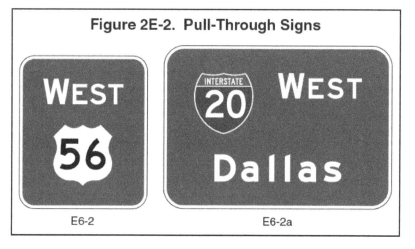

Figure 2E-2. Pull-Through Signs

Support:

03   Sections 2E.20 through 2E.24 contain information regarding the use of Overhead Arrow-per-Lane or Diagrammatic guide signs at multi-lane exits where there is a reduction in the number of through lanes and a through lane becomes an interior option lane for through or exiting traffic.

Section 2E.13 Designation of Destinations

Standard:

01   **The direction of a freeway and the major destinations or control cities along it shall be clearly identified through the use of appropriate destination legends (see Section 2D.37). Successive freeway guide signs shall provide continuity in destination names and consistency with available map information. At any decision point, a given destination shall be indicated by way of only one route.**

Guidance:

02   *Control city legends should be used in the following situations along a freeway:*
   A. *At interchanges between freeways;*
   B. *At separation points of overlapping freeway routes;*
   C. *On directional signs on intersecting routes, to guide traffic entering the freeway;*
   D. *On Pull-Through signs; and*
   E. *On the bottom line of post-interchange distance signs.*

Support:

03 Continuity of destination names is also useful on expressways serving long-distance or intrastate travel.

04 The determination of major destinations or control cities is important to the quality of service provided by the freeway. Control cities on freeway guide signs are selected by the States and are contained in the "Guidelines for the Selection of Supplemental Guide Signs for Traffic Generators Adjacent to Freeways, 4th Edition/Guide Signs, Part II: Guidelines for Airport Guide Signing/Guide Signs, Part III: List of Control Cities for Use in Guide Signs on Interstate Highways," published by and available from the American Association of State and Highway Transportation Officials (see Section 1A.11).

Section 2E.14 Size and Style of Letters and Signs

Standard:

01 **Except as provided in Section 2A.11, the sizes of freeway and expressway guide signs that have standardized designs shall be as shown in Table 2E-1.**

Support:

02 Section 2A.11 contains information regarding the applicability of the various columns in Table 2E-1.

Option:

03 Signs larger than those shown in Table 2E-1 may be used (see Section 2A.11).

Standard:

04 **For all freeway and expressway signs that do not have a standardized design, the message dimensions shall be determined first, and the outside sign dimensions secondarily. Word messages in the legend of expressway guide signs shall be in letters at least 8 inches high. Larger lettering shall be used for major guide signs at or in advance of interchanges and for all overhead signs. Minimum numeral and letter sizes for expressway guide signs according to interchange classification, type of sign, and component of sign legend shall be as shown in Tables 2E-2 and 2E-3. Minimum numeral and letter sizes for freeway guide signs according to interchange classification, type of sign, and component of sign legend shall be as shown in Tables 2E-4 and 2E-5. All names of places, streets, and highways on freeway and expressway guide signs shall be composed of lower-case letters with initial upper-case letters. The letters and the numerals used shall be Series E(M) of the "Standard Highway Signs and Markings" book (see Section 1A.11). The nominal loop height of the lower-case letters shall be 3/4 of the height of the initial upper-case letter (see Paragraph 2 of Section 2D.05 for additional information on the specification of letter heights). Other word legends shall be composed of upper-case letters. Interline and edge spacing shall be as provided in Section 2E.15.**

05 **Lettering size on freeway and expressway signs shall be the same for both rural and urban conditions.**

Support:

06 Sign size is determined primarily in terms of the length of the message and the size of the lettering necessary for proper legibility. Letter style and height, and arrow design have been standardized for freeway and expressway signs to assure uniform and effective application.

07 Designs for upper-case and lower-case alphabets together with Tables of recommended letter spacing, are shown in the "Standard Highway Signs and Markings" book (see Section 1A.11).

Guidance:

08 *Freeway lettering sizes (see Tables 2E-4 and 2E-5) should be used when expressway geometric design is comparable to freeway standards.*

09 *Other sign letter size requirements not specifically identified elsewhere in this Manual should be guided by these specifications. Abbreviations (see Section 2E.17) should be kept to a minimum.*

Support:

10 A sign mounted over a particular roadway lane to which it applies might have to be limited in horizontal dimension to the width of the lane, so that another sign can be placed over an adjacent lane. The necessity to maintain proper vertical clearance might also place a further limitation on the size of the overhead sign and the legend that can be accommodated.

Section 2E.15 Interline and Edge Spacing

Guidance:

01 *Interline spacing of upper-case letters should be approximately three-fourths the average of upper-case letter heights in adjacent lines of letters.*

02 *The spacings to the top and bottom borders should be equal to the average of the letter height of the adjacent line of letters. The lateral spacing to the vertical borders should be essentially the same as the height of the largest letter.*

Table 2E-1. Freeway or Expressway Guide Sign and Plaque Sizes (Sheet 1 of 2)

Sign or Plaque	Sign Designation	Section	Minimum Size
Exit Number (plaque)			
1-, 2-Digit Exit Number	E1-5P	2E.31	114 x 30
3-Digit Exit Number	E1-5P	2E.31	132 x 30
1-, 2-Digit Exit Number (with single letter suffix)	E1-5P	2E.31	138 x 30
3-Digit Exit Number (with single letter suffix)	E1-5P	2E.31	156 x 30
1-, 2-Digit Exit Number (with dual letter suffix)	E1-5P	2E.31	168 x 30
3-Digit Exit Number (with dual letter suffix)	E1-5P	2E.31	186 x 30
Left (plaque)	E1-5aP	2E.33	72 x 30
Left Exit Number (plaque)			
1-, 2-Digit Exit Number	E1-5bP	2E.31	114 x 54
3-Digit Exit Number	E1-5bP	2E.31	132 x 54
1-, 2-Digit Exit Number (with single letter suffix)	E1-5bP	2E.31	138 x 54
3-Digit Exit Number (with single letter suffix)	E1-5bP	2E.31	156 x 54
1-, 2-Digit Exit Number (with dual letter suffix)	E1-5bP	2E.31	168 x 54
3-Digit Exit Number (with dual letter suffix)	E1-5bP	2E.31	186 x 54
Next Exit XX Miles (1 line)	—	2E.34	Varies x 24
Next Exit XX Miles (2 lines)	—	2E.34	Varies x 36
Exit Gore (no exit number)	E5-1	2E.37	72 x 60
Exit Gore (with exit number)			
1-, 2-Digit Exit Number	E5-1a	2E.37	78 x 60
3-Digit Exit Number	E5-1a	2E.37	96 x 60
1-Digit Exit Number (with single letter suffix)	E5-1a	2E.37	90 x 60
2-Digit Exit Number (with single letter suffix)	E5-1a	2E.37	108 x 60
3-Digit Exit Number (with single letter suffix)	E5-1a	2E.37	126 x 60
1-Digit Exit Number (with dual letter suffix)	E5-1a	2E.37	120 x 60
2-Digit Exit Number (with dual letter suffix)	E5-1a	2E.37	138 x 60
3-Digit Exit Number (with dual letter suffix)	E5-1a	2E.37	156 x 60
Exit Number (plaque)			
1-, 2-Digit Exit Number	E5-1bP	2E.37	42 x 30
3-Digit Exit Number	E5-1bP	2E.37	60 x 30
1-Digit Exit Number (with single letter suffix)	E5-1bP	2E.37	48 x 30
1-Digit Exit Number (with dual letter suffix)	E5-1bP	2E.37	72 x 30
2-Digit Exit Number (with single or dual letter suffix)	E5-1bP	2E.37	72 x 30
3-Digit Exit Number (with single or dual letter suffix)	E5-1bP	2E.37	72 x 30
Narrow Exit Gore	E5-1c	2E.37	60 x 90*
Pull-Through	E6-2	2E.12	Varies x 120*
Pull-Through	E6-2a	2E.12	Varies x 90*
Exit Only (with arrow)	E11-1,1d	2E.24	174** x 36
Exit	E11-1a	2E.24	66 x 18
Only	E11-1b	2E.24	66 x 18
Exit Only	E11-1c	2E.24	120 x 18
Exit Only (with two arrows)	E11-1e,1f	2E.24	222** x 36
Left	E11-2	2E.40	60 x 18
Exit Gore Advisory Speed (plaque)	E13-1P	2E.37	72 x 24
Exit Direction Advisory Speed	E13-2	2E.36	162 x 24
Interstate Route Sign (1 or 2 digits)	M1-1	2E.27	36 x 36
Interstate Route Sign (3 digits)	M1-1	2E.27	45 x 36
Off-Interstate Route Sign (1 or 2 digits)	M1-2,3	2E.27	36 x 36
Off-Interstate Route Sign (3 digits)	M1-2,3	2E.27	45 x 36
U.S. Route Sign (1 or 2 digits)	M1-4	2E.27	36 x 36
U.S. Route Sign (3 digits)	M1-4	2E.27	45 x 36
State Route Sign (1 or 2 digits)	M1-5	2D.11	36 x 36

Table 2E-1. Freeway or Expressway Guide Sign and Plaque Sizes (Sheet 2 of 2)

Sign or Plaque	Sign Designation	Section	Minimum Size
State Route Sign (3 digits)	M1-5	2D.11	45 x 36
County Route Sign (1, 2, or 3 digits)	M1-6	2D.11	36 x 36
Forest Route (1, 2, or 3 digits)	M1-7	2D.11	36 x 36
Eisenhower Interstate System	M1-10,10a	2E.28	36 x 36
Junction	M2-1	2D.13	30 x 21
Combination Junction (2 route signs)	M2-2	2D.14	60 x 48*
Cardinal Direction	M3-1,2,3,4	2D.15	36 x 18
Alternate	M4-1,1a	2D.17	36 x 18
By-Pass	M4-2	2D.18	36 x 18
Business	M4-3	2D.19	36 x 18
Truck	M4-4	2D.20	36 x 18
To	M4-5	2D.21	36 x 18
End	M4-6	2D.22	36 x 18
Temporary	M4-7,7a	2D.24	36 x 18
Begin	M4-14	2D.23	36 x 18
Advance Turn Arrow	M5-1,2,3	2D.26	30 x 21
Lane Designation	M5-4,5,6	2D.27	36 x 24
Directional Arrow	M6-1,2,2a,3,4,5,6,7	2D.28	30 x 21
Destination (1 line)	D1-1	2D.37	Varies x 30
Destination and Distance (1 line)	D1-1a	2D.37	Varies x 30
Destination (2 lines)	D1-2	2D.37	Varies x 54
Destination and Distance (2 lines)	D1-2a	2D.37	Varies x 54
Destination (3 lines)	D1-3	2D.37	Varies x 72
Destination and Distance (3 lines)	D1-3a	2D.37	Varies x 72
Distance (1 line)	D2-1	2D.41	Varies x 30
Distance (2 lines)	D2-2	2D.41	Varies x 54
Distance (3 lines)	D2-3	2D.41	Varies x 72
Street Name	D3-1,1a	2D.43	Varies x 18
Advance Street Name (2 lines)	D3-2	2D.44	Varies x 42*
Advance Street Name (3 lines)	D3-2	2D.44	Varies x 66*
Advance Street Name (4 lines)	D3-2	2D.44	Varies x 84*
Park - Ride	D4-2	2D.48	36 x 48
National Scenic Byways	D6-4	2D.55	24 x 24
National Scenic Byways	D6-4a	2D.55	24 x 12
Weigh Station XX Miles	D8-1	2E.54	96 x 72 (F) 78 x 60 (E)
Weigh Station Next Right	D8-2	2E.54	108 x 90 (F) 84 x 72 (E)
Weigh Station (with arrow)	D8-3	2E.54	84 x 78 (F) 66 x 60 (E)
Crossover	D13-1,2	2D.54	78 x 42
Freeway Entrance	D13-3	2D.46	48 x 30
Freeway Entrance (with arrow)	D13-3a	2D.46	48 x 42
Combination Lane Use / Destination	D15-1	2D.33	Varies x 96
Next Truck Lane XX Miles	D17-1	2D.51	60 x 66
Truck Lane XX Miles	D17-2	2D.51	60 x 54
Slow Vehicle Turn-Out XX Miles	D17-7	2D.52	96 x 54

* The size shown is for a typical sign as illustrated in the figures in Chapters 2D and 2E. The size should be determined based on the amount of legend required for the sign.
** The width shown represents the minimum dimension. The width shall be increased as appropriate to match the width of the guide sign.

Notes: 1. Larger signs may be used when appropriate
2. Dimensions in inches are shown as width x height
3. Where two sizes are shown, the larger size is for freeways (F) and the smaller size is for expressways (E)

Table 2E-2. Minimum Letter and Numeral Sizes for Expressway Guide Signs According to Interchange Classification

Type of Sign	Type of Interchange (see Section 2E.32)				Overhead
	Major		Intermediate	Minor	
	Category a	Category b			
A. Advance Guide, Exit Direction, and Overhead Guide Signs					
Exit Number Plaques					
Words	10	10	10	8	10
Numerals & Letters	15	15	15	12	15
Interstate Route Signs					
Numerals	18	—	—	—	18
1- or 2-Digit Shields	36 x 36	—	—	—	36 x 36
3-Digit Shields	45 x 36	—	—	—	45 x 36
U.S. or State Route Signs					
Numerals	18	18	18	12	18
1- or 2-Digit Shields	36 x 36	36 x 36	36 x 36	24 x 24	36 x 36
3-Digit Shields	45 x 36	45 x 36	45 x 36	30 x 24	45 x 36
U.S. or State Route Text Identification (Example: US 56)					
Numerals & Letters	18	15	15	12	15
Cardinal Directions					
First Letters	18	15	12	10	15
Rest of Words	15	12	10	8	12
Auxiliary and Alternative Route Legends (Examples: JCT, TO, ALT, BUSINESS)					
Words	15	12	10	8	12
Names of Destinations					
Upper-Case Letters	20	16	13.33	10.67	16
Lower-Case Letters	15	12	10	8	12
Distance Numbers	18	15	12	10	15
Distance Fraction Numerals	12	10	10	8	10
Distance Words	12	10	10	8	10
Action Message Words	10	10	10	8	10
B. Gore Signs					
Words	10	10	10	8	—
Numerals & Letters	12	12	12	10	—

Note: Sizes are shown in inches and where applicable are shown as width x height

Table 2E-3. Minimum Letter and Numeral Sizes for Expressway Guide Signs According to Sign Type

Type of Sign	Minimum Size
A. Pull-Through Signs	
Destinations — Upper-Case Letters	13.33
Destinations — Lower-Case Letters	10
Route Signs	
1- or 2-Digit Shields	36 x 36
3-Digit Shields	45 x 36
Cardinal Directions — First Letters	12
Cardinal Directions — Rest of Word	10
B. Supplemental Guide Signs	
Exit Number — Words	8
Exit Number — Numerals and Letters	12
Place Names — Upper-Case Letters	10.67
Place Names — Lower-Case Letters	8
Action Messages	8
Route Signs	
Numerals	12
1- or 2-Digit Shield	24 x 24
3-Digit Shield	30 x 24
C. Interchange Sequence or Community Interchanges Identification Signs	
Words — Upper-Case Letters	10.67
Words — Lower-Case Letters	8
Numerals	10.67
Fraction Numerals	8
Route Signs	
Numerals	12
1- or 2-Digit Shield	24 x 24
3-Digit Shield	30 x 24
D. Next XX Exits Sign	
Place Names — Upper-Case Letters	10.67
Place Names — Lower-Case Letters	8
NEXT XX EXITS — Words	8
NEXT XX EXITS — Number	12
E. Distance Signs	
Words — Upper-Case Letters	8
Words — Lower-Case Letters	6
Numerals	8
Route Signs	
Numerals	9
1- or 2-Digit Shield	18 x 18
3-Digit Shield	22.5 x 18
F. General Services Signs (see Chapter 2I)	
Exit Number — Words	8
Exit Number — Numerals and Letters	12
Services	8
G. Rest Area, Scenic Area, and Roadside Area Signs (see Chapter 2I)	
Words	10
Distance Numerals	12
Distance Fraction Numerals	8
Distance Words	8
Action Message Words	10
H. Reference Location Signs (see Chapter 2H)	
Words	4
Numerals	10
I. Boundary and Orientation Signs (see Chapter 2H)	
Words — Upper-Case Letters	8
Words — Lower-Case Letters	6
J. Next Exit and Next Services Signs	
Words and Numerals	8
K. Exit Only Signs	
Words	12
L. Overhead Arrow-Per-Lane and Diagrammatic Signs	
See Table 2E-5	

Note: Sizes are shown in inches and where applicable are shown as width x height

Table 2E-4. Minimum Letter and Numeral Sizes for Freeway Guide Signs According to Interchange Classification

Type of Sign	Type of Interchange (see Section 2E.32)				Overhead
	Major		Intermediate	Minor	
	Category a	Category b			
A. Advance Guide, Exit Direction, and Overhead Guide Signs					
Exit Number Plaques					
Words	10	10	10	10	10
Numerals & Letters	15	15	15	15	15
Interstate Route Signs					
Numerals	24/18	—	—	—	18
1- or 2-Digit Shields	48 x 48/ 36 x 36	—	—	—	36 x 36
3-Digit Shields	60 x 48/ 45 x 36	—	—	—	45 x 36
U.S. or State Route Signs					
Numerals	24/18	18	18	12	18
1- or 2-Digit Shields	48 x 48/ 36 x 36	36 x 36	36 x 36	24 x 24	36 x 36
3-Digit Shields	60 x 48/ 45 x 36	45 x 36	45 x 36	30 x 24	45 x 36
U.S. or State Route Text Identification (Example: US 56)					
Numerals & Letters	18	18/15	15	12	15
Cardinal Directions					
First Letters	18	15	15	10	15
Rest of Words	15	12	12	8	12
Auxiliary and Alternative Route Legends (Examples: JCT, TO, ALT, BUSINESS)					
Words	15	12	12	8	12
Names of Destinations					
Upper-Case Letters	20	20	16	13.33	16
Lower-Case Letters	15	15	12	10	12
Distance Numbers	18	18/15	15	12	15
Distance Fraction Numerals	12	12/10	10	8	10
Distance Words	12	12/10	10	8	10
Action Message Words	12	12/10	10	8	10
B. Gore Signs					
Words	12	12	12	8	—
Numeral & Letters	18	18	18	12	—

Notes: 1. Sizes are shown in inches and where applicable are shown as width x height
2. Slanted line (/) signifies separation of desirable and minimum sizes

Table 2E-5. Minimum Letter and Numeral Sizes for Freeway Guide Signs According to Sign Type

Type of Sign	Minimum Size
A. Pull-Through Signs	
Destinations — Upper-Case Letters	16
Destinations — Lower-Case Letters	12
Route Signs	
1- or 2-Digit Shields	36 x 36
3-Digit Shields	45 x 36
Cardinal Directions — First Letter	15
Cardinal Directions — Rest of Word	12
B. Supplemental Guide Signs	
Exit Number Words	10
Exit Number Numerals and Letters	15
Place Names — Upper-Case Letters	13.33
Place Names — Lower-Case Letters	10
Action Messages	8
Route Signs	
Numerals	12
1- or 2-Digit Shield	24 x 24
3-Digit Shield	30 x 24
C. Interchange Sequence or Community Interchanges Identification Signs	
Words — Upper-Case Letters	13.33
Words — Lower-Case Letters	10
Numerals	13.33
Fraction Numerals	10
Route Signs	
Numerals	12
1- or 2-Digit Shield	24 x 24
3-Digit Shield	30 x 24
D. Next XX Exits Sign	
Place Names — Upper-Case Letters	13.33
Place Names — Lower-Case Letters	10
NEXT XX EXITS — Words	10
NEXT XX EXITS — Number	15
E. Distance Signs	
Words — Upper-Case Letters	8
Words — Lower-Case Letters	6
Numerals	8
Route Signs	
Numerals	9
1- or 2-Digit Shield	18 x 18
3-Digit Shield	22.5 x 18
F. General Services Signs (see Chapter 2I)	
Exit Number Words	10
Exit Number Numerals and Letters	15
Services	10

Type of Sign	Minimum Size
G. Rest Area, Scenic Area, and Roadside Area Signs (see Chapter 2I)	
Words	12
Distance Numerals	15
Distance Fraction Numerals	10
Distance Words	10
Action Message Words	12
H. Reference Location Signs (see Chapter 2H)	
Words	4
Numerals	10
I. Boundary and Orientation Signs (see Chapter 2H)	
Words — Upper-Case Letters	8
Words — Lower-Case Letters	6
J. Next Exit and Next Services Signs	
Words and Numerals	8
K. Exit Only Signs	
Words	12
L. Overhead Arrow-Per-Lane Signs	
Arrowhead (Type D Directional Arrow)	21.625
Arrow Shaft Width	8
Arrow Height	
Through	72
Left Only	48
Right Only	48
Optional-Diverge (Through with Left or Right)	72
Optional-Split (Left and Right)	66
Vertical Separator Width	2
Vertical Space between Vertical Separator and Top of Nearest Arrow	8
Horizontal Space between Vertical Separator and Top of Nearest Through Arrow	15
Horizontal Space between Arrow Shaft and EXIT and ONLY plaques	10
EXIT and ONLY Panels	60 x 18
M. Diagrammatic Signs	
Arrowhead (Type D Directional Arrow)	13.5*
Lane Widths	5
Lane Line Segments	1 x 6
Spacing between Lane Line Segments	6
Stem Height to Upper Point of Departure	30
Horizontal Space between Arrowhead and Route Shield or Destination	12

* The size shown is the arrowhead width per lane depicted on the corresponding arrow shaft

Note: Sizes are shown in inches and where applicable are shown as width x height

Section 2E.16 Sign Borders

Standard:

01 **Signs shall have a border of the same color as the legend in order to outline their distinctive shape and thereby give them easy recognition and a finished appearance.**

Guidance:

02 *For guide signs larger than 120 x 72 inches, the border should have a width of 2 inches. For smaller guide signs, a border width of 1.25 inches should be used, but the width should not exceed the stroke width of the lettering of the principal legend on the sign.*

03 *Corner radii of sign borders should be 1/8 of the minimum sign dimension on guide signs, except that the radii should not exceed 12 inches on any sign.*

Option:

04 The sign material in the area outside of the corner radius may be trimmed.

Section 2E.17 Abbreviations

Guidance:

01 *Abbreviations should be kept to a minimum; however, they are useful when complete destination messages produce excessively long signs. If used, abbreviations should be unmistakably recognized by road users (see Section 1A.15). Longer commonly used words that are not part of a proper name and are readily recognizable, such as Street, Boulevard, and Avenue, should be abbreviated to expedite recognition of the sign legend by reducing the amount and complexity of the legend.*

02 *Periods, apostrophes, question marks, ampersands, or other punctuation or characters that are not letters, numerals, or hyphens should not be used in abbreviations, unless necessary to avoid confusion.*

03 *The solidus (slanted line or forward slash) is intended to be used for fractions only and should not be used to separate words on the same line of legend. Instead, a hyphen should be used for this purpose, such as "CARS – TRUCKS."*

Standard:

04 **The words NORTH, SOUTH, EAST, and WEST shall not be abbreviated when used with route signs to indicate cardinal directions on guide signs.**

Section 2E.18 Symbols

Standard:

01 **Symbol designs shall be unmistakably like those shown in this Manual and in the "Standard Highway Signs and Markings" book (see Section 1A.11).**

Guidance:

02 *A special effort should be made to balance legend components for maximum legibility of the symbol with the rest of the sign.*

Option:

03 Educational plaques may be used below symbol signs where needed.

Section 2E.19 Arrows for Interchange Guide Signs

Standard:

01 **Arrows used on interchange guide signs shall be of the types shown in Figure 2D-2 and shall comply with the provisions of this Section and Section 2D.08.**

02 **Except on Overhead Arrow-per-Lane guide signs (see Section 2E.21) and on Exit Direction signs for lane drops (see Section 2E.24), and except as provided in Paragraphs 3 and 4, directional arrows on all overhead and post-mounted Exit Direction signs shall point diagonally upward and shall be located on the side of the sign consistent with the direction of the exiting movement.**

Option:

03 On post-mounted Exit Direction signs that are located where a directional arrow to the side of the legend farthest from the roadway might create an unusually wide sign that limits the road user's view of the arrow, the directional arrow may be placed at the bottom portion of the sign, centered under the legend.

Standard:

04 **Directional arrows on guide signs for multi-lane exits shall be positioned below the legend approximately over the center of each lane to which the arrow applies (see Figures 2E-4 and 2E-8).**

05 **On overhead signs where down arrows are used to indicate a lane to be followed, a down arrow shall be positioned approximately over the center of each lane and shall point vertically downward toward the**

approximate center of that lane. **Down arrows shall be used only on overhead guide signs that restrict the use of specific lanes to traffic bound for the destination(s) and/or route(s) indicated by these arrows. Down arrows shall not be used unless an arrow can be located over and pointed to the approximate center of each lane that can be used to reach the destination displayed on the sign.**

06 **If down arrows are used, having more than one down arrow pointing to the same lane on a single overhead sign (or on multiple signs on the same overhead sign structure) shall not be permitted.**

Support:

07 Directional and down arrows for use on guide signs are shown in Figure 2D-2. Detailed drawings and standardized sizes based on ranges of letter heights for these arrows are provided in the "Standard Highway Signs and Markings" book (see Section 1A.11). Information on the dimensions for arrows used in Overhead Arrow-per-Lane and Diagrammatic guide signing is also provided in the "Standard Highway Signs and Markings" book.

Section 2E.20 Signing for Option Lanes at Splits and Multi-Lane Exits

Support:

01 Some freeway and expressway splits or multi-lane exit interchanges contain an interior option lane serving both movements in which traffic can either leave the route or remain on the route, or choose either destination at a split, from the same lane.

Standard:

02 **On freeways and expressways, either the Overhead Arrow-per-Lane or Diagrammatic guide sign designs as provided in Sections 2E.21 and 2E.22 shall be used for all multi-lane exits at major interchanges (see Section 2E.32) that have an optional exit lane that also carries the through route (see Figures 2E-4, 2E-5, 2E-8, and 2E-9) and for all splits that include an option lane (see Figures 2E-6 and 2E-10). Overhead Arrow-per-Lane or Diagrammatic guide signs shall not be used on freeways and expressways for any other types of exits or splits, including single-lane exits and splits that do not have an option lane.**

Guidance:

03 *The Overhead Arrow-per-Lane guide sign design (see Section 2E.21) should also be considered for multi-lane exits with an option lane at intermediate interchanges (see Section 2E.32) based on such factors as the extent of the need to optimize the mainline operation by maximizing the usage of the option lane, the extent of the period(s) of the day during which the exiting volumes warrant the multi-lane exit arrangement, and the nature of the traffic that primarily uses the option lane during the high-volume periods.*

04 *Signing for multi-lane exits at minor interchanges (see Section 2E.32) that have an optional exit lane or at intermediate interchanges that have an optional exit lane at which it has been determined that the Overhead Arrow-per-Lane guide sign design is not warranted should use a combination of conventional guide signing and regulatory lane-use signing, in accordance with the provisions of Section 2E.23.*

Section 2E.21 Design of Overhead Arrow-per-Lane Guide Signs for Option Lanes

Support:

01 Overhead Arrow-per-Lane guide signs (see Figure 2E-3) are used where an option lane is present at freeway and expressway multi-lane exit interchanges and splits. They display an upward-pointing arrow above each lane that conveys the direction(s) of travel that the lane serves at the point of departure. At locations where an option lane is present at a multi-lane exit or split, Overhead Arrow-per-Lane guide signs have been shown to be superior to either conventional guide signs or Diagrammatic guide signs because they convey positive direction about which destination and direction each approach lane serves, particularly for the option lane, which is otherwise difficult to clearly sign.

Standard:

02 **Overhead Arrow-per-Lane guide signs shall be used on all new or reconstructed freeways and expressways as described in Section 2E.20.**

03 **Where used, the Overhead Arrow-per-Lane guide sign at the exit or split shall be located at or in the immediate vicinity of the point where the exiting lanes begin to diverge from the through lanes or, for a split, at the point where the approach lanes begin to diverge from one another, preserving the relation of the arrows displayed on the sign to their respective lanes. The Overhead Arrow-per-Lane guide sign at the exit shall not be located at or near the theoretical gore.**

Option:

04 At existing or non-reconstructed locations where Exit Direction and Pull-Through signs exist at the theoretical gore, the existing sign support structure may remain in place, continuing to use Exit Direction and Pull-Through signs, in conjunction with a replacement of the advance signs using the Overhead Arrow-per-Lane guide sign design.

Figure 2E-3. Overhead Arrow-per-Lane Guide Sign for a Multi-Lane Exit with an Option Lane

Standard:

05 If existing Exit Direction and Pull-Through signs are being retained at an interchange as provided in Paragraph 4, an Overhead Arrow-per-Lane guide sign shall not be used at the location of the Exit Direction and Pull-Through signs at or in the vicinity of the theoretical gore. New installations of Exit Direction and Pull-Through signs shall not be permitted in conjunction with Overhead Arrow-per-Lane guide signs on new or reconstructed facilities.

Guidance:

06 *Overhead Arrow-per-Lane guide signs should be located at approximately 1/2 mile and 1 mile in advance of the exit or split, and at approximately 2 miles in advance of the exit or split where space is available and conditions allow.*

Standard:

07 Overhead Arrow-per-Lane guide signs used on freeways and expressways shall include one arrow above each lane and shall be designed in accordance with the following criteria:
- A. The sign shall include an upward-pointing arrow for each lane of the approach to the split or exit, and the shaft of each arrow shall be located approximately over the center of the lane to which it applies.
- B. Arrows for continuing through lanes shall be vertically upward pointing (see Figure 2E-4) unless those lanes are on a significantly curved alignment beyond the theoretical gore, in which case the arrows for the continuing through lanes shall indicate the approximate degree of curvature (see Figure 2E-5).
- C. The arrow for a lane that must exit shall be curved in the direction of the exit and shall be accompanied by black-on-yellow EXIT (E11-1a) and ONLY (E11-1b) sign panels adjacent to the lower end of the arrow shaft. The E11-1a and E11-1b sign panels shall not be used for a split of two overlapping routes where neither of the diverging routes is designated as an exit. Where the through lanes curve and the exit continues on a straight alignment, upward-pointing vertical arrows shall be used for the exiting movement and curved arrows for the through movement.
- D. The arrow for an optional exit lane that also carries the through route shall have a single shaft that bifurcates into a vertically upward-pointing arrow and a curving arrow corresponding to the configuration of the through and exit lanes.
- E. For splits with an option lane, the arrow for the lane from which either direction of the split can be accessed shall have a single shaft that bifurcates into two upward-pointing curving arrows showing the approximate degrees of curvature of the two roadways beyond the theoretical gore (see Figure 2E-6).
- F. A vertical white line shall be used to separate the route shields and destinations for the two diverging movements from each other.

Figure 2E-4. Overhead Arrow-per-Lane Guide Signs for a Two-Lane Exit to the Right with an Option Lane

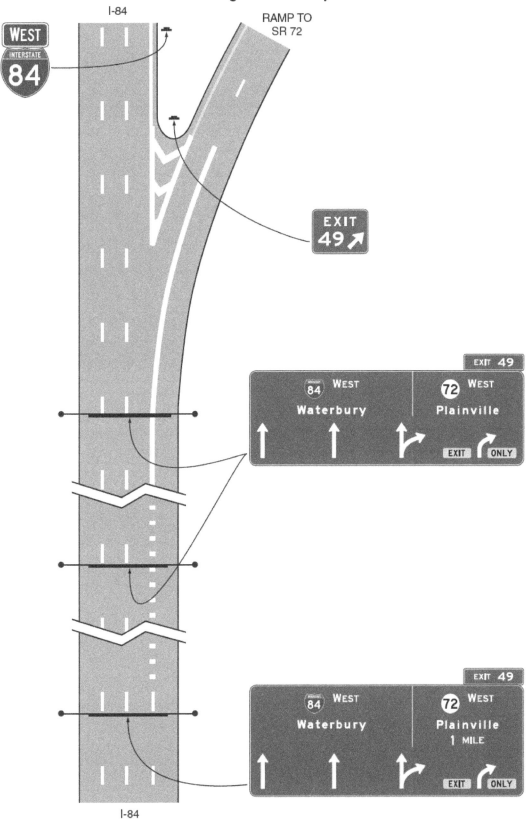

Figure 2E-5. Overhead Arrow-per-Lane Guide Signs for a Two-Lane Exit to the Right with an Option Lane (Through Lanes Curve to the Left)

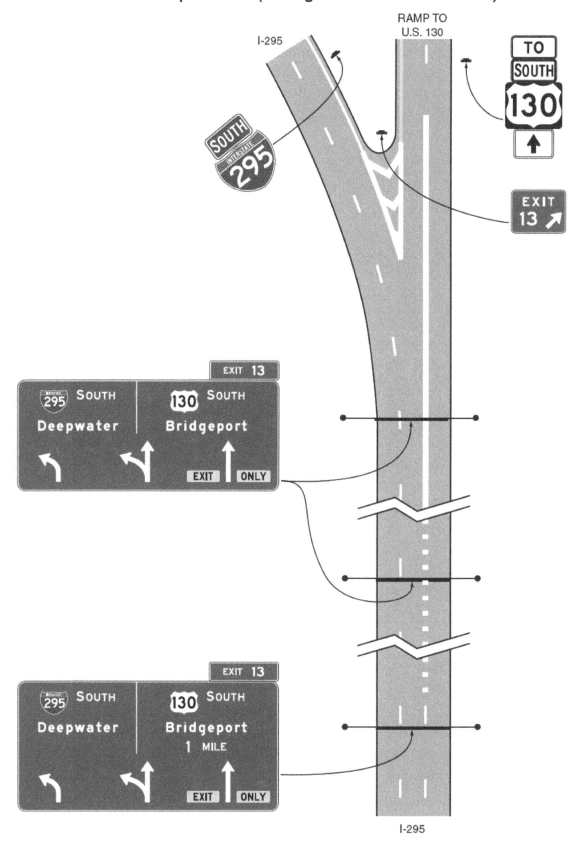

Figure 2E-6. Overhead Arrow-per-Lane Guide Signs for a Split with an Option Lane

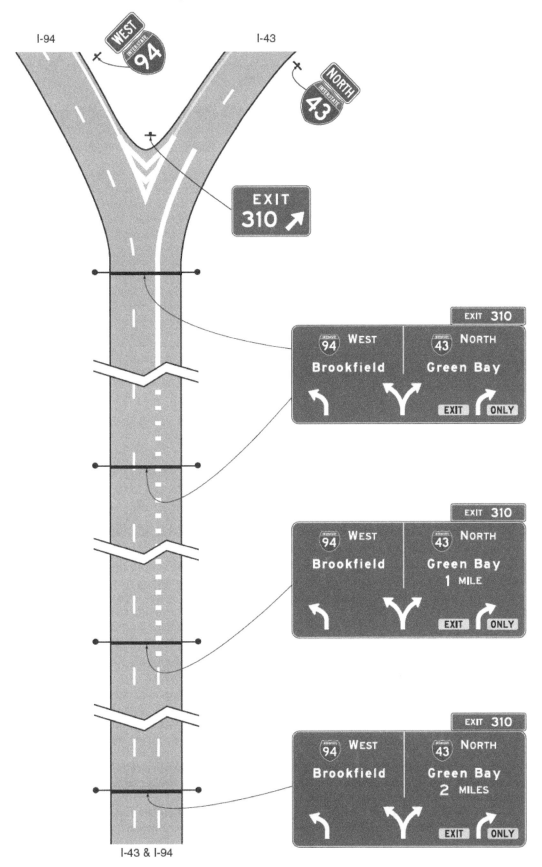

G. The distance to the exit or split shall be displayed below the off-movement destination on the advance signs at the 1-mile and 2-mile locations.
H. The number of lanes displayed on a sign shall correspond to the number of lanes at the location of that sign. An advance sign shall not depict lanes that are added downstream of a sign location.
I. For numbered exits, the Exit Number (E1-5P) or Left Exit Number (E1-5bP) plaque shall be used at the top of the sign in accordance with Section 2E.31. For unnumbered left exits, the LEFT (E1-5aP) plaque shall be used at the top left edge of the sign.

Guidance:

08 *Overhead Arrow-per-Lane guide signs used on freeways and expressways should be designed in accordance with the following additional criteria:*

A. No more than one destination should be displayed for each movement, and no more than two destinations should be displayed per sign.
B. The arrowhead(s) for the diverging movement should be positioned lower on the sign than the arrowhead(s) for the movement that continues straight ahead, independent of which movement carries the through route. Where the movements are freeway or expressway splits rather than exits, the arrowheads should be positioned at approximately the same height on the sign.
C. Route shields, cardinal directions, and destinations should be positioned on the sign such that they are clearly related to the arrowhead(s) for the movement to which they apply.
D. The cardinal direction should be placed adjacent to the route shield for exits or splits leading in a single cardinal direction.
E. The vertical white line that is used to separate the route shields and destinations for the two diverging movements from each other should not descend below the top of the arrowheads for the through lanes, and should be positioned approximately halfway between the diverging arrowheads for the optional movement lane (see Figure 2E-3).

Standard:

09 **Overhead Arrow-per-Lane guide signs shall not be used to depict a downstream split of an exit ramp on a sign located on the mainline.**

Support:

10 Specific guidelines for more detailed design of Overhead Arrow-per-Lane guide signs are contained in the "Standard Highway Signs and Markings" book (see Section 1A.11).

Option:

11 Where extra emphasis of an especially low advisory ramp speed is needed, an EXIT XX MPH (E13-2) sign panel (see Figure 2E-27) may be placed below the applicable destination legend to supplement, but not to replace, the exit or ramp advisory speed warning signs.

Section 2E.22 Design of Freeway and Expressway Diagrammatic Guide Signs for Option Lanes

Support:

01 Diagrammatic guide signs (see Figure 2E-7) are guide signs that show a simplified graphic view of the exit arrangement in relationship to the main highway. While the use of such guide signs might be helpful for the purpose of conveying relative direction of each movement, Diagrammatic guide signs have been shown to be less effective than conventional or Overhead Arrow-per-Lane guide signs at conveying the destination or direction(s) that each approach lane serves, regardless of whether dedicated or option lanes are present.

Standard:

02 **Diagrammatic guide signs used where an option lane is present at a freeway or expressway split or multi-lane exit shall be designed in accordance with the following criteria:**

A. The graphic legend shall be of a plan view showing the off-ramp arrangement.
B. No other symbols or route shields shall be used as a substitute for arrowheads.
C. They shall not be installed at the Exit Direction sign location (see Section 2E.36).
D. The EXIT ONLY sign panel shall not be used on diagrammatic guide signs in advance of the interchange.
E. For numbered exits, the Exit Number (E1-5P) or Left Exit Number (E1-5bP) plaque shall be used at the top of the sign in accordance with Section 2E.31. For unnumbered left exits, the LEFT (E1-5aP) plaque shall be used at the top left edge of the sign.
F. The EXIT ONLY (E11-1e or E11-1f) sign panels shall be used on the Exit Direction sign at the theoretical gore, except at splits of two overlapping routes where neither of the routes is designated as an exit.

Figure 2E-7. Diagrammatic Guide Sign for a Multi-Lane Exit with an Option Lane

Guidance:

03 *Diagrammatic guide signs used on freeways and expressways should be designed in accordance with the following additional criteria:*

   A. *The graphic should not depict deceleration lanes.*
   B. *No more than one destination should be displayed for each movement, and no more than two destinations should be displayed per sign.*
   C. *The arrowhead for the diverging movement should be positioned lower on the sign than the arrowhead for the movement that continues straight ahead, independent of which movement carries the through route (see Figures 2E-8 and 2E-9). Where the movements are freeway or expressway splits rather than exits, the arrowheads should be positioned at approximately the same height on the sign (see Figure 2E-10).*
   D. *Arrow shafts should contain lane lines.*
   E. *Route shields, cardinal directions, and destinations should be positioned on the sign such that they are clearly related to the arrowhead(s), and the arrowhead for the off movement should point toward the route shield for the off movement.*
   F. *For exits or splits leading in a single direction, the cardinal direction should be placed adjacent to the route shield, and the destination should be placed below the route shield and cardinal direction.*

Standard:

04 **Diagrammatic guide signs shall not be used at cloverleaf interchanges for the purpose of depicting successive departures from the mainline or separate downstream departures from a collector-distributor roadway. The use of Diagrammatic guide signs at cloverleaf interchanges shall be limited to the following cases:**

   A. **Where the outer (non-loop) exit ramp of the cloverleaf is a multi-lane exit having an optional exit lane that also carries the through route; and**
   B. **At cloverleaf interchanges that include collector-distributor roadways, such as those illustrated in Figure 2E-36, that are accessed from the mainline by a multi-lane exit having an optional exit lane that also carries the through route. In this case, the Diagrammatic guide sign shall only show the configuration of the lanes at the exit point to the collector-distributor roadway and not the entire interchange configuration.**

Support:

05 Specific guidelines for more detailed design of Diagrammatic guide signs are contained in the "Standard Highway Signs and Markings" book (see Section 1A.11).

Option:

06 Where extra emphasis of an especially low advisory ramp speed is needed, an EXIT XX MPH (E13-2) sign panel (see Figure 2E-27) may be placed below the applicable destination legend to supplement, but not to replace, the exit or ramp advisory speed warning signs.

Figure 2E-8. Diagrammatic Guide Signs for a Two-Lane Exit to the Right with an Option Lane

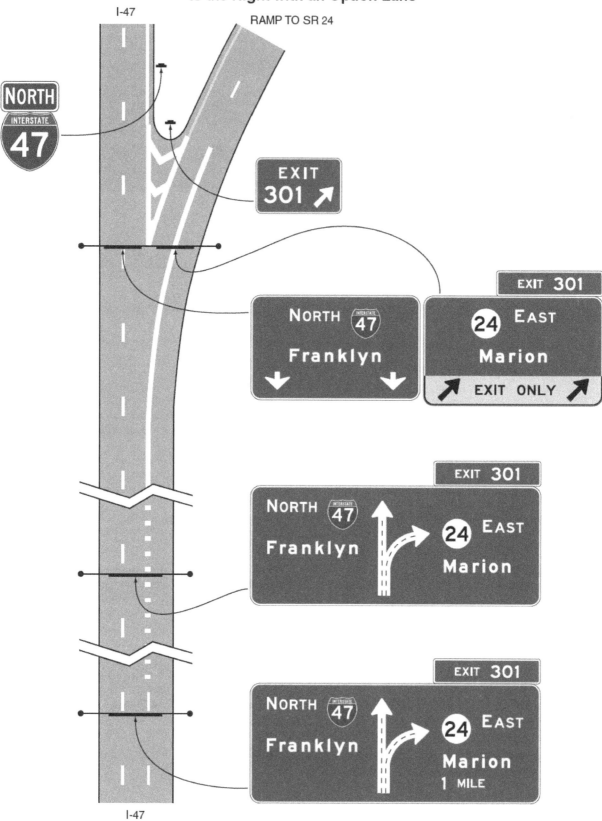

Figure 2E-9. Diagrammatic Guide Signs for a Two-Lane Exit to the Right with an Option Lane (Through Lanes Curve to the Left)

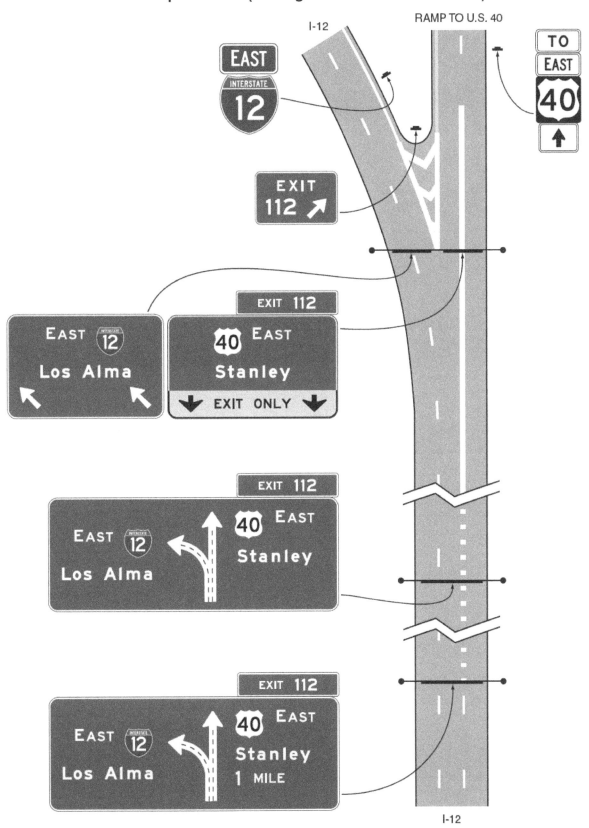

Figure 2E-10. Diagrammatic Guide Signs for a Split with an Option Lane

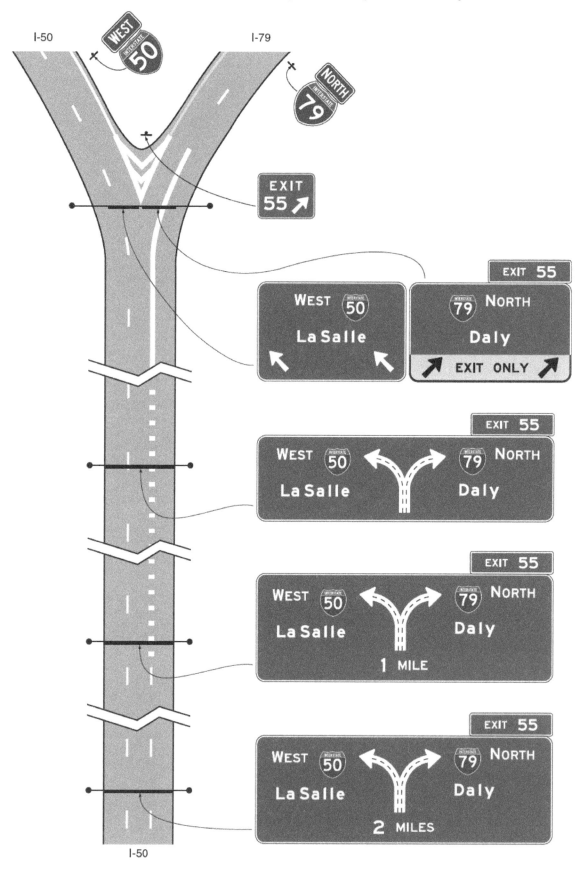

Section 2E.23 Signing for Intermediate and Minor Interchange Multi-Lane Exits with an Option Lane

Support:

01 Intermediate and minor multi-lane exits might have an operational need for the presence of an option lane for only the peak period during which excessive queues might otherwise develop if the option lane were not available. In such cases, the Overhead Arrow-per-Lane or Diagrammatic guide signing described for option lanes in Sections 2E.21 and 2E.22 might not be practical, depending on the level of use of the option lane and the spacing of nearby interchanges, particularly in non-rural areas.

Guidance:

02 *Signing for an intermediate or minor interchange that has a multi-lane exit with an option lane that also carries the through route should use the same basic principles as those for a conventional exit. In such cases, the option lane is not signed on the Advance Guide signs. For such exits that involve the addition of an auxiliary lane that is not present at the Advance Guide sign locations, but do not involve a lane drop (see Figure 2E-12), a sequence of post-mounted or overhead-mounted Advance Guide signs should be used, located in accordance with the interchange classification (see Section 2E.32). The Exit Direction sign should be located at the theoretical gore and display a diagonally upward-pointing directional arrow above each lane that departs from the mainline alignment. The Exit Direction sign should not contain the EXIT ONLY legend.*

03 *For such interchanges that also have a lane drop (see Figure 2E-11), the Advance Guide and Exit Direction signs should follow the provisions of Section 2E.24. The Exit Direction sign should be located at the theoretical gore and should contain the EXIT ONLY (E11-1e) sign panel.*

04 *The presence of the option lane should be conveyed by the use of post-mounted lane-use (R3-8 Series) signs (see Section 2B.22). When used, the R3-8 signs should be of an appropriate size for their application to optimize their conspicuity. The signs should be located in succession with the Advance Guide signs, where the option and exit lanes have developed (see Figure 2E-11). In cases where the exiting lane or lanes have not developed and the option lane is created by the addition of an auxiliary lane that exits, the R3-8 signs should be located only adjacent to where the lanes have been fully developed and not in advance of the lane or along its transition (see Figure 2E-12).*

Support:

05 The use of a down arrow on overhead freeway or expressway guide signs has been shown to be misinterpreted by road users as an indication of a dedicated lane.

Standard:

06 **Advance Guide signs that are mounted overhead shall not display a down arrow over an option lane.**

Section 2E.24 Signing for Interchange Lane Drops

Standard:

01 **The provisions of this Section shall only apply to lane drops at exits that do not have an optional exit lane. At exits that have an optional exit lane in addition to the dropped lane, the provisions of Sections 2E.20 through 2E.23 shall apply.**

02 **Major guide signs for all lane drops at interchanges shall be mounted overhead. An EXIT ONLY sign panel shall be used for all interchange lane drops at which the through route is carried on the mainline.**

03 **Except on Overhead Arrow-per-Lane and Diagrammatic guide signs (See Sections 2E.20 through 2E.22), the EXIT ONLY (down arrow) (E11-1 or E11-1f) sign panel (see Figure 2E-13) shall be used on all signing of lane drops on all overhead Advance Guide signs (see Figures 2E-14 through 2E-16). The number of arrows on each sign shall correspond to the number of dropped lanes at the location of each sign. Placement of the down arrow shall comply with the provisions of Section 2E.19.**

04 **For lane drops, the Exit Direction sign (see Section 2E.36 and Figure 2E-26) shall be of the format shown in Figures 2E-15 and 2E-16. The bottom portion of the Exit Direction sign shall be yellow with a black border and shall include a diagonally upward-pointing black directional arrow (left or right) for each lane dropped at the exit, with the sign designed and placed so that each arrow is located over the approximate center of each lane being dropped. The words EXIT and ONLY shall be positioned to the left and right, respectively, of the arrow on the E11-1d sign panel for a single-lane drop. For a two-lane drop, the words EXIT ONLY shall be located between the two arrows on the E11-1e sign panel. The number of arrows on the sign shall correspond to the number of dropped lanes at the location of the sign.**

Option:

05 EXIT ONLY messages of either the combination of E11-1a and E11-1b, or E11-1c formats may be used to retrofit existing signing to warn of a lane drop situation ahead.

Figure 2E-11. Example of Signing for a Two-Lane Intermediate or Minor Interchange Exit with an Option Lane and a Dropped Lane

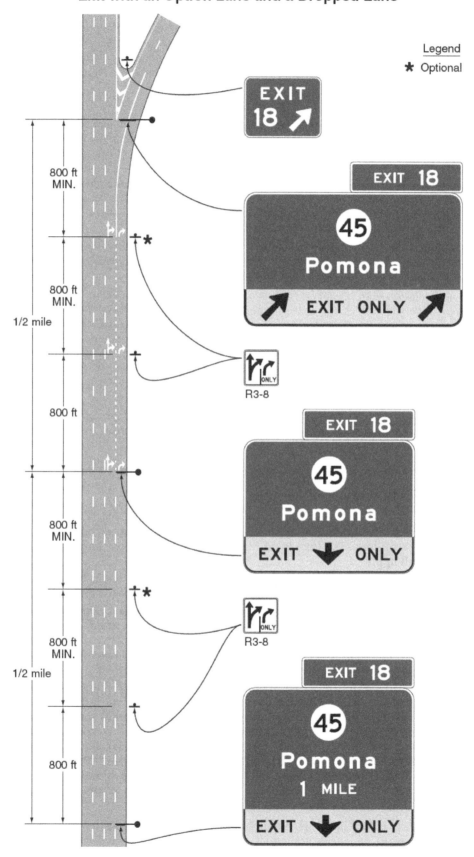

Figure 2E-12. Example of Signing for a Two-Lane Intermediate or Minor Interchange Exit with Option and Auxiliary Lanes

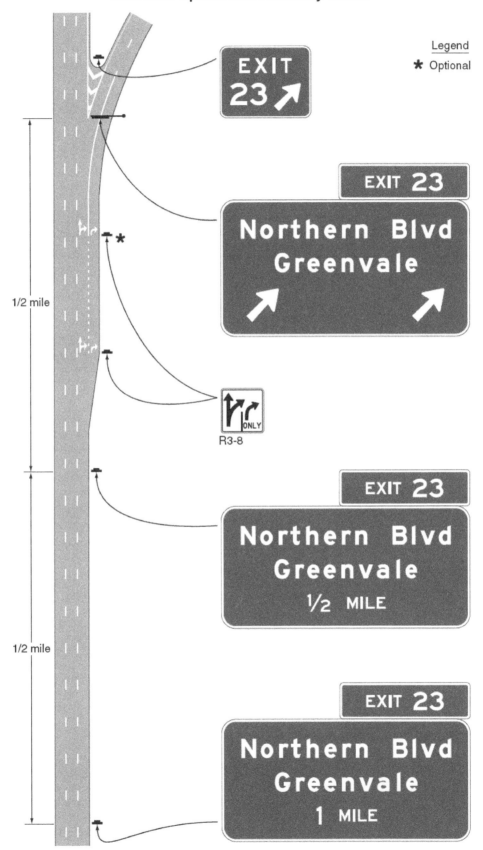

Figure 2E-13. EXIT ONLY and LEFT Sign Panels

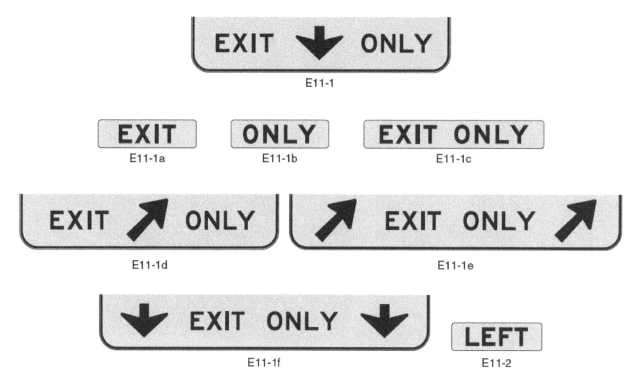

Standard:

06 **If used to retrofit an existing Advance Guide sign, the E11-1a and E11-1b sign panels (see Figure 2E-13) shall be placed on either side of a white down arrow. The E11-1c sign panel, if used to retrofit an existing sign, shall be placed between the lower destination message and the white down arrow.**

Guidance:

07 *Except as provided in Paragraph 8 for an auxiliary lane, Advance Guide signs for lane drops within 1 mile of the interchange should not contain the distance message.*

08 *Where the dropped lane is an auxiliary lane that is provided between successive entrance and exit ramps of two separate interchanges and the distance between the two ramps is less than 1 mile, the first Advance Guide sign in the sequence downstream from the entrance ramp should contain the distance message.*

09 *Wherever the dropped lane carries the through route, signs should be used without the EXIT ONLY sign panel.*

Support:

10 Sections 2E.20 through 2E.23 contain information on the signing of lane drops at exits that also have an option lane.

11 Section 2B.23 contains information regarding regulatory signs that can also be used for freeway lane drop situations and Section 2C.42 contains information regarding warning signs that can also be used for freeway lane drop situations.

Section 2E.25 Overhead Sign Installations

Support:

01 Specifications for the design and construction of structural supports for signs have been standardized by the American Association of State Highway and Transportation Officials (AASHTO). Overcrossing structures can often serve for the support of overhead signs, and might in some cases be the only practical location that will provide adequate viewing distance. Use of these structures as sign supports will eliminate the need for additional sign supports along the roadside. Factors justifying the installation of overhead signs are given in Section 2A.17. Vertical clearance of overhead signs is discussed in Section 2A.18.

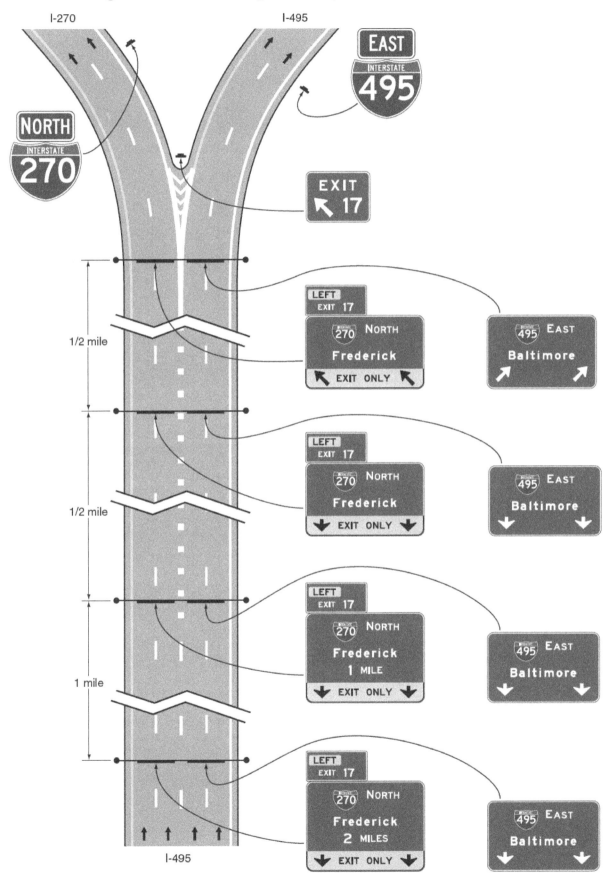

Figure 2E-14. Guide Signs for a Split with Dedicated Lanes

Figure 2E-15. Guide Signs for a Single-Lane Exit to the Left with a Dropped Lane

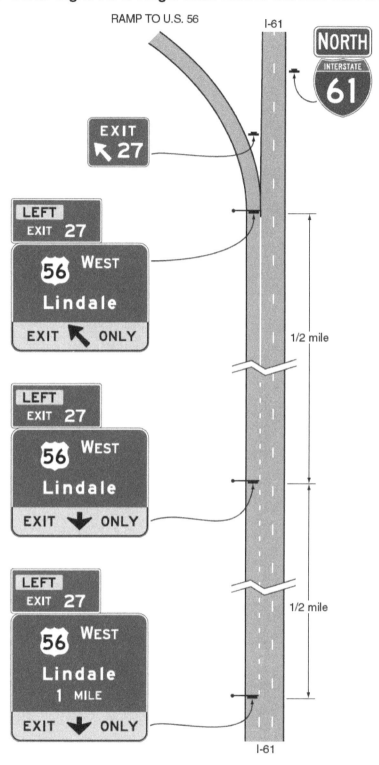

Figure 2E-16. Guide Signs for a Single-Lane Exit to the Right with a Dropped Lane

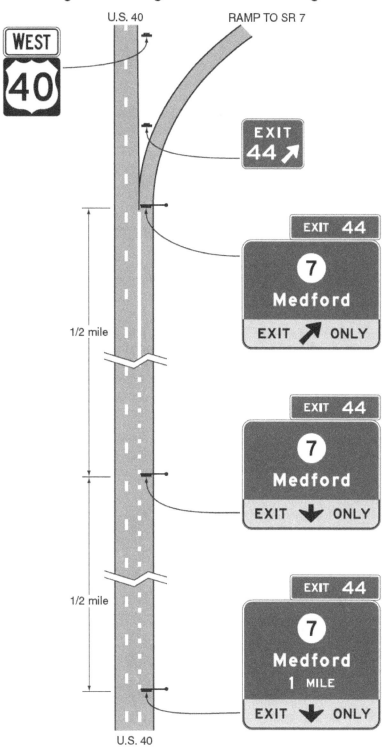

Section 2E.26 Lateral Offset
Standard:

01 **The minimum lateral offset outside the usable roadway shoulder for post-mounted freeway and expressway signs or for overhead sign supports, either to the right-hand or left-hand side of the roadway, shall be 6 feet. This minimum clearance shall also apply outside of a curb. If located within the clear zone, the signs shall be mounted on crashworthy supports or shielded by appropriate crashworthy barriers.**

Guidance:

02 *Where practical, a sign should not be less than 10 feet from the edge of the nearest traffic lane. Large guide signs especially should be farther removed, preferably 30 feet or more from the nearest traffic lane.*

03 *Where an expressway median is 12 feet or less in width, consideration should be given to spanning both roadways without a center support.*

04 *Where overhead sign supports cannot be placed sufficiently far away from the line of traffic or in an otherwise protected site, they should either be designed to minimize the impact forces, or be adequately shielded by a traffic barrier of suitable design.*

Standard:

05 **Butterfly-type sign supports and other overhead non-crashworthy sign supports shall not be installed in gores or other unshielded locations within the clear zone.**

Option:

06 Lesser clearances, but not generally less than 6 feet, may be used on connecting roadways or ramps at interchanges.

Section 2E.27 Route Signs and Trailblazer Assemblies
Standard:

01 **The official Route sign for the Interstate Highway System shall be the red, white, and blue retroreflective distinctive shield adopted by the American Association of State Highway and Transportation Officials (see Section 2D.11).**

Guidance:

02 *Route signs (see Figure 2E-17) should be incorporated as cut-out shields or other distinctive shapes on large directional guide signs. Where the Interstate shield is displayed in an assembly or on the face of a guide sign with U.S. or State Route signs, the Interstate numeral should be at least equal in size to the numerals on the other Route signs. The use of independent Route signs should be limited primarily to route confirmation assemblies.*

03 *Route signs and auxiliary signs showing junctions and turns should be used for guidance on approach roads, for route confirmation just beyond entrances and exits, and for reassurance along the freeway or expressway. When used along the freeway or expressway, the Route signs should be enlarged to a 36 x 36-inch minimum size for routes with one or two digits and to a 45 x 36-inch minimum size for routes with three digits as shown in the "Standard Highway Signs and Markings" book (see Section 1A.11). When independently mounted Route signs are used in place of Pull-Through signs, they should be located just beyond the exit.*

Figure 2E-17. Interstate, Off-Interstate, and U.S. Route Signs

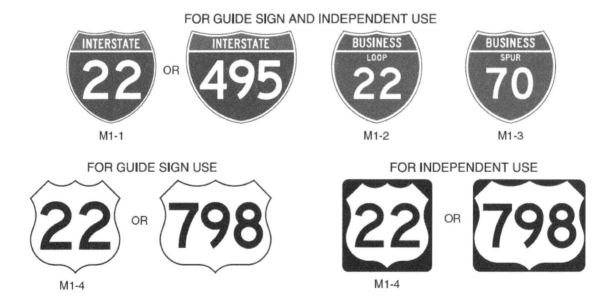

Option:

04 The standard Trailblazer Assembly (see Section 2D.35) may be used on roads leading to the freeway or expressway. Component messages of the Trailblazer Assembly may be included on a single sign in accordance with the provisions of Section 2D.12. Independently mounted Route signs may be used instead of Pull-Through signs (see Section 2E.12) as confirmation information.

Support:

05 Section 2H.07 contains information regarding the design of signs for Auto Tour Routes.

Option:

06 The commonly used name or trailblazer route sign for a toll highway (see Chapter 2F) may be displayed on non-toll sections of the Interstate Highway System at:
- A. The last exit before entering a toll Section of the Interstate Highway System;
- B. The interchange or connection with a toll highway, whether or not the toll highway is a part of the Interstate Highway System; and
- C. Other locations within a reasonable approach distance of toll highways when the name or trailblazer symbol for the toll highway would provide better guidance to road users unfamiliar with the area than would place names and route numbers.

07 The toll highway name or route sign may be included as a part of the guide sign installations on intersecting highways and approach roads to indicate the interchange with a toll Section of an Interstate route. Where needed for the proper direction of traffic, a trailblazer for a toll highway that is part of the Interstate Highway System may be displayed with the Interstate Trailblazer Assembly.

Support:

08 Chapter 2F contains additional information regarding signing for toll highways.

Section 2E.28 Eisenhower Interstate System Signs (M1-10, M1-10a)

Option:

01 The Eisenhower Interstate System (M1-10 and M1-10a) signs (see Figure 2E-18) may be used on Interstate highways at periodic intervals and in rest areas, scenic overlooks, or other similar roadside facilities on the Interstate Highway System.

Guidance:

02 *If used, the M1-10a sign should be used only in rest areas or other similar facilities where the sign can be viewed by occupants of parked vehicles or by pedestrians. The M1-10a sign should not be installed on Interstate highway mainlines, ramps, or other roadways where it can be viewed by vehicular traffic.*

Figure 2E-18. Eisenhower Interstate System Signs

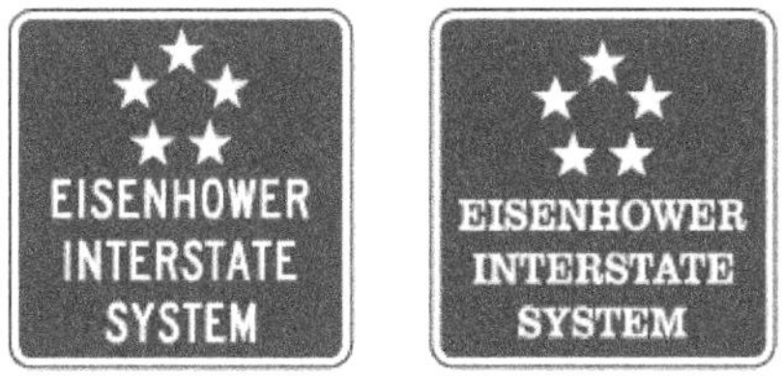

Standard:

03 **The M1-10 and M1-10a signs shall not be used as part of a Junction, Advance Route Turn, Directional, or Trailblazer Assembly or as part of a guide sign or similar assembly providing direction to a route or destination.**

Section 2E.29 Signs for Intersections at Grade

Guidance:

01 *If there are intersections at grade within the limits of an expressway, guide sign types provided in Chapter 2D should be used. However, such signs should be of a size compatible with the size of other signing on the expressway.*

Option:

02 Advance Guide signs for intersections at grade may take the form of diagrammatic layouts depicting the geometrics of the interSection along with essential directional information.

Section 2E.30 Interchange Guide Signs

Standard:

01 **The signs at interchanges and on their approaches shall include Advance Guide signs and Exit Direction signs. Consistent destination messages shall be displayed on these signs.**

Guidance:

02 *New destination information should not be introduced into the major sign sequence for one interchange, nor should destination information be dropped.*

03 *Reference should be made to Section 2E.11 and Sections 2E.33 through 2E.42 for a detailed description of the signs in the order that they should appear at the approach to and beyond each interchange. Guide signs placed in advance of an interchange deceleration lane should be spaced at least 800 feet apart.*

04 *Supplemental guide signing should be used sparingly as provided in Section 2E.35.*

Section 2E.31 Interchange Exit Numbering

Support:

01 Interchange exit numbering provides valuable orientation for the road user on a freeway or expressway. The feasibility of numbering interchanges or exits on an expressway will depend largely on the extent to which grade separations are provided. Where there is appreciable continuity of interchange facilities, interrupted only by an occasional interSection at grade, the numbering will be helpful to the expressway user.

Standard:

02 **Interchange numbering shall be used in signing each freeway interchange exit. Interchange exit numbers shall be displayed with each Advance Guide sign, Exit Direction sign, and Exit Gore sign. The exit number shall be displayed on a separate plaque at the top of the Advance Guide or Exit Direction sign. The exit number (E1-5P) plaque (see Figure 2E-22) shall be 30 inches in height and shall include the word EXIT and the appropriate exit number in a single-line format. Suffix letters shall be used for exit numbering at a multi-exit interchange. The suffix letter shall also be included on the exit number plaque and shall be separated from the exit number by a space having a width of between 1/2 and 3/4 of the height of the suffix letter. Exit numbers shall not include the cardinal initials corresponding to the directions of the cross route. Minimum numeral and letter sizes are given in Tables 2E-2 through 2E-5. If used, the interchange numbering system for expressways shall comply with the provisions prescribed for freeways.**

03 **At a multi-exit interchange where suffix letters are used for exit numbering, an exit of the same number without a suffix letter shall not be used on the same route in the same direction. For example, if an exit is designated as EXIT 256 A, then there shall not be an exit designated as EXIT 256 on the same route in the same direction.**

04 **Interchange exit numbering shall use the reference location sign exit numbering method. The consecutive exit numbering method shall not be used.**

Support:

05 Reference location sign exit numbering assists road users in determining their destination distances and travel mileage, and assists highway agencies because the exit numbering sequence does not have to be changed if new interchanges are added to a route.

Option:

06 Exit numbers may also be used with Supplemental Guide signs and Motorist Service signs.

Guidance:

07 *Exit number (E1-5P) plaques should be added to the top right-hand edge of the sign for an exit to the right.*

Standard:

08 **Because road users might not expect an exit to the left and might have difficulty in maneuvering to the left, a left exit number (E1-5bP) plaque (see Figure 2E-22) shall be added to the top left-hand edge of the sign for all left-hand exits (see Figures 2E-14 and 2E-15). The word LEFT on the E1-5bP plaque shall be a black legend on a yellow rectangular sign panel and shall be centered above the word EXIT.**

Support:

09 Example exit number plaque designs are shown in Figure 2E-22. Figures 2E-3, 2E-7, 2E-22, 2E-26, and 2E-27 illustrate the incorporation of exit number plaques on guide signs.

10 The general plan for numbering interchange exits is shown in Figures 2E-19 through 2E-21. Figure 2E-19 shows a circumferential route, which is a route that makes a complete circle around a city or town and usually has two interchanges (one on each side of the city or town) with each of the mainline routes that travel through the city or town. Figure 2E-20 shows a loop route, which is a route that departs from a mainline route and then rejoins the same mainline route at a subsequent point downstream, and a spur route, which is a route that departs from a mainline route and never rejoins the same mainline route. Figure 2E-21 shows two mainline routes that overlap each other.

Figure 2E-19. Example of Interchange Numbering for Mainline and Circumferential Routes

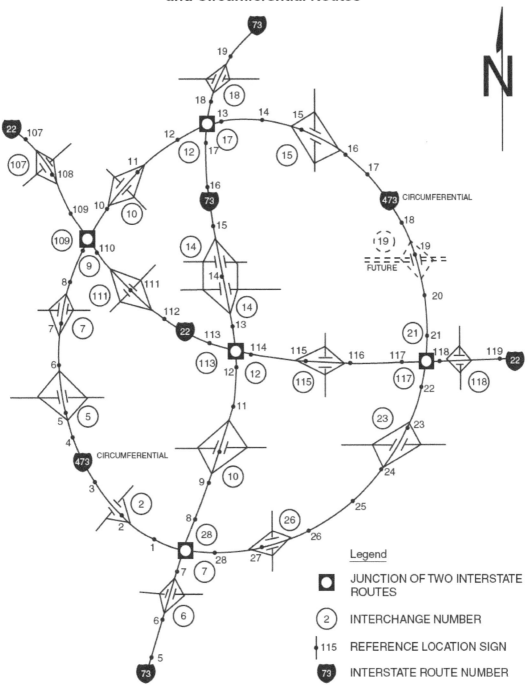

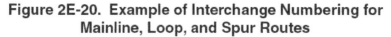

Figure 2E-20. Example of Interchange Numbering for Mainline, Loop, and Spur Routes

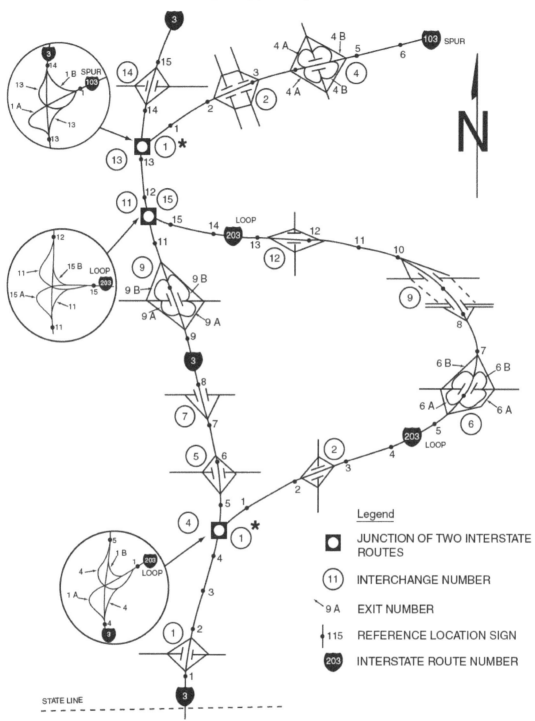

* The freeway/freeway interchange where the beginning of the loop or spur route intersects with the mainline route may be called either Exit 1 or Exit 0 on the loop or spur route.

Figure 2E-21. Example of Interchange Numbering for Overlapping Routes

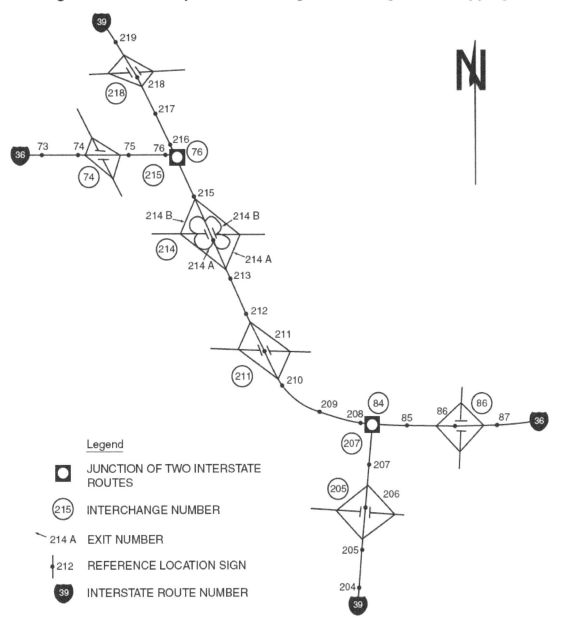

Standard:

11 **Regardless of whether a mainline route originates within a State or crosses into a State from another State, the southernmost or westernmost terminus within that State shall be the beginning point for interchange numbering.**

12 **For circumferential routes, interchange numbering shall be in a clockwise direction. The numbering shall begin with the first interchange west of the south end of an imaginary north-south line bisecting the circumferential route, at a radial freeway or other Interstate route, or some other conspicuous landmark in the circumferential route near a south polar location (see Figure 2E-19).**

13 **The interchange numbers on loop routes shall begin at the loop interchange nearest the south or west mainline junction and increase in magnitude toward the north or east mainline junction (see Figure 2E-20).**

14 **Spur route interchanges shall be numbered in ascending order starting at the interchange where the spur leaves the mainline route (see Figure 2E-20).**

15 **If a circumferential, loop, or spur route crosses State boundaries, the numbering sequence shall be coordinated by the States to provide continuous interchange numbering.**

16 **Where numbered routes overlap, continuity of interchange numbering shall be established for only one of the routes (see Figure 2E-21). If one of the routes is an Interstate and the other route is not an Interstate, the Interstate route shall maintain continuity of interchange numbering.**

Guidance:

17 *The route chosen for continuity of interchange numbering should also have reference location sign continuity (see Figure 2E-21).*

Section 2E.32 Interchange Classification

Support:

01 For signing purposes, interchanges are classified as major, intermediate, and minor. The minimum alphabet sizes contained in Tables 2E-2 and 2E-4 are based on this classification. Descriptions of these classifications are as follows:

- A. Major interchanges are subdivided into two categories: (a) interchanges with other expressways or freeways, or (b) interchanges with high-volume multi-lane highways, principal urban arterials, or major rural routes where the volume of interchanging traffic is heavy or includes many road users unfamiliar with the area.
- B. Intermediate interchanges are those with urban and rural routes not in the category of major or minor interchanges.
- C. Minor interchanges include those where traffic is local and very light, such as interchanges with land service access roads. Where the sum of exit volumes is estimated to be lower than 100 vehicles per day in the design year, the interchange is classified as minor.

Section 2E.33 Advance Guide Signs

Support:

01 An Advance Guide sign (see Figure 2E-22) gives notice well in advance of the exit point of the principal destinations served by the next interchange and the distance to that interchange.

Guidance:

02 *For major and intermediate interchanges (see Section 2E.32), Advance Guide signs should be placed at 1/2 mile and at 1 mile in advance of the exit with a third Advance Guide sign placed at 2 miles in advance of the exit if spacing permits. At minor interchanges, only one Advance Guide sign should be used. It should be located 1/2 to 1 mile from the exit gore. If the sign is located less than 1/2 mile from the exit, the distance displayed should be to the nearest 1/4 mile. Fractions of a mile, rather than decimals, should be displayed in all cases.*

Standard:

03 **For numbered exits to the left, a left exit number (E1-5bP) plaque (see Figure 2E-22) shall be added to the top left-hand edge of the sign.**

04 **For non-numbered exits to the left, a LEFT (E1-5aP) plaque (see Figure 2E-22) shall be added to the top left-hand edge of the sign.**

Support:

05 Section 2E.31 contains additional information regarding exit numbering.

Figure 2E-22. Examples of Interchange Advance Guide Signs, Exit Number Plaques, and LEFT Plaque

Note: Delete word EXIT(S) if exit number is used.

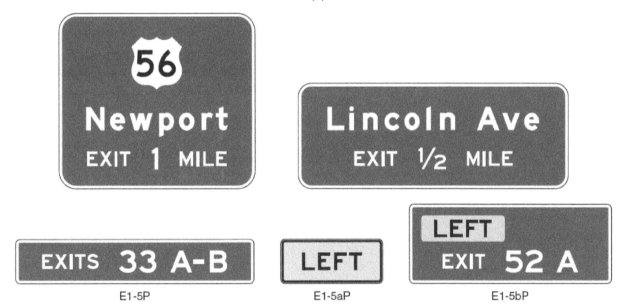

Standard:

06 Advance Guide signs for multi-lane exits having an optional exit lane that also carries the through route (see Figures 2E-4, 2E-5, 2E-8, and 2E-9) and for splits with an option lane (see Figures 2E-6 and 2E-10) shall be Overhead Arrow-per-Lane or diagrammatic signs designed in accordance with Sections 2E.20 through 2E.22.

07 Except as provided in Section 2E.24, Advance Guide signs, if used, shall contain the distance message. Except as provided in Paragraph 8 of this Section, the legend on the Advance Guide signs shall be the same as the legend on the Exit Direction sign, except that the last line shall read EXIT XX MILES. If the interchange has two or more exit roadways, the bottom line shall read EXITS XX MILES.

Guidance:

08 *Where interchange exit numbers are used, the word EXIT(S) should be omitted from the bottom line.*

Option:

09 Where the distance between interchanges is more than 1 mile, but less than 2 miles, the first Advance Guide sign may be closer than 2 miles, but not placed so as to overlap the signing for the previous exit. Duplicate Advance Guide signs or Interchange Sequence Series signs may be placed in the median on the opposite side of the roadway and are not included in the minimum requirements of interchange signing.

Guidance:

10 *Where there is less than 800 feet between interchanges, Interchange Sequence Series signs (see Section 2E.40) should be used instead of Advance Guide signs for the affected interchanges.*

11 *The Advance Guide signs for the last exit from a highway before it becomes a facility on which toll payments are required should include the LAST EXIT BEFORE TOLL (W16-16P) plaque (see Section 2F.10 and Figure 2F-3). The plaque should be installed above the Advance Guide signs.*

Option:

12 If there is insufficient space above the Advance Guide sign because of the presence of an exit number plaque, the W16-16P plaque may be installed below the Advance Guide sign.

Section 2E.34 Next Exit Plaques

Option:

01 Where the distance to the next interchange is unusually long, a Next Exit plaque (see Figure 2E-23) may be installed to inform road users of the distance to the next interchange.

Guidance:

02 *The Next Exit plaque should not be used unless the distance between successive interchanges is more than 5 miles.*

Standard:

03 **The Next Exit plaque shall carry the legend NEXT EXIT XX MILES. If the Next Exit plaque is used, it shall be placed below the Advance Guide sign nearest the interchange. It shall be mounted so as to not adversely affect the breakaway feature of the sign support structure.**

Option:

04 The legend for the Next Exit plaque may be displayed in either one or two lines as shown in Figure 2E-23.

Support:

05 The one-line message on the Next Exit plaque is the more desirable choice unless the message causes the sign to have a horizontal dimension greater than that of the Advance Guide sign.

Section 2E.35 Other Supplemental Guide Signs

Support:

01 Supplemental Guide signs can be used to provide information regarding destinations accessible from an interchange, other than places displayed on the standard interchange signing. However, such Supplemental Guide signing can reduce the effectiveness of other more important guide signing because of the possibility of overloading the road user's capacity to receive visual messages and make appropriate decisions. The AASHTO Guidelines for the Selection of Supplemental Guide Signs for Traffic Generators Adjacent to Freeways" is incorporated by reference in this Section (see Page i for AASHTO's address).

Guidance:

02 *No more than one Supplemental Guide sign should be used on each interchange approach.*

Figure 2E-23. Next Exit Plaques

⁰³ *A Supplemental Guide sign (see Figure 2E-24) should not list more than two destinations. Destination names should be followed by the interchange number (and suffix), or if interchanges are not numbered, by the legend NEXT RIGHT or SECOND RIGHT or both, as appropriate. The Supplemental Guide sign should be installed as an independent guide sign assembly.*

⁰⁴ *Where two or more Advance Guide signs are used, the Supplemental Guide sign should be installed approximately midway between two of the Advance Guide signs. If only one Advance Guide sign is used, the Supplemental Guide sign should follow it by at least 800 feet. If the interchanges are numbered, the interchange number should be used for the action message.*

⁰⁵ *States and other agencies should adopt an appropriate policy for installing supplemental signs using "The AASHTO Guidelines for the Selection of Supplemental Guide Signs for Traffic Generators Adjacent to Freeways." In developing policies for such signing, such items as population, amount of traffic generated, distance from the route, and the significance of the destination should be taken into account.*

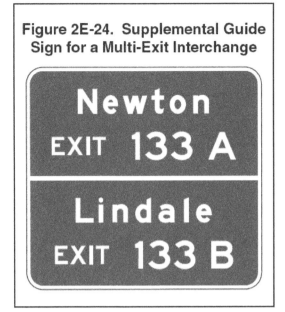

Figure 2E-24. Supplemental Guide Sign for a Multi-Exit Interchange

Standard:

⁰⁶ **Guide signs directing drivers to park - ride facilities shall be considered as Supplemental Guide signs (see Figure 2E-25).**

Option:

⁰⁷ A pictograph (see definition in Section 1A.13) may be used on a Supplemental Guide sign in conjunction with a destination that is associated with governmental agencies, military bases, universities, or other government-approved institutions.

Standard:

⁰⁸ **The maximum dimension (height or width) of a pictograph shall not exceed two times the upper-case letter height of the destination legend and shall not exceed the size of a route shield on the guide sign. If used, the pictograph shall be located to the left of the destination legend it represents, except as provided in Paragraph 9 for the park-ride Supplemental Guide sign.**

⁰⁹ **When a transit pictograph is displayed on the park-ride Supplemental Guide sign, it shall be located on the same line as the carpool symbol, if used, above the word legend.**

¹⁰ **A pictograph representing a State, county, or municipal corporation or other incorporated or unincorporated community shall not be displayed on a Supplemental Guide sign.**

¹¹ **Pictographs shall otherwise comply with the provisions of Section 2A.06.**

Figure 2E-25. Supplemental Guide Sign for a Park – Ride Facility

A – ROUTE WITHOUT EXIT NUMBERING

B – ROUTE WITH EXIT NUMBERING

Section 2E.36 Exit Direction Signs

Support:

01 The Exit Direction sign (see Figure 2E-26) repeats the route and destination information that was displayed on the Advance Guide sign(s) for the next exit, and thereby assures road users of the destination served and indicates whether they exit to the right or left for that destination.

Standard:

02 **Exit Direction signs shall be used at major and intermediate interchanges. Populations or other similar information shall not be displayed on Exit Direction signs.**

Guidance:

03 *Exit Direction signs should be used at minor interchanges.*

04 *post-mounted Exit Direction signs should be installed at the beginning of the deceleration lane. If there is less than 300 feet from the upstream end of the deceleration lane to the theoretical gore (see Figure 3B-8), the Exit Direction sign should be installed overhead over the exiting lane in the vicinity of the theoretical gore.*

Standard:

05 **Except where Overhead Arrow-per-Lane guide signs are used (see Section 2E.21 and Paragraph 6 of this Section), where a through lane is being terminated (dropped) at an exit, the Exit Direction sign shall be placed overhead at the theoretical gore (see Figures 2E-8 through 2E-11, and 2E-14 through 2E-16).**

06 **Except as provided in Paragraph 4 in Section 2E.21, where Overhead Arrow-per-Lane guide signs are used for the Advance Guide sign(s) for a multi-lane exit having an optional exit lane that also carries the through route or for a split with an option lane (see Section 2E.21), an Overhead Arrow-per-Lane guide sign shall also be used instead of the Exit Direction sign. This Overhead Arrow-per-Lane guide sign shall include the appropriate exit number (E1-5P or E1-5bP) plaque (if a numbered exit) and it shall be located near, but not downstream from, the point where the outside edge of the dropped lane begins to diverge from the mainline (see Figures 2E-4 through 2E-6).**

Figure 2E-26. Examples of Interchange Exit Direction Signs

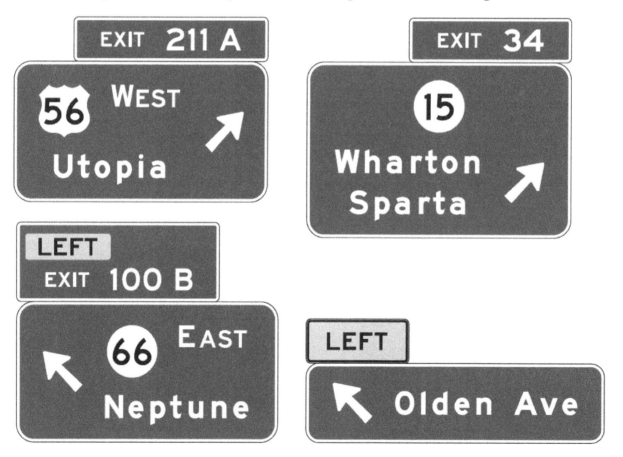

07 The following provisions shall govern the design and application of overhead Exit Direction signs:
   A. The sign shall carry the exit number (if exit numbering is used), the route number, cardinal direction, and destination, as applicable, with a diagonally upward-pointing directional arrow (see Figure 2E-26).
   B. The message EXIT ONLY in black on a yellow sign panel (E11-1d or E11-1e) shall be used on the overhead Exit Direction sign to advise road users of a lane drop situation (see Figures 2E-8 through 2E-11). The sign shall comply with the provisions of Section 2E.24.

Guidance:

08 *For numbered exits to the right, an exit number (E1-5P) plaque (see Figure 2E-22) should be added to the top right-hand edge of the sign.*

Standard:

09 **For numbered exits to the left, a left exit number (E1-5bP) plaque (see Figure 2E-22) shall be added to the top left-hand edge of the sign.**

10 **For non-numbered exits to the left, a LEFT (E1-5aP) plaque (see Figure 2E-22) shall be added to the top left-hand edge of the sign.**

Support:

11 Section 2E.31 contains additional information regarding exit numbering.

Option:

12 In some cases, principally in urban areas, where restricted sight distance because of structures or unusual alignment make it impossible to locate the Exit Direction sign without violating the required minimum spacing (see Section 2E.33) between major guide signs, Interchange Sequence signs (see Section 2E.40) may be substituted for an Advance Guide sign.

Guidance:

13 *At multi-exit interchanges, the Exit Direction sign should be located directly over the exiting lane for the first exit. At the same location, and normally over the right-hand through lane, an Advance Guide sign for the second exit should be located. Only for those conditions where the through movement is not evident should a confirmatory message (Pull-Through sign as shown in Figure 2E-2) be used over the left lane(s) to guide road users traveling through an interchange. In the interest of sign spreading, three signs on one structure should not be used. When the freeway or expressway is on an overpass, the Exit Direction sign should be installed on an overhead support over the exit lane in advance of the gore point.*

Option:

14 If the second exit is beyond an underpass, the Exit Direction sign may be mounted on the face of the overhead structure.

15 Where extra emphasis of an especially low advisory ramp speed is needed, an EXIT XX MPH (E13-2) sign panel (see Figure 2E-27) may be placed at the bottom of the Exit Direction sign to supplement, but not to replace, the exit or ramp advisory speed warning signs.

Guidance:

16 *At the last exit from a highway before it becomes a facility on which toll payments are required, the LAST EXIT BEFORE TOLL (W16-16P) plaque (see Section 2F.10 and Figure 2F-3) should be installed above the Exit Direction sign.*

Figure 2E-27. Interchange Exit Direction Sign with an Advisory Speed Panel

Exit Direction sign with E13-2 sign panel

OR

Exit Direction sign with E13-2 sign panel and flashing yellow beacons

Option:

17 If there is insufficient space above the Exit Direction sign because of the presence of an Exit Number (E1-5P) plaque, the W16-16P plaque may be mounted below the Exit Direction sign.

Section 2E.37 Exit Gore Signs (E5-1 Series)

Support:

01 The Exit Gore (E5-1 or E5-1a) sign (see Figure 2E-28) in the gore indicates the exiting point or the place of departure from the main roadway. Consistent application of this sign at each exit is important.

Standard:

02 **The gore shall be defined as the area located between the main roadway and the ramp just beyond where the ramp branches from the main roadway. The Exit Gore sign shall be located in the gore and shall carry the word EXIT or EXIT XX (if interchange numbering is used) and an appropriate upward slanting arrow. If suffix letters are used for exit numbering at a multi-exit interchange, the suffix letter shall also be included on the Exit Gore sign and shall be separated from the exit number by a space having a width of between 1/2 and 3/4 of the height of the suffix letter. Breakaway or yielding supports shall be used.**

Guidance:

03 *The arrow should be aligned to approximate the angle of departure. Each gore should be treated similarly, whether the interchange has one exit roadway or multiple exits.*

Option:

04 Where extra emphasis of an especially low advisory ramp speed is needed, an E13-1P plaque indicating the advisory speed may be mounted below the Exit Gore sign (see Figure 2E-28) to supplement, but not to replace, the exit or ramp advisory speed warning signs.

05 To improve the visibility of the gore for exiting drivers, a Type 1 object marker (see Chapter 2C) may be installed on each sign support below the Exit Gore sign.

06 An Exit Number (E5-1bP) plaque (see Figure 2E-22) may be installed above an existing Exit Gore (E5-1) sign when a non-numbered exit is converted to a numbered exit.

Standard:

07 **An Exit Gore (E5-1a) sign shall be used when the replacement of an existing assembly of an E5-1 sign and an E5-1bP plaque becomes necessary.**

Option:

08 The Narrow Exit Gore (E5-1c) sign may be used in gore areas of limited width where the width of the Exit Gore (E5-1a) sign would not permit sufficient lateral offset (see Section 2A.19), such as for ramp departures that are nearly parallel to the mainline roadway where the Exit Gore sign would be mounted on a narrow island or barrier. Where the E5-1c sign is mounted at a height of 14 feet or more from the roadway, the directional arrow may point diagonally downward.

Guidance:

09 *The E5-1c should not be used in gore areas where an E5-1a sign could be installed with sufficient lateral offset.*

Section 2E.38 Post-Interchange Signs

Guidance:

01 *If space between interchanges permits, as in rural areas, and where undue repetition of messages will not occur, a fixed sequence of signs should be displayed beginning 500 feet beyond the downstream end of the acceleration lane. At this point a Route sign assembly should be installed followed by a Speed Limit sign and a Distance sign, each at a spacing of 1,000 feet.*

02 *If space between interchanges does not permit placement of these three post-interchange signs without encroaching on or overlapping the Advance Guide signs necessary for the next interchange, or in rural areas where the interchanging traffic is primarily local, one or more of the post-interchange signs should be omitted.*

Figure 2E-28. Exit Gore Signs

Option:

03 Usually the Distance sign will be of less importance than the other two signs and may be omitted, especially if Interchange Sequence signs are used. If the sign for through traffic on an overhead assembly already contains the route sign, the post-interchange route sign assembly may also be omitted.

Section 2E.39 Post-Interchange Distance Signs
Standard:

01 **If used, the Post-Interchange Distance sign shall consist of a two- or three-line sign carrying the names of significant destination points and the distances to those points. The top line of the sign shall identify the next meaningful interchange with the name of the community near or through which the route passes, or if there is no community, the route number or name of the intersected highway (see Figure 2E-29).**

Support:

02 The minimum sizes of the route shields identifying a significant destination point are prescribed in Tables 2E-3 and 2E-5.

Option:

03 The text identification of a route may be displayed instead of a route shield, such as "US XX," "State Route XX," or "County Route XX."

Guidance:

04 *If a second line is used, it should be reserved for communities of general interest that are located on or immediately adjacent to the route or for major traffic generators along the route.*

Option:

05 The choice of names for the second line, if it is used, may be varied on successive Distance signs to give road users maximum information concerning communities served by the route.

Standard:

06 **The third, or bottom line, shall contain the name and distance to a control city (if any) that has national significance for travelers using the route.**

Guidance:

07 *Distances to the same destinations should not be shown more frequently than at 5-mile intervals. The distances displayed on these signs should be the actual distance to the destination points and not to the exit from the freeway or expressway. The distance displayed for each community should comply with the provisions of Section 2D.41.*

Section 2E.40 Interchange Sequence Signs
Option:

01 If interchanges are closely spaced, particularly through large urban areas, so that guide signs cannot be adequately spaced, Interchange Sequence signs identifying the next two or three interchanges may be used.

Guidance:

02 *If used, Interchange Sequence signs should be used over the entire length of a route in an urban area. Except as provided in Paragraph 3, they should not be used on a single interchange basis.*

03 *If there is less than 800 feet between interchanges, Interchange Sequence signs should be used instead of the Advance Guide signs for the affected interchanges.*

Support:

04 Interchange Sequence signs are generally supplemental to Advance Guide signs. Signing of this type is illustrated in Figures 2E-30 and 2E-31, and is compatible with the sign spreading concept described in Paragraph 3 of Section 2E.11.

05 These signs are installed in a series and display the next two or three interchanges by name or route number with distances to the nearest 1/4 mile.

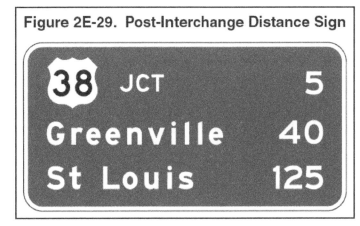

Figure 2E-29. Post-Interchange Distance Sign

Figure 2E-30. Example of Using an Interchange Sequence Sign for Closely-Spaced Interchanges

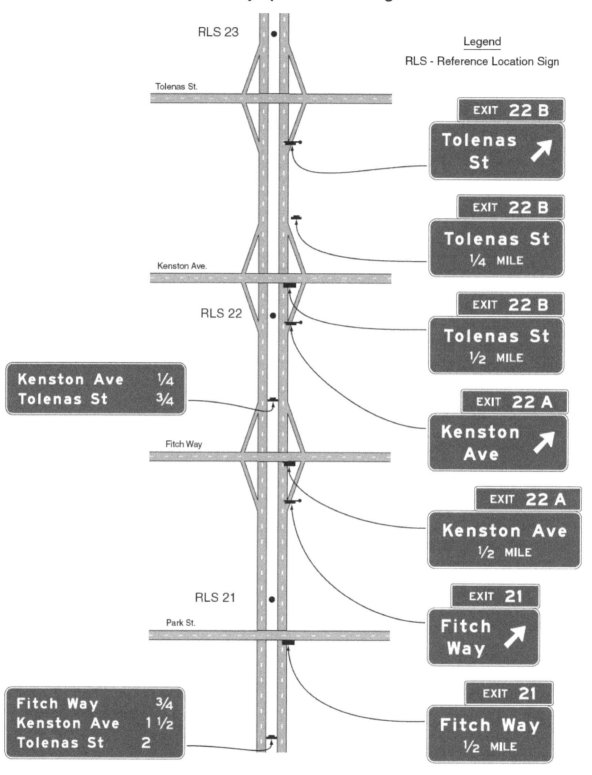

Standard:

06 **If used, the first sign in the series shall be located in advance of the first Advance Guide sign for the first interchange.**

07 **Where the exit direction is to the left, a LEFT (E11-2) sign panel (see Figure 2E-13) shall be displayed on the same line immediately to the right of the interchange name or route number.**

08 **Interchange Sequence signs shall not be substituted for Exit Direction signs.**

Guidance:

09 *Interchange Sequence signs should be located in the median. After the first of the series, Interchange Sequence signs should be placed approximately midway between interchanges.*

Standard:

10 **Interchange Sequence signs located in the median shall be installed at overhead sign height (see Section 2A.18).**

Option:

11 Interchange numbers may be displayed to the left of the interchange name or route number.

Section 2E.41 Community Interchanges Identification Signs

Support:

01 For suburban or rural communities served by two or three interchanges, Community Interchanges Identification signs are useful (see Figure 2E-32).

Guidance:

02 *In these cases, the name of the community followed by the word Exits should be displayed on the top line; the lines below should display the destination, road name or route number, and the corresponding distances to the nearest 1/4 mile.*

03 *The sign should be located in advance of the first Advance Guide sign for the first interchange within the community.*

Option:

04 If interchanges are not conveniently identifiable or if there are more than three interchanges to be identified, the NEXT XX EXITS sign (see Section 2E.42) may be used.

Section 2E.42 NEXT XX EXITS Sign

Support:

01 Many freeways or expressways pass through historical or recreational regions, or urban areas served by a succession of several interchanges.

Option:

02 Such regions or areas may be indicated by a NEXT XX EXITS sign (see Figure 2E-33) located in advance of the Advance Guide sign or signs for the first interchange.

Guidance:

03 *The sign legend should identify the region or area followed by the words NEXT XX EXITS.*

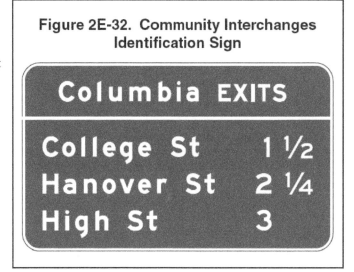

Figure 2E-31. Interchange Sequence Sign

Figure 2E-32. Community Interchanges Identification Sign

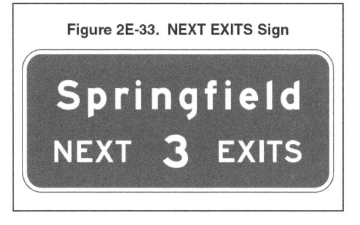

Figure 2E-33. NEXT EXITS Sign

Section 2E.43 Signing by Type of Interchange

Support:

01 Road users need signs to help identify the location of the exit, as well as to obtain route, direction, and destination information for specific exit ramps. Figures 2E-34 through 2E-40 show examples of guide signs for common types of interchanges. The interchange layouts shown in most of the figures illustrate only the major guide signs for one direction of traffic on the freeway and on the exit ramps. Section 2D.45 contains information regarding the signing of the crossroad approaches and connecting roadways to freeways and expressways.

Standard:

02 **Interchange guide signing shall be consistent for each type of interchange along a route.**

Guidance:

03 *The signing layout for all interchanges having only one exit ramp in the direction of travel should be similar, regardless of the interchange type. For the sake of uniform application, the significant features of the signing plan for each of the more frequent kinds of interchanges (illustrated in Figures 2E-34 through 2E-40) should be followed as closely as possible. Even when unusual geometric features exist, variations in signing layout should be held to a minimum.*

Section 2E.44 Freeway-to-Freeway Interchange

Support:

01 Freeway-to-freeway interchanges are major decision points where the effect of taking a wrong ramp cannot be easily corrected. Reversing direction on the connecting freeway or reentering to continue on the intended course is usually not possible. Figure 2E-34 shows examples of guide signs at a freeway-to-freeway interchange.

Guidance:

02 *The sign messages should contain only the route shield, cardinal direction, and the name of the next control city on the route. Arrows should point as indicated in Section 2D.08, except where Overhead Arrow-per-Lane or Diagrammatic signs are used in accordance with the provisions of Sections 2E.20 through 2E.22.*

Support:

03 At splits where the off-route movement is to the left or where there is an optional lane split, expectancy problems usually result.

Standard:

04 **At splits where the off-route movement is to the left, the Left Exit Number (E1-5bP) plaque shall be added at the top left-hand edge of the guide sign (see Section 2E.31). Overhead Arrow-per-Lane or Diagrammatic guide signs (see Sections 2E.21 and 2E.22) shall be used for freeway splits with an option lane and for multi-lane freeway-to-freeway exits having an option lane.**

05 **Overhead signs shall be used at a distance of 1 mile and at the theoretical gore of each connecting ramp. When Overhead Arrow-per-Lane or Diagrammatic guide signs are used, they shall comply with the provisions of Sections 2E.21 and 2E.22.**

Option:

06 Overhead signs may also be used at the 1/2-mile and 2-mile locations.

07 The arrow and/or the name of the control city may be omitted on signs that indicate the straight-ahead continuation of a route on a Pull-Through sign (see Section 2E.12).

08 An Advisory Exit Speed sign may be used where an engineering study shows that it is necessary to display a speed reduction message for ramp signing (see Section 2C.14).

09 Where extra emphasis of an especially low advisory ramp speed is needed, an EXIT XX MPH (E13-2) sign panel (see Figure 2E-27) may be placed at the bottom of the Exit Direction sign to supplement, but not to replace, the exit or ramp advisory speed warning signs.

Section 2E.45 Cloverleaf Interchange

Support:

01 A cloverleaf interchange has two exits for each direction of travel. The exits are closely spaced and have common Advance Guide signs. Examples of guide signs for cloverleaf interchanges are shown in Figure 2E-35.

Guidance:

02 *The Advance Guide signs should include two place names, one corresponding to each exit ramp, with the name of the place served by the first exit on the upper line.*

Figure 2E-34. Examples of Guide Signs for a Freeway-to-Freeway Interchange
(Sheet 1 of 2)

A - Example of Signing for a Two-Lane Exit Ramp with Two Dropped Lanes and a Bifurcation Beyond the Mainline Gore

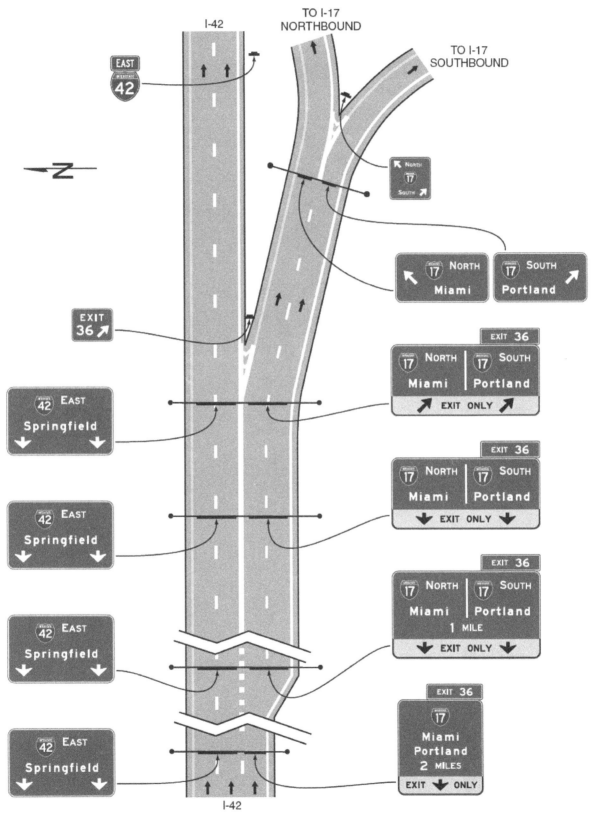

Figure 2E-34. Examples of Guide Signs for a Freeway-to-Freeway Interchange
(Sheet 2 of 2)

B - Example of Signing for Successive Exit Ramps with a Dropped Lane at the Second Exit

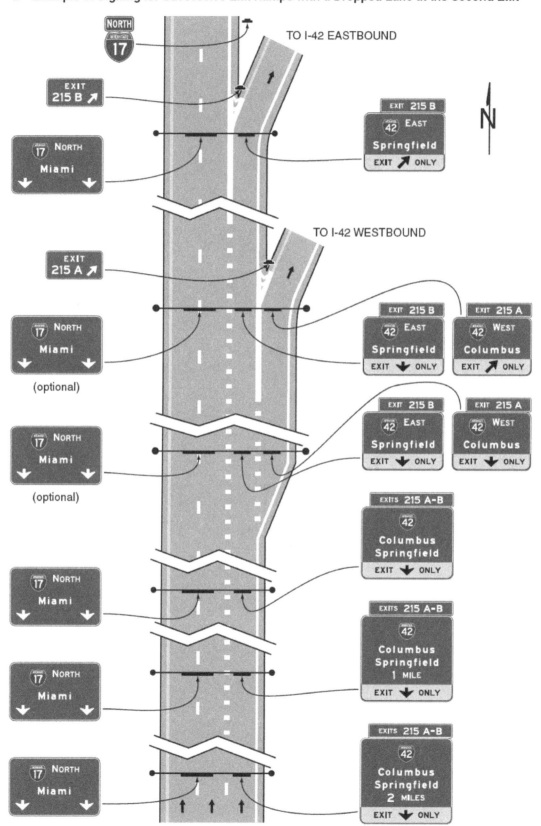

Figure 2E-35. Examples of Guide Signs for a Full Cloverleaf Interchange

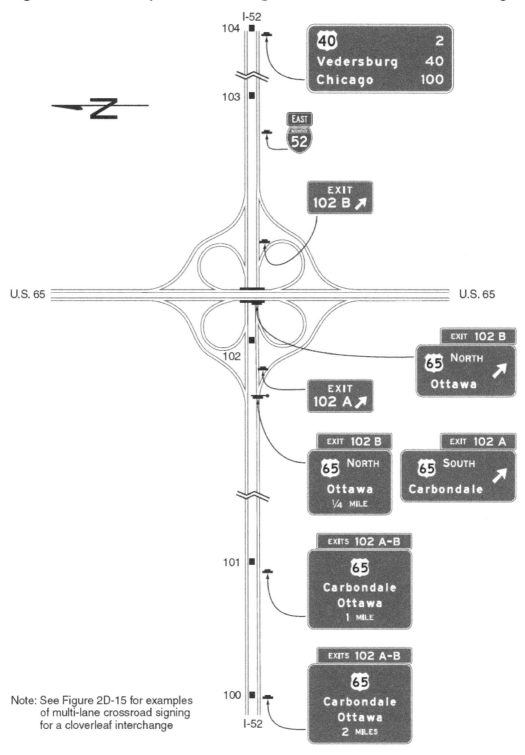

Note: See Figure 2D-15 for examples of multi-lane crossroad signing for a cloverleaf interchange

Standard:

03 **An overhead guide sign shall be placed at the theoretical gore of the first exit ramp, with a diagonally upward-pointing directional arrow on the Exit Direction sign for that exit and the message XX MILES, or EXIT XX MILES if interchange numbering is not used, on the Advance Guide sign for the second exit, as shown in Figure 2E-35. The second exit shall be indicated by an overhead Exit Direction sign over the auxiliary lane. An Exit Gore sign shall also be used at each gore (see Section 2E.37).**

04 **Interchanges with more than one exit from the main line shall be numbered as described in Section 2E.31 with an appropriate suffix.**

05 **Diagrammatic signs shall not be used for cloverleaf interchanges except as otherwise provided in Section 2E.22.**

Guidance:

06 *Where the mainline passes under the crossroad and the exit roadway is located beyond the overcrossing structure, the overhead Exit Direction sign for the second exit should be placed either on the overcrossing structure (see Figure 2E-35) or on a separate structure located immediately in front of the overcrossing structure.*

Section 2E.46 Cloverleaf Interchange with Collector-Distributor Roadways

Support:

01 Examples of guide signs for full cloverleaf interchanges with collector-distributor roadways are shown in Figure 2E-36.

Guidance:

02 *Signing on the collector-distributor roadways should be the same as the signing on the mainline of a cloverleaf interchange.*

Standard:

03 **Guide signs at exits from the collector-distributor roadways shall be overhead and located at the theoretical gore of the collector-distributor roadway and the exit ramp.**

Option:

04 Exits from the collector-distributor roadways may be numbered with an appropriate suffix. If the exits from a collector-distributor roadway are numbered with suffixes, the Advance Guide signs on the mainline may include two place names and their corresponding exit numbers with the plural EXITS. If only the exit from the mainline is numbered or if interchange numbering is not used, the Advance Guide signs on the mainline may use the singular EXIT.

Section 2E.47 Partial Cloverleaf Interchange

Support:

01 Examples of guide signs for partial cloverleaf interchanges are shown in Figure 2E-37.

Guidance:

02 *Where the mainline passes under the crossroad and the exit roadway is located beyond the overcrossing structure, the overhead Exit Direction sign should be placed either on the overcrossing structure (see Figure 2E-37) or on a separate structure located immediately in front of the overcrossing structure.*

Standard:

03 **A post-mounted Exit Gore sign shall also be installed in the ramp gore.**

Support:

04 Partial cloverleaf interchanges with successive exit ramps from the same direction of travel are signed the same as cloverleaf interchanges for that direction of travel (see Section 2E.45).

Section 2E.48 Diamond Interchange

Support:

01 Examples of guide signs for diamond interchanges are shown in Figure 2E-38.

Standard:

02 **For numbered exits, the singular message EXIT shall be used on the Exit Number plaques (see Section 2E.31) with the Advance Guide and Exit Direction signs. For non-numbered exits, the singular message EXIT shall be used as part of the distance message on the Advance Guide signs.**

Figure 2E-36. Examples of Guide Signs for a Full Cloverleaf Interchange with Collector-Distributor Roadways

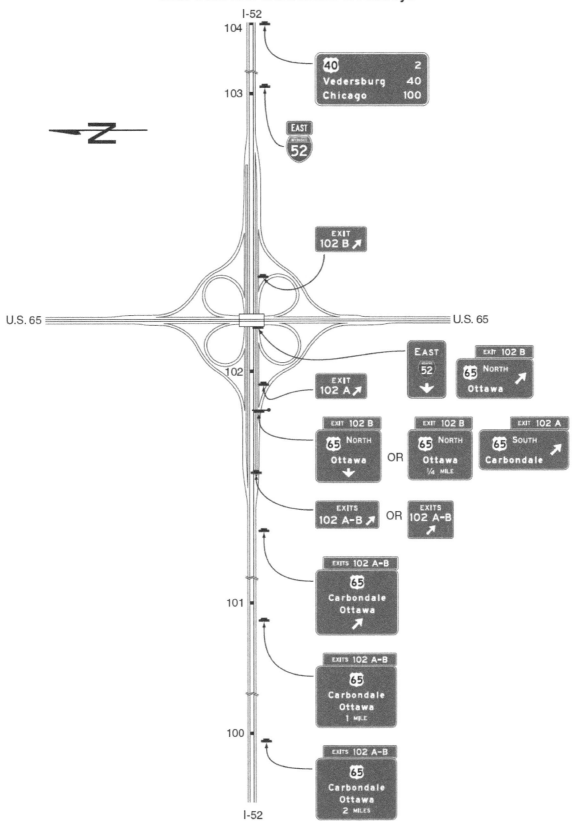

Note: See Figure 2D-15 for examples of multi-lane crossroad signing for a cloverleaf interchange

Figure 2E-37. Examples of Guide Signs for a Partial Cloverleaf Interchange

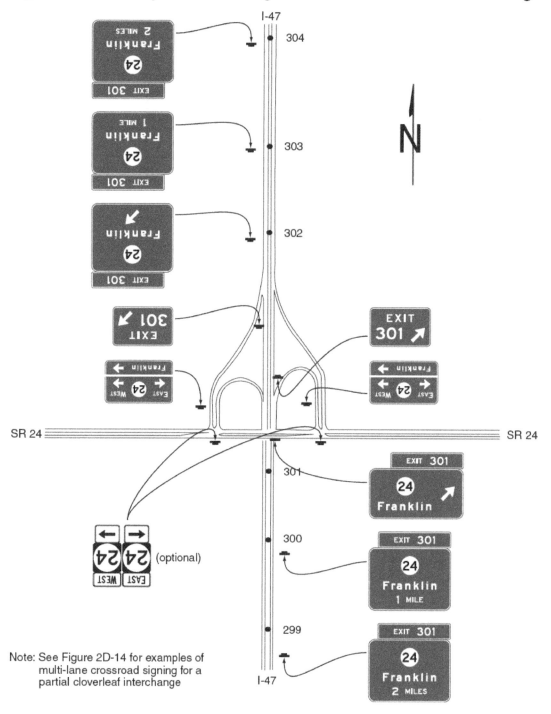

Note: See Figure 2D-14 for examples of multi-lane crossroad signing for a partial cloverleaf interchange

Figure 2E-38. Examples of Guide Signs for a Diamond Interchange

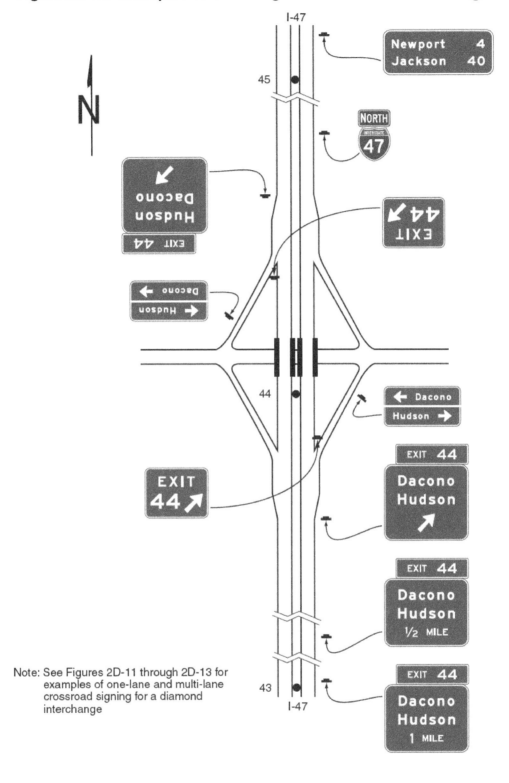

Note: See Figures 2D-11 through 2D-13 for examples of one-lane and multi-lane crossroad signing for a diamond interchange

Support:
⁰³ The typical diamond interchange ramp departs from the mainline roadway such that a speed reduction generally is not necessary in order for a driver to negotiate an exit maneuver from the mainline onto the ramp roadway.

Guidance:
⁰⁴ *When a speed reduction is not necessary, an exit speed sign should not be used.*

Option:
⁰⁵ An Advisory Exit Speed sign may be used where an engineering study shows that it is necessary to display a speed reduction message for ramp signing (see Section 2C.14).

Guidance:
⁰⁶ *The Advisory Exit Speed sign should be located along the deceleration lane or along the ramp such that it is visible to the driver far enough in advance to allow the driver to decelerate before reaching the curve associated with the exiting maneuver.*

Option:
⁰⁷ A Stop Ahead or Signal Ahead warning sign may be placed, where engineering judgment indicates a need, along the ramp in advance of the cross street, to give notice to the driver (see Section 2C.36).

Guidance:
⁰⁸ *When used on two-lane ramps, Stop Ahead or Signal Ahead signs should be used in pairs with one sign on each side of the ramp.*

Section 2E.49 Diamond Interchange in Urban Area

Support:
⁰¹ Examples of guide signs for diamond interchanges in an urban area are shown in Figure 2E-39. This example includes the use of the Community Interchanges Identification sign (see Section 2E.41), which might be useful if two or more interchanges serve the same community.
⁰² In urban areas, street names are often displayed as the principal message in destination signs.

Option:
⁰³ If interchanges are too closely spaced to properly locate the Advance Guide signs, they may be placed closer to the exit with the distances displayed adjusted accordingly.

Section 2E.50 Closely-Spaced Interchanges

Support:
⁰¹ Section 2E.11 contains information regarding sign spreading where the Exit Direction sign and the Advance Guide sign for the next interchange are mounted overhead. Sign spreading is particularly beneficial where interchanges are closely spaced and overhead signing is used in conjunction with Interchange Sequence signs as provided in Paragraph 2.

Guidance:
⁰² *Interchange Sequence signs (see Section 2E.40) should be used at closely-spaced interchanges. When used, they should identify and show street names and distances for the next two or three exits as shown in Figure 2E-30.*

Standard:
⁰³ **Advance Guide signs for closely-spaced interchanges shall show information for only one interchange.**

Section 2E.51 Minor Interchange

Option:
⁰¹ Less signing may be used for minor interchanges because such interchanges customarily serve low volumes of local traffic.

Support:
⁰² Examples of guide signs for minor interchanges are shown in Figure 2E-40.

Standard:
⁰³ **At least one Advance Guide sign and an Exit Gore sign shall be used at a minor interchange.**

Guidance:
⁰⁴ *An Exit Direction sign should also be used.*

Figure 2E-39. Examples of Guide Signs for a Diamond Interchange in an Urban Area

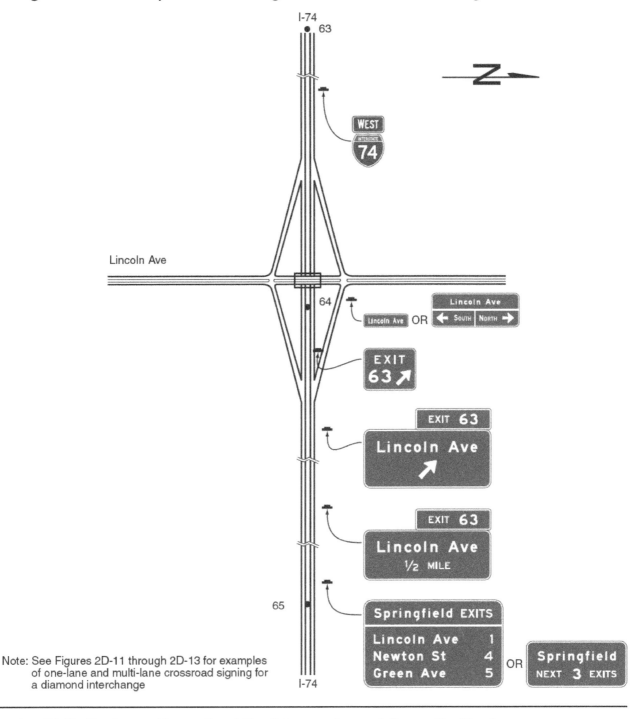

Section 2E.52 Signing on Conventional Road Approaches and Connecting Roadways
Support:
01 Section 2D.45 contains information regarding the signing on conventional roads on the approaches to interchanges and the signing on connecting roadways.

Section 2E.53 Wrong-Way Traffic Control at Interchange Ramps
Support:
01 Section 2B.41 contains information regarding the use of regulatory signs to deter wrong-way movements at intersections of freeway or expressway ramps with conventional roads, and in the area where entrance ramps intersect with the mainline lanes.

Figure 2E-40. Examples of Guide Signs for a Minor Interchange

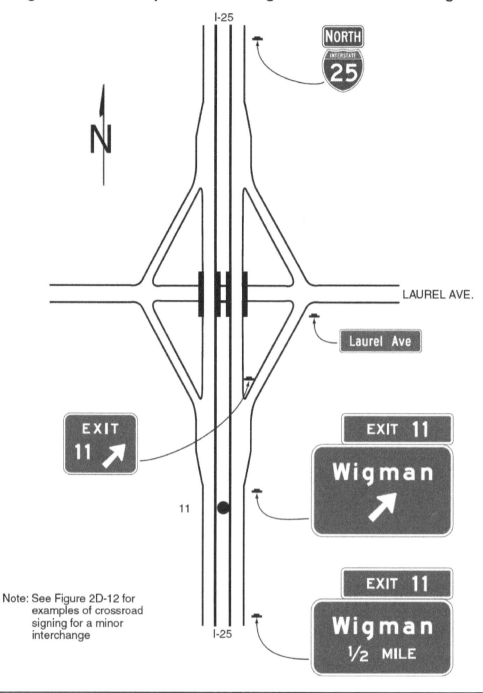

02 Section 2D.46 contains information regarding the use of a Directional assembly or a guide sign to mark the entrance to a freeway or expressway from a conventional road.

Section 2E.54 Weigh Station Signing
Standard:
01 **Weigh Station signing on freeways and expressways shall be the same as that provided in Section 2D.49, except for lettering size and the advance posting distance for the Exit Direction sign, which shall be located a minimum of 1,500 feet in advance of the gore.**
Support:
02 Weigh Station sign layouts for freeway and expressway applications are shown in the "Standard Highway Signs and Markings" book (see Section 1A.11).

CHAPTER 2F. TOLL ROAD SIGNS

Section 2F.01 Scope

Support:

01　Toll highways are typically limited-access freeway or expressway facilities. A portion of or an entire route might be a toll highway, or a bridge, tunnel, or other crossing point might be the only toll portion of a highway. A toll highway might be a conventional road. The general signing requirements for toll roads will depend on the type of facility and access (freeway, expressway, or conventional road). The provisions of Chapters 2D and 2E will generally apply for guide signs along the toll facility that direct road users within and off the facility where exit points and geometric configurations are not dependent specifically on the collection of tolls. The aspect of tolling and the presence of toll plazas or collection points necessitate additional considerations in the typical signing needs. The notification of the collection of tolls in advance of and at entry points to the toll highway also necessitate additional modifications to the typical signing.

02　The scope of this Section applies to a route or facility on which all lanes are tolled. Chapter 2G contains provisions for the signing of managed lanes within an otherwise non-toll facility that employ tolling or pricing as an operational strategy to manage congestion levels.

Standard:

03　**Except where specifically provided in this Chapter, the provisions of other Chapters in Part 2 shall apply to toll roads.**

Section 2F.02 Sizes of Toll Road Signs

Standard:

01　**Except as provided in Section 2A.11, the sizes of toll road signs that have standardized designs shall be as shown in Table 2F-1.**

Support:

02　Section 2A.11 contains information regarding the applicability of the various columns in Table 2F-1.

Option:

03　Signs larger than those shown in Table 2F-1 may be used (see Section 2A.11).

Table 2F-1. Toll Road Sign and Plaque Minimum Sizes

Sign or Plaque	Sign Designation	Section	Conventional Road Single Lane	Conventional Road Multi-Lane	Expressway	Freeway	Minimum	Oversized
Toll Rate	R3-28	2F.05	—	—	114 x 48	114 x 48	—	—
Pay Toll (plaque)	R3-29P	2F.05	—	—	24 x 18	24 x 18	—	—
Take Ticket (plaque)	R3-30P	2F.05	—	—	24 x 18	24 x 18	—	—
Pay Toll XX Miles Cars (price)	W9-6	2F.06	96 x 66	96 x 66	96 x 66	96 x 66	—	—
Pay Toll XX Miles Cars (price) (plaque)	W9-6P	2F.07	288* x 36	288* x 36	288* x 36	288* x 36	—	—
Stop Ahead Pay Toll Cars (price)	W9-6a	2F.08	114 x 66	114 x 66	114 x 66	114 x 66	—	—
Stop Ahead Pay Toll (plaque)	W9-6aP	2F.09	252* x 36	252* x 36	252* x 36	252* x 36	—	—
Last Exit Before Toll (plaque)	W16-16P	2F.10	—	—	252* x 36	252* x 36	—	—
Toll	M4-15	2F.11	24 x 12	24 x 12	36 x 18	36 x 18	24 x 12	36 x 18
No Cash	M4-16	2F.12	24 x 12	24 x 12	36 x 18	36 x 18	24 x 12	36 x 18
Toll Collector Symbol	M4-17	2F.13	—	—	48 x 48	48 x 48	—	—
Exact Change Symbol	M4-18	2F.13	—	—	48 x 48	48 x 48	—	—
ETC Only	M4-20	2F.12	24 x 24	24 x 24	36 x 36	36 x 36	24 x 24	36 x 36

* The width shown represents the minimum dimension. The width shall be increased as appropriate to match the width of the guide sign.

Notes: 1. Larger signs may be used when appropriate
　　　 2. Dimensions in inches are shown as width x height

Section 2F.03 Use of Purple Backgrounds and Underlay Panels with ETC Account Pictographs
Standard:

01 Use of the color purple on any sign shall comply with the provisions of Sections 1A.12 and 2A.10. Except as provided in Sections 2F.12 and 2F.16, purple as a background color shall be used only when the information associated with the appropriate ETC account is displayed on that portion of the sign. The background color of the remaining portion of such signs shall comply with the provisions of Sections 1A.12 and 2A.10 as appropriate for a regulatory, warning, or guide sign. Purple shall not be used as a background color to display a destination, action message, or other legend that is not a display of the requirement for all vehicles to have a registered ETC account.

02 If only vehicles with registered ETC accounts are allowed to use a highway lane, a toll plaza lane, an open-road tolling lane, or all lanes of a toll highway or connection, the signs for such lanes or highways shall incorporate the pictograph (see Chapter 2A) adopted by the toll facility's ETC payment system and the regulatory message ONLY. Except for ETC pictographs whose predominant background color is purple, if incorporated within the green background of a guide sign, the ETC pictograph shall be on a white rectangular or square panel set on a purple underlay panel with a white border. For rectangular ETC pictographs whose predominant background color is purple, a white border shall be used at the outer edges of the purple rectangle to provide contrast between the pictograph and the sign background color.

03 If an ETC pictograph is used on a separate plaque with a guide sign or on a header panel within a guide sign, the plaque or the header panel shall have a purple background with a white border and the ETC pictograph shall have a white border to provide contrast between the pictograph and the background of the plaque or header panel.

04 Purple underlay panels for ETC pictographs or purple backgrounds for plaques and header panels shall only be used in the manner described in Paragraphs 1 through 3 to convey the requirement of a registered ETC account on signs for lanes reserved exclusively for vehicles with such an account and on directional signs to an ETC account-only facility from a non-toll facility or from a toll facility that accepts multiple payment forms.

Support:

05 Figure 2F-1 shows examples of ETC account pictographs, their use with various background colors, and modifications involving underlay panels.

06 Section 2F.04 contains provisions regarding the size of pictographs for ETC accounts.

Section 2F.04 Size of ETC Pictographs
Standard:

01 The ETC pictograph (see Chapter 2A) shall be of a size that makes it a prominent feature of the sign legend as necessary for conspicuity for those road users with registered ETC accounts seeking such direction, as well as for those road users who do not have ETC accounts so that it is clear to them to avoid such direction when applicable.

Guidance:

02 *An ETC pictograph that is in the shape of a horizontal rectangle should have a minimum height between approximately 1.5 and 2 times the upper-case letter height of the principal legend on the sign. The width of an ETC pictograph in the shape of a horizontal rectangle should be between approximately two and three times the height of the pictograph. When the pictograph is the principal legend on the sign, such as for advance guide signs for open-road tolling lanes (see Section 2F.15), the minimum height of a horizontal rectangular ETC pictograph should be consistent with that of a route shield prescribed for the particular application and type of sign.*

03 *For ETC pictographs whose shape is square, circular, or otherwise similar in height and width, or is a vertical rectangle, the same basic principles for conspicuity and placement should be followed. ETC pictographs whose shape is not in that of a horizontal rectangle should be suitably sized to facilitate conspicuity as described in Paragraph 1 and should be of a similar approximate area as the horizontal rectangular pictographs designed in accordance with the height and width as provided in Paragraph 2.*

Section 2F.05 Regulatory Signs for Toll Plazas
Support:

01 Toll plaza operations often include lane-specific restrictions on vehicle type, forms of payment accepted, and speed limits or required stops. Vehicles are typically required to come to a stop to pay the toll or receive a toll ticket in the attended and exact change or automatic lanes. Electronic toll collection (ETC) lanes with favorable geometrics typically allow vehicles to move through the toll plaza without stopping, but usually within a set regulatory speed limit or advisory speed. In some ETC lanes and in most lanes that accommodate non-ETC vehicles, a stop might be required while the ETC payment is processed because of geometric or other conditions.

Figure 2F-1. Examples of ETC Account Pictographs and Use of Purple Backgrounds and Underlay Panels

A - PICTOGRAPH DESIGN WITH A PURPLE BACKGROUND AND A WHITE CONTRASTING BORDER

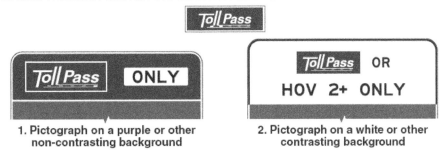

1. Pictograph on a purple or other non-contrasting background
2. Pictograph on a white or other contrasting background

B - PICTOGRAPH DESIGN WITH A BACKGROUND COLOR OTHER THAN PURPLE, SHOWN ON A PURPLE UNDERLAY PANEL WITH A WHITE CONTRASTING BORDER

1. Pictograph on a purple background
2. Pictograph with a purple underlay on a non-contrasting background
3. Pictograph with a purple underlay panel on a white or other contrasting background

Guidance:

02 *Regulatory signs applicable only to a particular lane or lanes should be located in a position that makes their applicability clear to road users approaching the toll plaza.*

03 *Regulatory signs, or regulatory panels within guide signs, indicating restrictions on vehicle type and forms of toll payment accepted at a specific toll plaza lane should be installed over the applicable lane either on the toll plaza canopy or on a separate structure immediately in advance of the canopy located in a manner such that each sign is clearly related to an individual toll lane.*

Support:

04 Section 2F.13 contains information regarding the incorporation of regulatory messages into guide signs for toll plazas.

05 Section 2F.16 contains information regarding the design and use of toll plaza canopy signs.

Guidance:

06 *One or more Speed Limit (R2-1) signs (see Section 2B.13) should be installed in the locations provided in Paragraph 8 for an ETC-Only lane at a toll plaza in which an enforceable regulatory speed limit is established for a lane in which it is intended that vehicles move through the toll plaza without stopping while toll payments requiring stops occur in other lanes at the toll plaza. The speed limit displayed on the signs should be based on an engineering study taking into account the geometry of the plaza and the lanes and other appropriate safety and operational factors.*

07 *A Speed Limit (R2-1) sign should not be installed for a toll plaza lane that is controlled by a STOP (R1-1) sign or where a stop is required.*

Option:

08 Speed limit signs may be installed over the applicable lane on the toll plaza canopy, on the approach end of the toll booth island, on the toll booth itself, or on a vertical element of the canopy structure. Down arrows or diagonally downward-pointing directional arrows may be used to supplement the speed limit signs if an engineering study or engineering judgment indicates that the arrow is needed to clarify the applicability of a sign to a specific lane or to improve compliance.

Standard:

09 A STOP (R1-1) sign shall not be installed for a toll plaza lane that is operated as an ETC-Only lane and that is designed for tolls to be collected while vehicles continue moving.

Option:

10 A STOP (R1-1) sign may be installed to require vehicles to come to a complete stop to pay a toll in an attended or exact change lane, even if that lane is also available for optional use by vehicles with registered ETC accounts. A PAY TOLL (R3-29P) or TAKE TICKET (R3-30P) plaque (see Figure 2F-2), as appropriate to the operation, may be installed directly under the STOP (R1-1) sign for a toll plaza lane, if needed.

11 The mounting height of the STOP sign and any supplemental plaque may be less than the normal mounting height requirements if constrained by the physical features of the toll island or toll plaza.

12 The lateral offset of a STOP or other regulatory sign located within a toll plaza island may be reduced to a minimum of 1 foot from the face of the toll island or raised barrier to the nearest edge of the sign.

Guidance:

13 *If used, a STOP (R1-1) sign for a toll plaza cash payment lane should be located in a longitudinal position as near as practical to the point where a vehicle is expected to stop to pay the toll or take a ticket.*

Option:

14 A Toll Rate (R3-28) sign (see Figure 2F-2) may be installed in advance of the toll plaza to indicate the toll applicable to the various vehicle types.

Guidance:

15 *If used, the Toll Rate (R3-28) sign should be located between the toll plaza and the first advance sign informing road users of the toll plaza.*

16 *The R3-28 sign should not contain more than three lines of legend. Each lines that shows a toll amount should display only a single toll amount.*

Option:

17 Additional toll rate information exceeding three lines of legend may be displayed on the toll booth adjacent to the payment window of an attended lane or the payment receptacle of an exact change or automatic lane where it is visible to a road user who has stopped to pay the toll, but is not visible to approaching road users who have not yet entered the toll lane.

Section 2F.06 Pay Toll Advance Warning Sign (W9-6)

Standard:

01 The Pay Toll Advance Warning (W9-6) sign shall be a horizontal rectangle with a black legend and border on a yellow background. The legend shall include the distance to the toll plaza and, except for toll-ticket facilities, the toll for passenger or 2-axle vehicles (see Figure 2F-3). Where the toll for passenger or 2-axle vehicles is variable by time of day, a changeable message element shall be incorporated into the W9-6 sign to display the toll in effect. For toll plazas where road users entering a toll-ticket facility are issued a toll ticket, the legend PAY TOLL shall be replaced with a suitable legend such as TAKE TICKET.

Guidance:

02 *The Pay Toll Advance Warning sign should be installed overhead at approximately 1 mile and 1/2 mile in advance of mainline toll plazas at which some or all lanes are required to come to a stop to pay a toll (see Sections 2F.14 and 2F.15).*

Option:

03 If there is insufficient space for the W9-6 sign at the 1-mile or 1/2-mile advance locations, the Pay Toll Advance Warning (W9-6P) plaque (see Section 2F.07) may be installed at those advance locations above the appropriate guide sign(s) that relate to toll payment types.

Figure 2F-2. Toll Plaza Regulatory Signs and Plaques

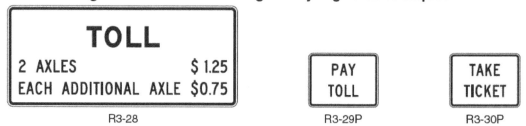

04 An additional W9-6 sign may be installed approximately 2 miles in advance of a mainline toll plaza. This sign may be either overhead or post-mounted.

05 If the visibility of a ramp toll plaza at which some or all lanes are required to come to a stop to pay a toll is limited, the W9-6 sign may also be installed in advance of the ramp toll plaza.

Section 2F.07 Pay Toll Advance Warning Plaque (W9-6P)

Option:

01 The Pay Toll Advance Warning (W9-6P) plaque (see Figure 2F-3) may be installed above the appropriate guide sign(s) relating to toll payment types at the 1-mile and/or 1/2-mile advance locations on the approach to a toll plaza if there is insufficient space for the W9-6 sign (see Section 2F.06) at those advance locations.

Standard:

02 **The W9-6P plaque shall be a horizontal rectangle with black legend and border on a yellow background. The legend shall include the distance to the toll plaza and, except for toll-ticket facilities, the toll for passenger or 2-axle vehicles. Where the toll for passenger or 2-axle vehicles is variable by time of day, a changeable message element shall be incorporated into the W9-6P plaque to display the toll in effect. For toll plazas where road users entering a toll-ticket facility are issued a toll ticket, the legend PAY TOLL shall be replaced with a suitable legend such as TAKE TICKET.**

Option:

03 The distance to the toll plaza may be omitted from the W9-6P plaque if the distance is displayed on the guide sign that the plaque accompanies.

04 The toll for passenger or 2-axle vehicles may be omitted from the W9-6P plaque if the toll information is displayed on the guide sign that the plaque accompanies.

Figure 2F-3. Toll Plaza Warning Signs and Plaques

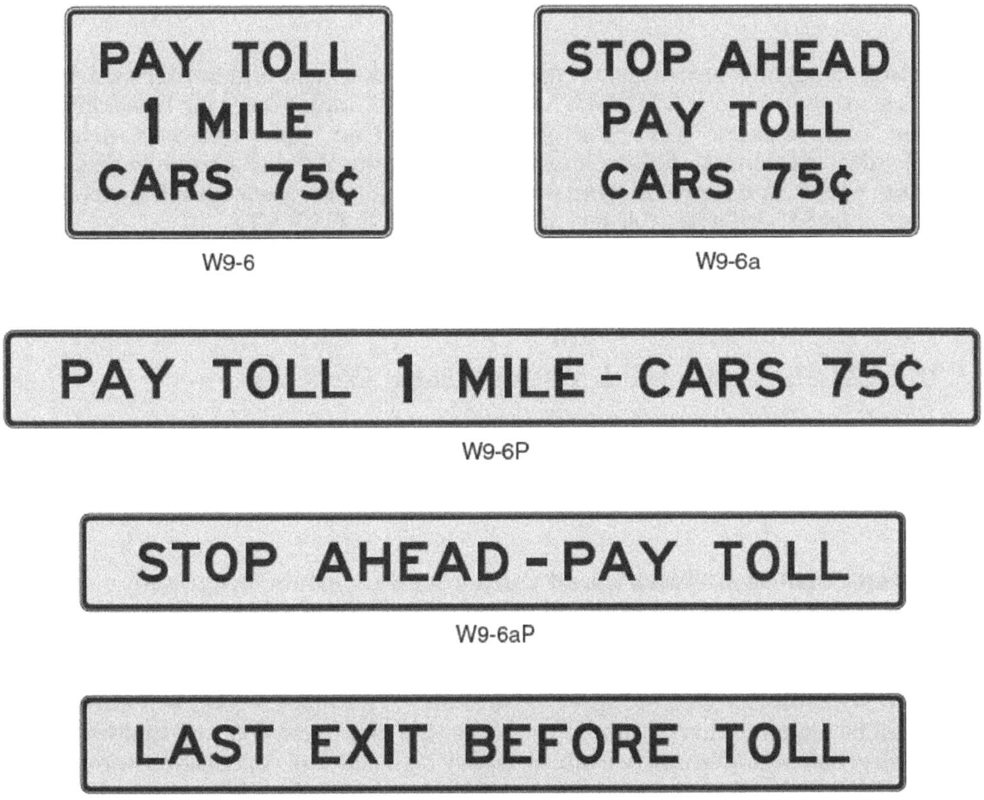

PAY TOLL
1 MILE
CARS 75¢

W9-6

STOP AHEAD
PAY TOLL
CARS 75¢

W9-6a

PAY TOLL 1 MILE – CARS 75¢

W9-6P

STOP AHEAD – PAY TOLL

W9-6aP

LAST EXIT BEFORE TOLL

W16-16P

Section 2F.08 Stop Ahead Pay Toll Warning Sign (W9-6a)
Standard:

01 **The Stop Ahead Pay Toll (W9-6a) sign shall be a horizontal rectangle with a black legend and border on a yellow background. The legend shall include STOP AHEAD PAY TOLL and, except for toll-ticket facilities, the toll for passenger or 2-axle vehicles (see Figure 2F-3). Where the toll for passenger or 2-axle vehicles is variable by time of day, a changeable message element shall be incorporated into the W9-6a sign to display the toll in effect. For toll plazas where road users entering a toll-ticket facility are issued a toll ticket, the legend PAY TOLL shall be replaced with a suitable legend such as TAKE TICKET.**

Guidance:

02 *The Stop Ahead Pay Toll sign should be installed overhead downstream from the W9-6 sign that is 1/2 mile in advance of a mainline toll plaza where some or all of the lanes are required to come to a stop to pay a toll (see Sections 2F.14 and 2F.15). The location of the overhead sign should coincide with the approximate location where the mainline lanes begin to widen on the approach to the toll plaza lanes.*

03 *Where open-road tolling is used in addition to a toll plaza at a particular location, the W9-6a sign should be located such that the message is clearly related to the lanes that access the toll plaza and not to the open-road tolling lanes.*

Option:

04 If there is insufficient space for the W9-6a sign at the recommended location, the Stop Ahead Pay Toll (W9-6aP) plaque (see Section 2F.09) may be installed at that location above the appropriate guide sign that relates to toll payment types.

05 If the visibility of a ramp toll plaza at which some or all lanes are required to come to a stop to pay a toll is limited, the W9-6a sign may also be installed in advance of the ramp toll plaza.

Section 2F.09 Stop Ahead Pay Toll Warning Plaque (W9-6aP)
Option:

01 The Stop Ahead Pay Toll (W9-6aP) plaque (see Figure 2F-3) may be installed above the appropriate guide sign at the location specified for the Stop Ahead Pay Toll (W9-6a) sign (see Section 2F.08) if there is insufficient space for the W9-6a sign at that location.

Standard:

02 **The W9-6aP plaque shall be a horizontal rectangle with black legend and border on a yellow background. The legend shall include STOP AHEAD PAY TOLL and, except for toll-ticket facilities, the toll for passenger or 2-axle vehicles. Where the toll for passenger or 2-axle vehicles is variable by time of day, a changeable message element shall be incorporated into the W9-6aP plaque to display the toll in effect. For toll plazas where road users entering a toll-ticket facility are issued a toll ticket, the legend PAY TOLL shall be replaced with a suitable legend such as TAKE TICKET.**

Option:

03 The toll for passenger or 2-axle vehicles may be omitted from the W9-6aP plaque if the toll information is displayed on the guide sign that the plaque accompanies.

Section 2F.10 LAST EXIT BEFORE TOLL Warning Plaque (W16-16P)
Guidance:

01 *The LAST EXIT BEFORE TOLL (W16-16P) plaque (see Figure 2F-3) should be used to notify road users of the last exit from a highway before it becomes a facility on which toll payments are required. The plaque should be installed above or below the appropriate guide signs for the exit (see Sections 2E.30 and 2E.33).*

Standard:

02 **The W16-16P plaque shall have a black legend and border on a yellow background.**

Section 2F.11 TOLL Auxiliary Sign (M4-15)
Standard:

01 **The TOLL (M4-15) auxiliary sign (see Figure 2F-4) shall have a black legend and border on a yellow background and shall be mounted directly above the route sign of a numbered toll highway or, if used, above the cardinal direction and alternative route auxiliary signs, in any route sign assembly providing directions from a non-toll highway to the toll highway or to a segment of a highway on which the payment of a toll is required.**

Figure 2F-4. ETC Account-Only Auxiliary Signs for Use in Route Sign Assemblies

NOTE: The ETC pictograph shown is an example only. The pictograph for the toll facility's adopted ETC system shall be used.

Example Route Sign Assembly

Section 2F.12 Electronic Toll Collection (ETC) Account-Only Auxiliary Signs (M4-16 and M4-20)
Standard:
01 In any route sign assembly providing directions from a non-toll highway to a toll facility, or to a tolled segment of a highway, where electronic toll collection (ETC) is the only payment method accepted and all vehicles are required to have a registered ETC account, the ETC Account-Only (M4-20) auxiliary sign (see Figure 2F-4) shall be mounted directly below the route sign of the numbered or named toll facility. The M4-20 auxiliary sign shall have a white border and purple background and incorporate the pictograph adopted by the toll facility's ETC payment system and the word ONLY in black letters on a white panel set on the purple background of the sign.

Option:
02 The NO CASH (M4-16) auxiliary sign (see Figure 2F-4) with a black legend and border on a white background may be used in a route sign assembly directly below the M4-20 auxiliary sign.

Section 2F.13 Toll Facility and Toll Plaza Guide Signs – General
Support:
01 Toll plazas are used on many toll highways, bridges, and tunnels for collection of tolls from road users. Electronic toll collection and/or open-road tolling might also be used on such facilities, either in addition to or in place of collecting toll payments at toll plazas.
02 Chapter 2G contains information regarding signs for preferential and managed lanes that are applicable to toll roads.
03 Chapter 3E contains information regarding pavement markings for certain toll plaza applications.

Standard:
04 **Directional assemblies for entrances to a toll highway or to a road leading directly to a toll highway with no opportunity to exit before paying or being charged a toll, shall clearly indicate that the facility is a toll facility. The TOLL (M4-15) auxiliary sign (see Section 2F.11) shall be used above the route sign of a numbered toll facility in any route sign assembly that provides directions to the toll route from another highway.**
05 **A rectangular panel with the black legend TOLL on a yellow background shall be incorporated into the guide signs leading road users to a toll highway (see Figure 2F-5).**

06 **Guide signs for toll highways, toll plazas, and tolled or priced managed lanes (see Chapter 2G) shall have white legends and borders on green backgrounds, except as specifically provided by Sections 2F.13 through 2F.16.**

Option:

07 Where conditions do not permit separate signs, or where it is important to associate a particular regulatory or warning message with specific guidance information, regulatory and/or warning messages may be combined with guide signs for toll plazas using plaques, header panels, or rectangular regulatory or warning panels incorporated within the guide signs, as long as the proper legend and background colors are preserved.

Standard:

08 **When regulatory messages are incorporated within a guide sign, they shall be on a rectangular panel with black legend on a white background. When warning messages are incorporated within a guide sign, they shall be on a rectangular panel with black legend on a yellow background.**

Support:

09 Figure 2F-5 shows examples of guide signs for entrances to various types of toll highways and for ETC account-only entrances to non-toll highways.

Standard:

10 **Signing for entrances to toll highways where ETC is employed only through license plate character recognition such that road users are not required to establish a toll account or register their vehicle equipment shall comply with the provisions of Paragraphs 4 and 5 (see Figure 2F-6).**

11 **If only vehicles with registered ETC accounts are allowed to use a toll highway, the guide signs for entrances to such facilities shall incorporate the pictograph adopted by the toll facility's ETC payment system and the regulatory message ONLY (see Figures 2F-1, 2F-5, and 2F-6). The use, size, and placement of the ETC pictograph shall comply with the provisions of Sections 2F.03 and 2F.04.**

Support:

12 Sections 2F.11, 2F.12, and 2F.17 contain additional provisions regarding signs for toll highways that only accept ETC payments.

13 Sections 2G.16 through 2G.18 contain additional provisions regarding signs for priced managed lanes that only accept ETC payments.

Option:

14 Where a toll highway on which tolls are collected only electronically also accepts payments from registered toll account users and those road users not registered in a toll account program are assessed a nominal surcharge in addition to the toll, or registered toll account users are assessed a discounted toll, such information may be displayed on a separate information sign near the entrance to such a facility (see Figure 2F-6).

Support:

15 Figure 2F-7 shows an example of guide signs for alternative toll and non-toll ramp connections to a non-toll highway.

16 Many different ETC payment systems are used by the various toll facility operators. Some of these systems accept payment from other systems' accounts.

Option:

17 Where a facility will accept payments from other systems' accounts in addition to its primary ETC-account payment system, such information may be displayed on a separate information sign near the entrances to such a facility or in advance of a toll plaza or open-road tolling lanes, as space allows between primary signs.

Guidance:

18 *Guide signs for toll plazas should be designed in accordance with the general principles of guide signs and the specific provisions of Chapter 2E.*

19 *Signs for toll plazas should systematically provide road users with advance and toll plaza lane-specific information regarding:*
   A. The amount of the toll, the types of payment accepted, and the type(s) of registered ETC accounts accepted for payment;
   B. Which lane or lanes are required or allowed to be used for each available payment type; and
   C. Restrictions on the use of a toll plaza lane or lanes by certain types of vehicles (such as cars only or no trucks).

Standard:

20 **Signs for attended lanes at toll plazas shall include word messages such as FULL SERVICE, CASH, CHANGE, or RECEIPTS (see Figures 2F-8 through 2F-11).**

2009 Edition   Page 245

Option:
21 Signs for Attended lanes at toll plazas may incorporate the Toll Taker (M4-17) symbol (see Figures 2F-8 and 2F-9), in a size that makes the symbol the predominant feature of the sign, to supplement the required word message.

Standard:
22 **Signs for Exact Change lanes at toll plazas shall incorporate an appropriate word message, such as EXACT CHANGE and the amount of the toll for passenger vehicles (see Figures 2F-8 through 2F-11).**

Option:
23 Signs for Exact Change lanes at toll plazas may include the Exact Change (M4-18) symbol (see Figures 2F-8 and 2F-9), in a size that makes the symbol the predominant feature of the sign, to supplement the required word message.

Figure 2F-5. Examples of Guide Signs for Entrances to Toll Highways or Ramps

A - ENTRANCE TO A TOLL HIGHWAY ON WHICH REGISTRATION IN A TOLL ACCOUNT PROGRAM IS NOT REQUIRED

B - ENTRANCE TO AN ETC ACCOUNT-ONLY TOLL HIGHWAY OR ENTRANCE TO A TOLL HIGHWAY VIA AN ETC ACCOUNT-ONLY RAMP

C - ENTRANCE TO A NON-TOLL HIGHWAY VIA AN ETC ACCOUNT-ONLY TOLL ENTRANCE RAMP

(the toll entrance is the only connection provided in the vicinity)

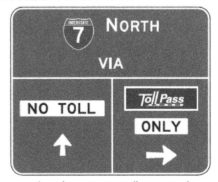
(an alternate non-toll entrance is provided in the vicinity)

Note: The ETC pictographs shown are examples only. The pictograph for the toll facility's adopted ETC system shall be used.

Page 246 — 2009 Edition

Figure 2F-6. Examples of Guide Signs for the Entrance to a Toll Highway on which Tolls are Collected Electronically Only

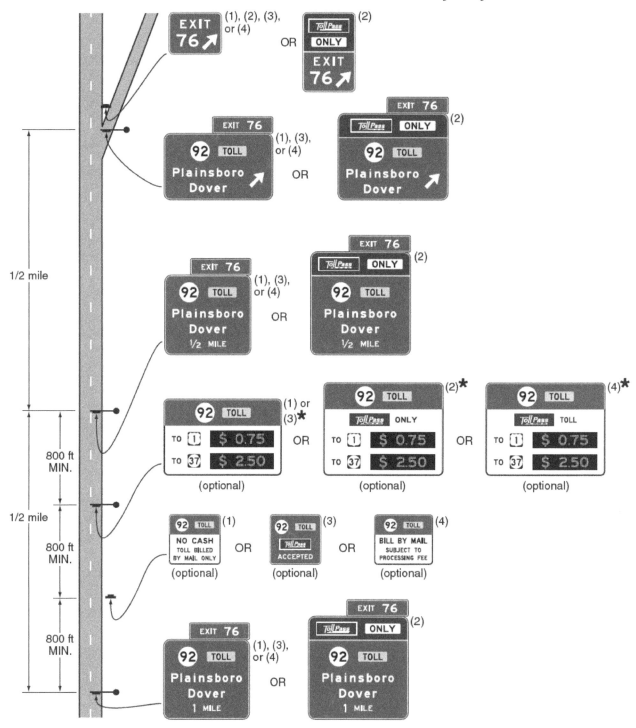

(1) All tolls are billed through license plate recognition only. A registered toll account or ETC device is not needed.
(2) All tolls are billed through registered toll accounts only. All vehicles must be registered in an ETC account program.
(3) Tolls are billed through license plate recognition in which registration in a toll account program is not required. Toll payments are also accepted from registered toll accounts. Registered toll accounts might receive a discount from the toll amount displayed on the signs.
(4) Tolls are billed through license plate character recognition or registered toll accounts. Vehicles not registered in a toll account program are assessed a nominal processing fee in addition to the toll amount displayed on the signs.
* For managed toll highways only (see Chapter 2G)

Sect. 2F.13

Figure 2F-7. Examples of Guide Signs for Alternative Toll and Non-Toll Ramp Connections to a Non-Toll Highway

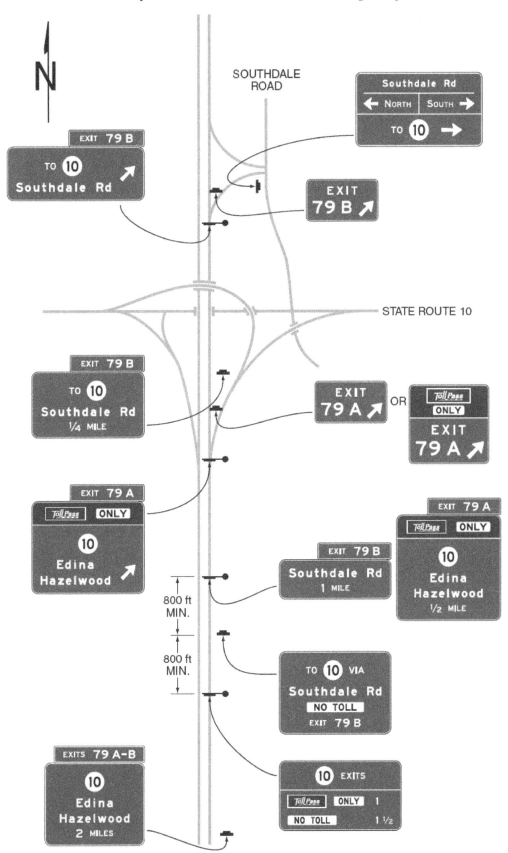

Figure 2F-8. Examples of Conventional Toll Plaza Advance Signs

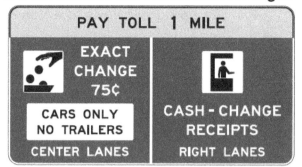

Notes:
1. The M4-17 symbol is optional for an attended lane.
2. The M4-18 symbol is optional for an exact change lane.
3. The ETC pictograph that is shown is only an example. The pictograph for the toll facility's adopted ETC system shall be used.

Figure 2F-9. Examples of Toll Plaza Canopy Signs

Attended Lane with an Optional M4-17 Toll Collector Symbol

Exact Change or ETC Account Lane with an Optional M4-18 Exact Change Symbol

ETC Account-Only Lane

★ Optional flashing yellow beacons that are separated from any lane-use control signals for the lane (see Section 2F.16)

★★ The ETC pictographs that are shown are only examples. The pictograph for the toll facility's adopted ETC system shall be used.

Standard:

24 **If used, the M4-17 and M4-18 symbols shall be used only as panels within guide signs that accompany the required word messages. The M4-17 and M4-18 symbols shall not be used as an independent sign or within a sign assembly.**

25 **If only vehicles with registered ETC accounts are allowed to use a toll plaza lane, the signs for such lanes shall incorporate the pictograph adopted by the toll facility's ETC payment system and the regulatory message ONLY (see Figures 2F-1, 2F-8, 2F-9, and 2F-11). The use, size, and placement of the ETC pictograph shall comply with the provisions of Sections 2F.03 and 2F.04.**

Option:

26 The ETC payment system's pictograph, without a purple underlay or purple header panel, may be used on signs for Exact Change or attended lanes at toll plazas to indicate that vehicles with registered ETC accounts may also use those lanes (see Figure 2F-9).

Section 2F.14 Advance Signs for Conventional Toll Plazas

Guidance:

01 *For conventional toll plazas (those without a divergence onto a separate alignment from mainline-aligned open-road tolling or ETC-Only lanes), one or more sets of overhead advance guide signs complying with the provisions of this Section should be provided. The advance guide signs for multi-lane toll plazas should provide information regarding which lanes to use for all of the toll payment methods accepted at the toll plaza. These signs should include toll plaza lane numbers (if used), or action messages or lane-use information such as LEFT LANE(S), CENTER LANE(S), RIGHT LANE(S), or down arrows over the approximate center of each applicable lane. These signs should also incorporate regulatory messages indicating any restrictions or prohibitions on the use of the lanes associated with the various types of payment methods by certain types of vehicles. For mainline toll plazas, these signs should be at least 1/2 mile in advance of the toll plaza, and farther if practical.*

02 *Additional guide signs with lane information for the toll payment types should be provided between approximately 1/4 mile and 800 feet in advance of the toll plaza at a location that avoids or minimizes obstruction of toll plaza canopy signs (see Section 2F.16) and lane-use control signals.*

03 *The number, mounting, and/or spacing of sets of advance signs for approaches to toll plazas on ramps, toll bridges, or tunnels, to accommodate a limited distance to the plaza from an intersection or from the start of the approach road to the bridge or tunnel, should be based on an engineering study or engineering judgment.*

Support:

04 Figure 2F-10 shows examples of advance signs for a conventional toll plaza.

Section 2F.15 Advance Signs for Toll Plazas on Diverging Alignments from Open-Road ETC Account-Only Lanes

Support:

01 Open-Road ETC lanes are sometimes located on the normal mainline alignment while the lanes for other toll payment methods are located at a toll plaza on a separate alignment (see Figure 2F-11). Since road users paying cash tolls must diverge from the mainline alignment, similar to a movement for an exit, it is important that the guide signs in advance of and at the point of divergence clearly indicate the required lane use and/or movements.

Guidance:

02 *For toll plazas located on a separate alignment that diverges from mainline-aligned Open-Road ETC lanes where vehicles are required to have a registered ETC account to use the Open-Road Tolling lanes, overhead advance signs should be provided at approximately 1 mile and 1/2 mile in advance of the divergence point. Both the 1-mile and 1/2-mile advance signs should include:*

  A. *The ETC (pictograph) Account-Only guide sign (see Figures 2F-8 and 2F-11) with a down arrow over the center of each lane that will become an Open-Road ETC lane;*
  B. *For the lane or lanes which will diverge to a toll plaza, guide signs conforming to the provisions of Section 2F.13, indicating which lane or lanes will diverge to the toll plaza for the various cash toll payment methods; and*
  C. *Regulatory signs, plaques, or panels within the guide signs, indicating any restrictions or prohibitions of certain types of vehicles from toll plaza lanes associated with the various types of payment methods.*

03 *At or near the theoretical gore of the divergence point, an additional set of overhead guide signs should be provided and should include:*

  A. *The ETC (pictograph) Account-Only guide sign (see Figures 2F-8 and 2F-11) with a down arrow over the center of each Open-Road ETC lane;*
  B. *Guide signs conforming to the provisions of Section 2F.13, with diagonally upward-pointing directional arrow(s) over the approximate center of each lane indicating the direction of the divergence, and providing lane information for all types of payment methods accepted at the toll plaza; and*
  C. *Regulatory signs, plaques, or panels within the guide signs, indicating any restrictions or prohibitions on the use of the toll plaza lanes associated with the various types of payment methods by certain types of vehicles.*

04 *Approximately 800 feet in advance of the toll plaza at a location that avoids or minimizes any obstruction of the toll plaza canopy signs (see Section 2F.16) and lane-use control signals, an additional set of overhead advance signs with lane information for the toll payment types should be provided.*

Standard:

05 **The use of down and directional arrows on the signs at the locations described in Paragraphs 2 through 4 shall comply with the provisions of Section 2D.08.**

Support:

06 Figure 2F-11 shows an example of advance signs for toll plazas on a diverging alignment from Open-Road ETC Account-Only Lanes.

07 Section 4K.02 contains information regarding the use of lane-use control signals for Open-Road ETC lanes for temporary lane closure purposes.

Figure 2F-10. Examples of Mainline Toll Plaza Approach and Canopy Signing

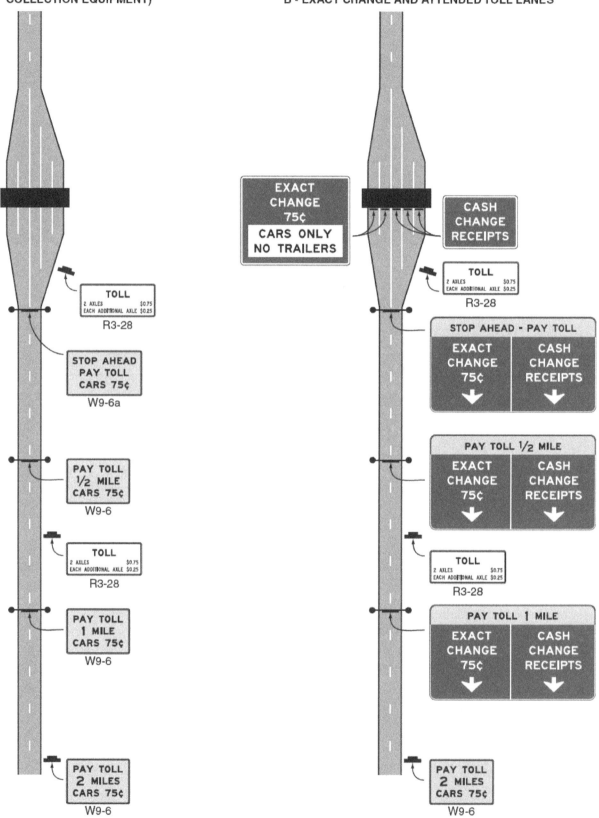

Figure 2F-11. Examples of Guide Signs for a Mainline Toll Plaza on a Diverging Alignment from Open-Road ETC Lanes

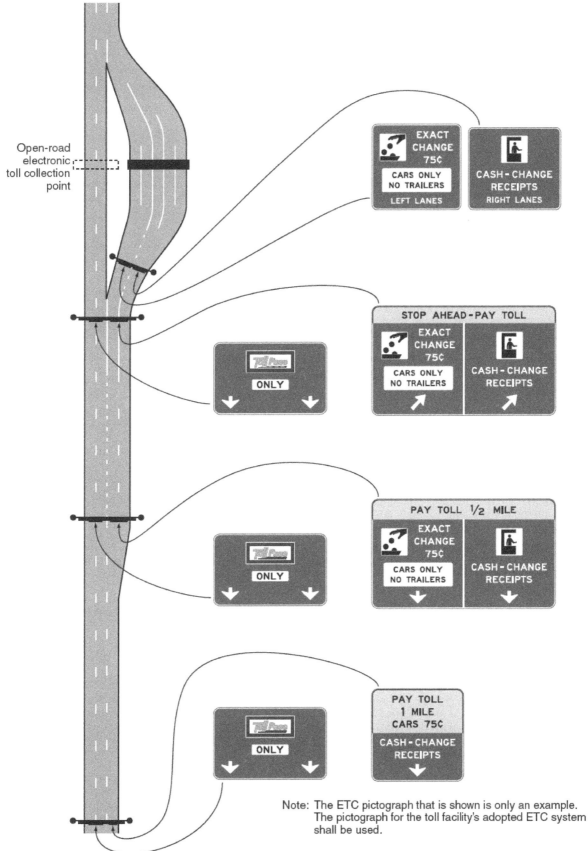

Note: The ETC pictograph that is shown is only an example. The pictograph for the toll facility's adopted ETC system shall be used.

Section 2F.16 Toll Plaza Canopy Signs

Standard:

01   A sign complying with the provisions of Section 2F.13 shall be provided above the center of each lane that is not an Open-Road ETC lane, mounted on or suspended from the toll plaza canopy, or on a separate structure immediately in advance of the plaza located such that each sign is clearly related to an individual toll lane, indicating the payment type(s) accepted in the lane and any restrictions or prohibitions of certain types of vehicles that apply to the lane. Except for toll-ticket systems, the toll for passenger or 2-axle vehicles shall be included on the canopy sign or on a separate sign mounted on the upstream side of the tollbooth.

02   The background color of a canopy sign for an ETC Account-Only toll plaza lane shall be purple (see Figure 2F-9).

Option:

03   Where vehicles are required to have a registered ETC account to use the lane, one or two flashing yellow beacons (see Section 4K.04) may supplement a canopy sign over an ETC Account-Only lane to call special attention to the location of the ETC Account-Only lane within the plaza.

04   The canopy sign for an ETC-Only toll plaza lane in which a regulatory speed limit is not posted and in which vehicles are not required to stop may display an advisory speed within a horizontal rectangular panel with a black legend and yellow background within the bottom portion of the canopy sign.

Standard:

05   Flashing beacons supplementing a canopy sign over an ETC Account-Only lane shall be mounted directly above or alongside the sign in a manner that is separated from any lane-use control signals for that lane (see Figure 2F-9).

06   For multi-lane toll plazas, lane-use control signals (see Section 4K.02) shall be provided above the center of each toll plaza lane that is not an Open-Road ETC lane to indicate the open or closed status of each lane. Lane-use control signals shall not be used to call attention to a lane for a specific toll payment type such as ETC Account-Only lanes.

Support:

07   Part 6 contains information regarding the closing of a lane for temporary traffic control purposes.

08   Figure 2F-9 shows examples of toll plaza canopy signs.

Section 2F.17 Guide Signs for Entrances to ETC Account-Only Facilities

Support:

01   Some toll highways, bridges, and tunnels are restricted to use only by vehicles with a specific registered ETC account.

Standard:

02   Where vehicles are required to have a registered ETC account to use an ETC Account-Only facility, guide signs for the facility shall comply with the applicable provisions of Chapter 2E and specifically with the applicable provisions of Section 2F.13.

03   Guide signs for the entrance ramps to such ETC Account-Only facilities shall incorporate the pictograph of the toll facility's ETC payment system and the word ONLY in a header panel or plaque designed in accordance with the provisions of Section 2F.13 (see Figure 2F-5).

Support:

04   Section 2F.12 contains information regarding ETC-Only auxiliary signs for use with route signs in route sign assemblies.

Section 2F.18 ETC Program Information Signs

Standard:

01   Except as provided in Paragraph 2, signs that inform road users of telephone numbers, Internet addresses, including domain names and uniform resource locators (URLs), or e-mail addresses for enrolling in an ETC program of a toll facility or managed lane, obtaining an ETC transponder, and/or obtaining ETC program information shall only be installed in rest areas, parking areas, or similar roadside facilities where the signs are viewed only by pedestrians or occupants of parked vehicles.

Option:

02   ETC program information signs displaying telephone numbers that have no more than four characters may be installed on roadways in locations where they will not obscure the road user's view of higher priority traffic control devices and that are removed from key decision points where the road user's view is more appropriately focused on other traffic control devices, roadway geometry, or traffic conditions, including exit and entrance ramps, intersections, toll plazas, temporary traffic control zones, and areas of limited sight distance.

CHAPTER 2G. PREFERENTIAL AND MANAGED LANE SIGNS

Section 2G.01 Scope
Support:
01　Preferential lanes are lanes designated for special traffic uses such as high-occupancy vehicles (HOVs), light rail, buses, taxis, or bicycles. Preferential lane treatments might be as simple as restricting a turning lane to a certain class of vehicles during peak periods, or as sophisticated as providing a separate roadway system within a highway corridor for certain vehicles.

02　Preferential lanes might be barrier-separated (on a separate alignment or physically separated from the other travel lanes by a barrier or median), buffer-separated (separated from the adjacent general-purpose lanes only by a narrow buffer area created with longitudinal pavement markings), or contiguous (separated from the adjacent general-purpose lanes only by a lane line). Preferential lanes might allow continuous access with the adjacent general-purpose lanes or restrict access only to designated locations. Preferential lanes might be operated in a constant direction or operated as reversible lanes. Some reversible preferential lanes on a divided highway might be operated counter-flow to the direction of traffic on the immediately adjacent general-purpose lanes.

03　Preferential lanes might be operated on a 24-hour basis, for extended periods of the day, during peak travel periods only, during special events, or during other activities.

04　Open-road tolling lanes and toll plaza lanes that segregate traffic based on payment method are not considered preferential lanes. Chapter 2F contains information regarding signing of open-road tolling lanes and toll plaza lanes.

05　Managed lanes typically restrict access with the adjacent general-purpose lanes to designated locations only.

06　Under certain operational strategies, such as the occupancy requirement of an HOV lane changing in response to actual congestion levels, a managed lane is a special type of preferential lane (see Sections 2G.03 through 2G.07).

07　A managed lane operated on a real-time basis in response to changing conditions might be operated as an HOV lane for a period of time as needed to manage congestion levels.

08　Sections 2G.16 through 2G.18 contain additional information regarding signs for managed lanes that use tolling or pricing as a management strategy.

09　Section 9B.04 contains information regarding Preferential Lane signs for bike lanes.

Section 2G.02 Sizes of Preferential and Managed Lane Signs
Standard:
01　**Except as provided in Section 2A.11, the sizes of preferential and managed lane signs that have standardized designs shall be as shown in Table 2G-1.**
Support:
02　Section 2A.11 contains information regarding the applicability of the various columns in Table 2G-1.
Option:
03　Signs larger than those shown in Table 2G-1 may be used (see Section 2A.11).

Section 2G.03 Regulatory Signs for Preferential Lanes – General
Standard:
01　**When a preferential lane is established, the Preferential Lane regulatory signs (see Figure 2G-1) and pavement markings (see Chapter 3D) for these lanes shall be used to advise road users.**
Support:
02　Preferential Lane (R3-10 series through R3-15 series) regulatory signs consist of several different general types of regulatory signs as follows (see Figure 2G-1):
   A. Vehicle Occupancy Definition signs define the vehicle occupancy requirements applicable to an HOV lane (such as "2 OR MORE PERSONS PER VEHICLE") or types of vehicles not meeting the minimum occupancy requirement (such as motorcycles or ILEVs) that are allowed to use an HOV lane (see Section 2G.04).
   B. Periods of Operation signs notify road users of the days and hours during which the preferential restrictions are in effect (see Section 2G.05).
   C. Preferential Lane Advance signs notify road users that a preferential lane restriction begins ahead (see Section 2G.06).
   D. Preferential Lane Ends signs notify users of the termination point of the preferential lane restrictions (see Section 2G.07).

Table 2G-1. Managed and Preferential Lanes Sign and Plaque Minimum Sizes

Sign or Plaque	Sign Designation	Section	Conventional Road		Expressway	Freeway	Oversized
			Single Lane	Multi-Lane			
Preferential Lane Vehicle Occupancy Definition (post-mounted)	R3-10,10a	2G.04	30 x 42	30 x 42	36 x 60	78 x 96	78 x 96
Preferential Lane Periods of Operation (post-mounted)	R3-11 series	2G.05	30 x 42	30 x 42	36 x 60	78 x 96	78 x 96
Motorcycles Allowed (plaque)	R3-11P	2G.03	30 x 15	30 x 15	36 x 18	78 x 36	78 x 36
Preferential Lane Ahead or Ends (post-mounted)	R3-12 series	2G.06	30 x 42	30 x 42	36 x 60	48 x 84	48 x 84
Preferential Lane Vehicle Occupancy Definition (overhead)	R3-13,13a	2G.04	66 x 36	66 x 36	84 x 48	144 x 78	144 x 78
HOV Lane Periods of Operation (overhead)	R3-14,14a,14b	2G.05	72 x 60	72 x 60	96 x 72	144 x 108	144 x 108
Preferential Lane Periods of Operation (overhead)	R3-14c	2G.05	90 x 60	90 x 60	108 x 72	156 x 102	168 x 102
HOV Lane Ahead (overhead)	R3-15	2G.06	66 x 36	66 x 36	84 x 48	102 x 60	102 x 60
HOV Lane Begins XX Miles (overhead)	R3-15a	2G.06	78 x 42	78 x 42	102 x 54	132 x 72	132 x 72
HOV Lane Ends (overhead)	R3-15b,15c	2G.07	66 x 36	66 x 36	84 x 48	102 x 60	102 x 60
Preferential Lane Ahead or Ends (overhead)	R3-15d,15e	2G.07	42 x 36	42 x 36	54 x 48	72 x 60	72 x 60
Priced Managed Lane Vehicle Occupancy Definition (post-mounted)	R3-40	2G.17	—	—	54 x 66	54 x 66	66 x 78
Priced Managed Lane Ends (post-mounted)	R3-42,42b	2G.17	—	—	48 x 60	48 x 60	60 x 78
Priced Managed Lane Ends Advance (post-mounted)	R3-42a,42c	2G.17	—	—	48 x 66	48 x 66	60 x 84
Priced Managed Lane Vehicle Occupancy Definition	R3-43	2G.17	—	—	138 x 66	138 x 66	—
Priced Managed Lane Periods of Operation (overhead)	R3-44	2G.17	—	—	90 x 84	90 x 84	—
Priced Managed Lane Periods of Operation (overhead)	R3-44a	2G.17	—	—	132 x 84	132 x 84	—
Priced Managed Lane Ends (overhead)	R3-45	2G.17	—	—	90 x 66	90 x 66	—
Priced Managed Lane Ends (overhead)	R3-45a	2G.17	—	—	114 x 66	114 x 66	—
Priced Managed Lane Toll Rate	R3-48	2G.17	—	—	Varies	Varies	—
Priced Managed Lane Toll Rate	R3-48a	2G.17	—	—	Varies	Varies	—
HOV (plaque)	W16-11P	2G.09	24 x 12	24 x 12	30 x 18	30 x 18	30 x 18
Preferential Lane Entrance Gore	E8-1	2G.10	—	—	48 x 96	48 x 96	—
Preferential Lane Intermediate Entrance Gore	E8-1a	2G.10	—	—	48 x 84	48 x 84	—
Preferential Lane Entrance Direction (overhead)	E8-2	2G.11	—	—	222 x 72	222 x 72	—
Preferential Lane Entrance Direction (post-mounted)	E8-2a	2G.11	—	—	186 x 108	186 x 108	—
Preferential Lane Entrance Advance	E8-3	2G.11	—	—	186 x 96	186 x 96	—
Preferential Lane Direct Exit Gore	E8-4	2G.15	—	—	60 x 78	60 x 78	—
Preferential Lane Intermediate Egress Direction	E8-5	2G.13	—	—	Varies x 90	Varies x 90	—
Preferential Lane Intermediate Egress Advance	E8-6	2G.13	—	—	Varies x 84	Varies x 84	—

Notes: 1. Larger signs may be used when appropriate
2. Dimensions in inches are shown as width x height

Figure 2G-1. Preferential Lane Regulatory Signs and Plaques (Sheet 1 of 2)

POST-MOUNTED PREFERENTIAL LANE SIGNS

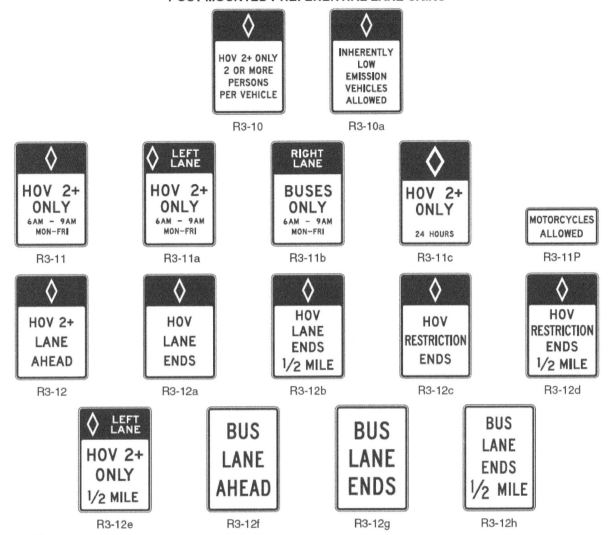

Notes:
1. The minimum vehicle occupancy requirement may vary for each facility (such as 2+, 3+, 4+).
2. The occupancy requirement may be added to the first line of the R3-12a, R3-12b, R3-12c, and R3-12d signs.
3. Some of the legends shown on these signs are for example purposes only. The specific legend for a particular application should be based upon local conditions, ordinances, and State statutes.

Standard:

03 **Regulatory signs applicable only to a preferential lane shall be distinguished from regulatory signs applicable to general-purpose lanes by the inclusion of the applicable symbol(s) and/or word(s) (see Figure 2G-1).**

Support:

04 The symbol and word message displayed on a particular Preferential Lane regulatory sign will vary based on the specific type of allowed traffic and on other related operational constraints that have been established for a particular lane, such as an HOV lane, a bus lane, or a taxi lane.

Option:

05 Changeable message signs may supplement, substitute for, or be incorporated into static Preferential Lane regulatory signs where travel conditions change or where multiple types of operational strategies (such as variable occupancy requirements or vehicle types) are used and varied throughout the day or week, or on a real-time basis, to manage the use of, control of, or access to preferential lanes.

Figure 2G-1. Preferential Lane Regulatory Signs and Plaques (Sheet 2 of 2)

OVERHEAD PREFERENTIAL LANE SIGNS

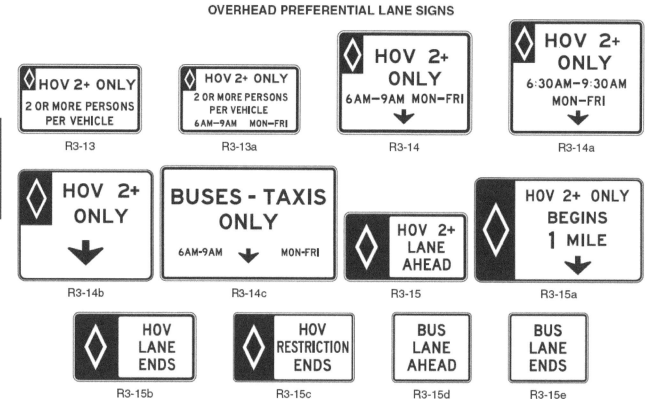

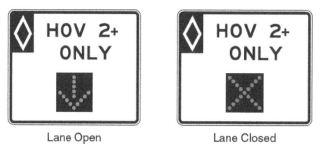

Notes:
1. The minimum vehicle occupancy requirement may vary for each facility (such as 2+, 3+, 4+).
2. The occupancy requirement may be added to the first line of the R3-15b and R3-15c signs.
3. Some of the legends shown on these signs are for example purposes only. The specific legend for a particular application should be based upon local conditions, ordinances, and State statutes.
4. Where sufficient median width is available, the R3-13 series and R3-15 series signs may be post-mounted.

Support:
06 Figure 2G-1 illustrates examples of changeable messages incorporated into static Preferential Lane regulatory signs.

Standard:

07 **When changeable message signs (see Chapter 2L) are used as regulatory signs for preferential lanes, they shall be the required sign size and shall display the required letter height and legend format that corresponds to the type of roadway facility and design speed.**

Guidance:

⁰⁸ *When Preferential Lane regulatory signs are used on conventional roads, the decision regarding whether to use a post-mounted or overhead version of a particular type of sign should be based on an engineering study that considers the available space, the existing signs for the adjacent general-purpose traffic lanes, roadway and traffic characteristics, the proximity to existing overhead signs, the ability to install overhead signs, and any other unique local factors.*

⁰⁹ *If overhead regulatory signs applicable only to a preferential lane are located in approximately the same longitudinal position along the highway as overhead signs applicable only to the general-purpose lanes, the signs for the preferential lane should be separated laterally from the signs for the general-purpose lanes to the maximum extent practical to minimize conflicting information, while maintaining their visual relationship to the lanes below necessitated by specific legend or arrows indicating lane assignment.*

Standard:

¹⁰ **If used, overhead Preferential Lane (R3-13 series, R3-14 series, and R3-15 series) regulatory signs shall be installed on the side of the roadway where the entrance to the preferential lane is located and any appropriate adjustments shall be made to the sign message.**

Option:

¹¹ Where a median of sufficient width is available, the R3-13 series and R3-15 series signs may be post-mounted.

Support:

¹² The sizes for Preferential Lane regulatory signs will differ to reflect the design speeds for each type of roadway facility. Table 2G-1 provides sizes for each type of roadway facility.

Guidance:

¹³ *The edges of Preferential Lane regulatory signs that are post-mounted on a median barrier should not project beyond the outer edges of the barrier, including in areas where lateral clearance is limited.*

Option:

¹⁴ Where lateral clearance is limited, Preferential Lane regulatory signs that are post-mounted on a median barrier and that are 72 inches or less in width may be skewed up to 45 degrees in order to fit within the barrier width or may be mounted higher, such that the vertical clearance to the bottom of the sign, light fixture, or structural support, whichever is lowest, is not less than 14 feet above any portion of the pavement and shoulders.

Standard:

¹⁵ **Where lateral clearance is limited, Preferential Lane regulatory signs that are post-mounted on a median barrier and that are wider than 72 inches shall be mounted with a vertical clearance that complies with the provisions of Section 2A.18 for overhead mounting.**

Guidance:

¹⁶ *On conventional roadways, Preferential Lane regulatory sign spacing should be determined by engineering judgment based on speed, block length, distances from adjacent intersections, and other site-specific considerations.*

Support:

¹⁷ Sections 2G.04 and 2G.05 contain provisions regarding the placement of Preferential Lane regulatory signs on freeways and expressways.

Standard:

¹⁸ **The signs illustrated in Figure 2G-1 that incorporate the diamond symbol shall be used exclusively with preferential lanes for high-occupancy vehicles to indicate the particular occupancy requirement and time restrictions applying to that lane. The signs illustrated in Figure 2G-1 that do not have a diamond symbol shall be used with preferential lanes that are not HOV lanes, but are designated for use by other types of vehicles (such as bus and/or taxi use).**

Option:

¹⁹ Agencies may select from either the HOV abbreviation or the diamond symbol, or use both, to reference the HOV lane designation.

Standard:

²⁰ **When the diamond symbol (or HOV abbreviation) is used without text on the post-mounted Preferential Lane (R3-10 series, R3-11 series, and R3-12 series) regulatory signs, it shall be centered on the top line of the sign. When the diamond symbol (or HOV abbreviation) is used with associated text on the post-mounted Preferential Lane (R3-10 series, R3-11 series, and R3-12 series) regulatory signs, it shall appear to the left of the associated text. When the diamond symbol is used on the overhead Preferential Lane (R3-13, R3-13a, R3-14, and R3-14a) regulatory signs, it shall appear in the top left quadrant. The diamond symbol for the R3-15, R3-15a, R3-15b, and R3-15c signs shall appear on the left side of the sign. The diamond symbol shall not be used on the bus, taxi, or bicycle Preferential Lane signs.**

21 **Vehicle Occupancy Definition, Periods of Operation, and Preferential Lane Advance regulatory signs for HOV lanes shall display the minimum allowable vehicle occupancy requirement established for each HOV lane, displayed immediately after the word message HOV or the diamond symbol.**

Support:

22 The agencies that own and operate HOV lanes have the authority and responsibility to determine how they are operated and the minimum occupancy requirements. Information about federal requirements for certain types of vehicles not meeting the minimum occupancy requirement to be eligible to use HOV lanes that receive Federal-aid program funding and about requirements associated with proposed significant changes to the operation of an existing HOV lane and certain vehicles are contained in the "Federal-Aid Highway Program Guidance on High Occupancy Vehicle (HOV) Lanes" (see Section 1A.11).

Standard:

23 **The provisions of Sections 2G.03 through 2G.07 regarding regulatory signs for Preferential lanes shall apply to managed lanes operated at all times or at certain times by varying vehicle occupancy requirements (HOV) or by using vehicle type restrictions as a congestion management strategy. Such managed lanes shall use changeable message signs or changeable message elements within static signs to display the appropriate regulatory sign messages only when they are in effect.**

24 **When certain types of vehicles (such as trucks) are prohibited from using a managed lane or when a managed lane is restricted to use by only certain types of vehicles during certain operational strategies, regulatory signs or regulatory panels within the appropriate guide signs that include changeable message elements shall be used to display the open/closed status of the managed lane for such vehicle types.**

25 **When the vehicle occupancy required for use of an HOV lane is varied as a part of a managed lane operational strategy, regulatory signs that include changeable message elements shall be used to display the required vehicle occupancy in effect.**

Support:

26 See Section 2G.17 for regulatory signs for managed lanes that use tolling or pricing as a congestion management strategy, either exclusively or with other management strategies.

27 Figures 2G-2 and 2G-3 illustrate the use of regulatory signs for the beginning, along the length, and at the end of contiguous or buffer-separated preferential lanes that provide continuous access with the adjacent general-purpose lanes.

Section 2G.04 Preferential Lane Vehicle Occupancy Definition Regulatory Signs (R3-10 Series and R3-13 Series)

Standard:

01 **The R3-10, R3-13, and R3-13a Vehicle Occupancy Definition signs (see Figure 2G-1) shall be used where agencies determine that it is appropriate to provide a sign that defines the minimum occupancy of vehicles that are allowed to use an HOV lane.**

Guidance:

02 *The Inherently Low Emission Vehicle (ILEV) (R3-10a) sign (see Figure 2G-1) should be used when it is permissible for a properly labeled and certified ILEV, regardless of the number of occupants, to use an HOV lane. When used, the ILEV signs should be post-mounted in advance of and at intervals along the HOV lane based upon engineering judgment and the placement of other Preferential Lane regulatory signs. The R3-10a sign is only applicable to HOV lanes and should not to be used with other preferential lane applications.*

Support:

03 ILEVs are defined by the Environmental Protection Agency (EPA) as vehicles having no fuel vapor (hydrocarbon) emissions and are certified by the EPA as meeting the emissions standards and requirements specified in 40 CFR 88-311-93 and 40 CFR 88.312-93(c).

Guidance:

04 *The legend format of the R3-10 and R3-13 signs should have the following sequence:*
   A. *Top Line: "HOV 2+ ONLY" (or 3+ or 4+ if appropriate)*
   B. *Bottom Lines: "2 OR MORE PERSONS PER VEHICLE" (or 3 or 4 if appropriate)*

05 *The legend format of the R3-13a sign should have the following sequence:*
   A. *Top Line: "HOV 2+ ONLY" (or 3+ or 4+ if appropriate)*
   B. *Middle Lines: "2 OR MORE PERSONS PER VEHICLE" (or 3 or 4 if appropriate)*
   C. *Bottom Lines: Times and days the occupancy restriction is in effect*

Support:

06 Section 2G.17 contains information regarding the legends of Vehicle Occupancy Definition signs for a priced managed lane that has an occupancy requirement for non-toll travel.

Standard:

07 **For barrier- or buffer-separated or contiguous preferential lanes where access between the preferential and general-purpose lanes is restricted to designated locations, an overhead Vehicle Occupancy Definition (R3-13 or R3-13a) sign shall be installed at least 1/2 mile in advance of the beginning of or initial entry point to an HOV lane. These signs shall only be displayed in advance of the beginning of or initial entry point to HOV lanes.**

Option:

08 For barrier-separated HOV lanes, the sequence of a post-mounted Periods of Operation (R3-11a) sign followed by a post-mounted Vehicle Occupancy Definition (R3-10) sign may be located at intervals of approximately 1/2 mile along the length of the HOV lane, at intermediate entry points, and at designated enforcement areas as defined by the operating agency.

Standard:

09 **For buffer-separated or contiguous HOV lanes where access is restricted to designated locations, the sequence of a post-mounted Periods of Operation (R3-11a) sign followed by a post-mounted Vehicle Occupancy Definition (R3-10) sign shall be located at intervals not greater than 1/2 mile along the length of the access-restricted HOV lane, at designated gaps where vehicles are allowed to legally access the HOV lane, and within designated enforcement areas as defined by the operating agency.**

10 **For buffer-separated or contiguous HOV lanes where continuous access with the adjacent general-purpose lanes is provided, the sequence of a post-mounted Periods of Operation (R3-11a) sign followed by a post-mounted Vehicle Occupancy Definition (R3-10) sign, and ILEV (R3-10a) signs if appropriate, shall be located at intervals not greater than 1/2 mile along the length of the HOV lane.**

Guidance:

11 *The signs within each Preferential Lane regulatory sign sequence should be separated by a minimum distance of 800 feet and a maximum distance of 1,000 feet.*

Standard:

12 **For all types of direct access ramps that provide access to or lead to HOV lanes, a post-mounted Vehicle Occupancy Definition (R3-10) sign, and an ILEV (R3-10a) sign if appropriate, shall be used at the beginning or initial entry point for the direct access ramp.**

Section 2G.05 Preferential Lane Periods of Operation Regulatory Signs (R3-11 Series and R3-14 Series)

Guidance:

01 *The sizes of post-mounted Periods of Operation (R3-11 series) signs should remain consistent to accommodate any manual addition or removal of a single line of text for each sign.*

Support:

02 Consistent sign sizes are beneficial for agencies when ordering sign materials, as well as when making text changes to existing signs if changes occur to operating times or occupancy restrictions in the future. For example, the R3-11c sign has space for one line located below "24 HOURS" if an agency determines that it is appropriate to display additional information (such as "MON – FRI"), yet the R3-11c sign has the same dimensions as the other R3-11 series signs.

Standard:

03 **When used, the post-mounted Periods of Operation (R3-11 series) signs shall be located adjacent to the preferential lane, and the overhead Periods of Operation (R3-14 series) signs shall be mounted directly over the lane.**

04 **The legend format of the post-mounted Periods of Operation (R3-11 series) signs shall have the following sequence:**
   A. **Top Lines: Lanes applicable, such as "RIGHT LANE" or "2 RIGHT LANES" or "THIS LANE"**
   B. **Middle Lines: Eligible uses, such as "HOV 2+ ONLY" (or 3+ or 4+ if appropriate) or "BUSES ONLY" or other applicable uses or eligible turning movements**
   C. **Bottom Lines: Applicable times and days, such as "7 AM – 9 AM" or "6:30 AM – 9:30 AM, MON-FRI"**

⁰⁵ **The legend format of the overhead Periods of Operation (R3-14 series) signs shall have the following sequence:**
  A. Top Line: Eligible uses, such as "HOV 2+ ONLY" (or 3+ or 4+ if appropriate) or "BUSES ONLY" or other applicable uses or eligible turning movements
  B. Bottom Lines: Applicable times and days, with the time and day placed above the down arrow, such as "7 AM – 9 AM" or "6:30 AM – 9:30 AM, MON-FRI" (When the operating periods exceed the available line width, the hours and days of the week shall be stacked as shown for the R3-14a sign in Figure 2G-1.)

⁰⁶ **For preferential lanes that are in effect on a full-time basis, either the full-time Periods of Operation (R3-11b and R3-14b) signs shall be used, or the legends of the part-time Periods of Operations (R3-11, R3-11a, R3-14, R3-14a) signs shall be modified to display the legend 24 HOURS.**

⁰⁷ **The full-time Periods of Operation (R3-14b) sign shall not be used where the preferential lane is in effect only on a part-time basis.**

Option:

⁰⁸ Where additional movements are permitted from a preferential lane on an approach to an intersection, the format and words used in the legend in the middle lines on the post-mounted Periods of Operation (R3-11 series) signs and on the top line of the overhead Periods of Operation (R3-14 series) signs may be modified to accommodate the permitted movements (such as "HOV 2+ AND RIGHT TURNS ONLY").

⁰⁹ A MOTORCYCLES ALLOWED (R3-11P) plaque may be used where motorcycles, regardless of the number of occupants, are allowed to use an HOV lane.

Standard:

¹⁰ **If used, the MOTORCYCLES ALLOWED plaque shall be mounted below a post-mounted Preferential Lane Periods of Operation (R3-11, R3-11a, or R3-11c) sign.**

¹¹ **For all barrier- or buffer-separated or contiguous preferential lanes where access is restricted to designated locations, an overhead Periods of Operation (R3-14 series) sign shall be used at the beginning or initial entry point, and at any intermediate entry points or gaps in the barrier where vehicles are allowed to legally access the access-restricted preferential lanes. For all barrier-separated and buffer-separated preferential lanes, post-mounted Periods of Operation (R3-11 series) signs shall be used only as a supplement to the overhead signs at the beginning or initial entry point, or at any intermediate entry points or gaps in the barrier or buffer.**

¹² **For buffer-separated or contiguous preferential lanes where continuous access with the adjacent general-purpose lanes is provided, including those where a preferential lane is added to the roadway (see Figure 2G-2 for HOV lanes) and those where a general-purpose lane transitions into a preferential lane (see Figure 2G-3 for HOV lanes), an overhead Periods of Operation (R3-14 series) sign shall be used at the beginning or initial entry point of the preferential lane.**

Guidance:

¹³ *Overhead (R3-14 series) or post-mounted (R3-11 series) Periods of Operation signs should be installed at periodic intervals along the length of a contiguous or buffer-separated preferential lane where continuous access with the adjacent general-purpose lanes is provided.*

Option:

¹⁴ Additional overhead (R3-14 series) or post-mounted (R3-11 series) Periods of Operation signs may be provided along the length of any type of preferential lane.

¹⁵ On conventional roads, the overhead Periods of Operation (R3-14 series) signs may be installed at the beginning or entry points and/or at intermediate points along preferential lanes in any geometric configuration.

Standard:

¹⁶ **For all types of direct access ramps that provide access to or lead to preferential lanes, a post-mounted Periods of Operation (R3-11 series) sign shall be used at the beginning or initial entry point of the direct access ramp.**

Option:

¹⁷ For direct access ramps to preferential lanes, an overhead Periods of Operation (R3-14 series) sign may be used at the beginning or initial entry point to supplement the required post-mounted signs.

¹⁸ Lane-use control signals (see Chapter 4M) may be used at access points to preferential lanes to indicate that a ramp or access roadway leading to the preferential lane or facility, or one or more specific lanes of the facility, are open or closed (see Figure 2G-14).

Figure 2G-2. Example of Signing for an Added Continuous-Access Contiguous or Buffer-Separated HOV Lane

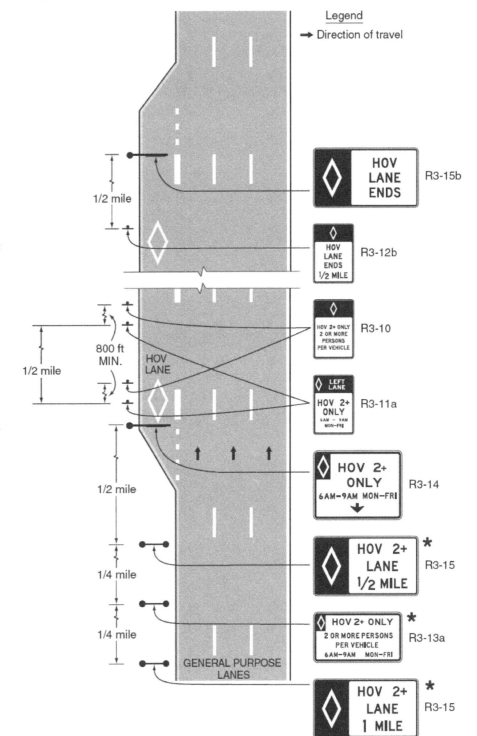

Notes:
1. The minimum vehicle occupancy requirement and hours of operation on the sign may vary for each facility
2. See Chapter 3D for pavement markings
3. Warning signs are not shown
4. Applicable to part-time or full-time HOV restriction
5. This roadway condition indicates the HOV lane will merge with the general purpose lanes upon termination
6. Sets of R3-10 and R3-11a signs should be placed following entrance ramps and at 1/2-mile intervals along the HOV lane

★ Where the median width is insufficient, post-mounted designs (R3-10, R3-11, and R3-12 series) may be used

Figure 2G-3. Example of Signing for a General-Purpose Lane that Becomes a Continuous-Access Contiguous or Buffer-Separated HOV Lane

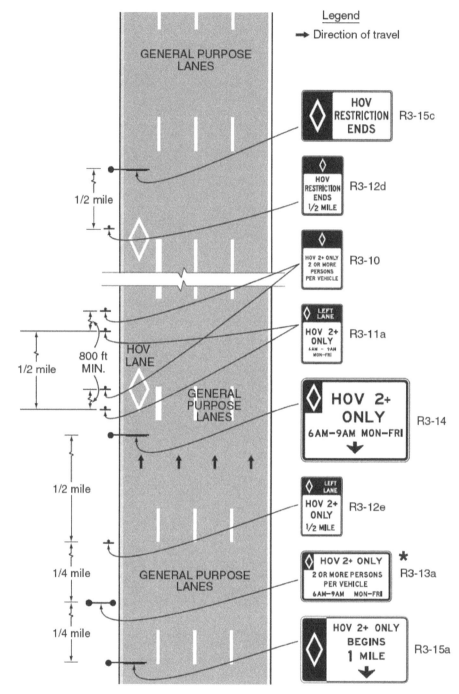

Notes:
1. The minimum vehicle occupancy requirement and hours of operation on the sign may vary for each facility
2. See Chapter 3D for pavement markings
3. Applicable to part-time or full-time HOV restriction
4. This roadway condition indicates the HOV lane will become a general purpose lane upon termination of the restriction
5. Sets of R3-10 and R3-11a signs should be placed following entrance ramps and at 1/2-mile intervals along the HOV lane
6. This signing scheme can also be used for an HOV lane on the right-hand side of the roadway

★ Where the median width is insufficient, this sign may be mounted overhead

Section 2G.06 Preferential Lane Advance Regulatory Signs (R3-12, R3-12e, R3-12f, R3-15, R3-15a, and R3-15d)

Guidance:

01 *The Preferential Lane Advance (R3-12, R3-12f, R3-15, and R3-15d) signs should be used for advance notification of a barrier-separated, buffer-separated, or contiguous preferential lane that is added to the general-purpose lanes (see Figure 2G-12).*

02 *The Preferential Lane Advance (R3-12e and R3-15a) signs should be used for advance notification of a general-purpose lane that becomes a preferential lane (see Figure 2G-13).*

Option:

03 The legends on the R3-12f and R3-15d signs may be modified to suit the type of preferential lane.

Guidance:

04 *On conventional roads, for general-purpose lanes that become preferential lanes, a post-mounted (R3-12e) or overhead (R3-15a) Preferential Lane Advance sign should be installed in advance of the beginning of or initial entry point to the preferential lane at a distance determined by engineering judgment based on speed, traffic characteristics, and other site-specific considerations. The distance selected should provide adequate opportunity for ineligible vehicles to vacate the lane prior to the beginning of the restriction.*

05 *On freeways and expressways, for general-purpose lanes that become preferential lanes, an overhead Preferential Lane Advance (R3-15a) sign should be installed at least 1 mile in advance of the beginning of the preferential lane restriction.*

Option:

06 Additional post-mounted or overhead Preferential Lane Advance signs may be placed farther in advance of or closer to the beginning or initial entry points to a preferential lane.

Section 2G.07 Preferential Lane Ends Regulatory Signs (R3-12a, R3-12b, R3-12c, R3-12d, R3-12g, R3-12h, R3-15b, R3-15c, and R3-15e)

Standard:

01 **A post-mounted Preferential Lane Ends (R3-12b or R3-12h) sign shall be installed at least 1/2 mile in advance of the termination of a preferential lane.**

02 **Except as provided in Paragraph 6, a post-mounted Preferential Lane Ends (R3-12a or R3-12g) sign shall be installed at the point where a preferential lane and restriction end and traffic must merge into the general-purpose lanes.**

03 **A post-mounted Preferential Lane Ends (R3-12d) sign shall be installed at least 1/2 mile in advance of the point where a preferential lane restriction ends and the lane becomes a general-purpose lane.**

04 **Except as provided in Paragraph 7, a post-mounted Preferential Lane Ends (R3-12c) sign shall be installed at the point where a preferential lane restriction ends and the lane becomes a general-purpose lane.**

Option:

05 The legends on the R3-12g and R3-15e signs may be modified to suit the type of preferential lane.

06 An overhead Preferential Lane Ends (R3-15b or R3-15e) sign may be installed instead of or in addition to a post-mounted R3-12a or R3-12g sign at the point where a preferential lane and restriction ends and traffic must merge into the general-purpose lanes.

07 An overhead Preferential Lane Ends (R3-15c) sign may be installed instead of or in addition to a post-mounted R3-12c sign at the point where the preferential lane restriction ends and the lane becomes a general-purpose lane.

Section 2G.08 Warning Signs on Median Barriers for Preferential Lanes

Option:

01 When a warning sign applicable only to a preferential lane is installed on a median barrier with limited lateral clearance to the adjacent travel lanes or shoulders, the warning sign may have a vertical rectangular shape. For a High Occupancy Vehicle lane, such signs may be used instead of using the HOV Plaque (W16-11P) (see Section 2G.09) with a standard diamond-shaped warning sign.

Standard:

02 **When a vertical rectangular-shaped warning sign applicable only to a preferential lane is installed on a median barrier, the top portion of the sign shall be comprised of a white symbol or legend denoting the type of preferential lane (such as the diamond symbol for HOV or the legend BUS LANE) on a black background with a white border, and the bottom portion of the sign shall be comprised of the standard word message or symbol of the standard warning sign as a black legend on a yellow background with a black border (see Figure 2G-4).**

Figure 2G-4. Examples of Warning Signs and Plaques Applicable Only to Preferential Lanes

A - BARRIER-MOUNTED RECTANGULAR WARNING SIGNS

W4-1L (modified)

W4-2L (modified)

W13-2 (modified)

B - WARNING PLAQUE FOR USE ABOVE STANDARD DIAMOND-SHAPED WARNING SIGNS

W16-11P

Note: An HOV lane example (diamond symbol) is illustrated. For other types of preferential lanes, the appropriate symbol or word message (see Section 2G.03) shall be displayed in white on the black background of the top portion of these signs.

Guidance:

03 *Where lateral clearance is limited, such as when a post-mounted warning sign applicable only to a preferential lane is installed on a median barrier, the edges of the sign should not project beyond the outer edges of the barrier.*

Option:

04 Where lateral clearance is limited, warning signs applicable only to a preferential lane that are post-mounted on a median barrier and that are 72 inches or less in width may be skewed up to 45 degrees in order to fit within the barrier width or may be mounted higher, such that the vertical clearance to bottom of the sign, light fixture, or its structural support, whichever is lowest, is not less than 14 feet above any portion of the pavement and shoulders.

Standard:

05 **Where lateral clearance is limited, Preferential Lane warning signs that are post-mounted on a median barrier and that that are wider than 72 inches shall be mounted with a vertical clearance that complies with the provisions of Section 2A.18 for overhead mounting.**

Section 2G.09 High-Occupancy Vehicle (HOV) Plaque (W16-11P)

Option:

01 In situations where there is a need to warn drivers in an HOV lane of a specific condition, a HOV (W16-11P) plaque (see Figure 2G-4) may be used above a warning sign. The HOV plaque may be used to differentiate a warning sign specific for HOV lanes when the sign is also visible to traffic on the adjacent general-purpose roadway. Among the warning signs that may be possible applications of the HOV plaque are the Advisory Exit Speed, Added Lane, and Merge signs.

02 The diamond symbol may be used instead of the word message HOV on the W16-11P plaque. When appropriate, the words LANE or ONLY may be used on this plaque.

Support:

03 Section 2G.08 contains information regarding warning signs that can be mounted on barriers for HOV or other types of preferential lanes.

Section 2G.10 Preferential Lane Guide Signs – General

Support:

01 Preferential lanes are used on freeways, expressways, and conventional roads. Except as otherwise provided, Sections 2G.10 through 2G.15 apply only to guide signs for preferential lanes on freeways and expressways.

Guidance:

02 *On conventional roads, guide signs applicable only to preferential lanes are ordinarily not needed, but if used they should comply with the provisions for guide signs in Chapter 2D and any principles for Preferential Lane guide signs in Sections 2G.10 through 2G.15 that engineering judgment finds to be appropriate for the conditions.*

Support:

03 Consistency in signs and pavement markings for preferential lanes plays a critical role in building public awareness, understanding, and acceptance, and makes enforcement more effective.

04 Additional guidance and standards related to the designation, operational considerations, signs, pavement markings, and other considerations for preferential lanes is provided in Sections 2G.03 through 2G.07, and 2G.09, and Chapter 3D.

Guidance:

05 *The appropriate combinations of pavement markings and standard overhead and post-mounted regulatory, warning, and guide signs for a specific preferential lane application should be selected based on an engineering study.*

06 *If overhead signs applicable only to a preferential lane are located in approximately the same longitudinal position along the highway as overhead signs applicable only to the general-purpose lanes, the signs for the preferential lane should be separated laterally from the signs for the general-purpose lanes to the maximum extent practical to minimize conflicting information.*

07 *The Preferential Lane signs should be designed and located to avoid overloading the road user. Based on the importance of the sign, regulatory signs should be given priority over guide signs. The order of priority of guide signs should be Advance Guide, Preferential Lane Entrance Direction, and finally Preferential Lane Exit Destination supplemental guide signs.*

Standard:

08 **Signs applicable only to a preferential lane shall be distinguished from signs applicable to general-purpose lanes by the inclusion of the applicable symbol(s) and/or word(s).**

Support:

09 The symbol and/or word message that appears on a particular guide sign applicable only to a preferential lane will vary based on the specific type of allowed traffic and on other related operational constraints that have been established for a particular lane, such as an HOV lane, a bus lane, or a taxi lane.

Standard:

10 **For HOV lanes, the diamond symbol shall appear on each Advance Guide sign, Preferential Lane Entrance Direction sign, and Preferential Lane Entrance Gore sign, as shown in Figures 2G-5 through 2G-7 for the designated entry and exit points for barrier- and buffer-separated geometric configurations and direct access ramps to or from such lanes. The diamond symbol shall not be used with preferential lanes for other types of traffic, such as bus lanes or taxi lanes.**

11 **Signing for an HOV lane that is managed by means of varying the occupancy requirement in response to changing conditions shall also comply with these provisions.**

12 **The diamond symbol shall be displayed in the legend of each Preferential Lane guide sign at the designated entry and exit points for all types of HOV lanes (including barrier- and buffer-separated, contiguous, and direct access ramps) in order to alert motorists that there is a minimum allowable vehicle occupancy requirement for vehicles to use the HOV lanes. Guide signs shall not display the occupancy requirement for the preferential lane.**

13 **A combination of guide and regulatory signs shall be used in advance of and at the initial entry point and all intermediate entry points from general-purpose lanes or facilities to contiguous, barrier-separated, and buffer-separated preferential lanes where access between the preferential and general-purpose lanes is restricted to designated locations. The regulatory signs shall comply with the provisions of Sections 2G.03 through 2G.07.**

¹⁴ **Regulatory signs alone shall be used in advance of, at the beginning of, and at periodic intervals along contiguous or buffer-separated preferential lanes that provide continuous access between the adjacent general-purpose lanes and the preferential lane (see Figures 2G-12 and 2G-13). The design and placement of the regulatory signs shall comply with the provisions of Sections 2G.03 through 2G.07.**

¹⁵ **Except as otherwise provided in Sections 2G.10 through 2G.13, guide signs applicable to a preferential lane with a vehicle occupancy requirement shall be distinguished from those applicable to general-purpose lanes by displaying the white diamond symbol on a black background at the left-hand edge of these signs.**

Option:

¹⁶ When post-mounted guide signs applicable only to a preferential lane are installed on a median barrier with limited lateral clearance to the adjacent travel lanes or shoulders, the guide signs may have a vertical rectangular shape.

Standard:

¹⁷ **When vertical rectangular shaped guide signs applicable only to a preferential lane are installed on a median barrier, the top portion of the signs shall be comprised of the applicable white symbol or white word message that identifies the type of preferential lane (such as the diamond symbol for an HOV lane) on a black background with a white border, and the bottom portion of the sign shall be comprised of the appropriate guide sign legend on a green background with a white border (see Figures 2G-3, 2G-6, and 2G-7).**

Guidance:

¹⁸ *Where lateral clearance is limited, such as when a post-mounted Preferential Lane guide sign is installed on a median barrier, the edges of the sign should not project beyond the outer edges of the barrier.*

Option:

¹⁹ Where lateral clearance is limited, Preferential Lane guide signs that are 72 inches or less in width may be skewed up to 45 degrees in order to fit within the barrier width or may be mounted higher, such that the vertical clearance to the bottom of the sign, light fixture, or its structural support, whichever is lowest, is not less than 14 feet above any portion of the pavement and shoulders.

Standard:

²⁰ **Where lateral clearance is limited, Preferential Lane guide signs that are post-mounted on a median barrier and that are wider than 72 inches shall be mounted with a vertical clearance that complies with the provisions of Section 2A.18 for overhead mounting.**

Option:

²¹ Lane-use control signals (see Chapter 4M) may be used at access points to preferential lanes to indicate that a ramp or access roadway leading to or from the preferential lane or facility, or one or more specific lanes of the facility, are open or closed.

²² Changeable message signs may supplement, substitute for, or be incorporated into static guide signs where travel conditions change or where multiple types of operational strategies (such as variable occupancy requirements, vehicle types, or pricing policies) are used and varied throughout the day or week to manage the use of, control of, or access to preferential lanes.

Standard:

²³ **When changeable message signs (see Chapter 2L) are used as guide signs for preferential lanes, they shall be the required sign size and shall display the required letter height and legend format that corresponds to the type of roadway facility and design speed.**

²⁴ **Advance Guide signs, Preferential Lane Entrance Direction signs, and Preferential Lane Entrance Gore signs for the initial entry point and intermediate entry points into a preferential lane from the general-purpose lanes on the same designated route shall not identify the entry point as an exit by using the word "EXIT" on the sign or on a plaque.**

Guidance:

²⁵ *Advance Guide signs and Preferential Lane Entrance Direction signs for initial and intermediate entry points into a preferential lane should use the word "ENTRANCE," such as "HOV LANE ENTRANCE" (see Figures 2G-5 and 2G-6) to convey the fact that vehicles are not leaving the designated route.*

²⁶ *Preferential Lane Entrance Gore signs (see Figure 2G-7) at the initial entry point to a preferential lane should use the word "ENTRANCE." Preferential Lane Entrance Gore signs at intermediate entry points to a barrier-separated preferential lane where the sign would be located immediately adjacent to and directly viewed by traffic in the preferential lane should not use the word "ENTRANCE."*

Figure 2G-5. Example of an Overhead Advance Guide Sign for a Preferential Lane Entrance

E8-3

Note: An example of an HOV Lane (diamond symbol) sign is illustrated. For other types of preferential lanes, the appropriate symbol or word message (see Section 2G.03) is displayed in white on the black background of the left-hand portion of this sign.

Figure 2G-6. Examples of Overhead or Post-Mounted Preferential Lane Entrance Direction Signs

E8-2
(overhead only)

E8-2a
(post-mounted only)

A changeable message sign may be incorporated into an overhead preferential lane guide sign to indicate the status of a reversible operation as shown in the following example:

Lane Open   Lane Closed

Note: Examples of HOV Lane (diamond symbol) signs are illustrated. For other types of preferential lanes, the appropriate symbol or word message (see Section 2G.03) is displayed in white on the black background of the top left-hand portion of these signs.

Standard:

27 When the entry point is on the left-hand side of the general-purpose lanes, a LEFT (E1-5aP) plaque (see Figure 2E-22) shall be added to the top left edge of the Advance Guide and Preferential Lane Entrance Direction signs. The LEFT plaque shall not be used on a preferential lane regulatory sign.

Section 2G.11 Guide Signs for Initial Entry Points to Preferential Lanes

Standard:

01 Except where a buffer-separated or contiguous preferential lane is added or where a general-purpose lane becomes a buffer-separated or contiguous preferential lane, and provides continuous access with the adjacent general-purpose lanes as illustrated in Figures 2G-2 and 2G-3, an Advance Guide sign shall be provided at least 1/2 mile prior to the initial entry point to all types of preferential lanes in any type of geometric configuration. A Preferential Lane Entrance Direction sign shall also be provided at the initial entry point. Advance Guide and Preferential Lane Entrance Direction signs for such entry points shall not include the word "EXIT" (see Section 2G.10).

Figure 2G-7. Entrance Gore Signs for Barrier-Separated Preferential Lanes

E8-1

E8-1a

Note: Examples of HOV Lane (diamond symbol) signs are illustrated. For other types of preferential lanes, the appropriate symbol or word message (see Section 2G.03) is displayed in white on the black background of the top portion of these signs.

Guidance:

02 *An Advance Guide sign should also be installed and located approximately 1 mile in advance of the initial entry point to a preferential lane that restricts access with the adjacent general-purpose lanes to designated locations.*

Option:

03 An Advance Guide sign may also be installed and located approximately 2 miles in advance of the initial entry point to a preferential lane that restricts access with the adjacent general-purpose lanes to designated locations.

Standard:

04 **For barrier-separated, buffer-separated, or contiguous preferential lanes where entry is restricted to only designated points, the Advance Guide and Preferential Lane Entrance Direction signs shall be mounted overhead.**

Guidance:

05 *Preferential Lane Exit Destination guide signs, identifying final destination and downstream exit locations accessible from the preferential lane (see Figures 2G-8, 2G-13, 2G-14, and 2G-16), should be installed in advance of the initial entry points to access-restricted preferential lanes (such as barrier- and buffer-separated). These signs should be located based on the priority of the message, the available space, the existing signs on adjacent general-purpose traffic lanes, roadway and traffic characteristics, the proximity to existing overhead signs, the ability to install overhead signs, and other unique local factors.*

Standard:

06 **Advance destination guide signs for preferential lanes shall include an upper section displaying a black legend that includes the type of preferential lane and the word "EXITS," such as "HOV EXITS," on a white background. For preferential lanes that incorporate a vehicle occupancy requirement, the white diamond symbol on a black background shall be displayed at the left edge of this upper section (see Figure 2G-8).**

Support:

07 Figure 2G-8 shows an example of signs for the initial entry point to a preferential lane.

Section 2G.12 Guide Signs for Intermediate Entry Points to Preferential Lanes

Standard:

01 **For barrier-separated, buffer-separated, and contiguous preferential lanes where entry is restricted only to designated points, an overhead Preferential Lane Entrance Direction sign shall be provided at intermediate entry points to the preferential lane from the general-purpose lanes.**

Guidance:

02 *For barrier- and buffer-separated preferential lanes where intermediate entry from the general-purpose lanes is provided via a separate lane or ramp (see Figure 2G-9), at least one Advance Guide sign should be provided in addition to the Preferential Lane Entrance Direction sign.*

03 *For access-restricted preferential lanes where intermediate entrance and egress are at the same designated access location, the Preferential Lane Entrance Direction sign should be located between 1/2 and 1/4 of the length of the designated entry area, as measured from the downstream end of the entry area (see Figure 2G-10).*

Figure 2G-8. Example of Signing for an Entrance to Access-Restricted HOV Lanes

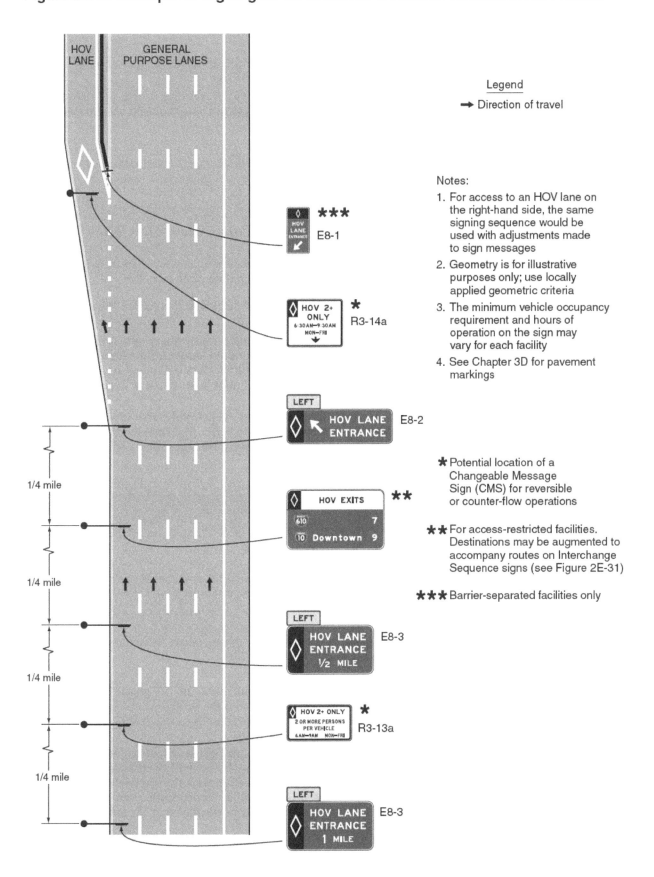

Standard:

04 **The Advance Guide signs, if used for intermediate entry points to a preferential lane from the general-purpose lanes, shall be overhead.Option:**

Option:

05 Advance Guide signs may be provided at approximately 1/2 mile, 1 mile, and 2 miles in advance of intermediate entry points from the general-purpose lanes to a preferential lane.

Standard:

06 **Advance Guide and Preferential Lane Entrance Direction signs for intermediate entry points shall not include the word "EXIT" (see Section 2G.10).**

Guidance:

07 *Exit Destination guide signs, identifying the final destination and downstream exit locations accessible from the preferential lane, should be installed in advance of intermediate entry points from the general-purpose lanes to access-restricted preferential lanes.*

Support:

08 Section 2G.11 contains information on the design and placement of Preferential Lane Exit Destination guide signs.

09 Figures 2G-9 and 2G-10 show examples of signs for various geometric configurations of intermediate entry to a barrier- or buffer-separated preferential lane where access is restricted to designated locations.

Section 2G.13 Guide Signs for Egress from Preferential Lanes to General-Purpose Lanes

Standard:

01 **For barrier-separated, buffer-separated, and contiguous preferential lanes where egress is restricted only to designated points, post-mounted Advance Guide and post-mounted Intermediate Egress Direction signs (see Figure 2G-11) shall be installed in the median or on median barriers that separate two directions of traffic prior to and at the intermediate exit points from the preferential lanes to the general-purpose lanes (see Figure 2G-9).**

02 **The legends of these signs shall refer to the next exit or exits from the general-purpose lanes by displaying the appropriate destination information, exit number(s), or both. The Intermediate Egress Direction signs for egress from the preferential lanes to the general-purpose lanes shall not refer to the egress as an exit.**

Support:

03 Section 2G.10 contains information on the design of post-mounted guide signs applicable to a preferential lane when installed on a median barrier. Figures 2G-9 and 2G-12 show examples of signs for various geometric configurations of intermediate egress from a barrier- or buffer-separated preferential lane where access is restricted to designated locations.

Guidance:

04 *Where two or more adjacent preferential lanes are present in a single direction, consideration should be given to the use of overhead guide signs to display the information related to egress from the preferential lanes.*

05 *For barrier-separated and buffer-separated preferential lanes where egress from a preferential lane to the general-purpose lanes is restricted only to designated points via a separate lane or ramp, the Advance Guide and Intermediate Egress Direction signs for the egress should be mounted overhead and a Pull-Through sign should be mounted with the Intermediate Egress Direction sign (see Figure 2G-12).*

Standard:

06 **For preferential lanes that incorporate a vehicle occupancy requirement, the design of the overhead Advance Guide and Egress Direction signs for intermediate egress from the preferential lanes to the general-purpose lanes shall display a white diamond symbol on a black background at the left-hand edge of the signs.**

07 **The design of Pull-Through signs when used in conjunction with an Egress Direction sign at an intermediate egress from the preferential lanes to the general-purpose lanes shall be distinguished from those applicable to general-purpose lanes by inclusion of an upper section with the applicable black legend on a white background, such as HOV LANE. For preferential lanes that incorporate a vehicle occupancy requirement, the white diamond symbol on a black background shall be displayed at the left-hand edge of this upper section.**

Figure 2G-9. Example of Signing for an Intermediate Entry to a Barrier- or Buffer-Separated HOV Lane

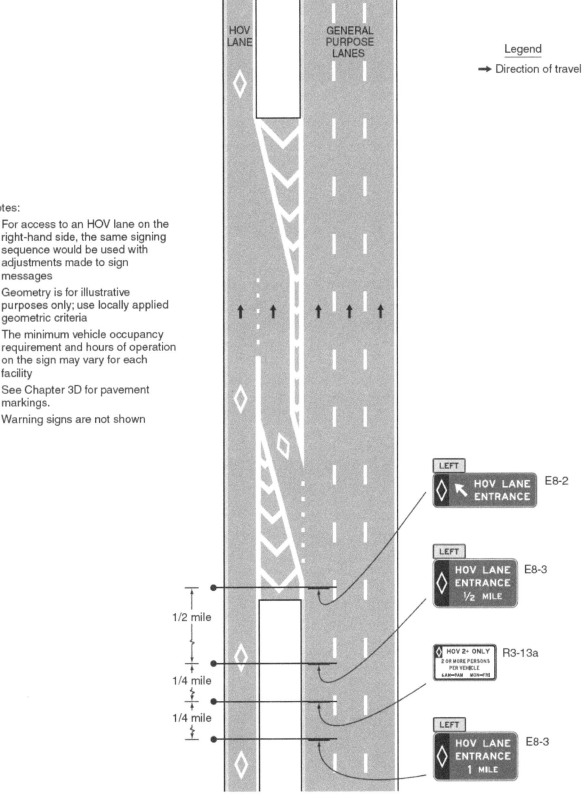

Figure 2G-10. Example of Signing for the Intermediate Entry to, Egress from, and End of Access-Restricted HOV Lanes

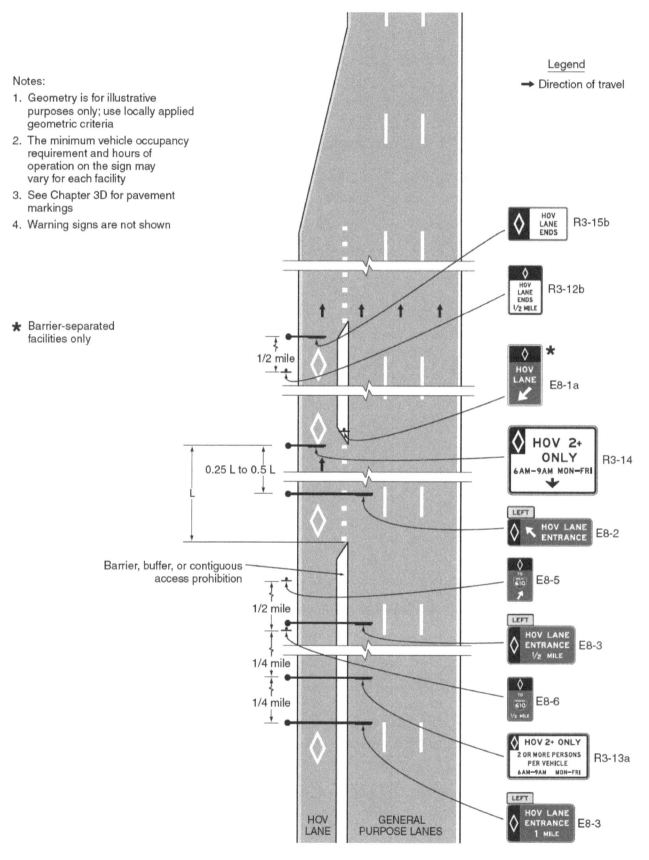

Figure 2G-11. Examples of Barrier-Mounted Guide Signs for an Intermediate Egress from Preferential Lanes

E8-5

E8-6

Note: Examples of HOV Lane (diamond symbol) signs are illustrated. For other types of preferential lanes, the appropriate symbol or word message (see Section 2G.03) is displayed in white on the black background of the top portion of these signs.

Section 2G.14 Guide Signs for Direct Entrances to Preferential Lanes from Another Highway

Standard:

01   **For direct access ramps to preferential lanes from a transit facility (such as a park - ride lot or a transit station or terminal) that is accessible from surface streets, advance guide signs shall be provided along the adjoining surface streets to direct traffic into and through the transit facility to the preferential lane (see Figure 2G-13).**

Support:

02   Figure 2G-14 provides examples of recommended uses and layouts of signs for HOV lanes for direct access ramps, park - ride lots, and access from surface streets.

Section 2G.15 Guide Signs for Direct Exits from Preferential Lanes to Another Highway

Standard:

01   **For contiguous preferential lanes on the left-hand side of the roadway, Advance Guide signs, Exit Direction signs, and Exit Gore signs (see Figure 2G-14) specifically applicable to the preferential lanes shall be used for exits to direct access ramps, such as HOV lane ramps (see Figure 2G-15) or ramps to park - ride facilities.**

02   **The design of Advance Guide, Exit Direction, and Pull-Through signs for direct exits from preferential lanes shall be distinguished from those applicable to general-purpose lanes by inclusion of an upper section with the applicable black legend on a white background, such as HOV LANE (for Pull-Through signs) or HOV EXIT (for Advance Guide and Exit Direction signs). For preferential lanes that incorporate a vehicle occupancy requirement, the white diamond symbol on a black background shall be displayed at the left-hand edge of this upper section (see Figures 2G-15 and 2G-16).**

Guidance:

03   *Advance Guide and Exit Direction signs for exits to direct access ramps from a preferential lane should be mounted overhead. A Pull-Through sign should be used with the Exit Direction sign at exits to direct access ramps.*

Standard:

04   **Post-mounted guide signs in a vertical rectangular shape installed on a median barrier shall not be used for the Advance Guide and Exit Direction signs for exits to direct access ramps.**

05   **Because direct access ramps for preferential lanes at interchanges connecting two freeways are typically left-hand side exits and typically have design speeds similar to the preferential lane, overhead Advance Guide signs and overhead Exit Direction signs shall be provided in advance of and at the entry point to each freeway-to-freeway preferential lane ramp (see Figure 2G-16).**

Guidance:

06   *The use of guide signs for preferential lanes at freeway interchanges should comply with the provisions for guide signs established in this Manual.*

Support:

07   Guide signs for direct access ramps for preferential lanes at interchanges connecting two freeways are similar to those for a connecting ramp between two freeway facilities.

Figure 2G-12. Examples of Signs for an Intermediate Egress from a Barrier- or Buffer-Separated HOV Lane

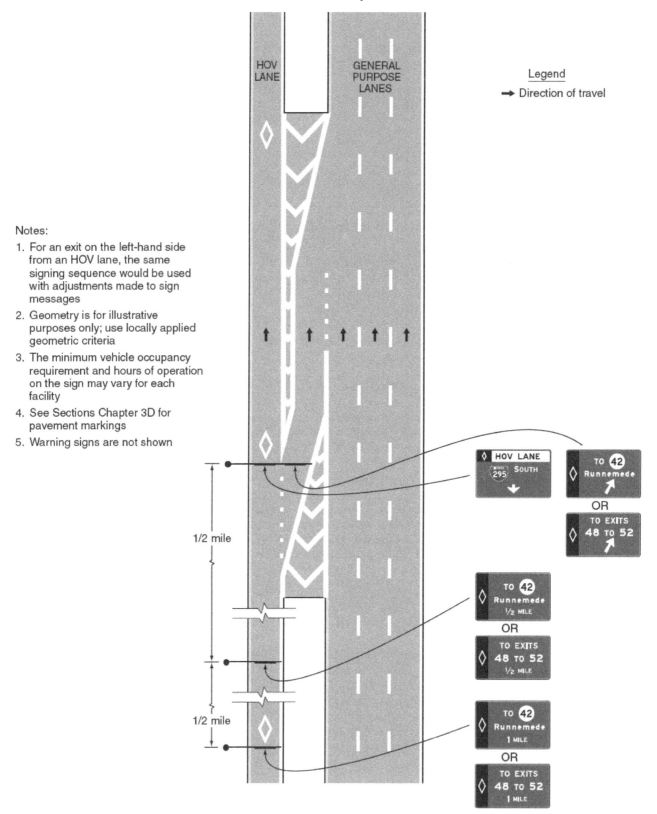

Figure 2G-13. Example of Signing for a Direct Entrance Ramp to an HOV Lane from a Park-and-Ride Facility and a Local Street

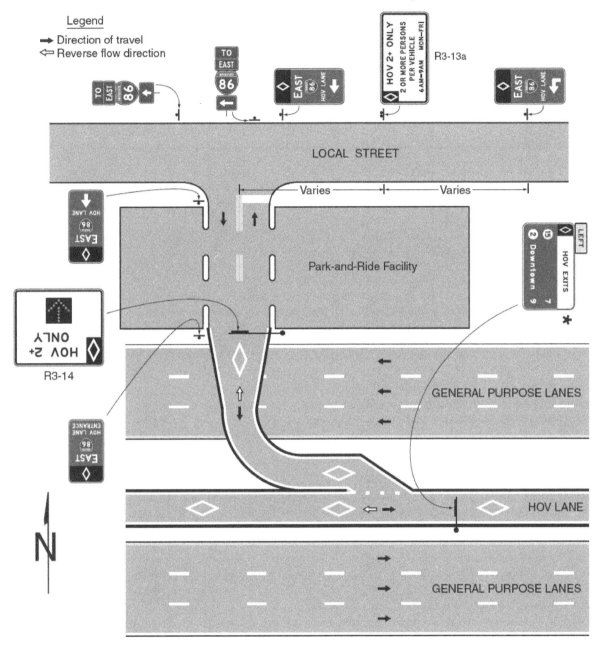

Notes:
1. The minimum vehicle occupancy requirement on the sign may vary for each facility
2. See Chapter 3D for pavement markings
3. Warning signs are not shown
4. Sign locations are approximate
5. Additional signs may be required to direct drivers from the surrounding streets into the park-and-ride lot and the HOV lane
6. Additional signs are required on the adjoining surface streets to inform non-HOVs that they should not enter the HOV facility
7. This figure illustrates a reversible HOV lane with a direct access ramp
8. The guide signs directing local street traffic to the HOV lane should include the word ENTRANCE when the direct access ramp does not traverse a park-and-ride facility

★ For access-restricted facilities; destinations may be augmented to accompany routes on Interchange Sequence signs (see Figure 2E-31)

Figure 2G-14. Exit Gore Sign for a Direct Exit from a Preferential Lane

E8-4

Note: An example of an HOV Lane (diamond symbol) sign is illustrated. For other types of preferential lanes, the appropriate symbol or word message (see Section 2G.03) is displayed in white on the black background of the top portion of this sign.

Section 2G.16 Signs for Priced Managed Lanes – General

Support:

01 A priced managed lane is a managed lane that employs tolling or pricing, typically through electronic toll collection, to manage congestion levels and maintain a certain level of service for users of the facility. A priced managed facility typically provides a less congested alternative to adjacent lanes along the same designated route, or to a nearby facility, that experience recurring congestion during peak periods. A priced managed lane might allow non-toll travel by certain vehicles based on occupancy or other criteria. A variety of operational management strategies might be used in conjunction with tolling or pricing.

02 The number and combination of operational strategies that are applied to a managed lane to manage congestion or improve efficiency might be practically limited by the amount of information that can be legibly displayed on signs or in signing sequences and still be readily comprehended by road users. Such factors to consider when evaluating alternatives for managed lanes are locations of signs for general-purpose interchanges and for other roadway conditions, the number of intermediate access points between the managed and general-purpose lanes and the need to repeat the operational information, and the distance over which a signing sequence that displays all of the eligibility requirements can be displayed.

03 Because managed lanes have the capability to employ a variety of operational strategies on a changing basis, it is not practical to assign a naming convention to such lanes for the purpose of signing based on the specific operational management strategies, as is more readily accomplished with other types of preferential lanes, such as HOV, Bus, or Bike lanes. Instead, the various requirements, restrictions, and eligibility criteria are more appropriately conveyed through a sequence of regulatory and guide signs with a more encompassing designation for the purpose of providing directional information.

04 As priced managed lanes become more prevalent as an operational strategy, it will be important to establish a uniform naming convention to distinguish those lanes that are an alternative to travel on adjacent general-purpose lanes on the same designated route to effectively communicate to motorists the range of basic requirements for similar facilities in different regions.

Standard:

05 **Priced managed lanes that are adjacent to general-purpose lanes along the same designated route shall be signed using the legend EXPRESS or EXPRESS LANE(S). This provision shall apply when any of the following operational strategies is used for a managed lane:**
   A. **All users of the managed lane are charged a fixed or variable toll;**
   B. **General-purpose traffic using the managed lane is charged a fixed or variable toll, but HOV traffic is allowed to travel without being charged a toll on either a full- or part-time basis;**
   C. **General-purpose traffic using the managed lane is charged a fixed or variable toll, but HOV traffic is offered a discounted toll on either a full- or part-time basis; or**
   D. **General-purpose traffic using the managed lane is charged a fixed or variable toll, but HOV traffic registered with a local program travels at a discounted toll or without being charged a toll on either a full- or part-time basis (a transponder or other identifier is typically required of HOVs to indicate registration in conjunction with electronic or visual enforcement and verification of vehicle occupancy).**

06 **The legends EXPRESS and EXPRESS LANE(S) shall not be used on signs for entrances to highways on which all lanes are managed and there are no adjacent general-purpose lanes on the same designated route. The legends EXPRESS and EXPRESS LANE(S) shall not be used on signs for a managed ramp connection that provides an alternative to a general-purpose ramp connection (see Figure 2F-7), except where the ramp leads directly to a managed lane as described in Section 2G.14. The legends EXPRESS and EXPRESS LANE(S) shall not be used on signs for open-road tolling lanes that bypass a conventional toll plaza (see Chapter 2F).**

Figure 2G-15. Examples of Guide Signs for Direct HOV Lane Entrance and Exit Ramps

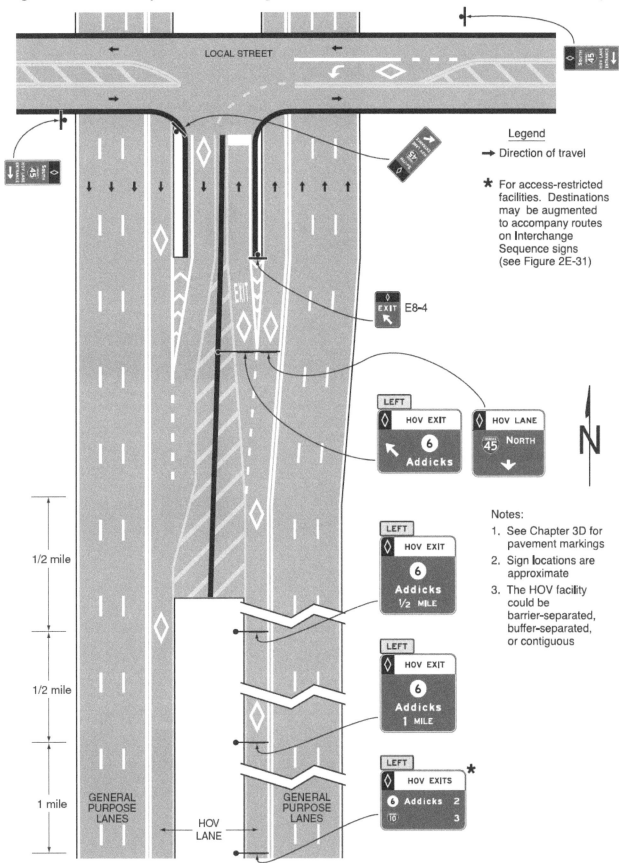

Figure 2G-16. Examples of Guide Signs for a Direct Access Ramp between HOV Lanes on Separate Freeways

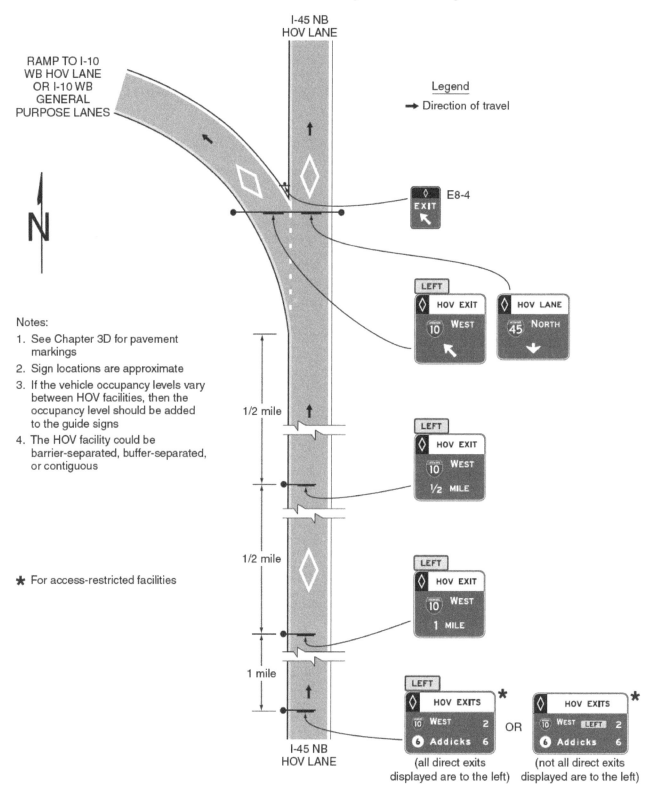

07 The diamond symbol shall be reserved exclusively for preferential lanes whose operational strategy is occupancy-based only (see Sections 2G.03 through 2G.14) and shall not be used to designate a managed lane in which other operational strategies, such as tolling and pricing, are employed to allow general-purpose traffic to use the lane.

Section 2G.17 Regulatory Signs for Priced Managed Lanes

Standard:

01 Except as otherwise provided in this Section, the provisions of Sections 2G.03 through 2G.07 regarding regulatory signs for Preferential lanes shall apply to priced managed lanes operated at all times or at certain times with a toll payment requirement of some or all vehicles to use the lane(s). Such managed lanes shall use changeable message signs or changeable message elements within static signs to display the appropriate regulatory sign messages only when they are in effect.

02 Regulatory signs for preferential lanes shall be appropriately modified for adaptation to a priced managed lane, where applicable, as shown in Figure 2G-17.

03 Regulatory signs shall be used to indicate the toll charged. If the toll varies, regulatory signs that include changeable message elements, such as the R3-48 and R3-48a signs that are shown in Figure 2G-17, shall be used to display the actual toll amount in effect at any given time.

04 When only vehicles with a registered ETC account are allowed to use a managed lane where some or all vehicles are charged a toll, regulatory signs to indicate such a restriction shall be provided and shall incorporate the pictograph adopted by the toll facility's ETC payment system and the word ONLY (see Section 2G.18 for the incorporation of such regulatory legends into the guide signs for the entrances to such facilities). The display of the ETC system pictograph shall comply with the provisions of Sections 2F.03 and 2F.04 as shown in Figures 2G-17 and 2G-18.

05 When HOV traffic is allowed to use a priced managed lane without paying a toll and registration in a local program is not required to receive the toll exemption, the Vehicle Occupancy Definition (R3-10 or R3-13) signs (see Section 2G.04) shall be modified to delete the diamond symbol to create priced managed lane Vehicle Occupancy Definition (R3-40 and R3-43) signs to indicate the minimum occupancy related to the management strategy (see Figure 2G-17).

06 A priced managed lane Periods of Operation (R3-44 or R3-44a) sign (see Figure 2G-17) shall be installed at the beginning or initial entry point, and at any intermediate entry points where vehicles are allowed to legally enter an access-restricted priced managed lane.

07 When the vehicle occupancy required for non-toll use of a managed lane is varied as a part of a priced managed lane operational strategy, regulatory signs that include changeable message elements shall be used to display the required vehicle occupancy in effect for non-toll travel.

Option:

08 Where registration in a local program or ETC account is required for HOV traffic to travel in a priced managed lane without being charged a toll or by being charged a discounted toll, such information may be displayed on a separate sign within the sequence of the required regulatory and guide signs.

Standard:

09 R3-42 Series and R3-45 Series signs (see Figure 2G-17) shall be installed in accordance with the provisions of Section 2G.07 to indicate the termination of a priced managed lane or restriction. The R3-42, R3-42a, and R3-45 signs shall be used only where the managed lane and restriction end and traffic must merge into the general-purpose lanes. The R3-42b, R3-42c, and R3-45a signs shall be used only where the managed lane restriction ends and the lane becomes a general-purpose lane.

Section 2G.18 Guide Signs for Priced Managed Lanes

Standard:

01 Except as otherwise provided in this Section, guide signs for barrier-separated, buffer-separated, and contiguous managed lanes shall follow the specific provisions for Preferential Lane guide signs contained in Sections 2G.10 through 2G.15. Except as otherwise provided in this Section, guide signs for highways on which all lanes are managed shall follow the general provisions for freeway and expressway guide signs as contained in Chapter 2E as a whole. Guide signs for highways on which all lanes are managed and tolling or pricing is used as a management strategy shall follow the applicable provisions for toll road guide signs as contained in Chapter 2F, in addition to the general provisions of Chapter 2E.

02 If fixed or variable tolls are used as an operational strategy for a managed lane, the guide signs shall comply with the provisions of Sections 2F.03, 2F.04, and 2F.17 regarding the use, size, and placement of ETC-account pictographs.

Figure 2G-17. Regulatory Signs for Managed Lanes

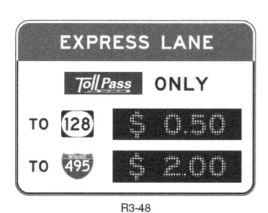

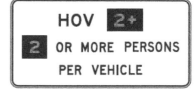

Notes:
1. The ETC pictograph shown is an example only. The pictograph for the toll facility's adopted ETC system shall be used.
2. Changeable message sign elements shall be used for the numerals displayed for the variable tolls.

Support:

03 Figure 2G-18 shows examples of Guide signs for entrances to priced managed lanes and other ETC account-only toll facilities that incorporate header panels with ETC account pictographs and regulatory legends.

Guidance:

04 *Exit Destination supplemental guide signs, identifying final destination and downstream exit locations accessible from the managed lane (see Figure 2G-19), should be installed in advance of the initial entry points to priced managed lanes. These signs should be located in accordance with the provisions of Paragraph 5 of Section 2G.11.*

05 *For managed lanes that are available as an alternative to travel on adjacent general-purpose lanes on the same designated route, changeable message signs indicating the comparative travel times or congestion levels using the managed lanes versus the general-purpose lanes (see Figure 2G-20) should be installed in advance of the initial and intermediate entry points to the managed lanes.*

Option:

06 Changeable message signs may also be used on non-managed highways to display comparative travel times or congestion levels for a nearby managed highway.

Standard:

07 **Guide signs at the initial and intermediate entry points to a priced managed lane in which all general-purpose passenger vehicles are allowed shall include the legend EXPRESS or EXPRESS LANE(S). The guide signs shall incorporate the pictograph of the ETC account system into a header panel within the guide sign in accordance with Sections 2F.03, 2F.04, and 2F.17. For a priced managed lane that allows non-toll travel by HOV traffic without registration in a local program, the header panel shall be modified to a regulatory format to display both the pictograph of the ETC account system and the minimum occupancy requirement for non-toll travel with a black legend on a white background (see Figure 2G-19).**

08 **Guide signs at the initial and intermediate entry points to a managed lane that allows only HOV traffic with either a fixed or variable occupancy requirement shall follow the provisions of Sections 2G.10 through 2G.12 and 2G.14.**

Support:

09 Figures 2G-21 through 2G-24 show examples of guide signs for various configurations of initial and intermediate entrances to a priced managed lane.

Figure 2G-18. Examples of Guide Signs for Entrances to Priced Managed Lanes

A - ENTRANCE TO A PRICED MANAGED LANE FROM A GENERAL PURPOSE LANE

B - DIRECT ENTRANCE TO A PRICED MANAGED LANE FROM A CROSSROAD

Note: 1. The ETC pictographs shown are examples only. The pictograph for the toll facility's adopted ETC system shall be used.
2. The examples shown are for facilities on which registration in a toll account program is required for toll payments.

Standard:

10 The use and locations of guide signs for intermediate egress locations and direct exits from a priced managed lane (see Figures 2G-24 through 2G-27) shall comply with the provisions of Sections 2G.13 and 2G.15. The signs shall be suitably modified to display header messages of white legend on a green background that relate the guide sign legends to the managed lane(s) as appropriate in accordance with the following:

   A. Post-mounted or overhead-mounted Advance Guide signs for intermediate egress to the general-purpose lanes shall include the legend LOCAL EXITS in a header panel within the guide signs, destination information or the exit number(s) for the next exit(s) accessible from the general-purpose lanes, and the appropriate distance information to the location of the egress (see Figures 2G-24 and 2G-25).
   B. Post-mounted or overhead-mounted Intermediate Egress Direction signs shall include the legend LOCAL EXITS in a header panel within the signs, the destination information or the exit number(s) of the next exit(s) accessible from the general-purpose lanes, and a diagonally upward-pointing directional arrow (see Figures 2G-24 and 2G-25).
   C. For direct exits to another roadway, the legend EXPRESS EXIT shall be used on the Advance Guide and Exit Direction signs (see Figure 2G-26).
   D. For pull-through signs, the legend EXPRESS LANE(S) shall be used, either as a header panel within the pull-through sign or as the principal legend of the sign without a header panel (see Figures 2G-25, 2G-26, and 2G-27).

Support:

11 Section 2G.13 contains information on the use of overhead-mounted guide signs for intermediate egress to the general-purpose lanes.

12 Figures 2G-28 and 2G-29 show examples of guide signing for direct entrances to a priced managed lane from a crossroad or surface street.

Figure 2G-19. Example of an Exit Destinations Sign for a Managed Lane

EXPRESS LANE EXITS

I-5 ½

Manchester Ave 2

Encinitas Blvd 6

Figure 2G-20. Example of a Comparative Travel Time Information Sign for Preferential or Managed Lanes

Notes:
1. The ETC pictograph shown is an example only. The pictograph for the toll facility's adopted ETC system shall be used.
2. CMS elements shall be used for the numerals displayed for the estimated travel times.

Figure 2G-21. Example of Signing for the Entrance to an Access-Restricted Priced Managed Lane

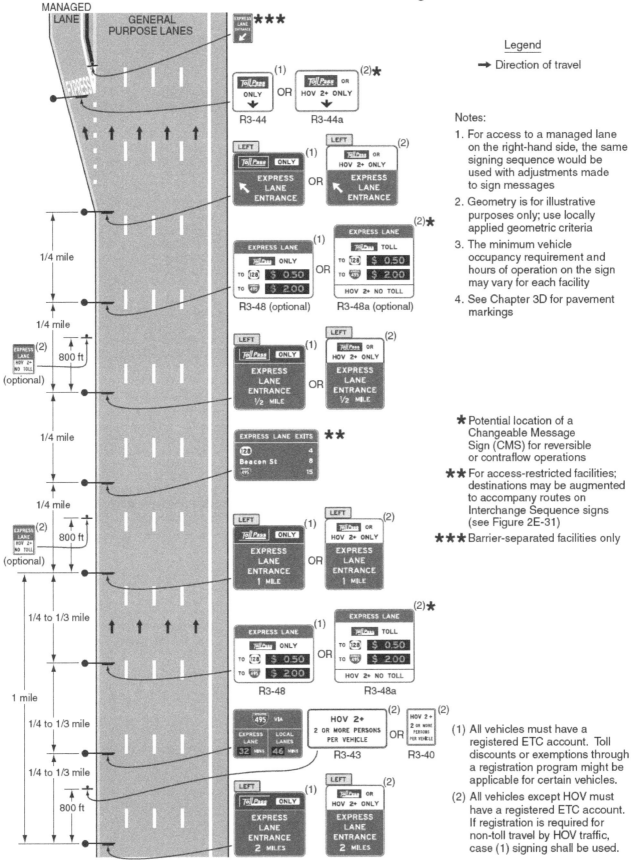

Figure 2G-22. Example of Signing for the Entrance to an Access-Restricted Priced Managed Lane Where a General-Purpose Lane Becomes the Managed Lane

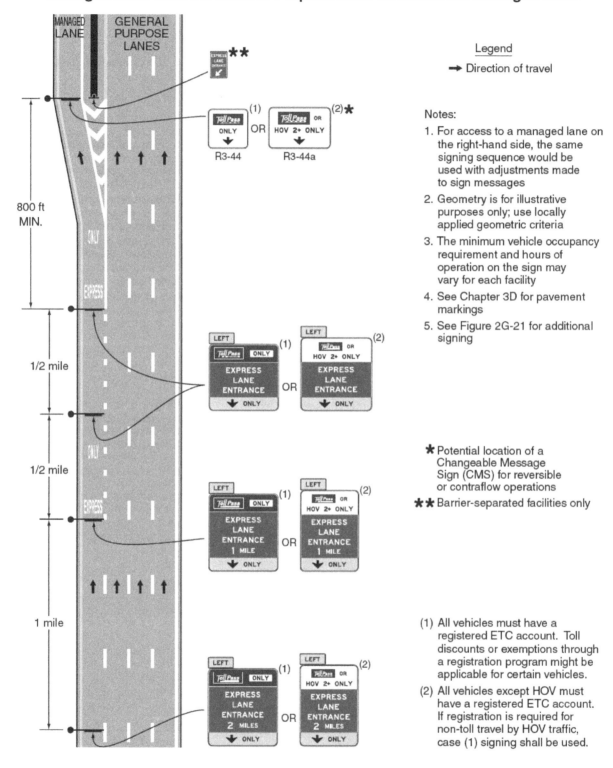

Figure 2G-23. Example of Signing for an Intermediate Entry to a Barrier- or Buffer-Separated Priced Managed Lane

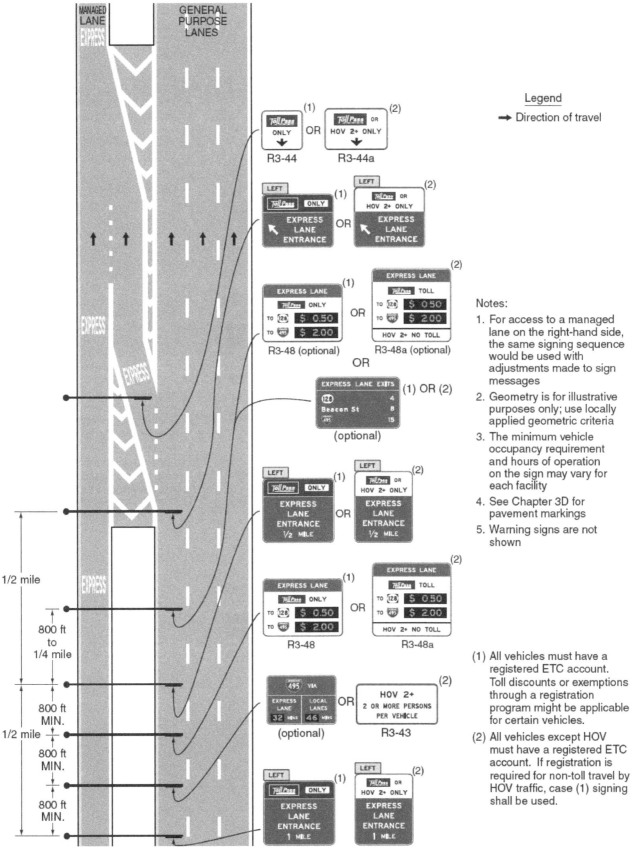

Figure 2G-24. Example of Signing for the Intermediate Entry to, Egress from, and End of Access-Restricted Priced Managed Lanes

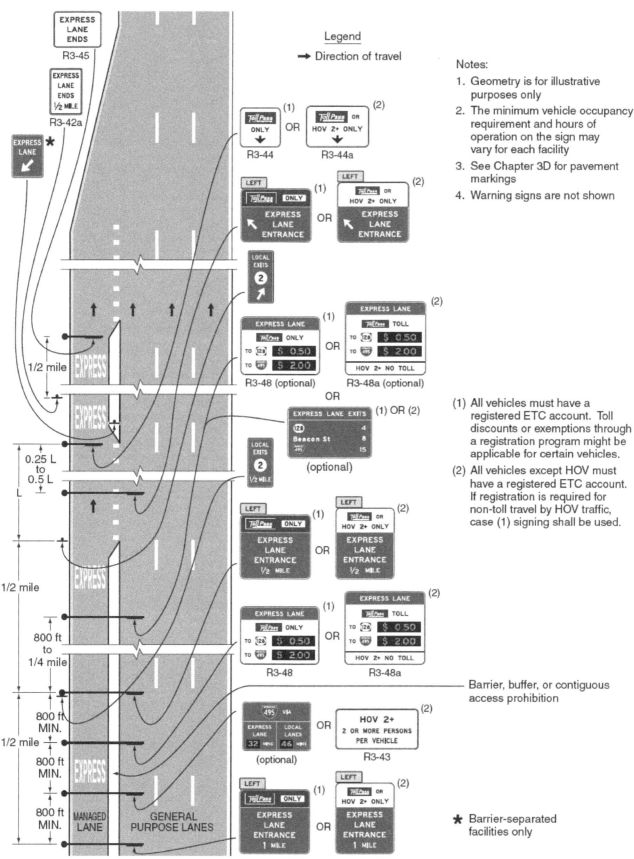

Figure 2G-25. Examples of Guide Signs for an Intermediate Egress from a Barrier- or Buffer-Separated Managed Lane

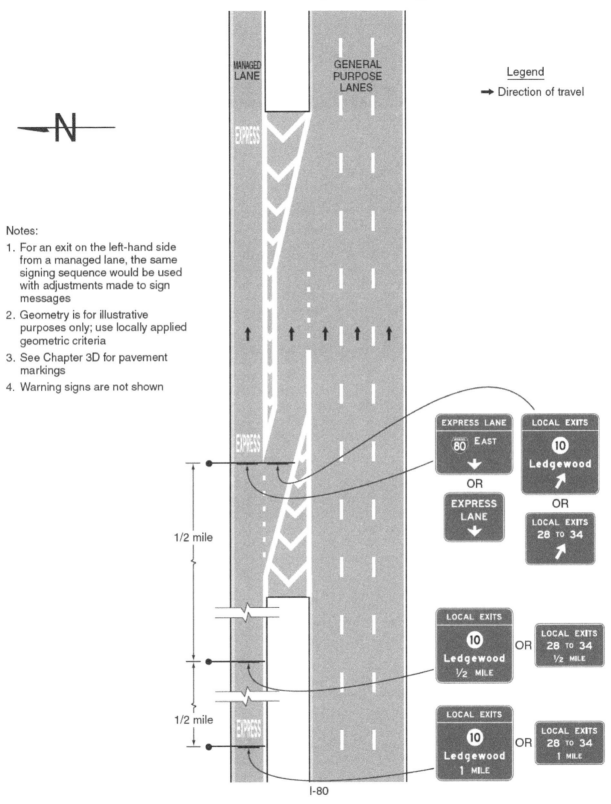

Figure 2G-26. Examples of Guide Signs for Direct Managed Lane Entrance and Exit Ramps

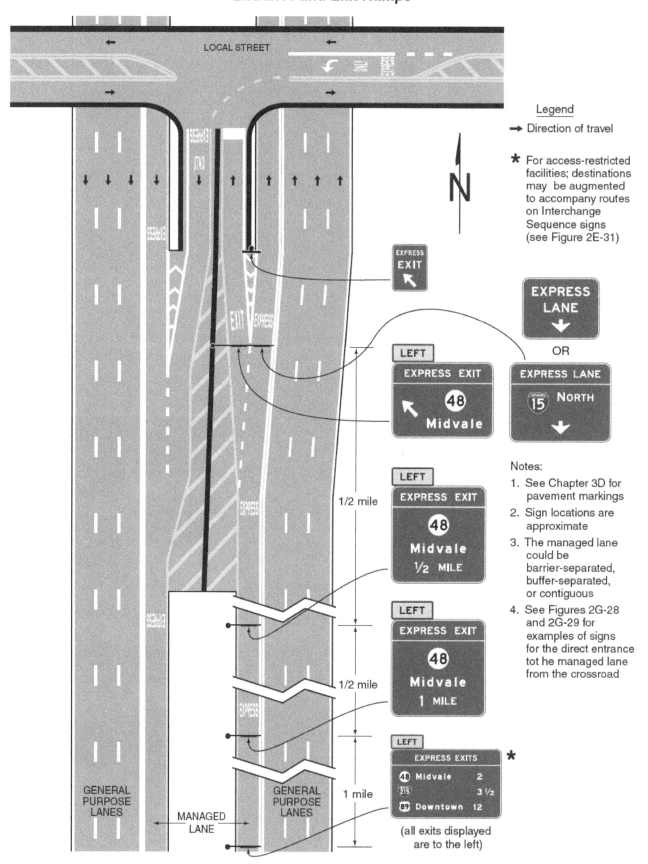

Figure 2G-27. Examples of Guide Signs for a Direct Access Ramp between Managed Lanes on Separate Freeways

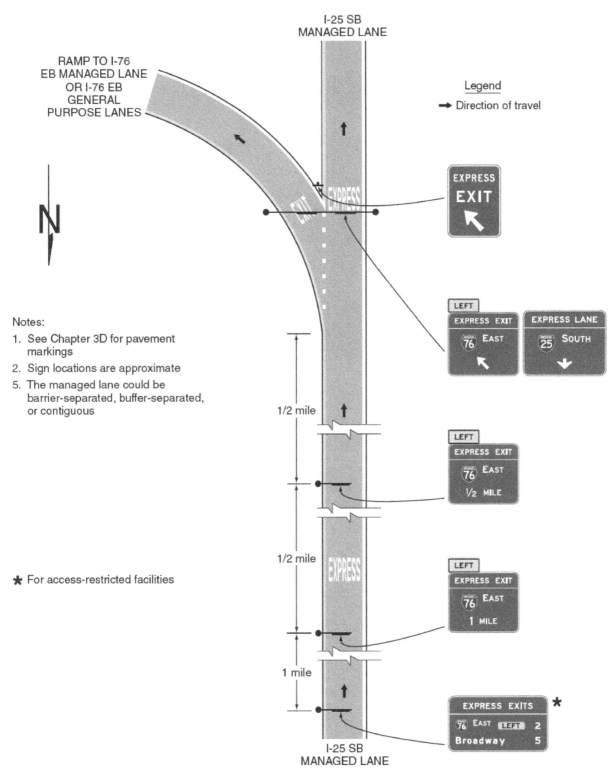

Figure 2G-28. Examples of Guide Signs for a Direct Entrance Ramp to a Priced Managed Lane and Trailblazing to a Nearby Entrance to the General-Purpose Lanes

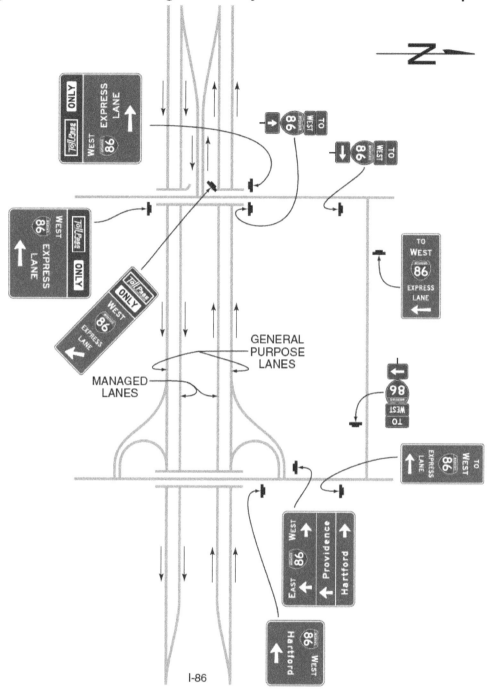

Figure 2G-29. Examples of Guide Signs for Separate Entrance Ramps to General-Purpose and Priced Managed Lanes from the Same Crossroad

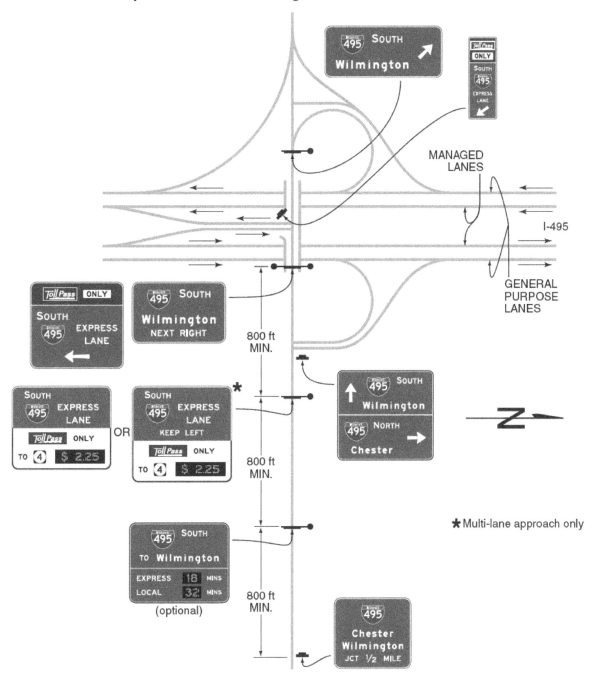

CHAPTER 2H. GENERAL INFORMATION SIGNS

Section 2H.01 Sizes of General Information Signs
Standard:
01 **Except as provided in Section 2A.11, the sizes of General Information signs that have a standardized design shall be as shown in Table 2H-1.**

Support:
02 Section 2A.11 contains information regarding the applicability of the various columns in Table 2H-1.

Option:
03 Signs larger than those shown in Table 2H-1 may be used (see Section 2A.11).

Section 2H.02 General Information Signs (I Series)
Support:
01 Of interest to the traveler, though not directly necessary for guidance, are numerous kinds of information that can properly be conveyed by General Information signs (see Figure 2H-1) or miscellaneous information signs (see Section 2H.04). They include such items as State lines, city limits, other political boundaries, time zones, stream names, elevations, landmarks, and similar items of geographical interest, and safety and transportation-related messages. Chapter 2M contains recreational and cultural interest area symbol signs that are sometimes used in combination with General Information signs.

Guidance:
02 *General Information signs should not be installed within a series of guide signs or at other equally critical locations, unless there are specific reasons for orienting the road user or identifying control points for activities that are clearly in the public interest. On all such signs, the designs should be simple and dignified, devoid of any advertising, and in general compliance with other guide signing.*

Table 2H-1. General Information Sign Sizes

Sign	Sign Designation	Section	Conventional Road	Freeway or Expressway
Reference Location (1 digit)	D10-1	2H.05	10 x 18	12 x 24
Intermediate Reference Location (2 digits)	D10-1a	2H.05	10 x 27	12 x 36
Reference Location (2 digits)	D10-2	2H.05	10 x 27	12 x 36
Intermediate Reference Location (3 digits)	D10-2a	2H.05	10 x 36	12 x 48
Reference Location (3 digits)	D10-3	2H.05	10 x 36	12 x 48
Intermediate Reference Location (4 digits)	D10-3a	2H.05	10 x 48	12 x 60
Enhanced Reference Location	D10-4	2H.06	18 x 54	18 x 54
Intermediate Enhanced Reference Location	D10-5	2H.06	18 x 60	18 x 60
Acknowledgement	D14-1	2H.08	36 x 30*	72 x 48*
Acknowledgement	D14-2	2H.08	36 x 30*	72 x 48*
Acknowledgement	D14-3	2H.08	42 x 24*	96 x 36*
Signals Set for XX MPH	I1-1	2H.03	24 x 36	—
Jurisdictional Boundary	I-2	2H.04	Varies x 18**	Varies x 36**
Geographical Features	I-3	2H.04	Varies x 18**	Varies x 36**
Airport	I-5	2H.02	24 x 24	30 x 30
Bus Station	I-6	2H.02	24 x 24	30 x 30
Train Station	I-7	2H.02	24 x 24	30 x 30
Library	I-8	2H.02	24 x 24	30 x 30
Vehicle Ferry Terminal	I-9	2H.02	24 x 24	30 x 30
Recycling Collection Center	I-11	2H.02	30 x 48	—
Light Rail Transit Station	I-12	2H.02	24 x 24	—

* The size shown is the maximum size for the corresponding roadway classification. The size of the sign and acknowledgement logo should be appropriately reduced where shorter legends are used.

** The size shown is for the typical sign illustrated in the figure. The size should be determined based on the amount of legend required for the sign.

Notes: 1. Larger signs may be used when appropriate, except for the D14 series signs
2. Dimensions in inches are shown as width x height

Standard:

03 **Except for political boundary signs, General Information signs shall have white legends and borders on green rectangular-shaped backgrounds.**

Option:

04 An information symbol sign (I-5 through I-9) may be used to identify a route leading to a transportation or general information facility, or to provide additional guidance to the facility. The symbol sign may be supplemented by an educational plaque where necessary; also, the name of the facility may be used if needed to distinguish between similar facilities.

05 The Advance Turn (M5 series) or Directional Arrow (M6 series) auxiliary signs shown in Figure 2H-1 with white arrows on green backgrounds may be used with General Information symbol signs to create a General Information Directional Assembly.

06 Guide signs for commercial service airports and non-carrier airports may be provided from the nearest Interstate, other freeway, or conventional highway intersection directly to the airport, normally not to exceed 15 miles. The Airport (I-5) symbol sign along with a supplemental plaque may be used to indicate the specific name of the airport. An Airport symbol sign, with or without a supplemental name plaque or the word AIRPORT, and an arrow may be used as a trailblazer.

Standard:

07 **Adequate trailblazer signs shall be in place prior to installing the airport guide signs.**

Support:

08 Location and placement of all airport guide signs depends upon the availability of longitudinal spacing on highways.

Option:

09 The Recycling Collection Center (I-11) symbol sign may be used to direct road users to recycling collection centers.

Guidance:

10 *The Recycling Collection Center symbol sign should not be used on freeways and expressways.*

Standard:

11 **If used on freeways or expressways, the Recycling Collection Center symbol sign shall be considered one of the supplemental sign destinations.**

Figure 2H-1. General Information and Miscellaneous Information Signs

Advance Turn and Directional Arrow Auxiliary Signs for use with General Information Signs

M5-1 M5-2 M6-1 M6-2 M6-3

12 **When a sign is used to display a safety or transportation-related message, the display format shall not be of a type that would be considered similar to advertising displays. Messages and symbols that resemble any official traffic control device shall not be used on safety or transportation-related message signs.**
Option:

13 The pictograph of a political jurisdiction (such as a State, county, or municipal corporation) may be displayed on a political boundary General Information sign.

Standard:

14 **If used, the height of a pictograph on a political boundary General Information sign shall not exceed two times the height of the upper-case letters of the principal legend on the sign. The pictograph shall comply with the provisions of Section 2A.06.**

Section 2H.03 Traffic Signal Speed Sign (I1-1)

Option:

01 The Traffic Signal Speed (I1-1) sign (see Figure 2H-1), reading SIGNALS SET FOR XX MPH, may be used to indicate a section of street or highway on which the traffic control signals are coordinated into a progressive system timed for a specified speed at all hours during which they are operated in a coordinated mode.

02 If different system progression speeds are set for different times of the day, a changeable message element may be used for the numerals of the Traffic Signal Speed (I1-1) sign. If the system is operated in coordinated mode only during certain times, a blank-out version of the Traffic Signal Speed (I1-1) sign may be used to display the message only during those times.

Guidance:

03 *If used, the sign should be mounted as near as practical to each intersection where the timed speed changes, and at intervals of several blocks throughout any section where the timed speed remains constant.*

Standard:

04 **The Traffic Signal Speed sign shall be a minimum of 24 x 36 inches with the longer dimension vertical. It shall have a white message and border on a green background.**

Section 2H.04 Miscellaneous Information Signs

Support:

01 Miscellaneous information are used to point out geographical features, such as rivers and summits, and other jurisdictional boundaries (see Section 2H.02). Figure 2H-1 shows examples of miscellaneous information (I-2 and I-3) signs.

Option:

02 Miscellaneous information signs may be used if they do not interfere with signing for interchanges or other critical points.

Guidance:

03 *Miscellaneous information signs should not be installed unless there are specific reasons for orienting the road users or identifying control points for activities that are clearly in the public interest. If Miscellaneous information signs are to be of value to the road user, they should be consistent with other guide signs in design and legibility. On all such signs, the design should be simple and dignified, devoid of any tendency toward flamboyant advertising, and in general compliance with other signing.*

Section 2H.05 Reference Location Signs (D10-1 through D10-3) and Intermediate Reference Location Signs (D10-1a through D10-3a)

Support:

01 There are two types of reference location signs:
  A. Reference Location (D10-1, 2, and 3) signs show an integer distance point along a highway, and
  B. Intermediate Reference Location (D10-1a, 2a, and 3a) signs also show a decimal between integer distance points along a highway.

Standard:

02 **Except when Enhanced Reference Location signs (see Section 2H.06) are used instead, Reference Location (D10-1 through D10-3) signs shall be placed on all expressway facilities that are located on a route where there is reference location sign continuity and on all freeway facilities to assist road users in estimating their progress, to provide a means for identifying the location of emergency incidents and traffic crashes, and to aid in highway maintenance and servicing.**

Option:

03 Reference Location (D10-1 to D10-3) signs (see Figure 2H-2) may be installed along any section of a highway route or ramp to assist road users in estimating their progress, to provide a means for identifying the location of emergency incidents and traffic crashes, and to aid in highway maintenance and servicing.

04 To augment the reference location sign system, Intermediate Reference Location (D10-1a to D10-3a) signs (see Figure 2H-3), which show the tenth of a mile with a decimal point, may be installed at one tenth of a mile intervals, or at some other regular spacing.

Standard:

05 **When Intermediate Reference Location (D10-1a to D10-3a) signs are used to augment the reference location sign system, the reference location sign at the integer mile point shall display a decimal point and a zero numeral.**

06 **When placed on freeways or expressways, reference location signs shall contain 10-inch white numerals on a 12-inch wide green background with a white border. The signs shall be 24, 36, or 48 inches in height for one, two, or three digits, respectively, and shall contain the word MILE in 4-inch white letters.**

07 **When placed on conventional roads, reference location signs shall contain 6-inch white numerals on a green background that is at least 10 inches wide with a white border. The signs shall contain the word MILE in 4-inch white letters.**

08 **Reference location signs shall have a minimum mounting height of 4 feet, measured vertically from the bottom of the sign to the elevation of the near edge of the roadway, and shall not be governed by the mounting height requirements prescribed in Section 2A.18.**

09 **The distance numbering shall be continuous for each route within a State, except where overlaps occur (see Section 2E.31). Where routes overlap, reference location sign continuity shall be established for only one of the routes. If one of the overlapping routes is an Interstate route, that route shall be selected for continuity of distance numbering.**

Guidance:

10 *The route selected for continuity of distance numbering should also have continuity in interchange exit numbering (see Section 2E.31).*

11 *On a route without reference location sign continuity, the first reference location sign beyond the overlap should indicate the total distance traveled on the route so that road users will have a means of correlating their travel distance between reference location signs with that shown on their odometer.*

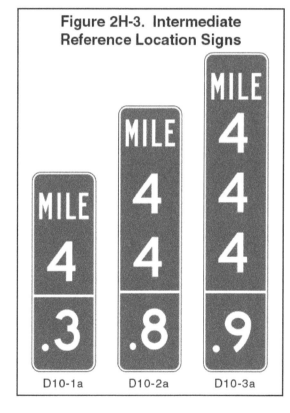

Figure 2H-2. Reference Location Signs

Figure 2H-3. Intermediate Reference Location Signs

Standard:

12 **For divided highways, the distance measurement shall be made on the northbound and eastbound roadways. The reference location signs for southbound or westbound roadways shall be set at locations directly opposite the reference location signs for the northbound or eastbound roadways.**

Guidance:

13 *Zero distance should begin at the south and west State lines, or at the south and west terminus points where routes begin within a State.*

Standard:

14 **Except as provided in Paragraph 15, reference location signs shall be installed on the right-hand side of the roadway.**

Option:

15 Where conditions limit or restrict the use of reference location signs on the right-hand side of the roadway, they may be installed in the median. On two-lane conventional roadways, reference location signs may be installed on one side of the roadway only and may be installed back-to-back. Reference location signs may be placed up to 30 feet from the edge of the pavement.

16 If a reference location sign cannot be installed in the correct location, it may be moved in either direction as much as 50 feet.

Guidance:

17 *If a reference location sign cannot be placed within 50 feet of the correct location, it should be omitted.*

Section 2H.06 Enhanced Reference Location Signs (D10-4, D10-5)

Support:

01 There are two types of enhanced reference location signs:
   A. Enhanced Reference Location signs (D10-4), and
   B. Intermediate Enhanced Reference Location signs (D10-5).

Option:

02 Enhanced Reference Location (D10-4) signs (see Figure 2H-4), which enhance the reference location sign system by identifying the route, may be placed on freeways or expressways (instead of Reference Location signs) or on conventional roads.

03 To augment an enhanced reference location sign system, Intermediate Enhanced Reference Location (D10-5) signs (see Figure 2H-4), which show the tenth of a mile with a decimal point, may be installed along any section of a highway route or ramp at one tenth of a mile intervals, or at some other regular spacing.

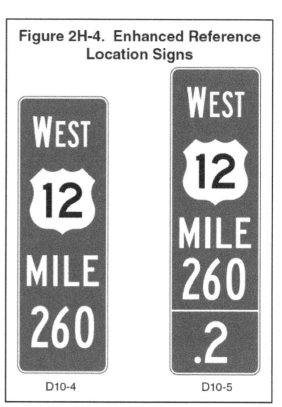

Figure 2H-4. Enhanced Reference Location Signs

Standard:

04 **If enhanced reference location signs are used, they shall be vertical signs having blue or green backgrounds with white numerals, letters, and borders, except for the route shield, which shall be the standard color and shape. The top line shall consist of the cardinal direction for the roadway. The second line shall consist of the applicable route shield for the roadway. The third line shall identify the mile reference for the location and the bottom line of the Intermediate Enhanced Reference Location sign shall give the tenth of a mile reference for the location. The bottom line of the Intermediate Enhanced Reference Location sign shall contain a decimal point. The height of the legend on enhanced reference location signs shall be a minimum of 6 inches. The height of the route shield on enhanced reference location signs shall be a minimum of 12 inches.**

05 **The background color shall be the same for all enhanced reference location signs within a jurisdiction.**

Support:

06 The provisions in Section 2H.05 regarding mounting height, distance numbering and measurements, sign continuity, and placement with respect to the right-hand shoulder and/or median for reference location signs also apply to enhanced reference location signs.

Section 2H.07 Auto Tour Route Signs

Support:
01   Auto Tour Route signs are informational signs, plaques, or shields designed to provide road users with route guidance in following an auto tour route of particular cultural, historical, or educational significance.
02   Signed auto tour routes are used in some cases to generally follow the historical route of a trail, such as the National Historic Trails administered by the National Park Service. Examples include auto tour routes that parallel the Lewis and Clark National Historic Trail, the Oregon National Historic Trail, and the Santa Fe National Historic Trail.

Guidance:
03   *If shields or other similar signs are used to provide route guidance in following an auto tour route, they should be designed in accordance with the sizes and other design principles for route signs, such as those described in Sections 2D.10 through 2D.12.*

Option:
04   Auto Tour Route signs may be installed on a highway if they have been approved by the appropriate transportation agency.

Standard:
05   **Auto Tour Route signs shall not be installed on freeways or expressways, except as necessary to provide continuity between discontinuous segments of conventional roadways that are designated as auto tour routes, for which the freeway or expressway provides the only connection between the segments. If installed on freeways or expressways, Auto Tour Route signs shall be installed as independent trailblazer assemblies (see Sections 2D.35 and 2E.27) and shall not be installed with other Route signs or confirmation assemblies or on guide signs. If installed on freeways or expressways, Auto Tour Route trailblazer assemblies shall be installed at less frequent intervals than route confirmation assemblies.**

Section 2H.08 Acknowledgment Signs

Support:
01   Acknowledgment signs are a way of recognizing a company, business, or volunteer group that provides a highway-related service. Acknowledgment signs include sponsorship signs for adopt-a-highway litter removal programs, maintenance of a parkway or interchange, and other highway maintenance or beautification sponsorship programs.

Guidance:
02   *A State or local highway agency that elects to have an acknowledgment sign program should develop an acknowledgment sign policy. The policy should require that eligible sponsoring organizations comply with State laws prohibiting discrimination based on race, religion, color, age, sex, national origin, and other applicable laws. The acknowledgment sign policy should include all of the provisions regarding sign placement and sign design that are described in this Section.*

Standard:
03   **Because regulatory, warning, and guide signs have a higher priority, acknowledgment signs shall only be installed where adequate spacing is available between the acknowledgment sign and other higher priority signs. Acknowledgment signs shall not be installed in a position where they would obscure the road users' view of other traffic control devices.**
04   **Acknowledgment signs shall not be installed at any of the following locations:**
   A. **On the front or back of, adjacent to, or around any other traffic control device, including traffic signs, highway traffic signals, and changeable message signs;**
   B. **On the front or back of, adjacent to, or around the supports or structures of other traffic control devices, or bridge piers; or**
   C. **At key decision points where a road user's attention is more appropriately focused on other traffic control devices, roadway geometry, or traffic conditions, including exit and entrance ramps, intersections, grade crossings, toll plazas, temporary traffic control zones, and areas of limited sight distance.**

Guidance:
05   *The minimum spacing between acknowledgment signs and any other traffic control signs, except parking regulation signs, should be:*
   A. *150 feet on roadways with speed limits of less than 30 mph,*
   B. *200 feet on roadways with speed limits of 30 to 45 mph, and*
   C. *500 feet on roadways with speed limits greater than 45 mph.*

If the placement of a newly-installed higher-priority traffic control device, such as a higher-priority sign, a highway traffic signal, or a temporary traffic control device, conflicts with an existing acknowledgment sign, the acknowledgment sign should be relocated, covered, or removed.

Option:

State or local highway agencies may develop their own acknowledgment sign designs and may also use their own pictograph (see definition in Section 1A.13) and/or a brief jurisdiction-wide program slogan as part of any portion of the acknowledgment sign, provided that the signs comply with the provisions for shape, color, and lettering style in this Chapter and in Chapter 2A.

Guidance:

Acknowledgment signs should clearly indicate the type of highway services provided by the sponsor.

Standard:

In addition to the general provisions for signs described in Chapter 2A and the sign design principles covered in the "Standard Highway Signs and Markings" book (see Section 1A.11), acknowledgment sign designs developed by State or local highway agencies shall comply with the following provisions:

- **A. Neither the sign design nor the sponsor acknowledgment logo shall contain any contact information, directions, slogans (other than a brief jurisdiction-wide program slogan, if used), telephone numbers, or Internet addresses, including domain names and uniform resource locators (URL);**
- **B. Except for the lettering, if any, on the sponsor acknowledgment logo, all of the lettering shall be in upper-case letters as provided in the "Standard Highway Signs and Markings" book (see Section 1A.11);**
- **C. In order to keep the main focus on the highway-related service and not on the sponsor acknowledgment logo, the area reserved for the sponsor acknowledgment logo shall not exceed 1/3 of the total area of the sign and shall be a maximum of 8 square feet, and shall not be located at the top of the sign;**
- **D. The entire sign display area shall not exceed 24 square feet;**
- **E. The sign shall not contain any messages, lights, symbols, or trademarks that resemble any official traffic control devices;**
- **F. The sign shall not contain any external or internal illumination, light-emitting diodes, luminous tubing, fiber optics, luminescent panels, or other flashing, moving, or animated features; and**
- **G. The sign shall not distract from official traffic control messages such as regulatory, warning, or guidance messages.**

Support:

Examples of acknowledgment sign designs are shown in Figure 2H-5.

Figure 2H-5. Examples of Acknowledgment Sign Designs

D14-1

D14-2

D14-3

CHAPTER 2I. GENERAL SERVICE SIGNS

Section 2I.01 Sizes of General Service Signs

Standard:

01 Except as provided in Section 2A.11, the sizes of General Service signs that have a standardized design shall be as shown in Table 2I-1.

Support:

02 Section 2A.11 contains information regarding the applicability of the various columns in Table 2I-1.

Option:

03 Signs larger than those shown in Table 2I-1 may be used (see Section 2A.11).

Table 2I-1. General Service Sign and Plaque Sizes (Sheet 1 of 2)

Sign or Plaque	Sign Designation	Section	Conventional Road	Freeway or Expressway
Rest Area XX Miles	D5-1	2I.05	66 x 36*	96 x 54*
Rest Area Next Right	D5-1a	2I.05	78 x 36*	120 x 60* (F) 114 x 48* (E)
Rest Area (with arrow)	D5-2	2I.05	66 x 36*	96 x 54*
Rest Area Gore	D5-2a	2I.05	42 x 48*	78 x 78* (F) 66 x 72* (E)
Rest Area (with horizontal arrow)	D5-5	2I.05	42 x 48*	—
Next Rest Area XX Miles	D5-6	2I.05	60 x 48*	90 x 72*
Rest Area Tourist Info Center XX Miles	D5-7	2I.08	90 x 72*	114 x 102* (F) 132 x 96* (E)
Rest Area Tourist Info Center (with arrow)	D5-8	2I.08	84 x 72*	120 x 102* (F) 120 x 96* (E)
Rest Area Tourist Info Center Next Right	D5-11	2I.08	90 x 72*	144 x 102* (F) 132 x 96* (E)
Interstate Oasis	D5-12	2I.04	—	156 x 78
Interstate Oasis (plaque)	D5-12P	2I.04	—	114 x 48
Brake Check Area XX Miles	D5-13	2I.06	84 x 48	126 x 72
Brake Check Area (with arrow)	D5-14	2I.06	78 x 60	96 x 72
Chain-Up Area XX Miles	D5-15	2I.07	66 x 48	96 x 72
Chain-Up Area (with arrow)	D5-16	2I.07	72 x 54	96 x 66
Telephone	D9-1	2I.02	24 x 24	30 x 30
Hospital	D9-2	2I.02	24 x 24	30 x 30
Camping	D9-3	2I.02	24 x 24	30 x 30
Trailer Camping	D9-3a	2I.02	24 x 24	30 x 30
Litter Container	D9-4	2I.02	24 x 30	36 x 48
Handicapped	D9-6	2I.02	24 x 24	30 x 30
Van Accessible (plaque)	D9-6P	2I.02	18 x 9	—
Gas	D9-7	2I.02	24 x 24	30 x 30
Food	D9-8	2I.02	24 x 24	30 x 30
Lodging	D9-9	2I.02	24 x 24	30 x 30
Tourist Information	D9-10	2I.02	24 x 24	30 x 30
Diesel Fuel	D9-11	2I.02	24 x 24	30 x 30
Alternative Fuel - Compressed Natural Gas	D9-11a	2I.02	24 x 24	30 x 30
Electric Vehicle Charging	D9-11b	2I.02	24 x 24	30 x 30
Electric Vehicle Charging (plaque)	D9-11bP	2I.02	24 x 18	30 x 24
Alternative Fuel - Ethanol	D9-11c	2I.02	24 x 24	30 x 30
RV Sanitary Station	D9-12	2I.02	24 x 24	30 x 30
Emergency Medical Services	D9-13	2I.02	24 x 24	30 x 30

Table 2I-1. General Service Sign and Plaque Sizes (Sheet 2 of 2)

Sign or Plaque	Sign Designation	Section	Conventional Road	Freeway or Expressway
Hospital (plaque)	D9-13aP	2I.02	24 x 12	30 x 12
Ambulance Station (plaque)	D9-13bP	2I.02	24 x 12	30 x 15
Emergency Medical Care (plaque)	D9-13cP	2I.02	24 x 18	30 x 24
Trauma Center (plaque)	D9-13dP	2I.02	24 x 12	30 x 15
Police	D9-14	2I.02	24 x 24	30 x 30
Propane Gas	D9-15	2I.02	24 x 24	30 x 30
Truck Parking	D9-16	2I.02	24 x 24	30 x 30
Next Services XX Miles (plaque)	D9-17P	2I.02	102 x 24	156 x 30
General Services (up to 6 symbols)	D9-18	2I.03	—	96 x 60
General Services	D9-18a	2I.03	—	96 x 60
General Services (up to 6 symbols) with Action or Exit Information	D9-18b	2I.03	108 x 84	132 x 114 (F) 132 x 108 (E)
General Services with Action or Exit Information	D9-18c	2I.03	72 x 60**	132 x 108** (F) 108 x 84** (E)
Pharmacy	D9-20	2I.02	24 x 24	30 x 30
24-Hour (plaque)	D9-20aP	2I.02	24 x 12	30 x 12
Telecommunication Device for the Deaf	D9-21	2I.05	24 x 24	30 x 30
Wireless Internet	D9-22	2I.05	24 x 24	30 x 30
Weather Information	D12-1	2I.09	84 x 48	132 x 84
Carpool Information	D12-2	2I.11	60 x 42	96 x 66
Channel 9 Monitored	D12-3	2I.09	84 x 48	132 x 84
Emergency Call 911	D12-4	2I.09	66 x 30	96 x 48
Travel Info Call 511 (pictograph)	D12-5	2I.10	42 x 60	66 x 78
Travel Info Call 511	D12-5a	2I.10	48 x 36	66 x 48

* The size shown is for a sign with a REST AREA and/or TOURIST INFO CENTER legend. The size should be appropriately adjusted if an alternate legend is used.
** The size shown is for a sign with four lines of services. The size should be appropriately adjusted depending on the amount of legend displayed.

Notes:
1. Larger signs may be used when appropriate
2. Dimensions in inches are shown as width x height
3. Where two sizes are shown, the larger size is for freeways (F) and the smaller size is for expressways (E)

Section 2I.02 General Service Signs for Conventional Roads

Support:

01 On conventional roads, commercial services such as gas, food, and lodging generally are within sight and are available to the road user at reasonably frequent intervals along the route. Consequently, on this class of road there usually is no need for special signs calling attention to these services. Moreover, General Service signing is usually not required in urban areas except for hospitals, law enforcement assistance, tourist information centers, and camping.

Option:

02 General Service signs (see Figure 2I-1) may be used where such services are infrequent and are found only on an intersecting highway or crossroad.

Standard:

03 **All General Service signs and supplemental sign panels shall have white letters, symbols, arrows, and borders on a blue background.**

Guidance:

04 *General Service signs should be installed at a suitable distance in advance of the turn-off point or intersecting highway.*

05 *States that elect to provide General Service signing should establish a statewide policy or warrant for its use, and criteria for the availability of services. Local jurisdictions electing to use such signing should follow State policy for the sake of uniformity.*

Option:

06 Individual States may sign for whatever alternative fuels are available at appropriate locations.

Figure 2I-1. General Service Signs and Plaques

D9-1 Telephone | D9-2 Hospital | D9-3 Camping | D9-3a Trailer Camping | D9-4 Litter Container | D9-6 Handicapped

D9-6P | D9-7 Gas | D9-8 Food | D9-9 Lodging | D9-10 Tourist Information | D9-11 Diesel Fuel

D9-11a Alternative Fuel- Compressed Natural Gas | D9-11b Electric Vehicle Charging | D9-11bP Electric Vehicle Charging | D9-11c Alternative Fuel- Ethanol | D9-12 RV Sanitary Station | D9-13 Emergency Medical Services

D9-13aP Hospital | D9-13bP Ambulance Station | D9-13cP Emergency Medical Care | D9-13dP Trauma Center | D9-14 Police | D9-15 Propane Gas

D9-16 Truck Parking | D9-20 Pharmacy / D9-20aP 24-Hour | D9-21 Telecommunication Device for the Deaf | D9-22 Wireless Internet

Advance Turn and Directional Arrow Auxiliary Signs for use with General Service Signs

M5-1 | M5-2 | M6-1 | M6-2 | M6-3

Example of directional assembly

Standard:

07 **General Service signs, if used at intersections, shall be accompanied by a directional message.**

Option:

08 The Advance Turn (M5 series) or Directional Arrow (M6 series) auxiliary signs with white arrows on blue backgrounds as shown in Figure 2I-1 may be used with General Service symbol signs to create a General Service Directional Assembly.

09 The General Service sign legends may be either symbols or word messages.

Standard:

10 **Symbols and word message General Service legends shall not be intermixed on the same sign. The Pharmacy (D9-20) sign shall only be used to indicate the availability of a pharmacy that is open, with a State-licensed pharmacist present and on duty, 24 hours per day, 7 days per week, and that is located within 3 miles of an interchange on the Federal-aid system. The D9-20 sign shall have a 24 HR (D9-20aP) plaque mounted below it.**

Support:

11 Formats for displaying different combinations of these services are described in Section 2I.03.

Option:

12 If the distance to the next point at which services are available is 10 miles or more, a NEXT SERVICES XX MILES (D9-17P) plaque (see Figure 2I-2) may be installed below the General Service sign.

Figure 2I-2. Example of Next Services Plaque

13 The International Symbol of Accessibility for the Handicapped (D9-6) sign may be used beneath General Service signs where paved ramps and rest room facilities accessible to, and usable by, the physically handicapped are provided.

Guidance:

14 *When the D9-6 sign is used in accordance with Paragraph 13, and van-accessible parking is available at the facility, a VAN ACCESSIBLE (D9-6P) plaque (see Figure 2I-1) should be mounted below the D9-6 sign.*

Option:

15 The Recreational Vehicle Sanitary Station (D9-12) sign may be used as needed to indicate the availability of facilities designed for the use of dumping wastes from recreational vehicle holding tanks.

16 The Litter Container (D9-4) sign may be placed in advance of roadside turnouts or rest areas, unless it distracts the driver's attention from other more important regulatory, warning, or directional signs.

17 The Emergency Medical Services (D9-13) symbol sign may be used to identify medical service facilities that have been included in the Emergency Medical Services system under a signing policy developed by the State and/or local highway agency.

Standard:

18 **The Emergency Medical Services symbol sign shall not be used to identify services other than qualified hospitals, ambulance stations, and qualified free-standing emergency medical treatment centers. If used, the Emergency Medical Services symbol sign shall be supplemented by a sign identifying the type of service provided.**

Option:

19 The Emergency Medical Services symbol sign may be used above the HOSPITAL (D9-13a) sign or Hospital (D9-2) symbol sign or above a sign with the legend AMBULANCE STATION (D9-13b), EMERGENCY MEDICAL CARE (D9-13c), or TRAUMA CENTER (D9-13d). The Emergency Medical Services symbol sign may also be used to supplement Telephone (D9-1), Channel 9 Monitored (D12-3), or POLICE (D9-14) signs.

Standard:

20 **The legend EMERGENCY MEDICAL CARE shall not be used for services other than qualified free-standing emergency medical treatment centers.**

Guidance:

21 *Each State should develop guidelines for the implementation of the Emergency Medical Services symbol sign.*

22 *The State should consider the following guidelines in the preparation of its policy:*

    A. AMBULANCE

        1. 24-hour service, 7 days per week.
        2. Staffed by two State-certified persons trained at least to the basic level.
        3. Vehicular communications with a hospital emergency department.
        4. Operator should have successfully completed an emergency-vehicle operator training course.

    B. HOSPITAL

        1. 24-hour service, 7 days per week.
        2. Emergency department facilities with a physician (or emergency care nurse on duty within the emergency department with a physician on call) trained in emergency medical procedures on duty.
        3. Licensed or approved for definitive medical care by an appropriate State authority.
        4. Equipped for radio voice communications with ambulances and other hospitals.

    C. Channel 9 Monitored

        1. Provided by either professional or volunteer monitors.
        2. Available 24 hours per day, 7 days per week.
        3. The service should be endorsed, sponsored, or controlled by an appropriate government authority to guarantee the level of monitoring.

Section 2I.03 General Service Signs for Freeways and Expressways

Support:

01   General Service (D9-18 series) signs (see Figure 2I-3) are generally not appropriate at major interchanges (see definition in Section 2E.32) and in urban areas.

Standard:

02   **General Service signs shall have white letters, symbols, arrows, and borders on a blue background. Letter and numeral sizes shall comply with the minimum requirements of Tables 2E-2 through 2E-5. All approved symbols shall be permitted as alternatives to word messages, but symbols and word service messages shall not be intermixed. If the services are not visible from the ramp of a single-exit interchange, the service signing shall be repeated in smaller size at the intersection of the exit ramp and the crossroad. Such service signs shall use arrows to indicate the direction to the services.**

Option:

03   For numbered interchanges, the exit number may be incorporated within the sign legend (D9-18b) or displayed on an Exit Number (E1-5P) plaque (see Section 2E.31).

Guidance:

04   *Distance to services should be displayed on General Service signs where distances are more than 1 mile.*

05   *General Service signing should only be provided at locations where the road user can return to the freeway or expressway and continue in the same direction of travel.*

06   *Only services that fulfill the needs of the road user should be displayed on General Service signs. If State or local agencies elect to provide General Service signing, there should be a statewide policy for such signing and criteria for the availability of the various types of services. The criteria should consider the following:*

    A. Gas, Diesel, LP Gas, EV Charging, and/or other alternative fuels if all of the following are available:

        1. Vehicle services such as gas, oil, and water;
        2. Modern sanitary facilities and drinking water;
        3. Continuous operations at least 16 hours per day, 7 days per week; and
        4. Public telephone.

    B. Food if all of the following are available:

        1. Licensing or approval, where required;
        2. Continuous operation to serve at least two meals per day, at least 6 days per week;
        3. Public telephone; and
        4. Modern sanitary facilities.

    C. Lodging if all of the following are available:

        1. Licensing or approval, where required;
        2. Adequate sleeping accommodations;
        3. Public telephone; and
        4. Modern sanitary facilities.

Figure 2I-3. Examples of General Service Signs with and without Exit Numbering

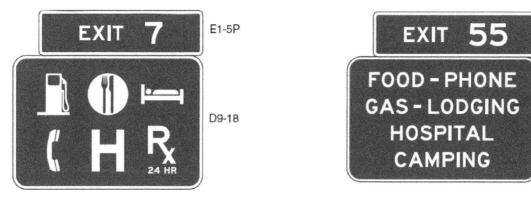

D. Public Telephone if continuous operation, 7 days per week is available.
E. Hospital if continuous emergency care capability, with a physician on duty 24 hours per day, 7 days per week is available. A physician on duty would include the following criteria and should be signed in accordance with the priority as follows:
  1. Physician on duty within the emergency department;
  2. Registered nurse on duty within the emergency department, with a physician in the hospital on call; or
  3. Registered nurse on duty within the emergency department, with a physician on call from office or home.
F. 24-Hour Pharmacy if a pharmacy is open, with a State-licensed pharmacist present and on duty, 24 hours per day, 7 days per week and is located within 3 miles of an interchange on the Federal-aid system.
G. Camping if all of the following are available:
  1. Licensing or approval, where required;
  2. Adequate parking accommodations; and
  3. Modern sanitary facilities and drinking water.

Standard:

07 **For any service that is operated on a seasonal basis only, the General Service signs shall be removed or covered during periods when the service is not available.**

08 **The General Service signs shall be mounted in an effective location, between the Advance Guide sign and the Exit Direction sign, in advance of the exit leading to the available services.**

Guidance:

09 *The General Service sign should contain the interchange number, if any, as shown in Figure 2I-3.*

Option:

10 If the distance to the next point where services are available is greater than 10 miles, a NEXT SERVICES XX MILES (D9-17P) plaque (see Figure 2I-2) may be installed below the Exit Direction sign.

Standard:

11 **Signs for services shall comply with the format for General Service signs (see Section 2I.02) and as provided in this Manual. No more than six general road user services shall be displayed on one sign, which includes any appended supplemental signs or plaques. General Service signs shall carry the legends for one or more of the following services: Food, Gas, Lodging, Camping, Phone, Hospital, 24-Hour Pharmacy, or Tourist Information.**

12 **The qualified services available shall be displayed at specific locations on the sign.**

13 **To provide flexibility for the future when the service might become available, the sign space normally reserved for a given service symbol or word shall be left blank when that service is not present.**

Guidance:

14 *The standard display of word messages should be FOOD and PHONE in that order on the top line, and GAS and LODGING on the second line. If used, HOSPITAL and CAMPING should be on separate lines (see Figure 2I-3).*

Option:

15 Signing for DIESEL, LP-Gas, or other alternative fuel services may be substituted for any of the general services or appended to such signs. The International Symbol of Accessibility for the Handicapped (D9-6) sign (see Figure 2I-1) may be used for facilities that qualify.

Guidance:

16 *When symbols are used for the road user services, they should be displayed as follows:*

  A. Six services:
    1. Top row—GAS, FOOD, and LODGING
    2. Bottom row—PHONE, HOSPITAL, and CAMPING
  B. Four services:
    1. Top row—GAS and FOOD
    2. Bottom row—LODGING and PHONE
  C. Three services:
    1. Top row—GAS, FOOD, and LODGING

Option:

17 Substitutions of other services for any of the services described in Paragraph 16 may be made by placing the substitution in the lower right (four or six services) or extreme right (three services) portion of the sign. An action message or an interchange number may be used for symbol signs in the same manner as they are used for word message signs. The Diesel Fuel (D9-11) symbol or the LP-Gas (D9-15) symbol may be substituted for the symbol representing fuel or appended to such assemblies. The Tourist Information (D9-10) symbol or the 24-Hour Pharmacy (D9-20 and D9-20aP) symbol may be substituted on any of the configurations provided in Paragraph 16.

18 At rural interchange areas where limited road user services are available and where it is unlikely that additional services will be provided within the near future, a supplemental plaque displaying one to three services (words or symbols) may be appended below a post-mounted interchange guide sign.

Standard:

19 **If more than three services become available at rural interchange areas where limited road user services were anticipated, the appended supplemental plaque described in Paragraph 18 shall be removed and replaced with an independently mounted General Service sign as described in this Section.**

Option:

20 A separate Telephone Service (D9-1) sign (see Figure 2I-1) may be installed if telephone facilities are located adjacent to the route at places where public telephones would not normally be expected.

Page 306   2009 Edition

21 The Recreational Vehicle Sanitary Station (D9-12) sign (see Figure 2I-1) may be used as needed to indicate the availability of facilities designed for dumping wastes from recreational vehicle holding tanks.

22 In some locations, signs may be used to indicate that services are not available.

23 A separate Truck Parking (D9-16) sign (see Figure 2I-1) may be mounted below the other general road user services to direct truck drivers to designated parking areas.

Section 2I.04 Interstate Oasis Signing

Support:

01 An Interstate Oasis is a facility near an Interstate highway that provides products and services to the public, 24-hour access to public restrooms, and parking for automobiles and heavy trucks. Interstate Oasis guide signs inform road users on Interstate highways as to the presence of an Interstate Oasis at an interchange and which businesses have been designated by the State within which they are traveling as having met the eligibility criteria of the Federal Highway Administration's Interstate Oasis policy. The FHWA's policy, which is dated October 18, 2006, and which can be viewed on the MUTCD website at http://mutcd.fhwa.dot.gov/res-policy.htm, provides a more detailed definition of an Interstate Oasis and specifies the eligibility criteria for an Interstate Oasis designation in compliance with the requirements of laws enacted by Congress.

Guidance:

02 *If a State elects to provide or allow Interstate Oasis signing (see Figure 2I-4), there should be a statewide policy, program, procedures, and criteria for the designation and signing of a facility as an Interstate Oasis that complies with FHWA's policy and with the provisions of this Section.*

03 *States electing to provide or allow Interstate Oasis signing should use the following signing practices on the freeway for any given exit to identify the availability of a designated Interstate Oasis:*

   A. *If adequate sign spacing allows, a separate Interstate Oasis (D5-12) sign should be installed in an effective location with spacing of at least 800 feet from other adjacent guide signs, including any Specific Service signs. This Interstate Oasis sign should be located upstream from the Advance Guide sign or between the Advance Guide sign and the Exit Direction sign for the exit leading to the Interstate Oasis. The Interstate Oasis sign should have a white legend with a letter height of at least 10 inches and a white border on a blue background and should contain the words INTERSTATE OASIS and the exit number or, for an unnumbered interchange, an action message such as NEXT RIGHT. The names or logos of the businesses designated as Interstate Oases should not be included on this sign.*

   B. *If the spacing of the other guide signs precludes the use of a separate sign as described in Item A, an INTERSTATE OASIS (D5-12P) supplemental plaque with a letter height of at least 10 inches and with a white legend and border on a blue background should be appended above or below an existing D9-18 series General Service sign for the interchange.*

04 *If a separate Interstate Oasis (D5-12) sign is installed, an Interstate Oasis sign panel should be incorporated into the design of the sign (see Figure 2I-4).*

Standard:

05 **The Interstate Oasis sign panel shall only be used on the separate Interstate Oasis sign where it is accompanied by the words INTERSTATE OASIS and shall not be used independently without the words.**

Figure 2I-4. Examples of Interstate Oasis Signs and Plaques

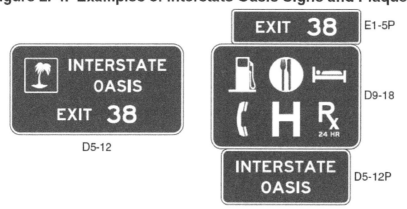

Option:
06 If Specific Service signing is provided at the interchange, a business designated as an Interstate Oasis and having a business logo sign panel on the Food and/or Gas Specific Service signs may use the bottom portion of the business logo sign panel to display the word OASIS.

Standard:
07 **If Specific Services signs containing the OASIS legend as a part of the business logo(s) are not used on the ramp and if the Interstate Oasis is not clearly visible and identifiable from the exit ramp, a sign with a white INTERSTATE OASIS legend with a letter height of at least 6 inches and a white border on a blue background shall be provided on the exit ramp to indicate the direction and distance to the Interstate Oasis.**

08 **If needed, additional trailblazer guide signs shall be used along the crossroad to guide road users to an Interstate Oasis.**

Section 2I.05 Rest Area and Other Roadside Area Signs

Standard:
01 **Rest Area signs (see Figure 2I-5) shall have a retroreflective white legend and border on a blue background.**

02 **Signs that include the legend REST AREA shall be used only where parking and restroom facilities are available.**

Guidance:
03 *A roadside area that does not contain restroom facilities should be signed to indicate the major road user service that is provided. For example, the sign legends for an area with only parking should use the words PARKING AREA instead of REST AREA. The sign legends for an area with only picnic tables and parking should use words such as PICNIC AREA, ROADSIDE TABLE, or ROADSIDE PARK instead of REST AREA.*

04 *Rest areas that have tourist information and welcome centers should be signed as discussed in Section 2I.08.*

05 *Scenic area signing should be consistent with that provided for rest areas, except that the legends should use words such as SCENIC AREA, SCENIC VIEW, or SCENIC OVERLOOK instead of REST AREA.*

06 *If a rest area or other roadside area is provided on a conventional road, a D5-1 and/or D5-1b sign should be installed in advance of the rest area or other roadside area to permit the driver to reduce speed in preparation for leaving the highway. A D5-5 sign (or a D5-2 sign if an exit ramp is provided) should be installed at the turnoff point where the driver needs to leave the highway to access the rest area or other roadside area.*

07 *If a rest area or other roadside area is provided on a freeway or expressway, a D5-1 sign should be placed 1 mile and/or 2 miles in advance of the rest area.*

Standard:
08 **A D5-2 sign shall be placed at the rest area or other roadside area exit gore.**

Figure 2I-5. Rest Area and Other Roadside Area Signs

D5-1   D5-1a   D5-2   D5-2a

D5-5   D5-6

NOTE: Alternate legends may be substituted for the REST AREA legend, such as PARKING AREA, PICNIC AREA, ROADSIDE TABLE, ROADSIDE PARK, SCENIC AREA, SCENIC VIEW, and SCENIC OVERLOOK.

Option:

09 A D5-1b sign may be placed between the D5-1 sign and the exit gore on a freeway or expressway. A second D5-1 sign may be used in place of the D5-1b sign with a distance to the nearest 1/2 or 1/4 mile displayed as a fraction rather than a decimal for distances of less than 1 mile.

10 To provide the road user with information on the location of succeeding rest areas, a NEXT REST AREA XX MILES (D5-6) sign (see Figure 2I-5) may be installed independently or as a supplemental sign mounted below one of the REST AREA advance guide signs.

Standard:

11 **All signs on freeways and expressways for rest and other roadside areas shall have letter and numeral sizes that comply with the minimum requirements of Tables 2E-2 through 2E-5. The sizes for General Service signs that have standardized designs shall be as shown in Table 2I-1.**

Option:

12 If the rest area has facilities for the physically impaired (see Section 2I.02), the International Symbol of Accessibility for the Handicapped (D9-6) sign (see Figure 2I-1) may be placed with or beneath the REST AREA advance guide sign.

13 If telecommunication devices for the deaf (TDD) are available at the rest area, the TDD (D9-21) symbol sign (see Figure 2I-1) may be used to supplement the advance guide signs for the rest area.

14 If wireless Internet services are available at the rest area, the Wi-Fi (D9-22) symbol sign (see Figure 2I-1) may be used to supplement the advance guide signs for the rest area.

Section 2I.06 Brake Check Area Signs (D5-13 and D5-14)

Guidance:

01 *If an area has been provided for drivers to check the brakes on their vehicle, a BRAKE CHECK AREA XX MILES (D5-13) sign (see Figure 2I-6) should be installed in advance of the brake check area, and a D5-14 sign (see Figure 2I-6) should be placed at the entrance to the brake check area.*

Section 2I.07 Chain-Up Area Signs (D5-15 and D5-16)

Guidance:

01 *If an area has been provided for drivers to pull off of the roadway to install chains on their tires, a CHAIN-UP AREA XX MILES (D5-15) sign (see Figure 2I-6) should be installed in advance of the chain-up area, and a D5-16 sign (see Figure 2I-6) should be placed at the entrance to the chain-up area.*

Section 2I.08 Tourist Information and Welcome Center Signs

Support:

01 Tourist information and welcome centers have been constructed within rest areas on freeways and expressways and are operated by either a State or a private organization. Others have been located within close proximity to these facilities and operated by civic clubs, chambers of commerce, or private enterprise.

Guidance:

02 *An excessive number of supplemental sign panels should not be installed with Tourist Information or Welcome Center signs so as not to overload the road user.*

Figure 2I-6. Brake Check Area and Chain-Up Area Signs

BRAKE CHECK AREA 1/2 MILE	BRAKE CHECK AREA ↗	CHAIN-UP AREA 1/2 MILE	CHAIN-UP AREA ↗
D5-13	D5-14	D5-15	D5-16

Standard:

⁰³ **Tourist Information or Welcome Center signs (see Figure 2I-7) shall have a white legend and border on a blue background. Continuously staffed or unstaffed operation at least 8 hours per day, 7 days per week, shall be required.**

⁰⁴ **If operated only on a seasonal basis, the Tourist Information or Welcome Center signs shall be removed or covered during the off seasons.**

Guidance:

⁰⁵ *For freeway or expressway rest area locations that also serve as tourist information or welcome centers, the following signing criteria should be used:*

   A. *The locations for tourist information and welcome center Advance Guide, Exit Direction, and Exit Gore signs should meet the General Service signing requirements described in Section 2I.03.*
   B. *If the signing for the tourist information or welcome center is to be accomplished in conjunction with the initial signing for the rest areas, the message on the Advance Guide (D5-7) sign should be REST AREA, TOURIST INFO CENTER, XX MILES or REST AREA, STATE NAME (optional), WELCOME CENTER XX MILES. On the Exit Direction (D5-8 or D5-11) sign the message should be REST AREA, TOURIST INFO CENTER with a diagonally upward-pointing directional arrow (or NEXT RIGHT), or REST AREA, STATE NAME (optional), WELCOME CENTER with a diagonally upward-pointing directional arrow (or NEXT RIGHT).*
   C. *If the initial rest area Advance Guide and Exit Direction signing is in place, these signs should include, on supplemental signs, the legend TOURIST INFO CENTER or STATE NAME (optional), WELCOME CENTER.*
   D. *The Exit Gore sign should contain only the legend REST AREA with the arrow and should not be supplemented with any legend pertaining to the tourist information center or welcome center.*

Option:

⁰⁶ An alternative to the supplemental TOURIST INFO CENTER legend is the Tourist Information (D9-10) sign (see Figure 2I-1), which may be appended beneath the REST AREA advance guide sign.

⁰⁷ The name of the State or local jurisdiction may appear on the Advance Guide and Exit Direction tourist information/welcome center signs if the jurisdiction controls the operation of the tourist information or welcome center and the center meets the operating criteria set forth in this Manual and is consistent with State policies.

Guidance:

⁰⁸ *For tourist information centers that are located off the freeway or expressway facility, additional signing criteria should be as follows:*

   A. *Each State should adopt a policy establishing the maximum distance that a tourist information center can be located from the interchange in order to be included on official signs.*
   B. *The location of signing should be in accordance with requirements pertaining to General Service signing (see Section 2I.03).*
   C. *Signing along the crossroad should be installed to guide the road user from the interchange to the tourist information center and back to the interchange.*

Option:

⁰⁹ As an alternative, the Tourist Information (D9-10) sign (see Figure 2I-1) may be appended to the guide signs for the exit that provides access to the tourist information center. As a second alternative, the Tourist Information sign may be combined with General Service signing.

Figure 2I-7. Examples of Tourist Information and Welcome Center Signs

D5-7

D5-8

D5-11

Note: Alternate legends may be substituted for the TOURIST INFO CENTER legend, such as WELCOME CENTER and (State Name) WELCOME CENTER.

Page 310   2009 Edition

Section 2I.09 Radio Information Signing

Option:

01   Radio-Weather Information (D12-1) signs (see Figure 2I-8) may be used in areas where difficult driving conditions commonly result from weather systems. Radio-Traffic Information signs may be used in conjunction with traffic management systems.

Standard:

02   Radio-Weather and Radio-Traffic Information signs shall have a white legend and border on a blue background. Only the numerical indication of the radio frequency shall be used to identify a station broadcasting travel-related weather or traffic information. No more than three frequencies shall be displayed on each sign. Only radio stations whose signal will be of value to the road user and who agree to broadcast either of the following two items shall be identified on Radio-Weather and Radio-Traffic Information signs:
  A. Periodic weather warnings at a rate of at least once every 15 minutes during periods of adverse weather; or
  B. Driving condition information (affecting the roadway being traveled) at a rate of at least once every 15 minutes, or when required, during periods of adverse traffic conditions, and when supplied by an official agency having jurisdiction.

Figure 2I-8. Radio, Telephone, and Carpool Information Signs

* The pictograph of the transportation agency or the travel information service or program may be used in place of the 511 pictograph (see Section 2I.08)

⁰³ **If a station to be considered operates only on a seasonal basis, its signs shall be removed or covered during the off season.**

Guidance:

⁰⁴ *The radio station should have a signal strength to adequately broadcast 70 miles along the route. Signs should be spaced as needed for each direction of travel at distances determined by an engineering study. The stations to be included on the signs should be selected in cooperation with the association(s) representing major broadcasting stations in the area to provide: (1) maximum coverage to all road users on both AM and FM frequencies; and (2) consideration of 24 hours per day, 7 days per week broadcast capability.*

Option:

⁰⁵ In roadway rest area locations, a smaller sign using a greater number of radio frequencies, but of the same general design, may be used.

Standard:

⁰⁶ **Radio-Weather and Radio-Traffic Information signs installed in rest areas shall be positioned such that they are not visible from the main roadway.**

Option:

⁰⁷ A Channel 9 Monitored (D12-3) sign (see Figure 2I-8) may be installed as needed. Official public agencies or their designees may be displayed as the monitoring agency on the sign.

Standard:

⁰⁸ **Only official public agencies or their designee shall be displayed as the monitoring agency on the Channel 9 Monitored sign.**

Option:

⁰⁹ An Emergency CALL XX (D12-4) sign (see Figure 2I-8), along with the appropriate number to call, may be used for cellular phone communications.

Section 2I.10 TRAVEL INFO CALL 511 Signs (D12-5 and D12-5a)

Option:

⁰¹ A TRAVEL INFO CALL 511 (D12-5) sign (see Figure 2I-8) may be installed if a 511 travel information services telephone number is available to road users for obtaining traffic, public transportation, weather, construction, or road condition information.

⁰² The pictograph of the transportation agency or the travel information service or program that is providing the travel information may be incorporated within the D12-5 sign either above or below the TRAVEL INFO CALL 511 legend.

Standard:

⁰³ **The logo of a commercial entity shall not be incorporated within the TRAVEL INFO CALL 511 sign.**

⁰⁴ **The TRAVEL INFO CALL 511 sign shall have a white legend and border on a blue background.**

Guidance:

⁰⁵ *If the pictograph of the transportation agency or the travel information service or program is used, the pictograph's maximum height should not exceed two times the letter height used in the legend of the sign.*

Section 2I.11 Carpool and Ridesharing Signing

Option:

⁰¹ In areas having carpool matching services, Carpool Information (D12-2) signs (see Figure 2I-8) may be provided adjacent to highways with preferential lanes or along any other highway.

⁰² Carpool Information signs may include an Internet domain name or telephone number of more than four characters within the legend.

Guidance:

⁰³ *Because this is an information sign related to road user services, the Carpool Information sign should have a white legend and border on a blue background.*

Standard:

⁰⁴ **If a local transit pictograph or carpool symbol is incorporated into the Carpool Information sign, the maximum vertical dimension of the logo or symbol shall not exceed 18 inches.**

CHAPTER 2J. SPECIFIC SERVICE SIGNS

Section 2J.01 Eligibility

Standard:

01 **Specific Service signs shall be defined as guide signs that provide road users with business identification and directional information for services and for eligible attractions. Eligible service categories shall be limited to gas, food, lodging, camping, attractions, and 24-hour pharmacies.**

Guidance:

02 *The use of Specific Service signs should be limited to areas primarily rural in character or to areas where adequate sign spacing can be maintained.*

Option:

03 Where an engineering study determines a need, Specific Service signs may be used on any class of highways.

Guidance:

04 *Specific Service signs should not be installed at an interchange where the road user cannot conveniently reenter the freeway or expressway and continue in the same direction of travel.*

Standard:

05 **Eligible service facilities shall comply with laws concerning the provisions of public accommodations without regard to race, religion, color, age, sex, or national origin, and laws concerning the licensing and approval of service facilities.**

06 **The attraction services shall include only facilities which have the primary purpose of providing amusement, historical, cultural, or leisure activities to the public.**

07 **Distances to eligible 24-hour pharmacies shall not exceed 3 miles in any direction of an interchange on the Federal-aid system.**

Guidance:

08 *Except as provided in Paragraph 9, distances to eligible services other than pharmacies should not exceed 3 miles in any direction.*

Option:

09 If, within the 3-mile limit, facilities for the services being considered other than pharmacies are not available or choose not to participate in the program, the limit of eligibility may be extended in 3-mile increments until one or more facilities for the services being considered chooses to participate, or until 15 miles is reached, whichever comes first.

Guidance:

10 *If State or local agencies elect to provide Specific Service signing, there should be a statewide policy for such signing and criteria for the availability of the various types of services. The criteria should consider the following:*

    A. *To qualify for a GAS logo sign panel, a business should have:*

        1. *Vehicle services including gas and/or alternative fuels, oil, and water;*
        2. *Continuous operation at least 16 hours per day, 7 days per week for freeways and expressways, and continuous operation at least 12 hours per day, 7 days per week for conventional roads;*
        3. *Modern sanitary facilities and drinking water; and*
        4. *Public telephone.*

    B. *To qualify for a FOOD logo sign panel, a business should have:*

        1. *Licensing or approval, where required;*
        2. *Continuous operations to serve at least two meals per day, at least 6 days per week;*
        3. *Modern sanitary facilities; and*
        4. *Public telephone.*

    C. *To qualify for a LODGING logo sign panel, a business should have:*

        1. *Licensing or approval, where required;*
        2. *Adequate sleeping accommodations;*
        3. *Modern sanitary facilities; and*
        4. *Public telephone.*

D. To qualify for a CAMPING logo sign panel, a business should have:
   1. Licensing or approval, where required;
   2. Adequate parking accommodations; and
   3. Modern sanitary facilities and drinking water.
E. To qualify for an ATTRACTION logo sign panel, a facility should have:
   1. Regional significance, in compliance with the provisions of Paragraph 6; and
   2. Adequate parking accommodations.

Standard:

11 **If State or local agencies elect to provide Specific Service signing for pharmacies, both of the following criteria shall be met for a pharmacy to qualify for signing:**
   A. **The pharmacy shall be continuously operated 24 hours per day, 7 days per week, and shall have a State-licensed pharmacist present and on duty at all times; and**
   B. **The pharmacy shall be located within 3 miles of an interchange on the Federal-aid system.**

Support:

12 Section 2I.04 contains information regarding the Interstate Oasis program.

Section 2J.02 Application

Standard:

01 **The number of Specific Service signs along an approach to an interchange or intersection, regardless of the number of service types displayed, shall be limited to a maximum of four. In the direction of traffic, successive Specific Service signs shall be for 24-hour pharmacy, attraction, camping, lodging, food, and gas services, in that order.**

02 **A Specific Service sign shall display the word message GAS, FOOD, LODGING, CAMPING, ATTRACTION, or 24-HOUR PHARMACY, an appropriate directional legend such as the word message EXIT XX, NEXT RIGHT, SECOND RIGHT, or directional arrows, and the related logo sign panels.**

03 **No more than three types of services shall be represented on any sign or sign assembly. If three types of services are displayed on one sign, then the logo sign panels shall be limited to two for each service type (for a total of six logo sign panels). If two types of services are displayed on one sign, then the logo sign panels shall be limited to either three for each service type (for a total of six logo sign panels) or four for one service type and two for the other service type (for a total of six logo sign panels). The legend and logo sign panels applicable to a service type shall be displayed such that the road user will not associate them with another service type on the same sign.**

04 **No service type shall appear on more than two signs (see Paragraph 6).**

05 **The signs shall have a blue background, a white border, and white legends of upper-case letters, numbers, and arrows.**

Guidance:

06 *Where a service type is displayed on two signs, the signs for that service should follow one another in succession.*

07 *The Specific Service signs should be located to take advantage of natural terrain, to have the least impact on the scenic environment, and to avoid visual conflict with other signs within the highway right-of-way.*

Option:

08 General Service signs (see Sections 2I.02 and 2I.03) may be used in conjunction with Specific Service signs for eligible types of services that are not represented by a Specific Service sign.

Support:

09 Examples of Specific Service signs are shown in Figure 2J-1. Examples of sign locations are shown in Figure 2J-2.

Section 2J.03 Logos and Logo Sign Panels

Standard:

01 **A logo shall be either an identification symbol/trademark or a word message. Each logo shall be placed on a separate logo sign panel that shall be attached to the Specific Service sign. Symbols or trademarks used alone for a logo shall be reproduced in the colors and general shape consistent with customary use, and any integral legend shall be in proportionate size. A logo that resembles an official traffic control device shall not be used.**

Figure 2J-1. Examples of Specific Service Signs

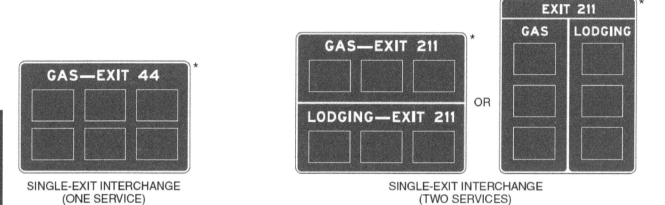

SINGLE-EXIT INTERCHANGE (ONE SERVICE)

SINGLE-EXIT INTERCHANGE (TWO SERVICES)

* See Section 2J.07 for option of displaying exit number on a separate plaque instead of on the sign

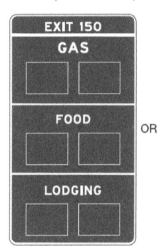

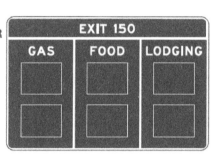

SINGLE-EXIT INTERCHANGE (THREE SERVICES)

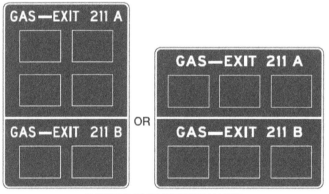

DOUBLE-EXIT INTERCHANGE

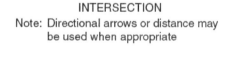

LOGO SIGN PANEL

INTERSECTION
Note: Directional arrows or distance may be used when appropriate

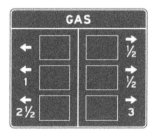 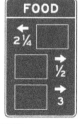 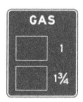

RAMP

Figure 2J-2. Examples of Specific Service Sign Locations

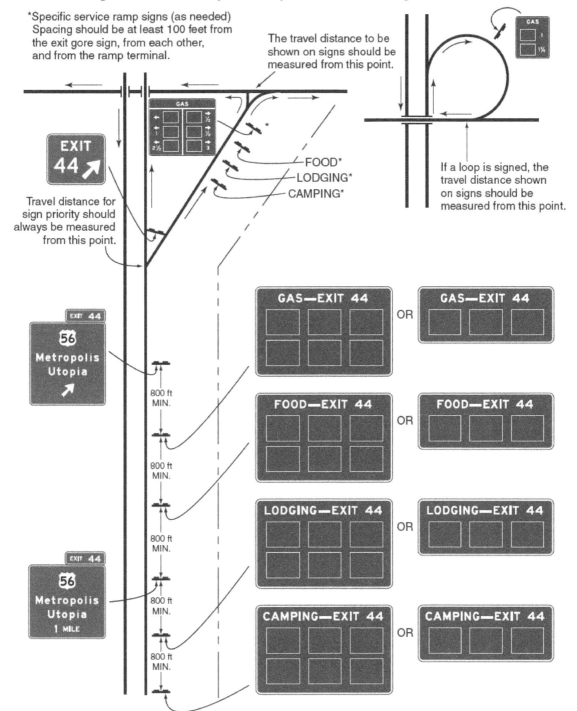

Guidance:

02 *A word message logo, not using a symbol or trademark, should have a blue background with white legend and border.*

Support:

03 Section 2J.05 contains information regarding the minimum letter heights for logo sign panels.

Option:

04 Where business identification symbols or trademarks are used alone for a logo, the border may be omitted from the logo sign panel.

05 A portion of a logo sign panel may be used to display a supplemental message horizontally along the bottom of the logo sign panel, provided that the message displays essential motorist information (see Figure 2J-3).

Standard:

06 **All supplemental messages shall be displayed within the logo sign panel and shall have letters and numerals that comply with the minimum height requirements shown in Table 2J-1.**

Guidance:

07 *A logo sign panel should not display more than one supplemental message.*

08 *The supplemental message should be displayed in a color to contrast effectively with the background of the business sign or separated from the other legend or logo by a divider bar.*

09 *State or local agencies that elect to allow supplemental messages on logo sign panels should develop a statewide policy for such messages.*

Support:

10 Typical supplemental messages might include DIESEL, 24 HOURS, CLOSED and the day of the week when the facility is closed, ALTERNATIVE FUELS (see Section 2I.03), and RV ACCESS.

Option:

11 The RV ACCESS supplemental message may be circular.

Standard:

12 **If the RV ACCESS supplemental message is circular, it shall be the abbreviation RV in black letters inside a yellow circle with a black border and it shall be displayed within the logo sign panel near the lower right-hand corner (see Figure 2J-4).**

Guidance:

13 *If the circular RV ACCESS supplemental message is used, the circle should have a diameter of 10 inches and the letters should have a height of 6 inches.*

14 *If a State or local agency elects to display the designation of businesses as providing on-premise accommodations for recreational vehicles with the RV ACCESS supplemental message or the RV Access circular message, there should be a statewide policy for such designation and criteria for qualifying businesses. The criteria should include such site conditions as access between the public roadway and the site, on-premise geometry, and parking.*

Option:

15 If a business designated as an Interstate Oasis (see Section 2I.04) has a business logo sign panel on the Food and/or Gas Specific Service signs, the word OASIS may be displayed on the bottom portion of the logo sign panel for that business.

Standard:

16 **A logo sign panel shall not display the symbol/trademark or name of more than one business.**

Figure 2J-3. Examples of Supplemental Messages on Logo Sign Panels

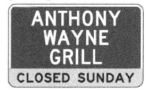

Table 2J-1. Minimum Letter and Numeral Sizes for Specific Service Signs According to Sign Type

Type of Sign	Freeway or Expressway	Conventional Road or Ramp
A. Specific Service Signs		
Service Categories	10	6
Exit Number Words	10	—
Exit Number Numerals and Letters	10	—
Action Message Words	10	6
Distance Numerals	—	6
Distance Fraction Numerals	—	4
B. Logo Sign Panels		
Logo Sign Panels	60 x 36	30 x 18
Words and Numerals (Non-Trademark/Graphic Logo)	8	4
Trademark/Graphic Logo	Proportional	Proportional
Supplemental Message Words and Numerals	5	2.5

Note: Sizes are shown in inches and where applicable are shown as width x height

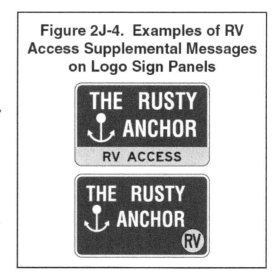

Figure 2J-4. Examples of RV Access Supplemental Messages on Logo Sign Panels

Section 2J.04 Number and Size of Signs and Logo Sign Panels

Guidance:

01 *Sign sizes should be determined by the amount and height of legend and the number and size of logo sign panels attached to the sign. All logo sign panels on a sign should be the same size.*

Standard:

02 **Each Specific Service sign or sign assembly shall be limited to no more than six logo sign panels.**

Option:

03 Where more than six businesses of a specific service type are eligible for logo sign panels at the same interchange, additional logo sign panels of that same specific service type may also be displayed in accordance with the provisions of Paragraph 4. The additional logo sign panels may be displayed either by placing more than one specific service type on the same sign (see Paragraph 3 of Section 2J.02) or by using a second Specific Service sign of that specific service type if the additional sign can be added without exceeding the limit of four Specific Service signs at an interchange or intersection approach (see Paragraph 6 of Section 2J.02).

Standard:

04 **Where logo sign panels for more than six businesses of a specific service type are displayed at the same interchange or intersection approach, the following provisions shall apply:**
   A. No more than 12 logo sign panels of a specific service type shall be displayed on no more than two Specific Service signs or sign assemblies;
   B. No more than six logo sign panels shall be displayed on a single Specific Service sign; and
   C. No more than four Specific Service signs shall be displayed on the approach.

Support:

05 Section 2J.08 contains information regarding Specific Service signs for double-exit interchanges.

Standard:

06 **Each logo sign panel attached to a Specific Service sign shall have a rectangular shape with a width longer than the height. A logo sign panel on signs for freeways and expressways shall not exceed 60 inches in width and 36 inches in height. A logo sign panel on signs for conventional roads and freeway and expressway ramps shall not exceed 30 inches in width and 18 inches in height. The vertical and horizontal spacing between logo sign panels shall not exceed 8 inches and 12 inches, respectively.**

Support:

07 Sections 2A.14, 2E.15, and 2E.16 contain information regarding borders, interline spacing, and edge spacing.

Section 2J.05 Size of Lettering

Standard:

01 **All Specific Service signs and logo sign panels shall have letter and numeral sizes that comply with the minimum requirements of Table 2J-1.**

Guidance:

02 *Any legend on a symbol/trademark should be proportional to the size of the symbol/trademark.*

Section 2J.06 Signs at Interchanges

Standard:

01 **The Specific Service signs shall be installed between the preceding interchange and at least 800 feet in advance of the Exit Direction sign at the interchange from which the services are available (see Figure 2J-2).**

Guidance:

02 *There should be at least an 800-foot spacing between the Specific Service signs, except for Specific Service ramp signs. However, excessive spacing is not desirable. Specific Service ramp signs should be spaced at least 100 feet from the Exit Gore sign, from each other, and from the ramp terminal.*

Section 2J.07 Single-Exit Interchanges

Standard:

01 **At numbered single-exit interchanges, the name of the service type followed by the exit number shall be displayed on one line above the logo sign panels. At unnumbered interchanges, the directional legend NEXT RIGHT (LEFT) shall be used.**

02 **At single-exit interchanges, Specific Service ramp signs shall be installed along the ramp or at the ramp terminal for facilities that have logo sign panels displayed along the main roadway if the facilities are not readily visible from the ramp terminal. Directions to the service facilities shall be indicated by arrows on the ramp signs. Logo sign panels on Specific Service ramp signs shall be duplicates of those displayed on the Specific Service signs located in advance of the interchange, but shall be reduced in size (see Paragraph 6 of Section 2J.04).**

Guidance:

03 *Specific Service ramp signs should include distances to the service facilities.*

Option:

04 An exit number plaque (see Section 2E.31) may be used instead of the exit number on the signs located in advance of an interchange.

Section 2J.08 Double-Exit Interchanges

Guidance:

01 *At double-exit interchanges, the Specific Service signs should consist of two sections, one for each exit (see Figure 2J-1).*

Standard:

02 **At a double-exit interchange, the top section shall display the logo sign panels for the first exit and the bottom section shall display the logo sign panels for the second exit. At numbered interchanges, the name of the service type and the exit number shall be displayed above the logo sign panels in each section. At unnumbered interchanges, the word message NEXT RIGHT (LEFT) and SECOND RIGHT (LEFT) shall be used in place of the exit number. The number of logo sign panels on the sign (total of both sections) or the sign assembly shall be limited to six.**

Guidance:

03 *At a double-exit interchange, where a service type is displayed on two Specific Service signs in accordance with the provisions of Section 2J.04, one of the signs should display the logo sign panels for that service type for the businesses that are accessible from one of the two exits and the other sign should display the logo sign panels for that service type for the businesses that are accessible from the other exit.*

Option:

04 At a double-exit interchange where there are four logo sign panels to be displayed for one of the exits and one or two logo sign panels to be displayed for the other exit, the logo sign panels may be arranged in three rows with two logo sign panels per row.

05 At a double-exit interchange, where a service is to be signed for only one exit, one section of the Specific Service sign may be omitted, or a single exit interchange sign may be used. Signs on ramps and crossroads as described in Section 2J.07 may be used at a double-exit interchange.

Section 2J.09 Specific Service Trailblazer Signs

Support:

01 Specific Service trailblazer signs (see Figure 2J-5) are guide signs with one to four logo sign panels that display business identification and directional information for services and for eligible attractions. Specific Service trailblazer signs are installed along crossroads for facilities that have logo sign panels displayed along the main roadway and ramp, and that require additional vehicle maneuvers.

Standard:

02 **Specific Service trailblazer signs shall be installed along crossroads where the route to the business requires a direction change, where it is questionable as to which roadway to follow, or where additional guidance is needed. Where it is not feasible or practical to install Specific Service trailblazer signs to such businesses, those businesses shall not be considered eligible for signing from the ramp and main roadway. A Specific Service trailblazer sign shall not be required at the point where the business is visible from the roadway and its access is readily apparent.**

Guidance:

03 *If used, a Specific Service trailblazer sign should be located a maximum of 500 feet in advance of any required turn.*

Standard:

04 **The location of other traffic control devices shall take precedence over the location of a Specific Service trailblazer sign.**

05 **When used, each Specific Service trailblazer sign or sign assembly shall be limited to no more than four logo sign panels. The logo sign panels on Specific Service trailblazer signs shall be duplicates of those displayed on the Specific Service ramp signs.**

06 **Appropriate legends, such as directional arrows or the word message NEXT RIGHT or SECOND RIGHT, shall be displayed with the logo sign panel to provide proper guidance. The directional legend and border shall be white and shall be displayed on a blue background.**

Figure 2J-5. Examples of Specific Service Trailblazer Signs

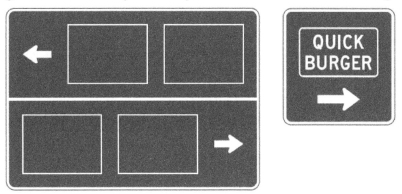

Option:
⁰⁷ Specific Service trailblazer signs may contain various types of services on a single sign or on a sign assembly.
⁰⁸ Specific Service trailblazer signs may be placed farther from the edge of the road than other traffic control signs.

Section 2J.10 Signs at Intersections
Standard:
⁰¹ **Where both tourist-oriented information (see Chapter 2K) and specific service information would be needed at the same intersection, the design of the tourist-oriented directional signs shall be used, and the needed specific service information shall be incorporated.**
Guidance:
⁰² *If Specific Service signs are used on conventional roads or at intersections on expressways, they should be installed between the previous interchange or intersection and at least 300 feet in advance of the intersection from which the services are available.*
⁰³ *The spacing between signs should be determined on the basis of an engineering study.*
⁰⁴ *Logo sign panels should not be displayed for a type of service for which a qualified facility is readily visible.*
Standard:
⁰⁵ **If Specific Service signs are used on conventional roads or at intersections on expressways, the name of each type of service shall be displayed above its logo sign panel(s), together with an appropriate legend, such as NEXT RIGHT (LEFT) or a directional arrow, either displayed on the same line as the name of the type of service or displayed below the logo sign panel(s).**
Option:
⁰⁶ Signs similar to Specific Service ramp signs as described in Section 2J.07 may be provided on the crossroad.

Section 2J.11 Signing Policy
Guidance:
⁰¹ *Each highway agency that elects to use Specific Service signs should establish a signing policy that includes, as a minimum, the guidelines of Section 2J.01 and at least the following criteria:*
  A. Selection of eligible businesses;
  B. Distances to eligible services;
  C. The use of logo sign panels, legends, and signs conforming with this Manual and State design requirements;
  D. Removal or covering of logo sign panels during off seasons for businesses that operate on a seasonal basis;
  E. The circumstances, if any, under which Specific Service signs are permitted to be used in non-rural areas; and
  F. Determination of the costs to businesses for initial permits, installations, annual maintenance, and removal of logo sign panels.

CHAPTER 2K. TOURIST-ORIENTED DIRECTIONAL SIGNS

Section 2K.01 Purpose and Application

Support:

01 Tourist-oriented directional signs are guide signs with one or more sign panels that display the business identification of and directional information for eligible business, service, and activity facilities.

Standard:

02 **A facility shall be eligible for tourist-oriented directional signs only if it derives its major portion of income or visitors during the normal business season from road users not residing in the area of the facility.**

Option:

03 Tourist-oriented directional signs may include businesses involved with seasonal agricultural products.

Standard:

04 **When used, tourist-oriented directional signs shall be used only on rural conventional roads and shall not be used on conventional roads in urban areas or at interchanges on freeways or expressways.**

05 **Where both tourist-oriented directional signs and Specific Service signs (see Chapter 2J) would be needed at the same intersection, the tourist-oriented directional signs shall incorporate the needed information from, and be used in place of, the Specific Service signs.**

Option:

06 Tourist-oriented directional signs may be used in conjunction with General Service signs (see Section 2I.02).

Support:

07 Section 2K.07 contains information on the adoption of a State policy for States that elect to use tourist-oriented directional signs.

Section 2K.02 Design

Standard:

01 **Tourist-oriented directional signs shall have one or more sign panels for the purpose of displaying the business identification of and directional information for eligible facilities. Each sign panel shall be rectangular in shape and shall have a white legend and border on a blue background.**

02 **The content of the legend on each sign panel shall be limited to the identification and directional information for no more than one eligible business, service, or activity facility. The legends shall not include promotional advertising.**

Guidance:

03 *Each sign panel should have a maximum of two lines of legend including no more than one symbol, a separate directional arrow, and the distance to the facility displayed beneath the arrow. Arrows pointing to the left or up should be at the extreme left of the sign panel. Arrows pointing to the right should be at the extreme right of the sign panel. Symbols, when used, should be to the left of the word legend or logo sign panel (see Paragraph 7).*

Option:

04 The General Service sign symbols (see Section 2I.02) and the symbols for recreational and cultural interest area signs (see Chapter 2M) may be used.

05 Logo sign panels (see Section 2J.03) for specific businesses, services, and activities may also be used. Based on engineering judgment, the hours of operation may be displayed on the sign panels.

Standard:

06 **When used, symbols and logo sign panels shall be an appropriate size (see Section 2K.04). Logos resembling official traffic control devices shall not be permitted.**

Option:

07 The tourist-oriented directional sign may display the word message TOURIST ACTIVITIES at the top of the sign.

Standard:

08 **The TOURIST ACTIVITIES word message shall have a white legend in all upper-case letters and a white border on a blue background. If used, it shall be placed above and in addition to the directional sign panels.**

Support:

09 Examples of tourist-oriented directional signs are shown in Figures 2K-1 and 2K-2.

Figure 2K-1. Examples of Tourist-Oriented Directional Signs

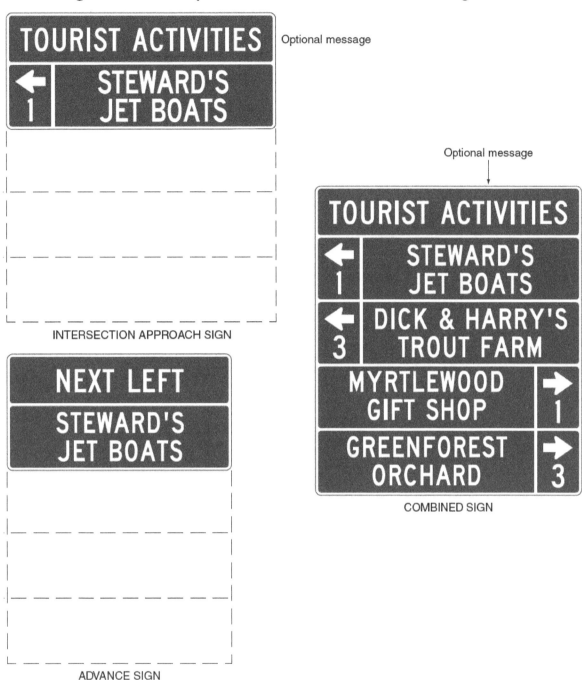

Figure 2K-2. Examples of Intersection Approach Signs and Advance Signs for Tourist-Oriented Directional Signs

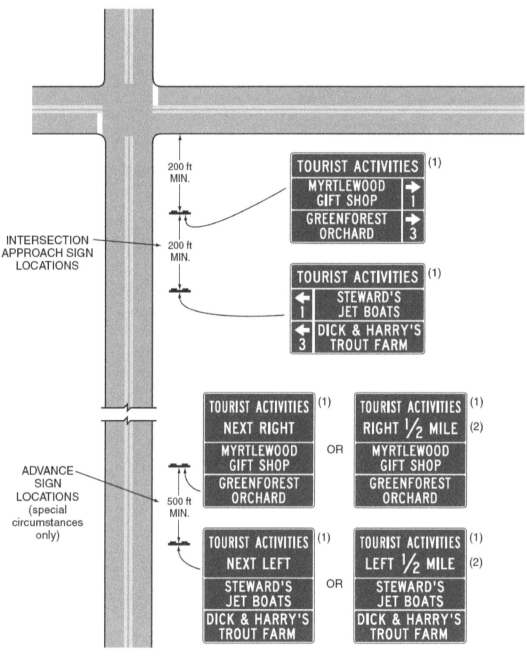

(1) Optional message
(2) Use if there is an intervening intersection

Section 2K.03 Style and Size of Lettering
Guidance:

01 *All letters and numbers on tourist-oriented directional signs, except on the logo sign panels, should be upper-case and at least 6 inches in height. Any legend on a logo should be proportional to the size of the logo.*

Standard:

02 **Design standards for letters, numerals, and spacing shall be as provided in the "Standard Highway Signs and Markings" book (see Section 1A.11).**

Section 2K.04 Arrangement and Size of Signs
Standard:

01 **The size of a tourist-oriented directional sign shall be limited to a maximum height of 6 feet. Additional height shall be allowed to accommodate the addition of the optional TOURIST ACTIVITIES message provided in Section 2K.02 and the action messages provided in Section 2K.05.**

Guidance:

02 *The number of intersection approach signs (one sign for tourist-oriented destinations to the left, one for destinations to the right, and one for destinations straight ahead) installed in advance of an intersection should not exceed three. The number of sign panels installed on each sign should not exceed four. The sign panels for right-turn, left-turn, and straight-through destinations should be on separate signs. The left-turn destination sign should be located farthest from the intersection, then the right-turn destination sign, with the straight-through destination sign located closest to the intersection (see Figure 2K-2). Signs for facilities in the straight-through direction should be considered only when there are signs for destinations in either the left or right direction.*

03 *If it has been determined to be appropriate to combine the left-turn and right-turn destination sign panels on a single sign, the left-turn destination sign panels should be above the right-turn destination sign panels (see Figure 2K-1). When there are multiple destinations in the same direction, they should be in order based on their distance from the intersection. Except as provided in Paragraph 5, a straight-through sign panel should not be combined with a sign displaying left- and/or right-turn destinations.*

04 *The sign panels should not exceed the size necessary to accommodate two lines of legend without crowding. Symbols and logo sign panels on a directional sign panel should not exceed the height of two lines of a word legend. All directional sign panels and other parts of the sign should be the same width, which should not exceed 6 feet.*

Option:

05 At intersection approaches where three or fewer facilities are displayed, the left-turn, right-turn, and straight-through destination sign panels may be combined on the same sign.

Section 2K.05 Advance Signs
Guidance:

01 *Advance signs should be limited to those situations where sight distance, intersection vehicle maneuvers, or other vehicle operating characteristics require advance notification of the destinations and their directions.*

02 *The design of the advance sign should be identical to the design of the intersection approach sign. However, the directional arrows and distances to the destinations should be omitted and the action messages NEXT RIGHT, NEXT LEFT, or AHEAD should be placed on the sign above the business identification sign panels. The action messages should have the same letter height as the other word messages on the directional sign panels (see Figures 2K-1 and 2K-2).*

Standard:

03 **The action message sign panels shall have a white legend in all upper-case letters and a white border on a blue background.**

Option:

04 The legend RIGHT 1/2 MILE or LEFT 1/2 MILE may be used on advance signs when there are intervening minor roads.

05 The height required to add the directional word messages recommended for the advance sign may be added to the maximum sign height of 6 feet.

Guidance:

06 *The optional TOURIST ACTIVITIES message, when used on an advance sign, and the action message should be combined on a single sign panel with TOURIST ACTIVITIES as the top line and the action message as the bottom line (see Figure 2K-2).*

Section 2K.06 Sign Locations

Guidance:

01 *If used, the intersection approach signs should be located at least 200 feet in advance of the intersection. Signs should be spaced at least 200 feet apart and at least 200 feet from other traffic control devices.*

02 *If used, advance signs should be located approximately 1/2 mile from the intersection with 500 feet between these signs. In the direction of travel, the order of advance sign placement should be to show the destinations to the left first, then destinations to the right, and last, the destinations straight ahead.*

03 *Position, height, and lateral offset of signs should be governed by Chapter 2A except as permitted in this Section.*

Option:

04 Tourist-oriented directional signs may be placed farther from the edge of the road than other traffic control signs.

Standard:

05 **The location of other traffic control devices shall take precedence over the location of tourist-oriented directional signs.**

Section 2K.07 State Policy

Standard:

01 **To be eligible for tourist-oriented directional signing, facilities shall comply with applicable State and Federal laws concerning the provisions of public accommodations without regard to race, religion, color, age, sex, or national origin, and with laws concerning the licensing and approval of service facilities. Each State that elects to use tourist-oriented directional signs shall adopt a policy that complies with these provisions.**

Guidance:

02 *The State policy should include:*

   A. A definition of tourist-oriented business, service, and activity facilities.
   B. Eligibility criteria for signs for facilities.
   C. Provision for incorporating Specific Service signs into the tourist-oriented directional signs as required by Paragraph 5 of Section 2K.01.
   D. Provision for covering signs during off seasons for facilities operated on a seasonal basis.
   E. Provisions for signs to facilities that are not located on the crossroad when such facilities are eligible for signs.
   F. A definition of the immediate area in compliance with the provisions of Paragraph 2 of Section 2K.01.
   G. Maximum distances to eligible facilities. The maximum distance should be 5 miles.
   H. Provision for information centers (plazas) when the number of eligible sign applicants exceeds the maximum permissible number of sign panel installations.
   I. Provision for limiting the number of signs when there are more applicants than the maximum number of signs permitted.
   J. Criteria for use at intersections on expressways.
   K. Provisions for controlling or excluding those businesses which have illegal signs as defined by the Highway Beautification Act of 1965 (23 U.S.C. 131).
   L. Provisions for States to charge fees to cover the cost of signs through a permit system.
   M. A definition of the conditions under which the time of operation is displayed.
   N. Provisions for determining if advance signs will be permitted, and the circumstances under which they will be installed.

CHAPTER 2L. CHANGEABLE MESSAGE SIGNS

Section 2L.01 Description of Changeable Message Signs

Support:

01 A changeable message sign (CMS) is a traffic control device that is capable of displaying one or more alternative messages. Some changeable message signs have a blank mode when no message is displayed, while others display multiple messages with only one of the messages displayed at a time (such as OPEN/CLOSED signs at weigh stations).

02 The provisions in this Chapter apply to both permanent and portable changeable message signs with electronic displays. Additional provisions that only apply to portable changeable message signs can be found in Section 6F.60. The provisions in this Chapter do not apply to changeable message signs with non-electronic displays that are changed either manually or electromechanically, such as a hinged-panel, rotating-drum, or back-lit curtain or scroll CMS.

Standard:

03 **Except as provided in Paragraph 2 of Section 2L.02, changeable message signs shall display only traffic operational, regulatory, warning, and guidance information. Advertising messages shall not be displayed on changeable message signs or its supports or other equipment.**

04 **The design of legends for non-electronic display changeable message signs shall comply with the provisions of Chapters 2A through 2K, 2M, and 2N of this Manual. All other changeable message signs shall comply with the design and application principles established in this Chapter and in Chapter 2A.**

Guidance:

05 *Blank-out signs that display only single-phase, predetermined electronic-display legends that are limited by their composition and arrangement of pixels or other illuminated forms in a fixed arrangement (such as a blank-out sign indicating a part-time turn prohibition, a blank-out or changeable lane-use sign, or a changeable OPEN/CLOSED sign for a weigh station) should comply with the provisions of the applicable Section for the specific type of sign, provided that the letter forms, symbols, and other legend elements are duplicates of the static messages as detailed in the "Standard Highway Signs and Markings" book (see Section 1A.11). Because such a sign is effectively an illuminated version of a static sign, the size of its legend elements, the overall size of the sign, and placement of the sign should comply with the applicable provisions for the static version of the sign.*

Section 2L.02 Applications of Changeable Message Signs

Support:

01 Changeable message signs have a large number of applications including, but not limited to, the following:
   A. Incident management and route diversion
   B. Warning of adverse weather conditions
   C. Special event applications associated with traffic control or conditions
   D. Control at crossing situations
   E. Lane, ramp, and roadway control
   F. Priced or other types of managed lanes
   G. Travel times
   H. Warning situations
   I. Traffic regulations
   J. Speed control
   K. Destination guidance

Option:

02 Changeable message signs may be used by State and local highway agencies to display safety messages, transportation-related messages, emergency homeland security messages, and America's Missing: Broadcast Emergency Response (AMBER) alert messages.

Guidance:

03 *State and local highway agencies should develop and establish a policy regarding the display of the types of messages provided in Paragraph 2. When changeable message signs are used at multiple locations to address a specific situation, the message displays should be consistent along the roadway corridor and adjacent corridors, which might necessitate coordination among different operating agencies.*

Support:

04 Examples of safety messages include "SEAT BELT BUCKLED?" and "DON'T DRINK AND DRIVE." Examples of transportation-related messages include "STADIUM EVENT SUNDAY, EXPECT DELAYS NOON TO 4 PM" and "OZONE ALERT CODE RED—USE TRANSIT."

Guidance:

05 *When a CMS is used to display a safety or transportation related message, the message should be simple, brief, legible, and clear. A CMS should not be used to display a safety or transportation-related message if doing so would adversely affect respect for the sign. "CONGESTION AHEAD" or other overly simplistic or vague messages should not be displayed alone. These messages should be supplemented with a message on the location or distance to the congestion or incident, delay and travel time, alternative route, or other similar messages.*

Standard:

06 **When a CMS is used to display a safety, transportation-related, emergency homeland security, or AMBER alert message, the display format shall not be of a type that could be considered similar to advertising displays.**

Support:

07 Section 2B.13 contains information regarding the design of changeable message signs that are used to display variable speed limits that change based on ambient or operational conditions, or that display the speed at which approaching drivers are traveling.

Section 2L.03 Legibility and Visibility of Changeable Message Signs

Support:

01 The maximum distance at which a driver can first correctly identify letters and words on a sign is called the legibility distance of the sign. Legibility distance is affected by the characteristics of the sign design and the visual capabilities of drivers. Visual capabilities, and thus legibility distances, vary among drivers.

02 For the more common types of changeable message signs, the longest measured legibility distances on sunny days occur during mid-day when the sun is overhead. Legibility distances are much shorter when the sun is behind the sign face, when the sun is on the horizon and shining on the sign face, or at night.

03 Visibility is the characteristic that enables a CMS to be seen. Visibility is associated with the point where the CMS is first detected, whereas legibility is the point where the message on the CMS can be read. Environmental conditions such as rain, fog, and snow impact the visibility of changeable message signs and can reduce the available legibility distances. During these conditions, there might not be enough viewing time for drivers to read the message.

Guidance:

04 *Changeable message signs used on roadways with speed limits of 55 mph or higher should be visible from 1/2 mile under both day and night conditions. The message should be designed to be legible from a minimum distance of 600 feet for nighttime conditions and 800 feet for normal daylight conditions. When environmental conditions that reduce visibility and legibility are present, or when the legibility distances stated in the previous sentences in this paragraph cannot be practically achieved, messages composed of fewer units of information should be used and consideration should be given to limiting the message to a single phase (see Section 2L.05 for information regarding the lengths of messages displayed on changeable message signs).*

Section 2L.04 Design Characteristics of Changeable Message Signs

Standard:

01 **Changeable message signs shall not include advertising, animation, rapid flashing, dissolving, exploding, scrolling, or other dynamic elements.**

Support:

02 Section 6F.61 contains information regarding the use of arrow boards that use flashing or sequential displays for lane closures.

Guidance:

03 *Except in the case of a limited-legend CMS (such as a blank-out or electronic-display changeable message regulatory sign) that is used in place of a static regulatory sign or an activated blank-out warning sign that supplements a static warning sign at a separate location, changeable message signs should be used as a supplement to and not as a substitute for conventional signs and markings.*

04 *CMS should be limited to no more than three lines, with no more than 20 characters per line.*

05 *The spacing between characters in a word should be between 25 to 40 percent of the letter height. The spacing between words in a message should be between 75 and 100 percent of the letter height. Spacing between the message lines should be between 50 and 75 percent of the letter height.*

06 *Except as provided in Paragraph 18, word messages on changeable message signs should be composed of all upper-case letters. The minimum letter height should be 18 inches for changeable message signs on roadways with speed limits of 45 mph or higher. The minimum letter height should be 12 inches for changeable message signs on roadways with speed limits of less than 45 mph.*

Support:

07 Using letter heights of more than 18 inches will not result in proportional increases in legibility distance.

Guidance:

08 *The width-to-height ratio of the sign characters should be between 0.7 and 1.0. The stroke width-to-height ratio should be 0.2.*

Support:

09 The width-to-height ratio is commonly accomplished using a minimum font matrix density of five pixels wide by seven pixels high.

Standard:

10 **Changeable message signs shall automatically adjust their brightness under varying light conditions to maintain legibility.**

Guidance:

11 *The luminance of changeable message signs should meet industry criteria for daytime and nighttime conditions. Luminance contrast should be between 8 and 12 for all conditions.*

12 *Contrast orientation of changeable message signs should always be positive, that is, with luminous characters on a dark or less luminous background.*

Support:

13 Legibility distances for negative-contrast changeable message signs are likely to be at least 25 percent shorter than those of positive-contrast messages. In addition, the increased light emitted by negative-contrast changeable message signs has not been shown to improve detection distances.

Standard:

14 **The colors used for the legends and backgrounds on changeable message signs shall be as provided in Table 2A-5.**

Guidance:

15 *If a black background is used, the color used for the legend on a changeable message sign should match the background color that would be used on a standard sign for that type of legend, such as white for regulatory, yellow for warning, orange for temporary traffic control, red for stop or yield, fluorescent pink for incident management, and fluorescent yellow-green for bicycle, pedestrian, and school warning.*

Standard:

16 **If a green background is used for a guide message on a CMS or if a blue background is used for a motorist services message on a CMS, the background color shall be provided by green or blue lighted pixels such that the entire CMS would be lighted, not just the white legend.**

Support:

17 Some CMS that employ newer technologies have the capability to display an exact duplicate of a standard sign or other sign legend using standard symbols, the Standard Alphabets and letter forms, route shields, and other typical sign legend elements with no apparent loss of resolution or recognition to the road user when compared with a static version of the same sign legend. Such signs are of the full-matrix type and can typically display full-color legends. Use of such technologies for new CMS is encouraged for greater legibility of their displays and enhanced recognition of the message as it pertains to regulatory, warning, or guidance information.

Guidance:

18 *If used, the CMS described in the preceding paragraph should not display symbols or route shields unless they can do so in the appropriate color combinations. For a single-phase message where the Standard Alphabets and other legend elements of standard designs are used, the lettering style, size, and line spacing should comply with the applicable provisions for the type of message displayed as provided elsewhere in this Manual. For two-phase messages, larger legend heights should be used as described previously in this Section because of the need for such messages to be legible at a greater distance. Regardless of the number of phases, the CMS should comply with the legibility and visibility provisions of Section 2L.03.*

Section 2L.05 Message Length and Units of Information

Guidance:

01 *The maximum length of a message should be dictated by the number of units of information contained in the message, in addition to the size of the CMS. A unit of information, which is a single answer to a single question that a driver can use to make a decision, should not be more than four words.*

Support:

02 In order to illustrate the concept of units of information, Table 2L-1 shows an example message that is comprised of four units of information.

03 The maximum allowable number of units of information in a CMS message is based on the principles described in this Section, the current highway operating speed, the legibility characteristics of the CMS, and the lighting conditions.

Standard:

04 **Each message shall consist of no more than two phases. A phase shall consist of no more than three lines of text. Each phase shall be understood by itself regardless of the sequence in which it is read. Messages shall be centered within each line of legend. Except for signs located on toll plaza structures or other facilities with a similar booth-lane arrangement, if more than one CMS is visible to road users, then only one sign shall display a sequential message at any given time.**

05 **Techniques of message display such as fading, rapid flashing, exploding, dissolving, or moving messages shall not be used. The text of the message shall not scroll or travel horizontally or vertically across the face of the sign.**

Guidance:

06 *When designing and displaying messages on changeable message signs, the following principles relative to message design should be used:*

  A. *The minimum time that an individual phase is displayed should be based on 1 second per word or 2 seconds per unit of information, whichever produces a lesser value. The display time for a phase should never be less than 2 seconds.*
  B. *The maximum cycle time of a two-phase message should be 8 seconds.*
  C. *The duration between the display of two phases should not exceed 0.3 seconds.*
  D. *No more than three units of information should be displayed on a phase of a message.*
  E. *No more than four units of information should be in a message when the traffic operating speeds are 35 mph or more.*
  F. *No more than five units of information should be in a message when the traffic operating speeds are less than 35 mph.*
  G. *Only one unit of information should appear on each line of the CMS.*
  H. *Compatible units of information should be displayed on the same message phase.*

Table 2L-1. Example of Units of Information

Question	Answer	Number of Information Units
What happened?	MAJOR CRASH	1
Where?	AT EXIT 12	1
Who is the advisory for?	Drivers Heading TO NEW YORK	1
What is advised?	USE ROUTE 46	1

Note: The following is an example of a two-phase message that could be developed from the four information units shown in this table:

```
MAJOR CRASH
AT EXIT 12
```
Phase 1

```
USE ROUTE 46
TO NEW YORK
```
Phase 2

Option:
07 A unit of information consisting of more than one word may be displayed on more than one line. An additional changeable message sign at a downstream location may be used for the purpose of allowing the entire message to be read twice.

Guidance:
08 *If more than two phases would be needed to display the necessary information, additional changeable message signs should be used to display this information as a series of two distinct, independent messages with a maximum of two phases at each location, in accordance with the provisions of Paragraph 4.*

09 *When the message on a CMS includes an abbreviation, the provisions of Section 1A.15 should be used.*

Section 2L.06 Installation of Permanent Changeable Message Signs

Guidance:
01 *A CMS that is used in place of a static sign (such as a blank-out or variable legend regulatory sign) should be located in accordance with the provisions of Chapter 2A. The following factors should be considered when installing other permanent changeable message signs:*
   A. *Changeable message signs should be located sufficiently upstream of known bottlenecks and high crash locations to enable road users to select an alternate route or take other appropriate action in response to a recurring condition.*
   B. *Changeable message signs should be located sufficiently upstream of major diversion decision points, such as interchanges, to provide adequate distance over which road users can change lanes to reach one destination or the other.*
   C. *Changeable message signs should not be located within an interchange except for toll plazas or managed lanes.*
   D. *Changeable message signs should not be positioned at locations where the information load on drivers is already high because of guide signs and other types of information.*
   E. *Changeable message signs should not be located in areas where drivers frequently perform lane-changing maneuvers in response to static guide sign information, or because of merging or weaving conditions.*

Support:
02 Information regarding the design and application of portable changeable message signs in temporary traffic control zones is contained in Section 6F.60.

CHAPTER 2M. RECREATIONAL AND CULTURAL INTEREST AREA SIGNS

Section 2M.01 Scope

Support:
01 Recreational or cultural interest areas are attractions or traffic generators that are open to the general public for the purpose of play, amusement, or relaxation. Recreational attractions include such facilities as parks, campgrounds, gaming facilities, and ski areas, while examples of cultural attractions include museums, art galleries, and historical buildings or sites.
02 The purpose of recreation and cultural interest area signs is to guide road users to a general area and then to specific facilities or activities within the area.

Option:
03 Recreational and cultural interest area guide signs directing road users to significant traffic generators may be used on freeways and expressways where there is direct access to these areas as provided in Section 2M.09.
04 Recreational and cultural interest area signs may be used off the road network, as appropriate.

Section 2M.02 Application of Recreational and Cultural Interest Area Signs

Support:
01 Provisions for signing recreational or cultural interest areas are subdivided into two different types of signs: (1) symbol signs and (2) destination guide signs.

Guidance:
02 *When highway agencies decide to provide recreational and cultural interest area signing, these agencies should have a policy for such signing. The policy should establish signing criteria for the eligibility of the various types of services, accommodations, and facilities. These signs should not be used where they might be confused with other traffic control signs.*

Option:
03 Recreational and cultural interest area guide signs may be used on any road to direct persons to facilities, structures, and places, and to identify various services available to the general public. These guide signs may also be used in recreational or cultural interest areas for signing non-vehicular events and amenities such as trails, structures, and facilities.

Support:
04 Section 2A.12 contains information regarding the use of recreational and cultural interest area symbols on other types of signs.

Section 2M.03 Regulatory and Warning Signs

Standard:
01 **All regulatory and warning signs installed on public roads and streets within recreational and cultural interest areas shall comply with the requirements of Chapters 2A, 2B, 2C, 7B, 8B, and 9B.**

Section 2M.04 General Design Requirements for Recreational and Cultural Interest Area Symbol Guide Signs

Standard:
01 **Recreational and cultural interest area symbol guide signs shall be square or rectangular in shape and shall have a white symbol or message and white border on a brown background. The symbols shall be grouped into the following usage and series categories:**
   A. **General Applications,**
   B. **Accommodations,**
   C. **Services,**
   D. **Land Recreation,**
   E. **Water Recreation, and**
   F. **Winter Recreation.**

Support:
02 Table 2M-1 contains a listing of the symbols within each series category. Drawings showing the design details for these symbols are found in the "Standard Highway Signs and Markings" book (see Section 1A.11).

Option:
03 Mirror images of symbols may be used where the reverse image will better convey the message.

Table 2M-1. Category Chart for Recreational and Cultural Interest Area Symbols

General	
Bear Viewing Area	RS-012
Bus Stop	RS-031
Campfires *	RS-042
Cans or Bottles *	RS-101
Cultural Interest Area	RS-142
Dam	RS-009
Deer Viewing Area	RS-011
Falling Rocks *	RS-008
Fire Extinguisher *	RS-090
Lighthouse	RS-007
Lookout Tower	RS-006
Nature Study Area	RS-141
Pets on Leash *	RS-017
Pick-Up Trucks	RS-140
Point of Interest	RS-080
Radios *	RS-103
Rattlesnakes *	RS-099
Recycling *	RS-200
Sea Plane	RS-115
Smoking *	RS-002
Snack Bar *	RS-102
Stay on Trail *	RS-123
Strollers *	RS-111
Tunnel	RS-005
Viewing Area	RS-036
Walk on Boardwalk *	RS-122
Wood Gathering *	RS-120

Accommodations	
Baby Changing Station (Men's Room)	RS-137
Baby Changing Station (Women's Room)	RS-138
Men's Restroom	RS-021
Parking	RS-034
Recreational Vehicle Site	RS-104
Restrooms	RS-022
Sleeping Shelter *	RS-037
Trailer Site	RS-040
Walk-In Camp	RS-148
Women's Restroom	RS-023

Services	
Drinking Water	RS-013
Electrical Hook-Up	RS-150
Firewood Cutting *	RS-112
First Aid	RS-024
Grocery Store	RS-020
Kennel	RS-045
Laundromat	RS-085
Litter Receptacle	RS-086
Lockers/Storage *	RS-030
Mechanic	RS-027
Picnic Shelter	RS-039
Picnic Site	RS-044
Post Office	RS-026
Ranger Station	RS-015
Sanitary Station	RS-041
Showers *	RS-035
Stable	RS-073
Theater	RS-109
Trail Shelter *	RS-043
Tramway	RS-071
Trash Dumpster	RS-091

Land Recreation	
All-Terrain Trail	RS-095
Amphitheater	RS-070
Archery	RS-116
Baseball *	RS-096
Climbing *	RS-082
Corral	RS-149
Driving Tour	RS-113
Exercise/Fitness	RS-097
Golfing *	RS-128
Hang Gliding	RS-126
Hiking Trail	RS-068
Horse Trail	RS-064
In-Line Skating	RS-125
Interpretive Trail	RS-114
Off-Road Vehicle Trail	RS-067
Rock Collecting *	RS-083
Skateboarding *	RS-098
Spelunking/Caves	RS-084
Technical Rock Climbing	RS-081
Tennis	RS-129
Wildlife Viewing	RS-076

Water Recreation	
Beach	RS-145
Boat Motor	RS-147
Boat Ramp	RS-054
Canoeing	RS-079
Diving	RS-062
Fish Cleaning *	RS-093
Fish Hatchery	RS-010
Fish Ladder *	RS-089
Fishing Area	RS-063
Fishing Pier	RS-119
Hand Launch/Small Boat Launch	RS-117
Jet Ski/Personal Watercraft	RS-121
Kayaking	RS-118
Lifejackets *	RS-094
Marina	RS-053
Motorboating	RS-055
Rafting	RS-146
Rowboating	RS-057
Sailing	RS-056
Scuba Diving	RS-060
Seal Viewing	RS-106
Surfing	RS-059
Swimming	RS-061
Tour Boat	RS-087
Wading	RS-088
Waterskiing	RS-058
Whale Viewing	RS-107
Wind Surfing	RS-108

Winter Recreation	
Chair Lift/Ski Lift	RS-105
Cross Country Skiing	RS-046
Dog Sledding	RS-143
Downhill Skiing	RS-047
Ice Fishing	RS-092
Ice Skating	RS-050
Ski Jumping	RS-048
Sledding	RS-049
Snow Tubing	RS-144
Snowboarding	RS-127
Snowmobiling	RS-052
Snowshoeing	RS-078
Winter Recreational Area	RS-077

* For non-road use only

Section 2M.05 Symbol Sign Sizes

Guidance:

01 *Recreational and cultural interest area symbol signs should be 24 x 24 inches. Where greater visibility or emphasis is needed, larger sizes should be used. Symbol sign enlargements should be in 6-inch increments.*

02 *Recreational and cultural interest area symbol signs should be 30 x 30 inches when used on guide signs on freeways or expressways.*

Option:

03 A smaller size of 18 x 18 inches may be used on low-speed, low-volume roadways and on non-road applications.

Section 2M.06 Use of Educational Plaques

Guidance:

01 *Educational plaques should accompany all initial installations of recreational and cultural interest area symbol signs. The educational plaque should remain in place for at least 3 years after the initial installation. If used, the educational plaque should be the same width as the symbol sign.*

Option:

02 Symbol signs that are readily recognizable by the public may be installed without educational plaques.

Support:

03 Figure 2M-1 illustrates some examples of the use of educational plaques.

Section 2M.07 Use of Prohibitive Circle and Diagonal Slash for Non-Road Applications

Standard:

01 **Where it is necessary to indicate a prohibition of an activity or an item within a recreational or cultural interest area for non-road use and a standard regulatory sign for such a prohibition is not provided in Chapter 2B, the appropriate recreational and cultural interest area symbol shall be used in combination with a red prohibitive circle and red diagonal slash. The recreational and cultural interest area symbol and the sign border shall be black and the sign background shall be white. The symbol shall be scaled proportionally to fit completely within the circle and the diagonal slash shall be oriented from the upper left to the lower right portions of the circle as shown in Figure 2M-1.**

02 **Requirements for retroreflection of the red circle and red diagonal slash shall be the same as those requirements for backgrounds, legends, symbols, arrows, and borders.**

Section 2M.08 Placement of Recreational and Cultural Interest Area Symbol Signs

Standard:

01 **If used, recreational and cultural interest area symbol signs shall be placed in accordance with the general requirements contained in Chapter 2A. The symbol(s) shall be placed as sign panels in the uppermost part of the sign and the directional information shall be placed below the symbol(s).**

02 **Except as provided in Paragraph 3, if the name of the recreational or cultural interest area facility or activity is displayed on a destination guide sign (see Section 2M.09) and a symbol is used, the symbol shall be placed below the name (see Figure 2M-2).**

Option:

03 When the legend Wildlife Viewing Area is displayed with the RS-076 symbol on a destination guide sign, the symbol may be placed to the left or right of the legend and the arrow may be placed below the symbol (see Figure 2M-2).

04 The symbols displayed with the facility or activity name may be placed below the destination guide sign as illustrated in Figure 2M-2 instead of as sign panels placed with the destination guide sign.

05 Secondary symbols of a smaller size (18 x 18 inches) may be placed beneath the primary symbols (see Drawing A in Figure 2M-1), where needed.

Standard:

06 **Recreational and cultural interest area symbols installed for non-road use shall be placed in accordance with the general sign position requirements of the authority having jurisdiction.**

Support:

07 Figure 2M-3 illustrates typical height and lateral mounting positions. Figure 2M-4 illustrates some examples of the placement of symbol signs within a recreational or cultural interest area. Figures 2M-5 through 2M-10 illustrate some of the symbols that can be used.

Figure 2M-1. Examples of Use of Arrows, Educational Plaques, and Prohibitive Slashes

A - DIRECTIONAL SIGNS

B - DIRECTIONAL ASSEMBLIES

C - DIRECTIONAL ASSEMBLY WITH EDUCATIONAL PLAQUE

D - PROHIBITED ACTIVITIES AND EDUCATIONAL PLAQUE FOR NON-ROAD USE*

* Standard regulatory signs shall be used where provided elsewhere in this Manual

Guidance:

08 *The number of symbols used in a single sign assembly should not exceed four.*

Option:

09 The Advance Turn (M5 series) or Directional Arrow (M6 series) auxiliary signs with white arrows on brown backgrounds shown in Figure 2D-5 may be used with Recreational and Cultural Area Interest symbol guide signs to create a Recreational and Cultural Interest Area Directional Assembly. The symbols may be used singularly, or in groups of two, three, or four on a single sign assembly (see Figures 2M-1, 2M-3, and 2M-4).

Section 2M.09 Destination Guide Signs

Guidance:

01 *When recreational or cultural interest area destinations are displayed on supplemental guide signs, the sign should be rectangular or trapezoidal in shape. The order of preference for use of shapes and colors should be as follows: (1) rectangular with a white legend and border on a green background; (2) rectangular with a white legend and border on a brown background; or (3) trapezoidal with a white legend and border on a brown background.*

Standard:

02 **Whenever the trapezoidal shape is used, the color combination shall be a white legend and border on a brown background.**

Figure 2M-2. Examples of Recreational and Cultural Interest Area Guide Signs

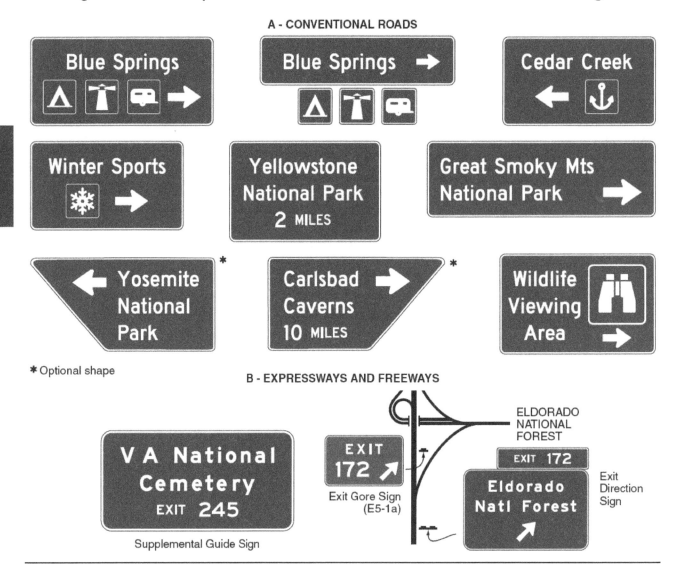

*Optional shape

Option:

03 Destination guide signs with a white legend and border on a brown background may be posted at the first point where an access or crossroad intersects a highway where recreational or cultural interest areas are a significant destination along conventional roads, expressways, or freeways. Supplemental guide signs with a white legend and border on a brown background may be used along conventional roads, expressways, or freeways to direct road users to recreational or cultural interest areas. Where access or crossroads lead exclusively to the recreational or cultural interest area, the advance guide sign and the exit direction sign may have a white legend and border on a brown background.

Standard:

04 **All Exit Gore (E5-1 and E5-1a) signs (see Section 2E.37) shall have a white legend and border on a green background. The background color of the interchange Exit Number (E1-5P and E1-5bP) plaque (see Section 2E.31) shall match the background color of the guide sign. Design characteristics of conventional road, expressway, or freeway guide signs shall comply with Chapter 2D or 2E except as provided in this Section for color combination.**

05 **The advance guide sign and the Exit Direction sign shall retain the white-on-green color combination where the crossroad leads to a destination other than a recreational or cultural interest area.**

Support:

06 Figure 2M-2 illustrates destination guide signs commonly used for identifying recreational or cultural interest areas or facilities.

Figure 2M-3. Arrangement, Height, and Lateral Position of Signs Located Within Recreational and Cultural Interest Areas

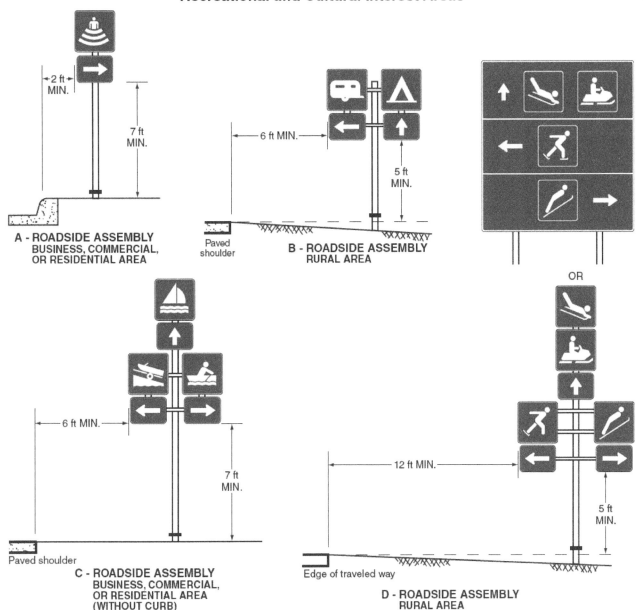

Figure 2M-4. Examples of Symbol and Destination Guide Signing Layout

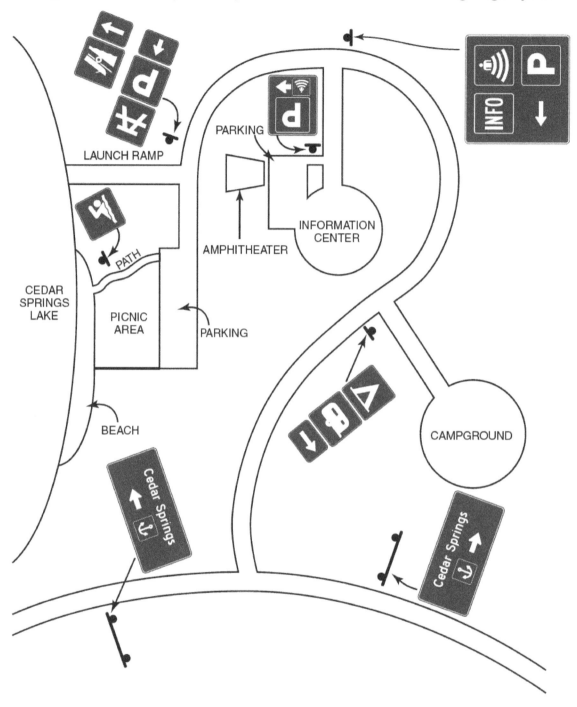

Figure 2M-5. Recreational and Cultural Interest Area Symbol Signs for General Applications

RS-002
Smoking

RS-005
Tunnel

RS-006
Lookout Tower

RS-007
Lighthouse

RS-008
Falling Rocks

RS-009
Dam

RS-011
Deer Viewing Area

RS-012
Bear Viewing Area

RS-017
Pets on Leash

RS-031
Bus Stop

RS-036
Viewing Area

RS-042
Campfires

RS-080
Point of Interest

RS-090
Fire Extinguisher

RS-099
Rattlesnakes

RS-101
Cans or Bottles

RS-102
Snack Bar

RS-103
Radios

RS-111
Strollers

RS-115
Sea Plane

RS-120
Wood Gathering

RS-122
Walk on Boardwalk

RS-123
Stay on Trail

RS-140
Pick-up Trucks

RS-141
Nature Study Area

RS-142
Cultural Interest Area

RS-200
Recycling

Figure 2M-6. Recreational and Cultural Interest Area Symbol Signs for Accommodations

RS-021
Men's Restroom

RS-022
Restrooms

RS-023
Women's Restroom

RS-034
Parking

RS-037
Sleeping Shelter

RS-040
Trailer Site

RS-104
Recreational Vehicle Site

RS-137
Baby Changing Station (Men's Room)

RS-138
Baby Changing Station (Women's Room)

RS-148
Walk-In Camp

Figure 2M-7. Recreational and Cultural Interest Area Symbol Signs for Services

RS-013
Drinking Water

RS-015
Ranger Station

RS-020
Grocery Store

RS-024
First Aid

RS-026
Post Office

RS-027
Mechanic

RS-030
Lockers/Storage

RS-035
Showers

RS-039
Picnic Shelter

RS-041
Sanitary Station

RS-043
Trail Shelter

RS-044
Picnic Site

RS-045
Kennel

RS-071
Tramway

RS-073
Stable

RS-085
Laundromat

RS-086
Litter Receptacle

RS-091
Trash Dumpster

RS-109
Theater

RS-112
Firewood Cutting

RS-114
Radiator Water

RS-150
Electrical Hook-Up

Figure 2M-8. Recreational and Cultural Interest Area Symbol Signs for Land Recreation

RS-064 Horse Trail | RS-067 Off-Road Vehicle Trail | RS-068 Hiking Trail | RS-070 Amphitheater | RS-076 Wildlife Viewing | RS-081 Technical Rock Climbing

RS-082 Climbing | RS-083 Rock Collecting | RS-084 Spelunking/Caves | RS-095 All-Terrain Trail | RS-096 Baseball | RS-097 Exercise/Fitness

RS-098 Skateboarding | RS-113 Driving Tour | RS-114 Interpretive Trail | RS-116 Archery | RS-125 In-Line Skating

RS-126 Hang Gliding | RS-128 Golfing | RS-129 Tennis | RS-149 Corral

Section 2M.10 Memorial or Dedication Signing

Support:

01 Legislative bodies will occasionally adopt an act or resolution memorializing or dedicating a highway, bridge, or other component of the highway.

Guidance:

02 *Such memorial or dedication names should not appear on or along a highway, or be placed on bridges or other highway components. If a route, bridge, or highway component is officially designated as a memorial or dedication, and if notification of the memorial or dedication is to be made on the highway right-of-way, such notification should consist of installing a memorial or dedication marker in a rest area, scenic overlook, recreational area, or other appropriate location where parking is provided with the signing inconspicuously located relative to vehicle operations along the highway.*

Option:

03 If the installation of a memorial or dedication marker off the main roadway is not practical, memorial or dedication signs may be installed on the mainline.

Guidance:

04 *Memorial or dedication signs should have a white legend and border on a brown background.*

Figure 2M-9. Recreational and Cultural Interest Area Symbol Signs for Water Recreation

RS-010
Fish Hatchery

RS-053
Marina

RS-054
Boat Ramp

RS-055
Motorboating

RS-056
Sailing

RS-057
Rowboating

RS-058
Waterskiing

RS-059
Surfing

RS-060
Scuba Diving

RS-061
Swimming

RS-062
Diving

RS-063
Fishing Area

RS-079
Canoeing

RS-087
Tour Boat

RS-088
Wading

RS-089
Fish Ladder

RS-093
Fish Cleaning

RS-094
Lifejackets

RS-106
Seal Viewing

RS-107
Whale Viewing

RS-108
Wind Surfing

RS-117
Hand Launch/
Small Boat Launch

RS-118
Kayaking

RS-119
Fishing Pier

RS-121
Jet Ski/Personal
Watercraft

RS-145
Beach

RS-146
Rafting

RS-147
Boat Motor

Figure 2M-10. Recreational and Cultural Interest Area Symbol Signs for Winter Recreation

RS-046 Cross Country Skiing | RS-047 Downhill Skiing | RS-048 Ski Jumping | RS-049 Sledding | RS-050 Ice Skating | RS-052 Snowmobiling

RS-077 Winter Recreational Area | RS-078 Snowshoeing | RS-092 Ice Fishing | RS-105 Chair Lift/Ski Lift | RS-127 Snowboarding

RS-143 Dog Sledding | RS-144 Snow Tubing

Standard:

05 Where such memorial or dedication signs are installed on the mainline, (1) memorial or dedication names shall not appear on directional guide signs, (2) memorial or dedication signs shall not interfere with the placement of any other necessary signing, and (3) memorial or dedication signs shall not compromise the safety or efficiency of traffic flow. The memorial or dedication signing shall be limited to one sign at an appropriate location in each route direction, each as an independent sign installation.

06 Memorial or dedication signs shall be rectangular in shape. The legend displayed on memorial or dedication signs shall be limited to the name of the person or entity being recognized and a simple message preceding or following the name, such as "Dedicated to" or "Memorial Parkway." Additional legend, such as biographical information, shall not be displayed on memorial or dedication signs. Decorative or graphical elements, pictographs, logos, or symbols shall not be displayed on memorial or dedication signs. All letters and numerals displayed on memorial or dedication signs shall be as provided in the "Standard Highway Signs and Markings" book (see Section 1A.11). The route number or officially mapped name of the highway shall not be displayed on the memorial or dedication sign.

07 Memorial or dedication names shall not appear on supplemental signs or on any other information sign on or along the highway or its intersecting routes.

Option:

08 The lettering for the name of the person or entity being recognized may be composed of a combination of lower-case letters with initial upper-case letters.

Guidance:

09 *Freeways and expressways should not be signed as memorial or dedicated highways.*

Support:

10 Named highways are officially designated and shown on official maps and serve the purpose of providing route guidance, primarily on unnumbered highways. A highway designated as a memorial or dedication is not considered to be a named highway. Section 2D.53 contains provisions for the signing of named highways.

CHAPTER 2N. EMERGENCY MANAGEMENT SIGNING

Section 2N.01 Emergency Management

Guidance:

01 *Contingency planning for an emergency evacuation should be considered by all State and local jurisdictions and should consider the use of all applicable roadways.*

02 *In the event of a disaster where highways that cannot be used will be closed, a successful contingency plan should account for the following elements: a controlled operation of certain designated highways, the establishment of traffic operations for the expediting of essential traffic, and the provision of emergency centers for civilian aid.*

Section 2N.02 Design of Emergency Management Signs

Standard:

01 **Emergency Management signs shall be used to guide and control highway traffic during an emergency.**

02 **Emergency Management signs shall not permanently displace any of the standard signs that are normally applicable.**

03 **Advance planning for transportation operations' emergencies shall be the responsibility of State and local authorities. The Federal Government shall provide guidance to the States as necessitated by changing circumstances.**

04 **Except as provided in Section 2A.11, the sizes for Emergency Management signs shall be as shown in Table 2N-1.**

Support:

05 Section 2A.11 contains information regarding the applicability of the various columns in Table 2N-1.

Option:

06 Signs larger than those shown in Table 2N-1 may be used (see Section 2A.11).

Guidance:

07 *As conditions permit, the Emergency Management signs should be replaced or augmented by standard signs.*

08 *The background of Emergency Management signs should be retroreflective.*

09 *Because Emergency Management signs might be needed in large numbers for temporary use during an emergency, consideration should be given to their fabrication from any light and economical material that can serve through the emergency period.*

Option:

10 Any Emergency Management sign that is used to mark an area that is contaminated by biological or chemical warfare agents or radioactive fallout may be accompanied by the standard symbol that is illustrated in the upper left corner of the EM-7c and EM-7d signs in Figure 2N-1.

Section 2N.03 Evacuation Route Signs (EM-1 and EM-1a)

Standard:

01 **The Evacuation Route (EM-1 and EM-1a) signs shall display a blue circular symbol on a white square sign without a border as shown in Figure 2N-1. The EM-1 sign shall include a white directional arrow (except as provided in Paragraph 3) and a white legend EVACUATION ROUTE within the blue circular symbol. The EM-1a sign shall include a white EVACUATION ROUTE legend and the tsunami symbol within the blue circular symbol. The EM-1 and EM-1a signs shall be retroreflective.**

02 **An Advance Turn Arrow (M5 series) or Directional Arrow (M6 series) auxiliary sign as shown in Figure 2D-5, but with a white arrow on a blue background instead of a black arrow on a white background, shall be installed below the EM-1a sign.**

Option:

03 Instead of including a directional arrow within the blue circular symbol on the EM-1 sign, an Advance Turn Arrow (M5 series) or Directional Arrow (M6 series) auxiliary sign as shown in Figure 2D-5, but with a white arrow on a blue background instead of a black arrow on a white background, may be installed below the EM-1 sign.

04 If desired, the word HURRICANE, or a word that describes some other type of evacuation route, may be added as a third line of text above the white EVACUATION ROUTE legend within the blue circular symbol on the EM-1 sign.

Table 2N-1. Emergency Management Sign Sizes

Sign or Plaque	Sign Designation	Section	Minimum Size
Evacuation Route	EM-1, EM-1a	2N.03	24 x 24*
Area Closed	EM-2	2N.04	30 x 24
Traffic Control Point	EM-3	2N.05	30 x 24
Maintain Top Safe Speed	EM-4	2N.06	24 x 30
Permit Required	EM-5	2N.07	24 x 30
Emergency Aid Center	EM-6a to EM-6d	2N.08	30 x 24
Shelter Directional	EM-7a to EM-7d	2N.09	30 x 24

* A minimum size of 18 x 18 may be used on low-volume roadways or roadways with speeds of 25 mph or less

Notes: 1. Larger signs may be used when appropriate
2. Dimensions in inches are shown as width x height

Figure 2N-1. Emergency Management Signs

* HURRICANE is an example of one type of evacuation route. Legends for other types may also be used, or this line of text may be omitted.

⁰⁵ An approved Emergency Management symbol with a diameter of 3.5 inches may appear near the bottom of an Evacuation Route sign.

Standard:

⁰⁶ The arrow designs, if used, on the EM-1 sign shall include a straight, vertical arrow pointing upward, a straight horizontal arrow pointing to the left or right, or a bent arrow pointing to the left or right for advance warning of a turn.

⁰⁷ If used, the Evacuation Route sign, with the appropriate arrow, shall be installed 150 to 300 feet in advance of, and at, any turn in an approved evacuation route. The sign shall also be installed elsewhere for straight-ahead confirmation where needed.

⁰⁸ If used in urban areas, the Evacuation Route sign shall be mounted at the right-hand side of the roadway, not less than 7 feet above the top of the curb, and at least 1 foot back from the face of the curb. If used in rural areas, the Evacuation Route sign shall be mounted at the right-hand side of the roadway, not less than 7 feet above the pavement and not less than 6 feet or more than 10 feet to the right of the right-hand roadway edge.

⁰⁹ Evacuation Route signs shall not be placed where they will conflict with other signs. Where conflict in placement would occur between the Evacuation Route sign and a standard regulatory sign, the regulatory sign shall take precedence.

Option:

¹⁰ In case of conflict with guide or warning signs, the Evacuation Route sign may take precedence.

Guidance:

¹¹ *Placement of Evacuation Route signs should be made under the supervision of the officials having jurisdiction over the placement of other traffic signs. Coordination with Emergency Management authorities and agreement between contiguous political entities should occur to assure continuity of routes.*

Section 2N.04 AREA CLOSED Sign (EM-2)

Standard:

⁰¹ The AREA CLOSED (EM-2) sign (see Figure 2N-1) shall be used to close a roadway in order to prohibit traffic from entering the area. It shall be installed on the shoulder as near as practical to the right-hand edge of the roadway, or preferably, on a portable mounting or barricade partly or entirely in the roadway.

Guidance:

⁰² *For best visibility, particularly at night, the sign height should not exceed 4 feet measured vertically from the pavement to the bottom of the sign. Unless adequate advance warning signs are used, it should not be placed to create a complete and unavoidable blocked route. Where feasible, the sign should be located at an intersection that provides a detour route.*

Section 2N.05 TRAFFIC CONTROL POINT Sign (EM-3)

Standard:

⁰¹ The TRAFFIC CONTROL POINT (EM-3) sign (see Figure 2N-1) shall be used to designate a location where an official traffic control point has been set up to impose such controls as are necessary to limit congestion, expedite emergency traffic, exclude unauthorized vehicles, or protect the public.

⁰² The sign shall be installed in the same manner as the AREA CLOSED sign (see Section 2N.04), and at the point where traffic must stop to be checked.

⁰³ The standard STOP (R1-1) sign shall be used in conjunction with the TRAFFIC CONTROL POINT sign. The TRAFFIC CONTROL POINT sign shall consist of a black legend and border on a retroreflectorized white background.

Guidance:

⁰⁴ *The TRAFFIC CONTROL POINT sign should be mounted directly below the STOP sign.*

Section 2N.06 MAINTAIN TOP SAFE SPEED Sign (EM-4)

Option:

⁰¹ The MAINTAIN TOP SAFE SPEED (EM-4) sign (see Figure 2N-1) may be used on highways where conditions are such that it is prudent to evacuate or traverse an area as quickly as possible.

⁰² Where an existing Speed Limit (R2-1) sign is in a suitable location, the MAINTAIN TOP SAFE SPEED sign may conveniently be mounted directly over the face of the speed limit sign that it supersedes.

Support:

01 Since any speed zoning would be impractical under such emergency conditions, no minimum speed limit can be prescribed by the MAINTAIN TOP SAFE SPEED sign in numerical terms. Where traffic is supervised by a traffic control point, official instructions will usually be given verbally, and the sign will serve as an occasional reminder of the urgent need for maintaining the proper speed.

Guidance:

04 *The sign should be installed as needed, in the same manner as other standard speed signs.*

Standard:

05 **If used in rural areas, the MAINTAIN TOP SAFE SPEED sign shall be mounted on the right-hand side of the road at a horizontal distance of not less than 6 feet or more than 10 feet from the roadway edge, and at a minimum height, measured vertically from the bottom of the sign to the elevation of the near edge of the traveled way, of 5 feet. If used in urban areas, the minimum height, measured vertically from the bottom of the sign to the top of the curb, or in the absence of curb, measured vertically from the bottom of the sign to the elevation of the near edge of the traveled way, shall be 7 feet, and the nearest edge of the sign shall be not less than 1 foot back from the face of the curb.**

Section 2N.07 ROAD (AREA) USE PERMIT REQUIRED FOR THRU TRAFFIC Sign (EM-5)

Support:

01 The intent of the ROAD (AREA) USE PERMIT REQUIRED FOR THRU TRAFFIC (EM-5) sign (see Figure 2N-1) is to notify road users of the presence of the traffic control point so that those who do not have priority permits issued by designated authorities can take another route, or turn back, without making a needless trip and without adding to the screening load at the post. Local traffic, without permits, can proceed as far as the traffic control post.

Standard:

02 **If used, the ROAD (AREA) USE PERMIT REQUIRED FOR THRU TRAFFIC (EM-5) sign shall be used at an intersection that is an entrance to a route on which a traffic control point is located.**

03 **If used, the sign shall be installed in a manner similar to that of the MAINTAIN TOP SAFE SPEED sign (see Section 2N.06).**

Section 2N.08 Emergency Aid Center Signs (EM-6 Series)

Standard:

01 **In the event of emergency, State and local authorities shall establish various centers for civilian relief, communication, medical service, and similar purposes. To guide the public to such centers a series of directional signs shall be used.**

02 **Emergency Aid Center (EM-6 series) signs (see Figure 2N-1) shall carry the designation of the center and an arrow indicating the direction to the center. They shall be installed as needed, at intersections and elsewhere, on the right-hand side of the roadway, in urban areas at a minimum height, measured vertically from the bottom of the sign to the top of the curb, or in the absence of curb, measured vertically from the bottom of the sign to the elevation of the near edge of the traveled way, of 7 feet, and not less than 1 foot back from the face of the curb, and in rural areas at a minimum height, measured vertically from the bottom of the sign to the elevation of the near edge of the traveled way, of 5 feet, and at a horizontal distance of not less than 6 feet or more than 10 feet from the roadway edge.**

03 **Emergency Aid Center signs shall carry one of the following legends, as appropriate, or others designating similar emergency facilities:**
   A. **MEDICAL CENTER (EM-6a),**
   B. **WELFARE CENTER (EM-6b),**
   C. **REGISTRATION CENTER (EM-6c), or**
   D. **DECONTAMINATION CENTER (EM-6d).**

04 **The Emergency Aid Center sign shall be a horizontal rectangle. Except as provided in Paragraph 5, the identifying word and the word CENTER, the directional arrow, and the border shall be black on a white background.**

Option:

05 When Emergency Aid Center signs are used in an incident situation, such as during the aftermath of a nuclear or biological attack, the background color may be fluorescent pink (see Chapter 6I).

Section 2N.09 Shelter Directional Signs (EM-7 Series)
Standard:

01 **Shelter Directional (EM-7 series) signs (see Figure 2N-1) shall be used to direct the public to selected shelters that have been licensed and marked for emergency use.**

02 **The installation of Shelter Directional signs shall comply with established signing standards. Where used, the signs shall not be installed in competition with other necessary highway guide, warning, and regulatory signs.**

03 **The Shelter Directional sign shall be a horizontal rectangle. Except as provided in Paragraph 4, the identifying word and the word SHELTER, the directional arrow, the distance to the shelter, and the border shall be black on a white background.**

Option:

04 When Shelter Directional signs are used in an incident situation, such as during the aftermath of a nuclear or biological attack, the background color may be fluorescent pink (see Chapter 6I).

05 The distance to the shelter may be omitted from the sign when appropriate.

06 Shelter Directional signs may carry one of the following legends, or others designating similar emergency facilities:
   A. EMERGENCY (EM-7a),
   B. HURRICANE (EM-7b),
   C. FALLOUT (EM-7c), or
   D. CHEMICAL (EM-7d).

07 If appropriate, the name of the facility may be used.

08 The Shelter Directional signs may be installed on the Interstate Highway System or any other major highway system when it has been determined that a need exists for such signs as part of a State or local shelter plan.

09 The Shelter Directional signs may be used to identify different routes to a shelter to provide for rapid movement of large numbers of persons.

Guidance:

10 *The Shelter Directional sign should be used sparingly and only in conjunction with approved plans of State and local authorities.*

11 *The Shelter Directional sign should not be posted more than 5 miles from a shelter.*

PART 3
MARKINGS

CHAPTER 3A. GENERAL

Section 3A.01 Functions and Limitations
Support:

01 Markings on highways and on private roads open to public travel have important functions in providing guidance and information for the road user. Major marking types include pavement and curb markings, delineators, colored pavements, channelizing devices, and islands. In some cases, markings are used to supplement other traffic control devices such as signs, signals, and other markings. In other instances, markings are used alone to effectively convey regulations, guidance, or warnings in ways not obtainable by the use of other devices.

02 Markings have limitations. Visibility of the markings can be limited by snow, debris, and water on or adjacent to the markings. Marking durability is affected by material characteristics, traffic volumes, weather, and location. However, under most highway conditions, markings provide important information while allowing minimal diversion of attention from the roadway.

Section 3A.02 Standardization of Application
Standard:

01 **Each standard marking shall be used only to convey the meaning prescribed for that marking in this Manual. When used for applications not described in this Manual, markings shall conform in all respects to the principles and standards set forth in this Manual.**

Guidance:

02 *Before any new highway, private road open to public travel (see definition in Section 1A.13), paved detour, or temporary route is opened to public travel, all necessary markings should be in place.*

Standard:

03 **Markings that must be visible at night shall be retroreflective unless ambient illumination assures that the markings are adequately visible. All markings on Interstate highways shall be retroreflective.**

04 **Markings that are no longer applicable for roadway conditions or restrictions and that might cause confusion for the road user shall be removed or obliterated to be unidentifiable as a marking as soon as practical.**

Option:

05 Until they can be removed or obliterated, markings may be temporarily masked with tape that is approximately the same color as the pavement.

Section 3A.03 Maintaining Minimum Pavement Marking Retroreflectivity

(This Section is reserved for future text based on FHWA rulemaking.)

Section 3A.04 Materials
Support:

01 Pavement and curb markings are commonly placed by using paints or thermoplastics; however, other suitable marking materials, including raised pavement markers and colored pavements, are also used. Delineators and channelizing devices are visibly placed in a vertical position similar to signs above the roadway.

02 Some marking systems consist of clumps or droplets of material with visible open spaces of bare pavement between the material droplets. These marking systems can function in a manner that is similar to the marking systems that completely cover the pavement surface and are suitable for use as pavement markings if they meet the other pavement marking requirements of the highway agency.

Guidance:

03 *The materials used for markings should provide the specified color throughout their useful life.*

04 *Consideration should be given to selecting pavement marking materials that will minimize tripping or loss of traction for road users, including pedestrians, bicyclists, and motorcyclists.*

05 *Delineators should not present a vertical or horizontal clearance obstacle for pedestrians.*

Section 3A.05 Colors

Standard:

01 Markings shall be yellow, white, red, blue, or purple. The colors for markings shall conform to the standard highway colors. Black in conjunction with one of the colors mentioned in the first sentence of this paragraph shall be a usable color.

02 When used, white markings for longitudinal lines shall delineate:
   A. The separation of traffic flows in the same direction, or
   B. The right-hand edge of the roadway.

03 When used, yellow markings for longitudinal lines shall delineate:
   A. The separation of traffic traveling in opposite directions,
   B. The left-hand edge of the roadways of divided highways and one-way streets or ramps, or
   C. The separation of two-way left-turn lanes and reversible lanes from other lanes.

04 When used, red raised pavement markers or delineators shall delineate:
   A. Truck escape ramps, or
   B. One-way roadways, ramps, or travel lanes that shall not be entered or used in the direction from which the markers are visible.

05 When used, blue markings shall supplement white markings for parking spaces for persons with disabilities.

06 When used, purple markings shall supplement lane line or edge line markings for toll plaza approach lanes that are restricted to use only by vehicles with registered electronic toll collection accounts.

Option:

07 Colors used for official route shield signs (see Section 2D.11) may be used as colors of symbol markings to simulate route shields on the pavement (see Section 3B.20.)

08 Black may be used in combination with the colors mentioned in the first sentence of Paragraph 1 where a light-colored pavement does not provide sufficient contrast with the markings.

Support:

09 When used in combination with other colors, black is not considered a marking color, but only a contrast-enhancing system for the markings.

Section 3A.06 Functions, Widths, and Patterns of Longitudinal Pavement Markings

Standard:

01 The general functions of longitudinal lines shall be:
   A. A double line indicates maximum or special restrictions,
   B. A solid line discourages or prohibits crossing (depending on the specific application),
   C. A broken line indicates a permissive condition, and
   D. A dotted line provides guidance or warning of a downstream change in lane function.

02 The widths and patterns of longitudinal lines shall be as follows:
   A. Normal line—4 to 6 inches wide.
   B. Wide line—at least twice the width of a normal line.
   C. Double line—two parallel lines separated by a discernible space.
   D. Broken line—normal line segments separated by gaps.
   E. Dotted line—noticeably shorter line segments separated by shorter gaps than used for a broken line. The width of a dotted line extension shall be at least the same as the width of the line it extends.

Support:

03 The width of the line indicates the degree of emphasis.

Guidance:

04 *Broken lines should consist of 10-foot line segments and 30-foot gaps, or dimensions in a similar ratio of line segments to gaps as appropriate for traffic speeds and need for delineation.*

Support:

05 Patterns for dotted lines depend on the application (see Sections 3B.04 and 3B.08.)

Guidance:

06 *A dotted line for line extensions within an intersection or taper area should consist of 2-foot line segments and 2- to 6-foot gaps. A dotted line used as a lane line should consist of 3-foot line segments and 9-foot gaps.*

CHAPTER 3B. PAVEMENT AND CURB MARKINGS

Section 3B.01 Yellow Center Line Pavement Markings and Warrants

Standard:

01 **Center line pavement markings, when used, shall be the pavement markings used to delineate the separation of traffic lanes that have opposite directions of travel on a roadway and shall be yellow.**

Option:

02 Center line pavement markings may be placed at a location that is not the geometric center of the roadway.

03 On roadways without continuous center line pavement markings, short sections may be marked with center line pavement markings to control the position of traffic at specific locations, such as around curves, over hills, on approaches to grade crossings, at grade crossings, and at bridges.

Standard:

04 **The center line markings on two-lane, two-way roadways shall be one of the following as shown in Figure 3B-1:**
   A. Two-direction passing zone markings consisting of a normal broken yellow line where crossing the center line markings for passing with care is permitted for traffic traveling in either direction;
   B. One-direction no-passing zone markings consisting of a double yellow line, one of which is a normal broken yellow line and the other is a normal solid yellow line, where crossing the center line markings for passing with care is permitted for the traffic traveling adjacent to the broken line, but is prohibited for traffic traveling adjacent to the solid line; or
   C. Two-direction no-passing zone markings consisting of two normal solid yellow lines where crossing the center line markings for passing is prohibited for traffic traveling in either direction.

05 **A single solid yellow line shall not be used as a center line marking on a two-way roadway.**

06 **The center line markings on undivided two-way roadways with four or more lanes for moving motor vehicle traffic always available shall be the two-direction no-passing zone markings consisting of a solid double yellow line as shown in Figure 3B-2.**

Guidance:

07 *On two-way roadways with three through lanes for moving motor vehicle traffic, two lanes should be designated for traffic in one direction by using one- or two-direction no-passing zone markings as shown in Figure 3B-3.*

Support:

08 Sections 11-301(c) and 11-311(c) of the "Uniform Vehicle Code (UVC)" contain information regarding left turns across center line no-passing zone markings and paved medians, respectively. The UVC can be obtained from the National Committee on Uniform Traffic Laws and Ordinances at the address shown on Page i.

Standard:

09 **Center line markings shall be placed on all paved urban arterials and collectors that have a traveled way of 20 feet or more in width and an ADT of 6,000 vehicles per day or greater. Center line markings shall also be placed on all paved two-way streets or highways that have three or more lanes for moving motor vehicle traffic.**

Guidance:

10 *Center line markings should be placed on paved urban arterials and collectors that have a traveled way of 20 feet or more in width and an ADT of 4,000 vehicles per day or greater. Center line markings should also be placed on all rural arterials and collectors that have a traveled way of 18 feet or more in width and an ADT of 3,000 vehicles per day or greater. Center line markings should also be placed on other traveled ways where an engineering study indicates such a need.*

11 *Engineering judgment should be used in determining whether to place center line markings on traveled ways that are less than 16 feet wide because of the potential for traffic encroaching on the pavement edges, traffic being affected by parked vehicles, and traffic encroaching into the opposing traffic lane.*

Option:

12 Center line markings may be placed on other paved two-way traveled ways that are 16 feet or more in width.

13 If a traffic count is not available, the ADTs described in this Section may be estimates that are based on engineering judgment.

Figure 3B-1. Examples of Two-Lane, Two-Way Marking Applications

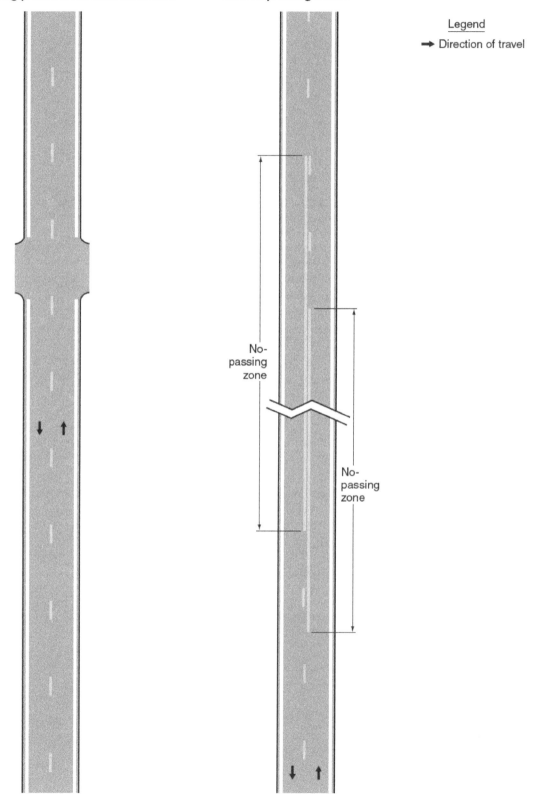

Figure 3B-2. Examples of Four-or-More Lane, Two-Way Marking Applications

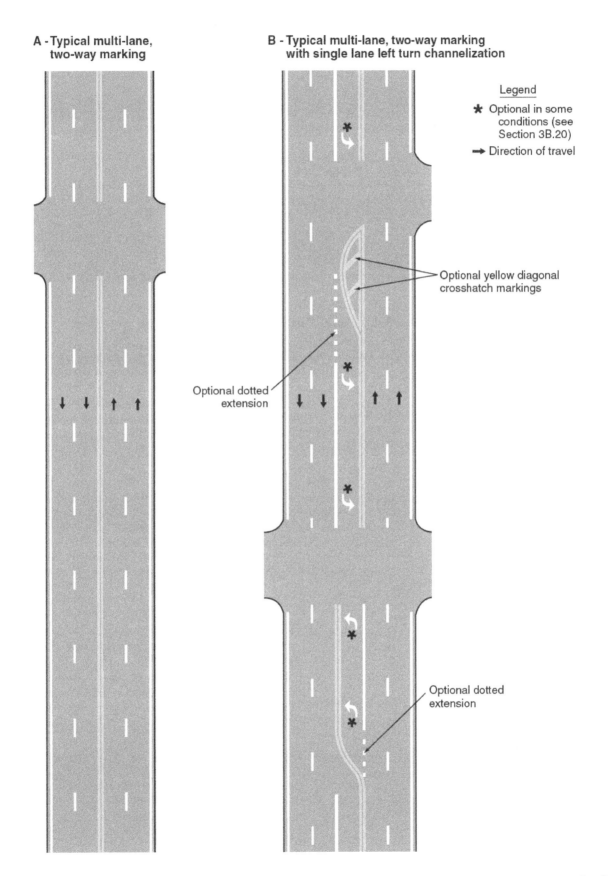

Figure 3B-3. Examples of Three-Lane, Two-Way Marking Applications

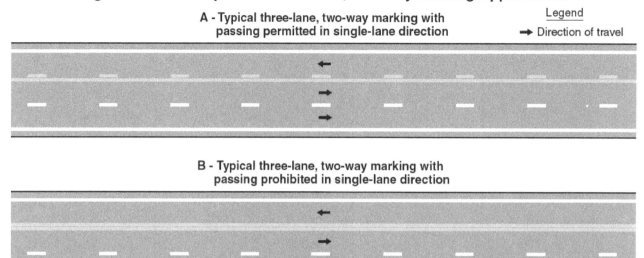

Section 3B.02 No-Passing Zone Pavement Markings and Warrants

Standard:

01 No-passing zones shall be marked by either the one direction no-passing zone pavement markings or the two-direction no-passing zone pavement markings described in Section 3B.01 and shown in Figures 3B-1 and 3B-3.

02 When center line markings are used, no-passing zone markings shall be used on two-way roadways at lane-reduction transitions (see Section 3B.09) and on approaches to obstructions that must be passed on the right (see Section 3B.10).

03 On two-way, two- or three-lane roadways where center line markings are installed, no-passing zones shall be established at vertical and horizontal curves and other locations where an engineering study indicates that passing must be prohibited because of inadequate sight distances or other special conditions.

04 On roadways with center line markings, no-passing zone markings shall be used at horizontal or vertical curves where the passing sight distance is less than the minimum shown in Table 3B-1 for the 85th-percentile speed or the posted or statutory speed limit. The passing sight distance on a vertical curve is the distance at which an object 3.5 feet above the pavement surface can be seen from a point 3.5 feet above the pavement (see Figure 3B-4). Similarly, the passing sight distance on a horizontal curve is the distance measured along the center line (or right-hand lane line of a three-lane roadway) between two points 3.5 feet above the pavement on a line tangent to the embankment or other obstruction that cuts off the view on the inside of the curve (see Figure 3B-4).

Support:

05 The upstream end of a no-passing zone at point "a" in Figure 3B-4 is that point where the sight distance first becomes less than that specified in Table 3B-1. The downstream end of the no-passing zone at point "b" in Figure 3B-4 is that point at which the sight distance again becomes greater than the minimum specified.

06 The values of the minimum passing sight distances that are shown in Table 3B-1 are for operational use in marking no-passing zones and are less than the values that are suggested for geometric design by the AASHTO Policy on Geometric Design of Streets and Highways (see Section 1A.11).

Table 3B-1. Minimum Passing Sight Distances for No-Passing Zone Markings

85th-Percentile or Posted or Statutory Speed Limit	Minimum Passing Sight Distance
25 mph	450 feet
30 mph	500 feet
35 mph	550 feet
40 mph	600 feet
45 mph	700 feet
50 mph	800 feet
55 mph	900 feet
60 mph	1,000 feet
65 mph	1,100 feet
70 mph	1,200 feet

Figure 3B-4. Method of Locating and Determining the Limits of No-Passing Zones at Curves

A - No-passing zone at VERTICAL CURVE

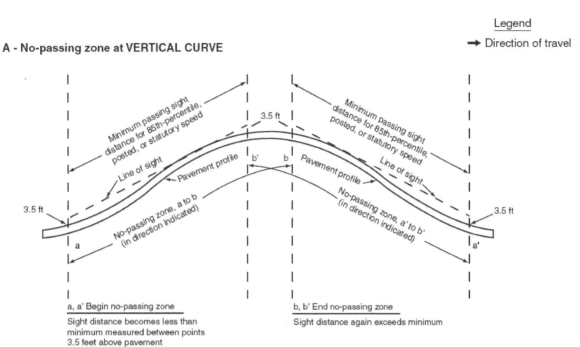

Profile View

Note: No-passing zones in opposite directions may or may not overlap, depending on alignment

B - No-passing zone at HORIZONTAL CURVE

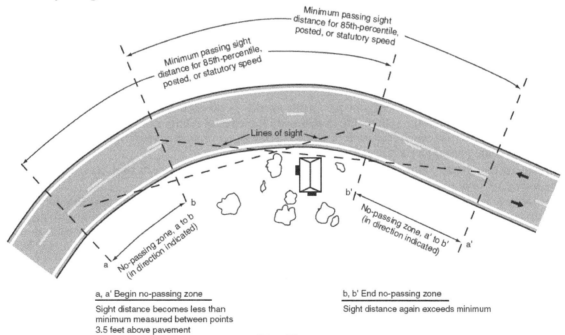

Plan View

Note: No-passing zones in opposite directions may or may not overlap, depending on alignment

Guidance:

07 *Where the distance between successive no-passing zones is less than 400 feet, no-passing markings should connect the zones.*

Standard:

08 **Where center line markings are used, no-passing zone markings shall be used on approaches to grade crossings in compliance with Section 8B.27.**

Option:

09 In addition to pavement markings, no-passing zone signs (see Sections 2B.28, 2B.29, and 2C.45) may be used to emphasize the existence and extent of a no-passing zone.

Support:

10 Section 11-307 of the "Uniform Vehicle Code (UVC)" contains further information regarding required road user behavior in no-passing zones. The UVC can be obtained from the National Committee on Uniform Traffic Laws and Ordinances at the address shown on Page i.

Standard:

11 **On three-lane roadways where the direction of travel in the center lane transitions from one direction to the other, a no-passing buffer zone shall be provided in the center lane as shown in Figure 3B-5. A lane-reduction transition (see Section 3B.09) shall be provided at each end of the buffer zone.**

12 **The buffer zone shall be a flush median island formed by two sets of double yellow center line markings that is at least 50 feet in length.**

Option:

13 Yellow diagonal crosshatch markings (see Section 3B.24) may be placed in the flush median area between the two sets of no-passing zone markings as shown in Figure 3B-5.

Guidance:

14 *For three-lane roadways having a posted or statutory speed limit of 45 mph or greater, the lane transition taper length should be computed by the formula $L = WS$. For roadways where the posted or statutory speed limit is less than 45 mph, the formula $L = WS^2/60$ should be used to compute the taper length.*

Support:

15 Under both formulas, L equals the taper length in feet, W equals the width of the center lane or offset distance in feet, and S equals the 85th-percentile speed or the posted or statutory speed limit, whichever is higher.

Guidance:

16 *The minimum lane transition taper length should be 100 feet in urban areas and 200 feet in rural areas.*

Section 3B.03 Other Yellow Longitudinal Pavement Markings

Standard:

01 **If reversible lanes are used, the lane line pavement markings on each side of reversible lanes shall consist of a normal broken double yellow line to delineate the edge of a lane in which the direction of travel is reversed from time to time, such that each of these markings serve as the center line markings of the roadway during some period (see Figure 3B-6).**

02 **Signs (see Section 2B.26), lane-use control signals (see Chapter 4M), or both shall be used to supplement reversible lane pavement markings.**

03 **If a two-way left-turn lane that is never operated as a reversible lane is used, the lane line pavement markings on each side of the two-way left-turn lane shall consist of a normal broken yellow line and a normal solid yellow line to delineate the edges of a lane that can be used by traffic in either direction as part of a left-turn maneuver. These markings shall be placed with the broken line toward the two-way left-turn lane and the solid line toward the adjacent traffic lane as shown in Figure 3B-7.**

Guidance:

04 *White two-way left-turn lane-use arrows (see Figure 3B-7), should be used in conjunction with the longitudinal two-way left-turn markings at the locations described in Section 3B.20.*

05 *Signs should be used in conjunction with the two-way left turn markings (see Section 2B.24).*

Standard:

06 **If a continuous flush median island formed by pavement markings separating travel in opposite directions is used, two sets of solid double yellow lines shall be used to form the island as shown in Figures 3B-2 and 3B-5. Other markings in the median island area shall also be yellow, except crosswalk markings which shall be white (see Section 3B.18).**

Figure 3B-5. Example of Application of Three-Lane, Two-Way Marking for Changing Direction of the Center Lane

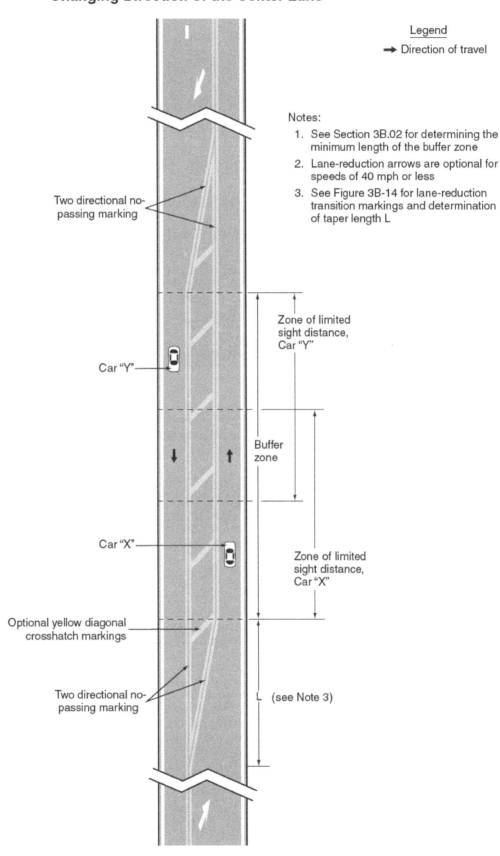

Figure 3B-6. Example of Reversible Lane Marking Application

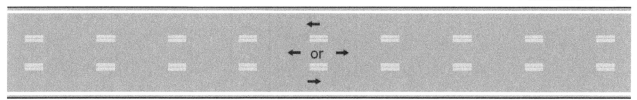

Section 3B.04 White Lane Line Pavement Markings and Warrants
Standard:
01   When used, lane line pavement markings delineating the separation of traffic lanes that have the same direction of travel shall be white.
02   Lane line markings shall be used on all freeways and Interstate highways.
Guidance:
03   *Lane line markings should be used on all roadways that are intended to operate with two or more adjacent traffic lanes in the same direction of travel, except as otherwise required for reversible lanes. Lane line markings should also be used at congested locations where the roadway will accommodate more traffic lanes with lane line markings than without the markings.*
Support:
04   Examples of lane line markings are shown in Figures 3B-2, 3B-3, and 3B-7 through 3B-13.
Standard:
05   Except as provided in Paragraph 6, where crossing the lane line markings with care is permitted, the lane line markings shall consist of a normal broken white line.
06   A dotted white line marking shall be used as the lane line to separate a through lane that continues beyond the interchange or intersection from an adjacent lane for any of the following conditions:
   A. A deceleration or acceleration lane,
   B. A through lane that becomes a mandatory exit or turn lane,
   C. An auxiliary lane 2 miles or less in length between an entrance ramp and an exit ramp, or
   D. An auxiliary lane 1 mile or less in length between two adjacent intersections.
07   For exit ramps with a parallel deceleration lane, a normal width dotted white lane line shall be installed from the upstream end of the full-width deceleration lane to the theoretical gore or to the upstream end of a solid white lane line, if used, that extends upstream from the theoretical gore as shown in Drawings A and C of Figure 3B-8.
Option:
08   For exit ramps with a parallel deceleration lane, a normal width dotted white line extension may be installed in the taper area upstream from the full-width deceleration lane as shown in Drawings A and C of Figure 3B-8.
09   For an exit ramp with a tapered deceleration lane, a normal width dotted white line extension may be installed from the theoretical gore through the taper area such that it meets the edge line at the upstream end of the taper as shown in Drawing B of Figure 3B-8.
Standard:
10   For entrance ramps with a parallel acceleration lane, a normal width dotted white lane line shall be installed from the theoretical gore or from the downstream end of a solid white lane line, if used, that extends downstream from the theoretical gore, to a point at least one-half the distance from the theoretical gore to the downstream end of the acceleration taper, as shown in Drawing A of Figure 3B-9.
Option:
11   For entrance ramps with a parallel acceleration lane, a normal width dotted white line extension may be installed from the downstream end of the dotted white lane line to the downstream end of the acceleration taper, as shown in Drawing A of Figure 3B-9.
12   For entrance ramps with a tapered acceleration lane, a normal width dotted white line extension may be installed from the downstream end of the channelizing line adjacent to the through lane to the downstream end of the acceleration taper, as shown in Drawings B and C of Figure 3B-9.

Figure 3B-7. Example of Two-Way Left-Turn Lane Marking Applications

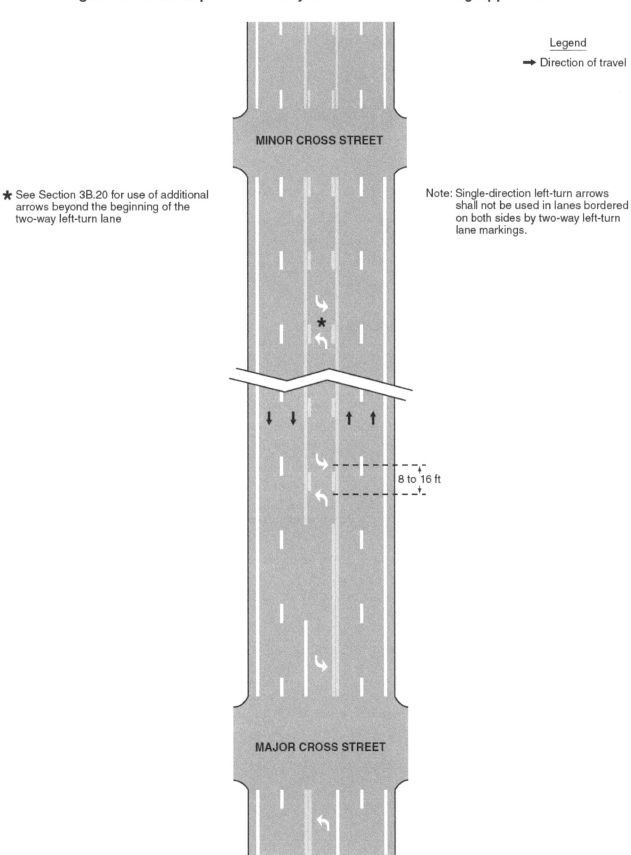

Figure 3B-8. Examples of Dotted Line and Channelizing Line Applications for Exit Ramp Markings (Sheet 1 of 2)

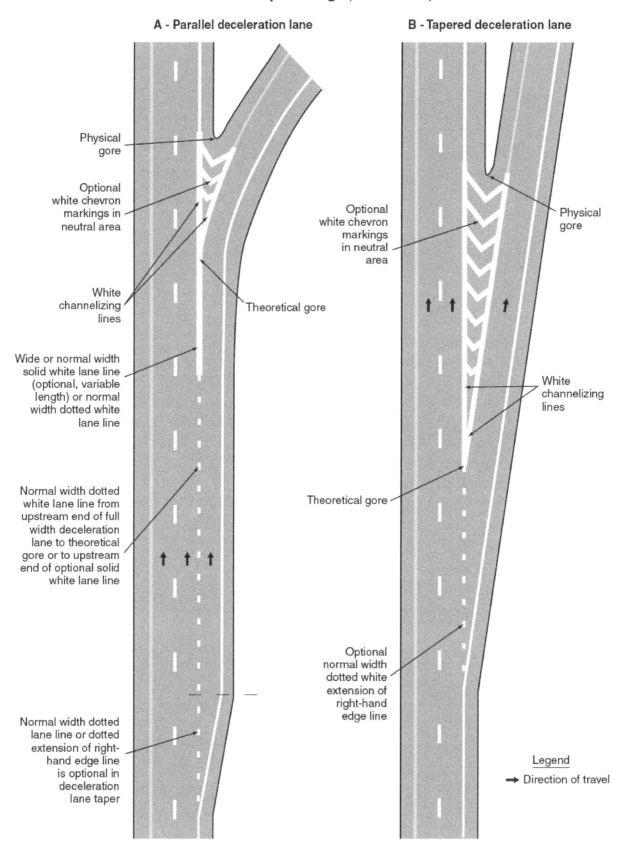

Figure 3B-8. Examples of Dotted Line and Channelizing Line Applications for Exit Ramp Markings (Sheet 2 of 2)

C – Parallel deceleration lane at a multi-lane exit ramp having an optional exit lane that also carries the through route

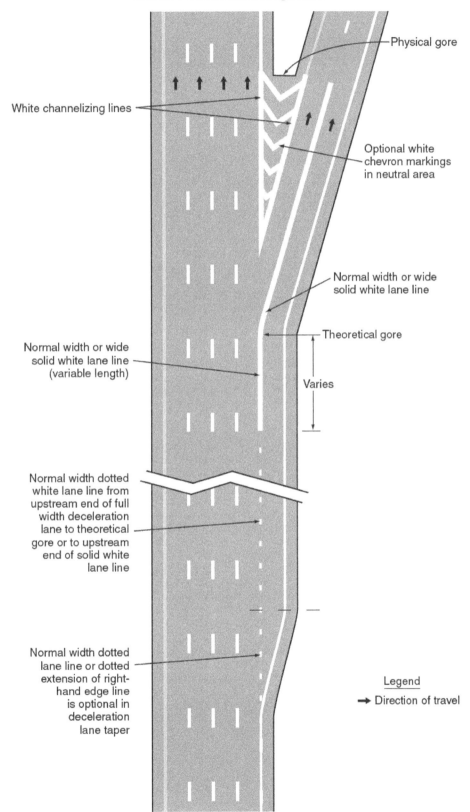

Figure 3B-9. Examples of Dotted Line and Channelizing Line Applications for Entrance Ramp Markings (Sheet 1 of 2)

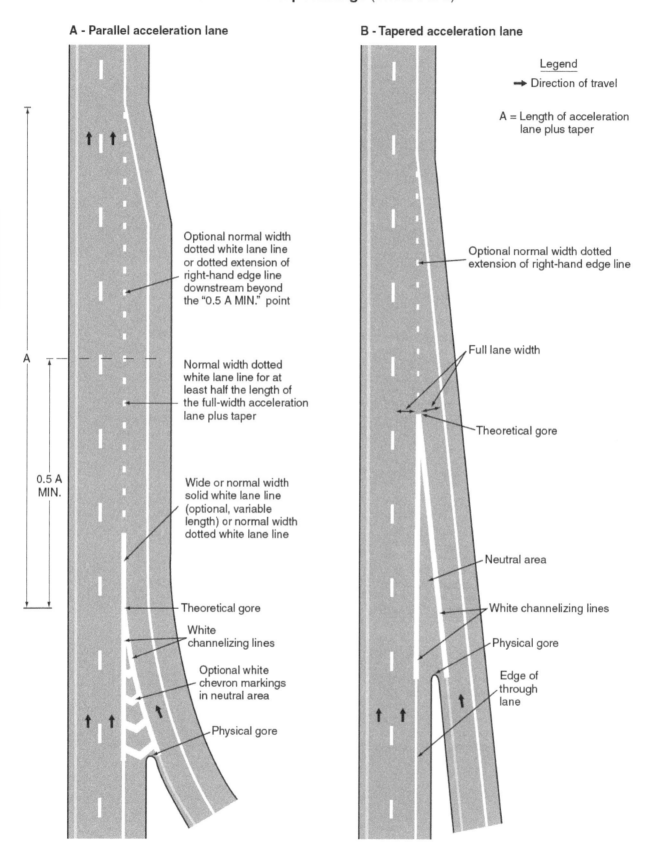

Figure 3B-9. Examples of Dotted Line and Channelizing Line Applications for Entrance Ramp Markings (Sheet 2 of 2)

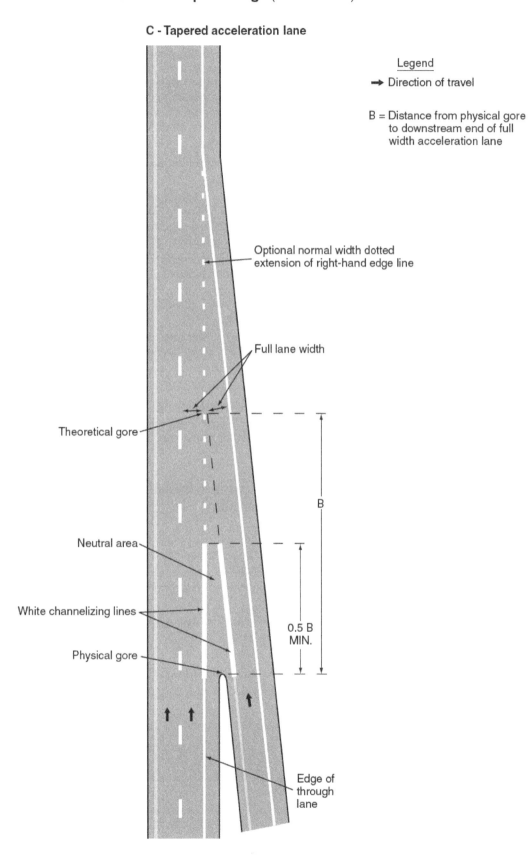

Standard:

13 **A wide dotted white lane line shall be used:**
   A. **As a lane drop marking in advance of lane drops at exit ramps to distinguish a lane drop from a normal exit ramp (see Drawings A, B, and C of Figure 3B-10),**
   B. **In advance of freeway route splits with dedicated lanes (see Drawing D of Figure 3B-10),**
   C. **To separate a through lane that continues beyond an interchange from an adjacent auxiliary lane between an entrance ramp and an exit ramp (see Drawing E of Figure 3B-10),**
   D. **As a lane drop marking in advance of lane drops at intersections to distinguish a lane drop from an intersection through lane (see Drawing A of Figure 3B-11), and**
   E. **To separate a through lane that continues beyond an intersection from an adjacent auxiliary lane between two intersections (see Drawing B of Figure 3B-11).**

Guidance:

14 *Lane drop markings used in advance of lane drops at freeway and expressway exit ramps should begin at least 1/2 mile in advance of the theoretical gore.*

15 *On the approach to a multi-lane exit ramp having an optional exit lane that also carries through traffic, lane line markings should be used as illustrated in Drawing B of Figure 3B-10. In this case, if the right-most exit lane is an added lane such as a parallel deceleration lane, the lane drop marking should begin at the upstream end of the full-width deceleration lane, as shown in Drawing C of Figure 3B-8.*

16 *Lane drop markings used in advance of lane drops at intersections should begin a distance in advance of the intersection that is determined by engineering judgment as suitable to enable drivers who do not desire to make the mandatory turn to move out of the lane being dropped prior to reaching the queue of vehicles that are waiting to make the turn. The lane drop marking should begin no closer to the intersection than the most upstream regulatory or warning sign associated with the lane drop.*

17 *The dotted white lane lines that are used for lane drop markings and that are used as a lane line separating through lanes from auxiliary lanes should consist of line segments that are 3 feet in length separated by 9-foot gaps.*

Support:

18 Section 3B.20 contains information regarding other markings that are associated with lane drops, such as lane-use arrow markings and ONLY word markings.

19 Section 3B.09 contains information about the lane line markings that are to be used for transition areas where the number of through lanes is reduced.

Standard:

20 **Where crossing the lane line markings is discouraged, the lane line markings shall consist of a normal or wide solid white line.**

Option:

21 Where it is intended to discourage lane changing on the approach to an exit ramp, a wide solid white lane line may extend upstream from the theoretical gore or, for multi-lane exits, as shown in Drawing B of Figure 3B-10, for a distance that is determined by engineering judgment.

22 Where lane changes might cause conflicts, a wide or normal solid white lane line may extend upstream from an intersection.

23 In the case of a lane drop at an exit ramp or intersection, such a solid white line may replace a portion, but not all of the length of the wide dotted white lane line.

Support:

24 Section 3B.09 contains information about the lane line markings that are to be used for transition areas where the number of through lanes is reduced.

Guidance:

25 *On approaches to intersections, a solid white lane line marking should be used to separate a through lane from an added mandatory turn lane.*

Option:

26 On approaches to intersections, solid white lane line markings may be used to separate adjacent through lanes or adjacent mandatory turn lanes from each other.

27 Where the median width allows the left-turn lanes to be separated from the through lanes to give drivers on opposing approaches a less obstructed view of opposing through traffic, white pavement markings may be used to form channelizing islands as shown in Figure 2B-17.

Figure 3B-10. Examples of Applications of Freeway and Expressway Lane-Drop Markings (Sheet 1 of 5)

A – Lane drop at a single lane exit ramp

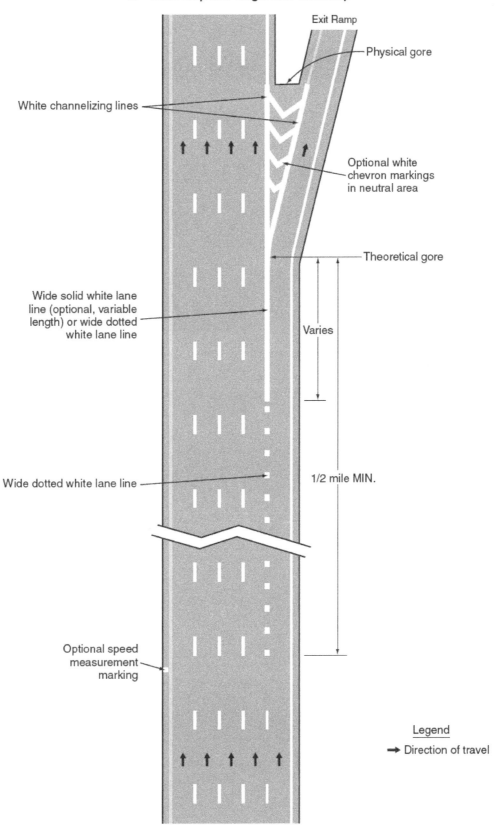

Figure 3B-10. Examples of Applications of Freeway and Expressway Lane-Drop Markings (Sheet 2 of 5)

B – Lane drop at a multi-lane exit ramp having an optional exit lane that also carries the through route

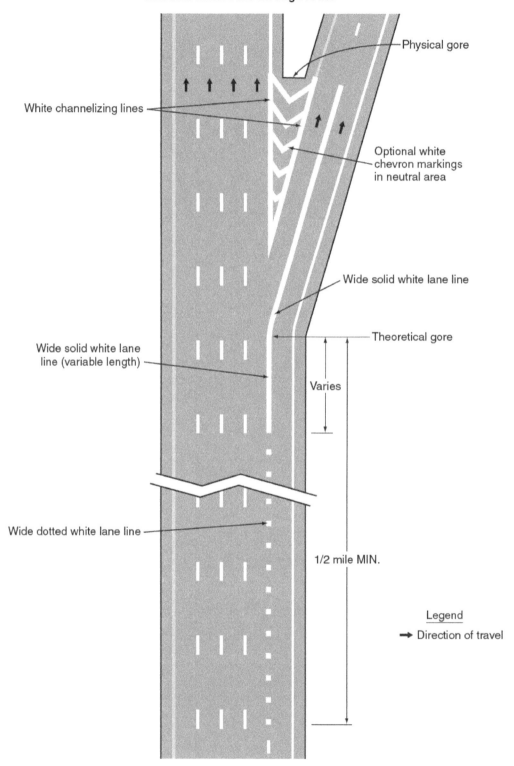

Figure 3B-10. Examples of Applications of Freeway and Expressway Lane-Drop Markings (Sheet 3 of 5)

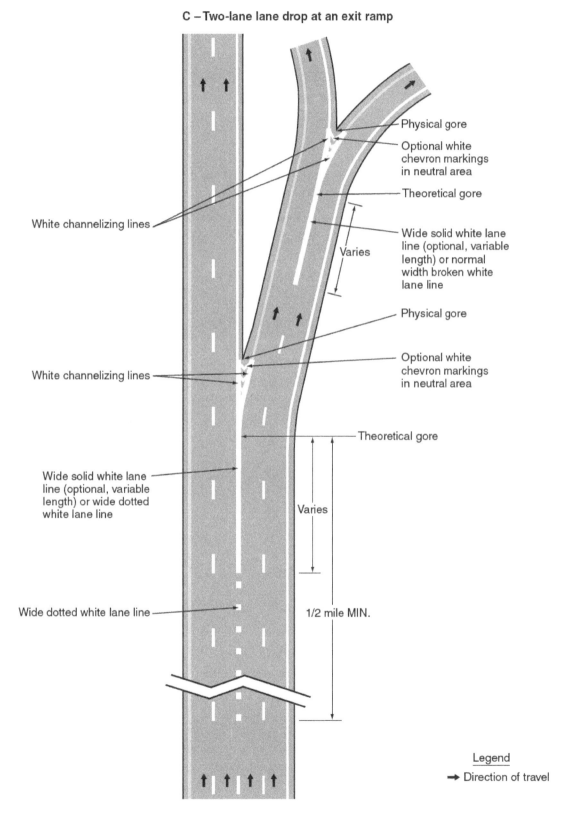

Figure 3B-10. Examples of Applications of Freeway and Expressway Lane-Drop Markings (Sheet 4 of 5)

D – Route split with dedicated lanes

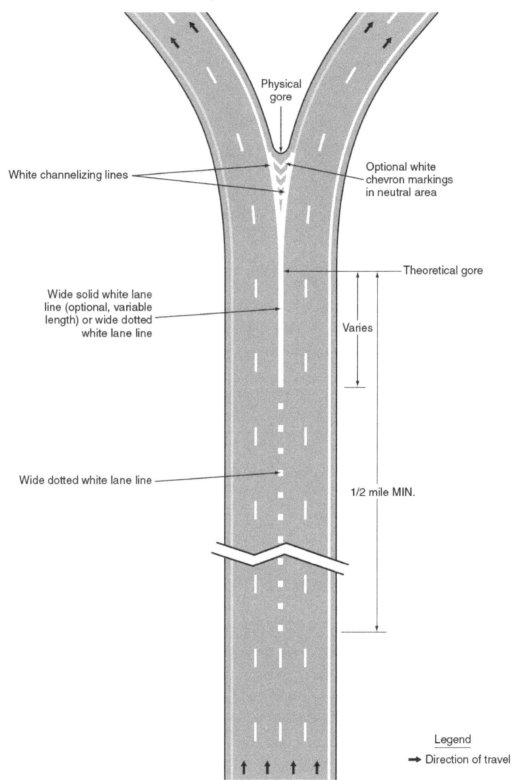

Figure 3B-10. Examples of Applications of Freeway and Expressway Lane-Drop Markings (Sheet 5 of 5)

E – Auxiliary lane, such as at a cloverleaf interchange

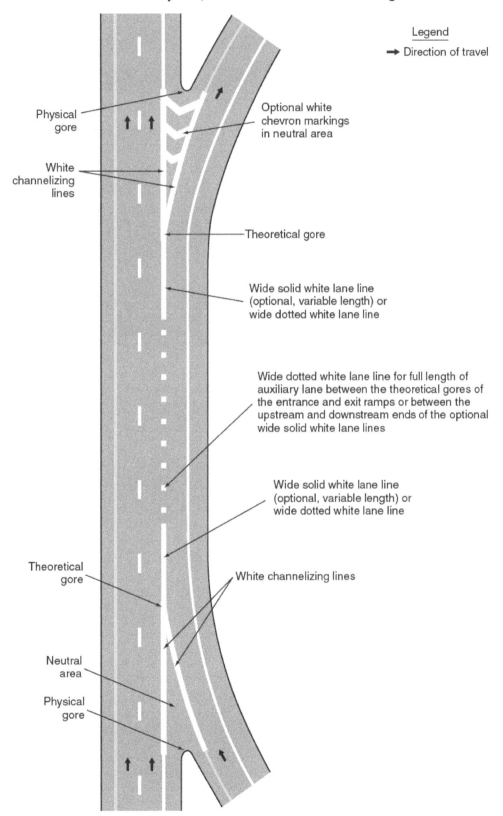

Figure 3B-11. Examples of Applications of Conventional Road Lane-Drop Markings
(Sheet 1 of 2)

A – Lane drop at an intersection

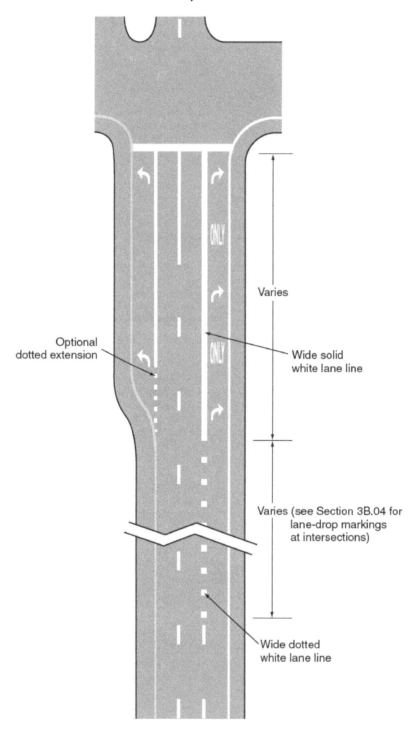

Figure 3B-11. Examples of Applications of Conventional Road Lane-Drop Markings
(Sheet 2 of 2)

B – Auxiliary lane between intersections

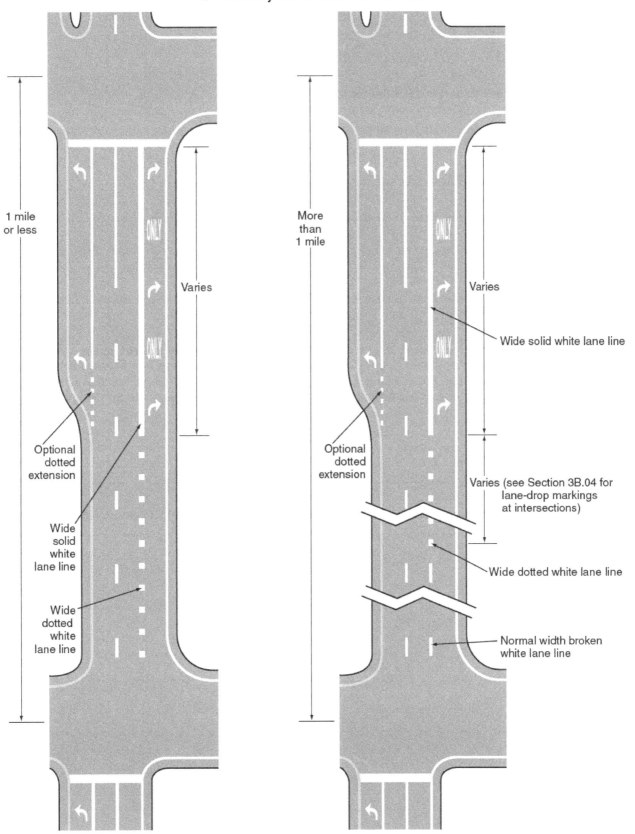

28 Solid white lane line markings may be used to separate through traffic lanes from auxiliary lanes, such as an added uphill truck lane or a preferential lane (see Section 3D.02).

29 Wide solid lane line markings may be used for greater emphasis.

Standard:

30 **Where crossing the lane line markings is prohibited, the lane line markings shall consist of a solid double white line (see Figure 3B-12).**

Section 3B.05 Other White Longitudinal Pavement Markings

Standard:

01 **A channelizing line shall be a wide or double solid white line.**

Option:

02 Channelizing lines may be used to form channelizing islands where traffic traveling in the same direction is permitted on both sides of the island.

Standard:

03 **Other pavement markings in the channelizing island area shall be white.**

Support:

04 Examples of channelizing line applications are shown in Figures 3B-8, 3B-9, and 3B-10, and in Drawing C of Figure 3B-15.

05 Channelizing lines at exit ramps as shown in Figures 3B-8 and 3B-10 define the neutral area, direct exiting traffic at the proper angle for smooth divergence from the main lanes into the ramp, and reduce the probability of colliding with objects adjacent to the roadway.

06 Channelizing lines at entrance ramps as shown in Figures 3B-9 and 3B-10 promote orderly and efficient merging with the through traffic.

Standard:

07 **For all exit ramps and for entrance ramps with parallel acceleration lanes, channelizing lines shall be placed on both sides of the neutral area (see Figures 3B-8 and 3B-10 and Drawing A of Figure 3B-9).**

08 **For entrance ramps with tapered acceleration lanes, channelizing lines shall be placed along both sides of the neutral area to a point at least one-half of the distance to the theoretical gore (see Drawing C of Figure 3B-9).**

Option:

09 For entrance ramps with tapered acceleration lanes, the channelizing lines may extend to the theoretical gore as shown in Drawing B of Figure 3B-9.

Figure 3B-12. Example of Solid Double White Lines Used to Prohibit Lane Changing

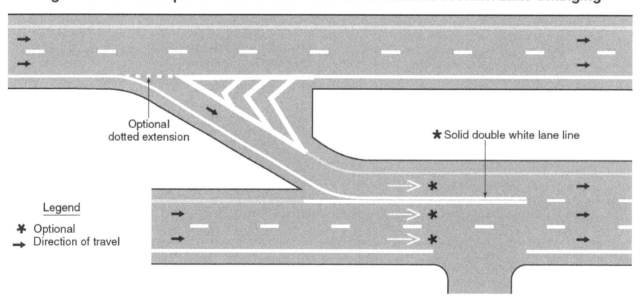

10 White chevron crosshatch markings (see Section 3B.24) may be placed in the neutral area of exit ramp and entrance ramp gores for special emphasis as shown in Figures 3B-8 and 3B-10 and Drawing A of Figure 3B-9. The channelizing lines and the optional chevron crosshatch markings at exit ramp and entrance ramp gores may be supplemented with white retroreflective or internally illuminated raised pavement markers (see Sections 3B.11 and 3B.13) for enhanced nighttime visibility.

Section 3B.06 Edge Line Pavement Markings
Standard:

01 **If used, edge line pavement markings shall delineate the right or left edges of a roadway.**

02 **Except for dotted edge line extensions (see Section 3B.08), edge line markings shall not be continued through intersections or major driveways.**

03 **If used on the roadways of divided highways or one-way streets, or on any ramp in the direction of travel, left edge line pavement markings shall consist of a normal solid yellow line to delineate the left-hand edge of a roadway or to indicate driving or passing restrictions left of these markings.**

04 **If used, right edge line pavement markings shall consist of a normal solid white line to delineate the right-hand edge of the roadway.**

Guidance:

05 *Edge line markings should not be broken for minor driveways.*

Support:

06 Edge line markings have unique value as visual references to guide road users during adverse weather and visibility conditions.

Option:

07 Wide solid edge line markings may be used for greater emphasis.

Section 3B.07 Warrants for Use of Edge Lines
Standard:

01 **Edge line markings shall be placed on paved streets or highways with the following characteristics:**
   A. Freeways,
   B. Expressways, and
   C. Rural arterials with a traveled way of 20 feet or more in width and an ADT of 6,000 vehicles per day or greater.

Guidance:

02 *Edge line markings should be placed on paved streets or highways with the following characteristics:*
   A. Rural arterials and collectors with a traveled way of 20 feet or more in width and an ADT of 3,000 vehicles per day or greater.
   B. At other paved streets and highways where an engineering study indicates a need for edge line markings.

03 *Edge line markings should not be placed where an engineering study or engineering judgment indicates that providing them is likely to decrease safety.*

Option:

04 Edge line markings may be placed on streets and highways with or without center line markings.

05 Edge line markings may be excluded, based on engineering judgment, for reasons such as if the traveled way edges are delineated by curbs, parking, or other markings.

06 If a bicycle lane is marked on the outside portion of the traveled way, the edge line that would mark the outside edge of the bicycle lane may be omitted.

07 Edge line markings may be used where edge delineation is desirable to minimize unnecessary driving on paved shoulders or on refuge areas that have lesser structural pavement strength than the adjacent roadway.

Section 3B.08 Extensions Through Intersections or Interchanges
Standard:

01 **Except as provided in Paragraph 2, pavement markings extended into or continued through an intersection or interchange area shall be the same color and at least the same width as the line markings they extend (see Figure 3B-13).**

Option:

02 A normal line may be used to extend a wide line through an intersection.

Figure 3B-13. Examples of Line Extensions through Intersections (Sheet 1 of 2)

A - Typical pavement markings with offset lane lines continued through the intersection and optional crosswalk lines and stop lines

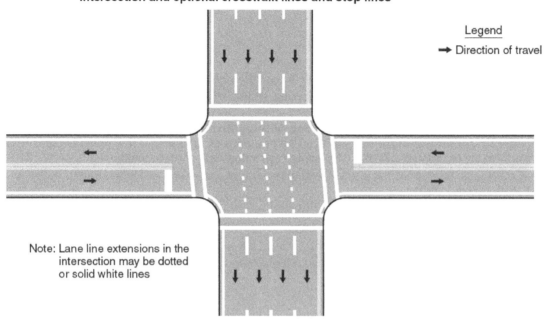

Legend
→ Direction of travel

Note: Lane line extensions in the intersection may be dotted or solid white lines

B - Typical pavement markings with double-turn lanes, lane-use turn arrows, and optional crosswalk lines, stop lines, and line extensions into intersection for double turns

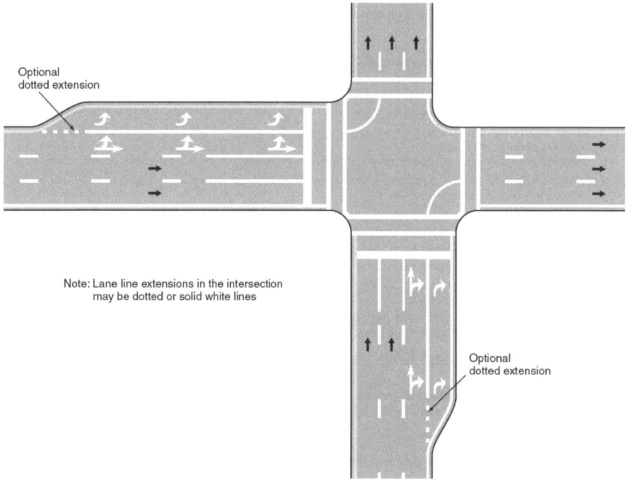

Optional dotted extension

Note: Lane line extensions in the intersection may be dotted or solid white lines

Optional dotted extension

Figure 3B-13. Examples of Line Extensions through Intersections (Sheet 2 of 2)

C - Typical dotted line markings to extend lane line markings into the intersection

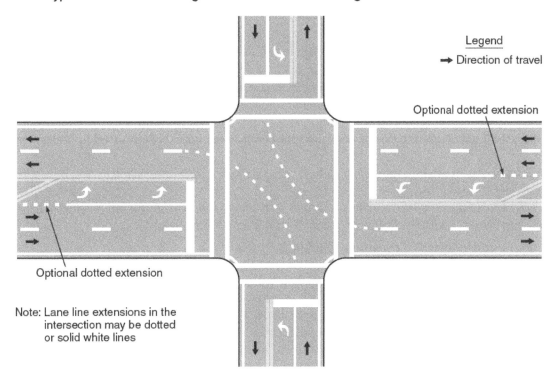

D - Typical dotted line markings to extend center line and lane line markings into the intersection

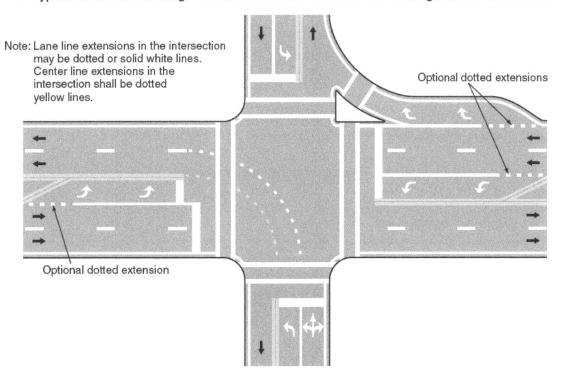

Guidance:

⁰³ *Where highway design or reduced visibility conditions make it desirable to provide control or to guide vehicles through an intersection or interchange, such as at offset, skewed, complex, or multi-legged intersections, on curved roadways, where multiple turn lanes are used, or where offset left turn lanes might cause driver confusion, dotted line extension markings consisting of 2-foot line segments and 2- to 6-foot gaps should be used to extend longitudinal line markings through an intersection or interchange area.*

Option:

⁰⁴ Dotted edge line extensions may be placed through intersections or major driveways.

Guidance:

⁰⁵ *Where greater restriction is required, solid lane lines or channelizing lines should be extended into or continued through intersections or major driveways.*

Standard:

⁰⁶ **Solid lines shall not be used to extend edge lines into or through intersections or major driveways.**

Guidance:

⁰⁷ *Where a double line is extended through an intersection, a single line of equal width to one of the lines of the double line should be used.*

⁰⁸ *To the extent possible, pavement marking extensions through intersections should be designed in a manner that minimizes potential confusion for drivers in adjacent or opposing lanes.*

Section 3B.09 Lane-Reduction Transition Markings

Support:

⁰¹ Lane-reduction transition markings are used where the number of through lanes is reduced because of narrowing of the roadway or because of a section of on-street parking in what would otherwise be a through lane. Lane-reduction transition markings are not used for lane drops.

Standard:

⁰² **Except as provided in Paragraph 3, where pavement markings are used, lane-reduction transition markings shall be used to guide traffic through transition areas where the number of through lanes is reduced, as shown in Figure 3B-14. On two-way roadways, no-passing zone markings shall be used to prohibit passing in the direction of the convergence, and shall continue through the transition area.**

Option:

⁰³ On low-speed urban roadways where curbs clearly define the roadway edge in the lane-reduction transition, or where a through lane becomes a parking lane, the edge line and/or delineators shown in Figure 3B-14 may be omitted as determined by engineering judgment.

Guidance:

⁰⁴ *For roadways having a posted or statutory speed limit of 45 mph or greater, the transition taper length for a lane-reduction transition should be computed by the formula $L = WS$. For roadways where the posted or statutory speed limit is less than 45 mph, the formula $L = WS^2/60$ should be used to compute the taper length.*

Support:

⁰⁵ Under both formulas, L equals the taper length in feet, W equals the width of the offset distance in feet, and S equals the 85th-percentile speed or the posted or statutory speed limit, whichever is higher.

Guidance:

⁰⁶ *Where observed speeds exceed posted or statutory speed limits, longer tapers should be used.*

Option:

⁰⁷ On new construction, where no posted or statutory speed limit has been established, the design speed may be used in the transition taper length formula.

Guidance:

⁰⁸ *Lane line markings should be discontinued one-quarter of the distance between the Lane Ends sign (see Section 2C.42) and the point where the transition taper begins.*

⁰⁹ *Except as provided in Paragraph 3 for low-speed urban roadways, the edge line markings shown in Figure 3B-14 should be installed from the location of the Lane Ends warning sign to beyond the beginning of the narrower roadway.*

Support:

¹⁰ Pavement markings at lane-reduction transitions supplement the standard signs. See Section 3B.20 for provisions regarding use of lane-reduction arrows.

Figure 3B-14. Examples of Applications of Lane-Reduction Transition Markings

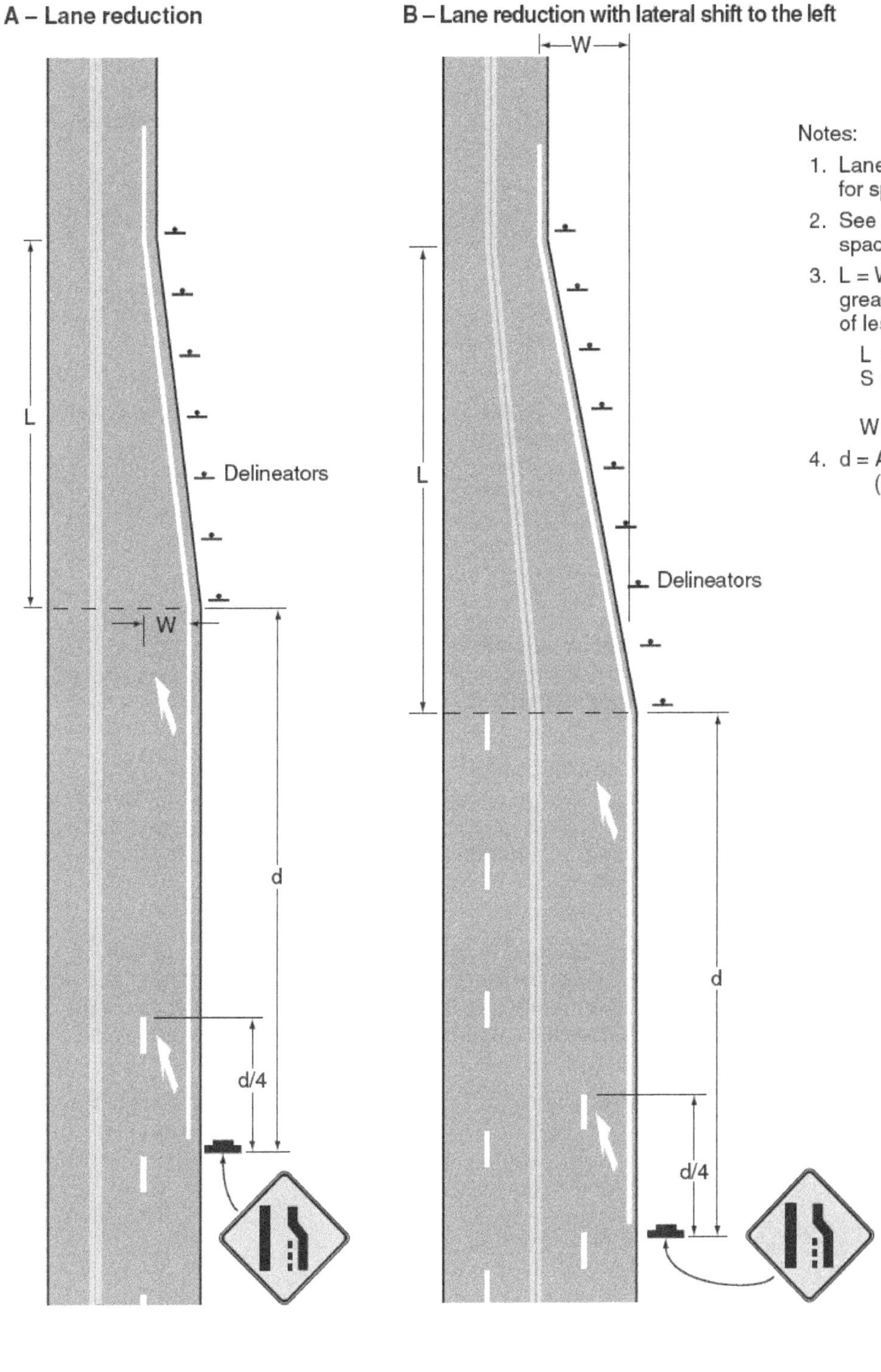

A – Lane reduction

B – Lane reduction with lateral shift to the left

Notes:
1. Lane-reduction arrows are optional for speeds of less than 45 mph
2. See Section 3F.04 for delineator spacing
3. $L = WS$ for speeds of 45 mph or greater and $L = WS^2/60$ for speeds of less than 45 mph, where:
   - L = Length of taper in feet
   - S = Posted, 85th-percentile, or statutory speed in mph
   - W = Offset in feet
4. d = Advance warning distance (see Section 2C.05)

Section 3B.10 Approach Markings for Obstructions

Standard:

01 **Pavement markings shall be used to guide traffic away from fixed obstructions within a paved roadway. Approach markings for bridge supports, refuge islands, median islands, toll plaza islands, and raised channelization islands shall consist of a tapered line or lines extending from the center line or the lane line to a point 1 to 2 feet to the right-hand side, or to both sides, of the approach end of the obstruction (see Figure 3B-15).**

Support:

02 See Chapter 3E for additional information on approach markings for toll plaza islands.

Guidance:

03 *For roadways having a posted or statutory speed limit of 45 mph or greater, the taper length of the tapered line markings should be computed by the formula $L = WS$. For roadways where the posted or statutory speed limit is less than 45 mph, the formula $L = WS^2/60$ should be used to compute the taper length.*

Support:

04 Under both formulas, L equals the taper length in feet, W equals the width of the offset distance in feet, and S equals the 85th-percentile speed or the posted or statutory speed limit, whichever is higher.

Guidance:

05 *The minimum taper length should be 100 feet in urban areas and 200 feet in rural areas.*

Support:

06 Examples of approach markings for obstructions in the roadway are shown in Figure 3B-15.

Standard:

07 **If traffic is required to pass only to the right of the obstruction, the markings shall consist of a two-direction no-passing zone marking at least twice the length of the diagonal portion as determined by the appropriate taper formula (see Drawing A of Figure 3B-15).**

Option:

08 If traffic is required to pass only to the right of the obstruction, yellow diagonal crosshatch markings (see Section 3B.24) may be placed in the flush median area between the no-passing zone markings as shown in Drawings A and B of Figure 3B-15. Other markings, such as yellow delineators, yellow channelizing devices, yellow raised pavement markers, and white crosswalk pavement markings, may also be placed in the flush median area.

Standard:

09 **If traffic can pass either to the right or left of the obstruction, the markings shall consist of two channelizing lines diverging from the lane line, one to each side of the obstruction. In advance of the point of divergence, a wide solid white line or normal solid double white line shall be extended in place of the broken lane line for a distance equal to the length of the diverging lines (see Drawing C of Figure 3B-15).**

Option:

10 If traffic can pass either to the right or left of the obstruction, additional white chevron crosshatch markings (see Section 3B.24) may be placed in the flush median area between the channelizing lines as shown in Drawing C of Figure 3B-15. Other markings, such as white delineators, white channelizing devices, white raised pavement markers, and white crosswalk markings may also be placed in the flush median area.

Section 3B.11 Raised Pavement Markers – General

Standard:

01 **The color of raised pavement markers under both daylight and nighttime conditions shall conform to the color of the marking for which they serve as a positioning guide, or for which they supplement or substitute.**

Option:

02 The side of a raised pavement marker that is visible to traffic proceeding in the wrong direction may be red (see Section 3A.05).

03 Retroreflective or internally illuminated raised pavement markers may be used in the roadway immediately adjacent to curbed approach ends of raised medians and curbs of islands, or on top of such curbs (see Section 3B.23).

Figure 3B-15. Examples of Applications of Markings for Obstructions in the Roadway
(Sheet 1 of 2)

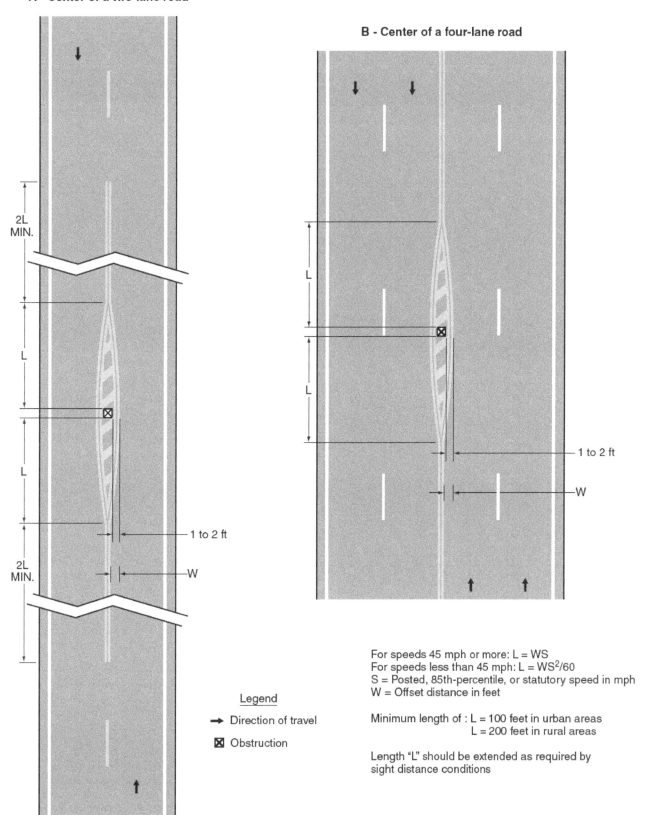

Figure 3B-15. Examples of Applications of Markings for Obstructions in the Roadway
(Sheet 2 of 2)

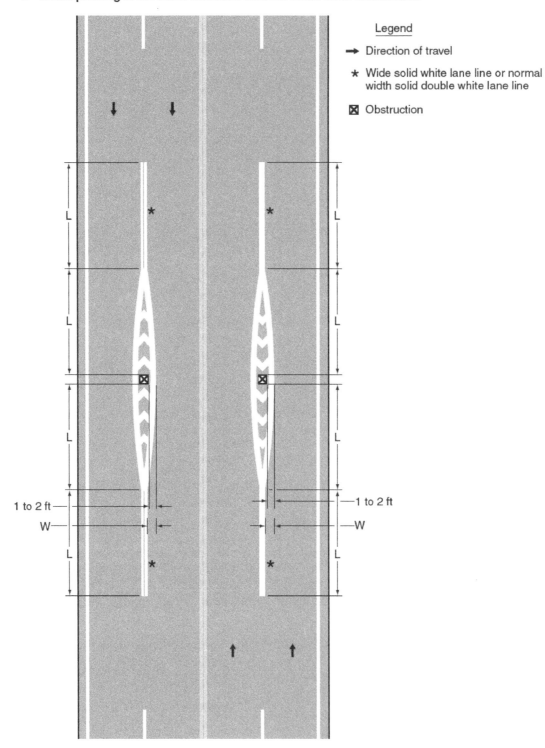

For speeds of 45 mph or more: L = WS
For speeds of less than 45 mph: L = WS^2/60
S = Posted, 85th-percentile, or statutory speed in mph
W = Offset distance in feet

Minimum length of: L = 100 feet in urban areas
L = 200 feet in rural areas

Length "L" should be extended as required by sight distance conditions

Support:

04 Retroreflective and internally illuminated raised pavement markers are available in mono-directional and bidirectional configurations. The bidirectional marker is capable of displaying the applicable color for each direction of travel.

05 Blue raised pavement markers are sometimes used in the roadway to help emergency personnel locate fire hydrants.

Standard:

06 **When used, internally illuminated raised pavement markers shall be steadily illuminated and shall not be flashed.**

Support:

07 Flashing raised pavement markers are considered to be In-Roadway Lights (see Chapter 4N).

Guidance:

08 *Non-retroreflective raised pavement markers should not be used alone, without supplemental retroreflective or internally illuminated markers, as a substitute for other types of pavement markings.*

09 *Directional configurations should be used to maximize correct information and to minimize confusing information provided to the road user. Directional configurations also should be used to avoid confusion resulting from visibility of markers that do not apply to the road user.*

10 *The spacing of raised pavement markers used to supplement or substitute for other types of longitudinal markings should correspond with the pattern of broken lines for which the markers supplement or substitute.*

Standard:

11 **The value of N cited in Sections 3B.12 through 3B.14 for the spacing of raised pavement markers shall equal the length of one line segment plus one gap of the broken lines used on the highway.**

Option:

12 For additional emphasis, retroreflective raised pavement markers may be spaced closer than described in Sections 3B.12 through 3B.14, as determined by engineering judgment or engineering study.

Support:

13 Figures 9-20 through 9-22 in the Traffic Control Devices Handbook" (see Section 1A.11) contain additional information regarding the spacing of raised pavement markers on longitudinal markings.

Section 3B.12 Raised Pavement Markers as Vehicle Positioning Guides with Other Longitudinal Markings

Option:

01 Retroreflective or internally illuminated raised pavement markers may be used as positioning guides with longitudinal line markings without necessarily conveying information to the road user about passing or lane-use restrictions. In such applications, markers may be positioned in line with or immediately adjacent to a single line marking, or positioned between the two lines of a double center line or double lane line marking.

Guidance:

02 *The spacing for such applications should be 2N, where N equals the length of one line segment plus one gap (see Section 3B.11).*

Option:

03 Where it is desired to alert the road user to changes in the travel path, such as on sharp curves or on transitions that reduce the number of lanes or that shift traffic laterally, the spacing may be reduced to N or less.

04 On freeways and expressways, the spacing may be increased to 3N for relatively straight and level roadway segments where engineering judgment indicates that such spacing will provide adequate delineation under wet night conditions.

Section 3B.13 Raised Pavement Markers Supplementing Other Markings

Guidance:

01 *The use of retroreflective or internally illuminated raised pavement markers for supplementing longitudinal line markings should comply with the following:*

   A. Lateral Positioning
   1. When supplementing double line markings, pairs of raised pavement markers placed laterally in line with or immediately outside of the two lines should be used.
   2. When supplementing wide line markings, pairs of raised pavement markers placed laterally adjacent to each other should be used.

B. *Longitudinal Spacing*
   1. *When supplementing solid line markings, raised pavement markers at a spacing no greater than N (see Section 3B.11) should be used, except that when supplementing channelizing lines or edge line markings, a spacing of no greater than N/2 should be used.*
   2. *When supplementing broken line markings, a spacing no greater than 3N should be used. However, when supplementing broken line markings identifying reversible lanes, a spacing of no greater than N should be used.*
   3. *When supplementing dotted lane line markings, a spacing appropriate for the application should be used.*
   4. *When supplementing longitudinal line extension markings through at-grade intersections, one raised pavement marker for each short line segment should be used.*
   5. *When supplementing line extensions through freeway interchanges, a spacing of no greater than N should be used.*

02 *Raised pavement markers should not supplement right-hand edge lines unless an engineering study or engineering judgment indicates the benefits of enhanced delineation of a curve or other location would outweigh possible impacts on bicycles using the shoulder, and the spacing of raised pavement markers on the right-hand edge is close enough to avoid misinterpretation as a broken line during wet night conditions.*

Option:

03 Raised pavement markers also may be used to supplement other markings such as channelizing islands, gore areas, approaches to obstructions, or wrong-way arrows.

04 To improve the visibility of horizontal curves, center lines may be supplemented with retroreflective or internally illuminated raised pavement markers for the entire curved section as well as for a distance in advance of the curve that approximates 5 seconds of travel time.

Section 3B.14 Raised Pavement Markers Substituting for Pavement Markings

Option:

01 Retroreflective or internally illuminated raised pavement markers, or non-retroreflective raised pavement markers supplemented by retroreflective or internally illuminated markers, may be substituted for markings of other types.

Guidance:

02 *If used, the pattern of the raised pavement markers should simulate the pattern of the markings for which they substitute.*

Standard:

03 **If raised pavement markers are used to substitute for broken line markings, a group of three to five markers equally spaced at a distance no greater than N/8 (see Section 3B.11) shall be used. If N is other than 40 feet, the markers shall be equally spaced over the line segment length (at 1/2 points for three markers, at 1/3 points for four markers, and at 1/4 points for five markers). At least one retroreflective or internally illuminated marker per group shall be used or a retroreflective or internally illuminated marker shall be installed midway in each gap between successive groups of non-retroreflective markers.**

04 **When raised pavement markers substitute for solid line markings, the markers shall be equally spaced at no greater than N/4, with retroreflective or internally illuminated units at a spacing no greater than N/2.**

Guidance:

05 *Raised pavement markers should not substitute for right-hand edge line markings unless an engineering study or engineering judgment indicates the benefits of enhanced delineation of a curve or other location would outweigh possible impacts on bicycles using the shoulder, and the spacing of raised pavement markers on the right-hand edge line is close enough to avoid misinterpretation as a broken line during wet night conditions.*

Standard:

06 **When raised pavement markers substitute for dotted lines, they shall be spaced at no greater than N/4, with not less than one raised pavement marker per dotted line segment. At least one raised marker every N shall be retroreflective or internally illuminated.**

Option:

07 When substituting for wide lines, raised pavement markers may be placed laterally adjacent to each other to simulate the width of the line.

Section 3B.15 Transverse Markings
Standard:

01 **Transverse markings, which include shoulder markings, word and symbol markings, arrows, stop lines, yield lines, crosswalk lines, speed measurement markings, speed reduction markings, speed hump markings, parking space markings, and others, shall be white unless otherwise provided in this Manual.**
Guidance:

02 *Because of the low approach angle at which pavement markings are viewed, transverse lines should be proportioned to provide visibility at least equal to that of longitudinal lines.*

Section 3B.16 Stop and Yield Lines
Guidance:

01 *Stop lines should be used to indicate the point behind which vehicles are required to stop in compliance with a traffic control signal.*

Option:

02 Stop lines may be used to indicate the point behind which vehicles are required to stop in compliance with a STOP (R1-1) sign, a Stop Here For Pedestrians (R1-5b or R1-5c) sign, or some other traffic control device that requires vehicles to stop, except YIELD signs that are not associated with passive grade crossings.

03 Yield lines may be used to indicate the point behind which vehicles are required to yield in compliance with a YIELD (R1-2) sign or a Yield Here To Pedestrians (R1-5 or R1-5a) sign.

Standard:

04 **Except as provided in Section 8B.28, stop lines shall not be used at locations where drivers are required to yield in compliance with a YIELD (R1-2) sign or a Yield Here To Pedestrians (R1-5 or R1-5a) sign or at locations on uncontrolled approaches where drivers are required by State law to yield to pedestrians.**

05 **Yield lines shall not be used at locations where drivers are required to stop in compliance with a STOP (R1-1) sign, a Stop Here For Pedestrians (R1-5b or R1-5c) sign, a traffic control signal, or some other traffic control device.**

06 **Stop lines shall consist of solid white lines extending across approach lanes to indicate the point at which the stop is intended or required to be made.**

07 **Yield lines (see Figure 3B-16) shall consist of a row of solid white isosceles triangles pointing toward approaching vehicles extending across approach lanes to indicate the point at which the yield is intended or required to be made.**

Guidance:

08 *Stop lines should be 12 to 24 inches wide.*

09 *The individual triangles comprising the yield line should have a base of 12 to 24 inches wide and a height equal to 1.5 times the base. The space between the triangles should be 3 to 12 inches.*

10 *If used, stop and yield lines should be placed a minimum of 4 feet in advance of the nearest crosswalk line at controlled intersections, except for yield lines at roundabouts as provided for in Section 3C.04 and at midblock crosswalks. In the absence of a marked crosswalk, the stop line or yield line should be placed at the desired stopping or yielding point, but should not be placed more than 30 feet or less than 4 feet from the nearest edge of the intersecting traveled way.*

11 *Stop lines at midblock signalized locations should be placed at least 40 feet in advance of the nearest signal indication (see Section 4D.14).*

12 *If yield or stop lines are used at a crosswalk that crosses an uncontrolled multi-lane approach, the yield lines or stop lines should be placed 20 to 50 feet in advance of the nearest crosswalk line, and parking should be prohibited in the area between the yield or stop line and the crosswalk (see Figure 3B-17).*

Standard:

13 **If yield (stop) lines are used at a crosswalk that crosses an uncontrolled multi-lane approach, Yield Here To (Stop Here For) Pedestrians (R1-5 series) signs (see Section 2B.11) shall be used.**

Guidance:

14 *Yield (stop) lines and Yield Here To (Stop Here For) Pedestrians signs should not be used in advance of crosswalks that cross an approach to or departure from a roundabout.*

Support:

15 When drivers yield or stop too close to crosswalks that cross uncontrolled multi-lane approaches, they place pedestrians at risk by blocking other drivers' views of pedestrians and by blocking pedestrians' views of vehicles approaching in the other lanes.

Figure 3B-16. Recommended Yield Line Layouts

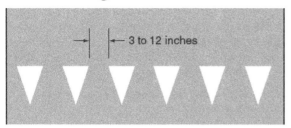
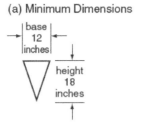
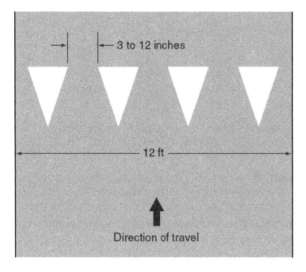
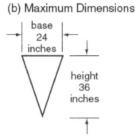

Notes:
Triangle height is equal to 1.5 times the base dimension.

Yield lines may be smaller than suggested when installed on much narrower, slow-speed facilities such as shared-use paths.

Option:
01 Stop and yield lines may be staggered longitudinally on a lane-by-lane basis (see Drawing D of Figure 3B-13).

Support:
02 Staggered stop lines and staggered yield lines can improve the driver's view of pedestrians, provide better sight distance for turning vehicles, and increase the turning radius for left-turning vehicles.

03 Section 8B.28 contains information regarding the use of stop lines and yield lines at grade crossings.

Section 3B.17 Do Not Block Intersection Markings

Option:
01 Do Not Block Intersection markings may be used to mark the edges of an intersection area that is in close proximity to a signalized intersection, railroad crossing, or other nearby traffic control that might cause vehicles to stop within the intersection and impede other traffic entering the intersection. If authorized by law, Do Not Block Intersection markings with appropriate signs may also be used at other locations.

Standard:
02 **If used, Do Not Block Intersection markings (see Figure 3B-18) shall consist of one of the following alternatives:**
   A. **Wide solid white lines that outline the intersection area that vehicles must not block;**
   B. **Wide solid white lines that outline the intersection area that vehicles must not block and a white word message such as DO NOT BLOCK or KEEP CLEAR;**
   C. **Wide solid white lines that outline the intersection area that vehicles must not block and white cross-hatching within the intersection area; or**
   D. **A white word message, such as DO NOT BLOCK or KEEP CLEAR, within the intersection area that vehicles must not block.**

03 **Do Not Block Intersection markings shall be accompanied by one or more DO NOT BLOCK INTERSECTION (DRIVEWAY) (CROSSING) (R10-7) signs (see Section 2B.53), one or more DO NOT STOP ON TRACKS (R8-8) signs (see Section 8B.09), or one or more similar signs.**

Figure 3B-17. Examples of Yield Lines at Unsignalized Midblock Crosswalks

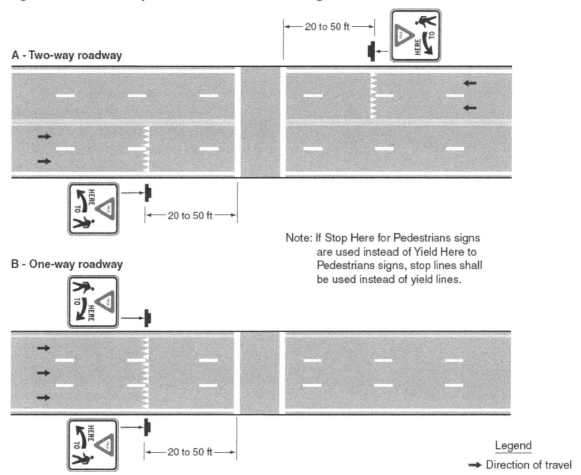

Section 3B.18 Crosswalk Markings

Support:

01 Crosswalk markings provide guidance for pedestrians who are crossing roadways by defining and delineating paths on approaches to and within signalized intersections, and on approaches to other intersections where traffic stops.

02 In conjunction with signs and other measures, crosswalk markings help to alert road users of a designated pedestrian crossing point across roadways at locations that are not controlled by traffic control signals or STOP or YIELD signs.

03 At non-intersection locations, crosswalk markings legally establish the crosswalk.

Standard:

04 **When crosswalk lines are used, they shall consist of solid white lines that mark the crosswalk. They shall not be less than 6 inches or greater than 24 inches in width.**

Guidance:

05 *If transverse lines are used to mark a crosswalk, the gap between the lines should not be less than 6 feet. If diagonal or longitudinal lines are used without transverse lines to mark a crosswalk, the crosswalk should be not less than 6 feet wide.*

06 *Crosswalk lines, if used on both sides of the crosswalk, should extend across the full width of pavement or to the edge of the intersecting crosswalk to discourage diagonal walking between crosswalks (see Figures 3B-17 and 3B-19).*

07 *At locations controlled by traffic control signals or on approaches controlled by STOP or YIELD signs, crosswalk lines should be installed where engineering judgment indicates they are needed to direct pedestrians to the proper crossing path(s).*

Figure 3B-18. Do Not Block Intersection Markings

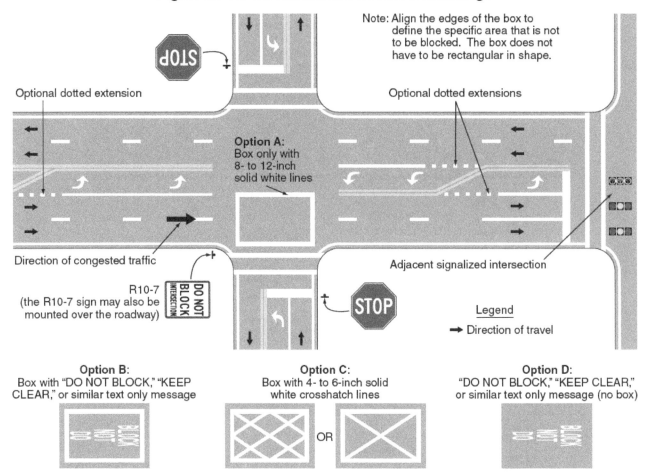

08 *Crosswalk lines should not be used indiscriminately. An engineering study should be performed before a marked crosswalk is installed at a location away from a traffic control signal or an approach controlled by a STOP or YIELD sign. The engineering study should consider the number of lanes, the presence of a median, the distance from adjacent signalized intersections, the pedestrian volumes and delays, the average daily traffic (ADT), the posted or statutory speed limit or 85th-percentile speed, the geometry of the location, the possible consolidation of multiple crossing points, the availability of street lighting, and other appropriate factors.*

09 *New marked crosswalks alone, without other measures designed to reduce traffic speeds, shorten crossing distances, enhance driver awareness of the crossing, and/or provide active warning of pedestrian presence, should not be installed across uncontrolled roadways where the speed limit exceeds 40 mph and either:*

A. *The roadway has four or more lanes of travel without a raised median or pedestrian refuge island and an ADT of 12,000 vehicles per day or greater; or*

B. *The roadway has four or more lanes of travel with a raised median or pedestrian refuge island and an ADT of 15,000 vehicles per day or greater.*

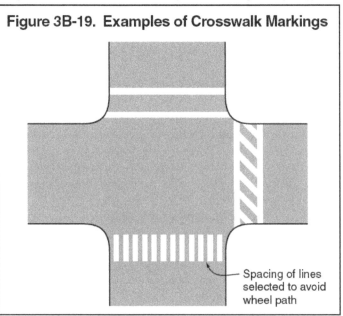

Figure 3B-19. Examples of Crosswalk Markings

2009 Edition   Page 385

Support:
10 Chapter 4F contains information on Pedestrian Hybrid Beacons. Section 4L.03 contains information regarding Warning Beacons to provide active warning of a pedestrian's presence. Section 4N.02 contains information regarding In-Roadway Warning Lights at crosswalks. Chapter 7D contains information regarding school crossing supervision.

Guidance:
11 *Because non-intersection pedestrian crossings are generally unexpected by the road user, warning signs (see Section 2C.50) should be installed for all marked crosswalks at non-intersection locations and adequate visibility should be provided by parking prohibitions.*

Support:
12 Section 3B.16 contains information regarding placement of stop line markings near crosswalk markings.

Option:
13 For added visibility, the area of the crosswalk may be marked with white diagonal lines at a 45-degree angle to the line of the crosswalk or with white longitudinal lines parallel to traffic flow as shown in Figure 3B-19.
14 When diagonal or longitudinal lines are used to mark a crosswalk, the transverse crosswalk lines may be omitted. This type of marking may be used at locations where substantial numbers of pedestrians cross without any other traffic control device, at locations where physical conditions are such that added visibility of the crosswalk is desired, or at places where a pedestrian crosswalk might not be expected.

Guidance:
15 *If used, the diagonal or longitudinal lines should be 12 to 24 inches wide and separated by gaps of 12 to 60 inches. The design of the lines and gaps should avoid the wheel paths if possible, and the gap between the lines should not exceed 2.5 times the width of the diagonal or longitudinal lines.*

Option:
16 When an exclusive pedestrian phase that permits diagonal crossing of an intersection is provided at a traffic control signal, a marking as shown in Figure 3B-20 may be used for the crosswalk.

Guidance:
17 *Crosswalk markings should be located so that the curb ramps are within the extension of the crosswalk markings.*

Support:
18 Detectable warning surfaces mark boundaries between pedestrian and vehicular ways where there is no raised curb. Detectable warning surfaces are required by 49 CFR, Part 37 and by the Americans with Disabilities Act (ADA) where curb ramps are constructed at the junction of sidewalks and the roadway, for marked and unmarked crosswalks. Detectable warning surfaces contrast visually with adjacent walking surfaces, either light-on-dark, or dark-on-light. The "Americans with Disabilities Act Accessibility Guidelines for Buildings and Facilities (ADAAG)" (see Section 1A.11) contains specifications for design and placement of detectable warning surfaces.

Section 3B.19 Parking Space Markings

Support:
01 Marking of parking space boundaries encourages more orderly and efficient use of parking spaces where parking turnover is substantial. Parking space markings tend to prevent encroachment into fire hydrant zones, bus stops, loading zones, approaches to intersections, curb ramps, and clearance spaces for islands and other zones where parking is restricted. Examples of parking space markings are shown in Figure 3B-21.

Standard:
02 **Parking space markings shall be white.**

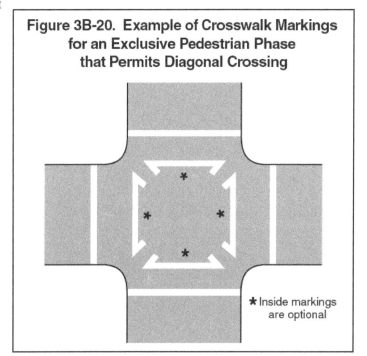

Figure 3B-20. Example of Crosswalk Markings for an Exclusive Pedestrian Phase that Permits Diagonal Crossing

★ Inside markings are optional

December 2009   Sect. 3B.18 to 3B.19

Figure 3B-21. Examples of Parking Space Markings

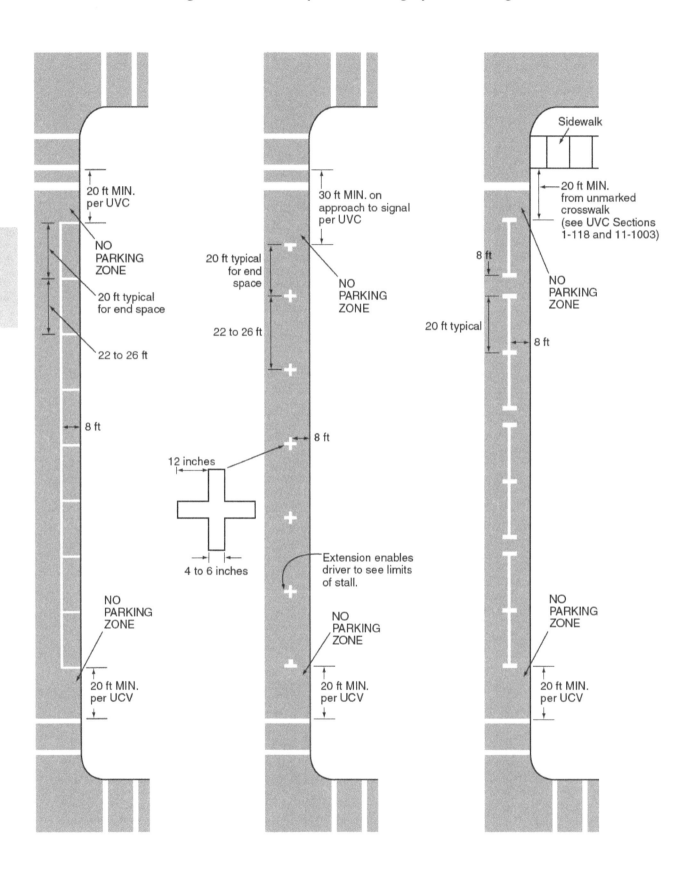

Option:

03 Blue lines may supplement white parking space markings of each parking space designated for use only by persons with disabilities.

Support:

04 Additional parking space markings for the purpose of designating spaces for use only by persons with disabilities are discussed in Section 3B.20 and illustrated in Figure 3B-22. The design and layout of accessible parking spaces for persons with disabilities is provided in the "Americans with Disabilities Act Accessibility Guidelines (ADAAG)" (see Section 1A.11).

Section 3B.20 Pavement Word, Symbol, and Arrow Markings

Support:

01 Word, symbol, and arrow markings on the pavement are used for the purpose of guiding, warning, or regulating traffic. These pavement markings can be helpful to road users in some locations by supplementing signs and providing additional emphasis for important regulatory, warning, or guidance messages, because the markings do not require diversion of the road user's attention from the roadway surface. Symbol messages are preferable to word messages. Examples of standard word and arrow pavement markings are shown in Figures 3B-23 and 3B-24.

Figure 3B-22. International Symbol of Accessibility Parking Space Marking

Height of symbol:
Minimum = 28 inches
Special = 41 inches

Width of symbol:
Minimum = 24 inches
Special = 36 inches

*Stroke width:
Minimum = 3 inches
Special = 4 inches

Note: Blue background and white border are optional

Figure 3B-23. Example of Elongated Letters for Word Pavement Markings

Figure 3B-24. Examples of Standard Arrows for Pavement Markings

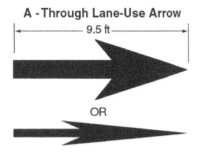

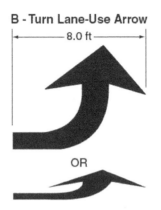

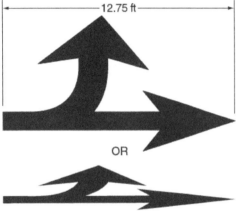

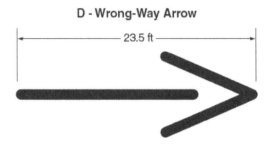

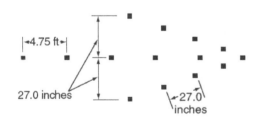

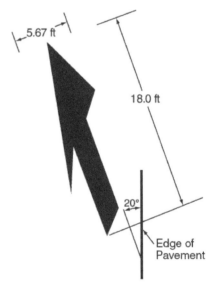

Notes:
1. Typical sizes for normal installation; sizes may be reduced approximately one-third for low-speed urban conditions; larger sizes may be needed for freeways, above average speeds, and other critical locations.
2. The narrow elongated arrow designs shown in Drawings A, B, and C are optional.
3. For proper proportion, see the Pavement Markings chapter of the "Standard Highway Signs and Markings" book (see Section 1A.11).

Option:

₀₂ Word, symbol, and arrow markings, including those contained in the "Standard Highway Signs and Markings" book (see Section 1A.11), may be used as determined by engineering judgment to supplement signs and/or to provide additional emphasis for regulatory, warning, or guidance messages. Among the word, symbol, and arrow markings that may be used are the following:

- A. Regulatory:
    1. STOP
    2. YIELD
    3. RIGHT (LEFT) TURN ONLY
    4. 25 MPH
    5. Lane-use and wrong-way arrows
    6. Diamond symbol for HOV lanes
    7. Other preferential lane word markings
- B. Warning:
    1. STOP AHEAD
    2. YIELD AHEAD
    3. YIELD AHEAD triangle symbol
    4. SCHOOL XING
    5. SIGNAL AHEAD
    6. PED XING
    7. SCHOOL
    8. R X R
    9. BUMP
    10. HUMP
    11. Lane-reduction arrows
- C. Guide:
    1. Route numbers (route shield pavement marking symbols and/or words such as I-81, US 40, STATE 135, or ROUTE 10)
    2. Cardinal directions (NORTH, SOUTH, EAST, or WEST)
    3. TO
    4. Destination names or abbreviations thereof

Standard:

₀₃ **Word, symbol, and arrow markings shall be white, except as otherwise provided in this Section.**

₀₄ **Pavement marking letters, numerals, symbols, and arrows shall be installed in accordance with the design details in the Pavement Markings chapter of the "Standard Highway Signs and Markings" book (see Section 1A.11).**

Guidance:

₀₅ *Letters and numerals should be 6 feet or more in height.*

₀₆ *Word and symbol markings should not exceed three lines of information.*

₀₇ *If a pavement marking word message consists of more than one line of information, it should read in the direction of travel. The first word of the message should be nearest to the road user.*

₀₈ *Except for the two opposing arrows of a two-way left-turn lane marking (see Figure 3B-7), the longitudinal space between word or symbol message markings, including arrow markings, should be at least four times the height of the characters for low-speed roads, but not more than ten times the height of the characters under any conditions.*

₀₉ *The number of different word and symbol markings used should be minimized to provide effective guidance and avoid misunderstanding.*

₁₀ *Except for the SCHOOL word marking (see Section 7C.03), pavement word, symbol, and arrow markings should be no more than one lane in width.*

₁₁ *Pavement word, symbol, and arrow markings should be proportionally scaled to fit within the width of the facility upon which they are applied.*

Option:

₁₂ On narrow, low-speed shared-use paths, the pavement words, symbols, and arrows may be smaller than suggested, but to the relative scale.

13 Pavement markings simulating Interstate, U.S., State, and other official highway route shield signs (see Figure 2D-3) with appropriate route numbers, but elongated for proper proportioning when viewed as a marking, may be used to guide road users to their destinations (see Figure 3B-25).

Standard:

14 **Except at the ends of aisles in parking lots, the word STOP shall not be used on the pavement unless accompanied by a stop line (see Section 3B.16) and STOP sign (see Section 2B.05). At the ends of aisles in parking lots, the word STOP shall not be used on the pavement unless accompanied by a stop line.**

15 **The word STOP shall not be placed on the pavement in advance of a stop line, unless every vehicle is required to stop at all times.**

Option:

16 A yield-ahead triangle symbol (see Figure 3B-26) or YIELD AHEAD word pavement marking may be used on approaches to intersections where the approaching traffic will encounter a YIELD sign at the intersection.

Standard:

17 **The yield-ahead triangle symbol or YIELD AHEAD word pavement marking shall not be used unless a YIELD sign (see Section 2B.08) is in place at the intersection. The yield-ahead symbol marking shall be as shown in Figure 3B-26.**

Guidance:

18 *The International Symbol of Accessibility parking space marking (see Figure 3B-22) should be placed in each parking space designated for use by persons with disabilities.*

Option:

19 A blue background with white border may supplement the wheelchair symbol as shown in Figure 3B-22.

Support:

20 Lane-use arrow markings (see Figure 3B-24) are used to indicate the mandatory or permissible movements in certain lanes (see Figure 3B-27) and in two-way left-turn lanes (see Figure 3B-7).

Guidance:

21 *Lane-use arrow markings (see Figure 3B-24) should be used in lanes designated for the exclusive use of a turning movement, including turn bays, except where engineering judgment determines that physical conditions or other markings (such as a dotted extension of the lane line through the taper into the turn bay) clearly discourage unintentional use of a turn bay by through vehicles. Lane-use arrow markings should also be used in lanes from which movements are allowed that are contrary to the normal rules of the road (see Drawing B of Figure 3B-13). When used in turn lanes, at least two arrows should be used, one at or near the upstream end of the full-width turn lane and one an appropriate distance upstream from the stop line or intersection (see Drawing A of Figure 3B-11).*

Figure 3B-25. Examples of Elongated Route Shields for Pavement Markings

A - Interstate Shield on dark or light pavement	B - U.S. Route Shield on dark pavement	C - U.S. Route Shield on light pavement	D - State Route Shield on dark pavement	E - State Route Shield on light pavement

Notes:
1. See the "Standard Highway Signs and Markings" book for other sizes and details
2. Colors and elongated shapes simulating State route shield signs may be used for route shield pavement markings where appropriate

Figure 3B-26. Yield Ahead Triangle Symbols

A - Posted or Statutory Speed Limit of 45 mph or greater

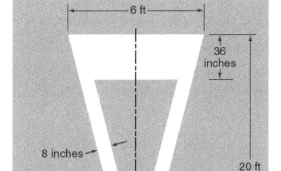

B - Posted or Statutory Speed Limit of less than 45 mph

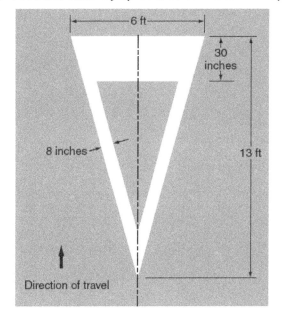

Option:

An additional arrow or arrows may be used in a turn lane. When arrows are used for a short turn lane, the second (downstream) arrow may be omitted based on engineering judgment.

Guidance:

Where opposing offset channelized left-turn lanes exist, lane-use arrow markings should be placed near the downstream terminus of the offset left-turn lanes to reduce wrong-way movements (see Figure 2B-17).

Support:

An arrow at the downstream end of a turn lane can help to prevent wrong way movements.

Standard:

Where through lanes approaching an intersection become mandatory turn lanes, lane-use arrow markings (see Figure 3B-24) shall be used and shall be accompanied by standard signs.

Guidance:

Where through lanes approaching an intersection become mandatory turn lanes, ONLY word markings (see Figure 3B-23) should be used in addition to the required lane-use arrow markings and signs (see Sections 2B.19 and 2B.20). These markings and signs should be placed well in advance of the turn and should be repeated as necessary to prevent entrapment and to help the road user select the appropriate lane in advance of reaching a queue of waiting vehicles (see Drawing A of Figure 3B-11).

Option:

On freeways or expressways where a through lane becomes a mandatory exit lane, lane-use arrow markings may be used on the approach to the exit in the dropped lane and in an adjacent optional through-or-exit lane if one exists.

Guidance:

A two-way left-turn lane-use arrow pavement marking, with opposing arrows spaced as shown in Figure 3B-7, should be used at or just downstream from the beginning of a two-way left-turn lane.

Option:

Additional two-way left-turn lane-use arrow markings may be used at other locations along a two-way left-turn lane where engineering judgment determines that such additional markings are needed to emphasize the proper use of the lane.

Figure 3B-27. Examples of Lane-Use Control Word and Arrow Pavement Markings

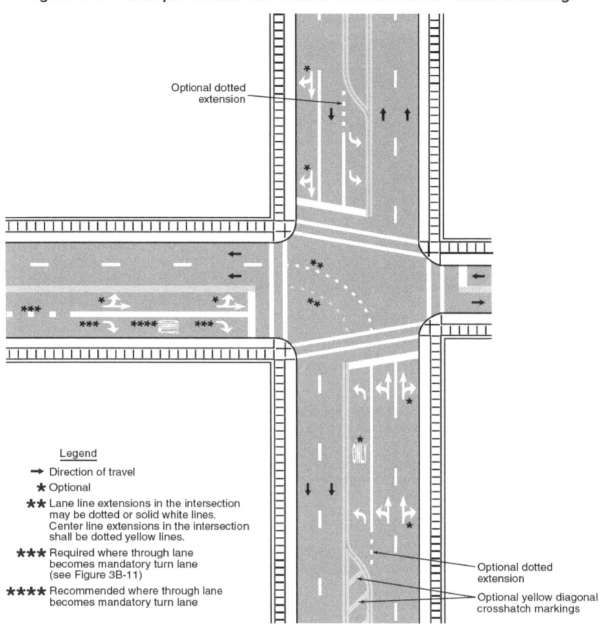

Standard:

A single-direction lane-use arrow shall not be used in a lane bordered on both sides by yellow two-way left-turn lane longitudinal markings.

Lane-use, lane-reduction, and wrong-way arrow markings shall be designed as shown in Figure 3B-24 and in the "Standard Highway Signs and Markings" book (see Section 1A.11).

Option:

The ONLY word marking (see Figure 3B-23) may be used to supplement the lane-use arrow markings in lanes that are designated for the exclusive use of a single movement (see Figure 3B-27) or to supplement a preferential lane word or symbol marking (see Section 3D.01).

Standard:

The ONLY word marking shall not be used in a lane that is shared by more than one movement.

Guidance:

34 *Where a lane-reduction transition occurs on a roadway with a speed limit of 45 mph or more, the lane-reduction arrow markings shown in Drawing F in Figure 3B-24 should be used (see Figure 3B-14). Except for acceleration lanes, where a lane-reduction transition occurs on a roadway with a speed limit of less than 45 mph, the lane-reduction arrow markings shown in Drawing F in Figure 3B-24 should be used if determined to be appropriate based on engineering judgment.*

Option:

35 Lane-reduction arrow markings may be used in long acceleration lanes based on engineering judgment.

Guidance:

36 *Where crossroad channelization or ramp geometrics do not make wrong-way movements difficult, the appropriate lane-use arrow should be placed in each lane of an exit ramp near the crossroad terminal where it will be clearly visible to a potential wrong-way road user (see Figure 2B-18).*

Option:

37 The wrong-way arrow markings shown in Drawing D in Figure 3B-24 may be placed near the downstream terminus of a ramp as shown in Figures 2B-18 and 2B-19, or at other locations where lane-use arrows are not appropriate, to indicate the correct direction of traffic flow and to discourage drivers from traveling in the wrong direction.

Section 3B.21 Speed Measurement Markings

Support:

01 A speed measurement marking is a transverse marking placed on the roadway to assist the enforcement of speed regulations.

Standard:

02 **Speed measurement markings, if used, shall be white, and shall not be greater than 24 inches in width.**

Option:

03 Speed measurement markings may extend 24 inches on either side of the center line or 24 inches on either side of edge line markings at 1/4-mile intervals over a 1-mile length of roadway. When paved shoulders of sufficient width are available, the speed measurement markings may be placed entirely on these shoulders (see Drawing A of Figure 3B-10). Advisory signs may be used in conjunction with these markings.

Section 3B.22 Speed Reduction Markings

Support:

01 Speed reduction markings (see Figure 3B-28) are transverse markings that are placed on the roadway within a lane (along both edges of the lane) in a pattern of progressively reduced spacing to give drivers the impression that their speed is increasing. These markings might be placed in advance of an unexpectedly severe horizontal or vertical curve or other roadway feature where drivers need to decelerate prior to reaching the feature and where the desired reduction in speeds has not been achieved by the installation of warning signs and/or other traffic control devices.

Guidance:

02 *If used, speed reduction markings should be reserved for unexpected curves and should not be used on long tangent sections of roadway or in areas frequented mainly by local or familiar drivers, (e.g., school zones). If used, speed reduction markings should supplement the appropriate warning signs and other traffic control devices and should not substitute for these devices.*

Standard:

03 **If used, speed reduction markings shall be a series of white transverse lines on both sides of the lane that are perpendicular to the center line, edge line, or lane line. The longitudinal spacing between the markings shall be progressively reduced from the upstream to the downstream end of the marked portion of the lane.**

Guidance:

04 *Speed reduction markings should not be greater than 12 inches in width, and should not extend more than 18 inches into the lane.*

Standard:

05 **Speed reduction markings shall not be used in lanes that do not have a longitudinal line (center line, edge line, or lane line) on both sides of the lane.**

Figure 3B-28. Example of the Application of Speed Reduction Markings

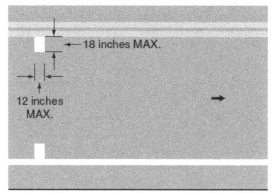

Section 3B.23 Curb Markings

Support:

01   Curb markings are most often used to indicate parking regulations or to delineate the curb.

Standard:

02   **Where curbs are marked to convey parking regulations in areas where curb markings are frequently obscured by snow and ice accumulation, signs shall be used with the curb markings except as provided in Paragraph 4.**

Guidance:

03   *Except as provided in Paragraph 4, when curb markings are used without signs to convey parking regulations, a legible word marking regarding the regulation (such as "No Parking" or "No Standing") should be placed on the curb.*

Option:

04   Curb markings without word markings or signs may be used to convey a general prohibition by statute of parking within a specified distance of a STOP sign, YIELD sign, driveway, fire hydrant, or crosswalk.

05   Local highway agencies may prescribe special colors for curb markings to supplement standard signs for parking regulation.

Support:

06   Since yellow and white curb markings are frequently used for curb delineation and visibility, it is advisable to establish parking regulations through the installation of standard signs (see Sections 2B.46 through 2B.48).

Standard:

07   **Where curbs are marked for delineation or visibility purposes, the colors shall comply with the general principles of markings (see Section 3A.05).**

Guidance:

08   *Retroreflective solid yellow markings should be placed on the approach ends of raised medians and curbs of islands that are located in the line of traffic flow where the curb serves to channel traffic to the right of the obstruction.*

09   *Retroreflective solid white markings should be used when traffic is permitted to pass on either side of the island.*

Support:

10   Where the curbs of the islands become parallel to the direction of traffic flow, it is not necessary to mark the curbs unless an engineering study indicates the need for this type of delineation.

11 Curbs at openings in a continuous median island need not be marked unless an engineering study indicates the need for this type of marking.

Option:

12 Retroreflective or internally illuminated raised pavement markers of the appropriate color may be placed on the pavement in front of the curb and/or on the top of curbed as of raised medians and curbs of islands, as a supplement to or substitute for retroreflective curb markings used for delineation.

Section 3B.24 Chevron and Diagonal Crosshatch Markings

Option:

01 Chevron and diagonal crosshatch markings may be used to discourage travel on certain paved areas, such as shoulders, gore areas, flush median areas between solid double yellow center line markings or between white channelizing lines approaching obstructions in the roadway (see Section 3B.10 and Figure 3B-15), between solid double yellow center line markings forming flush medians or channelized travel paths at intersections (see Figures 3B-2 and 3B-5), buffer spaces between preferential lanes and general-purpose lanes (see Figures 3D-2 and 3D-4), and at grade crossings (see Part 8).

Standard:

02 **When crosshatch markings are used in paved areas that separate traffic flows in the same general direction, they shall be white and they shall be shaped as chevron markings, with the point of each chevron facing toward approaching traffic, as shown in Figure 3B-8, Drawing A of Figure 3B-9, Figure 3B-10, and Drawing C of Figure 3B-15.**

03 **When crosshatch markings are used in paved areas that separate opposing directions of traffic, they shall be yellow diagonal markings that slant away from traffic in the adjacent travel lanes, as shown in Figures 3B-2 and 3B-5 and Drawings A and B of Figure 3B-15.**

04 **When crosshatch markings are used on paved shoulders, they shall be diagonal markings that slant away from traffic in the adjacent travel lane. The diagonal markings shall be yellow when used on the left-hand shoulders of the roadways of divided highways and on the left-hand shoulders of one-way streets or ramps. The diagonal markings shall be white when used on right-hand shoulders.**

Guidance:

05 *The chevrons and diagonal lines used for crosshatch markings should be at least 12 inches wide for roadways having a posted or statutory speed limit of 45 mph or greater, and at least 8 inches wide for roadways having posted or statutory speed limit of less than 45 mph. The longitudinal spacing of the chevrons or diagonal lines should be determined by engineering judgment considering factors such as speeds and desired visual impacts. The chevrons and diagonal lines should form an angle of approximately 30 to 45 degrees with the longitudinal lines that they intersect.*

Section 3B.25 Speed Hump Markings

Standard:

01 **If speed hump markings are used, they shall be a series of white markings placed on a speed hump to identify its location. If markings are used for a speed hump that does not also function as a crosswalk or speed Table, the markings shall comply with Option A, B, or C shown in Figure 3B-29. If markings are used for a speed hump that also functions as a crosswalk or speed Table, the markings shall comply with Option A or B shown in Figure 3B-30.**

Section 3B.26 Advance Speed Hump Markings

Option:

01 Advance speed hump markings (see Figure 3B-31) may be used in advance of speed humps or other engineered vertical roadway deflections such as dips where added visibility is desired or where such deflection is not expected.

02 Advance pavement wording such as BUMP or HUMP (see Section 3B.20) may be used on the approach to a speed hump either alone or in conjunction with advance speed hump markings. Appropriate advance warning signs may be used in compliance with Section 2C.29.

Standard:

03 **If advance speed hump markings are used, they shall be a series of eight white 12-inch transverse lines that become longer and are spaced closer together as the vehicle approaches the speed hump or other deflection. If advance markings are used, they shall comply with the detailed design shown in Figure 3B-31.**

Guidance:

04 *If used, advance speed hump markings should be installed in each approach lane.*

Figure 3B-29. Pavement Markings for Speed Humps without Crosswalks

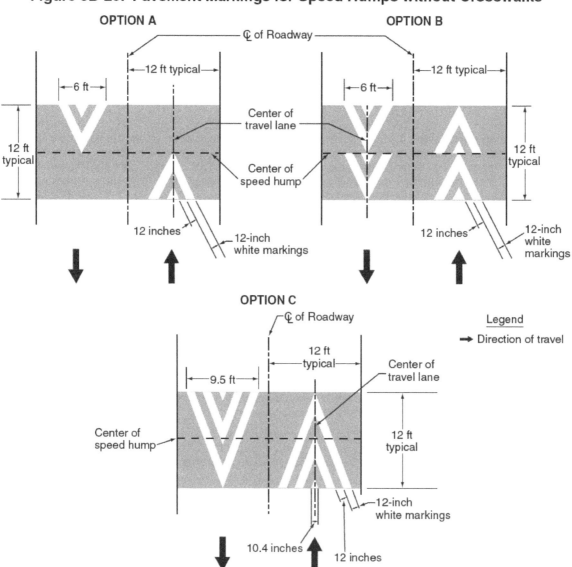

Figure 3B-30. Pavement Markings for Speed Tables or Speed Humps with Crosswalks

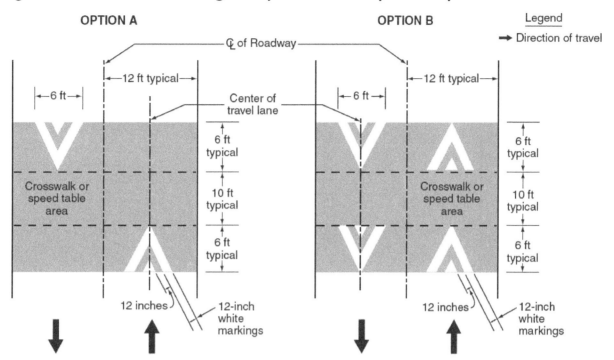

Note: Optional crosswalk lines are not shown in this figure

Figure 3B-31. Advance Warning Markings for Speed Humps

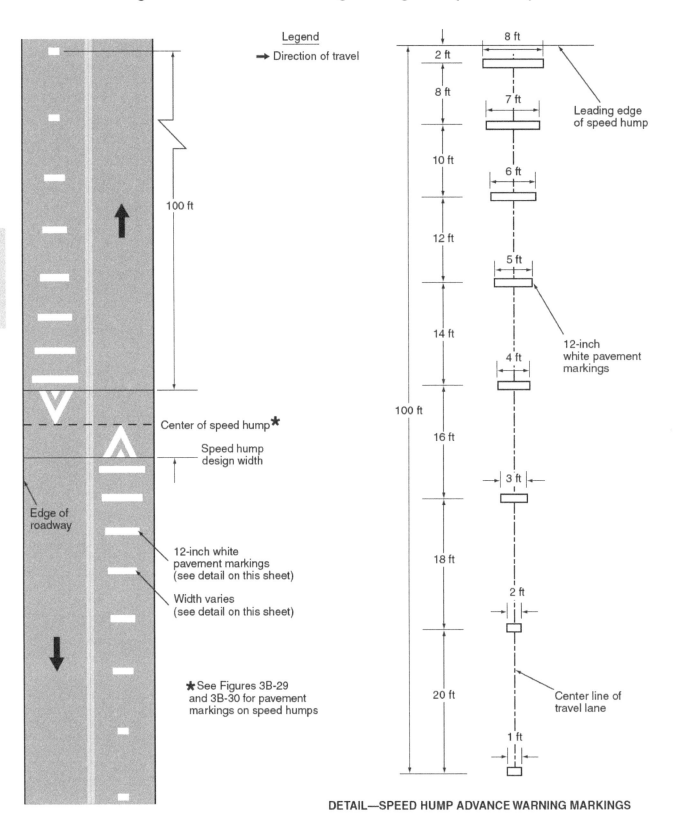

DETAIL—SPEED HUMP ADVANCE WARNING MARKINGS

CHAPTER 3C. ROUNDABOUT MARKINGS

Section 3C.01 General

Support:

01 A roundabout (see definition in Section 1A.13) is a specific type of circular intersection designed to control speeds and having specific traffic control features.

Guidance:

02 *Pavement markings and signing for a roundabout should be integrally designed to correspond to the geometric design and intended lane use of a roundabout.*

03 *Markings on the approaches to a roundabout and on the circular roadway should be compatible with each other to provide a consistent message to road users and should facilitate movement through the roundabout such that vehicles do not have to change lanes within the circulatory roadway in order to exit the roundabout in a given direction.*

Support:

04 Figure 3C-1 provides an example of the pavement markings for approach and circulatory roadways at a roundabout. Figure 3C-2 shows the options that are available for lane-use pavement marking arrows on approaches to roundabouts. Figures 3C-3 through 3C-14 illustrate examples of markings for roundabouts of various geometric and lane-use configurations.

05 Traffic control signals or pedestrian hybrid beacons (see Part 4) are sometimes used at roundabouts to facilitate the crossing of pedestrians or to meter traffic.

06 Section 8C.12 contains information about roundabouts that contain or are in close proximity to grade crossings.

Figure 3C-1. Example of Markings for Approach and Circulatory Roadways at a Roundabout

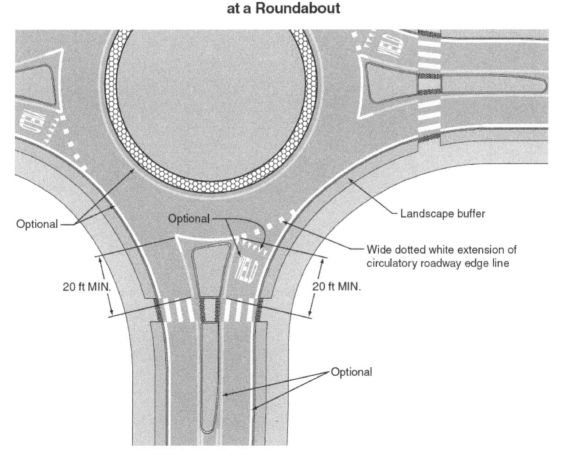

Figure 3C-2. Lane-Use Arrow Pavement Marking Options for Roundabout Approaches

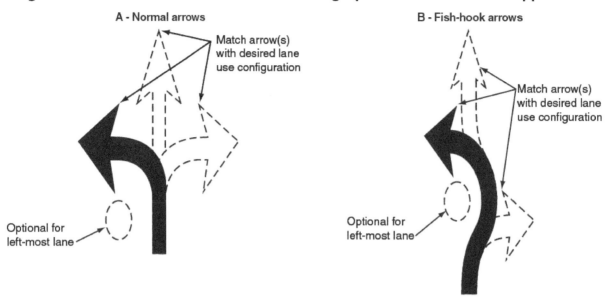

Figure 3C-3. Example of Markings for a One-Lane Roundabout

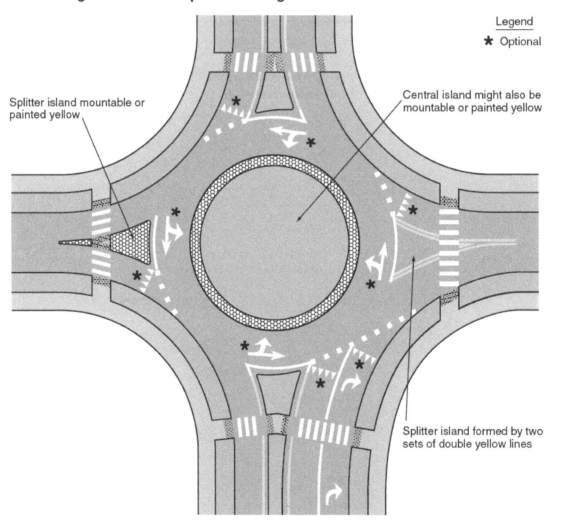

Figure 3C-4. Example of Markings for a Two-Lane Roundabout with One- and Two-Lane Approaches (Sheet 1 of 2)

A – Unextended central island

Figure 3C-4. Example of Markings for a Two-Lane Roundabout with One- and Two-Lane Approaches (Sheet 2 of 2)

B – Central island extended by pavement markings

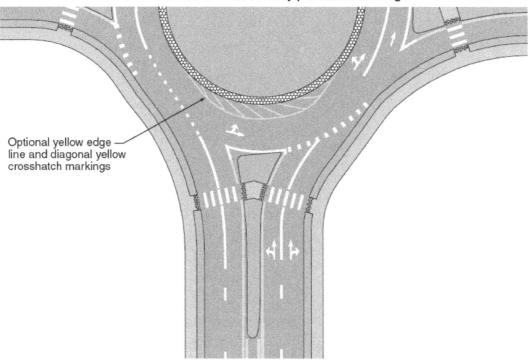

Optional yellow edge line and diagonal yellow crosshatch markings

C – Central island extended by a truck apron

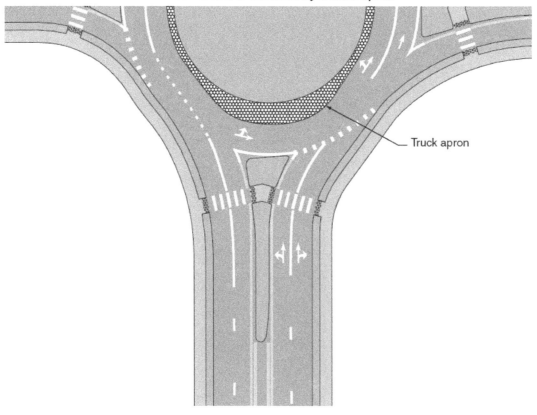

Truck apron

Figure 3C-5. Example of Markings for a Two-Lane Roundabout with One-Lane Exits

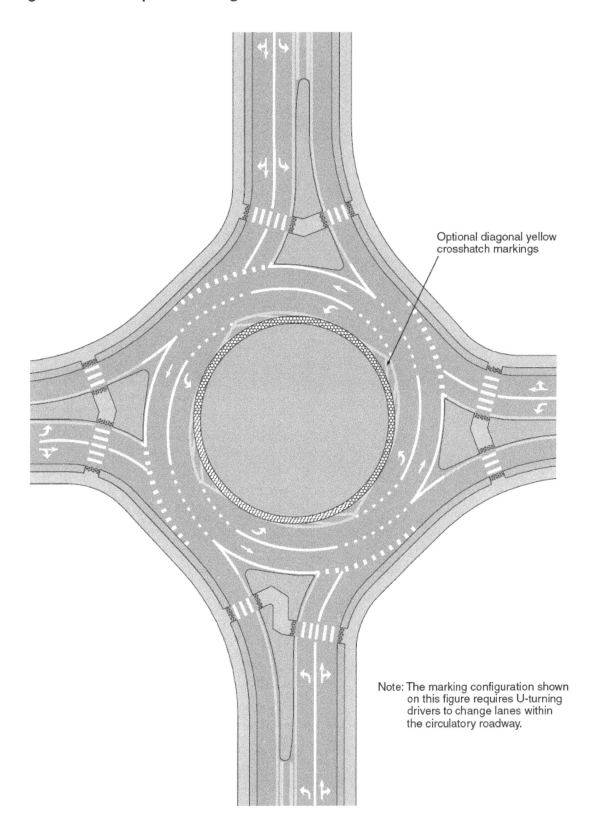

Optional diagonal yellow crosshatch markings

Note: The marking configuration shown on this figure requires U-turning drivers to change lanes within the circulatory roadway.

Figure 3C-6. Example of Markings for a Two-Lane Roundabout with Two-Lane Exits

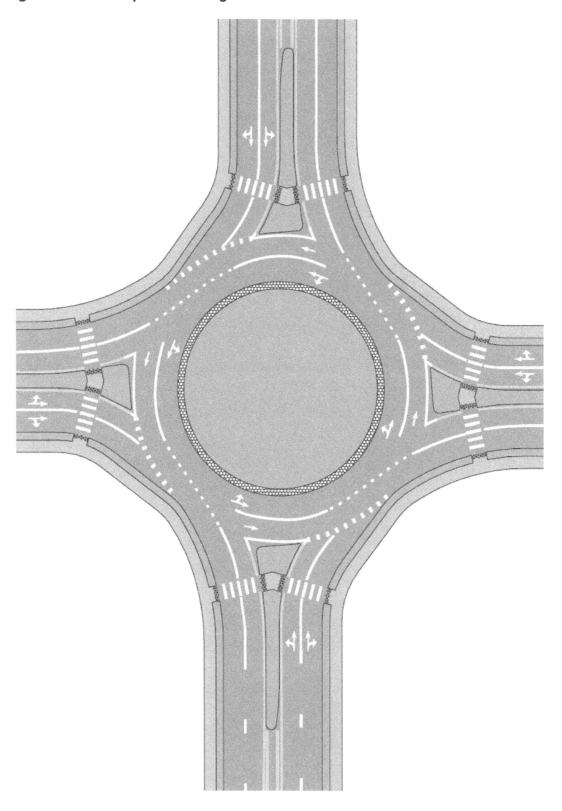

Figure 3C-7. Example of Markings for a Two-Lane Roundabout with a Double Left Turn

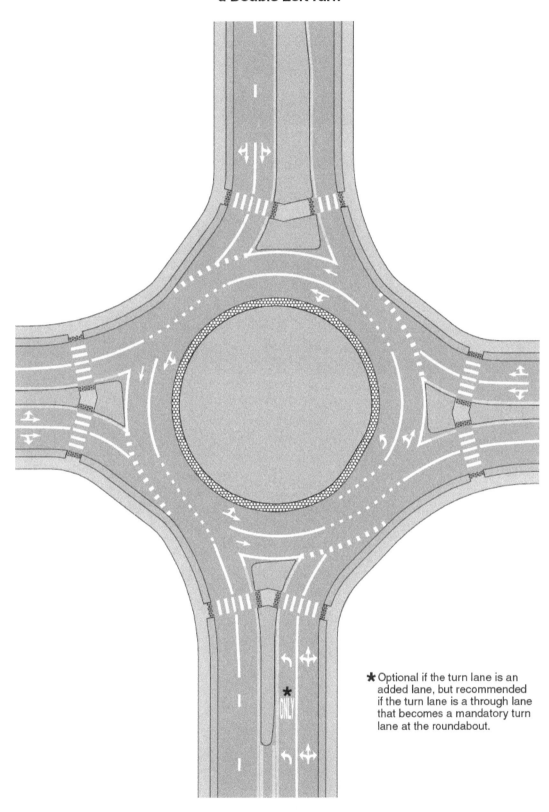

★ Optional if the turn lane is an added lane, but recommended if the turn lane is a through lane that becomes a mandatory turn lane at the roundabout.

Figure 3C-8. Example of Markings for a Two-Lane Roundabout with a Double Right Turn

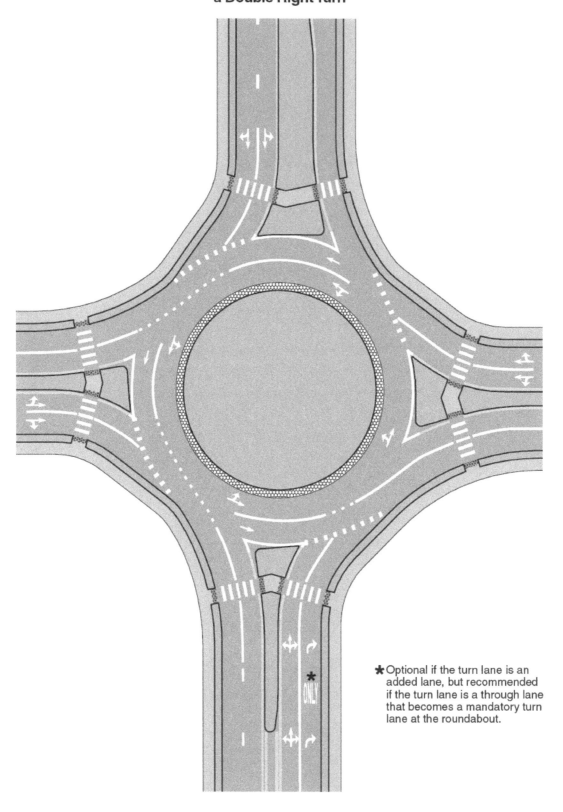

* Optional if the turn lane is an added lane, but recommended if the turn lane is a through lane that becomes a mandatory turn lane at the roundabout.

Figure 3C-9. Example of Markings for a Two-Lane Roundabout with Consecutive Double Left Turns

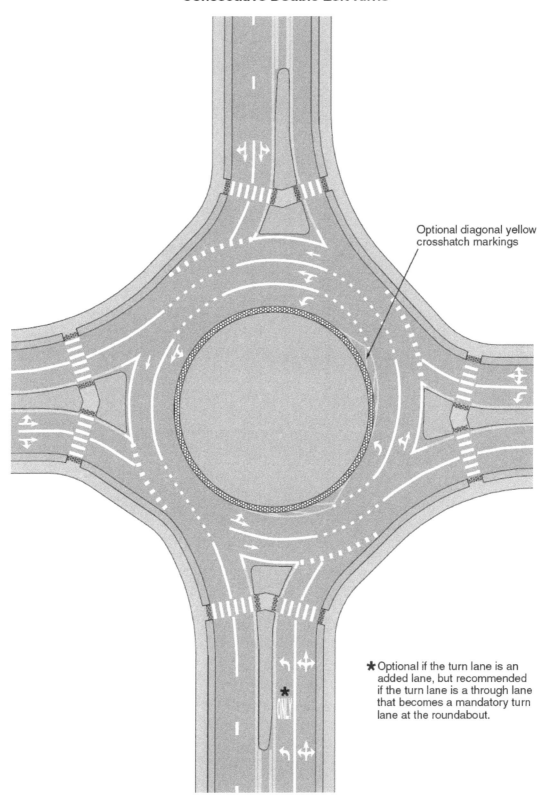

Figure 3C-10. Example of Markings for a Three-Lane Roundabout with Two- and Three-Lane Approaches

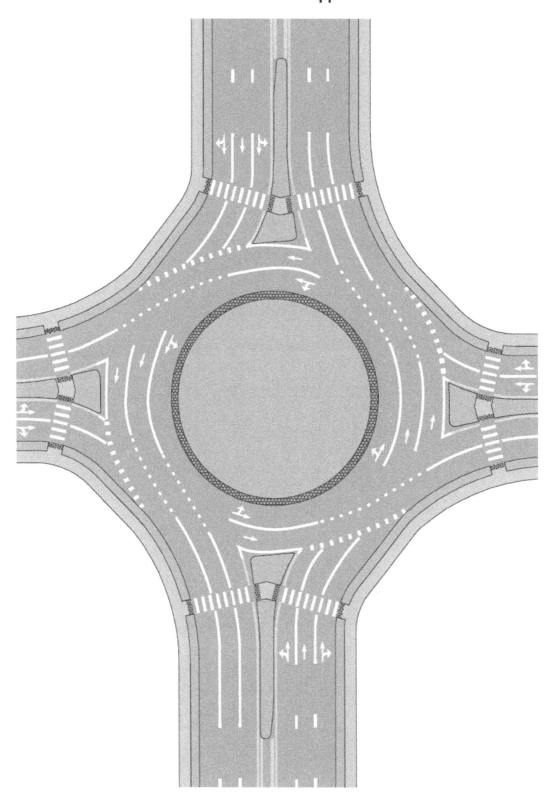

Figure 3C-11. Example of Markings for a Three-Lane Roundabout with Three-Lane Approaches

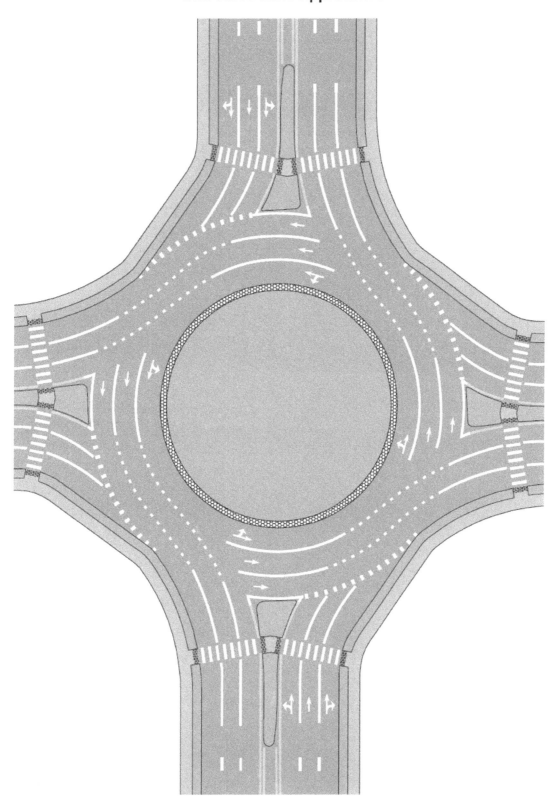

Figure 3C-12. Example of Markings for a Three-Lane Roundabout with Two-Lane Exits

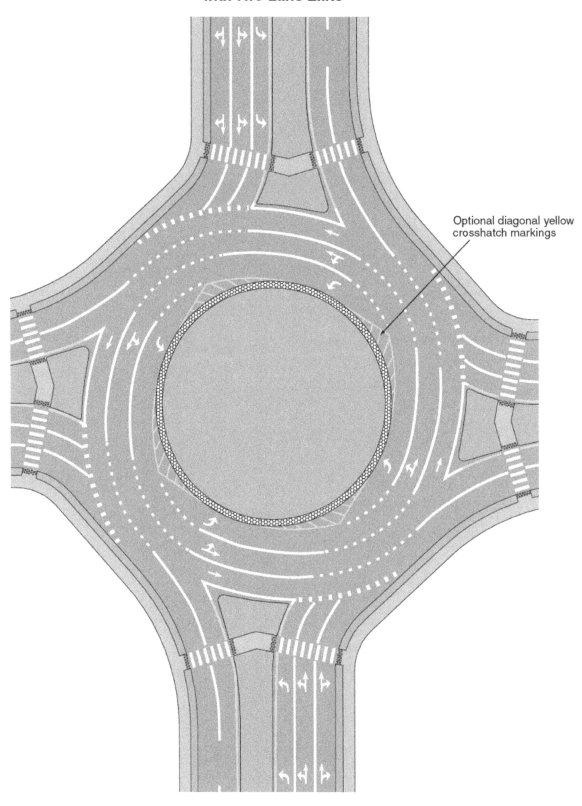

Figure 3C-13. Example of Markings for Two Linked Roundabouts

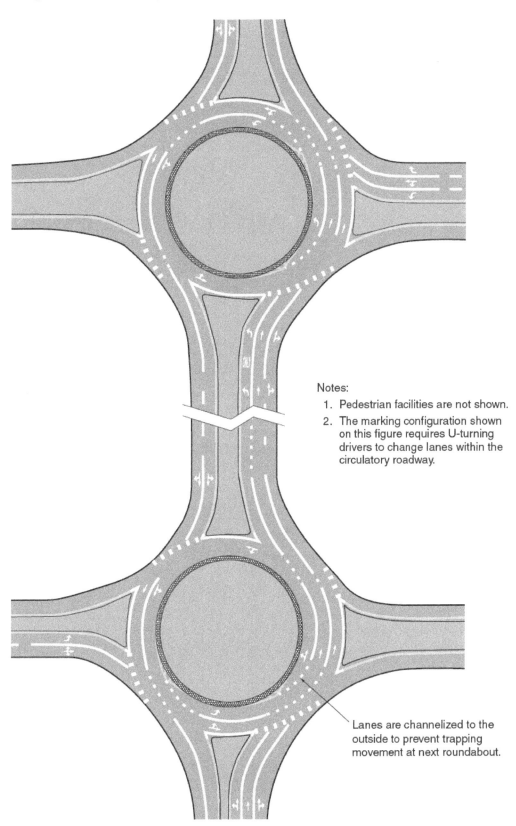

Notes:
1. Pedestrian facilities are not shown.
2. The marking configuration shown on this figure requires U-turning drivers to change lanes within the circulatory roadway.

Lanes are channelized to the outside to prevent trapping movement at next roundabout.

Figure 3C-14. Example of Markings for a Diamond Interchange with Two Circular-Shaped Roundabout Ramp Terminals

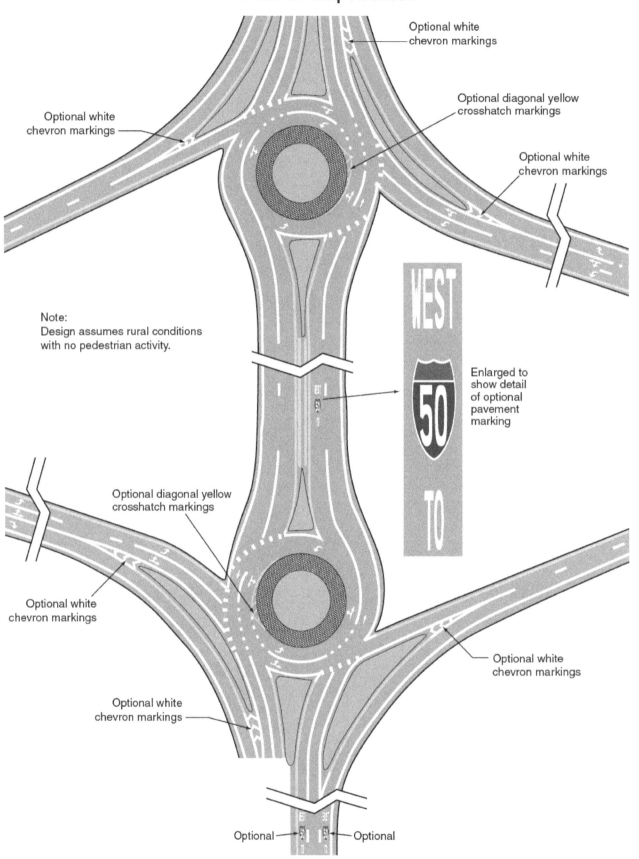

Section 3C.02 White Lane Line Pavement Markings for Roundabouts

Standard:

01 **Multi-lane approaches to roundabouts shall have lane lines.**

02 **A through lane on a roadway that becomes a dropped lane (mandatory turn lane) at a roundabout shall be marked with a dotted white lane line in accordance with Section 3B.04.**

Guidance:

03 *Multi-lane roundabouts should have lane line markings within the circulatory roadway to channelize traffic to the appropriate exit lane.*

Standard:

04 **Continuous concentric lane lines shall not be used within the circulatory roadway of roundabouts.**

Support:

05 Section 9C.04 contains information regarding bicycle lane markings at roundabouts.

Section 3C.03 Edge Line Pavement Markings for Roundabout Circulatory Roadways

Guidance:

01 *A white edge line should be used on the outer (right-hand) side of the circulatory roadway.*

02 *Where a white edge line is used for the circulatory roadway, it should be as follows (see Figure 3C-1):*
   A. *A solid line adjacent to the splitter island, and*
   B. *A wide dotted line across the lane(s) entering the roundabout.*

Standard:

03 **Edge lines and edge line extensions shall not be placed across the exits from the circulatory roadway at roundabouts.**

Option:

04 A yellow edge line may be placed around the inner (left-hand) edge of the circulatory roadway (see Figure 3C-1) and may be used to channelize traffic (see Drawing B of Figure 3C-4).

Section 3C.04 Yield Lines for Roundabouts

Option:

01 A yield line (see Section 3B.16) may be used to indicate the point behind which vehicles are required to yield at the entrance to a roundabout (see Figure 3C-1).

Section 3C.05 Crosswalk Markings at Roundabouts

Standard:

01 **Pedestrian crosswalks shall not be marked to or from the central island of roundabouts.**

Guidance:

02 *If pedestrian facilities are provided, crosswalks (see Section 3B.18) should be marked across roundabout entrances and exits to indicate where pedestrians are intended to cross.*

03 *Crosswalks should be a minimum of 20 feet from the edge of the circulatory roadway.*

Support:

04 Various arrangements of crosswalks at roundabouts are illustrated in the figures in this Chapter.

Section 3C.06 Word, Symbol, and Arrow Pavement Markings for Roundabouts

Option:

01 Lane-use arrows may be used on any approach to and within the circulatory roadway of any roundabout.

02 YIELD (word) and YIELD AHEAD (symbol or word) pavement markings (see Figure 3C-1) may be used on approaches to roundabouts.

03 Word and/or route shield pavement markings may be used on an approach to or within the circulatory roadway of a roundabout to provide route and/or destination guidance information to road users (see Figure 3C-14).

Guidance:

04 *Within the circulatory roadway of multi-lane roundabouts, normal lane-use arrows (see Section 3B.20 and Figure 3B-24) should be used.*

05 *On multi-lane approaches with double left-turn and/or double right-turn lanes, lane-use arrows as shown in Figures 3C-7 and 3C-8 should be used.*

Option:

06 If used on approaches to a roundabout, lane-use arrows may be either normal or fish-hook arrows, either with or without an oval symbolizing the central island, as shown in Figure 3C-2.

Section 3C.07 Markings for Other Circular Intersections

Support:

01 Other circular intersections include, but are not limited to, rotaries, traffic circles, and residential traffic calming designs.

Option:

02 The markings shown in this Chapter may be used at other circular intersections if engineering judgment indicates that their presence will benefit drivers, pedestrians, or other road users.

CHAPTER 3D. MARKINGS FOR PREFERENTIAL LANES

Section 3D.01 Preferential Lane Word and Symbol Markings

Support:

01   Preferential lanes are established for one or more of a wide variety of special uses, including, but not limited to, high-occupancy vehicle (HOV) lanes, ETC lanes, high-occupancy toll (HOT) lanes, bicycle lanes, bus only lanes, taxi only lanes, and light rail transit only lanes.

Standard:

02   **When a lane is assigned full or part time to a particular class or classes of vehicles, the preferential lane word and symbol markings described in this Section and the preferential lane longitudinal markings described in Section 3D.02 shall be used.**

03   **All longitudinal pavement markings, as well as word and symbol pavement markings, associated with a preferential lane shall end where the Preferential Lane Ends (R3-12a or R3-12c) sign (see Section 2G.07) designating the downstream end of the preferential only lane restriction is installed.**

04   **Static or changeable message regulatory signs (see Sections 2G.03 to 2G.07) shall be used with preferential lane word or symbol markings.**

05   **All preferential lane word and symbol markings shall be white and shall be positioned laterally in the center of the preferential lane.**

06   **Where a preferential lane use exists contiguous to a general-purpose lane or is separated from a general-purpose lane by a flush buffered space that can be traversed by motor vehicles, the preferential lane shall be marked with one or more of the following symbol or word markings for the preferential lane use specified:**
   A. **HOV lane—the preferential lane-use marking for high-occupancy vehicle lanes shall consist of white lines formed in a diamond shape symbol or the word message HOV. The diamond shall be at least 2.5 feet wide and 12 feet in length. The lines shall be at least 6 inches in width.**
   B. **HOT lane or ETC Account-Only lane—except as provided in Paragraph 8, the preferential lane-use marking for a HOT lane or an ETC Account-Only lane shall consist of a word marking using the name of the ETC payment system required for use of the lane, such as E-Z PASS ONLY.**
   C. **Bicycle lane—the preferential lane-use marking for a bicycle lane shall consist of a bicycle symbol or the word marking BIKE LANE (see Chapter 9C and Figures 9C-1 and 9C-3 through 9C-6).**
   D. **Bus only lane—the preferential lane-use marking for a bus only lane shall consist of the word marking BUS ONLY.**
   E. **Taxi only lane—the preferential lane-use marking for a taxi only lane shall consist of the word marking TAXI ONLY.**
   F. **Light rail transit lane—the preferential lane-use marking for a light rail transit lane shall consist of the word marking LRT ONLY.**
   G. **Other type of preferential lane—the preferential lane-use markings shall consist of a word marking appropriate to the restriction.**

07   **If two or more preferential lane uses are permitted in a single lane, the symbol or word marking for each preferential lane use shall be installed.**

Option:

08   Preferential lane-use symbol or word markings may be omitted at toll plazas where physical conditions preclude the use of the markings (see Section 3E.01).

Guidance:

09   *The spacing of the markings should be based on engineering judgment that considers the prevailing speed, block lengths, distance from intersections, and other factors that affect clear communication to the road user.*

Support:

10   Markings spaced as close as 80 feet apart might be appropriate on city streets, while markings spaced as far as 1,000 feet apart might be appropriate for freeways.

Guidance:

11   *In addition to a regular spacing interval, the preferential lane marking should be placed at strategic locations such as major decision points, direct exit ramp departures from the preferential lane, and along access openings to and from adjacent general-purpose lanes. At decision points, the preferential lane marking should be placed on all applicable lanes and should be visible to approaching traffic for all available departures. At direct exits from preferential lanes where extra emphasis is needed, the use of word markings (such as "EXIT" or "EXIT ONLY") in the deceleration lane for the direct exit and/or on the direct exit ramp itself just beyond the exit gore should be considered.*

Option:

¹² A numeral indicating the vehicle occupancy requirements established for a high-occupancy vehicle lane may be included in sequence after the diamond symbol or HOV word message.

Guidance:

¹³ *Engineering judgment should determine the need for supplemental devices such as tubular markers, traffic cones, or other channelizing devices (see Chapter 3H).*

Section 3D.02 Preferential Lane Longitudinal Markings for Motor Vehicles

Support:

⁰¹ Preferential lanes can take many forms depending on the level of usage and the design of the facility. They might be barrier-separated or buffer-separated from the adjacent general-purpose lanes, or they might be contiguous with the adjacent general-purpose lanes. Barrier-separated preferential lanes might be operated in a constant direction or be operated as reversible lanes. Some reversible preferential lanes on a divided highway might be operated counter-flow to the direction of traffic on the immediately adjacent general-purpose lanes. See Section 1A.13 for definitions of terms.

⁰² Preferential lanes might be operated full-time (24 hours per day on all days), for extended periods of the day, part-time (restricted usage during specific hours on specified days), or on a variable basis (such as a strategy for a managed lane).

Standard:

⁰³ **Longitudinal pavement markings for preferential lanes shall be as follows (these same requirements are presented in tabular form in Table 3D-1):**

   A. **Barrier-separated, non-reversible preferential lane**—the longitudinal pavement markings for preferential lanes that are physically separated from the other travel lanes by a barrier or median shall consist of a normal solid single yellow line at the left-hand edge of the travel lane(s), and a normal solid single white line at the right-hand edge of the travel lane(s) (see Drawing A in Figure 3D-1).

   B. **Barrier-separated, reversible preferential lane**—the longitudinal pavement markings for reversible preferential lanes that are physically separated from the other travel lanes by a barrier or median shall consist of a normal solid single white line at both edges of the travel lane(s) (see Drawing B in Figure 3D-1).

   C. **Buffer-separated (left-hand side) preferential lane**—the longitudinal pavement markings for a full-time or part-time preferential lane on the left-hand side of and separated from the other travel lanes by a neutral buffer space shall consist of a normal solid single yellow line at the left-hand edge of the preferential travel lane(s) and one of the following at the right-hand edge of the preferential travel lane(s):

   1. A wide solid double white line along both edges of the buffer space where crossing the buffer space is prohibited (see Drawing A in Figure 3D-2).
   2. A wide solid single white line along both edges of the buffer space where crossing the buffer space is discouraged (see Drawing B in Figure 3D-2).
   3. A wide broken single white line along both edges of the buffer space, or a wide broken single white lane line within the allocated buffer space (resulting in wider lanes), where crossing the buffer space is permitted (see Drawing C in Figure 3D-2).

   D. **Buffer-separated (right-hand side) preferential lane**—the longitudinal pavement markings for a full-time or part-time preferential lane on the right-hand side of and separated from the other travel lanes by a neutral buffer space shall consist of a normal solid single white line at the right-hand edge of the preferential travel lane(s) if warranted (see Section 3B.07) and one of the following at the left-hand edge of the preferential travel lane(s) (see Drawing D in Figure 3D-2):

   1. A wide solid double white line along both edges of the buffer space where crossing the buffer space is prohibited.
   2. A wide solid single white line along both edges of the buffer space where crossing of the buffer space is discouraged.
   3. A wide broken single white line along both edges of the buffer space, or a wide broken single white line within the allocated buffer space (resulting in wider lanes), where crossing the buffer space is permitted.
   4. A wide dotted single white lane line within the allocated buffer space (resulting in wider lanes) where crossing the buffer space is permitted for any vehicle to perform a right-turn maneuver.

Table 3D-1. Standard Edge Line and Lane Line Markings for Preferential Lanes

Type of Preferential Lane	Left-Hand Edge Line	Right-Hand Edge Line
Barrier-Separated, Non-Reversible	A normal solid single yellow line	A normal solid single white line (see Drawing A of Figure 3D-1)
Barrier-Separated, Reversible	A normal solid single white line	A normal solid single white line (see Drawing B of Figure 3D-1)
Buffer-Separated, Left-Hand Side	A normal solid single yellow line	A wide solid double white line along both edges of the buffer space where crossing is prohibited (see Drawing A of Figure 3D-2) A wide solid single white line along both edges of the buffer space where crossing is discouraged (see Drawing B of Figure 3D-2) A wide broken single white line along both edges of the buffer space, or a wide broken single white line within the buffer space (resulting in wider lanes), where crossing is permitted (see Drawing C of Figure 3D-2)
Buffer-Separated, Right-Hand Side	A wide solid double white line along both edges of the buffer space where crossing is prohibited (see Drawing D of Figure 3D-2) A wide solid single white line along both edges of the buffer space where crossing is discouraged (see Drawing D of Figure 3D-2) A wide broken single white line along both edges of the buffer space, or a wide broken single white line within the buffer space (resulting in wider lanes), where crossing is permitted (see Drawing D of Figure 3D-2) A wide dotted single white line within the buffer space (resulting in wider lanes) where crossing is permitted for any vehicle to perform a right-turn maneuver (see Drawing D of Figure 3D-2)	A normal solid single white line (if warranted)
Contiguous, Left-Hand Side	A normal solid single yellow line	A wide solid double white line where crossing is prohibited (see Drawing A of Figure 3D-3) A wide solid single white line where crossing is discouraged (see Drawing B of Figure 3D-3) A wide broken single white line where crossing is permitted (see Drawing C of Figure 3D-3)
Contiguous, Right-Hand Side	A wide solid double white line where crossing is prohibited (see Drawing D of Figure 3D-3) A wide solid single white line where crossing is discouraged (see Drawing D of Figure 3D-3) A wide broken single white line where crossing is permitted (see Drawing D of Figure 3D-3) A wide dotted single white line where crossing is permitted for any vehicle to perform a right-turn maneuver (see Drawing D of Figure 3D-3)	A normal solid single white line

Notes: 1. If there are two or more preferential lanes, the lane lines between the preferential lanes shall be normal broken white lines.
2. The standard lane markings listed in this table are provided in a tabular format for reference.
3. This information is also described in Paragraph 3 of Section 3D.02.

    E. **Contiguous (left-hand side) preferential lane**—the longitudinal pavement markings for a full-time or part-time preferential lane on the left-hand side of and contiguous to the other travel lanes shall consist of a normal solid single yellow line at the left-hand edge of the preferential travel lane(s) and one of the following at the right-hand edge of the preferential travel lane(s):
       1. A wide solid double white lane line where crossing is prohibited (see Drawing A in Figure 3D-3).
       2. A wide solid single white lane line where crossing is discouraged (see Drawing B in Figure 3D 3).
       3. A wide solid single white lane line where crossing is permitted (see Drawing C in Figure 3D-3).

    F. **Contiguous (right-hand side) preferential lane**—the longitudinal pavement markings for a full-time or part-time preferential lane on the right-hand side of and contiguous to the other travel lanes shall consist of a normal solid single white line at the right-hand edge of the preferential travel lane(s) if warranted (see Section 3B.07) and one of the following at the left-hand edge of the preferential travel lane(s) (see Drawing D in Figure 3D-3):
       1. A wide solid double white lane line where crossing is prohibited.
       2. A wide solid single white lane line where crossing is discouraged.
       3. A wide broken single white lane line where crossing is permitted.
       4. A wide dotted single white lane line where crossing is permitted for any vehicle to perform a right-turn maneuver.

Figure 3D-1. Markings for Barrier-Separated Preferential Lanes

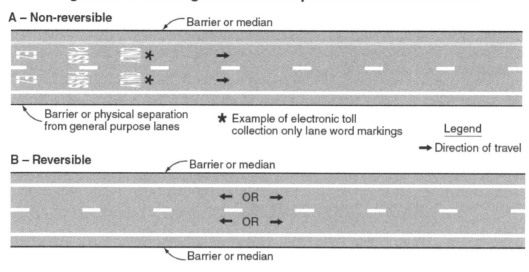

Figure 3D-2. Markings for Buffer-Separated Preferential Lanes (Sheet 1 of 2)

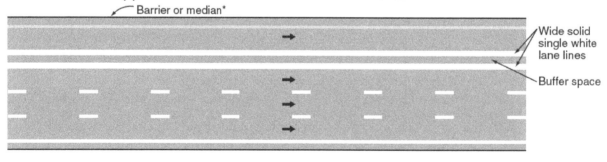

Figure 3D-2. Markings for Buffer-Separated Preferential Lanes (Sheet 2 of 2)

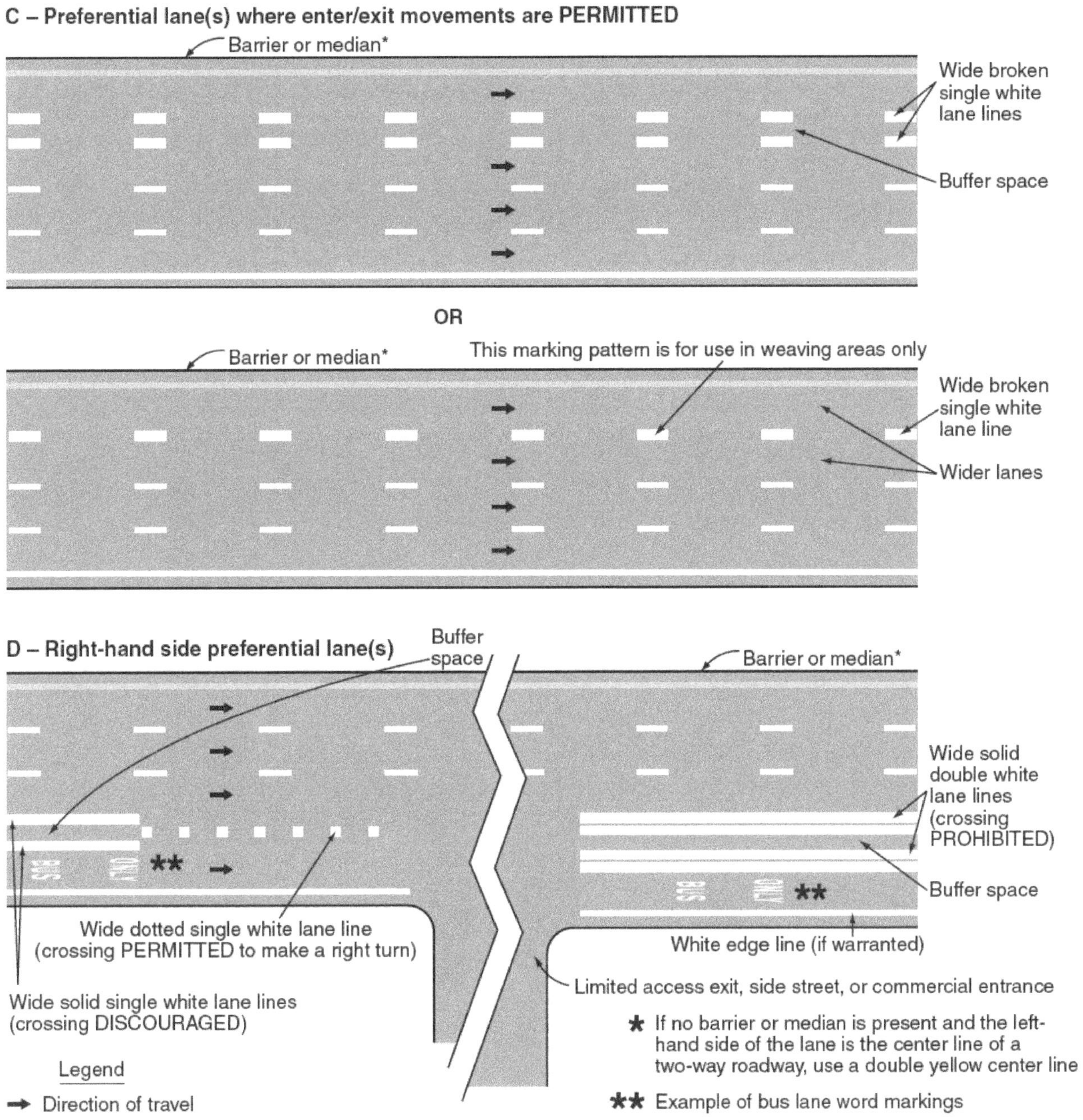

C – Preferential lane(s) where enter/exit movements are PERMITTED

- Barrier or median*
- Wide broken single white lane lines
- Buffer space

OR

This marking pattern is for use in weaving areas only

- Barrier or median*
- Wide broken single white lane line
- Wider lanes

D – Right-hand side preferential lane(s)

- Buffer space
- Barrier or median*
- Wide solid double white lane lines (crossing PROHIBITED)
- Buffer space
- White edge line (if warranted)
- Wide dotted single white lane line (crossing PERMITTED to make a right turn)
- Wide solid single white lane lines (crossing DISCOURAGED)
- Limited access exit, side street, or commercial entrance

Legend
→ Direction of travel

* If no barrier or median is present and the left-hand side of the lane is the center line of a two-way roadway, use a double yellow center line

** Example of bus lane word markings

Figure 3D-3. Markings for Contiguous Preferential Lanes

A – Full-time preferential lane(s) where enter/exit movements are PROHIBITED

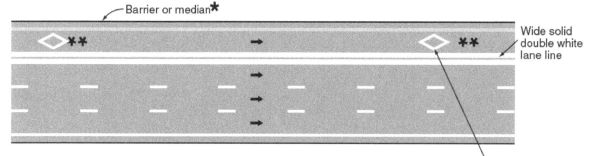

B – Preferential lane(s) where enter/exit movements are DISCOURAGED Space at 1/4-mile intervals

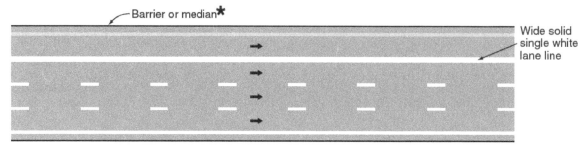

C – Preferential lane(s) where enter/exit movements are PERMITTED

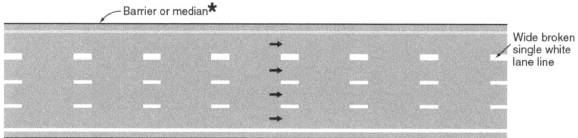

D – Right-hand side preferential lane(s)

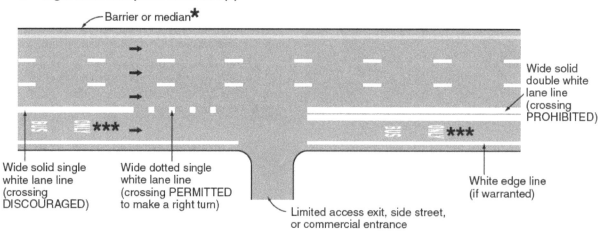

Legend
→ Direction of travel

★ If no barrier or median is present and the left-hand side of the lane is the center line of a two-way roadway, use a double yellow center line

★★ Example of HOV only lane symbol markings

★★★ Example of bus lane word markings

Guidance:

04 *Where preferential lanes and other travel lanes are separated by a buffer space wider than 4 feet and crossing the buffer space is prohibited, chevron markings (see Section 3B.24) should be placed in the buffer area (see Drawing A in Figure 3D-2). The chevron spacing should be 100 feet or greater.*

Option:

05 If a full-time or part-time contiguous preferential lane is separated from the other travel lanes by a wide broken single white line (see Drawing C in Figure 3D-3), the spacing or skip pattern of the line may be reduced and the width of the line may be increased.

Standard:

06 **If there are two or more preferential lanes for traffic moving in the same direction, the lane lines between the preferential lanes shall be normal broken white lines.**

07 **Preferential lanes for motor vehicles shall also be marked with the appropriate word or symbol pavement markings in accordance with Section 3D.01 and shall have appropriate regulatory signs in accordance with Sections 2G.03 through 2G.07.**

Guidance:

08 *At direct exits from a preferential lane, dotted white line markings should be used to separate the tapered or parallel deceleration lane for the direct exit (including the taper) from the adjacent continuing preferential through lane, to reduce the chance of unintended exit maneuvers.*

Standard:

09 **On a divided highway, a part-time counter-flow preferential lane that is contiguous to the travel lanes in the opposing direction shall be separated from the opposing direction lanes by the standard reversible lane longitudinal marking, a normal width broken double yellow line (see Section 3B.03 and Drawing A of Figure 3D-4). If a buffer space is provided between the part-time counter-flow preferential lane and the opposing direction lanes, a normal width broken double yellow line shall be placed along both edges of the buffer space (see Drawing B of Figure 3D-4). Signs (see Section 2B.26), lane-use control signals (see Chapter 4M), or both shall be used to supplement the reversible lane markings.**

10 **On a divided highway, a full-time counter-flow preferential lane that is contiguous to the travel lanes in the opposing direction shall be separated from the opposing direction lanes by a solid double yellow center line marking (see Drawing C of Figure 3D-4). If a buffer space is provided between the full-time counter-flow preferential lane and the opposing direction lanes, a normal width solid double yellow line shall be placed along both edges of the buffer space (see Drawing D of Figure 3D-4).**

Option:

11 Cones, tubular markers, or other channelizing devices (see Chapter 3H) may also be used to separate the opposing lanes when a counter-flow preferential lane operation is in effect.

Figure 3D-4. Markings for Counter-Flow Preferential Lanes on Divided Highways

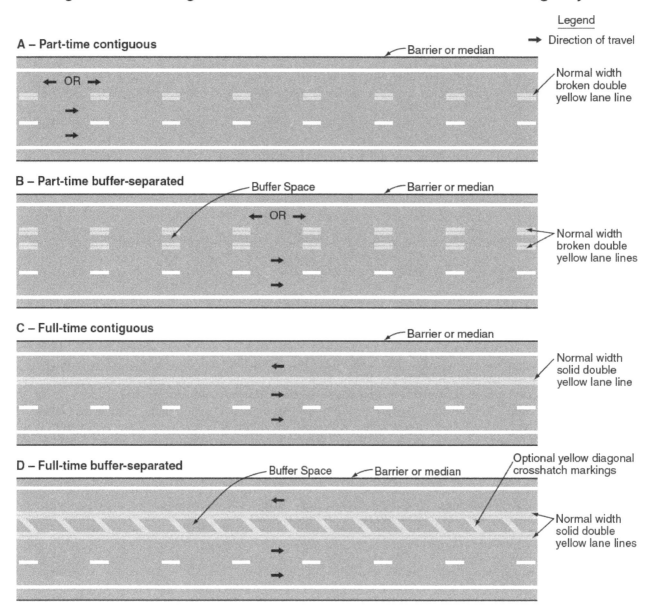

CHAPTER 3E. MARKINGS FOR TOLL PLAZAS

Section 3E.01 Markings for Toll Plazas

Support:

01 At toll plazas, pavement markings help road users identify the proper lane(s) to use for the type of toll payment they plan to use, to channelize movements into the various lanes, and to delineate obstructions in the roadway.

Standard:

02 **When a lane on the approach to a toll plaza is restricted to use only by vehicles with registered ETC accounts, the ETC Account-Only lane word markings described in Section 3D.01 and the preferential lane longitudinal markings described in Section 3D.02 shall be used. When one or more ORT lanes that are restricted to use only by vehicles with registered ETC accounts bypass a mainline toll plaza on a separate alignment, these word markings and longitudinal markings shall be used on the approach to the point where the ORT lanes diverge from the lanes destined for the mainline toll plaza.**

Option:

03 Preferential lane-use symbol or word markings may be omitted at toll plazas where physical conditions preclude the use of the markings.

Guidance:

04 *If an ORT lane that is immediately adjacent to a mainline toll plaza is not separated from adjacent cash payment toll plaza lanes by a curb or barrier, then channelizing devices (see Section 3H.01), and/or longitudinal pavement markings that discourage or prohibit lane changing should be used to separate the ORT lane from the adjacent cash payment lane. This separation should begin on the approach to the mainline toll plaza at approximately the point where the vehicle speeds in the adjacent cash lanes drop below 30 mph during off-peak periods and should extend downstream beyond the toll plaza approximately to the point where the vehicles departing the toll plaza in the adjacent cash lanes have accelerated to 30 mph.*

Option:

05 For a toll plaza approach lane that is restricted to use only by vehicles with registered ETC accounts, the solid white lane line or edge line on the right-hand side of the ETC Account-Only lane and the solid white lane line or solid yellow edge line on the left-hand side of the ETC Account-Only lane may be supplemented with purple solid longitudinal markings placed contiguous to the inside edges of the lines defining the lane.

Standard:

06 **If used, the purple solid longitudinal marking described in the previous paragraph shall be a minimum of 3 inches in width and a maximum width equal to the width of the line it supplements, and ETC Account-Only preferential lane word markings (see Section 3D.01) shall be installed within the lane.**

07 **Toll booths and the islands on which they are located are considered to be obstructions in the roadway and they shall be provided with markings that comply with the provisions of Section 3B.10 and Chapter 3G.**

Option:

08 Longitudinal pavement markings may be omitted alongside toll booth islands between the approach markings and any departure markings.

CHAPTER 3F. DELINEATORS

Section 3F.01 Delineators
Support:
01 Delineators are particularly beneficial at locations where the alignment might be confusing or unexpected, such as at lane-reduction transitions and curves. Delineators are effective guidance devices at night and during adverse weather. An important advantage of delineators in certain locations is that they remain visible when the roadway is wet or snow covered.

02 Delineators are considered guidance devices rather than warning devices.

Option:
03 Delineators may be used on long continuous sections of highway or through short stretches where there are changes in horizontal alignment.

Section 3F.02 Delineator Design
Standard:
01 **Delineators shall consist of retroreflective devices that are capable of clearly retroreflecting light under normal atmospheric conditions from a distance of 1,000 feet when illuminated by the high beams of standard automobile lights.**

02 **Retroreflective elements for delineators shall have a minimum dimension of 3 inches.**

Support:
03 Within a series of delineators along a roadway, delineators for a given direction of travel at a specific location are referred to as single delineators if they have one retroreflective element for that direction, double delineators if they have two identical retroreflective elements for that direction mounted together, or vertically elongated delineators if they have a single retroreflective element with an elongated vertical dimension to approximate the vertical dimension of two separate single delineators.

Option:
04 A vertically elongated delineator of appropriate size may be used in place of a double delineator.

Section 3F.03 Delineator Application
Standard:
01 **The color of delineators shall comply with the color of edge lines stipulated in Section 3B.06.**

02 **A series of single delineators shall be provided on the right-hand side of freeways and expressways and on at least one side of interchange ramps, except when either Condition A or Condition B is met, as follows:**
   A. **On tangent sections of freeways and expressways when both of the following conditions are met:**
      1. **Raised pavement markers are used continuously on lane lines throughout all curves and on all tangents to supplement pavement markings, and**
      2. **Roadside delineators are used to lead into all curves.**
   B. **On sections of roadways where continuous lighting is in operation between interchanges.**

Option:
03 Delineators may be provided on other classes of roads. A series of single delineators may be provided on the left-hand side of roadways.

Standard:
04 **Delineators on the left-hand side of a two-way roadway shall be white (see Figure 3F-1).**

Guidance:
05 *A series of single delineators should be provided on the outside of curves on interchange ramps.*

06 *Where median crossovers are provided for official or emergency use on divided highways and where these crossovers are to be marked, a double yellow delineator should be placed on the left-hand side of the through roadway on the far side of the crossover for each roadway.*

07 *Double or vertically elongated delineators should be installed at 100-foot intervals along acceleration and deceleration lanes.*

08 *A series of delineators should be used wherever guardrail or other longitudinal barriers are present along a roadway or ramp.*

Figure 3F-1. Examples of Delineator Placement

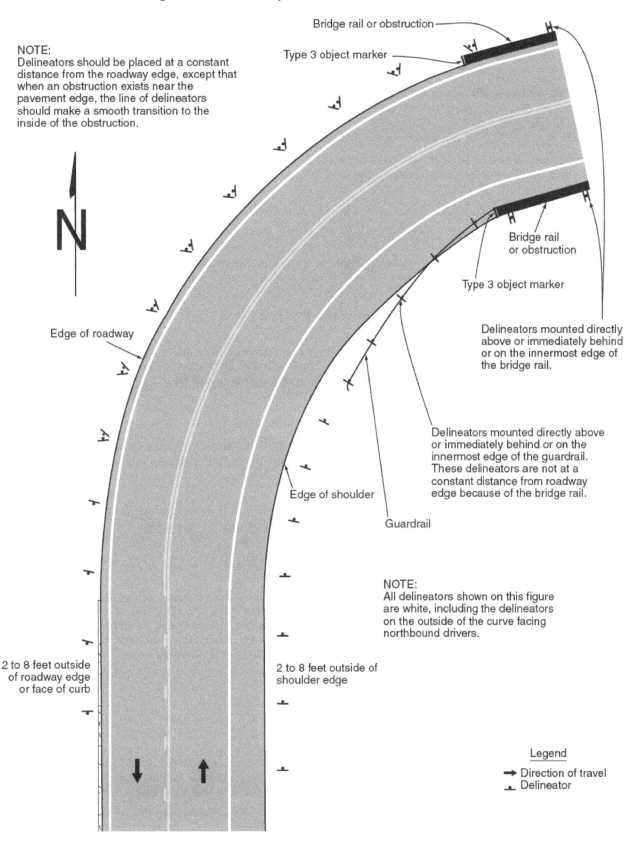

Option:

09 Red delineators may be used on the reverse side of any delineator where it would be viewed by a road user traveling in the wrong direction on that particular ramp or roadway.

10 Delineators of the appropriate color may be used to indicate a lane-reduction transition where either an outside or inside lane merges into an adjacent lane.

Guidance:

11 *When used for lane-reduction transitions, the delineators should be installed adjacent to the lane or lanes reduced for the full length of the transition and should be so placed and spaced to show the reduction (see Figure 3B-14).*

Support:

12 Delineators are not necessary for traffic moving in the direction of a wider pavement or on the side of the roadway where the alignment is not affected by the lane-reduction transition.

Guidance:

13 *On a highway with continuous delineation on either or both sides, delineators should be carried through transitions.*

Option:

14 On a highway with continuous delineation on either or both sides, the spacing between a series of delineators may be closer.

Standard:

15 **When used on a truck escape ramp, delineators shall be red.**

Guidance:

16 *Red delineators should be placed on both sides of truck escape ramps. The delineators should be spaced at 50-foot intervals for a distance sufficient to identify the ramp entrance. Delineator spacing beyond the ramp entrance should be adequate for guidance according to the length and design of the escape ramp.*

Section 3F.04 Delineator Placement and Spacing

Guidance:

01 *Delineators should be mounted on suitable supports at a mounting height, measured vertically from the bottom of the lowest retroreflective device to the elevation of the near edge of the roadway, of approximately 4 feet.*

Option:

02 When mounted on the face of or on top of guardrails or other longitudinal barriers, delineators may be mounted at a lower elevation than the normal delineator height recommended in Paragraph 1.

Guidance:

03 *Delineators should be placed 2 to 8 feet outside the outer edge of the shoulder, or if appropriate, in line with the roadside barrier that is 8 feet or less outside the outer edge of the shoulder.*

04 *Delineators should be placed at a constant distance from the edge of the roadway, except that where an obstruction intrudes into the space between the pavement edge and the extension of the line of the delineators, the delineators should be transitioned to be in line with or inside the innermost edge of the obstruction. If the obstruction is a guardrail or other longitudinal barrier, the delineators should be transitioned to be just behind, directly above (in line with), or on the innermost edge of the guardrail or longitudinal barrier.*

05 *Delineators should be spaced 200 to 530 feet apart on mainline tangent sections. Delineators should be spaced 100 feet apart on ramp tangent sections.*

Support:

06 Examples of delineator installations are shown in Figure 3F-1.

Option:

07 When uniform spacing is interrupted by such features as driveways and intersections, delineators which would ordinarily be located within the features may be relocated in either direction for a distance not exceeding one quarter of the uniform spacing. Delineators still falling within such features may be eliminated.

08 Delineators may be transitioned in advance of a lane transition or obstruction as a guide for oncoming traffic.

Guidance:

09 *The spacing of delineators should be adjusted on approaches to and throughout horizontal curves so that several delineators are always simultaneously visible to the road user. The approximate spacing shown in Table 3F-1 should be used.*

Option:

10 When needed for special conditions, delineators of the appropriate color may be mounted in a closely-spaced manner on the face of or on top of guardrails or other longitudinal barriers to form a continuous or nearly continuous "ribbon" of delineation.

Table 3F-1. Approximate Spacing for Delineators on Horizontal Curves

Radius (R) of Curve	Approximate Spacing (S) on Curve
50 feet	20 feet
115 feet	25 feet
180 feet	35 feet
250 feet	40 feet
300 feet	50 feet
400 feet	55 feet
500 feet	65 feet
600 feet	70 feet
700 feet	75 feet
800 feet	80 feet
900 feet	85 feet
1,000 feet	90 feet

Notes:
1. Spacing for specific radii may be interpolated from table.
2. The minimum spacing should be 20 feet.
3. The spacing on curves should not exceed 300 feet.
4. In advance of or beyond a curve, and proceeding away from the end of the curve, the spacing of the first delineator is 2S, the second 3S, and the third 6S, but not to exceed 300 feet.
5. S refers to the delineator spacing for specific radii computed from the formula $S = 3\sqrt{R-50}$.
6. The distances for S shown in the table above were rounded to the nearest 5 feet.

CHAPTER 3G. COLORED PAVEMENTS

Section 3G.01 General

Support:

01 Colored pavements consist of differently colored road paving materials, such as colored asphalt or concrete, or paint or other marking materials applied to the surface of a road or island to simulate a colored pavement.

02 If non-retroreflective colored pavement, including bricks and other types of patterned surfaces, is used as a purely aesthetic treatment and is not intended to communicate a regulatory, warning, or guidance message to road users, the colored pavement is not considered to be a traffic control device, even if it is located between the lines of a crosswalk.

Standard:

03 **If colored pavement is used within the traveled way, on flush or raised islands, or on shoulders to regulate, warn, or guide traffic or if retroreflective colored pavement is used, the colored pavement is considered to be a traffic control device and shall be limited to the following colors and applications:**

    A. Yellow pavement color shall be used only for flush or raised median islands separating traffic flows in opposite directions or for left-hand shoulders of roadways of divided highways or one-way streets or ramps.

    B. White pavement color shall be used for flush or raised channelizing islands where traffic passes on both sides in the same general direction or for right-hand shoulders.

04 **Colored pavements shall not be used as a traffic control device, unless the device is applicable at all times.**

Guidance:

05 *Colored pavements used as traffic control devices should be used only where they contrast significantly with adjoining paved areas.*

06 *Colored pavement located between crosswalk lines should not use colors or patterns that degrade the contrast of white crosswalk lines, or that might be mistaken by road users as a traffic control application.*

CHAPTER 3H. CHANNELIZING DEVICES USED FOR EMPHASIS OF PAVEMENT MARKING PATTERNS

Section 3H.01 Channelizing Devices

Option:

01 Channelizing devices, as described in Sections 6F.63 through 6F.73, and 6F.75, and as shown in Figure 6F-7, such as cones, tubular markers, vertical panels, drums, lane separators, and raised islands, may be used for general traffic control purposes such as adding emphasis to reversible lane delineation, channelizing lines, or islands. Channelizing devices may also be used along a center line to preclude turns or along lane lines to preclude lane changing, as determined by engineering judgment.

Standard:

02 **Except for color, the design of channelizing devices, including but not limited to retroreflectivity, minimum dimensions, and mounting height, shall comply with the provisions of Chapter 6F.**

03 **The color of channelizing devices used outside of temporary traffic control zones shall be either orange or the same color as the pavement marking that they supplement, or for which they are substituted.**

04 **For nighttime use, channelizing devices shall be retroreflective (as described in Part 6) or internally illuminated. On channelizing devices used outside of temporary traffic control zones, retroreflective sheeting or bands shall be white if the devices separate traffic flows in the same direction and shall be yellow if the devices separate traffic flows in the opposite direction or are placed along the left-hand edge line of a one-way roadway or ramp.**

Guidance:

05 *Channelizing devices should be kept clean and bright to maximize target value.*

CHAPTER 3I. ISLANDS

Section 3I.01 General

Support:

01 This Chapter addresses the characteristics of islands (see definition in Section 1A.13) as traffic-control devices. Criteria for the design of islands are set forth in "A Policy on Geometric Design of Highways and Streets" (see Section 1A.11).

Option:

02 An island may be designated by curbs, pavement edges, pavement markings, channelizing devices, or other devices.

Section 3I.02 Approach-End Treatment

Guidance:

01 *The ends of islands first approached by traffic should be preceded by diverging longitudinal pavement markings on the roadway surface, to guide vehicles into desired paths of travel along the island edge.*

Support:

02 The neutral area between approach-end markings that can be readily crossed even at considerable speed sometimes contains slightly raised (usually less than 1 inch high) sections of coarse aggregate or other suitable materials to create rumble sections that provide increased visibility of the marked areas and that produce an audible warning to road users traveling across them. For additional discouragement to driving in the neutral area, bars or buttons projecting 1 to 3 inches above the pavement surface are sometimes placed in the neutral area. These bars or buttons are designed so that any wheel encroachment within the area will be obvious to the vehicle operator, but will result in only minimal effects on control of the vehicle. Such bars or buttons are sometimes preceded by rumble sections or their height is gradually increased as approached by traffic.

Guidance:

03 *When raised bars or buttons are used in these neutral areas, they should be marked with white or yellow retroreflective materials, as determined by the direction or directions of travel they separate.*

Standard:

04 **Channelizing devices, when used in advance of islands having raised curbs, shall not be placed in such a manner as to constitute an unexpected obstacle.**

Option:

05 Pavement markings may be used with raised bars to better designate the island area.

Section 3I.03 Island Marking Application

Standard:

01 **Markings, as related to islands, shall consist only of pavement and curb markings, channelizing devices, and delineators.**

Guidance:

02 *Pavement markings as described in Section 3B.10 for the approach to an obstruction may be omitted on the approach to a particular island based on engineering judgment.*

Section 3I.04 Island Marking Colors

Guidance:

01 *Islands outlined by curbs or pavement markings should be marked with retroreflective white or yellow material as determined by the direction or directions of travel they separate (see Section 3A.05).*

02 *The retroreflective area should be of sufficient length to denote the general alignment of the edge of the island along which vehicles travel, including the approach end, when viewed from the approach to the island.*

Option:

03 On long islands, curb retroreflection may be discontinued such that it does not extend for the entire length of the curb, especially if the island is illuminated or marked with delineators or edge lines.

Section 3I.05 Island Delineation
Standard:
01   **Delineators installed on islands shall be the same colors as the related edge lines except that, when facing wrong-way traffic, they shall be red (see Section 3F.03).**

02   **Each roadway through an intersection shall be considered separately in positioning delineators to assure maximum effectiveness.**

Option:

03   Retroreflective or internally illuminated raised pavement markers of the appropriate color may be placed on the pavement in front of the curb and/or on the top of curbed approach ends of raised medians and curbs of islands, as a supplement to or as a substitute for retroreflective curb markings.

Section 3I.06 Pedestrian Islands and Medians
Support:

01   Raised islands or medians of sufficient width that are placed in the center area of a street or highway can serve as a place of refuge for pedestrians who are attempting to cross at a midblock or intersection location. Center islands or medians allow pedestrians to find an adequate gap in one direction of traffic at a time, as the pedestrians are able to stop, if necessary, in the center island or median area and wait for an adequate gap in the other direction of traffic before crossing the second half of the street or highway. The minimum widths for accessible refuge islands and for design and placement of detectable warning surfaces are provided in the "Americans with Disabilities Act Accessibility Guidelines for Buildings and Facilities (ADAAG)" (see Section 1A.11).

CHAPTER 3J. RUMBLE STRIP MARKINGS

Section 3J.01 Longitudinal Rumble Strip Markings

Support:

01 Longitudinal rumble strips consist of a series of rough-textured or slightly raised or depressed road surfaces intended to alert inattentive drivers through vibration and sound that their vehicle has left the travel lane. Shoulder rumble strips are typically installed along the shoulder near the travel lane. On divided highways, rumble strips are sometimes installed on the median side (left-hand side) shoulder as well as on the outside (right-hand side) shoulder. On two-way roadways, rumble strips are sometimes installed along the center line.

02 This Manual contains no provisions regarding the design and placement of longitudinal rumble strips. The provisions in this Manual address the use of markings in combination with a longitudinal rumble strip.

Option:

03 An edge line or center line may be located over a longitudinal rumble strip to create a rumble stripe.

Standard:

04 **The color of an edge line or center line associated with a longitudinal rumble stripe shall be in accordance with Section 3A.05.**

05 **An edge line shall not be used in addition to a rumble stripe that is located along a shoulder.**

Support:

06 Figure 3J-1 illustrates markings used with or near longitudinal rumble strips.

Section 3J.02 Transverse Rumble Strip Markings

Support:

01 Transverse rumble strips consist of intermittent narrow, transverse areas of rough-textured or slightly raised or depressed road surface that extend across the travel lanes to alert drivers to unusual vehicular traffic conditions. Through noise and vibration, they attract the attention of road users to features such as unexpected changes in alignment and conditions requiring a reduction in speed or a stop.

02 This Manual contains no provisions regarding the design and placement of transverse rumble strips that approximate the color of the pavement. The provisions in this Manual address the use of markings in combination with a transverse rumble strip.

Standard:

03 **Except as otherwise provided in Section 6F.87 for TTC zones, if the color of a transverse rumble strip used within a travel lane is not the color of the pavement, the color of the transverse rumble strip shall be either black or white.**

Guidance:

04 *White transverse rumble strips used in a travel lane should not be placed in locations where they could be confused with other transverse markings such as stop lines or crosswalks.*

Figure 3J-1. Examples of Longitudinal Rumble Strip Markings

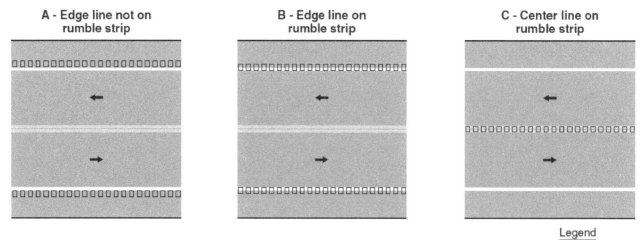

Note: Edge line may be located alongside the rumble strip (Option A) or on the rumble strip (Option B). Center line markings may also be located on a center line rumble strip (Option C).

PART 4
HIGHWAY TRAFFIC SIGNALS

CHAPTER 4A. GENERAL

Section 4A.01 Types

Support:

01 The following types and uses of highway traffic signals are discussed in Part 4: traffic control signals; pedestrian signals; hybrid beacons; emergency-vehicle signals; traffic control signals for one-lane, two-way facilities; traffic control signals for freeway entrance ramps; traffic control signals for movable bridges; toll plaza traffic signals; flashing beacons; lane-use control signals; and in-roadway lights.

Section 4A.02 Definitions Relating to Highway Traffic Signals

Support:

01 Definitions and acronyms pertaining to Part 4 are provided in Sections 1A.13 and 1A.14.

CHAPTER 4B. TRAFFIC CONTROL SIGNALS—GENERAL

Section 4B.01 General

Support:

01 Words such as pedestrians and bicyclists are used redundantly in selected Sections of Part 4 to encourage sensitivity to these elements of "traffic."

02 Standards for traffic control signals are important because traffic control signals need to attract the attention of a variety of road users, including those who are older, those with impaired vision, as well as those who are fatigued or distracted, or who are not expecting to encounter a signal at a particular location.

Section 4B.02 Basis of Installation or Removal of Traffic Control Signals

Guidance:

01 *The selection and use of traffic control signals should be based on an engineering study of roadway, traffic, and other conditions.*

Support:

02 A careful analysis of traffic operations, pedestrian and bicyclist needs, and other factors at a large number of signalized and unsignalized locations, coupled with engineering judgment, has provided a series of signal warrants, described in Chapter 4C, that define the minimum conditions under which installing traffic control signals might be justified.

Guidance:

03 *Engineering judgment should be applied in the review of operating traffic control signals to determine whether the type of installation and the timing program meet the current requirements of all forms of traffic.*

04 *If changes in traffic patterns eliminate the need for a traffic control signal, consideration should be given to removing it and replacing it with appropriate alternative traffic control devices, if any are needed.*

05 *If the engineering study indicates that the traffic control signal is no longer justified, and a decision is made to remove the signal, removal should be accomplished using the following steps:*

 A. *Determine the appropriate traffic control to be used after removal of the signal.*
 B. *Remove any sight-distance restrictions as necessary.*
 C. *Inform the public of the removal study.*
 D. *Flash or cover the signal heads for a minimum of 90 days, and install the appropriate stop control or other traffic control devices.*
 E. *Remove the signal if the engineering data collected during the removal study period confirms that the signal is no longer needed.*

Option:

06 Because Items C, D, and E in Paragraph 5 are not relevant when a temporary traffic control signal (see Section 4D.32) is removed, a temporary traffic control signal may be removed immediately after Items A and B are completed.

07 Instead of total removal of a traffic control signal, the poles, controller cabinet, and cables may remain in place after removal of the signal heads for continued analysis.

Section 4B.03 Advantages and Disadvantages of Traffic Control Signals

Support:

01 When properly used, traffic control signals are valuable devices for the control of vehicular and pedestrian traffic. They assign the right-of-way to the various traffic movements and thereby profoundly influence traffic flow.

02 Traffic control signals that are properly designed, located, operated, and maintained will have one or more of the following advantages:

 A. They provide for the orderly movement of traffic.
 B. They increase the traffic-handling capacity of the intersection if:
 1. Proper physical layouts and control measures are used, and
 2. The signal operational parameters are reviewed and updated (if needed) on a regular basis (as engineering judgment determines that significant traffic flow and/or land use changes have occurred) to maximize the ability of the traffic control signal to satisfy current traffic demands.
 C. They reduce the frequency and severity of certain types of crashes, especially right-angle collisions.
 D. They are coordinated to provide for continuous or nearly continuous movement of traffic at a definite speed along a given route under favorable conditions.
 E. They are used to interrupt heavy traffic at intervals to permit other traffic, vehicular or pedestrian, to cross.

₀₃ Traffic control signals are often considered a panacea for all traffic problems at intersections. This belief has led to traffic control signals being installed at many locations where they are not needed, adversely affecting the safety and efficiency of vehicular, bicycle, and pedestrian traffic.

₀₄ Traffic control signals, even when justified by traffic and roadway conditions, can be ill-designed, ineffectively placed, improperly operated, or poorly maintained. Improper or unjustified traffic control signals can result in one or more of the following disadvantages:
- A. Excessive delay,
- B. Excessive disobedience of the signal indications,
- C. Increased use of less adequate routes as road users attempt to avoid the traffic control signals, and
- D. Significant increases in the frequency of collisions (especially rear-end collisions).

Section 4B.04 Alternatives to Traffic Control Signals

Guidance:

₀₁ *Since vehicular delay and the frequency of some types of crashes are sometimes greater under traffic signal control than under STOP sign control, consideration should be given to providing alternatives to traffic control signals even if one or more of the signal warrants has been satisfied.*

Option:

₀₂ These alternatives may include, but are not limited to, the following:
- A. Installing signs along the major street to warn road users approaching the intersection;
- B. Relocating the stop line(s) and making other changes to improve the sight distance at the intersection;
- C. Installing measures designed to reduce speeds on the approaches;
- D. Installing a flashing beacon at the intersection to supplement STOP sign control;
- E. Installing flashing beacons on warning signs in advance of a STOP sign controlled intersection on major- and/or minor-street approaches;
- F. Adding one or more lanes on a minor-street approach to reduce the number of vehicles per lane on the approach;
- G. Revising the geometrics at the intersection to channelize vehicular movements and reduce the time required for a vehicle to complete a movement, which could also assist pedestrians;
- H. Revising the geometrics at the intersection to add pedestrian median refuge islands and/or curb extensions;
- I. Installing roadway lighting if a disproportionate number of crashes occur at night;
- J. Restricting one or more turning movements, perhaps on a time-of-day basis, if alternate routes are available;
- K. If the warrant is satisfied, installing multi-way STOP sign control;
- L. Installing a pedestrian hybrid beacon (see Chapter 4F) or In-Roadway Warning Lights (see Chapter 4N) if pedestrian safety is the major concern;
- M. Installing a roundabout; and
- N. Employing other alternatives, depending on conditions at the intersection.

Section 4B.05 Adequate Roadway Capacity

Support:

₀₁ The delays inherent in the alternating assignment of right-of-way at intersections controlled by traffic control signals can frequently be reduced by widening the major roadway, the minor roadway, or both roadways. Widening the minor roadway often benefits the operations on the major roadway, because it reduces the green time that must be assigned to minor-roadway traffic. In urban areas, the effect of widening can be achieved by eliminating parking on intersection approaches. It is desirable to have at least two lanes for moving traffic on each approach to a signalized location. Additional width on the departure side of the intersection, as well as on the approach side, will sometimes be needed to clear traffic through the intersection effectively.

Guidance:

₀₂ *Adequate roadway capacity should be provided at a signalized location. Before an intersection is widened, the additional green time pedestrians need to cross the widened roadways should be considered to determine if it will exceed the green time saved through improved vehicular flow.*

₀₃ *Other methods of increasing the roadway capacity at signalized locations that do not involve roadway widening, such as revisions to the pavement markings and the careful evaluation of proper lane-use assignments (including varying the lane use by time of day), should be considered where appropriate. Such consideration should include evaluation of any impacts that changes to pavement markings and lane assignments will have on bicycle travel.*

CHAPTER 4C. TRAFFIC CONTROL SIGNAL NEEDS STUDIES

Section 4C.01 Studies and Factors for Justifying Traffic Control Signals

Standard:

01   An engineering study of traffic conditions, pedestrian characteristics, and physical characteristics of the location shall be performed to determine whether installation of a traffic control signal is justified at a particular location.

02   The investigation of the need for a traffic control signal shall include an analysis of factors related to the existing operation and safety at the study location and the potential to improve these conditions, and the applicable factors contained in the following traffic signal warrants:

    Warrant 1, Eight-Hour Vehicular Volume
    Warrant 2, Four-Hour Vehicular Volume
    Warrant 3, Peak Hour
    Warrant 4, Pedestrian Volume
    Warrant 5, School Crossing
    Warrant 6, Coordinated Signal System
    Warrant 7, Crash Experience
    Warrant 8, Roadway Network
    Warrant 9, Intersection Near a Grade Crossing

03   The satisfaction of a traffic signal warrant or warrants shall not in itself require the installation of a traffic control signal.

Support:

04   Sections 8C.09 and 8C.10 contain information regarding the use of traffic control signals instead of gates and/or flashing-light signals at highway-rail grade crossings and highway-light rail transit grade crossings, respectively.

Guidance:

05   *A traffic control signal should not be installed unless one or more of the factors described in this Chapter are met.*

06   *A traffic control signal should not be installed unless an engineering study indicates that installing a traffic control signal will improve the overall safety and/or operation of the intersection.*

07   *A traffic control signal should not be installed if it will seriously disrupt progressive traffic flow.*

08   *The study should consider the effects of the right-turn vehicles from the minor-street approaches. Engineering judgment should be used to determine what, if any, portion of the right-turn traffic is subtracted from the minor-street traffic count when evaluating the count against the signal warrants listed in Paragraph 2.*

09   *Engineering judgment should also be used in applying various traffic signal warrants to cases where approaches consist of one lane plus one left-turn or right-turn lane. The site-specific traffic characteristics should dictate whether an approach is considered as one lane or two lanes. For example, for an approach with one lane for through and right-turning traffic plus a left-turn lane, if engineering judgment indicates that it should be considered a one-lane approach because the traffic using the left-turn lane is minor, the total traffic volume approaching the intersection should be applied against the signal warrants as a one-lane approach. The approach should be considered two lanes if approximately half of the traffic on the approach turns left and the left-turn lane is of sufficient length to accommodate all left-turn vehicles.*

10   *Similar engineering judgment and rationale should be applied to a street approach with one through/left-turn lane plus a right-turn lane. In this case, the degree of conflict of minor-street right-turn traffic with traffic on the major street should be considered. Thus, right-turn traffic should not be included in the minor-street volume if the movement enters the major street with minimal conflict. The approach should be evaluated as a one-lane approach with only the traffic volume in the through/left-turn lane considered.*

11   *At a location that is under development or construction and where it is not possible to obtain a traffic count that would represent future traffic conditions, hourly volumes should be estimated as part of an engineering study for comparison with traffic signal warrants. Except for locations where the engineering study uses the satisfaction of Warrant 8 to justify a signal, a traffic control signal installed under projected conditions should have an engineering study done within 1 year of putting the signal into stop-and-go operation to determine if the signal is justified. If not justified, the signal should be taken out of stop-and-go operation or removed.*

12   *For signal warrant analysis, a location with a wide median, even if the median width is greater than 30 feet, should be considered as one intersection.*

Option:

13　At an intersection with a high volume of left-turn traffic from the major street, the signal warrant analysis may be performed in a manner that considers the higher of the major-street left-turn volumes as the "minor-street" volume and the corresponding single direction of opposing traffic on the major street as the "major-street" volume.

14　For signal warrants requiring conditions to be present for a certain number of hours in order to be satisfied, any four sequential 15-minute periods may be considered as 1 hour if the separate 1-hour periods used in the warrant analysis do not overlap each other and both the major-street volume and the minor-street volume are for the same specific one-hour periods.

15　For signal warrant analysis, bicyclists may be counted as either vehicles or pedestrians.

Support:

16　When performing a signal warrant analysis, bicyclists riding in the street with other vehicular traffic are usually counted as vehicles and bicyclists who are clearly using pedestrian facilities are usually counted as pedestrians.

Option:

17　Engineering study data may include the following:
   A. The number of vehicles entering the intersection in each hour from each approach during 12 hours of an average day. It is desirable that the hours selected contain the greatest percentage of the 24-hour traffic volume.
   B. Vehicular volumes for each traffic movement from each approach, classified by vehicle type (heavy trucks, passenger cars and light trucks, public-transit vehicles, and, in some locations, bicycles), during each 15-minute period of the 2 hours in the morning and 2 hours in the afternoon during which total traffic entering the intersection is greatest.
   C. Pedestrian volume counts on each crosswalk during the same periods as the vehicular counts in Item B and during hours of highest pedestrian volume. Where young, elderly, and/or persons with physical or visual disabilities need special consideration, the pedestrians and their crossing times may be classified by general observation.
   D. Information about nearby facilities and activity centers that serve the young, elderly, and/or persons with disabilities, including requests from persons with disabilities for accessible crossing improvements at the location under study. These persons might not be adequately reflected in the pedestrian volume count if the absence of a signal restrains their mobility.
   E. The posted or statutory speed limit or the 85th-percentile speed on the uncontrolled approaches to the location.
   F. A condition diagram showing details of the physical layout, including such features as intersection geometrics, channelization, grades, sight-distance restrictions, transit stops and routes, parking conditions, pavement markings, roadway lighting, driveways, nearby railroad crossings, distance to nearest traffic control signals, utility poles and fixtures, and adjacent land use.
   G. A collision diagram showing crash experience by type, location, direction of movement, severity, weather, time of day, date, and day of week for at least 1 year.

18　The following data, which are desirable for a more precise understanding of the operation of the intersection, may be obtained during the periods described in Item B of Paragraph 17:
   A. Vehicle-hours of stopped time delay determined separately for each approach.
   B. The number and distribution of acceptable gaps in vehicular traffic on the major street for entrance from the minor street.
   C. The posted or statutory speed limit or the 85th-percentile speed on controlled approaches at a point near to the intersection but unaffected by the control.
   D. Pedestrian delay time for at least two 30-minute peak pedestrian delay periods of an average weekday or like periods of a Saturday or Sunday.
   E. Queue length on stop-controlled approaches.

Section 4C.02 Warrant 1, Eight-Hour Vehicular Volume

Support:

01　The Minimum Vehicular Volume, Condition A, is intended for application at locations where a large volume of intersecting traffic is the principal reason to consider installing a traffic control signal.

02　The Interruption of Continuous Traffic, Condition B, is intended for application at locations where Condition A is not satisfied and where the traffic volume on a major street is so heavy that traffic on a minor intersecting street suffers excessive delay or conflict in entering or crossing the major street.

03　It is intended that Warrant 1 be treated as a single warrant. If Condition A is satisfied, then Warrant 1 is satisfied and analyses of Condition B and the combination of Conditions A and B are not needed. Similarly, if Condition B is satisfied, then Warrant 1 is satisfied and an analysis of the combination of Conditions A and B is not needed.

Standard:

04 **The need for a traffic control signal shall be considered if an engineering study finds that one of the following conditions exist for each of any 8 hours of an average day:**
   A. **The vehicles per hour given in both of the 100 percent columns of Condition A in Table 4C-1 exist on the major-street and the higher-volume minor-street approaches, respectively, to the intersection; or**
   B. **The vehicles per hour given in both of the 100 percent columns of Condition B in Table 4C-1 exist on the major-street and the higher-volume minor-street approaches, respectively, to the intersection.**

In applying each condition the major-street and minor-street volumes shall be for the same 8 hours. On the minor street, the higher volume shall not be required to be on the same approach during each of these 8 hours.

Option:

05 If the posted or statutory speed limit or the 85th-percentile speed on the major street exceeds 40 mph, or if the intersection lies within the built-up area of an isolated community having a population of less than 10,000, the traffic volumes in the 70 percent columns in Table 4C-1 may be used in place of the 100 percent columns.

Guidance:

06 *The combination of Conditions A and B is intended for application at locations where Condition A is not satisfied and Condition B is not satisfied and should be applied only after an adequate trial of other alternatives that could cause less delay and inconvenience to traffic has failed to solve the traffic problems.*

Standard:

07 **The need for a traffic control signal shall be considered if an engineering study finds that both of the following conditions exist for each of any 8 hours of an average day:**
   A. **The vehicles per hour given in both of the 80 percent columns of Condition A in Table 4C-1 exist on the major-street and the higher-volume minor-street approaches, respectively, to the intersection; and**
   B. **The vehicles per hour given in both of the 80 percent columns of Condition B in Table 4C-1 exist on the major-street and the higher-volume minor-street approaches, respectively, to the intersection.**

These major-street and minor-street volumes shall be for the same 8 hours for each condition; however, the 8 hours satisfied in Condition A shall not be required to be the same 8 hours satisfied in Condition B. On the minor street, the higher volume shall not be required to be on the same approach during each of the 8 hours.

Table 4C-1. Warrant 1, Eight-Hour Vehicular Volume

Condition A—Minimum Vehicular Volume

Number of lanes for moving traffic on each approach		Vehicles per hour on major street (total of both approaches)				Vehicles per hour on higher-volume minor-street approach (one direction only)			
Major Street	Minor Street	100%[a]	80%[b]	70%[c]	56%[d]	100%[a]	80%[b]	70%[c]	56%[d]
1	1	500	400	350	280	150	120	105	84
2 or more	1	600	480	420	336	150	120	105	84
2 or more	2 or more	600	480	420	336	200	160	140	112
1	2 or more	500	400	350	280	200	160	140	112

Condition B—Interruption of Continuous Traffic

Number of lanes for moving traffic on each approach		Vehicles per hour on major street (total of both approaches)				Vehicles per hour on higher-volume minor-street approach (one direction only)			
Major Street	Minor Street	100%[a]	80%[b]	70%[c]	56%[d]	100%[a]	80%[b]	70%[c]	56%[d]
1	1	750	600	525	420	75	60	53	42
2 or more	1	900	720	630	504	75	60	53	42
2 or more	2 or more	900	720	630	504	100	80	70	56
1	2 or more	750	600	525	420	100	80	70	56

[a] Basic minimum hourly volume
[b] Used for combination of Conditions A and B after adequate trial of other remedial measures
[c] May be used when the major-street speed exceeds 40 mph or in an isolated community with a population of less than 10,000
[d] May be used for combination of Conditions A and B after adequate trial of other remedial measures when the major-street speed exceeds 40 mph or in an isolated community with a population of less than 10,000

Option:

08 If the posted or statutory speed limit or the 85th-percentile speed on the major street exceeds 40 mph, or if the intersection lies within the built-up area of an isolated community having a population of less than 10,000, the traffic volumes in the 56 percent columns in Table 4C-1 may be used in place of the 80 percent columns.

Section 4C.03 Warrant 2, Four-Hour Vehicular Volume

Support:

01 The Four-Hour Vehicular Volume signal warrant conditions are intended to be applied where the volume of intersecting traffic is the principal reason to consider installing a traffic control signal.

Standard:

02 **The need for a traffic control signal shall be considered if an engineering study finds that, for each of any 4 hours of an average day, the plotted points representing the vehicles per hour on the major street (total of both approaches) and the corresponding vehicles per hour on the higher-volume minor-street approach (one direction only) all fall above the applicable curve in Figure 4C-1 for the existing combination of approach lanes. On the minor street, the higher volume shall not be required to be on the same approach during each of these 4 hours.**

Option:

03 If the posted or statutory speed limit or the 85th-percentile speed on the major street exceeds 40 mph, or if the intersection lies within the built-up area of an isolated community having a population of less than 10,000, Figure 4C-2 may be used in place of Figure 4C-1.

Section 4C.04 Warrant 3, Peak Hour

Support:

01 The Peak Hour signal warrant is intended for use at a location where traffic conditions are such that for a minimum of 1 hour of an average day, the minor-street traffic suffers undue delay when entering or crossing the major street.

Standard:

02 **This signal warrant shall be applied only in unusual cases, such as office complexes, manufacturing plants, industrial complexes, or high-occupancy vehicle facilities that attract or discharge large numbers of vehicles over a short time.**

03 **The need for a traffic control signal shall be considered if an engineering study finds that the criteria in either of the following two categories are met:**
  A. **If all three of the following conditions exist for the same 1 hour (any four consecutive 15-minute periods) of an average day:**
    1. **The total stopped time delay experienced by the traffic on one minor-street approach (one direction only) controlled by a STOP sign equals or exceeds: 4 vehicle-hours for a one-lane approach or 5 vehicle-hours for a two-lane approach; and**
    2. **The volume on the same minor-street approach (one direction only) equals or exceeds 100 vehicles per hour for one moving lane of traffic or 150 vehicles per hour for two moving lanes; and**
    3. **The total entering volume serviced during the hour equals or exceeds 650 vehicles per hour for intersections with three approaches or 800 vehicles per hour for intersections with four or more approaches.**
  B. **The plotted point representing the vehicles per hour on the major street (total of both approaches) and the corresponding vehicles per hour on the higher-volume minor-street approach (one direction only) for 1 hour (any four consecutive 15-minute periods) of an average day falls above the applicable curve in Figure 4C-3 for the existing combination of approach lanes.**

Option:

04 If the posted or statutory speed limit or the 85th-percentile speed on the major street exceeds 40 mph, or if the intersection lies within the built-up area of an isolated community having a population of less than 10,000, Figure 4C-4 may be used in place of Figure 4C-3 to evaluate the criteria in the second category of the Standard.

05 If this warrant is the only warrant met and a traffic control signal is justified by an engineering study, the traffic control signal may be operated in the flashing mode during the hours that the volume criteria of this warrant are not met.

Guidance:

06 *If this warrant is the only warrant met and a traffic control signal is justified by an engineering study, the traffic control signal should be traffic-actuated.*

Figure 4C-1. Warrant 2, Four-Hour Vehicular Volume

*Note: 115 vph applies as the lower threshold volume for a minor-street approach with two or more lanes and 80 vph applies as the lower threshold volume for a minor-street approach with one lane.

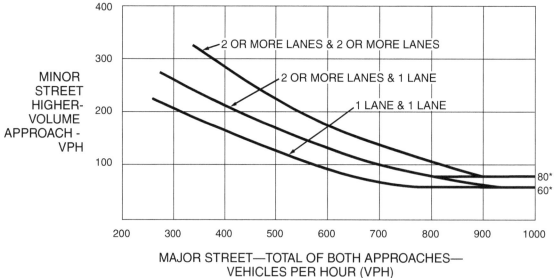

Figure 4C-2. Warrant 2, Four-Hour Vehicular Volume (70% Factor)

(COMMUNITY LESS THAN 10,000 POPULATION OR ABOVE 40 MPH ON MAJOR STREET)

*Note: 80 vph applies as the lower threshold volume for a minor-street approach with two or more lanes and 60 vph applies as the lower threshold volume for a minor-street approach with one lane.

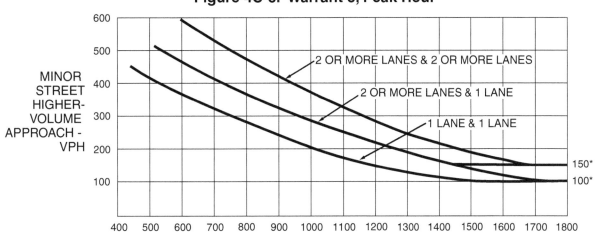

Figure 4C-3. Warrant 3, Peak Hour

*Note: 150 vph applies as the lower threshold volume for a minor-street approach with two or more lanes and 100 vph applies as the lower threshold volume for a minor-street approach with one lane.

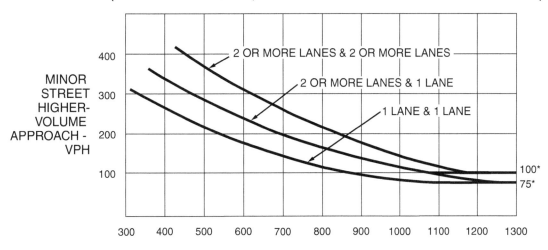

Figure 4C-4. Warrant 3, Peak Hour (70% Factor)

(COMMUNITY LESS THAN 10,000 POPULATION OR ABOVE 40 MPH ON MAJOR STREET)

*Note: 100 vph applies as the lower threshold volume for a minor-street approach with two or more lanes and 75 vph applies as the lower threshold volume for a minor-street approach with one lane.

Section 4C.05 Warrant 4, Pedestrian Volume

Support:

01　The Pedestrian Volume signal warrant is intended for application where the traffic volume on a major street is so heavy that pedestrians experience excessive delay in crossing the major street.

Standard:

02　**The need for a traffic control signal at an intersection or midblock crossing shall be considered if an engineering study finds that one of the following criteria is met:**

　　A. For each of any 4 hours of an average day, the plotted points representing the vehicles per hour on the major street (total of both approaches) and the corresponding pedestrians per hour crossing the major street (total of all crossings) all fall above the curve in Figure 4C-5; or

　　B. For 1 hour (any four consecutive 15-minute periods) of an average day, the plotted point representing the vehicles per hour on the major street (total of both approaches) and the corresponding pedestrians per hour crossing the major street (total of all crossings) falls above the curve in Figure 4C-7.

Option:

03　If the posted or statutory speed limit or the 85th-percentile speed on the major street exceeds 35 mph, or if the intersection lies within the built-up area of an isolated community having a population of less than 10,000, Figure 4C-6 may be used in place of Figure 4C-5 to evaluate Criterion A in Paragraph 2, and Figure 4C-8 may be used in place of Figure 4C-7 to evaluate Criterion B in Paragraph 2.

Standard:

04　**The Pedestrian Volume signal warrant shall not be applied at locations where the distance to the nearest traffic control signal or STOP sign controlling the street that pedestrians desire to cross is less than 300 feet, unless the proposed traffic control signal will not restrict the progressive movement of traffic.**

05　**If this warrant is met and a traffic control signal is justified by an engineering study, the traffic control signal shall be equipped with pedestrian signal heads complying with the provisions set forth in Chapter 4E.**

Guidance:

06　*If this warrant is met and a traffic control signal is justified by an engineering study, then:*

　　A. If it is installed at an intersection or major driveway location, the traffic control signal should also control the minor-street or driveway traffic, should be traffic-actuated, and should include pedestrian detection.

　　B. If it is installed at a non-intersection crossing, the traffic control signal should be installed at least 100 feet from side streets or driveways that are controlled by STOP or YIELD signs, and should be pedestrian-actuated. If the traffic control signal is installed at a non-intersection crossing, at least one of the signal faces should be over the traveled way for each approach, parking and other sight obstructions should be prohibited for at least 100 feet in advance of and at least 20 feet beyond the crosswalk or site accommodations should be made through curb extensions or other techniques to provide adequate sight distance, and the installation should include suitable standard signs and pavement markings.

　　C. Furthermore, if it is installed within a signal system, the traffic control signal should be coordinated.

Option:

07　The criterion for the pedestrian volume crossing the major street may be reduced as much as 50 percent if the 15th-percentile crossing speed of pedestrians is less than 3.5 feet per second.

08　A traffic control signal may not be needed at the study location if adjacent coordinated traffic control signals consistently provide gaps of adequate length for pedestrians to cross the street.

Section 4C.06 Warrant 5, School Crossing

Support:

01　The School Crossing signal warrant is intended for application where the fact that schoolchildren cross the major street is the principal reason to consider installing a traffic control signal. For the purposes of this warrant, the word "schoolchildren" includes elementary through high school students.

Standard:

02　**The need for a traffic control signal shall be considered when an engineering study of the frequency and adequacy of gaps in the vehicular traffic stream as related to the number and size of groups of schoolchildren at an established school crossing across the major street shows that the number of adequate gaps in the traffic stream during the period when the schoolchildren are using the crossing is less than the number of minutes in the same period (see Section 7A.03) and there are a minimum of 20 schoolchildren during the highest crossing hour.**

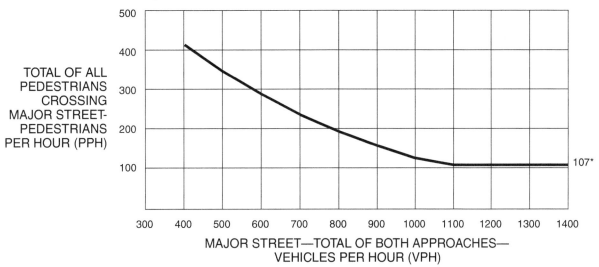

Figure 4C-5. Warrant 4, Pedestrian Four-Hour Volume

*Note: 107 pph applies as the lower threshold volume.

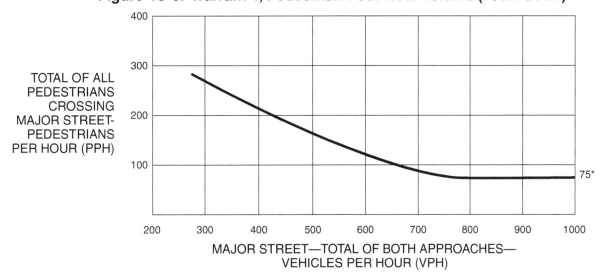

Figure 4C-6. Warrant 4, Pedestrian Four-Hour Volume (70% Factor)

*Note: 75 pph applies as the lower threshold volume.

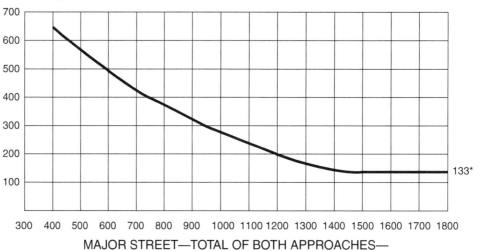

Figure 4C-7. Warrant 4, Pedestrian Peak Hour

*Note: 133 pph applies as the lower threshold volume.

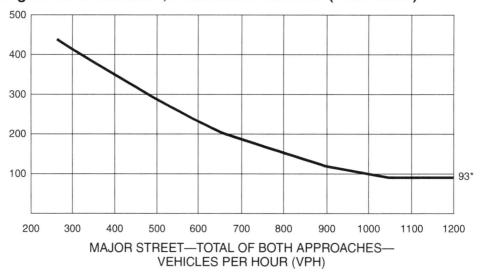

Figure 4C-8. Warrant 4, Pedestrian Peak Hour (70% Factor)

*Note: 93 pph applies as the lower threshold volume.

03 **Before a decision is made to install a traffic control signal, consideration shall be given to the implementation of other remedial measures, such as warning signs and flashers, school speed zones, school crossing guards, or a grade-separated crossing.**

04 **The School Crossing signal warrant shall not be applied at locations where the distance to the nearest traffic control signal along the major street is less than 300 feet, unless the proposed traffic control signal will not restrict the progressive movement of traffic.**

Guidance:

05 *If this warrant is met and a traffic control signal is justified by an engineering study, then:*

   A. If it is installed at an intersection or major driveway location, the traffic control signal should also control the minor-street or driveway traffic, should be traffic-actuated, and should include pedestrian detection.

   B. If it is installed at a non-intersection crossing, the traffic control signal should be installed at least 100 feet from side streets or driveways that are controlled by STOP or YIELD signs, and should be pedestrian-actuated. If the traffic control signal is installed at a non-intersection crossing, at least one of the signal faces should be over the traveled way for each approach, parking and other sight obstructions should be prohibited for at least 100 feet in advance of and at least 20 feet beyond the crosswalk or site accommodations should be made through curb extensions or other techniques to provide adequate sight distance, and the installation should include suitable standard signs and pavement markings.

   C. Furthermore, if it is installed within a signal system, the traffic control signal should be coordinated.

Section 4C.07 Warrant 6, Coordinated Signal System

Support:

01 Progressive movement in a coordinated signal system sometimes necessitates installing traffic control signals at intersections where they would not otherwise be needed in order to maintain proper platooning of vehicles.

Standard:

02 **The need for a traffic control signal shall be considered if an engineering study finds that one of the following criteria is met:**

   A. On a one-way street or a street that has traffic predominantly in one direction, the adjacent traffic control signals are so far apart that they do not provide the necessary degree of vehicular platooning.

   B. On a two-way street, adjacent traffic control signals do not provide the necessary degree of platooning and the proposed and adjacent traffic control signals will collectively provide a progressive operation.

Guidance:

03 *The Coordinated Signal System signal warrant should not be applied where the resultant spacing of traffic control signals would be less than 1,000 feet.*

Section 4C.08 Warrant 7, Crash Experience

Support:

01 The Crash Experience signal warrant conditions are intended for application where the severity and frequency of crashes are the principal reasons to consider installing a traffic control signal.

Standard:

02 **The need for a traffic control signal shall be considered if an engineering study finds that all of the following criteria are met:**

   A. Adequate trial of alternatives with satisfactory observance and enforcement has failed to reduce the crash frequency; and

   B. Five or more reported crashes, of types susceptible to correction by a traffic control signal, have occurred within a 12-month period, each crash involving personal injury or property damage apparently exceeding the applicable requirements for a reportable crash; and

   C. For each of any 8 hours of an average day, the vehicles per hour (vph) given in both of the 80 percent columns of Condition A in Table 4C-1 (see Section 4C.02), or the vph in both of the 80 percent columns of Condition B in Table 4C-1 exists on the major-street and the higher-volume minor-street approach, respectively, to the intersection, or the volume of pedestrian traffic is not less than 80 percent of the requirements specified in the Pedestrian Volume warrant. These major-street and minor-street volumes shall be for the same 8 hours. On the minor street, the higher volume shall not be required to be on the same approach during each of the 8 hours.

Option:

01 If the posted or statutory speed limit or the 85th-percentile speed on the major street exceeds 40 mph, or if the intersection lies within the built-up area of an isolated community having a population of less than 10,000, the traffic volumes in the 56 percent columns in Table 4C-1 may be used in place of the 80 percent columns.

Section 4C.09 Warrant 8, Roadway Network

Support:

01 Installing a traffic control signal at some intersections might be justified to encourage concentration and organization of traffic flow on a roadway network.

Standard:

02 **The need for a traffic control signal shall be considered if an engineering study finds that the common intersection of two or more major routes meets one or both of the following criteria:**
   A. The intersection has a total existing, or immediately projected, entering volume of at least 1,000 vehicles per hour during the peak hour of a typical weekday and has 5-year projected traffic volumes, based on an engineering study, that meet one or more of Warrants 1, 2, and 3 during an average weekday; or
   B. The intersection has a total existing or immediately projected entering volume of at least 1,000 vehicles per hour for each of any 5 hours of a non-normal business day (Saturday or Sunday).

03 **A major route as used in this signal warrant shall have at least one of the following characteristics:**
   A. It is part of the street or highway system that serves as the principal roadway network for through traffic flow.
   B. It includes rural or suburban highways outside, entering, or traversing a city.
   C. It appears as a major route on an official plan, such as a major street plan in an urban area traffic and transportation study.

Section 4C.10 Warrant 9, Intersection Near a Grade Crossing

Support:

01 The Intersection Near a Grade Crossing signal warrant is intended for use at a location where none of the conditions described in the other eight traffic signal warrants are met, but the proximity to the intersection of a grade crossing on an intersection approach controlled by a STOP or YIELD sign is the principal reason to consider installing a traffic control signal.

Guidance:

02 *This signal warrant should be applied only after adequate consideration has been given to other alternatives or after a trial of an alternative has failed to alleviate the safety concerns associated with the grade crossing. Among the alternatives that should be considered or tried are:*
   A. Providing additional pavement that would enable vehicles to clear the track or that would provide space for an evasive maneuver, or
   B. Reassigning the stop controls at the intersection to make the approach across the track a non-stopping approach.

Standard:

03 **The need for a traffic control signal shall be considered if an engineering study finds that both of the following criteria are met:**
   A. A grade crossing exists on an approach controlled by a STOP or YIELD sign and the center of the track nearest to the intersection is within 140 feet of the stop line or yield line on the approach; and
   B. During the highest traffic volume hour during which rail traffic uses the crossing, the plotted point representing the vehicles per hour on the major street (total of both approaches) and the corresponding vehicles per hour on the minor-street approach that crosses the track (one direction only, approaching the intersection) falls above the applicable curve in Figure 4C-9 or 4C-10 for the existing combination of approach lanes over the track and the distance D, which is the clear storage distance as defined in Section 1A.13.

Guidance:

04 *The following considerations apply when plotting the traffic volume data on Figure 4C-9 or 4C-10:*
   A. Figure 4C-9 should be used if there is only one lane approaching the intersection at the track crossing location and Figure 4C-10 should be used if there are two or more lanes approaching the intersection at the track crossing location.

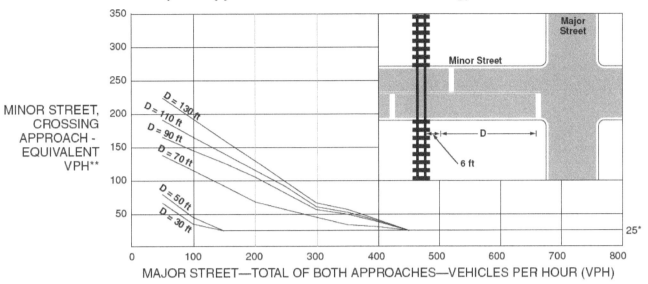

Figure 4C-9. Warrant 9, Intersection Near a Grade Crossing (One Approach Lane at the Track Crossing)

* 25 vph applies as the lower threshold volume
** VPH after applying the adjustment factors in Tables 4C-2, 4C-3, and/or 4C-4, if appropriate

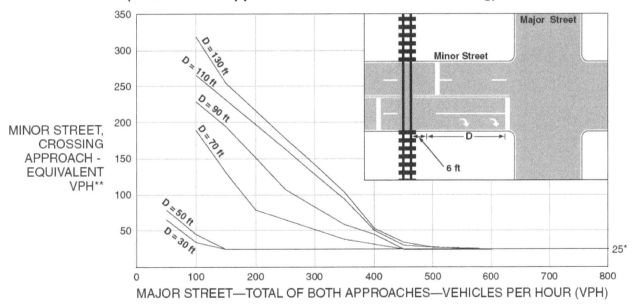

Figure 4C-10. Warrant 9, Intersection Near a Grade Crossing (Two or More Approach Lanes at the Track Crossing)

* 25 vph applies as the lower threshold volume
** VPH after applying the adjustment factors in Tables 4C-2, 4C-3, and/or 4C-4, if appropriate

B. After determining the actual distance D, the curve for the distance D that is nearest to the actual distance D should be used. For example, if the actual distance D is 95 feet, the plotted point should be compared to the curve for D = 90 feet.

C. If the rail traffic arrival times are unknown, the highest traffic volume hour of the day should be used.

Option:

05 The minor-street approach volume may be multiplied by up to three adjustment factors as provided in Paragraphs 6 through 8.

06 Because the curves are based on an average of four occurrences of rail traffic per day, the vehicles per hour on the minor-street approach may be multiplied by the adjustment factor shown in Table 4C-2 for the appropriate number of occurrences of rail traffic per day.

07 Because the curves are based on typical vehicle occupancy, if at least 2% of the vehicles crossing the track are buses carrying at least 20 people, the vehicles per hour on the minor-street approach may be multiplied by the adjustment factor shown in Table 4C-3 for the appropriate percentage of high-occupancy buses.

08 Because the curves are based on tractor-trailer trucks comprising 10% of the vehicles crossing the track, the vehicles per hour on the minor-street approach may be multiplied by the adjustment factor shown in Table 4C-4 for the appropriate distance and percentage of tractor-trailer trucks.

Standard:

09 **If this warrant is met and a traffic control signal at the intersection is justified by an engineering study, then:**
   A. The traffic control signal shall have actuation on the minor street;
   B. Preemption control shall be provided in accordance with Sections 4D.27, 8C.09, and 8C.10; and
   C. The grade crossing shall have flashing-light signals (see Chapter 8C).

Guidance:

10 *If this warrant is met and a traffic control signal at the intersection is justified by an engineering study, the grade crossing should have automatic gates (see Chapter 8C).*

Table 4C-2. Warrant 9, Adjustment Factor for Daily Frequency of Rail Traffic

Rail Traffic per Day	Adjustment Factor
1	0.67
2	0.91
3 to 5	1.00
6 to 8	1.18
9 to 11	1.25
12 or more	1.33

Table 4C-3. Warrant 9, Adjustment Factor for Percentage of High-Occupancy Buses

% of High-Occupancy Buses* on Minor-Street Approach	Adjustment Factor
0%	1.00
2%	1.09
4%	1.19
6% or more	1.32

* A high-occupancy bus is defined as a bus occupied by at least 20 people.

Table 4C-4. Warrant 9, Adjustment Factor for Percentage of Tractor-Trailer Trucks

% of Tractor-Trailer Trucks on Minor-Street Approach	Adjustment Factor	
	D less than 70 feet	D of 70 feet or more
0% to 2.5%	0.50	0.50
2.6% to 7.5%	0.75	0.75
7.6% to 12.5%	1.00	1.00
12.6% to 17.5%	2.30	1.15
17.6% to 22.5%	2.70	1.35
22.6% to 27.5%	3.28	1.64
More than 27.5%	4.18	2.09

CHAPTER 4D. TRAFFIC CONTROL SIGNAL FEATURES

Section 4D.01 General

Support:

01 The features of traffic control signals of interest to road users are the location, design, and meaning of the signal indications. Uniformity in the design features that affect the traffic to be controlled, as set forth in this Manual, is especially important for the safety and efficiency of operations.

02 Traffic control signals can be operated in pretimed, semi-actuated, or full-actuated modes. For isolated (non-interconnected) signalized locations on rural high-speed highways, full-actuated mode with advance vehicle detection on the high-speed approaches is typically used. These features are designed to reduce the frequency with which the onset of the yellow change interval is displayed when high-speed approaching vehicles are in the "dilemma zone" such that the drivers of these high-speed vehicles find it difficult to decide whether to stop or proceed.

Standard:

03 **When a traffic control signal is not in operation, such as before it is placed in service, during seasonal shutdowns, or when it is not desirable to operate the traffic control signal, the signal faces shall be covered, turned, or taken down to clearly indicate that the traffic control signal is not in operation.**

Support:

04 Seasonal shutdown is a condition in which a permanent traffic signal is turned off or otherwise made non-operational during a particular season when its operation is not justified. This might be applied in a community where tourist traffic during most of the year justifies the permanent signalization, but a seasonal shutdown of the signal during an annual period of lower tourist traffic would reduce delays; or where a major traffic generator, such as a large factory, justifies the permanent signalization, but the large factory is shut down for an annual factory vacation for a few weeks in the summer.

Standard:

05 **A traffic control signal shall control traffic only at the intersection or midblock location where the signal faces are placed.**

06 **Midblock crosswalks shall not be signalized if they are located within 300 feet from the nearest traffic control signal, unless the proposed traffic control signal will not restrict the progressive movement of traffic.**

Guidance:

07 *A midblock crosswalk location should not be controlled by a traffic control signal if the crosswalk is located within 100 feet from side streets or driveways that are controlled by STOP signs or YIELD signs.*

08 *Engineering judgment should be used to determine the proper phasing and timing for a traffic control signal. Since traffic flows and patterns change, phasing and timing should be reevaluated regularly and updated if needed.*

09 *Traffic control signals within 1/2 mile of one another along a major route or in a network of intersecting major routes should be coordinated, preferably with interconnected controller units. Where traffic control signals that are within 1/2 mile of one another along a major route have a jurisdictional boundary or a boundary between different signal systems between them, coordination across the boundary should be considered.*

Support:

10 Signal coordination need not be maintained between control sections that operate on different cycle lengths.

11 For coordination with grade crossing signals and movable bridge signals, see Sections 4D.27, 4J.03, 8C.09, and 8C.10.

Section 4D.02 Responsibility for Operation and Maintenance

Guidance:

01 *Prior to installing any traffic control signal, the responsibility for the maintenance of the signal and all of the appurtenances, hardware, software, and the timing plan(s) should be clearly established. The responsible agency should provide for the maintenance of the traffic control signal and all of its appurtenances in a competent manner.*

02 *To this end the agency should:*

  A. *Keep every controller assembly in effective operation in accordance with its predetermined timing schedule; check the operation of the controller assembly frequently enough to verify that it is operating in accordance with the predetermined timing schedule; and establish a policy to maintain a record of all timing changes and that only authorized persons are permitted to make timing changes;*

B. Clean the optical system of the signal sections and replace the light sources as frequently as experience proves necessary;
C. Clean and service equipment and other appurtenances as frequently as experience proves necessary;
D. Provide for alternate operation of the traffic control signal during a period of failure, using flashing mode or manual control, or manual traffic direction by proper authorities as might be required by traffic volumes or congestion, or by erecting other traffic control devices;
E. Have properly skilled maintenance personnel available without undue delay for all signal malfunctions and signal indication failures;
F. Provide spare equipment to minimize the interruption of traffic control signal operation as a result of equipment failure;
G. Provide for the availability of properly skilled maintenance personnel for the repair of all components; and
H. Maintain the appearance of the signal displays and equipment.

Section 4D.03 Provisions for Pedestrians

Support:

01 Chapter 4E contains additional information regarding pedestrian signals and Chapter 4F contains additional information regarding pedestrian hybrid beacons.

Standard:

02 **The design and operation of traffic control signals shall take into consideration the needs of pedestrian as well as vehicular traffic.**

03 **If engineering judgment indicates the need for provisions for a given pedestrian movement, signal faces conveniently visible to pedestrians shall be provided by pedestrian signal heads (see Chapter 4E) or a vehicular signal face(s) for a concurrent vehicular movement.**

Guidance:

04 *Accessible pedestrian signals (see Sections 4E.09 through 4E.13) that provide information in non-visual formats (such as audible tones, speech messages, and/or vibrating surfaces) should be provided where determined appropriate by engineering judgment.*

05 *Where pedestrian movements regularly occur, pedestrians should be provided with sufficient time to cross the roadway by adjusting the traffic control signal operation and timing to provide sufficient crossing time every cycle or by providing pedestrian detectors.*

06 *If it is necessary or desirable to prohibit certain pedestrian movements at a traffic control signal location, No Pedestrian Crossing (R9-3) signs (see Section 2B.51) should be used if it is not practical to provide a barrier or other physical feature to physically prevent the pedestrian movements.*

Section 4D.04 Meaning of Vehicular Signal Indications

Support:

01 The "Uniform Vehicle Code" (see Section 1A.11) is the primary source for the standards for the meaning of vehicular signal indications to both vehicle operators and pedestrians as provided in this Section, and the standards for the meaning of separate pedestrian signal head indications as provided in Section 4E.02.

02 The physical area that is defined as being "within the intersection" is dependent upon the conditions that are described in the definition of intersection in Section 1A.13.

Standard:

03 **The following meanings shall be given to highway traffic signal indications for vehicles and pedestrians:**
   A. **Steady green signal indications shall have the following meanings:**
      1. **Vehicular traffic facing a CIRCULAR GREEN signal indication is permitted to proceed straight through or turn right or left or make a U-turn movement except as such movement is modified by lane-use signs, turn prohibition signs, lane markings, roadway design, separate turn signal indications, or other traffic control devices.**
         Such vehicular traffic, including vehicles turning right or left or making a U-turn movement, shall yield the right-of-way to:
         (a) Pedestrians lawfully within an associated crosswalk, and
         (b) Other vehicles lawfully within the intersection.
         In addition, vehicular traffic turning left or making a U-turn movement to the left shall yield the right-of-way to other vehicles approaching from the opposite direction so closely as to constitute an immediate hazard during the time when such turning vehicle is moving across or within the intersection.

2. Vehicular traffic facing a GREEN ARROW signal indication, displayed alone or in combination with another signal indication, is permitted to cautiously enter the intersection only to make the movement indicated by such arrow, or such other movement as is permitted by other signal indications displayed at the same time.

   Such vehicular traffic, including vehicles turning right or left or making a U-turn movement, shall yield the right-of-way to:
   (a) Pedestrians lawfully within an associated crosswalk, and
   (b) Other vehicles lawfully within the intersection.
3. Pedestrians facing a CIRCULAR GREEN signal indication, unless otherwise directed by a pedestrian signal indication or other traffic control device, are permitted to proceed across the roadway within any marked or unmarked associated crosswalk. The pedestrian shall yield the right-of-way to vehicles lawfully within the intersection or so close as to create an immediate hazard at the time that the green signal indication is first displayed.
4. Pedestrians facing a GREEN ARROW signal indication, unless otherwise directed by a pedestrian signal indication or other traffic control device, shall not cross the roadway.

B. Steady yellow signal indications shall have the following meanings:
1. Vehicular traffic facing a steady CIRCULAR YELLOW signal indication is thereby warned that the related green movement or the related flashing arrow movement is being terminated or that a steady red signal indication will be displayed immediately thereafter when vehicular traffic shall not enter the intersection. The rules set forth concerning vehicular operation under the movement(s) being terminated shall continue to apply while the steady CIRCULAR YELLOW signal indication is displayed.
2. Vehicular traffic facing a steady YELLOW ARROW signal indication is thereby warned that the related GREEN ARROW movement or the related flashing arrow movement is being terminated. The rules set forth concerning vehicular operation under the movement(s) being terminated shall continue to apply while the steady YELLOW ARROW signal indication is displayed.
3. Pedestrians facing a steady CIRCULAR YELLOW or YELLOW ARROW signal indication, unless otherwise directed by a pedestrian signal indication or other traffic control device shall not start to cross the roadway.

C. Steady red signal indications shall have the following meanings:
1. Vehicular traffic facing a steady CIRCULAR RED signal indication, unless entering the intersection to make another movement permitted by another signal indication, shall stop at a clearly marked stop line; but if there is no stop line, traffic shall stop before entering the crosswalk on the near side of the intersection; or if there is no crosswalk, then before entering the intersection; and shall remain stopped until a signal indication to proceed is displayed, or as provided below.

   Except when a traffic control device is in place prohibiting a turn on red or a steady RED ARROW signal indication is displayed, vehicular traffic facing a steady CIRCULAR RED signal indication is permitted to enter the intersection to turn right, or to turn left from a one-way street into a one-way street, after stopping. The right to proceed with the turn shall be subject to the rules applicable after making a stop at a STOP sign.
2. Vehicular traffic facing a steady RED ARROW signal indication shall not enter the intersection to make the movement indicated by the arrow and, unless entering the intersection to make another movement permitted by another signal indication, shall stop at a clearly marked stop line; but if there is no stop line, before entering the crosswalk on the near side of the intersection; or if there is no crosswalk, then before entering the intersection; and shall remain stopped until a signal indication or other traffic control device permitting the movement indicated by such RED ARROW is displayed.

   When a traffic control device is in place permitting a turn on a steady RED ARROW signal indication, vehicular traffic facing a steady RED ARROW signal indication is permitted to enter the intersection to make the movement indicated by the arrow signal indication, after stopping. The right to proceed with the turn shall be limited to the direction indicated by the arrow and shall be subject to the rules applicable after making a stop at a STOP sign.
3. Unless otherwise directed by a pedestrian signal indication or other traffic control device, pedestrians facing a steady CIRCULAR RED or steady RED ARROW signal indication shall not enter the roadway.

D. A flashing green signal indication has no meaning and shall not be used.

E. Flashing yellow signal indications shall have the following meanings:
1. Vehicular traffic, on an approach to an intersection, facing a flashing CIRCULAR YELLOW signal indication is permitted to cautiously enter the intersection to proceed straight through or turn right or left or make a U-turn except as such movement is modified by lane-use signs, turn prohibition signs, lane markings, roadway design, separate turn signal indications, or other traffic control devices.

    Such vehicular traffic, including vehicles turning right or left or making a U-turn, shall yield the right-of-way to:
    (a) Pedestrians lawfully within an associated crosswalk, and
    (b) Other vehicles lawfully within the intersection.

    In addition, vehicular traffic turning left or making a U-turn to the left shall yield the right-of-way to other vehicles approaching from the opposite direction so closely as to constitute an immediate hazard during the time when such turning vehicle is moving across or within the intersection.
2. Vehicular traffic, on an approach to an intersection, facing a flashing YELLOW ARROW signal indication, displayed alone or in combination with another signal indication, is permitted to cautiously enter the intersection only to make the movement indicated by such arrow, or other such movement as is permitted by other signal indications displayed at the same time.

    Such vehicular traffic, including vehicles turning right or left or making a U-turn, shall yield the right-of-way to:
    (a) Pedestrians lawfully within an associated crosswalk, and
    (b) Other vehicles lawfully within the intersection.

    In addition, vehicular traffic turning left or making a U-turn to the left shall yield the right-of-way to other vehicles approaching from the opposite direction so closely as to constitute an immediate hazard during the time when such turning vehicle is moving across or within the intersection.
3. Pedestrians facing any flashing yellow signal indication at an intersection, unless otherwise directed by a pedestrian signal indication or other traffic control device, are permitted to proceed across the roadway within any marked or unmarked associated crosswalk. Pedestrians shall yield the right-of-way to vehicles lawfully within the intersection at the time that the flashing yellow signal indication is first displayed.
4. When a flashing CIRCULAR YELLOW signal indication(s) is displayed as a beacon (see Chapter 4L) to supplement another traffic control device, road users are notified that there is a need to pay extra attention to the message contained thereon or that the regulatory or warning requirements of the other traffic control device, which might not be applicable at all times, are currently applicable.

F. Flashing red signal indications shall have the following meanings:
1. Vehicular traffic, on an approach to an intersection, facing a flashing CIRCULAR RED signal indication shall stop at a clearly marked stop line; but if there is no stop line, before entering the crosswalk on the near side of the intersection; or if there is no crosswalk, at the point nearest the intersecting roadway where the driver has a view of approaching traffic on the intersecting roadway before entering the intersection. The right to proceed shall be subject to the rules applicable after making a stop at a STOP sign.
2. Vehicular traffic, on an approach to an intersection, facing a flashing RED ARROW signal indication if intending to turn in the direction indicated by the arrow shall stop at a clearly marked stop line; but if there is no stop line, before entering the crosswalk on the near side of the intersection; or if there is no crosswalk, at the point nearest the intersecting roadway where the driver has a view of approaching traffic on the intersecting roadway before entering the intersection. The right to proceed with the turn shall be limited to the direction indicated by the arrow and shall be subject to the rules applicable after making a stop at a STOP sign.
3. Pedestrians facing any flashing red signal indication at an intersection, unless otherwise directed by a pedestrian signal indication or other traffic control device, are permitted to proceed across the roadway within any marked or unmarked associated crosswalk. Pedestrians shall yield the right-of-way to vehicles lawfully within the intersection at the time that the flashing red signal indication is first displayed.
4. When a flashing CIRCULAR RED signal indication(s) is displayed as a beacon (see Chapter 4L) to supplement another traffic control device, road users are notified that there is a need to pay extra attention to the message contained thereon or that the regulatory requirements of the other traffic control device, which might not be applicable at all times, are currently applicable. Use of this signal indication shall be limited to supplementing STOP (R1-1), DO NOT ENTER (R5-1), or WRONG WAY (R5-1a) signs, and to applications where compliance with the supplemented traffic control device requires a stop at a designated point.

Section 4D.05 Application of Steady Signal Indications

Standard:

01   When a traffic control signal is being operated in a steady (stop-and-go) mode, at least one indication in each signal face shall be displayed at any given time.

02   A signal face(s) that controls a particular vehicular movement during any interval of a cycle shall control that same movement during all intervals of the cycle.

03   Steady signal indications shall be applied as follows:

   A. A steady CIRCULAR RED signal indication:
   1. Shall be displayed when it is intended to prohibit traffic, except pedestrians directed by a pedestrian signal head, from entering the intersection or other controlled area. Turning after stopping is permitted as stated in Item C.1 in Paragraph 3 of Section 4D.04.
   2. Shall be displayed with the appropriate GREEN ARROW signal indications when it is intended to permit traffic to make a specified turn or turns, and to prohibit traffic from proceeding straight ahead through the intersection or other controlled area, except in protected only mode operation (see Sections 4D.19 and 4D.23), or in protected/permissive mode operation with separate turn signal faces (see Sections 4D.20 and 4D.24).

   B. A steady CIRCULAR YELLOW signal indication:
   1. Shall be displayed following a CIRCULAR GREEN or straight-through GREEN ARROW signal indication in the same signal face.
   2. Shall not be displayed in conjunction with the change from the CIRCULAR RED signal indication to the CIRCULAR GREEN signal indication.
   3. Shall be followed by a CIRCULAR RED signal indication except that, when entering preemption operation, the return to the previous CIRCULAR GREEN signal indication shall be permitted following a steady CIRCULAR YELLOW signal indication (see Section 4D.27).
   4. Shall not be displayed to an approach from which drivers are turning left permissively or making a U-turn to the left permissively unless one of the following conditions exists:
      (a) A steady CIRCULAR YELLOW signal indication is also simultaneously being displayed to the opposing approach;
      (b) An engineering study has determined that, because of unique intersection conditions, the condition described in Item (a) cannot reasonably be implemented without causing significant operational or safety problems and that the volume of impacted left-turning or U-turning traffic is relatively low, and those left-turning or U-turning drivers are advised that a steady CIRCULAR YELLOW signal indication is not simultaneously being displayed to the opposing traffic if this operation occurs continuously by the installation near the left-most signal head of a W25-1 sign (see Section 2C.48) with the legend ONCOMING TRAFFIC HAS EXTENDED GREEN; or
      (c) Drivers are advised of the operation if it occurs only occasionally, such as during a preemption sequence, by the installation near the left-most signal head of a W25-2 sign (see Section 2C.48) with the legend ONCOMING TRAFFIC MAY HAVE EXTENDED GREEN.

   C. A steady CIRCULAR GREEN signal indication shall be displayed only when it is intended to permit traffic to proceed in any direction that is lawful and practical.

   D. A steady RED ARROW signal indication shall be displayed when it is intended to prohibit traffic, except pedestrians directed by a pedestrian signal head, from entering the intersection or other controlled area to make the indicated turn. Except as described in Item C.2 in Paragraph 3 of Section 4D.04, turning on a steady RED ARROW signal indication shall not be permitted.

   E. A steady YELLOW ARROW signal indication:
   1. Shall be displayed in the same direction as a GREEN ARROW signal indication following a GREEN ARROW signal indication in the same signal face, unless:
      (a) The GREEN ARROW signal indication and a CIRCULAR GREEN (or straight-through GREEN ARROW) signal indication terminate simultaneously in the same signal face, or
      (b) The green arrow is a straight-through GREEN ARROW (see Item B.1).
   2. Shall be displayed in the same direction as a flashing YELLOW ARROW signal indication or flashing RED ARROW signal indication following a flashing YELLOW ARROW signal indication or flashing RED ARROW signal indication in the same signal face, when the flashing arrow indication is displayed as part of a steady mode operation, if the signal face will subsequently display a steady red signal indication.

3. Shall not be displayed in conjunction with the change from a steady RED ARROW, flashing RED ARROW, or flashing YELLOW ARROW signal indication to a GREEN ARROW signal indication, except when entering preemption operation as provided in Item 5(a).
4. Shall not be displayed when any conflicting vehicular movement has a green or yellow signal indication (except for the situation regarding U-turns to the left provided in Paragraph 4) or any conflicting pedestrian movement has a WALKING PERSON (symbolizing WALK) or flashing UPRAISED HAND (symbolizing DONT WALK) signal indication, except that a steady left-turn (or U-turn to the left) YELLOW ARROW signal indication used to terminate a flashing left-turn (or U-turn to the left) YELLOW ARROW or a flashing left-turn (or U-turn to the left) RED ARROW signal indication in a signal face controlling a permissive left-turn (or U-turn to the left) movement as described in Sections 4D.18 and 4D.20 shall be permitted to be displayed when a CIRCULAR YELLOW signal indication is displayed for the opposing through movement. Vehicles departing in the same direction shall not be considered in conflict if, for each turn lane with moving traffic, there is a separate departing lane, and pavement markings or raised channelization clearly indicate which departure lane to use.
5. Shall not be displayed to terminate a flashing arrow signal indication on an approach from which drivers are turning left permissively or making a U-turn to the left permissively unless one of the following conditions exists:
   (a) A steady CIRCULAR YELLOW signal indication is also simultaneously being displayed to the opposing approach;
   (b) An engineering study has determined that, because of unique intersection conditions, the condition described in Item (a) cannot reasonably be implemented without causing significant operational or safety problems and that the volume of impacted left-turning or U-turning traffic is relatively low, and those left-turning or U-turning drivers are advised that a steady CIRCULAR YELLOW signal indication is not simultaneously being displayed to the opposing traffic if this operation occurs continuously by the installation near the left-most signal head of a W25-1 sign (see Section 2C.48) with the legend ONCOMING TRAFFIC HAS EXTENDED GREEN; or
   (c) Drivers are advised of the operation if it occurs only occasionally, such as during a preemption sequence, by the installation near the left-most signal head of a W25-2 sign (see Section 2C.48) with the legend ONCOMING TRAFFIC MAY HAVE EXTENDED GREEN.
6. Shall be terminated by a RED ARROW signal indication for the same direction or a CIRCULAR RED signal indication except:
   (a) When entering preemption operation, the display of a GREEN ARROW signal indication or a flashing arrow signal indication shall be permitted following a steady YELLOW ARROW signal indication.
   (b) When the movement controlled by the arrow is to continue on a permissive mode basis during an immediately following CIRCULAR GREEN or flashing YELLOW ARROW signal indication.

F. A steady GREEN ARROW signal indication:
1. Shall be displayed only to allow vehicular movements, in the direction indicated, that are not in conflict with other vehicles moving on a green or yellow signal indication and are not in conflict with pedestrians crossing in compliance with a WALKING PERSON (symbolizing WALK) or flashing UPRAISED HAND (symbolizing DONT WALK) signal indication. Vehicles departing in the same direction shall not be considered in conflict if, for each turn lane with moving traffic, there is a separate departing lane, and pavement markings or raised channelization clearly indicate which departure lane to use.
2. Shall be displayed on a signal face that controls a left-turn movement when said movement is not in conflict with other vehicles moving on a green or yellow signal indication (except for the situation regarding U-turns provided in Paragraph 4) and is not in conflict with pedestrians crossing in compliance with a WALKING PERSON (symbolizing WALK) or flashing UPRAISED HAND (symbolizing DONT WALK) signal indication. Vehicles departing in the same direction shall not be considered in conflict if, for each turn lane with moving traffic, there is a separate departing lane, and pavement markings or raised channelization clearly indicate which departure lane to use.
3. Shall not be required on the stem of a T-intersection or for turns from a one-way street.

Option:

04 If U-turns are permitted from the approach and a right-turn GREEN ARROW signal indication is simultaneously being displayed to road users making a right turn from the conflicting approach to the left, road users making a U-turn may be advised of the operation by the installation near the left-turn signal face of a U-TURN YIELD TO RIGHT TURN (R10-16) sign (see Section 2B.53).

05 If not otherwise prohibited, a steady straight-through green arrow signal indication may be used instead of a circular green signal indication in a signal face on an approach intersecting a one-way street to discourage wrong-way turns.

06 If not otherwise prohibited, steady red, yellow, and green turn arrow signal indications may be used instead of steady circular red, yellow, and green signal indications in a signal face on an approach where all traffic is required to turn or where the straight-through movement is not physically possible.

Support:

07 Section 4D.25 contains information regarding the signalization of approaches that have a shared left-turn/right-turn lane and no through movement.

Standard:

08 **If supplemental signal faces are used, the following limitations shall apply:**
   A. Left-turn arrows and U-turn arrows to the left shall not be used in near-right signal faces.
   B. Right-turn arrows and U-turn arrows to the right shall not be used in far-left signal faces. A far-side median-mounted signal face shall be considered a far-left signal for this application.

09 **A straight-through RED ARROW signal indication or a straight-through YELLOW ARROW signal indication shall not be displayed on any signal face, either alone or in combination with any other signal indication.**

10 **The following combinations of signal indications shall not be simultaneously displayed on any one signal face:**
   A. CIRCULAR RED with CIRCULAR YELLOW;
   B. CIRCULAR GREEN with CIRCULAR RED; or
   C. Straight-through GREEN ARROW with CIRCULAR RED;

11 **Additionally, the above combinations shall not be simultaneously displayed on an approach as a result of the combination of displays from multiple signal faces unless the display is created by a signal face(s) devoted exclusively to the control of a right-turning movement and:**
   A. The signal face(s) controlling the right-turning movement is visibility-limited from the adjacent through movement or positioned to minimize potential confusion to approaching road users, or
   B. A RIGHT TURN SIGNAL (R10-10) sign (see Sections 4D.21 through 4D.24) is mounted adjacent to the signal face(s) controlling the right-turning movement.

12 **The following combinations of signal indications shall not be simultaneously displayed on any one signal face or as a result of the combination of displays from multiple signal faces on an approach:**
   A. CIRCULAR GREEN with CIRCULAR YELLOW;
   B. Straight-through GREEN ARROW with CIRCULAR YELLOW;
   C. GREEN ARROW with YELLOW ARROW pointing in the same direction;
   D. RED ARROW with YELLOW ARROW pointing in the same direction; or
   E. GREEN ARROW with RED ARROW pointing in the same direction.

13 **Except as otherwise provided in Sections 4F.03 and 4G.04, the same signal section shall not be used to display both a flashing yellow and a steady yellow indication during steady mode operation. Except as otherwise provided in Sections 4D.18, 4D.20, 4D.22, and 4D.24, the same signal section shall not be used to display both a flashing red and a steady red indication during steady mode operation.**

Guidance:

14 *No movement that creates an unexpected crossing of pathways of moving vehicles or pedestrians should be allowed during any green or yellow interval, except when all three of the following conditions are met:*
   A. The movement involves only slight conflict, and
   B. Serious traffic delays are substantially reduced by permitting the conflicting movement, and
   C. Drivers and pedestrians subjected to the unexpected conflict are effectively warned thereof by a sign.

Section 4D.06 Signal Indications – Design, Illumination, Color, and Shape

Standard:

01 Each signal indication, except those used for pedestrian signal heads and lane-use control signals, shall be circular or arrow.

02 Letters or numbers (including those associated with countdown displays) shall not be displayed as part of a vehicular signal indication.

03 Strobes shall not be used within or adjacent to any signal indication.

04 Except for the flashing signal indications and the pre-emption confirmation lights that are expressly allowed by the provisions of this Chapter, flashing displays shall not be used within or adjacent to any signal indications.

05 Each circular signal indication shall emit a single color: red, yellow, or green.

06 Each arrow signal indication shall emit a single color: red, yellow, or green except that the alternate display (dual-arrow signal section) of a GREEN ARROW and a YELLOW ARROW signal indication, both pointing in the same direction, shall be permitted, provided that they are not displayed simultaneously.

07 The arrow, which shall show only one direction, shall be the only illuminated part of an arrow signal indication.

08 Arrows shall be pointed:
   A. Vertically upward to indicate a straight-through movement, or
   B. Horizontally in the direction of the turn to indicate a turn at approximately or greater than a right angle, or
   C. Upward with a slope at an angle approximately equal to that of the turn if the angle of the turn is substantially less than a right angle, or
   D. In a manner that directs the driver through the turn if a U-turn arrow is used (see Figure 4D-1).

09 Except as provided in Paragraph 10, the requirements of the publication entitled "Vehicle Traffic Control Signal Heads" (see Section 1A.11) that pertain to the aspects of the signal head design that affect the display of the signal indications shall be met.

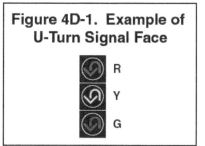

Figure 4D-1. Example of U-Turn Signal Face

Guidance:

10 *The intensity and distribution of light from each illuminated signal lens should comply with the publications entitled "Vehicle Traffic Control Signal Heads" and "Traffic Signal Lamps" (see Section 1A.11).*

Standard:

11 References to signal lenses in this section shall not be used to limit signal optical units to incandescent lamps within optical assemblies that include lenses.

Support:

12 Research has resulted in signal optical units that are not lenses, such as, but not limited to, light emitting diode (LED) traffic signal modules. Some units are practical for all signal indications, and some are practical for specific types such as visibility-limited signal indications.

Guidance:

13 *If a signal indication is so bright that it causes excessive glare during nighttime conditions, some form of automatic dimming should be used to reduce the brilliance of the signal indication.*

Section 4D.07 Size of Vehicular Signal Indications

Standard:

01 There shall be two nominal diameter sizes for vehicular signal indications: 8 inches and 12 inches.

02 Except as provided in Paragraph 3 below, 12-inch signal indications shall be used for all signal sections in all new signal faces.

Option:

03 Eight-inch circular signal indications may be used in new signal faces only for:
   A. The green or flashing yellow signal indications in an emergency-vehicle traffic control signal (see Section 4G.02);
   B. The circular indications in signal faces controlling the approach to the downstream location where two adjacent signalized locations are close to each other and it is not practical because of factors such as high approach speeds, horizontal or vertical curves, or other geometric factors to install visibility-limited signal faces for the downstream approach;

C. The circular indications in a signal face that is located less than 120 feet from the stop line on a roadway with a posted or statutory speed limit of 30 mph or less;
D. The circular indications in a supplemental near-side signal face:
E. The circular indications in a supplemental signal face installed for the sole purpose of controlling pedestrian movements (see Section 4D.03) rather than vehicular movements; and
F. The circular indications in a signal face installed for the sole purpose of controlling a bikeway or a bicycle movement.

04 Existing 8-inch circular signal indications that are not included in Items A through F in Paragraph 3 may be retained for the remainder of their useful service life.

Section 4D.08 Positions of Signal Indications Within a Signal Face – General

Support:

01 Standardization of the number and arrangements of signal sections in vehicular traffic control signal faces enables road users who are color vision deficient to identify the illuminated color by its position relative to other signal sections.

Standard:

02 **Unless otherwise provided in this Manual for a particular application, each signal face at a signalized location shall have three, four, or five signal sections. Unless otherwise provided in this Manual for a particular application, if a vertical signal face includes a cluster (see Section 4D.09), the signal face shall have at least three vertical positions.**

03 **A single-section signal face shall be permitted at a traffic control signal if it consists of a continuously-displayed GREEN ARROW signal indication that is being used to indicate a continuous movement.**

04 **The signal sections in a signal face shall be arranged in a vertical or horizontal straight line, except as otherwise provided in Section 4D.09.**

05 **The arrangement of adjacent signal sections in a signal face shall follow the relative positions listed in Sections 4D.09 or 4D.10, as applicable.**

06 **If a signal section that displays a CIRCULAR YELLOW signal indication is used, it shall be located between the signal section that displays the red signal indication and all other signal sections.**

07 **If a U-turn arrow signal section is used in a signal face for a U-turn to the left, its position in the signal face shall be the same as stated in Sections 4D.09 and 4D.10 for a left-turn arrow signal section of the same color. If a U-turn arrow signal section is used in a signal face for a U-turn to the right, its position in the signal face shall be the same as stated in Sections 4D.09 and 4D.10 for a right-turn arrow signal section of the same color.**

08 **A U-turn arrow signal indication pointing to the left shall not be used in a signal face that also contains a left-turn arrow signal indication. A U-turn arrow signal indication pointing to the right shall not be used in a signal face that also contains a right-turn arrow signal indication.**

Option:

09 Within a signal face, two identical CIRCULAR RED or RED ARROW signal indications may be displayed immediately horizontally adjacent to each other in a vertical or horizontal signal face (see Figure 4D-2) for emphasis.

10 Horizontally-arranged and vertically-arranged signal faces may be used on the same approach provided they are separated to meet the lateral separation spacing required in Section 4D.13.

Support:

11 Figure 4D-2 illustrates some of the typical arrangements of signal sections in signal faces that do not control separate turning movements. Figures 4D-6 through 4D-12 illustrate the typical arrangements of signal sections in left-turn signal faces. Figures 4D-13 through 4D-19 illustrate the typical arrangements of signal sections in right-turn signal faces.

Section 4D.09 Positions of Signal Indications Within a Vertical Signal Face

Standard:

01 **In each vertically-arranged signal face, all signal sections that display red signal indications shall be located above all signal sections that display yellow and green signal indications.**

02 **In vertically-arranged signal faces, each signal section that displays a YELLOW ARROW signal indication shall be located above the signal section that displays the GREEN ARROW signal indication to which it applies.**

Figure 4D-2. Typical Arrangements of Signal Sections in Signal Faces That Do Not Control Turning Movements

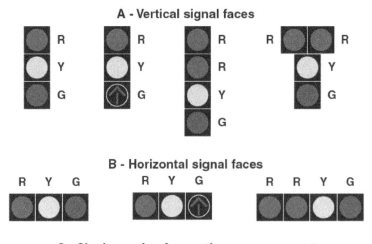

03   The relative positions of signal sections in a vertically-arranged signal face, from top to bottom, shall be as follows:
   CIRCULAR RED
   Steady and/or flashing left-turn RED ARROW
   Steady and/or flashing right-turn RED ARROW
   CIRCULAR YELLOW
   CIRCULAR GREEN
   Straight-through GREEN ARROW
   Steady left-turn YELLOW ARROW
   Flashing left-turn YELLOW ARROW
   Left-turn GREEN ARROW
   Steady right-turn YELLOW ARROW
   Flashing right-turn YELLOW ARROW
   Right-turn GREEN ARROW

04   If a dual-arrow signal section (capable of alternating between the display of a GREEN ARROW and a YELLOW ARROW signal indication) is used in a vertically-arranged signal face, the dual-arrow signal section shall occupy the same position relative to the other sections as the signal section that displays the GREEN ARROW signal indication in a vertically-arranged signal face would occupy.

Option:

05   In a vertically-arranged signal face, signal sections that display signal indications of the same color may be arranged horizontally adjacent to each other at right angles to the basic straight line arrangement to form a clustered signal face (see Figures 4D-2, 4D-9, 4D-11, 4D-16, and 4D-18).

Standard:

06   Such clusters shall be limited to the following:
   A. Two identical signal sections,
   B. Two or three different signal sections that display signal indications of the same color, or
   C. For only the specific case described in Section 4D.25 (see Drawing B of Figure 4D-20), two signal sections, one of which displays a GREEN ARROW signal indication and the other of which displays a flashing YELLOW ARROW signal indication.

07   The signal section that displays a flashing yellow signal indication during steady mode operation:
   A. Shall not be placed in the same vertical position as the signal section that displays a steady yellow signal indication, and
   B. Shall be placed below the signal section that displays a steady yellow signal indication.

Support:

08 Sections 4F.02 and 4G.04 contain exceptions to the provisions of this Section that are applicable to hybrid beacons.

Section 4D.10 Positions of Signal Indications Within a Horizontal Signal Face

Standard:

01 In each horizontally-arranged signal face, all signal sections that display red signal indications shall be located to the left of all signal sections that display yellow and green signal indications.

02 In horizontally-arranged signal faces, each signal section that displays a YELLOW ARROW signal indication shall be located to the left of the signal section that displays the GREEN ARROW signal indication to which it applies.

03 The relative positions of signal sections in a horizontally-arranged signal face, from left to right, shall be as follows:
CIRCULAR RED
Steady and/or flashing left-turn RED ARROW
Steady and/or flashing right-turn RED ARROW
CIRCULAR YELLOW
Steady left-turn YELLOW ARROW
Flashing left-turn YELLOW ARROW
Left-turn GREEN ARROW
CIRCULAR GREEN
Straight-through GREEN ARROW
Steady right-turn YELLOW ARROW
Flashing right-turn YELLOW ARROW
Right-turn GREEN ARROW

04 If a dual-arrow signal section (capable of alternating between the display of a GREEN ARROW and a YELLOW ARROW signal indication) is used in a horizontally-arranged signal face, the signal section that displays the dual left-turn arrow signal indication shall be located immediately to the right of the signal section that displays the CIRCULAR YELLOW signal indication, the signal section that displays the straight-through GREEN ARROW signal indication shall be located immediately to the right of the signal section that displays the CIRCULAR GREEN signal indication, and the signal section that displays the dual right-turn arrow signal indication shall be located to the right of all other signal sections.

05 The signal section that displays a flashing yellow signal indication during steady mode operation:
  A. Shall not be placed in the same horizontal position as the signal section that displays a steady yellow signal indication, and
  B. Shall be placed to the right of the signal section that displays a steady yellow signal indication.

Section 4D.11 Number of Signal Faces on an Approach

Standard:

01 The signal faces for each approach to an intersection or a midblock location shall be provided as follows:
  A. If a signalized through movement exists on an approach, a minimum of two primary signal faces shall be provided for the through movement. If a signalized through movement does not exist on an approach, a minimum of two primary signal faces shall be provided for the signalized turning movement that is considered to be the major movement from the approach (also see Section 4D.25).
  B. See Sections 4D.17 through 4D.20 for left-turn (and U-turn to the left) signal faces.
  C. See Sections 4D.21 through 4D.24 for right-turn (and U-turn to the right) signal faces.

Option:

02 Where a movement (or a certain lane or lanes) at the intersection never conflicts with any other signalized vehicular or pedestrian movement, a continuously-displayed single-section GREEN ARROW signal indication may be used to inform road users that the movement is free-flow and does not need to stop.

Support:

03 In some circumstances where the through movement never conflicts with any other signalized vehicular or pedestrian movement at the intersection, such as at T-intersections with appropriate geometrics and/or pavement markings and signing, an engineering study might determine that the through movement (or certain lanes of the through movement) can be free-flow and not signalized.

Page 460   2009 Edition

Guidance:

04   *If two or more left-turn lanes are provided for a separately controlled protected only mode left-turn movement, or if a left-turn movement represents the major movement from an approach, two or more primary left-turn signal faces should be provided.*

05   *If two or more right-turn lanes are provided for a separately controlled right-turn movement, or if a right-turn movement represents the major movement from an approach, two or more primary right-turn signal faces should be provided.*

Support:

06   Locating primary signal faces overhead on the far side of the intersection has been shown to provide safer operation by reducing intersection entries late in the yellow interval and by reducing red signal violations, as compared to post-mounting signal faces at the roadside or locating signal faces overhead within the intersection on a diagonally-oriented mast arm or span wire. On approaches with two or more lanes for the through movement, one signal face per through lane, centered over each through lane, has also been shown to provide safer operation.

Guidance:

07   *If the posted or statutory speed limit or the 85th-percentile speed on an approach to a signalized location is 45 mph or higher, signal faces should be provided as follows for all new or reconstructed signal installations (see Figure 4D-3):*

   A. The minimum number and location of primary (non-supplemental) signal faces for through traffic should be provided in accordance with Table 4D-1.

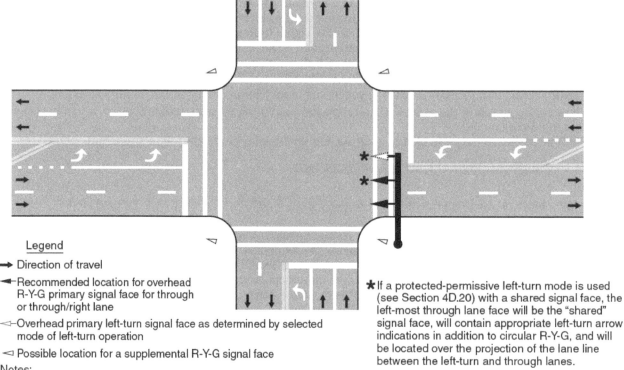

Figure 4D-3. Recommended Vehicular Signal Faces for Approaches with Posted, Statutory, or 85th-Percentile Speed of 45 mph or Higher

Notes:

1. Signal faces for only one direction and only one possible set of geometrics (number of lanes, etc.) are illustrated. If there are fewer or more than two through lanes on the approach, see Table 4D-2.
2. Any primary left-turn and/or right-turn signal faces, as determined by Sections 4D.17 through 4D.24, should be overhead for each exclusive turn lane.
3. One or more pole-mounted or overhead supplemental faces should be considered, based on the geometrics of the approach, to maximize visibility for approaching traffic.
4. All signal faces should have backplates.

Sect. 4D.11

Table 4D-1. Recommended Minimum Number of Primary Signal Faces for Through Traffic on Approaches with Posted, Statutory, or 85th-Percentile Speed of 45 mph or Higher

Number of Through Lanes on Approach	Total Number of Primary Through Signal Faces for Approach*	Minimum Number of Overhead-Mounted Primary Through Signal Faces for Approach
1	2	1
2	2	1
3	3	2**
4 or more	4 or more	3**

NOTES: *A minimum of two through signal faces is always required (See Section 4D.11). These recommended numbers of through signal faces may be exceeded. Also, see cone of vision requirements otherwise indicated in Section 4D.13.

** If practical, all of the recommended number of primary through signal faces should be located overhead.

B. *If the number of overhead primary signal faces for through traffic is equal to the number of through lanes on an approach, one overhead signal face should be located approximately over the center of each through lane.*
C. *Except for shared left-turn and right-turn signal faces, any primary signal face required by Sections 4D.17 through 4D.25 for an exclusive turn lane should be located overhead approximately over the center of each exclusive turn lane.*
D. *All primary signal faces should be located on the far side of the intersection.*
E. *In addition to the primary signal faces, one or more supplemental pole-mounted or overhead signal faces should be considered to provide added visibility for approaching traffic that is traveling behind large vehicles.*
F. *All signal faces should have backplates.*

08 *This layout of signal faces should also be considered for any major urban or suburban arterial street with four or more lanes and for other approaches with speeds of less than 45 mph.*

Section 4D.12 Visibility, Aiming, and Shielding of Signal Faces

Standard:

01 The primary consideration in signal face placement, aiming, and adjustment shall be to optimize the visibility of signal indications to approaching traffic.

02 Road users approaching a signalized intersection or other signalized area, such as a midblock crosswalk, shall be given a clear and unmistakable indication of their right-of-way assignment.

03 The geometry of each intersection to be signalized, including vertical grades, horizontal curves, and obstructions as well as the lateral and vertical angles of sight toward a signal face, as determined by typical driver-eye position, shall be considered in determining the vertical, longitudinal, and lateral position of the signal face.

Guidance:

04 *The two primary signal faces required as a minimum for each approach should be continuously visible to traffic approaching the traffic control signal, from a point at least the minimum sight distance provided in Table 4D-2 in advance of and measured to the stop line. This range of continuous visibility should be provided unless precluded by a physical obstruction or unless another signalized location is within this range.*

Table 4D-2. Minimum Sight Distance for Signal Visibility

85th-Percentile Speed	Minimum Sight Distance
20 mph	175 feet
25 mph	215 feet
30 mph	270 feet
35 mph	325 feet
40 mph	390 feet
45 mph	460 feet
50 mph	540 feet
55 mph	625 feet
60 mph	715 feet

Note: Distances in this table are derived from stopping sight distance plus an assumed queue length for shorter cycle lengths (60 to 75 seconds).

05 *There should be legal authority to prohibit the display of any unauthorized sign, signal, marking, or device that interferes with the effectiveness of any official traffic control device (see Section 11-205 of the "Uniform Vehicle Code").*

06 *At signalized midblock crosswalks, at least one of the signal faces should be over the traveled way for each approach.*

Standard:

07 **If approaching traffic does not have a continuous view of at least two signal faces for at least the minimum sight distance shown in Table 4D-2, a sign (see Section 2C.36) shall be installed to warn approaching traffic of the traffic control signal.**

Option:

08 If a sign is installed to warn approaching road users of the traffic control signal, the sign may be supplemented by a Warning Beacon (see Section 4L.03).

09 A Warning Beacon used in this manner may be interconnected with the traffic signal controller assembly in such a manner as to flash yellow during the period when road users passing this beacon at the legal speed for the roadway might encounter a red signal indication (or a queue resulting from the display of the red signal indication) upon arrival at the signalized location.

10 If the sight distance to the signal faces for an approach is limited by horizontal or vertical alignment, supplemental signal faces aimed at a point on the approach at which the signal indications first become visible may be used.

Guidance:

11 *Supplemental signal faces should be used if engineering judgment has shown that they are needed to achieve intersection visibility both in advance and immediately before the signalized location.*

12 *If supplemental signal faces are used, they should be located to provide optimum visibility for the movement to be controlled.*

Standard:

13 **In cases where irregular street design necessitates placing signal faces for different street approaches with a comparatively small angle between their respective signal indications, each signal indication shall, to the extent practical, be visibility-limited by signal visors, signal louvers, or other means so that an approaching road user's view of the signal indication(s) controlling movements on other approaches is minimized.**

14 **Signal visors exceeding 12 inches in length shall not be used on free-swinging signal faces.**

Guidance:

15 *Signal visors should be used on signal faces to aid in directing the signal indication specifically to approaching traffic, as well as to reduce "sun phantom," which can result when external light enters the lens.*

16 *The use of signal visors, or the use of signal faces or devices that direct the light without a reduction in intensity, should be considered as an alternative to signal louvers because of the reduction in light output caused by signal louvers.*

Option:

17 Special signal faces, such as visibility-limited signal faces, may be used such that the road user does not see signal indications intended for other approaches before seeing the signal indications for their own approach, if simultaneous viewing of both signal indications could cause the road user to be misdirected.

Guidance:

18 *If the posted or statutory speed limit or the 85th-percentile speed on an approach to a signalized location is 45 mph or higher, signal backplates should be used on all of the signal faces that face the approach. Signal backplates should also be considered for use on signal faces on approaches with posted or statutory speed limits or 85th-percentile speeds of less than 45 mph where sun glare, bright sky, and/or complex or confusing backgrounds indicate a need for enhanced signal face target value.*

Support:

19 The use of backplates enhances the contrast between the traffic signal indications and their surroundings for both day and night conditions, which is also helpful to older drivers.

Standard:

20 **The inside of signal visors (hoods), the entire surface of louvers and fins, and the front surface of backplates shall have a dull black finish to minimize light reflection and to increase contrast between the signal indication and its background.**

Option:

A yellow retroreflective strip with a minimum width of 1 inch and a maximum width of 3 inches may be placed along the perimeter of the face of a signal backplate to project a rectangular appearance at night.

Section 4D.13 Lateral Positioning of Signal Faces

Standard:

At least one and preferably both of the minimum of two primary signal faces required for the through movement (or the major turning movement if there is no through movement) on the approach shall be located between two lines intersecting with the center of the approach at a point 10 feet behind the stop line, one making an angle of approximately 20 degrees to the right of the center of the approach extended, and the other making an angle of approximately 20 degrees to the left of the center of the approach extended. The signal face that satisfies this requirement shall simultaneously satisfy the longitudinal placement requirement described in Section 4D.14 (see Figure 4D-4).

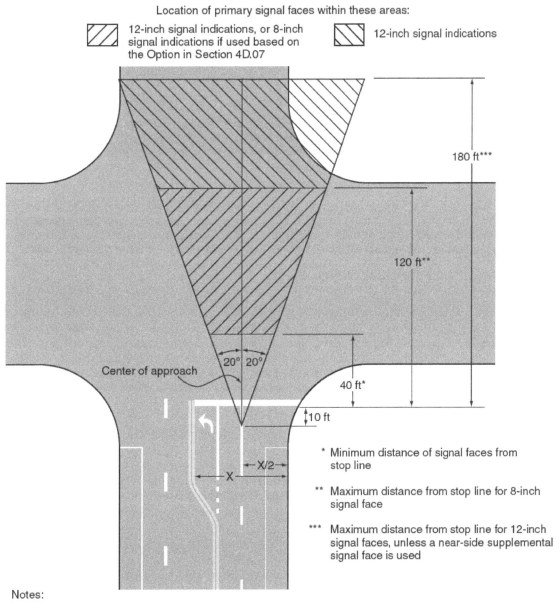

Figure 4D-4. Lateral and Longitudinal Location of Primary Signal Faces

* Minimum distance of signal faces from stop line

** Maximum distance from stop line for 8-inch signal face

*** Maximum distance from stop line for 12-inch signal faces, unless a near-side supplemental signal face is used

Notes:
1. See Section 4D.11 for approaches with posted, statutory, or 85th-percentile speeds of 45 mph or higher
2. See Section 4D.13 regarding location of signal faces that display a CIRCULAR GREEN signal indication for a permissive left-turn movement on approaches with an exclusive left-turn lane or lanes

02 **If both of the minimum of two primary signal faces required for the through movement (or the major turning movement if there is no through movement) on the approach are post-mounted, they shall both be on the far side of the intersection, one on the right and one on the left of the approach lane(s).**

03 **The required signal faces for through traffic on an approach shall be located not less than 8 feet apart measured horizontally perpendicular to the approach between the centers of the signal faces.**

04 **If more than one separate turn signal face is provided for a turning movement and if one or both of the separate turn signal faces are located over the roadway, the signal faces shall be located not less than 8 feet apart measured horizontally perpendicular to the approach between the centers of the signal faces.**

Guidance:

05 *If a signal face controls a specific lane or lanes of an approach, its position should make it readily visible to road users making that movement.*

Support:

06 Section 4D.11 contains additional provisions regarding lateral positioning of signal faces for approaches having a posted or statutory speed limit or an 85th-percentile speed of 45 mph or higher.

Standard:

07 **If an exclusive left-turn, right-turn, or U-turn lane is present on an approach and if a primary separate turn signal face controlling that lane is mounted over the roadway, the primary separate turn signal face shall not be positioned any further to the right than the extension of the right-hand edge of the exclusive turn lane or any further to the left than the extension of the left-hand edge of the exclusive turn lane.**

08 **Supplemental turn signal faces mounted over the roadway shall not be subject to the positioning requirements in the previous paragraph.**

Guidance:

09 *For new or reconstructed signal installations, on an approach with an exclusive turn lane(s) for a left-turn (or U-turn to the left) movement and with opposing vehicular traffic, signal faces that display a CIRCULAR GREEN signal indication should not be post-mounted on the far-side median or mounted overhead above the exclusive turn lane(s) or the extension of the lane(s).*

Standard:

10 **If supplemental post-mounted signal faces are used, the following limitations shall apply:**
   A. Left-turn arrows and U-turn arrows to the left shall not be used in near-right signal faces.
   B. Right-turn arrows and U-turn arrows to the right shall not be used in far-left signal faces. A far-side median-mounted signal face shall be considered a far-left signal for this application.

Section 4D.14 Longitudinal Positioning of Signal Faces

Standard:

01 **Except where the width of an intersecting roadway or other conditions make it physically impractical, the signal faces for each approach to an intersection or a midblock location shall be provided as follows:**
   A. A signal face installed to satisfy the requirements for primary left-turn signal faces (see Sections 4D.17 through 4D.20) and primary right-turn signal faces (see Sections 4D.21 through 4D.24), and at least one and preferably both of the minimum of two primary signal faces required for the through movement (or the major turning movement if there is no through movement) on the approach shall be located:
   1. No less than 40 feet beyond the stop line,
   2. No more than 180 feet beyond the stop line unless a supplemental near-side signal face is provided, and
   3. As near as practical to the line of the driver's normal view, if mounted over the roadway.
   The primary signal face that satisfies this requirement shall simultaneously satisfy the lateral placement requirement described in Section 4D.13 (see Figure 4D-4).
   B. Where the nearest signal face is located between 150 and 180 feet beyond the stop line, engineering judgment of the conditions, including the worst-case visibility conditions, shall be used to determine if the provision of a supplemental near-side signal face would be beneficial.

Support:

02 Section 4D.11 contains additional provisions regarding longitudinal positioning of signal faces for approaches having a posted or 85th-percentile speed of 45 mph or higher.

Guidance:

03 *Supplemental near-side signal faces should be located as near as practical to the stop line.*

Section 4D.15 Mounting Height of Signal Faces

Standard:

01　The top of the signal housing of a vehicular signal face located over any portion of a highway that can be used by motor vehicles shall not be more than 25.6 feet above the pavement.

02　For viewing distances between 40 and 53 feet from the stop line, the maximum mounting height to the top of the signal housing shall be as shown in Figure 4D-5.

03　The bottom of the signal housing and any related attachments to a vehicular signal face located over any portion of a highway that can be used by motor vehicles shall be at least 15 feet above the pavement.

04　The bottom of the signal housing (including brackets) of a vehicular signal face that is vertically arranged and not located over a roadway:
   A. Shall be a minimum of 8 feet and a maximum of 19 feet above the sidewalk or, if there is no sidewalk, above the pavement grade at the center of the roadway.
   B. Shall be a minimum of 4.5 feet and a maximum of 19 feet above the median island grade of a center median island if located on the near side of the intersection.

05　The bottom of the signal housing (including brackets) of a vehicular signal face that is horizontally arranged and not located over a roadway:
   A. Shall be a minimum of 8 feet and a maximum of 22 feet above the sidewalk or, if there is no sidewalk, above the pavement grade at the center of the roadway.
   B. Shall be a minimum of 4.5 feet and a maximum of 22 feet above the median island grade of a center median island if located on the near side of the intersection.

Section 4D.16 Lateral Offset (Clearance) of Signal Faces

Standard:

01　Signal faces mounted at the side of a roadway with curbs at less than 15 feet from the bottom of the housing and any related attachments shall have a horizontal offset of not less than 2 feet from the face of a vertical curb, or if there is no curb, not less than 2 feet from the edge of a shoulder.

Section 4D.17 Signal Indications for Left-Turn Movements – General

Standard:

01　In Sections 4D.17 through 4D.20, provisions applicable to left-turn movements and left-turn lanes shall also apply to signal indications for U-turns to the left that are provided at locations where left turns are prohibited or not geometrically possible.

Figure 4D-5. Maximum Mounting Height of Signal Faces Located Between 40 Feet and 53 Feet from Stop Line

Support:

02 Left-turning traffic is controlled by one of four modes as follows:
   A. Permissive Only Mode—turns made on a CIRCULAR GREEN signal indication, a flashing left-turn YELLOW ARROW signal indication, or a flashing left-turn RED ARROW signal indication after yielding to pedestrians, if any, and/or opposing traffic, if any.
   B. Protected Only Mode—turns made only when a left-turn GREEN ARROW signal indication is displayed.
   C. Protected/Permissive Mode—both modes can occur on an approach during the same cycle.
   D. Variable Left-Turn Mode—the operating mode changes among the protected only mode and/or the protected/permissive mode and/or the permissive only mode during different periods of the day or as traffic conditions change.

Option:

03 In areas having a high percentage of older drivers, special consideration may be given to the use of protected only mode left-turn phasing, when appropriate.

Standard:

04 **During a permissive left-turn movement, the signal faces for through traffic on the opposing approach shall simultaneously display green or steady yellow signal indications. If pedestrians crossing the lane or lanes used by the permissive left-turn movement to depart the intersection are controlled by pedestrian signal heads, the signal indications displayed by those pedestrian signal heads shall not be limited to any particular display during the permissive left-turn movement.**

05 **During a protected left-turn movement, the signal faces for through traffic on the opposing approach shall simultaneously display steady CIRCULAR RED signal indications. If pedestrians crossing the lane or lanes used by the protected left-turn movement to depart the intersection are controlled by pedestrian signal heads, the pedestrian signal heads shall display a steady UPRAISED HAND (symbolizing DONT WALK) signal indication during the protected left-turn movement.**

06 **A protected only mode left-turn movement that does not begin and terminate at the same time as the adjacent through movement shall not be provided on an approach unless an exclusive left-turn lane exists.**

07 **A yellow change interval for the left-turn movement shall not be displayed when the status of the left-turn operation is changing from permissive to protected within any given signal sequence.**

08 **If the operating mode changes among the protected only mode and/or the protected/permissive mode and/or the permissive only mode during different periods of the day or as traffic conditions change, the requirements in Sections 4D.18 through 4D.20 that are appropriate to that mode of operation shall be met, subject to the following:**
   A. **The CIRCULAR GREEN and CIRCULAR YELLOW signal indications shall not be displayed when operating in the protected only mode.**
   B. **The left-turn GREEN ARROW and left-turn YELLOW ARROW signal indications shall not be displayed when operating in the permissive only mode.**

Option:

09 Additional static signs or changeable message signs may be used to meet the requirements for the variable left-turn mode or to inform drivers that left-turn green arrows will not be available during certain times of the day.

Support:

10 Sections 4D.17 through 4D.20 describe the use of the following two types of signal faces for controlling left-turn movements:
   A. Shared signal face – This type of signal face controls both the left-turn movement and the adjacent movement (usually the through movement) and can serve as one of the two required primary signal faces for the adjacent movement. A shared signal face always displays the same color of circular indication that is displayed by the signal face or faces for the adjacent movement. If a shared signal face that provides protected/permissive mode left turns is mounted overhead at the intersection, it is usually positioned over or slightly to the right of the extension of the lane line separating the left-turn lane from the adjacent lane.
   B. Separate left-turn signal face – This type of signal face controls only the left-turn movement and cannot serve as one of the two required primary signal faces for the adjacent movement (usually the through movement) because it displays signal indications that are applicable only to the left-turn movement. If a separate left-turn signal face is mounted overhead at the intersection, it is positioned over the extension of the left-turn lane. In a separate left-turn signal face, a flashing left-turn YELLOW ARROW signal indication or a flashing left-turn RED ARROW signal indication is used to control permissive left-turning movements.

¹¹ Section 4D.13 contains provisions regarding the lateral positioning of signal faces that control left-turn movements.

¹² It is not necessary that the same mode of left-turn operation or same type of left-turn signal face be used on every approach to a signalized location. Selecting different modes and types of left-turn signal faces for the various approaches to the same signalized location is acceptable.

Option:

¹³ A signal face that is shared by left-turning and right-turning traffic may be provided for a shared left-turn/right-turn lane on an approach that has no through traffic (see Section 4D.25).

Section 4D.18 Signal Indications for Permissive Only Mode Left-Turn Movements

Standard:

⁰¹ If a shared signal face is provided for a permissive only mode left turn, it shall meet the following requirements (see Figure 4D-6):
  A. It shall be capable of displaying the following signal indications: steady CIRCULAR RED, steady CIRCULAR YELLOW, and CIRCULAR GREEN. Only one of the three indications shall be displayed at any given time.
  B. During the permissive left-turn movement, a CIRCULAR GREEN signal indication shall be displayed.
  C. A permissive only shared signal face, regardless of where it is positioned and regardless of how many adjacent through signal faces are provided, shall always simultaneously display the same color of circular indication that the adjacent through signal face or faces display.
  D. If the permissive only mode is not the only left-turn mode used for the approach, the signal face shall be the same shared signal face that is used for the protected/permissive mode (see Section 4D.20) except that the left-turn GREEN ARROW and left-turn YELLOW ARROW signal indications shall not be displayed when operating in the permissive only mode.

⁰² If a separate left-turn signal face is being operated in a permissive only left-turns mode, a CIRCULAR GREEN signal indication shall not be used in that face.

⁰³ If a separate left-turn signal face is being operated in a permissive only left-turn mode and a flashing left-turn YELLOW ARROW signal indication is provided, it shall meet the following requirements (see Figure 4D-7):
  A. It shall be capable of displaying the following signal indications: steady left-turn RED ARROW, steady left-turn YELLOW ARROW, and flashing left-turn YELLOW ARROW. Only one of the three indications shall be displayed at any given time.

Figure 4D-6. Typical Position and Arrangements of Shared Signal Faces for Permissive Only Mode Left Turns

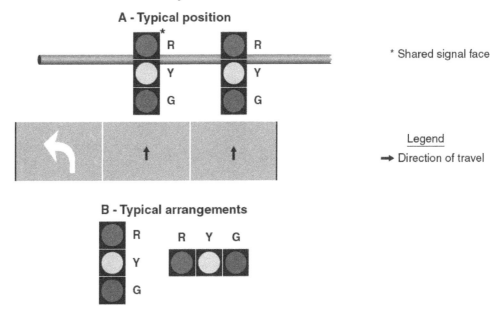

Figure 4D-7. Typical Position and Arrangements of Separate Signal Faces with Flashing Yellow Arrow for Permissive Only Mode Left Turns

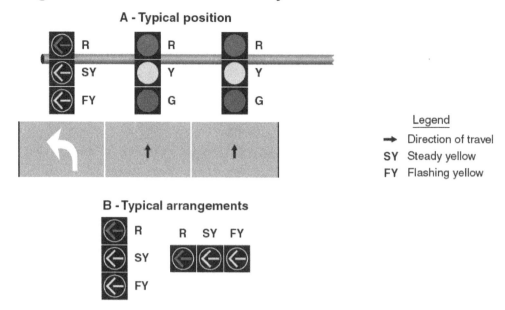

- B. During the permissive left-turn movement, a flashing left-turn YELLOW ARROW signal indication shall be displayed.
- C. A steady left-turn YELLOW ARROW signal indication shall be displayed following the flashing left-turn YELLOW ARROW signal indication.
- D. It shall be permitted to display a flashing left-turn YELLOW ARROW signal indication for a permissive left-turn movement while the signal faces for the adjacent through movement display steady CIRCULAR RED signal indications and the opposing left-turn signal faces display left-turn GREEN ARROW signal indications for a protected left-turn movement.
- E. During steady mode (stop-and-go) operation, the signal section that displays the steady left-turn YELLOW ARROW signal indication during change intervals shall not be used to display the flashing left-turn YELLOW ARROW signal indication for permissive left turns.
- F. During flashing mode operation (see Section 4D.30), the display of a flashing left-turn YELLOW ARROW signal indication shall be only from the signal section that displays a steady left-turn YELLOW ARROW signal indication during steady mode (stop-and-go) operation.
- G. If the permissive only mode is not the only left-turn mode used for the approach, the signal face shall be the same separate left-turn signal face with a flashing YELLOW ARROW signal indication that is used for the protected/permissive mode (see Section 4D.20) except that the left-turn GREEN ARROW signal indication shall not be displayed when operating in the permissive only mode.

Option:

04 A separate left-turn signal face with a flashing left-turn RED ARROW signal indication during the permissive left-turn movement may be used for unusual geometric conditions, such as wide medians with offset left-turn lanes, but only when an engineering study determines that each and every vehicle must successively come to a full stop before making a permissive left turn.

Standard:

05 If a separate left-turn signal face is being operated in a permissive only left-turn mode and a flashing left-turn RED ARROW signal indication is provided, it shall meet the following requirements (see Figure 4D-8):
- A. It shall be capable of displaying the following signal indications: steady or flashing left-turn RED ARROW, steady left-turn YELLOW ARROW, and left-turn GREEN ARROW. Only one of the three indications shall be displayed at any given time. The GREEN ARROW indication is required in order to provide a three-section signal face, but shall not be displayed during the permissive only mode.
- B. During the permissive left-turn movement, a flashing left-turn RED ARROW signal indication shall be displayed, thus indicating that each and every vehicle must successively come to a full stop before making a permissive left turn.

Figure 4D-8. Typical Position and Arrangements of Separate Signal Faces with Flashing Red Arrow for Permissive Only Mode and Protected/Permissive Mode Left Turns

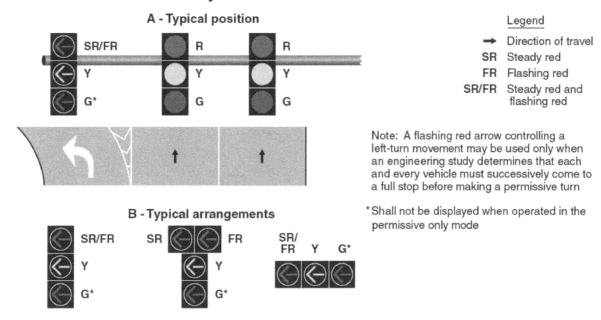

C. A steady left-turn YELLOW ARROW signal indication shall be displayed following the flashing left-turn RED ARROW signal indication.
D. It shall be permitted to display a flashing left-turn RED ARROW signal indication for a permissive left-turn movement while the signal faces for the adjacent through movement display steady CIRCULAR RED signal indications and the opposing left-turn signal faces display left-turn GREEN ARROW signal indications for a protected left-turn movement.
E. A supplementary sign shall not be required. If used, it shall be a LEFT TURN YIELD ON FLASHING RED ARROW AFTER STOP (R10-27) sign (see Figure 2B-27).

Option:

06 The requirements of Item A in Paragraph 5 may be met by a vertically-arranged signal face with a horizontal cluster of two left-turn RED ARROW signal indications, the left-most of which displays a steady indication and the right-most of which displays a flashing indication (see Figure 4D-8).

Section 4D.19 Signal Indications for Protected Only Mode Left-Turn Movements

Standard:

01 A shared signal face shall not be used for protected only mode left turns unless the CIRCULAR GREEN and left-turn GREEN ARROW signal indications always begin and terminate together. If a shared signal face is provided for a protected only mode left turn, it shall meet the following requirements (see Figure 4D-9):
   A. It shall be capable of displaying the following signal indications: steady CIRCULAR RED, steady CIRCULAR YELLOW, CIRCULAR GREEN, and left-turn GREEN ARROW. Only one of the three colors shall be displayed at any given time.
   B. During the protected left-turn movement, the shared signal face shall simultaneously display both a CIRCULAR GREEN signal indication and a left-turn GREEN ARROW signal indication.
   C. The shared signal face shall always simultaneously display the same color of circular indication that the adjacent through signal face or faces display.
   D. If the protected only mode is not the only left-turn mode used for the approach, the signal face shall be the same shared signal face that is used for the protected/permissive mode (see Section 4D.20).

Option:

02 A straight-through GREEN ARROW signal indication may be used instead of the CIRCULAR GREEN signal indication in Items A and B in Paragraph 1 on an approach where right turns are prohibited and a straight-through GREEN ARROW signal indication is also used instead of a CIRCULAR GREEN signal indication in the other signal face(s) for through traffic.

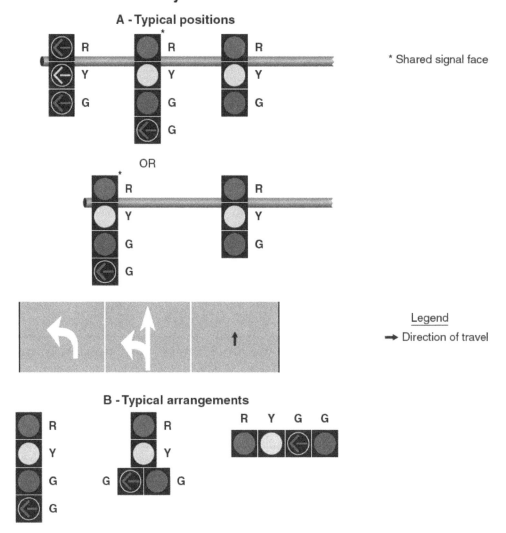

Figure 4D-9. Typical Positions and Arrangements of Shared Signal Faces for Protected Only Mode Left Turns

Standard:

03 If a separate left-turn signal face is provided for a protected only mode left turn, it shall meet the following requirements (see Figure 4D-10):
   A. It shall be capable of displaying, the following signal indications: steady left-turn RED ARROW, steady left-turn YELLOW ARROW, and left-turn GREEN ARROW. Only one of the three indications shall be displayed at any given time. A signal instruction sign shall not be required with this set of signal indications. If used, it shall be a LEFT ON GREEN ARROW ONLY (R10-5) sign (see Figure 2B-27).
   B. During the protected left-turn movement, a left-turn GREEN ARROW signal indication shall be displayed.
   C. A steady left-turn YELLOW ARROW signal indication shall be displayed following the left-turn GREEN ARROW signal indication.
   D. If the protected only mode is not the only left-turn mode used for the approach, the signal face shall be the same separate left-turn signal face that is used for the protected/permissive mode (see Section 4D.20 and Figures 4D-8 and 4D-12) except that the flashing left-turn YELLOW ARROW or flashing left-turn RED ARROW signal indication shall not be displayed when operating in the protected only mode.

Figure 4D-10. Typical Position and Arrangements of Separate Signal Faces for Protected Only Mode Left Turns

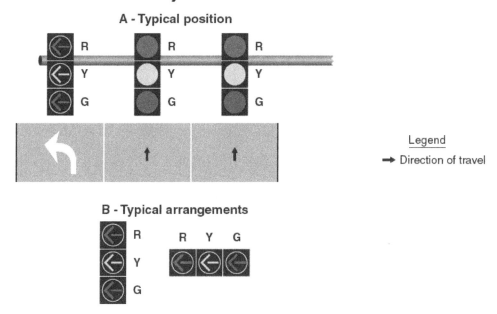

Section 4D.20 Signal Indications for Protected/Permissive Mode Left-Turn Movements

Standard:

01   If a shared signal face is provided for a protected/permissive mode left turn, it shall meet the following requirements (see Figure 4D-11):
   A. It shall be capable of displaying the following signal indications: steady CIRCULAR RED, steady CIRCULAR YELLOW, CIRCULAR green, steady left-turn YELLOW ARROW, and left-turn GREEN ARROW. Only one of the three circular indications shall be displayed at any given time. Only one of the two arrow indications shall be displayed at any given time. If the left-turn GREEN ARROW signal indication and the CIRCULAR GREEN signal indication(s) for the adjacent through movement are always terminated together, the steady left-turn YELLOW ARROW signal indication shall not be required.
   B. During the protected left-turn movement, the shared signal face shall simultaneously display a left-turn GREEN ARROW signal indication and a circular signal indication that is the same color as the signal indication for the adjacent through lane on the same approach as the protected left turn.
   C. A steady left-turn YELLOW ARROW signal indication shall be displayed following the left-turn GREEN ARROW signal indication, unless the left-turn GREEN ARROW signal indication and the CIRCULAR GREEN signal indication(s) for the adjacent through movement are being terminated together. When the left-turn GREEN ARROW and CIRCULAR GREEN signal indications are being terminated together, the required display following the left-turn GREEN ARROW signal indication shall be either the display of a CIRCULAR YELLOW signal indication alone or the simultaneous display of the CIRCULAR YELLOW and left-turn YELLOW ARROW signal indications.
   D. During the permissive left-turn movement, the shared signal face shall display only a CIRCULAR GREEN signal indication.
   E. A protected/permissive shared signal face, regardless of where it is positioned and regardless of how many adjacent through signal faces are provided, shall always simultaneously display the same color of circular indication that the adjacent through signal face or faces display.
   F. A supplementary sign shall not be required. If used, it shall be a LEFT TURN YIELD ON GREEN (symbolic circular green) (R10-12) sign (see Figure 2B-27).

02   If a separate left-turn signal face is being operated in a protected/permissive left-turn mode, a CIRCULAR GREEN signal indication shall not be used in that face.

Figure 4D-11. Typical Position and Arrangements of Shared Signal Faces for Protected/Permissive Mode Left Turns

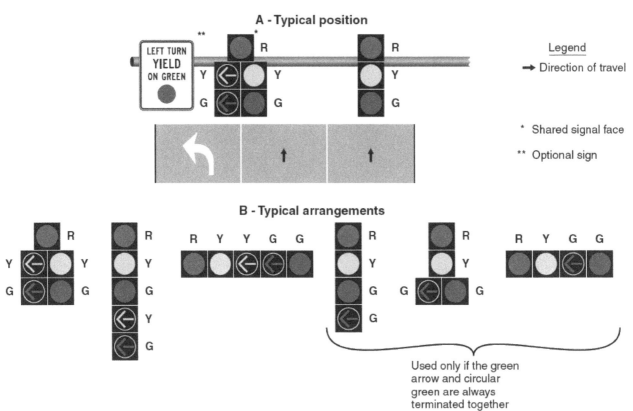

03   If a separate left-turn signal face is being operated in a protected/permissive left-turn mode and a flashing left-turn yellow arrow signal indication is provided, it shall meet the following requirements (see Figure 4D-12):

   A. It shall be capable of displaying the following signal indications: steady left-turn RED ARROW, steady left-turn YELLOW ARROW, flashing left-turn YELLOW ARROW, and left-turn GREEN ARROW. Only one of the four indications shall be displayed at any given time.
   B. During the protected left-turn movement, a left-turn GREEN ARROW signal indication shall be displayed.
   C. A steady left-turn YELLOW ARROW signal indication shall be displayed following the left-turn GREEN ARROW signal indication.
   D. During the permissive left-turn movement, a flashing left-turn YELLOW ARROW signal indication shall be displayed.
   E. A steady left-turn YELLOW ARROW signal indication shall be displayed following the flashing left-turn YELLOW ARROW signal indication if the permissive left-turn movement is being terminated and the separate left-turn signal face will subsequently display a steady left-turn RED ARROW indication.
   F. It shall be permitted to display a flashing left-turn YELLOW ARROW signal indication for a permissive left-turn movement while the signal faces for the adjacent through movement display steady CIRCULAR RED signal indications and the opposing left-turn signal faces display left-turn GREEN ARROW signal indications for a protected left-turn movement.
   G. When a permissive left-turn movement is changing to a protected left-turn movement, a left-turn GREEN ARROW signal indication shall be displayed immediately upon the termination of the flashing left-turn YELLOW ARROW signal indication. A steady left-turn YELLOW ARROW signal indication shall not be displayed between the display of the flashing left-turn YELLOW ARROW signal indication and the display of the steady left-turn GREEN ARROW signal indication.

Figure 4D-12. Typical Position and Arrangements of Separate Signal Faces with Flashing Yellow Arrow for Protected/Permissive Mode and Protected Only Mode Left Turns

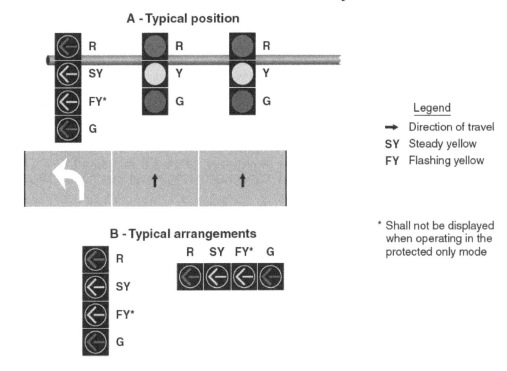

H. The display shall be a four-section signal face except that a three-section signal face containing a dual-arrow signal section shall be permitted where signal head height limitations (or lateral positioning limitations for a horizontally-mounted signal face) will not permit the use of a four-section signal face. The dual-arrow signal section, where used, shall display a GREEN ARROW for the protected left-turn movement and a flashing YELLOW ARROW for the permissive left-turn movement.

I. During steady mode (stop-and-go) operation, the signal section that displays the steady left-turn YELLOW ARROW signal indication during change intervals shall not be used to display the flashing left-turn YELLOW ARROW signal indication for permissive left turns.

J. During flashing mode operation (see Section 4D.30), the display of a flashing left-turn YELLOW ARROW signal indication shall be only from the signal section that displays a steady left-turn YELLOW ARROW signal indication during steady mode (stop-and-go) operation.

Option:

04 A separate left-turn signal face with a flashing left-turn RED ARROW signal indication during the permissive left-turn movement may be used for unusual geometric conditions, such as wide medians with offset left-turn lanes, but only when an engineering study determines that each and every vehicle must successively come to a full stop before making a permissive left turn.

Standard:

05 If a separate left-turn signal face is being operated in a protected/permissive left-turn mode and a flashing left-turn RED arrow signal indication is provided, it shall meet the following requirements (see Figure 4D-8):

A. It shall be capable of displaying the following signal indications: steady or flashing left-turn RED ARROW, steady left-turn YELLOW ARROW, and left-turn GREEN ARROW. Only one of the three indications shall be displayed at any given time.

B. During the protected left-turn movement, a left-turn GREEN ARROW signal indication shall be displayed.

C. A steady left-turn YELLOW ARROW signal indication shall be displayed following the left-turn GREEN ARROW signal indication.

D. During the permissive left-turn movement, a flashing left-turn RED ARROW signal indication shall be displayed.

E. A steady left-turn YELLOW ARROW signal indication shall be displayed following the flashing left-turn RED ARROW signal indication if the permissive left-turn movement is being terminated and the separate left-turn signal face will subsequently display a steady left-turn RED ARROW indication.
F. When a permissive left-turn movement is changing to a protected left-turn movement, a left-turn GREEN ARROW signal indication shall be displayed immediately upon the termination of the flashing left-turn RED ARROW signal indication. A steady left-turn YELLOW ARROW signal indication shall not be displayed between the display of the flashing left-turn RED ARROW signal indication and the display of the steady left-turn GREEN ARROW signal indication.
G. It shall be permitted to display a flashing left-turn RED ARROW signal indication for a permissive left-turn movement while the signal faces for the adjacent through movement display steady CIRCULAR RED signal indications and the opposing left-turn signal faces display left-turn GREEN ARROW signal indications for a protected left-turn movement.
H. A supplementary sign shall not be required. If used, it shall be a LEFT TURN YIELD ON FLASHING RED ARROW AFTER STOP (R10-27) sign (see Figure 2B-27).

Option:

06 The requirements of Item A in Paragraph 5 may be met by a vertically-arranged signal face with a horizontal cluster of two left-turn RED ARROW signal indications, the left-most of which displays a steady indication and the right-most of which displays a flashing indication (see Figure 4D-8).

Section 4D.21 Signal Indications for Right-Turn Movements – General

Standard:

01 **In Sections 4D.21 through 4D.24, provisions applicable to right-turn movements and right-turn lanes shall also apply to signal indications for U-turns to the right that are provided at locations where right turns are prohibited or not geometrically possible.**

Support:

02 Right-turning traffic is controlled by one of four modes as follows:
A. Permissive Only Mode—turns made on a CIRCULAR GREEN signal indication, a flashing right-turn YELLOW ARROW signal indication, or a flashing right-turn RED ARROW signal indication after yielding to pedestrians, if any.
B. Protected Only Mode—turns made only when a right-turn GREEN ARROW signal indication is displayed.
C. Protected/Permissive Mode—both modes occur on an approach during the same cycle.
D. Variable Right-Turn Mode—the operating mode changes among the protected only mode and/or the protected/permissive mode and/or the permissive only mode during different periods of the day or as traffic conditions change.

Standard:

03 **During a permissive right-turn movement, the signal faces, if any, that exclusively control U-turn traffic that conflicts with the permissive right-turn movement (see Item F.1 in Section 4D.05) shall simultaneously display steady U-turn RED ARROW signal indications. If pedestrians crossing the lane or lanes used by the permissive right-turn movement to depart the intersection are controlled by pedestrian signal heads, the signal indications displayed by those pedestrian signal heads shall not be limited to any particular display during the permissive right-turn movement.**

04 **During a protected right-turn movement, the signal faces for left-turn traffic, if any, on the opposing approach shall not simultaneously display a steady left-turn GREEN ARROW or steady left-turn YELLOW ARROW signal indication, and signal faces, if any, that exclusively control U-turn traffic that conflicts with the protected right-turn movement (see Item F.1 in Section 4D.05) shall simultaneously display steady U-turn RED ARROW signal indications. If pedestrians crossing the lane or lanes used by the protected right-turn movement to depart the intersection are controlled by pedestrian signal heads, the pedestrian signal heads shall display a steady UPRAISED HAND (symbolizing DONT WALK) signal indication during the protected right-turn movement.**

05 **A protected only mode right-turn movement that does not begin and terminate at the same time as the adjacent through movement shall not be provided on an approach unless an exclusive right-turn lane exists.**

06 **A yellow change interval for the right-turn movement shall not be displayed when the status of the right-turn operation is changing from permissive to protected within any given signal sequence.**

07 If the operating mode changes among the protected only mode and/or the protected/permissive mode and/or the permissive only mode during different periods of the day or as traffic conditions change, the requirements in Sections 4D.22 through 4D.24 that are appropriate to that mode of operation shall be met, subject to the following:
   A. The CIRCULAR GREEN and CIRCULAR YELLOW signal indications shall not be displayed when operating in the protected only mode.
   B. The right-turn GREEN ARROW and right-turn YELLOW ARROW signal indications shall not be displayed when operating in the permissive only mode.

Option:

08 Additional static signs or changeable message signs may be used to meet the requirements for the variable right-turn mode or to inform drivers that right-turn green arrows will not be available during certain times of the day.

Support:

09 Sections 4D.21 through 4D.24 describe the use of the following two types of signal faces for controlling right-turn movements:
   A. Shared signal face – This type of signal face controls both the right-turn movement and the adjacent movement (usually the through movement) and can serve as one of the two required primary signal faces for the adjacent movement. A shared signal face always displays the same color of circular indication that is displayed by the signal face or faces for the adjacent movement.
   B. Separate right-turn signal face – This type of signal face controls only the right-turn movement and cannot serve as one of the two required primary signal faces for the adjacent movement (usually the through movement) because it displays signal indications that are applicable only to the right-turn movement. If a separate right-turn signal face is mounted overhead at the intersection, it is positioned over the extension of the right-turn lane. In a separate right-turn signal face, a flashing right-turn YELLOW ARROW signal indication or a flashing right-turn RED ARROW signal indication is used to control permissive right-turning movements.

10 Section 4D.13 contains provisions regarding the lateral positioning of signal faces that control right-turn movements.

11 It is not necessary that the same mode of right-turn operation or same type of right-turn signal face be used on every approach to a signalized location. Selecting different modes and types of right-turn signal faces for the various approaches to the same signalized location is acceptable.

Option:

12 A signal face that is shared by left-turning and right-turning traffic may be provided for a shared left-turn/right-turn lane on an approach that has no through traffic (see Section 4D.25).

Section 4D.22 Signal Indications for Permissive Only Mode Right-Turn Movements
Standard:

01 If a shared signal face is provided for a permissive only mode right turn, it shall meet the following requirements (see Figure 4D-13):
   A. It shall be capable of displaying the following signal indications: steady CIRCULAR RED, steady CIRCULAR YELLOW, and CIRCULAR GREEN. Only one of the three indications shall be displayed at any given time.
   B. During the permissive right-turn movement, a CIRCULAR GREEN signal indication shall be displayed.
   C. A permissive only shared signal face, regardless of where it is positioned and regardless of how many adjacent through signal faces are provided, shall always simultaneously display the same color of circular indication that the adjacent through signal face or faces display.
   D. If the permissive only mode is not the only right-turn mode used for the approach, the signal face shall be the same shared signal face that is used for the protected/permissive mode (see Section 4D.24) except that the right-turn GREEN ARROW and right-turn YELLOW ARROW signal indications shall not be displayed when operating in the permissive only mode.

02 If a separate right-turn signal face is being operated in a permissive only right-turn mode, a CIRCULAR GREEN signal indication shall not be used in that face.

Figure 4D-13. Typical Positions and Arrangements of Shared Signal Faces for Permissive Only Mode Right Turns

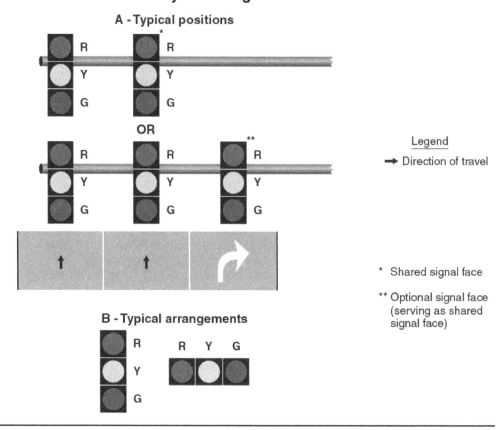

03 If a separate right-turn signal face is being operated in a permissive only right-turn mode and a flashing right-turn yellow arrow signal indication is provided, it shall meet the following requirements (see Figure 4D-14):

    A. It shall be capable of displaying one of the following sets of signal indications:
        1. Steady right-turn RED ARROW, steady right-turn YELLOW ARROW, and flashing right-turn YELLOW ARROW. Only one of the three indications shall be displayed at any given time.
        2. Steady CIRCULAR RED, steady right-turn YELLOW ARROW, and flashing right-turn YELLOW ARROW. Only one of the three indications shall be displayed at any given time. If the CIRCULAR RED signal indication is sometimes displayed when the signal faces for the adjacent through lane(s) are not displaying a CIRCULAR RED signal indication, a RIGHT TURN SIGNAL (R10-10R) sign (see Figure 2B-27) shall be used unless the CIRCULAR RED signal indication in the separate right-turn signal face is shielded, hooded, louvered, positioned, or designed such that it is not readily visible to drivers in the through lane(s).
    B. During the permissive right-turn movement, a flashing right-turn YELLOW ARROW signal indication shall be displayed.
    C. A steady right-turn YELLOW ARROW signal indication shall be displayed following the flashing right-turn YELLOW ARROW signal indication.
    D. When the separate right-turn signal face is providing a message to stop and remain stopped, a steady right-turn RED ARROW signal indication shall be displayed if it is intended that right turns on red not be permitted (except when a traffic control device is in place permitting a turn on a steady RED ARROW signal indication) or a steady CIRCULAR RED signal indication shall be displayed if it is intended that right turns on red be permitted.
    E. It shall be permitted to display a flashing right-turn YELLOW ARROW signal indication for a permissive right-turn movement while the signal faces for the adjacent through movement display steady CIRCULAR RED signal indications.
    F. During steady mode (stop-and-go) operation, the signal section that displays the steady right-turn YELLOW ARROW signal indication during change intervals shall not be used to display the flashing right-turn YELLOW ARROW signal indication for permissive right turns.

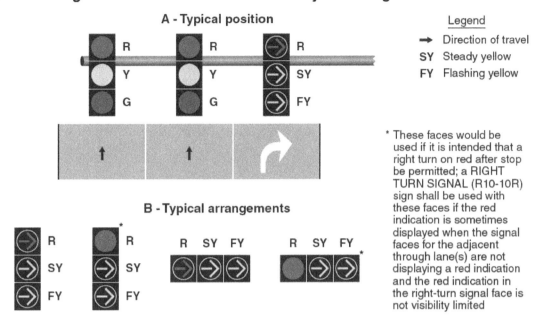

Figure 4D-14. Typical Position and Arrangements of Separate Signal Faces with Flashing Yellow Arrow for Permissive Only Mode Right Turns

G. During flashing mode operation (see Section 4D.30), the display of a flashing right-turn YELLOW ARROW signal indication shall be only from the signal section that displays a steady right-turn YELLOW ARROW signal indication during steady mode (stop-and-go) operation.

H. If the permissive only mode is not the only right-turn mode used for the approach, the signal face shall be the same separate right-turn signal face with a flashing YELLOW ARROW signal indication that is used for the protected/permissive mode (see Section 4D.24) except that the right-turn GREEN ARROW signal indication shall not be displayed when operating in the permissive only mode.

Option:

04 When an engineering study determines that each and every vehicle must successively come to a full stop before making a permissive right turn, a separate right-turn signal face with a flashing right-turn RED ARROW signal indication during the permissive right-turn movement may be used.

Standard:

05 If a separate right-turn signal face is being operated in a permissive only right-turn mode and a flashing right-turn RED arrow signal indication is provided, it shall meet the following requirements (see Figure 4D-15):

A. It shall be capable of displaying one of the following sets of signal indications:
1. Steady or flashing right-turn RED ARROW, steady right-turn YELLOW ARROW, and right-turn GREEN ARROW. Only one of the three indications shall be displayed at any given time. The GREEN ARROW indication is required in order to provide a three-section signal face, but shall not be displayed during permissive only mode.
2. Steady CIRCULAR RED on the left and steady right-turn RED ARROW on the right of the top position, steady right-turn YELLOW ARROW in the middle position, and right-turn GREEN ARROW in the bottom position. Only one of the four indications shall be displayed at any given time. The GREEN ARROW indication is required in order to provide three vertical positions, but shall not be displayed during permissive only mode. If the CIRCULAR RED signal indication is sometimes displayed when the signal faces for the adjacent through lane(s) are not displaying a CIRCULAR RED signal indication, a RIGHT TURN SIGNAL (R10-10R) sign (see Figure 2B-27) shall be used unless the CIRCULAR RED signal indication in the separate right-turn signal face is shielded, hooded, louvered, positioned, or designed such that it is not readily visible to drivers in the through lane(s).

B. During the permissive right-turn movement, a flashing right-turn RED ARROW signal indication shall be displayed, thus indicating that each and every vehicle must successively come to a full stop before making a permissive right turn.

Figure 4D-15. Typical Position and Arrangements of Separate Signal Faces with Flashing Red Arrow for Permissive Only Mode and Protected/Permissive Mode Right Turns

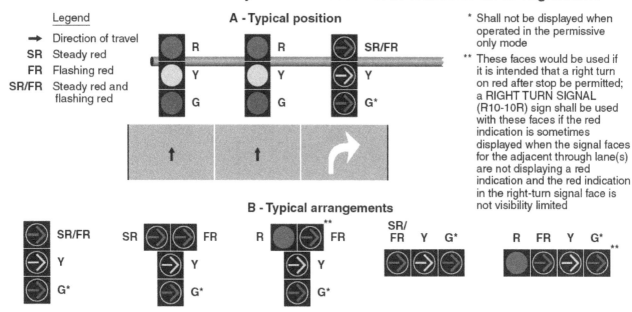

Note: A flashing red arrow controlling a right-turn movement may be used only when an engineering study determines that each and every vehicle must successively come to a full stop before making a permissive turn

---

C. A steady right-turn YELLOW ARROW signal indication shall be displayed following the flashing right-turn RED ARROW signal indication.

D. When the separate right-turn signal face is providing a message to stop and remain stopped, a steady right-turn RED ARROW signal indication shall be displayed if it is intended that right turns on red not be permitted (except when a traffic control device is in place permitting a turn on a steady RED ARROW signal indication) or a steady CIRCULAR RED signal indication shall be displayed if it is intended that right turns on red be permitted.

E. The display of a flashing right-turn RED ARROW signal indication for a permissive right-turn movement while the signal faces for the adjacent through movement display steady CIRCULAR RED signal indications and the opposing left-turn signal faces display left-turn GREEN ARROW signal indications for a protected left-turn movement shall be permitted.

F. A supplementary sign shall not be required. If used, it shall be a RIGHT TURN YIELD ON FLASHING RED ARROW AFTER STOP (R10-27) sign (see Figure 2B-27).

Option:

06 The requirements of Item A.1 in Paragraph 5 may be met by a vertically-arranged signal face with a horizontal cluster of two right-turn RED ARROW signal indications, the left-most of which displays a steady indication and the right-most of which displays a flashing indication (see Figure 4D-15).

Section 4D.23 Signal Indications for Protected Only Mode Right-Turn Movements

Standard:

01 A shared signal face shall not be used for protected only mode right turns unless the CIRCULAR GREEN and right-turn GREEN ARROW signal indications always begin and terminate together. If a shared signal face is provided for a protected only right turn, it shall meet the following requirements (see Figure 4D-16):

A. It shall be capable of displaying the following signal indications: steady CIRCULAR RED, steady CIRCULAR YELLOW, CIRCULAR GREEN, and right-turn GREEN ARROW. Only one of the three colors shall be displayed at any given time.

B. During the protected right-turn movement, the shared signal face shall simultaneously display both a CIRCULAR GREEN signal indication and a right-turn GREEN ARROW signal indication.

C. The shared signal face shall always simultaneously display the same color of circular indication that the adjacent through signal face or faces display.

D. If the protected only mode is not the only right-turn mode used for the approach, the signal face shall be the same shared signal face that is used for the protected/permissive mode (see Section 4D.24).

Option:

02 A straight-through GREEN ARROW signal indication may be used instead of the CIRCULAR GREEN signal indication in Items A and B in Paragraph 1 on an approach where left turns are prohibited and a straight-through GREEN ARROW signal indication is also used instead of a CIRCULAR GREEN signal indication in the other signal face(s) for through traffic.

Standard:

03 If a separate right-turn signal face is provided for a protected only mode right turn, it shall meet the following requirements (see Figure 4D-17):

   A. It shall be capable of displaying one of the following sets of signal indications:
      1. Steady right-turn RED ARROW, steady right-turn YELLOW ARROW, and right-turn GREEN ARROW. Only one of the three indications shall be displayed at any given time. A signal instruction sign shall not be required with this set of signal indications. If used, it shall be a RIGHT ON GREEN ARROW ONLY (R10-5a) sign (see Figure 2B-27).

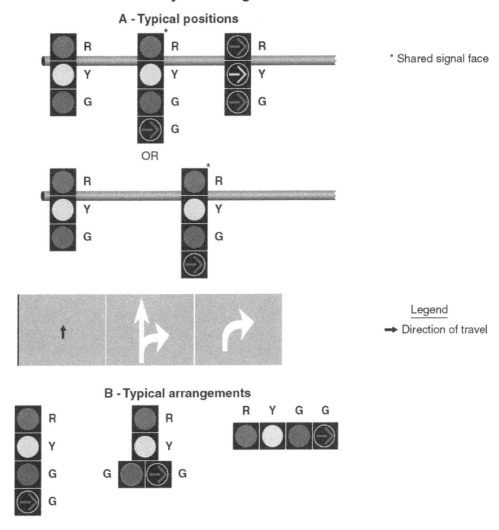

Figure 4D-16. Typical Positions and Arrangements of Shared Signal Faces for Protected Only Mode Right Turns

Note: Shared signal faces shall only be used for a protected-only mode right turn if the circular green and green right-turn arrow indications always begin and terminate together

Figure 4D-17. Typical Position and Arrangements of Separate Signal Faces for Protected Only Mode Right Turns

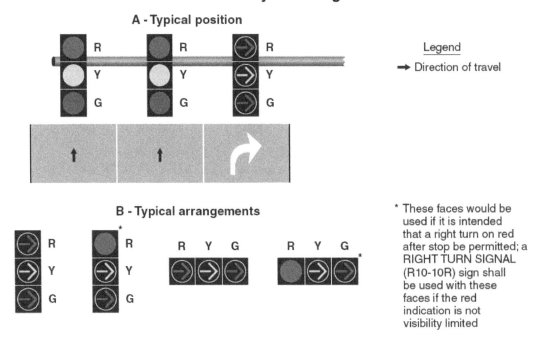

2. Steady CIRCULAR RED, steady right-turn YELLOW ARROW, and right-turn GREEN ARROW. Only one of three indications shall be displayed at any given time. If the CIRCULAR RED signal indication is sometimes displayed when the signal faces for the adjacent through lane(s) are not displaying a CIRCULAR RED signal indication, a RIGHT TURN SIGNAL (R10-10R) sign (see Figure 2B-27) shall be used unless the CIRCULAR RED signal indication is shielded, hooded, louvered, positioned, or designed such that it is not readily visible to drivers in the through lane(s).
B. During the protected right-turn movement, a right-turn GREEN ARROW signal indication shall be displayed.
C. A steady right-turn YELLOW ARROW signal indication shall be displayed following the right-turn GREEN ARROW signal indication.
D. When the separate signal face is providing a message to stop and remain stopped, a steady right-turn RED ARROW signal indication shall be displayed if it is intended that right turns on red not be permitted (except when a traffic control device is in place permitting a turn on a steady RED ARROW signal indication) or a steady CIRCULAR RED signal indication shall be displayed if it is intended that right turns on red be permitted.
E. If the protected only mode is not the only right-turn mode used for the approach, the signal face shall be the same separate right-turn signal face that is used for the protected/permissive mode (see Section 4D.24 and Figure 4D-19) except that a flashing right-turn YELLOW ARROW or flashing right-turn RED ARROW signal indication shall not be displayed when operating in the protected only mode.

Section 4D.24 Signal Indications for Protected/Permissive Mode Right-Turn Movements

Standard:

01 If a shared signal face is provided for a protected/permissive mode right turn, it shall meet the following requirements (see Figure 4D-18):
A. It shall be capable of displaying the following signal indications: steady CIRCULAR RED, steady CIRCULAR YELLOW, CIRCULAR green, steady right-turn YELLOW ARROW, and right-turn GREEN ARROW. Only one of the three circular indications shall be displayed at any given time. Only one of the two arrow indications shall be displayed at any given time. If the right-turn GREEN ARROW signal indication and the CIRCULAR GREEN signal indication(s) for the adjacent through movement are always terminated together, the steady right-turn YELLOW ARROW signal indication shall not be required.

Figure 4D-18. Typical Positions and Arrangements of Shared Signal Faces for Protected/Permissive Mode Right Turns

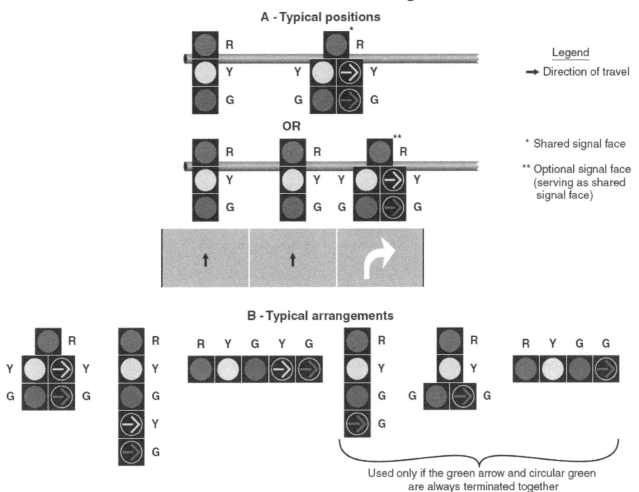

B. During the protected right-turn movement, the shared signal face shall simultaneously display a right-turn GREEN ARROW signal indication and a circular signal indication that is the same color as the signal indication for the adjacent through lane on the same approach as the protected right turn.

C. A steady right-turn YELLOW ARROW signal indication shall be displayed following the right-turn GREEN ARROW signal indication, unless the right-turn GREEN ARROW signal indication and the CIRCULAR GREEN signal indication(s) for the adjacent through movement are being terminated together. When the right-turn GREEN ARROW and CIRCULAR GREEN signal indications are being terminated together, the required display following the right-turn GREEN ARROW signal indication shall be either the display of a CIRCULAR YELLOW signal indication alone or the simultaneous display of the CIRCULAR YELLOW and right-turn YELLOW ARROW signal indications.

D. During the permissive right-turn movement, the shared signal face shall display only a CIRCULAR GREEN signal indication.

E. A protected/permissive shared signal face, regardless of where it is positioned and regardless of how many adjacent through signal faces are provided, shall always simultaneously display the same color of circular indication that the adjacent through signal face or faces display.

02 If a separate right-turn signal face is being operated in a protected/permissive right-turn mode, a CIRCULAR GREEN signal indication shall not be used in that face.

03 If a separate right-turn signal face is being operated in a protected/permissive right-turn mode and a flashing right-turn yellow arrow signal indication is provided, it shall meet the following requirements (see Figure 4D-19):
   A. It shall be capable of displaying one of the following sets of signal indications:
      1. Steady right-turn RED ARROW, steady right-turn YELLOW ARROW, flashing right-turn YELLOW ARROW, and right-turn GREEN ARROW. Only one of the four indications shall be displayed at any given time.
      2. Steady CIRCULAR RED, steady right-turn YELLOW ARROW, flashing right-turn YELLOW ARROW, and right-turn GREEN ARROW. Only one of the four indications shall be displayed at any given time. If the CIRCULAR RED signal indication is sometimes displayed when the signal faces for the adjacent through lane(s) are not displaying a CIRCULAR RED signal indication, a RIGHT TURN SIGNAL (R10-10R) sign (see Figure 2B-27) shall be used unless the CIRCULAR RED signal indication in the separate right-turn signal face is shielded, hooded, louvered, positioned, or designed such that it is not readily visible to drivers in the through lane(s).
   B. During the protected right-turn movement, a right-turn GREEN ARROW signal indication shall be displayed.
   C. A steady right-turn YELLOW ARROW signal indication shall be displayed following the right-turn GREEN ARROW signal indication.
   D. During the permissive right-turn movement, a flashing right-turn YELLOW ARROW signal indication shall be displayed.
   E. A steady right-turn YELLOW ARROW signal indication shall be displayed following the flashing right-turn YELLOW ARROW signal indication if the permissive right-turn movement is being terminated and the separate right-turn signal face will subsequently display a steady red indication.
   F. When a permissive right-turn movement is changing to a protected right-turn movement, a right-turn GREEN ARROW signal indication shall be displayed immediately upon the termination of the flashing right-turn YELLOW ARROW signal indication. A steady right-turn YELLOW ARROW signal indication shall not be displayed between the display of the flashing right-turn YELLOW ARROW signal indication and the display of the steady right-turn GREEN ARROW signal indication.

Figure 4D-19. Typical Position and Arrangements of Separate Signal Faces with Flashing Yellow Arrow for Protected/Permissive Mode and Protected Only Mode Right Turns

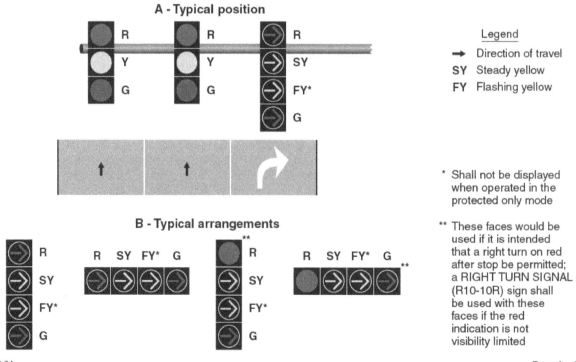

G. When the separate right-turn signal face is providing a message to stop and remain stopped, a steady right-turn RED ARROW signal indication shall be displayed if it is intended that right turns on red not be permitted (except when a traffic control device is in place permitting a turn on a steady RED ARROW signal indication) or a steady CIRCULAR RED signal indication shall be displayed if it is intended that right turns on red be permitted.
H. It shall be permitted to display a flashing right-turn YELLOW ARROW signal indication for a permissive right-turn movement while the signal faces for the adjacent through movement display steady CIRCULAR RED signal indications.
I. A signal face containing a dual-arrow signal section in place of separate flashing right-turn YELLOW ARROW and right-turn GREEN ARROW signal sections shall be permitted where signal head height limitations (or lateral positioning limitations for a horizontally-mounted signal face) are a concern. The dual-arrow signal section, where used, shall display a GREEN ARROW for the protected right-turn movement and a flashing YELLOW ARROW for the permissive right-turn movement.
J. During steady mode (stop-and-go) operation, the signal section that displays the steady right-turn YELLOW ARROW signal indication during change intervals shall not be used to display the flashing right-turn YELLOW ARROW signal indication for permissive right turns.
K. During flashing mode operation (see Section 4D.30), the display of a flashing right-turn YELLOW ARROW signal indication shall be only from the signal section that displays a steady right-turn YELLOW ARROW signal indication during steady mode (stop-and-go) operation.

Option:

04 When an engineering study determines that each and every vehicle must successively come to a full stop before making a permissive right turn, a separate signal face that has a flashing right-turn RED ARROW signal indication during the permissive right-turn movement may be used.

Standard:

05 If a separate right-turn signal face is being operated in a protected/permissive right-turn mode and a flashing right-turn RED arrow signal indication is provided, it shall meet the following requirements (see Figure 4D-15):
   A. It shall be capable of displaying one of the following sets of signal indications:
      1. Steady or flashing right-turn RED ARROW, steady right-turn YELLOW ARROW, and right-turn GREEN ARROW. Only one of the three indications shall be displayed at any given time.
      2. Steady CIRCULAR RED on the left and steady or flashing right-turn RED ARROW on the right of the top position, steady right-turn YELLOW ARROW in the middle position, and right-turn GREEN ARROW in the bottom position. Only one of the four indications shall be displayed at any given time. If the CIRCULAR RED signal indication is sometimes displayed when the signal faces for the adjacent through lane(s) are not displaying a CIRCULAR RED signal indication, a RIGHT TURN SIGNAL (R10-10R) sign (see Figure 2B-27) shall be used unless the CIRCULAR RED signal indication in the separate right-turn signal face is shielded, hooded, louvered, positioned, or designed such that it is not readily visible to drivers in the through lane(s).
   B. During the protected right-turn movement, a right-turn GREEN ARROW signal indication shall be displayed.
   C. A steady right-turn YELLOW ARROW signal indication shall be displayed following the right-turn GREEN ARROW signal indication.
   D. During the permissive right-turn movement, the separate right-turn signal face shall display a flashing right-turn RED ARROW signal indication.
   E. A steady right-turn YELLOW ARROW signal indication shall be displayed following the flashing right-turn RED ARROW signal indication if the permissive right-turn movement is being terminated and the separate right-turn signal face will subsequently display a steady red indication.
   F. When a permissive right-turn movement is changing to a protected right-turn movement, a right-turn GREEN ARROW signal indication shall be displayed immediately upon the termination of the flashing right-turn RED ARROW signal indication. A steady right-turn YELLOW ARROW signal indication shall not be displayed between the display of the flashing right-turn RED ARROW signal indication and the display of the steady right-turn GREEN ARROW signal indication.

G. When the separate right-turn signal face is providing a message to stop and remain stopped, a steady right-turn RED ARROW signal indication shall be displayed if it is intended that right turns on red not be permitted (except when a traffic control device is in place permitting a turn on a steady RED ARROW signal indication) or a steady CIRCULAR RED signal indication shall be displayed if it is intended that right turns on red be permitted.
H. It shall be permitted to display a flashing right-turn RED ARROW signal indication for a permissive right-turn movement while the signal faces for the adjacent through movement display steady CIRCULAR RED signal indications and the opposing left-turn signal faces display left-turn GREEN ARROW signal indications for a protected left-turn movement.
I. A supplementary sign shall not be required. If used, it shall be a RIGHT TURN YIELD ON FLASHING RED ARROW AFTER STOP (R10-27) sign (see Figure 2B-27).

Option:

06   The requirements of Item A.1 in Paragraph 5 may be met by a vertically-arranged signal face with a horizontal cluster of two right-turn RED ARROW signal indications, the left-most of which displays a steady indication and the right-most of which displays a flashing indication (see Figure 4D-15).

Section 4D.25 Signal Indications for Approaches With Shared Left-Turn/Right-Turn Lanes and No Through Movement

Support:

01   A lane that is shared by left-turn and right-turn movements is sometimes provided on an approach that has no through movement, such as the stem of a T-intersection or where the opposite approach is a one-way roadway in the opposing direction.

Standard:

02   **When a shared left-turn/right-turn lane exists on a signalized approach, the left-turn and right-turn movements shall start and terminate simultaneously and the red signal indication used in each of the signal faces on the approach shall be a CIRCULAR RED.**

Support:

03   This requirement for the use of CIRCULAR RED signal indications in signal faces for approaches having a shared lane for left-turn and right-turn movements is a specific exception to other provisions in this Chapter that would otherwise require the use of RED ARROW signal indications.

Standard:

04   **The signal faces provided for an approach with a shared left-turn/right-turn lane and no through movement shall be one of the following:**
   A. **Two or more signal faces, each capable of displaying CIRCULAR RED, CIRCULAR YELLOW, and CIRCULAR GREEN signal indications, shall be provided for the approach. This display shall be permissible regardless of number of exclusive left-turn and/or right-turn lanes that exist on the approach in addition to the shared left-turn/right-turn lane and regardless of whether or not there are pedestrian or opposing vehicular movements that conflict with the left-turn or right-turn movements. However, if there is an opposing approach and the signal phasing protects the left-turn movement on the approach with the shared left-turn/right-turn lane from conflicts with the opposing vehicular movements and any signalized pedestrian movements, a left-turn GREEN ARROW signal indication shall also be included in the left-most signal face and shall be displayed simultaneously with the CIRCULAR GREEN signal indication.**
   B. **If the approach has one or more exclusive turn lanes in addition to the shared left-turn/right-turn lane and there is no conflict with a signalized vehicular or pedestrian movement, and GREEN ARROW signal indications are used in place of CIRCULAR GREEN signal indications on the approach, the signal faces for the approach shall be:**
      1. **A signal face(s) capable of displaying CIRCULAR RED, YELLOW ARROW, and GREEN ARROW signal indications for the exclusive turn lane(s), with the arrows pointing in the direction of the turn, and**
      2. **A shared left-turn/right-turn signal face capable of displaying CIRCULAR RED, left-turn YELLOW ARROW, left-turn GREEN ARROW, right-turn YELLOW ARROW, and right-turn GREEN ARROW signal indications, in an arrangement of signal sections that complies with the provisions of Section 4D.09 or 4D.10.**
   C. **If the approach has one or more exclusive turn lanes in addition to the shared left-turn/right-turn lane and there is a conflict with a signalized vehicular or pedestrian movement, and flashing YELLOW ARROW signal indications are used in place of CIRCULAR GREEN signal indications**

on the approach, the signal faces for the approach shall be as described in Items B.1 and B.2, except that flashing YELLOW ARROW signal indications shall be used in place of the GREEN ARROW signal indications for the turning movement(s) that conflicts with the signalized vehicular or pedestrian movement.

Support:

05 Figure 4D-20 illustrates application of these Standards on approaches that have only a shared left-turn/right-turn lane, and on approaches that have one or more exclusive turn lanes in addition to the shared left-turn/right-turn lane.

Option:

06 If the lane-use regulations on an approach are variable such that at certain times all of the lanes on the approach are designated as exclusive turn lanes and no lane is designated as a shared left-turn/right-turn lane:

A. During the times that no lane is designated as a shared left-turn/right-turn lane, the left-turn and right-turn movements may start and terminate independently, and the left-turn and right-turn movements may be operated in one or more of the modes of operation as described in Sections 4D.17 through 4D.24; and

B. If a protected-permissive mode is used, the shared left-turn/right-turn signal face provided in Paragraph 4 may be modified to include a dual-arrow signal section capable of displaying both a GREEN ARROW signal indication and a flashing YELLOW ARROW signal indication for a turn movement(s) in order to not exceed the maximum of five sections per signal face provided in Section 4D.08.

Section 4D.26 Yellow Change and Red Clearance Intervals

Standard:

01 **A steady yellow signal indication shall be displayed following every CIRCULAR GREEN or GREEN ARROW signal indication and following every flashing YELLOW ARROW or flashing RED ARROW signal indication displayed as a part of a steady mode operation. This requirement shall not apply when a CIRCULAR GREEN, a flashing YELLOW ARROW, or a flashing RED ARROW signal indication is followed immediately by a GREEN ARROW signal indication.**

02 **The exclusive function of the yellow change interval shall be to warn traffic of an impending change in the right-of-way assignment.**

03 **The duration of the yellow change interval shall be determined using engineering practices.**

Support:

04 Section 4D.05 contains provisions regarding the display of steady CIRCULAR YELLOW signal indications to approaches from which drivers are allowed to make permissive left turns.

Guidance:

05 *When indicated by the application of engineering practices, the yellow change interval should be followed by a red clearance interval to provide additional time before conflicting traffic movements, including pedestrians, are released.*

Standard:

06 **When used, the duration of the red clearance interval shall be determined using engineering practices.**

Support:

07 Engineering practices for determining the duration of yellow change and red clearance intervals can be found in ITE's "Traffic Control Devices Handbook" and in ITE's "Manual of Traffic Signal Design" (see Section 1A.11).

Standard:

08 **The durations of yellow change intervals and red clearance intervals shall be consistent with the determined values within the technical capabilities of the controller unit.**

09 **The duration of a yellow change interval shall not vary on a cycle-by-cycle basis within the same signal timing plan.**

10 **Except as provided in Paragraph 12, the duration of a red clearance interval shall not be decreased or omitted on a cycle-by-cycle basis within the same signal timing plan.**

Option:

11 The duration of a red clearance interval may be extended from its predetermined value for a given cycle based upon the detection of a vehicle that is predicted to violate the red signal indication.

12 When an actuated signal sequence includes a signal phase for permissive/protected (lagging) left-turn movements in both directions, the red clearance interval may be shown during those cycles when the lagging left-turn signal phase is skipped and may be omitted during those cycles when the lagging left-turn signal phase is shown.

Figure 4D-20. Signal Indications for Approaches with a Shared Left-Turn/Right-Turn Lane and No Through Movement (Sheet 1 of 3)

A - No conflicting vehicular or pedestrian movements

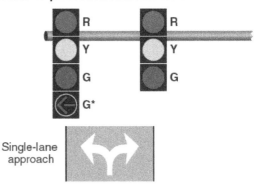

* Left-turn GREEN ARROW section shall be included if there is an opposing one-way approach and the signal phasing eliminates conflicts.

Notes:
1. Horizontally-aligned signal faces may also be used.
2. Shared signal faces may also be 5 sections in a vertical straight line instead of a cluster.

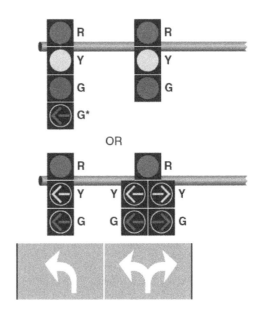

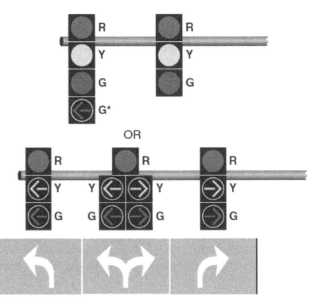

Figure 4D-20. Signal Indications for Approaches with a Shared Left-Turn/Right-Turn Lane and No Through Movement (Sheet 2 of 3)

B - Pedestrian or vehicular conflict with one turn movement

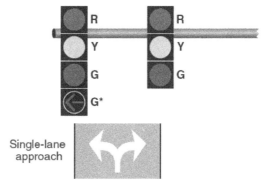

* Left-turn GREEN ARROW section shall be included if there is an opposing one-way approach and the signal phasing eliminates conflicts.

Notes:
1. A conflict with the right-turn movement is illustrated.
2. Horizontally-aligned signal faces may also be used.
3. Shared signal faces may also be 5 sections in a vertical straight line instead of a cluster.

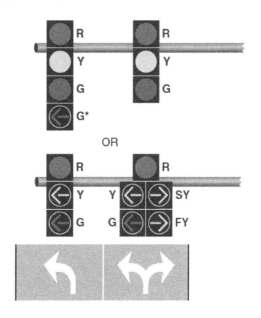

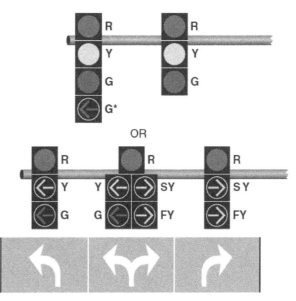

Figure 4D-20. Signal Indications for Approaches with a Shared Left-Turn/Right-Turn Lane and No Through Movement (Sheet 3 of 3)

C - Pedestrian or vehicular conflicts with both turn movements

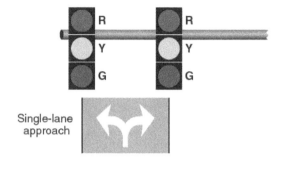

Notes:
1. Horizontally-aligned signal faces may also be used.
2. Shared signal faces may also be 5 sections in a vertical straight line instead of a cluster.

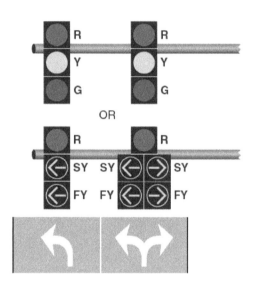

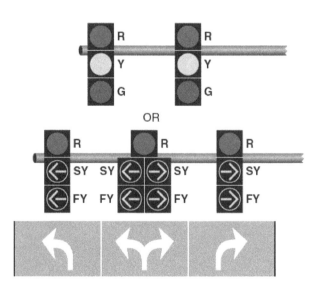

¹³ The duration of a yellow change interval or a red clearance interval may be different in different signal timing plans for the same controller unit.

Guidance:

¹⁴ *A yellow change interval should have a minimum duration of 3 seconds and a maximum duration of 6 seconds. The longer intervals should be reserved for use on approaches with higher speeds.*

¹⁵ *Except when clearing a one-lane, two-way facility (see Section 4H.02) or when clearing an exceptionally wide intersection, a red clearance interval should have a duration not exceeding 6 seconds.*

Standard:

¹⁶ **Except for warning beacons mounted on advance warning signs on the approach to a signalized location (see Section 2C.36), signal displays that are intended to provide a "pre-yellow warning" interval, such as flashing green signal indications, vehicular countdown displays, or other similar displays, shall not be used at a signalized location.**

Support:

¹⁷ The use of signal displays (other than warning beacons mounted on advance warning signs) that convey a "pre-yellow warning" have been found by research to increase the frequency of crashes.

Section 4D.27 Preemption and Priority Control of Traffic Control Signals

Option:

⁰¹ Traffic control signals may be designed and operated to respond to certain classes of approaching vehicles by altering the normal signal timing and phasing plan(s) during the approach and passage of those vehicles. The alternative plan(s) may be as simple as extending a currently displayed green interval or as complex as replacing the entire set of signal phases and timing.

Support:

⁰² Preemption control (see definition in Section 1A.13) is typically given to trains, boats, emergency vehicles, and light rail transit.

⁰³ Examples of preemption control include the following:
  A. The prompt displaying of green signal indications at signalized locations ahead of fire vehicles, law enforcement vehicles, ambulances, and other official emergency vehicles;
  B. A special sequence of signal phases and timing to expedite and/or provide additional clearance time for vehicles to clear the tracks prior to the arrival of rail traffic; and
  C. A special sequence of signal phases to display a steady red indication to prohibit turning movements toward the tracks during the approach or passage of rail traffic.

⁰⁴ Priority control (see definition in Section 1A.13) is typically given to certain non-emergency vehicles such as light-rail transit vehicles operating in a mixed-use alignment and buses.

⁰⁵ Examples of priority control include the following:
  A. The displaying of early or extended green signal indications at an intersection to assist public transit vehicles in remaining on schedule, and
  B. Special phasing to assist public transit vehicles in entering the travel stream ahead of the platoon of traffic.

⁰⁶ Some types or classes of vehicles supersede others when a traffic control signal responds to more than one type or class. In general, a vehicle that is more difficult to control supersedes a vehicle that is easier to control.

Option:

⁰⁷ Preemption or priority control of traffic control signals may also be a means of assigning priority right-of-way to specified classes of vehicles at certain non-intersection locations such as on approaches to one-lane bridges and tunnels, movable bridges, highway maintenance and construction activities, metered freeway entrance ramps, and transit operations.

Standard:

⁰⁸ **During the transition into preemption control:**
  A. The yellow change interval, and any red clearance interval that follows, shall not be shortened or omitted.
  B. The shortening or omission of any pedestrian walk interval and/or pedestrian change interval shall be permitted.
  C. The return to the previous green signal indication shall be permitted following a steady yellow signal indication in the same signal face, omitting the red clearance interval, if any.

09 **During preemption control and during the transition out of preemption control:**
   A. **The shortening or omission of any yellow change interval, and of any red clearance interval that follows, shall not be permitted.**
   B. **A signal indication sequence from a steady yellow signal indication to a green signal indication shall not be permitted.**
10 **During priority control and during the transition into or out of priority control:**
   A. **The shortening or omission of any yellow change interval, and of any red clearance interval that follows, shall not be permitted.**
   B. **The shortening of any pedestrian walk interval below that time described in Section 4E.06 shall not be permitted.**
   C. **The omission of a pedestrian walk interval and its associated change interval shall not be permitted unless the associated vehicular phase is also omitted or the pedestrian phase is exclusive.**
   D. **The shortening or omission of any pedestrian change interval shall not be permitted.**
   E. **A signal indication sequence from a steady yellow signal indication to a green signal indication shall not be permitted.**

Guidance:

11 *Except for traffic control signals interconnected with light rail transit systems, traffic control signals with railroad preemption or coordinated with flashing-light signal systems should be provided with a back-up power supply.*

12 *When a traffic control signal that is returning to a steady mode from a dark mode (typically upon restoration from a power failure) receives a preemption or priority request, care should be exercised to minimize the possibility of vehicles or pedestrians being misdirected into a conflict with the vehicle making the request.*

Option:

13 During the change from a dark mode to a steady mode under a preemption or priority request, the display of signal indications that could misdirect road users may be prevented by one or more of the following methods:
   A. Having the traffic control signal remain in the dark mode,
   B. Having the traffic control signal remain in the flashing mode,
   C. Altering the flashing mode,
   D. Executing the normal start-up routine before responding, or
   E. Responding directly to initial or dwell period.

Guidance:

14 *If a traffic control signal is installed near or within a grade crossing or if a grade crossing with active traffic control devices is within or near a signalized highway intersection, Chapter 8C should be consulted.*

15 *Traffic control signals operating under preemption control or under priority control should be operated in a manner designed to keep traffic moving.*

16 *Traffic control signals that are designed to respond under preemption or priority control to more than one type or class of vehicle should be designed to respond in the relative order of importance or difficulty in stopping the type or class of vehicle. The order of priority should be: train, boat, heavy vehicle (fire vehicle, emergency medical service), light vehicle (law enforcement), light rail transit, rubber-tired transit.*

Option:

17 A distinctive indication may be provided at the intersection to show that an emergency vehicle has been given control of the traffic control signal (see Section 11-106 of the "Uniform Vehicle Code"). In order to assist in the understanding of the control of the traffic signal, a common distinctive indication may be used where drivers from different agencies travel through the same intersection when responding to emergencies.

18 If engineering judgment indicates that light rail transit signal indications would reduce road user confusion that might otherwise occur if standard traffic signal indications were used to control these movements, light rail transit signal indications complying with Section 8C.11 and as illustrated in Figure 8C-3 may be used for preemption or priority control of the following exclusive movements at signalized intersections:
   A. Public transit buses in "queue jumper" lanes, and
   B. Bus rapid transit in semi-exclusive or mixed-use alignments.

Section 4D.28 Flashing Operation of Traffic Control Signals – General

Standard:

01   The light source of a flashing signal indication shall be flashed continuously at a rate of not less than 50 or more than 60 times per minute.

02   The displayed period of each flash shall be a minimum of 1/2 and a maximum of 2/3 of the total flash cycle.

03   Flashing signal indications shall comply with the requirements of other Sections of this Manual regarding visibility-limiting or positioning of conflicting signal indications, except that flashing yellow signal indications for through traffic shall not be required to be visibility-limited or positioned to minimize visual conflict for road users in separately controlled turn lanes.

04   Each traffic control signal shall be provided with an independent flasher mechanism that operates in compliance with this Section.

05   The flashing operation shall not be terminated by removal or turn off of the controller unit or of the conflict monitor (malfunction management unit) or both.

06   A manual switch, a conflict monitor (malfunction management unit) circuit, and, if appropriate, automatic means shall be provided to initiate the flashing mode.

Option:

07   Based on engineering study or engineering judgment, traffic control signals may be operated in the flashing mode on a scheduled basis during one or more periods of the day rather than operated continuously in the steady (stop-and-go) mode.

Support:

08   Sections 4E.06 and 4E.09 contain information regarding the operation of pedestrian signal heads and accessible pedestrian signal detector pushbutton locator tones, respectively, during flashing operation.

Section 4D.29 Flashing Operation – Transition Into Flashing Mode

Standard:

01   The transition from steady (stop-and-go) mode to flashing mode, if initiated by a conflict monitor (malfunction management unit) or by a manual switch, shall be permitted to be made at any time.

02   Programmed changes from steady (stop-and-go) mode to flashing mode shall be made under either of the following circumstances:
   A. At the end of the common major-street red interval (such as just prior to the start of the green in both directions on the major street), or
   B. Directly from a CIRCULAR GREEN signal indication to a flashing CIRCULAR YELLOW signal indication, or from a GREEN ARROW signal indication to a flashing YELLOW ARROW signal indication, or from a flashing YELLOW ARROW signal indication (see Sections 4D.17 to 4D.24) to a flashing YELLOW ARROW signal indication in a different signal section.

03   During programmed changes into flashing mode, no green signal indication or flashing yellow signal indication shall be terminated and immediately followed by a steady red or flashing red signal indication without first displaying the steady yellow signal indication.

Section 4D.30 Flashing Operation – Signal Indications During Flashing Mode

Guidance:

01   *When a traffic control signal is operated in the flashing mode, a flashing yellow signal indication should be used for the major street and a flashing red signal indication should be used for the other approaches unless flashing red signal indications are used on all approaches.*

Standard:

02   When a traffic control signal is operated in the flashing mode, all of the green signal indications at the signalized location shall be dark (non-illuminated) and shall not be displayed in either a steady or flashing manner, except for single-section GREEN ARROW signal indications as provided elsewhere in this Section.

03   Flashing yellow signal indications shall be used on more than one approach to a signalized location only if those approaches do not conflict with each other.

04   Except as provided in Paragraph 5, when a traffic control signal is operated in the flashing mode, one and only one signal indication in every signal face at the signalized location shall be flashed.

Option:

05   If a signal face has two identical CIRCULAR RED or RED ARROW signal indications (see Section 4D.08), both of those identical signal indications may be flashed simultaneously.

Standard:

06 No steady indications, other than a single-section signal face consisting of a continuously-displayed GREEN ARROW signal indication that is used alone to indicate a continuous movement in the steady (stop-and-go) mode, shall be displayed at the signalized location during the flashing mode. A single-section GREEN ARROW signal indication shall remain continuously-displayed when the traffic control signal is operated in the flashing mode.

07 If a signal face includes both circular and arrow signal indications of the color that is to be flashed, only the circular signal indication shall be flashed.

08 All signal faces that are flashed on an approach shall flash the same color, either yellow or red, except that separate turn signal faces (see Sections 4D.17 and 4D.21) shall be permitted to flash a RED ARROW signal indication when the adjacent through movement signal indications are flashed yellow. Shared signal faces (see Sections 4D.17 and 4D.21) for turn movements shall not be permitted to flash a CIRCULAR RED signal indication when the adjacent through movement signal indications are flashed yellow.

09 The appropriate RED ARROW or YELLOW ARROW signal indication shall be flashed when a signal face consists entirely of arrow indications. A signal face that consists entirely of arrow indications and that provides a protected only turn movement during the steady (stop-and-go) mode or that provides a flashing yellow arrow or flashing red arrow signal indication for a permissive turn movement during the steady (stop-and-go) mode shall be permitted to flash the YELLOW ARROW signal indication during the flashing mode if the adjacent through movement signal indications are flashed yellow and if it is intended that a permissive turn movement not requiring a full stop by each turning vehicle be provided during the flashing mode.

Section 4D.31 Flashing Operation – Transition Out of Flashing Mode

Standard:

01 All changes from flashing mode to steady (stop-and-go) mode shall be made under one of the following procedures:
   A. Yellow-red flashing mode: Changes from flashing mode to steady (stop-and-go) mode shall be made at the beginning of the major-street green interval (when a green signal indication is displayed to through traffic in both directions on the major street), or if there is no common major-street green interval, at the beginning of the green interval for the major traffic movement on the major street.
   B. Red-red flashing mode: Changes from flashing mode to steady (stop-and-go) mode shall be made by changing the flashing red indications to steady red indications followed by appropriate green indications to begin the steady mode cycle. These green indications shall be the beginning of the major-street green interval (when a green signal indication is displayed to through traffic in both directions on the major street) or if there is no common major-street green interval, at the beginning of the green interval for the major traffic movement on the major street.

Guidance:

02 *The steady red clearance interval provided during the change from red-red flashing mode to steady (stop-and-go) mode should have a duration of 6 seconds.*

03 *When changing from the yellow-red flashing mode to steady (stop-and-go) mode, if there is no common major-street green interval, the provision of a steady red clearance interval for the other approaches before changing from a flashing yellow or a flashing red signal indication to a green signal indication on the major approach should be considered.*

Standard:

04 During programmed changes out of flashing mode, no flashing yellow signal indication shall be terminated and immediately followed by a steady red or flashing red signal indication without first displaying the steady yellow signal indication.

Option:

05 Because special midblock signals that rest in flashing circular yellow in the position normally occupied by the green signal indication do not have a green signal indication in the signal face, these signals may go directly from flashing circular yellow (in the position normally occupied by the green signal indication) to steady yellow without going first to a green signal indication.

Section 4D.32 Temporary and Portable Traffic Control Signals

Support:

01 A temporary traffic control signal is generally installed using methods that minimize the costs of installation, relocation, and/or removal. Typical temporary traffic control signals are for specific purposes, such as for one-lane, two-way facilities in temporary traffic control zones (see Chapter 4H), for a haul-road intersection, or for access to a site that will have a permanent access point developed at another location in the near future.

Standard:

02　**Advance signing shall be used when employing a temporary traffic control signal.**

03　**A temporary traffic control signal shall:**
　　A. Meet the physical display and operational requirements of a conventional traffic control signal.
　　B. Be removed when no longer needed.
　　C. Be placed in the flashing mode when not being used if it will be operated in the steady mode within 5 working days; otherwise, it shall be removed.
　　D. Be placed in the flashing mode during periods when it is not desirable to operate the signal, or the signal heads shall be covered, turned, or taken down to indicate that the signal is not in operation.

Guidance:

04　*A temporary traffic control signal should be used only if engineering judgment indicates that installing the signal will improve the overall safety and/or operation of the location.*

05　*The use of temporary traffic control signals by a work crew on a regular basis in their work area should be subject to the approval of the jurisdiction having authority over the roadway.*

06　*A temporary traffic control signal should not operate longer than 30 days unless associated with a longer-term temporary traffic control zone project.*

07　*For use of temporary traffic control signals in temporary traffic control zones, reference should be made to Section 6F.84.*

Section 4D.33 Lateral Offset of Signal Supports and Cabinets

Guidance:

01　*The following items should be considered when placing signal supports and cabinets:*
　　A. Reference should be made to the American Association of State Highway and Transportation Officials (AASHTO) "Roadside Design Guide" (see Section 1A.11) and to the "Americans with Disabilities Act Accessibility Guidelines for Buildings and Facilities (ADAAG)" (see Section 1A.11).
　　B. Signal supports should be placed as far as practical from the edge of the traveled way without adversely affecting the visibility of the signal indications.
　　C. Where supports cannot be located based on the recommended AASHTO clearances, consideration should be given to the use of appropriate safety devices.
　　D. No part of a concrete base for a signal support should extend more than 4 inches above the ground level at any point. This limitation does not apply to the concrete base for a rigid support.
　　E. In order to minimize hindrance to the passage of persons with physical disabilities, a signal support or controller cabinet should not obstruct the sidewalk, or access from the sidewalk to the crosswalk.
　　F. Controller cabinets should be located as far as practical from the edge of the roadway.
　　G. On medians, the minimum clearances provided in Items A through E for signal supports should be obtained if practical.

Section 4D.34 Use of Signs at Signalized Locations

Support:

01　Traffic signal signs are sometimes used at highway traffic signal locations to instruct or guide pedestrians, bicyclists, or motorists. Among the signs typically used at or on the approaches to signalized locations are movement prohibition signs (see Section 2B.18), lane control signs (see Sections 2B.19 to 2B.22), pedestrian crossing signs (see Section 2B.51), pedestrian actuation signs (see Section 2B.52), traffic signal signs (see Sections 2B.53 and 2C.48), Signal Ahead warning signs (see Section 2C.36), Street Name signs (see Section 2D.43), and Advance Street Name signs (see Section 2D.44).

Guidance:

02　*Regulatory, warning, and guide signs should be used at traffic control signal locations as provided in Part 2 and as specifically provided elsewhere in Part 4.*

03　*Traffic signal signs should be located adjacent to the signal face to which they apply.*

Support:

04　Section 2B.19 contains information regarding the use of overhead lane control signs on signalized approaches where lane drops, multiple-lane turns involving shared through-and-turn lanes, or other lane-use regulations that would be unexpected by unfamiliar road users are present.

Standard:

05 If used, illuminated traffic signal signs shall be designed and mounted in such a manner as to avoid glare and reflections that seriously detract from the signal indications. Traffic control signal faces shall be given dominant position and brightness to maximize their priority in the overall display.

06 The minimum vertical clearance and horizontal offset of the total assembly of traffic signal signs (see Section 2B.53) shall comply with the provisions of Sections 4D.15 and 4D.16.

07 STOP signs shall not be used in conjunction with any traffic control signal operation, except in either of the following cases:
   A. If the signal indication for an approach is a flashing red at all times, or
   B. If a minor street or driveway is located within or adjacent to the area controlled by the traffic control signal, but does not require separate traffic signal control because an extremely low potential for conflict exists.

Section 4D.35 Use of Pavement Markings at Signalized Locations

Support:

01 Pavement markings (see Part 3) that clearly communicate the operational plan of an intersection to road users play an important role in the effective operation of traffic control signals. By designating the number of lanes, the use of each lane, the length of additional lanes on the approach to an intersection, and the proper stopping points, the engineer can design the signal phasing and timing to best match the goals of the operational plan.

Guidance:

02 *Pavement markings should be used at traffic control signal locations as provided in Part 3. If the road surface will not retain pavement markings, signs should be installed to provide the needed road user information.*

CHAPTER 4E. PEDESTRIAN CONTROL FEATURES

Section 4E.01 Pedestrian Signal Heads

Support:
01 Pedestrian signal heads provide special types of traffic signal indications exclusively intended for controlling pedestrian traffic. These signal indications consist of the illuminated symbols of a WALKING PERSON (symbolizing WALK) and an UPRAISED HAND (symbolizing DONT WALK).

Guidance:
02 *Engineering judgment should determine the need for separate pedestrian signal heads (see Section 4D.03) and accessible pedestrian signals (see Section 4E.09).*

Support:
03 Chapter 4F contains information regarding the use of pedestrian hybrid beacons and Chapter 4N contains information regarding the use of In-Roadway Warning Lights at unsignalized marked crosswalks.

Section 4E.02 Meaning of Pedestrian Signal Head Indications

Standard:
01 **Pedestrian signal head indications shall have the following meanings:**
   A. **A steady WALKING PERSON (symbolizing WALK) signal indication means that a pedestrian facing the signal indication is permitted to start to cross the roadway in the direction of the signal indication, possibly in conflict with turning vehicles. The pedestrian shall yield the right-of-way to vehicles lawfully within the intersection at the time that the WALKING PERSON (symbolizing WALK) signal indication is first shown.**
   B. **A flashing UPRAISED HAND (symbolizing DONT WALK) signal indication means that a pedestrian shall not start to cross the roadway in the direction of the signal indication, but that any pedestrian who has already started to cross on a steady WALKING PERSON (symbolizing WALK) signal indication shall proceed to the far side of the traveled way of the street or highway, unless otherwise directed by a traffic control device to proceed only to the median of a divided highway or only to some other island or pedestrian refuge area.**
   C. **A steady UPRAISED HAND (symbolizing DONT WALK) signal indication means that a pedestrian shall not enter the roadway in the direction of the signal indication.**
   D. **A flashing WALKING PERSON (symbolizing WALK) signal indication has no meaning and shall not be used.**

Section 4E.03 Application of Pedestrian Signal Heads

Standard:
01 **Pedestrian signal heads shall be used in conjunction with vehicular traffic control signals under any of the following conditions:**
   A. **If a traffic control signal is justified by an engineering study and meets either Warrant 4, Pedestrian Volume or Warrant 5, School Crossing (see Chapter 4C);**
   B. **If an exclusive signal phase is provided or made available for pedestrian movements in one or more directions, with all conflicting vehicular movements being stopped;**
   C. **At an established school crossing at any signalized location; or**
   D. **Where engineering judgment determines that multi-phase signal indications (as with split-phase timing) would tend to confuse or cause conflicts with pedestrians using a crosswalk guided only by vehicular signal indications.**

Guidance:
02 *Pedestrian signal heads should be used under any of the following conditions:*
   A. *If it is necessary to assist pedestrians in deciding when to begin crossing the roadway in the chosen direction or if engineering judgment determines that pedestrian signal heads are justified to minimize vehicle-pedestrian conflicts;*
   B. *If pedestrians are permitted to cross a portion of a street, such as to or from a median of sufficient width for pedestrians to wait, during a particular interval but are not permitted to cross the remainder of the street during any part of the same interval; and/or*
   C. *If no vehicular signal indications are visible to pedestrians, or if the vehicular signal indications that are visible to pedestrians starting a crossing provide insufficient guidance for them to decide when to begin crossing the roadway in the chosen direction, such as on one-way streets, at T-intersections, or at multi-phase signal operations.*

Option:

01 Pedestrian signal heads may be used under other conditions based on engineering judgment.

Section 4E.04 Size, Design, and Illumination of Pedestrian Signal Head Indications

Standard:

01 All new pedestrian signal head indications shall be displayed within a rectangular background and shall consist of symbolized messages (see Figure 4E-1), except that existing pedestrian signal head indications with lettered or outline style symbol messages shall be permitted to be retained for the remainder of their useful service life. The symbol designs that are set forth in the "Standard Highway Signs and Markings" book (see Section 1A.11) shall be used. Each pedestrian signal head indication shall be independently displayed and emit a single color.

02 If a two-section pedestrian signal head is used, the UPRAISED HAND (symbolizing DONT WALK) signal section shall be mounted directly above the WALKING PERSON (symbolizing WALK) signal section. If a one-section pedestrian signal head is used, the symbols shall be either overlaid upon each other or arranged side-by-side with the UPRAISED HAND symbol to the left of the WALKING PERSON symbol, and a light source that can display each symbol independently shall be used.

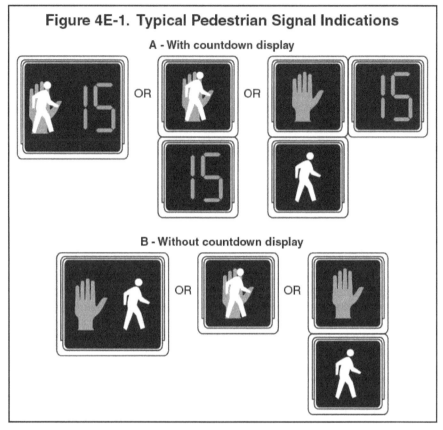

Figure 4E-1. Typical Pedestrian Signal Indications

03 The WALKING PERSON (symbolizing WALK) signal indication shall be white, conforming to the publication entitled "Pedestrian Traffic Control Signal Indications" (see Section 1A.11), with all except the symbol obscured by an opaque material.

04 The UPRAISED HAND (symbolizing DONT WALK) signal indication shall be Portland orange, conforming to the publication entitled "Pedestrian Traffic Control Signal Indications" (see Section 1A.11), with all except the symbol obscured by an opaque material.

05 When not illuminated, the WALKING PERSON (symbolizing WALK) and UPRAISED HAND (symbolizing DONT WALK) symbols shall not be readily visible to pedestrians at the far end of the crosswalk that the pedestrian signal head indications control.

06 For pedestrian signal head indications, the symbols shall be at least 6 inches high.

07 The light source of a flashing UPRAISED HAND (symbolizing DONT WALK) signal indication shall be flashed continuously at a rate of not less than 50 or more than 60 times per minute. The displayed period of each flash shall be a minimum of 1/2 and a maximum of 2/3 of the total flash cycle.

Guidance:

08 *Pedestrian signal head indications should be conspicuous and recognizable to pedestrians at all distances from the beginning of the controlled crosswalk to a point 10 feet from the end of the controlled crosswalk during both day and night.*

09 *For crosswalks where the pedestrian enters the crosswalk more than 100 feet from the pedestrian signal head indications, the symbols should be at least 9 inches high.*

10 *If the pedestrian signal indication is so bright that it causes excessive glare in nighttime conditions, some form of automatic dimming should be used to reduce the brilliance of the signal indication.*

Option:

An animated eyes symbol may be added to a pedestrian signal head in order to prompt pedestrians to look for vehicles in the intersection during the time that the WALKING PERSON (symbolizing WALK) signal indication is displayed.

Standard:

If used, the animated eyes symbol shall consist of an outline of a pair of white steadily-illuminated eyes with white eyeballs that scan from side to side at a rate of approximately once per second. The animated eyes symbol shall be at least 12 inches wide with each eye having a width of at least 5 inches and a height of at least 2.5 inches. The animated eyes symbol shall be illuminated at the start of the walk interval and shall terminate at the end of the walk interval.

Section 4E.05 Location and Height of Pedestrian Signal Heads

Standard:

01 Pedestrian signal heads shall be mounted with the bottom of the signal housing including brackets not less than 7 feet or more than 10 feet above sidewalk level, and shall be positioned and adjusted to provide maximum visibility at the beginning of the controlled crosswalk.

02 If pedestrian signal heads are mounted on the same support as vehicular signal heads, there shall be a physical separation between them.

Section 4E.06 Pedestrian Intervals and Signal Phases

Standard:

01 At intersections equipped with pedestrian signal heads, the pedestrian signal indications shall be displayed except when the vehicular traffic control signal is being operated in the flashing mode. At those times, the pedestrian signal indications shall not be displayed.

02 When the pedestrian signal heads associated with a crosswalk are displaying either a steady WALKING PERSON (symbolizing WALK) or a flashing UPRAISED HAND (symbolizing DONT WALK) signal indication, a steady or a flashing red signal indication shall be shown to any conflicting vehicular movement that is approaching the intersection or midblock location perpendicular or nearly perpendicular to the crosswalk.

03 When pedestrian signal heads are used, a WALKING PERSON (symbolizing WALK) signal indication shall be displayed only when pedestrians are permitted to leave the curb or shoulder.

04 A pedestrian change interval consisting of a flashing UPRAISED HAND (symbolizing DONT WALK) signal indication shall begin immediately following the WALKING PERSON (symbolizing WALK) signal indication. Following the pedestrian change interval, a buffer interval consisting of a steady UPRAISED HAND (symbolizing DONT WALK) signal indication shall be displayed for at least 3 seconds prior to the release of any conflicting vehicular movement. The sum of the time of the pedestrian change interval and the buffer interval shall not be less than the calculated pedestrian clearance time (see Paragraphs 7 through 16). The buffer interval shall not begin later than the beginning of the red clearance interval, if used.

Option:

05 During the yellow change interval, the UPRAISED HAND (symbolizing DON'T WALK) signal indication may be displayed as either a flashing indication, a steady indication, or a flashing indication for an initial portion of the yellow change interval and a steady indication for the remainder of the interval.

Support:

06 Figure 4E-2 illustrates the pedestrian intervals and their possible relationships with associated vehicular signal phase intervals.

Guidance:

07 *Except as provided in Paragraph 8, the pedestrian clearance time should be sufficient to allow a pedestrian crossing in the crosswalk who left the curb or shoulder at the end of the WALKING PERSON (symbolizing WALK) signal indication to travel at a walking speed of 3.5 feet per second to at least the far side of the traveled way or to a median of sufficient width for pedestrians to wait.*

Option:

08 A walking speed of up to 4 feet per second may be used to evaluate the sufficiency of the pedestrian clearance time at locations where an extended pushbutton press function has been installed to provide slower pedestrians an opportunity to request and receive a longer pedestrian clearance time. Passive pedestrian detection may also be used to automatically adjust the pedestrian clearance time based on the pedestrian's actual walking speed or actual clearance of the crosswalk.

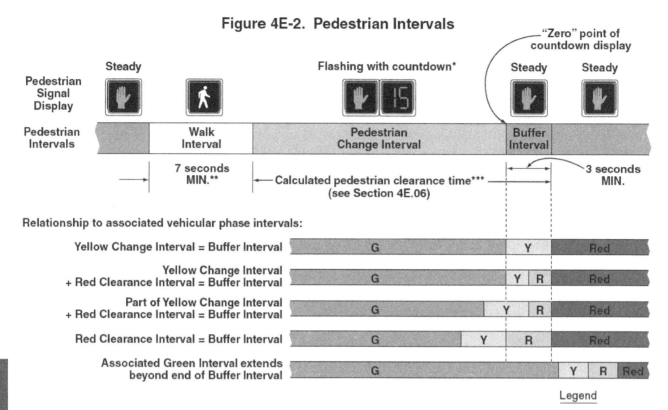

Figure 4E-2. Pedestrian Intervals

*   The countdown display is optional for Pedestrian Change Intervals of 7 seconds or less.
** The Walk Interval may be reduced under some conditions (see Section 4E.06).
*** The Buffer Interval, which shall always be provided and displayed, may be used to help satisfy the calculated pedestrian clearance time, or may begin after the calculated pedestrian clearance time has ended.

Legend
G = Green Interval
Y = Yellow Change Interval (of at least 3 seconds)
R = Red Clearance Interval
Red = Red because conflicting traffic has been released

09 The additional time provided by an extended pushbutton press to satisfy pedestrian clearance time needs may be added to either the walk interval or the pedestrian change interval.

Guidance:

10 *Where pedestrians who walk slower than 3.5 feet per second, or pedestrians who use wheelchairs, routinely use the crosswalk, a walking speed of less than 3.5 feet per second should be considered in determining the pedestrian clearance time.*

11 *Except as provided in Paragraph 12, the walk interval should be at least 7 seconds in length so that pedestrians will have adequate opportunity to leave the curb or shoulder before the pedestrian clearance time begins.*

Option:

12 If pedestrian volumes and characteristics do not require a 7-second walk interval, walk intervals as short as 4 seconds may be used.

Support:

13 The walk interval is intended for pedestrians to start their crossing. The pedestrian clearance time is intended to allow pedestrians who started crossing during the walk interval to complete their crossing. Longer walk intervals are often used when the duration of the vehicular green phase associated with the pedestrian crossing is long enough to allow it.

Guidance:

14 *The total of the walk interval and pedestrian clearance time should be sufficient to allow a pedestrian crossing in the crosswalk who left the pedestrian detector (or, if no pedestrian detector is present, a location 6 feet from the face of the curb or from the edge of the pavement) at the beginning of the WALKING PERSON (symbolizing WALK) signal indication to travel at a walking speed of 3 feet per second to the far side of the traveled way being crossed or to the median if a two-stage pedestrian crossing sequence is used. Any additional time that is required to satisfy the conditions of this paragraph should be added to the walk interval.*

Option:

¹⁵ On a street with a median of sufficient width for pedestrians to wait, a pedestrian clearance time that allows the pedestrian to cross only from the curb or shoulder to the median may be provided.

Standard:

¹⁶ **Where the pedestrian clearance time is sufficient only for crossing from the curb or shoulder to a median of sufficient width for pedestrians to wait, median-mounted pedestrian signals (with pedestrian detectors if actuated operation is used) shall be provided (see Sections 4E.08 and 4E.09) and signing such as the R10-3d sign (see Section 2B.52) shall be provided to notify pedestrians to cross only to the median to await the next WALKING PERSON (symbolizing WALK) signal indication.**

Guidance:

¹⁷ *Where median-mounted pedestrian signals and detectors are provided, the use of accessible pedestrian signals (see Sections 4E.09 through 4E.13) should be considered.*

Option:

¹⁸ During the transition into preemption, the walk interval and the pedestrian change interval may be shortened or omitted as described in Section 4D.27.

¹⁹ At intersections with high pedestrian volumes and high conflicting turning vehicle volumes, a brief leading pedestrian interval, during which an advance WALKING PERSON (symbolizing WALK) indication is displayed for the crosswalk while red indications continue to be displayed to parallel through and/or turning traffic, may be used to reduce conflicts between pedestrians and turning vehicles.

Guidance:

²⁰ *If a leading pedestrian interval is used, the use of accessible pedestrian signals (see Sections 4E.09 through 4E.13) should be considered.*

Support:

²¹ If a leading pedestrian interval is used without accessible features, pedestrians who are visually impaired can be expected to begin crossing at the onset of the vehicular movement when drivers are not expecting them to begin crossing.

Guidance:

²² *If a leading pedestrian interval is used, it should be at least 3 seconds in duration and should be timed to allow pedestrians to cross at least one lane of traffic or, in the case of a large corner radius, to travel far enough for pedestrians to establish their position ahead of the turning traffic before the turning traffic is released.*

²³ *If a leading pedestrian interval is used, consideration should be given to prohibiting turns across the crosswalk during the leading pedestrian interval.*

Support:

²⁴ At intersections with pedestrian volumes that are so high that drivers have difficulty finding an opportunity to turn across the crosswalk, the duration of the green interval for a parallel concurrent vehicular movement is sometimes intentionally set to extend beyond the pedestrian clearance time to provide turning drivers additional green time to make their turns while the pedestrian signal head is displaying a steady UPRAISED HAND (symbolizing DONT WALK) signal indication after pedestrians have had time to complete their crossings.

Section 4E.07 Countdown Pedestrian Signals

Standard:

⁰¹ **All pedestrian signal heads used at crosswalks where the pedestrian change interval is more than 7 seconds shall include a pedestrian change interval countdown display in order to inform pedestrians of the number of seconds remaining in the pedestrian change interval.**

Option:

⁰² Pedestrian signal heads used at crosswalks where the pedestrian change interval is 7 seconds or less may include a pedestrian change interval countdown display in order to inform pedestrians of the number of seconds remaining in the pedestrian change interval.

Standard:

⁰³ **Where countdown pedestrian signals are used, the countdown shall always be displayed simultaneously with the flashing UPRAISED HAND (symbolizing DONT WALK) signal indication displayed for that crosswalk.**

⁰⁴ **Countdown pedestrian signals shall consist of Portland orange numbers that are at least 6 inches in height on a black opaque background. The countdown pedestrian signal shall be located immediately adjacent to the associated UPRAISED HAND (symbolizing DONT WALK) pedestrian signal head indication (see Figure 4E-1).**

⁰⁵ **The display of the number of remaining seconds shall begin only at the beginning of the pedestrian change interval (flashing UPRAISED HAND). After the countdown displays zero, the display shall remain dark until the beginning of the next countdown.**

⁰⁶ **The countdown pedestrian signal shall display the number of seconds remaining until the termination of the pedestrian change interval (flashing UPRAISED HAND). Countdown displays shall not be used during the walk interval or during the red clearance interval of a concurrent vehicular phase.**

Guidance:

⁰⁷ *If used with a pedestrian signal head that does not have a concurrent vehicular phase, the pedestrian change interval (flashing UPRAISED HAND) should be set to be approximately 4 seconds less than the required pedestrian clearance time (see Section 4E.06) and an additional clearance interval (during which a steady UPRAISED HAND is displayed) should be provided prior to the start of the conflicting vehicular phase.*

⁰⁸ *For crosswalks where the pedestrian enters the crosswalk more than 100 feet from the countdown pedestrian signal display, the numbers should be at least 9 inches in height.*

⁰⁹ *Because some technology includes the countdown pedestrian signal logic in a separate timing device that is independent of the timing in the traffic signal controller, care should be exercised by the engineer when timing changes are made to pedestrian change intervals.*

¹⁰ *If the pedestrian change interval is interrupted or shortened as a part of a transition into a preemption sequence (see Section 4E.06), the countdown pedestrian signal display should be discontinued and go dark immediately upon activation of the preemption transition.*

Section 4E.08 Pedestrian Detectors

Option:

⁰¹ Pedestrian detectors may be pushbuttons or passive detection devices.

Support:

⁰² Passive detection devices register the presence of a pedestrian in a position indicative of a desire to cross, without requiring the pedestrian to push a button. Some passive detection devices are capable of tracking the progress of a pedestrian as the pedestrian crosses the roadway for the purpose of extending or shortening the duration of certain pedestrian timing intervals.

⁰³ The provisions in this Section place pedestrian pushbuttons within easy reach of pedestrians who are intending to cross each crosswalk and make it obvious which pushbutton is associated with each crosswalk. These provisions also position pushbutton poles in optimal locations for installation of accessible pedestrian signals (see Sections 4E.09 through 4E.13). Information regarding reach ranges can be found in the "Americans with Disabilities Act Accessibility Guidelines for Buildings and Facilities (ADAAG)" (see Section 1A.11).

Guidance:

⁰⁴ *If pedestrian pushbuttons are used, they should be capable of easy activation and conveniently located near each end of the crosswalks. Except as provided in Paragraphs 5 and 6, pedestrian pushbuttons should be located to meet all of the following criteria (see Figure 4E-3):*
   A. *Unobstructed and adjacent to a level all-weather surface to provide access from a wheelchair;*
   B. *Where there is an all-weather surface, a wheelchair accessible route from the pushbutton to the ramp;*
   C. *Between the edge of the crosswalk line (extended) farthest from the center of the intersection and the side of a curb ramp (if present), but not greater than 5 feet from said crosswalk line;*
   D. *Between 1.5 and 6 feet from the edge of the curb, shoulder, or pavement;*
   E. *With the face of the pushbutton parallel to the crosswalk to be used; and*
   F. *At a mounting height of approximately 3.5 feet, but no more than 4 feet, above the sidewalk.*

⁰⁵ *Where there are physical constraints that make it impractical to place the pedestrian pushbutton adjacent to a level all-weather surface, the surface should be as level as feasible.*

⁰⁶ *Where there are physical constraints that make it impractical to place the pedestrian pushbutton between 1.5 and 6 feet from the edge of the curb, shoulder, or pavement, it should not be farther than 10 feet from the edge of curb, shoulder, or pavement.*

⁰⁷ *Except as provided in Paragraph 8, where two pedestrian pushbuttons are provided on the same corner of a signalized location, the pushbuttons should be separated by a distance of at least 10 feet.*

Option:

⁰⁸ Where there are physical constraints on a particular corner that make it impractical to provide the 10-foot separation between the two pedestrian pushbuttons, the pushbuttons may be placed closer together or on the same pole.

Figure 4E-3. Pushbutton Location Area

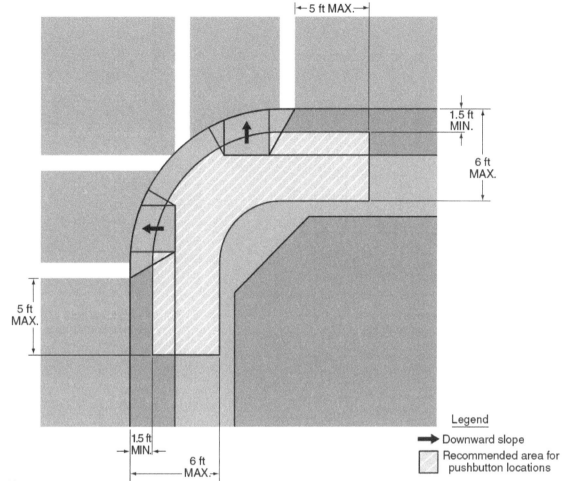

Notes:
1. Where there are constraints that make it impractical to place the pedestrian pushbutton between 1.5 feet and 6 feet from the edge of the curb, shoulder, or pavement, it should not be further than 10 feet from the edge of curb, shoulder, or pavement.
2. Two pedestrian pushbuttons on a corner should be separated by 10 feet.
3. This figure is not drawn to scale.
4. Figure 4E-4 shows typical pushbutton locations.

Support:
09   Figure 4E-4 shows typical pedestrian pushbutton locations for a variety of situations.

Standard:

10   **Signs (see Section 2B.52) shall be mounted adjacent to or integral with pedestrian pushbuttons, explaining their purpose and use.**

Option:

11   At certain locations, a supplemental sign in a more visible location may be used to call attention to the pedestrian pushbutton.

Standard:

12   **The positioning of pedestrian pushbuttons and the legends on the pedestrian pushbutton signs shall clearly indicate which crosswalk signal is actuated by each pedestrian pushbutton.**

13   **If the pedestrian clearance time is sufficient only to cross from the curb or shoulder to a median of sufficient width for pedestrians to wait and the signals are pedestrian actuated, an additional pedestrian detector shall be provided in the median.**

Figure 4E-4. Typical Pushbutton Locations (Sheet 1 of 2)

A - Parallel ramps with wide sidewalk

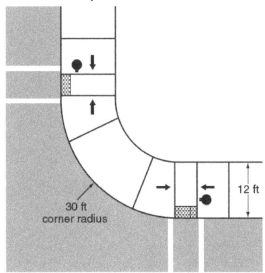

B - Parallel ramps with narrow sidewalk

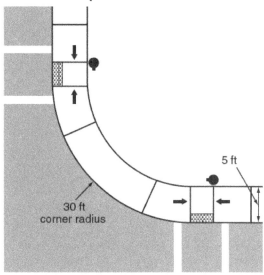

C - Parallel ramps with narrow sidewalk and tight corner radius

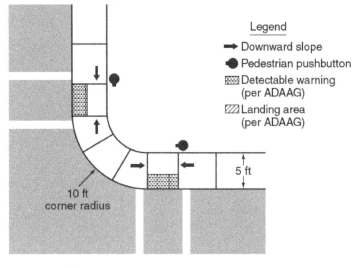

D - Perpendicular ramps with crosswalks far apart

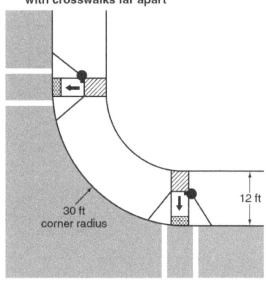

Legend
→ Downward slope
● Pedestrian pushbutton
▓ Detectable warning (per ADAAG)
▨ Landing area (per ADAAG)

Notes:
1. This figure is not drawn to scale.
2. These drawings are intended to describe the typical locations for pedestrian pushbutton installations. They are not intended to be a guide for the design of curb cut ramps.
3. Figure 4E-3 shows the recommended area for pushbutton locations.

Guidance:

The use of additional pedestrian detectors on islands or medians where a pedestrian might become stranded should be considered.

If used, special purpose pushbuttons (to be operated only by authorized persons) should include a housing capable of being locked to prevent access by the general public and do not need an instructional sign.

Standard:

If used, a pilot light or other means of indication installed with a pedestrian pushbutton shall not be illuminated until actuation. Once it is actuated, the pilot light shall remain illuminated until the pedestrian's green or WALKING PERSON (symbolizing WALK) signal indication is displayed.

Figure 4E-4. Typical Pushbutton Locations (Sheet 2 of 2)

E - Perpendicular ramps with crosswalks close together

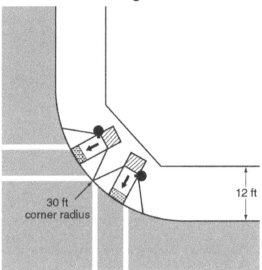

F - Perpendicular ramps with sidewalk set back from road with crosswalks far apart

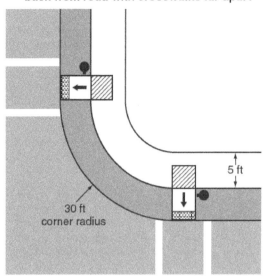

G - Perpendicular ramps with sidewalk set back from road with crosswalks close together

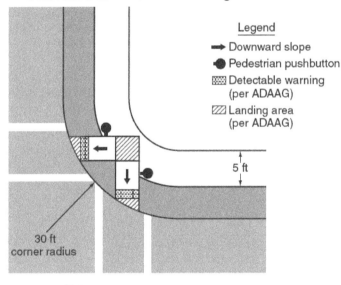

H - Perpendicular ramps with sidewalk set back from road with continuous sidewalk between ramps

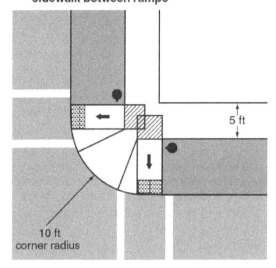

Legend
- → Downward slope
- ● Pedestrian pushbutton
- ▒ Detectable warning (per ADAAG)
- ▨ Landing area (per ADAAG)

Notes:
1. This figure is not drawn to scale.
2. Tese drawings are intended to describe the typical locations for pedestrian pushbutton installations. They are not intended to be a guide for the design of curb cut ramps.
3. Figure 4E-3 shows the recommended area for pushbutton locations.

---

17 If a pilot light is used at an accessible pedestrian signal location (see Sections 4E.09 through 4E.13), each actuation shall be accompanied by the speech message "wait."

Option:

18 At signalized locations with a demonstrated need and subject to equipment capabilities, pedestrians with special needs may be provided with additional crossing time by means of an extended pushbutton press.

Standard:

19 **If additional crossing time is provided by means of an extended pushbutton press, a PUSH BUTTON FOR 2 SECONDS FOR EXTRA CROSSING TIME (R10-32P) plaque (see Figure 2B-26) shall be mounted adjacent to or integral with the pedestrian pushbutton.**

Section 4E.09 Accessible Pedestrian Signals and Detectors – General

Support:

01 Accessible pedestrian signals and detectors provide information in non-visual formats (such as audible tones, speech messages, and/or vibrating surfaces).

02 The primary technique that pedestrians who have visual disabilities use to cross streets at signalized locations is to initiate their crossing when they hear the traffic in front of them stop and the traffic alongside them begin to move, which often corresponds to the onset of the green interval. The existing environment is often not sufficient to provide the information that pedestrians who have visual disabilities need to cross a roadway at a signalized location.

Guidance:

03 *If a particular signalized location presents difficulties for pedestrians who have visual disabilities to cross the roadway, an engineering study should be conducted that considers the needs of pedestrians in general, as well as the information needs of pedestrians with visual disabilities. The engineering study should consider the following factors:*
   A. *Potential demand for accessible pedestrian signals;*
   B. *A request for accessible pedestrian signals;*
   C. *Traffic volumes during times when pedestrians might be present, including periods of low traffic volumes or high turn-on-red volumes;*
   D. *The complexity of traffic signal phasing (such as split phases, protected turn phases, leading pedestrian intervals, and exclusive pedestrian phases); and*
   E. *The complexity of intersection geometry.*

Support:

04 The factors that make crossing at a signalized location difficult for pedestrians who have visual disabilities include: increasingly quiet cars, right turn on red (which masks the beginning of the through phase), continuous right-turn movements, complex signal operations, traffic circles, and wide streets. Furthermore, low traffic volumes might make it difficult for pedestrians who have visual disabilities to discern signal phase changes.

05 Local organizations, providing support services to pedestrians who have visual and/or hearing disabilities, can often act as important advisors to the traffic engineer when consideration is being given to the installation of devices to assist such pedestrians. Additionally, orientation and mobility specialists or similar staff also might be able to provide a wide range of advice. The U.S. Access Board (www.access-board.gov) provides technical assistance for making pedestrian signal information available to persons with visual disabilities (see Page i for the address for the U.S. Access Board).

Standard:

06 **When used, accessible pedestrian signals shall be used in combination with pedestrian signal timing. The information provided by an accessible pedestrian signal shall clearly indicate which pedestrian crossing is served by each device.**

07 **Under stop-and-go operation, accessible pedestrian signals shall not be limited in operation by the time of day or day of week.**

Option:

08 Accessible pedestrian signal detectors may be pushbuttons or passive detection devices.

09 At locations with pretimed traffic control signals or non-actuated approaches, pedestrian pushbuttons may be used to activate the accessible pedestrian signals.

Support:

10 Accessible pedestrian signals are typically integrated into the pedestrian detector (pushbutton), so the audible tones and/or messages come from the pushbutton housing. They have a pushbutton locator tone and tactile arrow, and can include audible beaconing and other special features.

Option:

11 The name of the street to be crossed may also be provided in accessible format, such as Braille or raised print. Tactile maps of crosswalks may also be provided.

Support:

12 Specifications regarding the use of Braille or raised print for traffic control devices can be found in the "Americans with Disabilities Act Accessibility Guidelines for Buildings and Facilities (ADAAG)" (see Section 1A.11).

Standard:

13   **At accessible pedestrian signal locations where pedestrian pushbuttons are used, each pushbutton shall activate both the walk interval and the accessible pedestrian signals.**

Section 4E.10 Accessible Pedestrian Signals and Detectors – Location

Support:

01   Accessible pedestrian signals that are located as close as possible to pedestrians waiting to cross the street provide the clearest and least ambiguous indication of which pedestrian crossing is served by a device.

Guidance:

02   *Pushbuttons for accessible pedestrian signals should be located in accordance with the provisions of Section 4E.08 and should be located as close as possible to the crosswalk line furthest from the center of the intersection and as close as possible to the curb ramp.*

Standard:

03   **If two accessible pedestrian pushbuttons are placed less than 10 feet apart or on the same pole, each accessible pedestrian pushbutton shall be provided with the following features (see Sections 4E.11 through 4E.13):**
       A. **A pushbutton locator tone,**
       B. **A tactile arrow,**
       C. **A speech walk message for the WALKING PERSON (symbolizing WALK) indication, and**
       D. **A speech pushbutton information message.**

04   **If the pedestrian clearance time is sufficient only to cross from the curb or shoulder to a median of sufficient width for pedestrians to wait and accessible pedestrian detectors are used, an additional accessible pedestrian detector shall be provided in the median.**

Section 4E.11 Accessible Pedestrian Signals and Detectors – Walk Indications

Support:

01   Technology that provides different sounds for each non-concurrent signal phase has frequently been found to provide ambiguous information. Research indicates that a rapid tick tone for each crossing coming from accessible pedestrian signal devices on separated poles located close to each crosswalk provides unambiguous information to pedestrians who are blind or visually impaired. Vibrotactile indications provide information to pedestrians who are blind and deaf and are also used by pedestrians who are blind or who have low vision to confirm the walk signal in noisy situations.

Standard:

02   **Accessible pedestrian signals shall have both audible and vibrotactile walk indications.**

03   **Vibrotactile walk indications shall be provided by a tactile arrow on the pushbutton (see Section 4E.12) that vibrates during the walk interval.**

04   **Accessible pedestrian signals shall have an audible walk indication during the walk interval only. The audible walk indication shall be audible from the beginning of the associated crosswalk.**

05   **The accessible walk indication shall have the same duration as the pedestrian walk signal except when the pedestrian signal rests in walk.**

Guidance:

06   *If the pedestrian signal rests in walk, the accessible walk indication should be limited to the first 7 seconds of the walk interval. The accessible walk indication should be recalled by a button press during the walk interval provided that the crossing time remaining is greater than the pedestrian change interval.*

Standard:

07   **Where two accessible pedestrian signals are separated by a distance of at least 10 feet, the audible walk indication shall be a percussive tone. Where two accessible pedestrian signals on one corner are not separated by a distance of at least 10 feet, the audible walk indication shall be a speech walk message.**

08   **Audible tone walk indications shall repeat at eight to ten ticks per second. Audible tones used as walk indications shall consist of multiple frequencies with a dominant component at 880 Hz.**

Guidance:

09   *The volume of audible walk indications and pushbutton locator tones (see Section 4E.12) should be set to be a maximum of 5 dBA louder than ambient sound, except when audible beaconing is provided in response to an extended pushbutton press.*

Standard:

10 **Automatic volume adjustment in response to ambient traffic sound level shall be provided up to a maximum volume of 100 dBA.**

Guidance:

11 *The sound level of audible walk indications and pushbutton locator tones should be adjusted to be low enough to avoid misleading pedestrians who have visual disabilities when the following conditions exist:*
   A. *Where there is an island that allows unsignalized right turns across a crosswalk between the island and the sidewalk.*
   B. *Where multi-leg approaches or complex signal phasing require more than two pedestrian phases, such that it might be unclear which crosswalk is served by each audible tone.*
   C. *At intersections where a diagonal pedestrian crossing is allowed, or where one street receives a WALKING PERSON (symbolizing WALK) signal indication simultaneously with another street.*

Option:

12 An alert tone, which is a very brief burst of high-frequency sound at the beginning of the audible walk indication that rapidly decays to the frequency of the walk tone, may be used to alert pedestrians to the beginning of the walk interval.

Support:

13 An alert tone can be particularly useful if the walk tone is not easily audible in some traffic conditions.

14 Speech walk messages communicate to pedestrians which street has the walk interval. Speech messages might be either directly audible or transmitted, requiring a personal receiver to hear the message. To be a useful system, the words and their meaning need to be correctly understood by all users in the context of the street environment where they are used. Because of this, tones are the preferred means of providing audible walk indications except where two accessible pedestrian signals on one corner are not separated by a distance of at least 10 feet.

15 If speech walk messages are used, pedestrians have to know the names of the streets that they are crossing in order for the speech walk messages to be unambiguous. In getting directions to travel to a new location, pedestrians with visual disabilities do not always get the name of each street to be crossed. Therefore, it is desirable to give users of accessible pedestrian signals the name of the street controlled by the pushbutton. This can be done by means of a speech pushbutton information message (see Section 4D.13) during the flashing or steady UPRAISED HAND intervals, or by raised print and Braille labels on the pushbutton housing.

16 By combining the information from the pushbutton message or Braille label, the tactile arrow aligned in the direction of travel on the relevant crosswalk, and the speech walk message, pedestrians with visual disabilities are able to correctly respond to speech walk messages even if there are two pushbuttons on the same pole.

Standard:

17 **If speech walk messages are used to communicate the walk interval, they shall provide a clear message that the walk interval is in effect, as well as to which crossing it applies. Speech walk messages shall be used only at intersections where it is technically infeasible to install two accessible pedestrian signals at one corner separated by a distance of at least 10 feet.**

18 **Speech walk messages that are used at intersections having pedestrian phasing that is concurrent with vehicular phasing shall be patterned after the model: "Broadway. Walk sign is on to cross Broadway."**

19 **Speech walk messages that are used at intersections having exclusive pedestrian phasing shall be patterned after the model: "Walk sign is on for all crossings."**

20 **Speech walk messages shall not contain any additional information, except they shall include designations such as "Street" or "Avenue" where this information is necessary to avoid ambiguity at a particular location.**

Guidance:

21 *Speech walk messages should not state or imply a command to the pedestrian, such as "Cross Broadway now." Speech walk messages should not tell pedestrians that it is "safe to cross," because it is always the pedestrian's responsibility to check actual traffic conditions.*

Standard:

22 **A speech walk message is not required at times when the walk interval is not timing, but, if provided:**
   A. **It shall begin with the term "wait."**
   B. **It need not be repeated for the entire time that the walk interval is not timing.**

23 **If a pilot light (see Section 4E.08) is used at an accessible pedestrian signal location, each actuation shall be accompanied by the speech message "wait."**

Option:

24   Accessible pedestrian signals that provide speech walk messages may provide similar messages in languages other than English, if needed, except for the terms "walk sign" and "wait."

Standard:

25   **Following the audible walk indication, accessible pedestrian signals shall revert to the pushbutton locator tone (see Section 4E.12) during the pedestrian change interval.**

Section 4E.12 Accessible Pedestrian Signals and Detectors – Tactile Arrows and Locator Tones

Standard:

01   **To enable pedestrians who have visual disabilities to distinguish and locate the appropriate pushbutton at an accessible pedestrian signal location, pushbuttons shall clearly indicate by means of tactile arrows which crosswalk signal is actuated by each pushbutton. Tactile arrows shall be located on the pushbutton, have high visual contrast (light on dark or dark on light), and shall be aligned parallel to the direction of travel on the associated crosswalk.**

02   **An accessible pedestrian pushbutton shall incorporate a locator tone.**

Support:

03   A pushbutton locator tone is a repeating sound that informs approaching pedestrians that a pushbutton to actuate pedestrian timing or receive additional information exists, and that enables pedestrians with visual disabilities to locate the pushbutton.

Standard:

04   **Pushbutton locator tones shall have a duration of 0.15 seconds or less, and shall repeat at 1-second intervals.**

05   **Pushbutton locator tones shall be deactivated when the traffic control signal is operating in a flashing mode. This requirement shall not apply to traffic control signals or pedestrian hybrid beacons that are activated from a flashing or dark mode to a stop-and-go mode by pedestrian actuations.**

06   **Pushbutton locator tones shall be intensity responsive to ambient sound, and be audible 6 to 12 feet from the pushbutton, or to the building line, whichever is less.**

Support:

07   Section 4E.11 contains additional provisions regarding the volume and sound level of pushbutton locator tones.

Section 4E.13 Accessible Pedestrian Signals and Detectors – Extended Pushbutton Press Features

Option:

01   Pedestrians may be provided with additional features such as increased crossing time, audible beaconing, or a speech pushbutton information message as a result of an extended pushbutton press.

Standard:

02   **If an extended pushbutton press is used to provide any additional feature(s), a pushbutton press of less than one second shall actuate only the pedestrian timing and any associated accessible walk indication, and a pushbutton press of one second or more shall actuate the pedestrian timing, any associated accessible walk indication, and any additional feature(s).**

03   **If additional crossing time is provided by means of an extended pushbutton press, a PUSH BUTTON FOR 2 SECONDS FOR EXTRA CROSSING TIME (R10-32P) plaque (see Figure 2B-26) shall be mounted adjacent to or integral with the pedestrian pushbutton.**

Support:

04   Audible beaconing is the use of an audible signal in such a way that pedestrians with visual disabilities can home in on the signal that is located on the far end of the crosswalk as they cross the street.

05   Not all crosswalks at an intersection need audible beaconing; audible beaconing can actually cause confusion if used at all crosswalks at some intersections. Audible beaconing is not appropriate at locations with channelized turns or split phasing, because of the possibility of confusion.

Guidance:

06   *Audible beaconing should only be considered following an engineering study at:*
   A. Crosswalks longer than 70 feet, unless they are divided by a median that has another accessible pedestrian signal with a locator tone;
   B. Crosswalks that are skewed;
   C. Intersections with irregular geometry, such as more than four legs;
   D. Crosswalks where audible beaconing is requested by an individual with visual disabilities; or
   E. Other locations where a study indicates audible beaconing would be beneficial.

Option:

07 Audible beaconing may be provided in several ways, any of which are initiated by an extended pushbutton press.

Standard:

08 **If audible beaconing is used, the volume of the pushbutton locator tone during the pedestrian change interval of the called pedestrian phase shall be increased and operated in one of the following ways:**
   A. **The louder audible walk indication and louder locator tone comes from the far end of the crosswalk, as pedestrians cross the street,**
   B. **The louder locator tone comes from both ends of the crosswalk, or**
   C. **The louder locator tone comes from an additional speaker that is aimed at the center of the crosswalk and that is mounted on a pedestrian signal head.**

Option:

09 Speech pushbutton information messages may provide intersection identification, as well as information about unusual intersection signalization and geometry, such as notification regarding exclusive pedestrian phasing, leading pedestrian intervals, split phasing, diagonal crosswalks, and medians or islands.

Standard:

10 **If speech pushbutton information messages are made available by actuating the accessible pedestrian signal detector, they shall only be actuated when the walk interval is not timing. They shall begin with the term "Wait," followed by intersection identification information modeled after: "Wait to cross Broadway at Grand." If information on intersection signalization or geometry is also given, it shall follow the intersection identification information.**

Guidance:

11 *Speech pushbutton information messages should not be used to provide landmark information or to inform pedestrians with visual disabilities about detours or temporary traffic control situations.*

Support:

12 Additional information on the structure and wording of speech pushbutton information messages is included in ITE's "Electronic Toolbox for Making Intersections More Accessible for Pedestrians Who Are Blind or Visually Impaired," which is available at ITE's website (see Page i).

CHAPTER 4F. PEDESTRIAN HYBRID BEACONS

Section 4F.01 Application of Pedestrian Hybrid Beacons

Support:
01 A pedestrian hybrid beacon is a special type of hybrid beacon used to warn and control traffic at an unsignalized location to assist pedestrians in crossing a street or highway at a marked crosswalk.

Option:
02 A pedestrian hybrid beacon may be considered for installation to facilitate pedestrian crossings at a location that does not meet traffic signal warrants (see Chapter 4C), or at a location that meets traffic signal warrants under Sections 4C.05 and/or 4C.06 but a decision is made to not install a traffic control signal.

Standard:
03 **If used, pedestrian hybrid beacons shall be used in conjunction with signs and pavement markings to warn and control traffic at locations where pedestrians enter or cross a street or highway. A pedestrian hybrid beacon shall only be installed at a marked crosswalk.**

Guidance:
04 *If one of the signal warrants of Chapter 4C is met and a traffic control signal is justified by an engineering study, and if a decision is made to install a traffic control signal, it should be installed based upon the provisions of Chapters 4D and 4E.*

05 *If a traffic control signal is not justified under the signal warrants of Chapter 4C and if gaps in traffic are not adequate to permit pedestrians to cross, or if the speed for vehicles approaching on the major street is too high to permit pedestrians to cross, or if pedestrian delay is excessive, the need for a pedestrian hybrid beacon should be considered on the basis of an engineering study that considers major-street volumes, speeds, widths, and gaps in conjunction with pedestrian volumes, walking speeds, and delay.*

06 *For a major street where the posted or statutory speed limit or the 85th-percentile speed is 35 mph or less, the need for a pedestrian hybrid beacon should be considered if the engineering study finds that the plotted point representing the vehicles per hour on the major street (total of both approaches) and the corresponding total of all pedestrians crossing the major street for 1 hour (any four consecutive 15-minute periods) of an average day falls above the applicable curve in Figure 4F-1 for the length of the crosswalk.*

07 *For a major street where the posted or statutory speed limit or the 85th-percentile speed exceeds 35 mph, the need for a pedestrian hybrid beacon should be considered if the engineering study finds that the plotted point representing the vehicles per hour on the major street (total of both approaches) and the corresponding total of all pedestrians crossing the major street for 1 hour (any four consecutive 15-minute periods) of an average day falls above the applicable curve in Figure 4F-2 for the length of the crosswalk.*

08 *For crosswalks that have lengths other than the four that are specifically shown in Figures 4F-1 and 4F-2, the values should be interpolated between the curves.*

Section 4F.02 Design of Pedestrian Hybrid Beacons

Standard:
01 **Except as otherwise provided in this Section, a pedestrian hybrid beacon shall meet the provisions of Chapters 4D and 4E.**

02 **A pedestrian hybrid beacon face shall consist of three signal sections, with a CIRCULAR YELLOW signal indication centered below two horizontally aligned CIRCULAR RED signal indications (see Figure 4F-3).**

03 **When an engineering study finds that installation of a pedestrian hybrid beacon is justified, then:**
   A. At least two pedestrian hybrid beacon faces shall be installed for each approach of the major street,
   B. A stop line shall be installed for each approach to the crosswalk,
   C. A pedestrian signal head conforming to the provisions set forth in Chapter 4E shall be installed at each end of the marked crosswalk, and
   D. The pedestrian hybrid beacon shall be pedestrian actuated.

Guidance:
04 *When an engineering study finds that installation of a pedestrian hybrid beacon is justified, then:*
   A. The pedestrian hybrid beacon should be installed at least 100 feet from side streets or driveways that are controlled by STOP or YIELD signs,

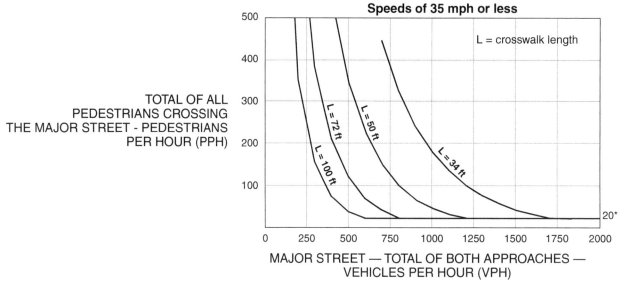

Figure 4F-1. Guidelines for the Installation of Pedestrian Hybrid Beacons on Low-Speed Roadways

* Note: 20 pph applies as the lower threshold volume

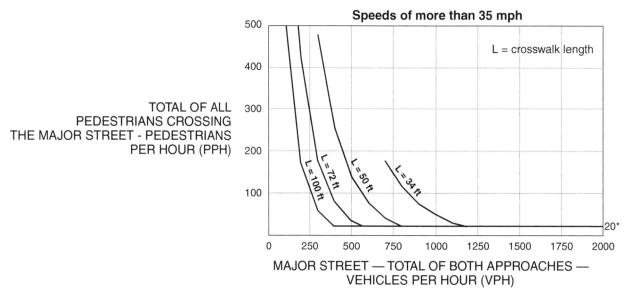

Figure 4F-2. Guidelines for the Installation of Pedestrian Hybrid Beacons on High-Speed Roadways

* Note: 20 pph applies as the lower threshold volume

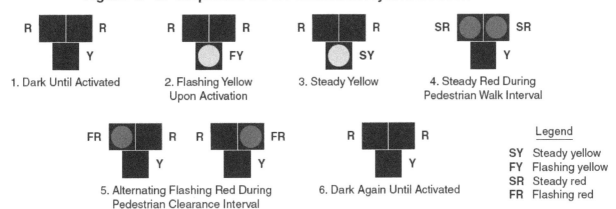

Figure 4F-3. Sequence for a Pedestrian Hybrid Beacon

B. Parking and other sight obstructions should be prohibited for at least 100 feet in advance of and at least 20 feet beyond the marked crosswalk, or site accommodations should be made through curb extensions or other techniques to provide adequate sight distance,
C. The installation should include suitable standard signs and pavement markings, and
D. If installed within a signal system, the pedestrian hybrid beacon should be coordinated.

05 On approaches having posted or statutory speed limits or 85th-percentile speeds in excess of 35 mph and on approaches having traffic or operating conditions that would tend to obscure visibility of roadside hybrid beacon face locations, both of the minimum of two pedestrian hybrid beacon faces should be installed over the roadway.

06 On multi-lane approaches having a posted or statutory speed limits or 85th-percentile speeds of 35 mph or less, either a pedestrian hybrid beacon face should be installed on each side of the approach (if a median of sufficient width exists) or at least one of the pedestrian hybrid beacon faces should be installed over the roadway.

07 A pedestrian hybrid beacon should comply with the signal face location provisions described in Sections 4D.11 through 4D.16.

Standard:

08 **A CROSSWALK STOP ON RED (symbolic circular red) (R10-23) sign (see Section 2B.53) shall be mounted adjacent to a pedestrian hybrid beacon face on each major street approach. If an overhead pedestrian hybrid beacon face is provided, the sign shall be mounted adjacent to the overhead signal face.**

Option:

09 A Pedestrian (W11-2) warning sign (see Section 2C.50) with an AHEAD (W16-9P) supplemental plaque may be placed in advance of a pedestrian hybrid beacon. A warning beacon may be installed to supplement the W11-2 sign.

Guidance:

10 *If a warning beacon supplements a W11-2 sign in advance of a pedestrian hybrid beacon, it should be programmed to flash only when the pedestrian hybrid beacon is not in the dark mode.*

Standard:

11 **If a warning beacon is installed to supplement the W11-2 sign, the design and location of the warning beacon shall comply with the provisions of Sections 4L.01 and 4L.03.**

Section 4F.03 Operation of Pedestrian Hybrid Beacons

Standard:

01 **Pedestrian hybrid beacon indications shall be dark (not illuminated) during periods between actuations.**

02 **Upon actuation by a pedestrian, a pedestrian hybrid beacon face shall display a flashing CIRCULAR yellow signal indication, followed by a steady CIRCULAR yellow signal indication, followed by both steady CIRCULAR RED signal indications during the pedestrian walk interval, followed by alternating flashing CIRCULAR RED signal indications during the pedestrian clearance interval (see Figure 4F-3). Upon termination of the pedestrian clearance interval, the pedestrian hybrid beacon faces shall revert to a dark (not illuminated) condition.**

03 **Except as provided in Paragraph 4, the pedestrian signal heads shall continue to display a steady UPRAISED HAND (symbolizing DONT WALK) signal indication when the pedestrian hybrid beacon faces are either dark or displaying flashing or steady CIRCULAR yellow signal indications. The pedestrian signal heads shall display a WALKING PERSON (symbolizing WALK) signal indication when the pedestrian hybrid beacon faces are displaying steady CIRCULAR RED signal indications. The pedestrian signal heads shall display a flashing UPRAISED HAND (symbolizing DONT WALK) signal indication when the pedestrian hybrid beacon faces are displaying alternating flashing CIRCULAR RED signal indications. Upon termination of the pedestrian clearance interval, the pedestrian signal heads shall revert to a steady UPRAISED HAND (symbolizing DONT WALK) signal indication.**

Option:

04 Where the pedestrian hybrid beacon is installed adjacent to a roundabout to facilitate crossings by pedestrians with visual disabilities and an engineering study determines that pedestrians without visual disabilities can be allowed to cross the roadway without actuating the pedestrian hybrid beacon, the pedestrian signal heads may be dark (not illuminated) when the pedestrian hybrid beacon faces are dark.

Guidance:

05 *The duration of the flashing yellow interval should be determined by engineering judgment.*

Standard:

06 **The duration of the steady yellow change interval shall be determined using engineering practices.**

Guidance:

07 *The steady yellow interval should have a minimum duration of 3 seconds and a maximum duration of 6 seconds (see Section 4D.26). The longer intervals should be reserved for use on approaches with higher speeds.*

CHAPTER 4G. TRAFFIC CONTROL SIGNALS AND HYBRID BEACONS FOR EMERGENCY-VEHICLE ACCESS

Section 4G.01 Application of Emergency-Vehicle Traffic Control Signals and Hybrid Beacons

Support:

01 An emergency-vehicle traffic control signal is a special traffic control signal that assigns the right-of-way to an authorized emergency vehicle.

Option:

02 An emergency-vehicle traffic control signal may be installed at a location that does not meet other traffic signal warrants such as at an intersection or other location to permit direct access from a building housing the emergency vehicle.

03 An emergency-vehicle hybrid beacon may be installed instead of an emergency-vehicle traffic control signal under conditions described in Section 4G.04.

Guidance:

04 *If a traffic control signal is not justified under the signal warrants of Chapter 4C and if gaps in traffic are not adequate to permit the timely entrance of emergency vehicles, or the stopping sight distance for vehicles approaching on the major street is insufficient for emergency vehicles, installing an emergency-vehicle traffic control signal should be considered. If one of the signal warrants of Chapter 4C is met and a traffic control signal is justified by an engineering study, and if a decision is made to install a traffic control signal, it should be installed based upon the provisions of Chapter 4D.*

05 *The sight distance determination should be based on the location of the visibility obstruction for the critical approach lane for each street or drive and the posted or statutory speed limit or 85th-percentile speed on the major street, whichever is higher.*

Section 4G.02 Design of Emergency-Vehicle Traffic Control Signals

Standard:

01 **Except as otherwise provided in this Section, an emergency-vehicle traffic control signal shall meet the requirements of this Manual.**

02 **An Emergency Vehicle (W11-8) sign (see Section 2C.49) with an EMERGENCY SIGNAL AHEAD (W11-12P) supplemental plaque shall be placed in advance of all emergency-vehicle traffic control signals. If a warning beacon is installed to supplement the W11-8 sign, the design and location of the beacon shall comply with the Standards of Sections 4L.01 and 4L.03.**

Guidance:

03 *At least one of the two required signal faces for each approach on the major street should be located over the roadway.*

04 *The following size signal indications should be used for emergency-vehicle traffic control signals: 12-inch diameter for steady red and steady yellow circular signal indications and any arrow indications, and 8-inch diameter for green or flashing yellow circular signal indications.*

Standard:

05 **An EMERGENCY SIGNAL (R10-13) sign shall be mounted adjacent to a signal face on each major street approach (see Section 2B.53). If an overhead signal face is provided, the EMERGENCY SIGNAL sign shall be mounted adjacent to the overhead signal face.**

Option:

06 An approach that only serves emergency vehicles may be provided with only one signal face consisting of one or more signal sections.

07 Besides using an 8-inch diameter signal indication, other appropriate means to reduce the flashing yellow light output may be used.

Section 4G.03 Operation of Emergency-Vehicle Traffic Control Signals

Standard:

01 **Right-of-way for emergency vehicles at signalized locations operating in the steady (stop-and-go) mode shall be obtained as provided in Section 4D.27.**

02 **As a minimum, the signal indications, sequence, and manner of operation of an emergency-vehicle traffic control signal installed at a midblock location shall be as follows:**

A. The signal indication, between emergency-vehicle actuations, shall be either green or flashing yellow. If the flashing yellow signal indication is used instead of the green signal indication, it shall be displayed in the normal position of the green signal indication, while the steady red and steady yellow signal indications shall be displayed in their normal positions.
B. When an emergency-vehicle actuation occurs, a steady yellow change interval followed by a steady red interval shall be displayed to traffic on the major street.
C. A yellow change interval is not required following the green interval for the emergency-vehicle driveway.

03 Emergency-vehicle traffic control signals located at intersections shall either be operated in the flashing mode between emergency-vehicle actuations (see Sections 4D.28 and 4D.30) or be full-actuated or semi-actuated to accommodate normal vehicular and pedestrian traffic on the streets.

04 Warning beacons, if used with an emergency-vehicle traffic control signal, shall be flashed only:
A. For an appropriate time in advance of and during the steady yellow change interval for the major street; and
B. During the steady red interval for the major street.

Guidance:

05 *The duration of the steady red interval for traffic on the major street should be determined by on-site test-run time studies, but should not exceed 1.5 times the time required for the emergency vehicle to clear the path of conflicting vehicles.*

Option:

06 An emergency-vehicle traffic control signal sequence may be initiated manually from a local control point such as a fire station or law enforcement headquarters or from an emergency vehicle equipped for remote operation of the signal.

Section 4G.04 Emergency-Vehicle Hybrid Beacons

Standard:

01 **Emergency-vehicle hybrid beacons shall be used only in conjunction with signs to warn and control traffic at an unsignalized location where emergency vehicles enter or cross a street or highway. Emergency-vehicle hybrid beacons shall be actuated only by authorized emergency or maintenance personnel.**

Guidance:

02 *Emergency-vehicle hybrid beacons should only be used when all of the following criteria are satisfied:*
   A. The conditions justifying an emergency-vehicle traffic control signal (see Section 4G.01) are met; and
   B. An engineering study, considering the road width, approach speeds, and other pertinent factors, determines that emergency-vehicle hybrid beacons can be designed and located in compliance with the requirements contained in this Section and in Section 4L.01, such that they effectively warn and control traffic at the location; and
   C. The location is not at or within 100 feet from an intersection or driveway where the side road or driveway is controlled by a STOP or YIELD sign.

Standard:

03 **Except as otherwise provided in this Section, an emergency-vehicle hybrid beacon shall meet the requirements of this Manual.**

04 **An emergency-vehicle hybrid beacon face shall consist of three signal sections, with a CIRCULAR YELLOW signal indication centered below two horizontally aligned CIRCULAR RED signal indications (see Figure 4G-1).**

05 **Emergency-vehicle hybrid beacons shall be placed in a dark mode (no indications displayed) during periods between actuations.**

06 **Upon actuation by authorized emergency personnel, the emergency-vehicle hybrid beacon faces shall each display a flashing yellow signal indication, followed by a steady yellow change interval, prior to displaying two CIRCULAR RED signal indications in an alternating flashing array for a duration of time adequate for egress of the emergency vehicles. The alternating flashing red signal indications shall only be displayed when it is required that drivers on the major street stop and then proceed subject to the rules applicable after making a stop at a STOP sign. Upon termination of the flashing red signal indications, the emergency-vehicle hybrid beacons shall revert to a dark mode (no indications displayed) condition.**

Guidance:

07 *The duration of the flashing yellow interval should be determined by engineering judgment.*

Figure 4G-1. Sequence for an Emergency-Vehicle Hybrid Beacon

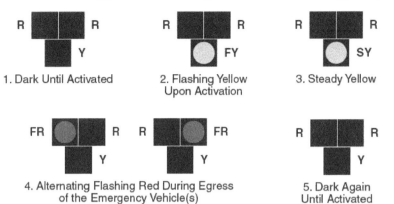

Standard:

08 **The duration of the steady yellow change interval shall be determined using engineering practices.**

Guidance:

09 *The steady yellow change interval should have a minimum duration of 3 seconds and a maximum duration of 6 seconds (see Section 4D.26). The longer intervals should be reserved for use on approaches with higher speeds.*

Option:

10 A steady red clearance interval may be used after the steady yellow change interval.

11 Emergency-vehicle hybrid beacons may be equipped with a light or other display visible to the operator of the egressing emergency vehicle to provide confirmation that the beacons are operating.

12 Emergency-vehicle hybrid beacons may be supplemented with an advance warning sign, which may also be supplemented with a Warning Beacon (see Section 4L.03).

Guidance:

13 *If a Warning Beacon is used to supplement the advance warning sign, it should be programmed to flash only when the emergency-vehicle hybrid beacon is not in the dark mode.*

Standard:

14 **At least two emergency-vehicle hybrid beacon faces shall be installed for each approach of the major street and a stop line shall be installed for each approach of the major street.**

Guidance:

15 *On approaches having posted or statutory speed limits or 85th-percentile speeds in excess of 40 mph, and on approaches having traffic or operating conditions that would tend to obscure visibility of roadside beacon faces, both of the minimum of two emergency-vehicle hybrid beacon faces should be installed over the roadway.*

16 *On multi-lane approaches having posted or statutory speed limits or 85th-percentile speeds of 40 mph or less, either an emergency-vehicle hybrid beacon face should be installed on each side of the approach (if a median of sufficient width exists) or at least one of the emergency-vehicle hybrid beacon faces should be installed over the roadway.*

17 *An emergency-vehicle hybrid beacon should comply with the signal face location provisions described in Sections 4D.11 through 4D.16.*

Standard:

18 **Stop lines and EMERGENCY SIGNAL—STOP WHEN FLASHING RED (R10-14 or R10-14a) signs (see Section 2B.53) shall be used with emergency-vehicle hybrid beacons.**

Option:

19 If needed for extra emphasis, a STOP HERE ON FLASHING RED (R10-14b) sign (see Section 2B.53) may be installed with an emergency-vehicle hybrid beacon.

CHAPTER 4H. TRAFFIC CONTROL SIGNALS FOR ONE-LANE, TWO-WAY FACILITIES

Section 4H.01 Application of Traffic Control Signals for One-Lane, Two-Way Facilities

Support:

01 A traffic control signal at a narrow bridge, tunnel, or roadway section is a special signal that assigns the right-of-way for vehicles passing over a bridge or through a tunnel or roadway section that is not of sufficient width for two opposing vehicles to pass.

02 Temporary traffic control signals (see Sections 4D.32 and 6F.84) are the most frequent application of one-lane, two-way facilities.

Guidance:

03 *Sight distance across or through the one-lane, two-way facility should be considered as well as the approach speed and sight distance approaching the facility when determining whether traffic control signals should be installed.*

Option:

04 At a narrow bridge, tunnel, or roadway section where a traffic control signal is not justified under the conditions of Chapter 4C, a traffic control signal may be used if gaps in opposing traffic do not permit the flow of traffic through the one-lane section of roadway.

Section 4H.02 Design of Traffic Control Signals for One-Lane, Two-Way Facilities

Standard:

01 **The provisions of Chapter 4D shall apply to traffic control signals for one-lane, two-way facilities, except that:**
   A. **Durations of red clearance intervals shall be adequate to clear the one-lane section of conflicting vehicles.**
   B. **Adequate means, such as interconnection, shall be provided to prevent conflicting signal indications, such as green and green, at opposite ends of the section.**

Section 4H.03 Operation of Traffic Control Signals for One-Lane, Two-Way Facilities

Standard:

01 **Traffic control signals at one-lane, two-way facilities shall operate in a manner consistent with traffic requirements.**

02 **When in the flashing mode, the signal indications shall flash red.**

Guidance:

03 *Adequate time should be provided to allow traffic to clear the narrow facility before opposing traffic is allowed to move. Engineering judgment should be used to determine the proper timing for the signal.*

CHAPTER 4I. TRAFFIC CONTROL SIGNALS FOR FREEWAY ENTRANCE RAMPS

Section 4I.01 Application of Freeway Entrance Ramp Control Signals

Support:

01 Ramp control signals are traffic control signals that control the flow of traffic entering the freeway facility. This is often referred to as "ramp metering."

02 Freeway entrance ramp control signals are sometimes used if controlling traffic entering the freeway could reduce the total expected delay to traffic in the freeway corridor, including freeway ramps and local streets.

Guidance:

03 *The installation of ramp control signals should be preceded by an engineering study of the physical and traffic conditions on the highway facilities likely to be affected. The study should include the ramps and ramp connections and the surface streets that would be affected by the ramp control, as well as the freeway section concerned.*

Support:

04 Information on conditions that might justify freeway entrance ramp control signals, factors to be evaluated in traffic engineering studies for ramp control signals, design of ramp control signals, and operation of ramp control signals can be found in the FHWA's "Ramp Management and Control Handbook" (see Section 1A.11).

Section 4I.02 Design of Freeway Entrance Ramp Control Signals

Standard:

01 **Ramp control signals shall meet all of the standard design specifications for traffic control signals, except as otherwise provided in this Section.**

02 **The signal face for freeway entrance ramp control signals shall be either a two-section signal face containing red and green signal indications or a three-section signal face containing red, yellow, and green signal indications.**

03 **If only one lane is present on an entrance ramp or if more than one lane is present on an entrance ramp and the ramp control signals are operated such that green signal indications are always displayed simultaneously to all of the lanes on the ramp, then a minimum of two signal faces per ramp shall face entering traffic.**

04 **If more than one lane is present on an entrance ramp and the ramp control signals are operated such that green signal indications are not always displayed simultaneously to all of the lanes on the ramp, then one signal face shall be provided over the approximate center of each separately-controlled lane.**

Guidance:

05 *Additional side-mounted signal faces should be considered for ramps with two or more separately-controlled lanes.*

Standard:

06 **Ramp control signals shall be located and designed to minimize their viewing by mainline freeway traffic.**

Option:

07 Ramp control signals may be placed in the dark mode (no indications displayed) when not in use.

08 Ramp control signals may be used to control some, but not all, lanes on a ramp, such as when non-metered HOV bypass lanes are provided on a ramp.

09 The required signal faces, if located at the side of the ramp roadway, may be mounted such that the height above the pavement grade at the center of the ramp roadway to the bottom of the signal housing of the lowest signal face is between 4.5 and 6 feet.

10 For entrance ramps with only one controlled lane, the two required signal faces may both be mounted at the side of the roadway on a single pole, with one face at the normal mounting height and one face mounted lower as provided in Paragraph 9, as a specific exception to the normal 8-foot minimum lateral separation of signal faces required by Section 4D.13.

Guidance:

11 *Regulatory signs with legends appropriate to the control, such as XX VEHICLE(S) PER GREEN or XX VEHICLE(S) PER GREEN EACH LANE (see Section 2B.56), should be installed adjacent to the ramp control signal faces. When ramp control signals are installed on a freeway-to-freeway ramp, special consideration should be given to assuring adequate visibility of the ramp control signals, and multiple advance warning signs with flashing warning beacons should be installed to warn road users of the metered operation.*

Section 4I.03 Operation of Freeway Entrance Ramp Control Signals

Guidance:

01 *Operational strategies for ramp control signals, such as periods of operation, metering rates and algorithms, and queue management, should be determined by the operating agency prior to the installation of the ramp control signals and should be closely monitored and adjusted as needed thereafter.*

02 *When the ramp control signals are operated only during certain periods of the day, a RAMP METERED WHEN FLASHING (W3-8) sign (see Section 2C.37) should be installed in advance of the ramp control signal near the entrance to the ramp, or on the arterial on the approach to the ramp, to alert road users to the presence and operation of ramp meters.*

Standard:

03 **The RAMP METERED WHEN FLASHING sign shall be supplemented with a warning beacon (see Section 4L.03) that flashes when the ramp control signal is in operation.**

CHAPTER 4J. TRAFFIC CONTROL FOR MOVABLE BRIDGES

Section 4J.01 Application of Traffic Control for Movable Bridges

Support:

01 Traffic control signals for movable bridges are a special type of highway traffic signal installed at movable bridges to notify road users to stop because of a road closure rather than alternately giving the right-of-way to conflicting traffic movements. The signals are operated in coordination with the opening and closing of the movable bridge, and with the operation of movable bridge warning and resistance gates, or other devices and features used to warn, control, and stop traffic.

02 Movable bridge warning gates installed at movable bridges decrease the likelihood of vehicles and pedestrians passing the stop line and entering an area where potential hazards exist because of bridge operations.

03 A movable bridge resistance gate is sometimes used at movable bridges and located downstream of the movable bridge warning gate. A movable bridge resistance gate provides a physical deterrent to road users when placed in the appropriate position. The movable bridge resistance gates are considered a design feature and not a traffic control device; requirements for them are contained in AASHTO's "Standard Specifications for Movable Highway Bridges" (see Page i for AASHTO's address).

Standard:

04 **Traffic control at movable bridges shall include both signals and gates, except in the following cases:**
  A. Neither is required if other traffic control devices or measures considered appropriate are used under either of the following conditions:
    1. On low-volume roads (roads of less than 400 vehicles average daily traffic), or
    2. At manually operated bridges if electric power is not available.
  B. Only signals are required in urban areas if intersecting streets or driveways make gates ineffective.
  C. Only movable bridge warning gates are required if a traffic control signal that is controlled as part of the bridge operations exists within 500 feet of the movable bridge resistance gates and no intervening traffic entrances exist.

Section 4J.02 Design and Location of Movable Bridge Signals and Gates

Standard:

01 **The signal faces and mountings of movable bridge signals shall comply with the provisions of Chapter 4D except as provided in this Section.**

02 **Signal faces with 12-inch diameter signal indications shall be used for all new movable bridge signals.**

Option:

03 Existing signal faces with 8-inch diameter lenses may be retained for the remainder of their useful service life.

Standard:

04 **Since movable bridge operations cover a variable range of time periods between openings, the signal faces shall be one of the following types:**
  A. Three-section signal faces with red, yellow, and green signal indications; or
  B. Two one-section signal faces with red signal indications in a vertical array separated by a STOP HERE ON RED (R10-6) sign (see Section 2B.53).

05 **Regardless of which signal type is selected, at least two signal faces shall be provided for each approach to the movable span and a stop line (see Section 3B.16) shall be installed to indicate the point behind which vehicles are required to stop.**

Guidance:

06 *If movable bridge operation is frequent, the use of three-section signal faces should be considered.*

07 *Insofar as practical, the height and lateral placement of signal faces should comply with the requirements for other traffic control signals in accordance with Chapter 4D. They should be located no more than 50 feet in advance of the movable bridge warning gate.*

Option:

08 Movable bridge signals may be supplemented with audible warning devices to provide additional warning to drivers and pedestrians.

Standard:

09 **A DRAW BRIDGE (W3-6) sign (see Section 2C.39) shall be used in advance of movable bridge signals and gates to give warning to road users, except in urban conditions where such signing would not be practical.**

Standard:

10   If physical conditions prevent a road user from having a continuous view of at least two signal indications for the distance specified in Table 4D-2, an auxiliary device (either a supplemental signal face or the mandatory DRAW BRIDGE (W3-6) sign to which has been added a warning beacon that is interconnected with the movable bridge controller unit) shall be provided in advance of movable bridge signals and gates.

Option:

11   The DRAW BRIDGE (W3-6) sign may be supplemented by a Warning Beacon (see Section 4L.03).

Standard:

12   If two sets of gates (both a warning and a resistance gate) are used for a single direction, highway traffic signals shall not be required to accompany the resistance gate nearest the span opening.

13   Movable bridge warning gates, if used, shall be at least standard railroad size, striped with 16-inch alternate vertical, fully reflectorized red and white stripes. Flashing red lights in accordance with the Standards for those on railroad gates (see Section 8C.04) shall be included on the gate arm and they shall only be operated if the gate is closed or in the process of being opened or closed. In the horizontal position, the top of the gate shall be approximately 4 feet above the pavement.

Guidance:

14   *Movable bridge warning gates should be of lightweight construction. In its normal upright position, the gate arm should provide adequate lateral clearance.*

Option:

15   The movable bridge resistance gates may be delineated, if practical, in a manner similar to the movable bridge warning gate.

Standard:

16   Movable bridge warning gates, if used, shall extend at least across the full width of the approach lanes if movable bridge resistance gates are used. On divided highways in which the roadways are separated by a barrier median, movable bridge warning gates, if used, shall extend across all roadway lanes approaching the span openings.

Guidance:

17   *If movable bridge resistance gates are not used on undivided highways, movable bridge warning gates, if used, should extend across the full width of the roadway.*

Option:

18   A single full-width gate or two half-width gates may be used.

Support:

19   The locations of movable bridge signals and gates are determined by the location of the movable bridge resistance gate (if used) rather than by the location of the movable spans. The movable bridge resistance gates for high-speed highways are preferably located 50 feet or more from the span opening except for bascule and lift bridges, where they are often attached to, or are a part of, the structure.

Standard:

20   Except where physical conditions make it impractical, movable bridge warning gates shall be located 100 feet or more from the movable bridge resistance gates or, if movable bridge resistance gates are not used, 100 feet or more from the movable span.

Guidance:

21   *On bridges or causeways that cross a long reach of water and that might be hit by large marine vessels, within the limits of practicality, traffic should not be halted on a section of the bridge or causeway that is subject to impact.*

22   *In cases where it is not practical to halt traffic on a span that is not subject to impact, traffic should be halted at least one span from the opening. If traffic is halted by signals and gates more than 330 feet from the movable bridge warning gates (or from the span opening if movable bridge warning gates are not used), a second set of gates should be installed approximately 100 feet from the gate or span opening.*

23   *If the movable bridge is close to a grade crossing and traffic might possibly be stopped on the crossing as a result of the bridge opening, a traffic control device should notify the road users to not stop on the railroad tracks.*

Section 4J.03 Operation of Movable Bridge Signals and Gates

Standard:

01 Traffic control devices at movable bridges shall be coordinated with the movable spans, so that the signals, gates, and movable spans are controlled by the bridge tender through an interlocked control.

02 If the three-section type of signal face is used, the green signal indication shall be displayed at all times between bridge openings, except that if the bridge is not expected to open during continuous periods in excess of 5 hours, a flashing yellow signal indication shall be permitted to be used. The signal shall display a steady red signal indication when traffic is required to stop. The duration of the yellow change interval between the display of the green and steady red signal indications, or flashing yellow and steady red signal indications, shall be determined using engineering practices (see Section 4D.26).

03 If the vertical array of red signal indications is the type of signal face selected, the red signal indications shall flash alternately only when traffic is required to stop.

Guidance:

04 *The yellow change interval should have a minimum duration of 3 seconds and a maximum duration of 6 seconds. The longer intervals should be reserved for use on approaches with higher speeds.*

05 *Traffic control signals on adjacent streets and highways should be interconnected with the drawbridge control if indicated by engineering judgment. When such interconnection is provided, the traffic control signals at adjacent intersections should be preempted by the operation of the movable bridge in the manner described in Section 4D.27.*

CHAPTER 4K. HIGHWAY TRAFFIC SIGNALS AT TOLL PLAZAS

Section 4K.01 Traffic Signals at Toll Plazas
Standard:

01 **Traffic control signals or devices that closely resemble traffic control signals that use red or green circular indications shall not be used at toll plazas to indicate the open or closed status of the toll plaza lanes.**

Guidance:

02 *Traffic control signals or devices that closely resemble traffic control signals that use red or green circular indications should not be used for new or reconstructed installations at toll plazas to indicate the success or failure of electronic toll payments or to alternately direct drivers making cash toll payments to stop and then proceed.*

Section 4K.02 Lane-Use Control Signals at or Near Toll Plazas
Standard:

01 **Lane-use control signals used at toll plazas shall comply with the provisions of Chapter 4M except as otherwise provided in this Section.**

02 **At toll plazas with multiple lanes where one or more lanes is sometimes closed to traffic, a lane-use control signal shall be installed above the center of each toll plaza lane to indicate the open or closed status of the controlled lane.**

Option:

03 The bottom of the signal housing of a lane-use control signal above a toll plaza lane having a canopy may be mounted lower than 15 feet above the pavement, but not lower than the vertical clearance of the canopy structure.

04 Lane-use control signals may also be used to indicate the open or closed status of an Open-Road ETC lane as a supplement to other devices used for the temporary closure of a lane (see Part 6).

Section 4K.03 Warning Beacons at Toll Plazas
Standard:

01 **Warning Beacons used at toll plazas shall comply with the provisions of Chapter 4L except as otherwise provided in this Section.**

Guidance:

02 *Warning Beacons, if used with a toll plaza canopy sign (see Section 2F.16) to assist drivers of such vehicles in locating the dedicated ETC Account-Only lane(s), should be installed in a manner such that the beacons are distinctly separate from the lane-use control signals (see Section 4M.01) for the toll plaza lane.*

Option:

03 Warning Beacons that are mounted on toll plaza islands, behind impact attenuators in front of toll plaza islands, and/or on toll booth pylons (ramparts) to identify them as objects in the roadway may be mounted at a height that is appropriate for viewing in a toll plaza context, even if that height is lower than the normal minimum of 8 feet above the pavement.

CHAPTER 4L. FLASHING BEACONS

Section 4L.01 General Design and Operation of Flashing Beacons

Support:

01 A Flashing Beacon is a highway traffic signal with one or more signal sections that operates in a flashing mode. It can provide traffic control when used as an intersection control beacon (see Section 4L.02) or it can provide warning when used in other applications (see Sections 4L.03, 4L.04, and 4L.05).

Standard:

02 **Flashing Beacon units and their mountings shall comply with the provisions of Chapter 4D, except as otherwise provided in this Chapter.**

03 **Beacons shall be flashed at a rate of not less than 50 or more than 60 times per minute. The illuminated period of each flash shall be a minimum of 1/2 and a maximum of 2/3 of the total cycle.**

04 **A beacon shall not be included within the border of a sign except for SCHOOL SPEED LIMIT sign beacons (see Sections 4L.04 and 7B.15).**

Guidance:

05 *If used to supplement a warning or regulatory sign, the edge of the beacon signal housing should normally be located no closer than 12 inches outside of the nearest edge of the sign.*

Option:

06 An automatic dimming device may be used to reduce the brilliance of flashing yellow signal indications during night operation.

Section 4L.02 Intersection Control Beacon

Standard:

01 **An Intersection Control Beacon shall consist of one or more signal faces directed toward each approach to an intersection. Each signal face shall consist of one or more signal sections of a standard traffic signal face, with flashing CIRCULAR YELLOW or CIRCULAR RED signal indications in each signal face. They shall be installed and used only at an intersection to control two or more directions of travel.**

02 **Application of Intersection Control Beacon signal indications shall be limited to the following:**
   A. **Yellow on one route (normally the major street) and red for the remaining approaches, and**
   B. **Red for all approaches (if the warrant described in Section 2B.07 for a multi-way stop is satisfied).**

03 **Flashing yellow signal indications shall not face conflicting vehicular approaches.**

04 **A STOP sign shall be used on approaches to which a flashing red signal indication is displayed on an Intersection Control Beacon (see Section 2B.04).**

05 **If two horizontally aligned red signal indications are used on an approach for an Intersection Control Beacon, they shall be flashed simultaneously to avoid being confused with grade crossing flashing-light signals. If two vertically aligned red signal indications are used on an approach for an Intersection Control Beacon, they shall be flashed alternately.**

Guidance:

06 *An Intersection Control Beacon should not be mounted on a pedestal in the roadway unless the pedestal is within the confines of a traffic or pedestrian island.*

Option:

07 Supplemental signal indications may be used on one or more approaches in order to provide adequate visibility to approaching road users.

08 Intersection Control Beacons may be used at intersections where traffic or physical conditions do not justify conventional traffic control signals but crash rates indicate the possibility of a special need.

09 An Intersection Control Beacon is generally located over the center of an intersection; however, it may be used at other suitable locations.

Section 4L.03 Warning Beacon

Support:

01 Typical applications of Warning Beacons include the following:
   A. At obstructions in or immediately adjacent to the roadway;
   B. As supplemental emphasis to warning signs;
   C. As emphasis for midblock crosswalks;

D. As supplemental emphasis to regulatory signs, except STOP, DO NOT ENTER, WRONG WAY, and SPEED LIMIT signs; and
E. In conjunction with a regulatory or warning sign that includes the phrase WHEN FLASHING in its legend to indicate that the regulation is in effect or that the condition is present only at certain times.

Standard:

02 **A Warning Beacon shall consist of one or more signal sections of a standard traffic signal face with a flashing CIRCULAR YELLOW signal indication in each signal section.**

03 **A Warning Beacon shall be used only to supplement an appropriate warning or regulatory sign or marker.**

04 **Warning Beacons, if used at intersections, shall not face conflicting vehicular approaches.**

05 **If a Warning Beacon is suspended over the roadway, the clearance above the pavement shall be a minimum of 15 feet and a maximum of 19 feet.**

Guidance:

06 *The condition or regulation justifying Warning Beacons should largely govern their location with respect to the roadway.*

07 *If an obstruction is in or adjacent to the roadway, illumination of the lower portion or the beginning of the obstruction or a sign on or in front of the obstruction, in addition to the beacon, should be considered.*

08 *Warning Beacons should be operated only during those periods or times when the condition or regulation exists.*

Option:

09 Warning Beacons that are actuated by pedestrians, bicyclists, or other road users may be used as appropriate to provide additional warning to vehicles approaching a crossing or other location.

10 If Warning Beacons have more than one signal section, they may be flashed either alternately or simultaneously.

11 A flashing yellow beacon interconnected with a traffic signal controller assembly may be used with a traffic signal warning sign (see Section 2C.36).

Section 4L.04 Speed Limit Sign Beacon

Standard:

01 **A Speed Limit Sign Beacon shall be used only to supplement a Speed Limit sign.**

02 **A Speed Limit Sign Beacon shall consist of one or more signal sections of a standard traffic control signal face, with a flashing CIRCULAR YELLOW signal indication in each signal section. The signal indications shall have a nominal diameter of not less than 8 inches. If two signal indications are used, they shall be vertically aligned, except that they shall be permitted to be horizontally aligned if the Speed Limit (R2-1) sign is longer horizontally than vertically. If two signal indications are used, they shall be alternately flashed.**

Option:

03 A Speed Limit Sign Beacon may be used with a fixed or variable Speed Limit sign. If applicable, a flashing Speed Limit Sign Beacon (with an appropriate accompanying sign) may be used to indicate that the displayed speed limit is in effect.

04 A Speed Limit Sign Beacon may be included within the border of a School Speed Limit (S5-1) sign (see Section 7B.15).

Section 4L.05 Stop Beacon

Standard:

01 **A Stop Beacon shall be used only to supplement a STOP sign, a DO NOT ENTER sign, or a WRONG WAY sign.**

02 **A Stop Beacon shall consist of one or more signal sections of a standard traffic signal face with a flashing CIRCULAR RED signal indication in each signal section. If two horizontally aligned signal indications are used for a Stop Beacon, they shall be flashed simultaneously to avoid being confused with grade crossing flashing-light signals. If two vertically aligned signal indications are used for a Stop Beacon, they shall be flashed alternately.**

03 **The bottom of the signal housing of a Stop Beacon shall be not less than 12 inches or more than 24 inches above the top of a STOP sign, a DO NOT ENTER sign, or a WRONG WAY sign.**

CHAPTER 4M. LANE-USE CONTROL SIGNALS

Section 4M.01 Application of Lane-Use Control Signals

Support:

01　Lane-use control signals are special overhead signals that permit or prohibit the use of specific lanes of a street or highway or that indicate the impending prohibition of their use. Lane-use control signals are distinguished by placement of special signal faces over a certain lane or lanes of the roadway and by their distinctive shapes and symbols. Supplementary signs are sometimes used to explain their meaning and intent.

02　Lane-use control signals are most commonly used for reversible-lane control, but are also used in certain non-reversible lane applications and for toll plaza lanes (see Section 4K.02).

Guidance:

03　*An engineering study should be conducted to determine whether a reversible-lane operation can be controlled satisfactorily by static signs (see Section 2B.26) or whether lane-use control signals are necessary. Lane-use control signals should be used to control reversible-lane operations if any of the following conditions are present:*

　　A. *More than one lane is reversed in direction;*
　　B. *Two-way or one-way left turns are allowed during peak-period reversible operations, but those turns are from a different lane than used during off-peak periods;*
　　C. *Other unusual or complex operations are included in the reversible-lane pattern;*
　　D. *Demonstrated crash experience occurring with reversible-lane operation controlled by static signs that can be corrected by using lane-use control signals at the times of transition between peak and off-peak patterns; and/or*
　　E. *An engineering study indicates that the safety and efficiency of the traffic operations of a reversible-lane system would be improved by lane-use control signals.*

Standard:

04　**Pavement markings (see Section 3B.03) shall be used in conjunction with reversible-lane control signals.**

Option:

05　Lane-use control signals may also be used if there is no intent or need to reverse lanes, but there is a need to indicate the open or closed status of one or more lanes, such as:

　　A. On a freeway, if it is desired to close certain lanes at certain hours to facilitate the merging of traffic from a ramp or other freeway;
　　B. On a freeway, near its terminus, to indicate a lane that ends;
　　C. On a freeway or long bridge, to indicate that a lane may be temporarily blocked by a crash, breakdown, construction or maintenance activities, or similar temporary conditions; and
　　D. On a conventional road or driveway, at access or egress points to or from a facility, such as a parking garage, where one or more lanes of the access or egress are opened or closed at various times.

Section 4M.02 Meaning of Lane-Use Control Signal Indications

Standard:

01　**The meanings of lane-use control signal indications shall be as follows:**

　　A. **A steady DOWNWARD GREEN ARROW signal indication shall mean that a road user is permitted to drive in the lane over which the arrow signal indication is located.**
　　B. **A steady YELLOW X signal indication shall mean that a road user is to prepare to vacate the lane over which the signal indication is located because a lane control change is being made to a steady RED X signal indication.**
　　C. **A steady WHITE TWO-WAY LEFT-TURN ARROW signal indication (see Figure 4M-1) shall mean that a road user is permitted to use a lane over which the signal indication is located for a left turn, but not for through travel, with the understanding that common use of the lane by oncoming road users for left turns is also permitted.**
　　D. **A steady WHITE ONE WAY LEFT-TURN ARROW signal indication (see Figure 4M-1) shall mean that a road user is permitted to use a lane over which the signal indication is located for a left turn (without opposing turns in the same lane), but not for through travel.**
　　E. **A steady RED X signal indication shall mean that a road user is not permitted to use the lane over which the signal indication is located and that this signal indication shall modify accordingly the meaning of other traffic controls present.**

Figure 4M-1. Left-Turn Lane-Use Control Signals

Two-way left-turn arrow

One-way left-turn arrow

White arrows on an opaque 30 x 30-inch background

Section 4M.03 Design of Lane-Use Control Signals

Standard:

01   All lane-use control signal indications shall be in units with rectangular signal faces and shall have opaque backgrounds. Nominal minimum height and width of each DOWNWARD GREEN ARROW, YELLOW X, and RED X signal face shall be 18 inches for typical applications. The WHITE TWO-WAY LEFT-TURN ARROW and WHITE ONE WAY LEFT-TURN ARROW signal faces shall have a nominal minimum height and width of 30 inches.

02   Each lane to be reversed or closed shall have signal faces with a DOWNWARD GREEN ARROW and a RED X symbol.

03   Each reversible lane that also operates as a two-way or one-way left-turn lane during certain periods shall have signal faces that also include the applicable WHITE TWO-WAY LEFT-TURN ARROW or WHITE ONE WAY LEFT-TURN ARROW symbol.

04   Each non-reversible lane immediately adjacent to a reversible lane shall have signal indications that display a DOWNWARD GREEN ARROW to traffic traveling in the permitted direction and a RED X to traffic traveling in the opposite direction.

05   If in separate signal sections, the relative positions, from left to right, of the signal indications shall be RED X, YELLOW X, DOWNWARD GREEN ARROW, WHITE TWO-WAY LEFT-TURN ARROW, WHITE ONE WAY LEFT-TURN ARROW.

06   The color of lane-use control signal indications shall be clearly visible for 2,300 feet at all times under normal atmospheric conditions, unless otherwise physically obstructed.

07   Lane-use control signal faces shall be located approximately over the center of the lane controlled.

08   If the area to be controlled is more than 2,300 feet in length, or if the vertical or horizontal alignment is curved, intermediate lane-use control signal faces shall be located over each controlled lane at frequent intervals. This location shall be such that road users will at all times be able to see at least one signal indication and preferably two along the roadway, and will have a definite indication of the lanes specifically reserved for their use.

09   All lane-use control signal faces shall be located in a straight line across the roadway approximately at right angles to the roadway alignment.

10   On roadways having intersections controlled by traffic control signals, the lane-use control signal face shall be located sufficiently far in advance of or beyond such traffic control signals to prevent them from being misconstrued as traffic control signals.

11   Except as provided in Paragraph 12, the bottom of the signal housing of any lane-use control signal face shall be a minimum of 15 feet and a maximum of 19 feet above the pavement grade.

Option:

12   The bottom of a lane-use control signal housing may be lower than 15 feet above the pavement if it is mounted on a canopy or other structure over the pavement, but not lower than the vertical clearance of the structure.

13   Except for lane-use control signals at toll plazas (see Section 4K.02), in areas with minimal visual clutter and with speeds of less than 40 mph, lane-use control signal faces with nominal height and width of 12 inches may be used for the DOWNWARD GREEN ARROW, YELLOW X, and RED X signal faces, and lane-use control signal faces with nominal height and width of 18 inches may be used for the WHITE TWO-WAY LEFT-TURN ARROW and WHITE ONE-WAY LEFT-TURN ARROW signal faces.

14 Other sizes of lane-use control signal faces larger than 18 inches with message recognition distances appropriate to signal spacing may be used for the DOWNWARD GREEN ARROW, YELLOW X, and RED X signal faces.

15 Non-reversible lanes not immediately adjacent to a reversible lane on any street so controlled may also be provided with signal indications that display a DOWNWARD GREEN ARROW to traffic traveling in the permitted direction and a RED X to traffic traveling in the opposite direction.

16 The signal indications provided for each lane may be in separate signal sections or may be superimposed in the same signal section.

Section 4M.04 Operation of Lane-Use Control Signals

Standard:

01 All lane-use control signals shall be coordinated so that all the signal indications along the controlled section of roadway are operated uniformly and consistently. The lane-use control signal system shall be designed to reliably guard against showing any prohibited combination of signal indications to any traffic at any point in the controlled lanes.

02 For reversible-lane control signals, the following combination of signal indications shall not be simultaneously displayed over the same lane to both directions of travel:
   A. DOWNWARD GREEN ARROW in both directions,
   B. YELLOW X in both directions,
   C. WHITE ONE WAY LEFT-TURN ARROW in both directions,
   D. DOWNWARD GREEN ARROW in one direction and YELLOW X in the other direction,
   E. WHITE TWO-WAY LEFT-TURN ARROW or WHITE ONE WAY LEFT-TURN ARROW in one direction and DOWNWARD GREEN ARROW in the other direction,
   F. WHITE TWO-WAY LEFT-TURN ARROW in one direction and WHITE ONE WAY LEFT-TURN ARROW in the other direction, and
   G. WHITE ONE WAY LEFT-TURN ARROW in one direction and YELLOW X in the other direction.

03 A moving condition in one direction shall be terminated either by the immediate display of a RED X signal indication or by a YELLOW X signal indication followed by a RED X signal indication. In either case, the duration of the RED X signal indication shall be sufficient to allow clearance of the lane before any moving condition is allowed in the opposing direction.

04 Whenever a DOWNWARD GREEN ARROW signal indication is changed to a WHITE TWO-WAY LEFT-TURN ARROW signal indication, the RED X signal indication shall continue to be displayed to the opposite direction of travel for an appropriate duration to allow traffic time to vacate the lane being converted to a two-way left-turn lane.

05 If an automatic control system is used, a manual control to override the automatic control shall be provided.

Guidance:

06 *The type of control provided for reversible-lane operation should be such as to permit either automatic or manual operation of the lane-use control signals.*

Standard:

07 If used, lane-use control signals shall be operated continuously, except that lane-use control signals that are used only for special events or other infrequent occurrences and lane-use control signals on non-reversible freeway lanes shall be permitted to be darkened when not in operation. The change from normal operation to non-operation shall occur only when the lane-use control signals display signal indications that are appropriate for the lane use that applies when the signals are not operated. The lane-use control signals shall display signal indications that are appropriate for the existing lane use when changed from non-operation to normal operations. Also, traffic control devices shall clearly indicate the proper lane use when the lane control signals are not in operation.

Support:

08 Section 2B.26 contains additional information concerning considerations involving left-turn prohibitions in conjunction with reversible-lane operations.

CHAPTER 4N. IN-ROADWAY LIGHTS

Section 4N.01 Application of In-Roadway Lights

Support:

01   In-Roadway Lights are special types of highway traffic signals installed in the roadway surface to warn road users that they are approaching a condition on or adjacent to the roadway that might not be readily apparent and might require the road users to slow down and/or come to a stop. This includes situations warning of marked school crosswalks, marked midblock crosswalks, marked crosswalks on uncontrolled approaches, marked crosswalks in advance of roundabouts as described in Chapter 3C, and other roadway situations involving pedestrian crossings.

Standard:

02   **In-Roadway Lights shall not be used for any application that is not described in this Chapter.**

03   **If used, In-Roadway Lights shall not exceed a height of 3/4 inch above the roadway surface.**

04   **When used, In-Roadway Lights shall be flashed and shall not be steadily illuminated.**

Support:

05   Steadily illuminated lights installed in the roadway surface are considered to be internally illuminated raised pavement markers (see Section 3B.11).

Option:

06   In-Roadway Lights may be flashed in a manner that includes a continuous flash of varying intensity and time duration that is repeated to provide a flickering effect (see Section 4N.02).

Section 4N.02 In-Roadway Warning Lights at Crosswalks

Option:

01   In-roadway lights may be installed at certain marked crosswalks, based on an engineering study or engineering judgment, to provide additional warning to road users.

Standard:

02   **If used, In-Roadway Warning Lights at crosswalks shall be installed only at marked crosswalks with applicable warning signs. They shall not be used at crosswalks controlled by YIELD signs, STOP signs, or traffic control signals.**

03   **If In-Roadway Warning Lights are used at a crosswalk, the following requirements shall apply:**
   A. **Except as provided in Paragraphs 7 and 8, they shall be installed along both sides of the crosswalk and shall span its entire length.**
   B. **They shall initiate operation based on pedestrian actuation and shall cease operation at a predetermined time after the pedestrian actuation or, with passive detection, after the pedestrian clears the crosswalk.**
   C. **They shall display a flashing yellow light when actuated. The flash rate shall be at least 50, but no more than 60, flash periods per minute. If they are flashed in a manner that includes a continuous flash of varying intensity and time duration that is repeated to provide a flickering effect, the flickers or pulses shall not repeat at a rate that is between 5 and 30 per second to avoid frequencies that might cause seizures.**
   D. **They shall be installed in the area between the outside edge of the crosswalk line and 10 feet from the outside edge of the crosswalk.**
   E. **They shall face away from the crosswalk if unidirectional, or shall face away from and across the crosswalk if bidirectional.**

04   **If used on one-lane, one-way roadways, a minimum of two In-Roadway Warning Lights shall be installed on the approach side of the crosswalk. If used on two-lane roadways, a minimum of three In-Roadway Warning Lights shall be installed along both sides of the crosswalk. If used on roadways with more than two lanes, a minimum of one In-Roadway Warning Light per lane shall be installed along both sides of the crosswalk.**

Guidance:

05   *If used, In-Roadway Warning Lights should be installed in the center of each travel lane, at the center line of the roadway, at each edge of the roadway or parking lanes, or at other suitable locations away from the normal tire track paths.*

06   *The location of the In-Roadway Warning Lights within the lanes should be based on engineering judgment.*

Option:

07 On one-way streets, In-Roadway Warning Lights may be omitted on the departure side of the crosswalk.

08 Based on engineering judgment, the In-Roadway Warning Lights on the departure side of the crosswalk on the left side of a median may be omitted.

09 Unidirectional In-Roadway Warning Lights installed at crosswalk locations may have an optional, additional yellow light indication in each unit that is visible to pedestrians in the crosswalk to indicate to pedestrians in the crosswalk that the In-Roadway Warning Lights are in fact flashing as they cross the street. These yellow lights may flash with and at the same flash rate as the light module in which each is installed.

Guidance:

10 *If used, the period of operation of the In-Roadway Warning Lights following each actuation should be sufficient to allow a pedestrian crossing in the crosswalk to leave the curb or shoulder and travel at a walking speed of 3.5 feet per second to at least the far side of the traveled way or to a median of sufficient width for pedestrians to wait. Where pedestrians who walk slower than 3.5 feet per second, or pedestrians who use wheelchairs, routinely use the crosswalk, a walking speed of less than 3.5 feet per second should be considered in determining the period of operation.*

Standard:

11 **If pedestrian pushbuttons are used to actuate the in-roadway lights, a Push Button To Turn On Warning Lights (with pushbutton symbol) (R10-25) sign (see Figure 2B-26) shall be mounted adjacent to or integral with each pedestrian pushbutton.**

12 **Where the period of operation is sufficient only for crossing from a curb or shoulder to a median of sufficient width for pedestrians to wait, median-mounted pedestrian actuators shall be provided.**

(This page left intentionally blank)

PART 5
TRAFFIC CONTROL DEVICES FOR LOW-VOLUME ROADS

CHAPTER 5A. GENERAL

Section 5A.01 Function
Standard:

01 **A low-volume road shall be defined for this Part of the Manual as follows:**
   A. **A low-volume road shall be a facility lying outside of built-up areas of cities, towns, and communities, and it shall have a traffic volume of less than 400 AADT.**
   B. **A low-volume road shall not be a freeway, an expressway, an interchange ramp, a freeway service road, a road on a designated State highway system, or a residential street in a neighborhood. In terms of highway classification, it shall be a variation of a conventional road or a special purpose road as defined in Section 1A.13.**
   C. **A low-volume road shall be classified as either paved or unpaved.**

Support:

02 Low-volume roads typically include agricultural, recreational, resource management and development such as mining and logging and grazing, and local roads in rural areas.

Guidance:

03 *The needs of unfamiliar road users for occasional, recreational, and commercial transportation purposes should be considered.*

Support:

04 At some locations on low-volume roads, the use of traffic control devices might be needed to provide the road user limited, but essential, information regarding regulation, guidance, and warning.

05 Other Parts of this Manual contain provisions applicable to all low-volume roads; however, Part 5 specifically supplements and references the provisions for traffic control devices commonly used on low-volume roads.

Section 5A.02 Application
Support:

01 It is possible, in many cases, to provide essential information to road users on low-volume roads with a limited number of traffic control devices. The focus might be on devices that:
   A. Warn of conditions not normally encountered,
   B. Prohibit unsafe movements, or
   C. Provide minimal destination guidance.

Standard:

02 **The provisions contained in Part 5 shall not prohibit the installation or the full application of traffic control devices on a low-volume road where conditions justify their use.**

Guidance:

03 *Additional traffic control devices and provisions contained in other Parts of the Manual should be considered for use on low-volume roads.*

Support:

04 Section 1A.09 contains information regarding the assistance that is available to jurisdictions that do not have engineers on their staffs who are trained and/or experienced in traffic control devices.

Section 5A.03 Design
Standard:

01 **Traffic control devices for use on low-volume roads shall be designed in accordance with the provisions contained in Part 5, and where required, in other applicable Parts of this Manual.**

02 **The typical sizes for signs and plaques installed on low-volume roads shall be as shown in Table 5A-1. The sizes in the minimum column shall only be used on low-volume roads where the 85th-percentile speed or posted speed limit is less than 35 mph.**

Guidance:

03 *The sizes in the oversized column should be used where engineering judgment indicates a need based on high vehicle operating speeds, driver expectancy, traffic operations, or roadway conditions.*

Option:

04 Signs and plaques larger than those shown in Table 5A-1 may be used (see Section 2A.11).

Table 5A-1. Sign and Plaque Sizes on Low-Volume Roads (Sheet 1 of 2)

Sign or Plaque	Sign Designation	Section	Sign Sizes Typical	Sign Sizes Minimum	Sign Sizes Oversized
Stop	R1-1	5B.02	30 x 30	—	36 x 36
Yield	R1-2	5B.02	30 x 30 x 30	—	36 x 36 x 36
Speed Limit (English)	R2-1	5B.03	24 x 30	18 x 24	36 x 48
Do Not Pass	R4-1	5B.04	24 x 30	—	36 x 48
Pass With Care	R4-2	5B.04	24 x 30	18 x 24	36 x 48
Keep Right	R4-7	5B.04	24 x 30	18 x 24	36 x 48
Do Not Enter	R5-1	5B.04	30 x 30	—	36 x 36
No Trucks	R5-2	5B.04	24 x 24	—	30 x 30
One Way	R6-2	5B.04	18 x 24	—	24 x 30
No Parking (symbol)	R8-3	5B.05	24 x 24	18 x 18	30 x 30
No Parking	R8-3a	5B.05	18 x 24	—	24 x 30
No Parking (plaque)	R8-3cP,3dP	5B.05	24 x 18	18 x 12	30 x 24
Road Closed	R11-2	5B.04	48 x 30	—	—
Road Closed, Local Traffic Only	R11-3a	5B.04	60 x 30	—	—
Bridge Out, Local Traffic Only	R11-3b	5B.04	60 x 30	—	—
Road Closed to Thru Traffic	R11-4	5B.04	60 x 30	—	—
Weight Limit	R12-1	5B.04	24 x 30	—	36 x 48
Grade Crossing (Crossbuck)	R15-1	5F.02	48 x 9	—	—
Number of Tracks (plaque)	R15-2P	5F.02	27 x 18	—	—
Horizontal Alignment	W1-1,2,3,4,5	5C.02	30 x 30	—	36 x 36
One-Direction Large Arrow	W1-6	5C.02	36 x 18	—	48 x 24
Two-Direction Large Arrow	W1-7	5C.02	36 x 18	—	48 x 24
Chevron Alignment	W1-8	5C.02	12 x 18	—	18 x 24
Intersection Warning	W2-1,2,3,4,5,6	5C.03	30 x 30	—	36 x 36
Stop Ahead	W3-1	5C.04	30 x 30	—	36 x 36
Yield Ahead	W3-2	5C.04	30 x 30	—	36 x 36
Be Prepared to Stop	W3-4	5G.05	36 x 36	—	48 x 48
Narrow Bridge	W5-2	5C.05	30 x 30	—	36 x 36
One Lane Bridge	W5-3	5C.06	30 x 30	—	36 x 36
Hill	W7-1	5C.07	30 x 30	—	36 x 36
XX % Grade (plaque)	W7-3P	5C.07	24 x 18	—	30 x 24
Next XX Miles (plaque)	W7-3aP	5C.09	24 x 18	—	30 x 24
Pavement Ends	W8-3	5C.08	30 x 30	—	36 x 36
Truck Crossing	W8-6	5C.09	30 x 30	—	36 x 36
Loose Gravel	W8-7	5G.05	30 x 30	—	36 x 36
Rough Road	W8-8	5G.05	30 x 30	—	36 x 36
Road May Flood	W8-18	5G.05	30 x 30	—	36 x 36
Grade Crossing Advance Warning	W10-1	5F.03	30 Dia.	—	36 Dia.
Grade Crossing Advance Warning	W10-2,3,4	5F.03	30 x 30	—	36 x 36
Trains May Exceed 80 mph	W10-8	5F.06	30 x 30	—	36 x 36
Storage Space Symbol	W10-11	5F.06	30 x 30	—	36 x 36
Skewed Crossing	W10-12	5F.06	30 x 30	—	36 x 36
Entering/Crossing	W11 Series	5C.09	30 x 30	—	36 x 36
Advisory Speed (plaque)	W13-1P	5C.10	18 x 18	—	24 x 24
Dead End/No Outlet	W14-1,2	5C.11	30 x 30	—	36 x 36
Dead End/No Outlet	W14-1a,2a	5C.11	36 x 9	24 x 6	—

Table 5A-1. Sign and Plaque Sizes on Low-Volume Roads (Sheet 2 of 2)

Sign or Plaque	Sign Designation	Section	Sign Sizes Typical	Sign Sizes Minimum	Sign Sizes Oversized
No Passing Zone (pennant)	W14-3	5G.05	40 x 40 x 30	—	48 x 48 x 36
Supplemental Distance (plaque)	W16-2P	5C.09	24 x 18	18 x 12	30 x 24
Diagonal Arrow (plaque)	W16-7P	5C.09	24 x 12	—	30 x 18
Ahead (plaque)	W16-9P	5C.09	24 x 12	—	30 x 18
No Traffic Signs	W18-1	5C.12	30 x 30	24 x 24	36 x 36
Road Work (with distance)	W20-1	5G.05	36 x 36	—	48 x 48
Road Closed (with distance)	W20-3	5G.05	36 x 36	—	48 x 48
One Lane Road (with distance)	W20-4	5G.05	36 x 36	—	48 x 48
Flagger	W20-7	5G.05	36 x 36	—	48 x 48
Workers	W21-1	5G.05	36 x 36	—	48 x 48
Fresh Oil	W21-2	5G.05	30 x 30	—	48 x 48
Road Machinery Ahead	W21-3	5G.05	30 x 30	—	48 x 48
Shoulder Work	W21-5	5G.05	36 x 36	—	48 x 48
Survey Crew	W21-6	5G.05	36 x 36	—	48 x 48
Utility Work (with distance)	W21-7	5G.05	36 x 36	—	48 x 48

Notes: 1. Larger sizes may be used when appropriate
2. Dimensions are shown in inches and are shown as width x height

Standard:

05 All signs shall be retroreflective or illuminated to show the same shape and similar color both day and night, unless specifically stated otherwise in other applicable Parts of this Manual. The requirements for sign illumination shall not be considered to be satisfied by street, highway, or strobe lighting.

06 All markings shall be visible at night and shall be retroreflective unless ambient illumination provides adequate visibility of the markings.

Section 5A.04 Placement

Standard:

01 Except as provided in Paragraph 3, the traffic control devices used on low-volume roads shall be placed and positioned in accordance with the lateral, longitudinal, and vertical placement provisions contained in Part 2 and other applicable Sections of this Manual.

Guidance:

02 *The placement of warning signs should comply with the guidance contained in Section 2C.05 and other applicable Sections of this Manual.*

Option:

03 A lateral offset of not less than 2 feet from the roadway edge to the roadside edge of a sign may be used where roadside features such as terrain, shrubbery, and/or trees prevent lateral placement in accordance with Section 2A.19.

Standard:

04 If located within a clear zone, post-mounted sign supports shall be yielding, breakaway, or shielded with a longitudinal barrier or crash cushion as required in Section 2A.19.

CHAPTER 5B. REGULATORY SIGNS

Section 5B.01 Introduction
Support:
01 The purpose of a regulatory sign is to inform highway users of traffic laws or regulations, and to indicate the applicability of legal requirements that would not otherwise be apparent.
02 The provisions for regulatory signs are contained in Chapter 2B and in other Sections of this Manual. Provisions for regulatory signs that are specific to low-volume roads are contained in this Chapter.

Section 5B.02 STOP and YIELD Signs (R1-1 and R1-2)
Guidance:
01 *STOP (R1-1) and YIELD (R1-2) signs (see Figure 5B-1) should be considered for use on low-volume roads where engineering judgment or study, consistent with the provisions of Sections 2B.04 to 2B.10, indicates that either of the following conditions applies:*
   A. *An intersection of a less-important road with a main road where application of the normal right-of-way rule might not be readily apparent.*
   B. *An intersection that has restricted sight distance for the prevailing vehicle speeds.*

Section 5B.03 Speed Limit Signs (R2 Series)
Standard:
01 **If used, Speed Limit (R2 series) signs (see Figure 5B-1) shall display the speed limit established by law, ordinance, regulation, or as adopted by the authorized agency following an engineering study. The displayed speed limits shall be in multiples of 5 mph.**
02 **Speed limits shall be established in accordance with Section 2B.13.**
Option:
03 Speed limit signs may be used on low-volume roads that carry traffic from, onto, or adjacent to higher-volume roads that have posted speed limits.

Figure 5B-1. Regulatory Signs on Low-Volume Roads

Section 5B.04 Traffic Movement and Prohibition Signs (R3, R4, R5, R6, R9, R10, R11, R12, R13, and R14 Series)

Support:

01   The regulatory signs (see Figure 5B-1) in these series inform road users of required, permitted, or prohibited traffic movements involving turn, alignment, exclusion, and pedestrians.

Standard:

02   **If used, signs for traffic prohibitions or restrictions shall be placed in advance of the prohibition or restriction so that traffic can use an alternate route or turn around.**

Guidance:

03   *Signs should be used on low-volume roads to indicate traffic prohibitions and restrictions such as road closures and weight restrictions.*

Option:

04   Signs for traffic prohibitions or restrictions may be used on a low-volume road near and at the intersections or the connections with a higher class of road, and where the regulatory message is essential for transition from the low-volume road to the higher-class facility or vice versa.

Section 5B.05 Parking Signs (R8 Series)

Option:

01   Parking signs (see Figure 5B-2) may be installed selectively on low-volume roads with due consideration of enforcement.

Section 5B.06 Other Regulatory Signs

Standard:

01   **Other regulatory signs used on low-volume roads that are not discussed in Part 5 shall comply with the provisions contained in other Parts of this Manual.**

Figure 5B-2. Parking Signs and Plaques on Low-Volume Roads

CHAPTER 5C. WARNING SIGNS

Section 5C.01 Introduction

Support:

01 The purpose of a warning sign is to provide advance warning to the road user of unexpected conditions on or adjacent to the roadway that might not be readily apparent.

02 The provisions for warning signs are contained in Chapter 2C and in other Sections of this Manual. Provisions for warning signs that are specific to low-volume roads are contained in this Chapter.

Section 5C.02 Horizontal Alignment Signs (W1-1 through W1-8)

Support:

01 Horizontal Alignment signs (see Sections 2C.06 through 2C.12 and Figure 5C-1) include turn, curve, reverse turn, reverse curve, winding road, large arrow, and chevron alignment signs.

Option:

02 Horizontal Alignment signs may be used where engineering judgment indicates a need to inform the road user of a change in the horizontal alignment of the roadway.

Figure 5C-1. Horizontal Alignment and Intersection Warning Signs and Plaques and Object Markers on Low-Volume Roads

Type 1 Object Markers (obstructions within the roadway)

OM1-1   OM1-2   OM1-3

Type 2 Object Markers (obstructions adjacent to the roadway)

OM2-1V   OM2-2V   OM2-1H   OM2-2H

Type 3 Object Markers (obstructions adjacent to or within the roadway)

OM3-L   OM3-C   OM3-R

Type 4 Object Markers (end of roadway)

OM4-1   OM4-2   OM4-3

Section 5C.03 Intersection Warning Signs (W2-1 through W2-6)

Support:

01　Intersection signs (see Figure 5C-1) include the crossroad, side road, T-symbol, Y-symbol, and circular intersection signs.

Option:

02　Intersection signs may be used where engineering judgment indicates a need to inform the road user in advance of an intersection.

Section 5C.04 Stop Ahead and Yield Ahead Signs (W3-1, W3-2)

Standard:

01　**A Stop Ahead (W3-1) sign (see Figure 5C-2) shall be used where a STOP sign is not visible for a sufficient distance to permit the road user to bring the vehicle to a stop at the STOP sign.**

02　**A Yield Ahead (W3-2) sign (see Figure 5C-2) shall be used where a YIELD sign is not visible for a sufficient distance to permit the road user to bring the vehicle to a stop, if necessary, at the YIELD sign.**

Section 5C.05 NARROW BRIDGE Sign (W5-2)

Option:

01　The NARROW BRIDGE (W5-2) sign (see Figure 5C-2) may be used on an approach to a bridge or culvert that has a clear width less than that of the approach roadway.

Section 5C.06 ONE LANE BRIDGE Sign (W5-3)

Guidance:

01　*A ONE LANE BRIDGE (W5-3) sign (see Figure 5C-2) should be used on low-volume two-way roadways in advance of any bridge or culvert:*
  A. Having a clear roadway width of less than 16 feet, or
  B. Having a clear roadway width of less than 18 feet when commercial vehicles constitute a high proportion of the traffic, or
  C. Having a clear roadway width of 18 feet or less where the approach sight distance is limited on the approach to the structure.

Option:

02　Roadway alignment and additional warning may be provided on the approach to a bridge or culvert by the use of object markers and/or delineators.

Section 5C.07 Hill Sign (W7-1)

Option:

01　An engineering study of vehicles and road characteristics, such as percent grade and length of grade, may be conducted to determine hill signing requirements.

Section 5C.08 PAVEMENT ENDS Sign (W8-3)

Option:

01　A PAVEMENT ENDS (W8-3) sign (see Figure 5C-2) may be used to warn road users where a paved surface changes to a gravel or earth road surface.

Section 5C.09 Vehicular Traffic Warning and Non-Vehicular Warning Signs (W11 Series and W8-6)

Guidance:

01　*Vehicular Traffic Warning signs (see Figure 5C-2) should be used to alert road users to locations where frequent unexpected entries into the roadway by trucks, bicyclists, farm vehicles, fire trucks, and other vehicles might occur. Such signs should be used only at locations where the road user's sight distance is restricted or the condition, activity, or entering traffic would be unexpected.*

Option:

02　Non-Vehicular Warning signs (see Figure 5C-2) may be used to alert road users in advance of locations where unexpected entries into the roadway or shared use by pedestrians, large animals, or other crossing activities might occur.

03　A W7-3aP, W16-2P, or W16-9P supplemental plaque (see Figure 5C-2), with the legend NEXT XX MILES, XX FEET, or AHEAD may be installed below a Vehicular Traffic Warning or Non-Vehicular Warning sign (see Sections 2C.49 and 2C.50) to inform road users that they are approaching a portion of the roadway or a point where crossing activity might occur.

Figure 5C-2. Other Warning Signs and Plaques on Low-Volume Roads

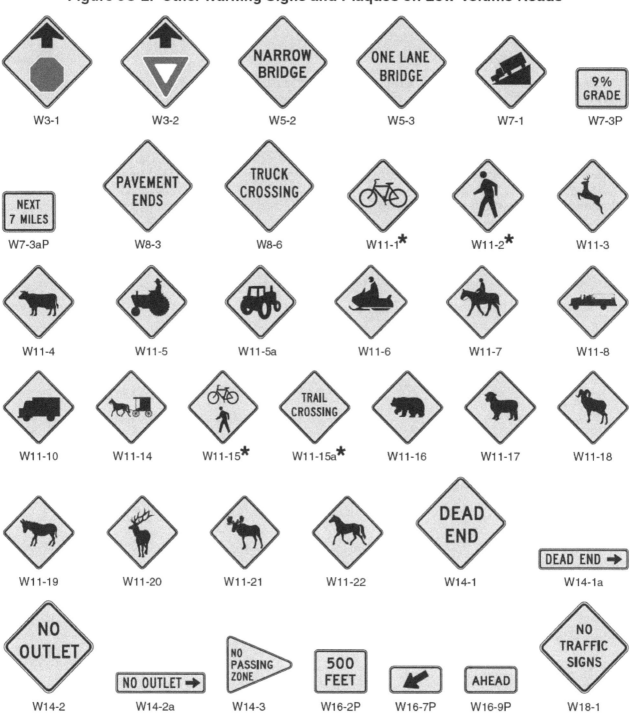

* A fluorescent yellow-green background color may be used for this sign or plaque

Standard:

04 **When a Non-Vehicular Warning sign is placed at the location of the crossing point, a diagonal downward pointing arrow (W16-7P) plaque (see Figure 5C-2) shall be mounted below the sign.**

Guidance:

05 *If the activity is seasonal or temporary, the sign should be removed or covered when the condition or activity does not exist.*

Section 5C.10 Advisory Speed Plaque (W13-1P)

Option:

01 An Advisory Speed (W13-1P) plaque (see Figure 5C-1) may be mounted below a warning sign when the condition requires a reduced speed.

Section 5C.11 DEAD END or NO OUTLET Signs (W14-1, W14-1a, W14-2, W14-2a)

Option:

01 The DEAD END (W14-1) and NO OUTLET (W14-2) signs (see Figure 5C-2) and the DEAD END (W14-1a) and NO OUTLET (W14-2a) signs (see Figure 5C-2) may be used to warn road users of a road that has no outlet or that terminates in a dead end or cul-de-sac.

Guidance:

02 *If used, these signs should be placed at a location that gives drivers of large commercial or recreational vehicles an opportunity to select a different route or turn around.*

Section 5C.12 NO TRAFFIC SIGNS Sign (W18-1)

Option:

01 A W18-1 warning sign (see Figure 5C-2) with the legend NO TRAFFIC SIGNS may be used only on unpaved, low-volume roads to advise users that no signs are installed along the distance of the road. If used, the sign may be installed at the point where road users would enter the low-volume road or where, based on engineering judgment, the road user might need this information.

02 A W7-3aP, W16-2P, or W16-9P supplemental plaque (see Figure 5C-2) with the legend NEXT XX MILES, XX FEET, or AHEAD may be installed below the W18-1 sign when appropriate.

Section 5C.13 Other Warning Signs

Standard:

01 **Other warning signs used on low-volume roads that are not discussed in Part 5, but are in this Manual, shall comply with the provisions contained in other Parts of this Manual. Warning signs that are not provided in this Manual shall comply with the provisions in Sections 2C.02 and 2C.03.**

Section 5C.14 Object Markers and Barricades

Support:

01 The purpose of object markers is to mark obstructions located within or adjacent to the roadway, such as bridge abutments, drainage structures, and other physical objects.

Guidance:

02 *The end of a low-volume road should be marked with a Type 4 object marker in compliance with Section 2C.66.*

Option:

03 A Type 3 Barricade may be used where engineering studies or judgment indicates a need for a more visible end-of-roadway treatment (see Section 2B.67).

Standard:

04 **Barricades used on low-volume roads shall comply with the provisions contained in Section 2B.67.**

CHAPTER 5D. GUIDE SIGNS

Section 5D.01 Introduction

Support:
01 The purpose of a guide sign is to inform road users regarding positions, directions, destinations, and routes.
02 The provisions for guide signs, in general, are contained in Chapters 2D through 2N and in other Sections of this Manual. Provisions for guide signs that are specific to low-volume roads are contained in this Chapter.

Guidance:
03 *The familiarity of the road users with the road should be considered in determining the need for guide signs on low-volume roads.*

Support:
04 Low-volume roads generally do not require guide signs to the extent that they are needed on higher classes of roads. Because guide signs are typically only beneficial as a navigational aid for road users who are unfamiliar with a low-volume road, guide signs might not be needed on low-volume roads that serve only local traffic.

Guidance:
05 *If used, destination names should be as specific and descriptive as possible. Destinations such as campgrounds, ranger stations, recreational areas, and the like should be clearly indicated so that they are not interpreted to be communities or locations with road user services.*

Option:
06 Guide signs may be used at intersections to provide information for road users returning to a higher class of roads.

CHAPTER 5E. MARKINGS

Section 5E.01 Introduction

Support:

01 The purpose of markings on highways is to provide guidance and information for road users regarding roadway conditions and restrictions.

02 The provisions for markings and delineators, in general, are contained in Part 3 and in other Sections of this Manual. Provisions for markings that are specific to low-volume roads are contained in this Chapter.

Section 5E.02 Center Line Markings

Standard:

01 **Where center line markings are installed, no-passing zone markings in compliance with Section 3B.02 shall also be installed.**

Guidance:

02 *Center line markings should be used on paved low-volume roads consistent with the principles of this Manual and with the policies and practices of the road agency and on the basis of either an engineering study or the application of engineering judgment.*

Option:

03 Center line markings may be placed on highways with or without edge line markings.

Section 5E.03 Edge Line Markings

Support:

01 The purpose of edge line markings is to delineate the left-hand or right-hand edge of the roadway.

Guidance:

02 *Edge line markings should be considered for use on paved low-volume roads based on engineering judgment or an engineering study.*

Option:

03 Edge line markings may be placed on highways with or without center line markings.

04 Edge line markings may be placed on paved low-volume roads for roadway features such as horizontal curves, narrow bridges, pavement width transitions, curvilinear alignment, and at other locations based on engineering judgment or an engineering study.

Section 5E.04 Delineators

Support:

01 The purpose of delineators is to enhance driver safety where it is desirable to call attention to a changed or changing condition such as abrupt roadway narrowing or curvature.

Option:

02 Delineators may be used on low-volume roads based on engineering judgment, such as for curves, T-intersections, and abrupt changes in the roadway width. In addition, they may be used to mark the location of driveways or other minor roads entering the low-volume road.

Section 5E.05 Other Markings

Standard:

01 **Other markings, such as stop lines, crosswalks, pavement legends, channelizing devices, and islands, used on low-volume roads shall comply with the provisions contained in this Manual.**

CHAPTER 5F. TRAFFIC CONTROL FOR HIGHWAY-RAIL GRADE CROSSINGS

Section 5F.01 Introduction
Support:
01 The provisions for highway-rail grade crossing traffic control devices are contained in Part 8 and in other Sections of this Manual.
02 Traffic control for highway-rail grade crossings includes all signs, signals, markings, illumination, and other warning devices and their supports along roadways either approaching or at highway-rail grade crossings. The purpose of this traffic control is to promote a safer and more efficient operation of both rail and highway traffic at highway-rail grade crossings.

Section 5F.02 Grade Crossing (Crossbuck) Sign and Number of Tracks Plaque (R15-1, R15-2P)
Support:
01 In most States, the Grade Crossing (Crossbuck) (R15-1) sign (see Figure 5F-1) requires road users to yield the right-of-way to rail traffic at a highway-rail grade crossing.
Standard:
02 **The Crossbuck (R15-1) sign shall be used at all highway-rail grade crossings, except as otherwise provided in Section 8B.03. For all low-volume roads, Crossbuck signs shall be used on the right-hand side of each approach. If there are two or more tracks, the supplemental Number of Tracks (R15-2P) plaque (see Figure 5F-1) shall display the number of tracks and shall be installed below the Crossbuck sign.**
03 **A strip of retroreflective white material not less than 2 inches in width shall be used on the back of each blade of each Crossbuck sign for the length of each blade, at all highway-rail grade crossings, except those where Crossbuck signs have been installed back-to-back.**
04 **A vertical strip of retroreflective white material, not less than 2 inches in width, shall be used on each support at passive highway-rail grade crossings for the full length of the front and back of the support from the Crossbuck sign or Number of Tracks plaque to within 2 feet above the ground, except on the side of those supports where a STOP (R1-1) or YIELD (R1-2) sign or flashing lights have been installed or on the back side of supports for Crossbuck signs installed on one-way streets.**

Section 5F.03 Grade Crossing Advance Warning Signs (W10 Series)
Standard:
01 **Except as provided in Paragraph 2, a Grade Crossing Advance Warning (W10-1) sign (see Figure 5F-1) shall be used on all low-volume roads in advance of every highway-rail grade crossing.**
Option:
02 The Grade Crossing Advance Warning sign may be omitted for highway-rail grade crossings that are flagged by train crews.
03 The W10-2, W10-3, and W10-4 signs (see Figure 5F-1) may be used on low-volume roads that run parallel to railroad tracks to warn road users making a turn that they will encounter a highway-rail grade crossing soon after making the turn.

Figure 5F-1. Highway-Rail Grade Crossing Signs and Plaques for Low-Volume Roads

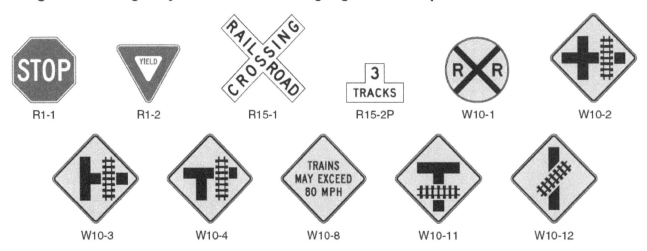

Section 5F.04 STOP and YIELD Signs (R1-1, R1-2)
Standard:
01 **The use and application at passive highway-rail grade crossings on low-volume roads of Crossbuck Assemblies with YIELD (R1-2) signs or STOP (R1-1) signs shall comply with the provisions of Section 8B.04.**

02 **At all highway-rail grade crossings where YIELD or STOP signs are installed, Yield Ahead (W3-2) or Stop Ahead (W3-1) signs shall also be installed if the criteria for their installation in Section 2C.36 is met.**

Section 5F.05 Pavement Markings
Guidance:

01 *Pavement markings at highway-rail grade crossings should be used on paved low-volume roads, particularly if they are already deployed at most other highway-rail grade crossings within the immediate vicinity, or when the roadway has center line markings.*

Section 5F.06 Other Traffic Control Devices
Standard:

01 **Other traffic control devices that are used at highway-rail grade crossings on low-volume roads, such as other signs, signals, and illumination that are not in this Chapter, shall comply with the provisions contained in Part 8 and other applicable Parts of this Manual.**

CHAPTER 5G. TEMPORARY TRAFFIC CONTROL ZONES

Section 5G.01 Introduction

Guidance:

01 *The safety of road users, including pedestrians and bicyclists, as well as personnel in work zones, should be an integral and high priority element of every project in the planning, design, maintenance, and construction phases. Part 6 should be reviewed for additional criteria, specific details, and more complex temporary traffic control zone requirements. The following principles should be applied to temporary traffic control zones:*

  A. *Traffic movement should be disrupted as little as possible.*
  B. *Road users should be guided in a clear and positive manner while approaching and within construction, maintenance, and utility work areas.*
  C. *Routine inspection and maintenance of traffic control elements should be performed both day and night.*
  D. *Both the contracting agency and the contractor should assign at least one person on each project to have day-to-day responsibility for assuring that the traffic control elements are operating effectively and any needed operational changes are brought to the attention of their supervisors.*

02 *Traffic control in temporary traffic control zones should be designed on the assumption that road users will only reduce their speeds if they clearly perceive a need to do so, and then only in small increments of speed. Temporary traffic control zones should not present a surprise to the road user. Frequent and/or abrupt changes in geometrics and other features should be avoided. Transitions should be well delineated and long enough to accommodate driving conditions at the speeds vehicles are realistically expected to travel.*

03 *A temporary traffic control plan (see Section 6C.01) should be used for a temporary traffic control zone on a low-volume road to specify particular traffic control devices and features, or to reference typical drawings such as those contained in Part 6.*

Support:

04 Applications of speed reduction countermeasures and enforcement can be effective in reducing traffic speeds in temporary traffic control zones.

Section 5G.02 Applications

Guidance:

01 *Planned work phasing and sequencing should be the basis for the use of traffic control devices for temporary traffic control zones. Part 6 should be consulted for specific traffic control requirements and examples where construction or maintenance work is planned.*

Support:

02 Maintenance activities might not require extensive temporary traffic control if the traffic volumes and speeds are low.

Option:

03 The traffic applications shown in Figures 6H-1, 6H-10, 6H-11, 6H-13, 6H-15, 6H-16, and 6H-18 of Part 6 are among those that may be used on low-volume roads.

Support:

04 Table 6H-3 provides distances for the advance placement of the traffic control devices shown in the typical applications.

Option:

05 For low-volume roadways with speeds of 30 miles per hour or less, a minimum distance of 100 feet may be used for the advance placement distance and the distance between signs shown in the typical applications.

06 For temporary traffic control zones on low-volume roads that require flaggers, a single flagger may be adequate if the flagger is visible to approaching traffic from all appropriate directions.

Section 5G.03 Channelization Devices

Standard:

01 **Channelization devices for nighttime use shall have the same retroreflective requirements as specified for higher-volume roadways.**

Option:

02 To alert, guide, and direct road users through temporary traffic control zones on low-volume roads, tapers may be used to move a road user out of the traffic lane and around the work space using the spacing of devices that is described in Section 6F.63.

Section 5G.04 Markings
Guidance:

01 *Pavement markings should be considered for temporary traffic control zones on paved low-volume roads, especially roads that had existing pavement markings or that have a surfaced detour or temporary roadway.*

Option:

02 Interim pavement markings may be omitted in a temporary traffic control zone if they are not needed based on the criteria for these markings in Section 6F.78.

Section 5G.05 Other Traffic Control Devices
Standard:

01 **Other traffic control devices, such as other signs, signals, and illumination that are used on low-volume roads in temporary traffic control zones, but are not described in Part 5, shall comply with the provisions contained in other Parts of this Manual.**

Support:

02 Some of the signs that might be applicable in a temporary traffic control zone on a low-volume road are shown in Figure 5G-1.

Figure 5G-1. Temporary Traffic Control Signs and Plaques on Low-Volume Roads

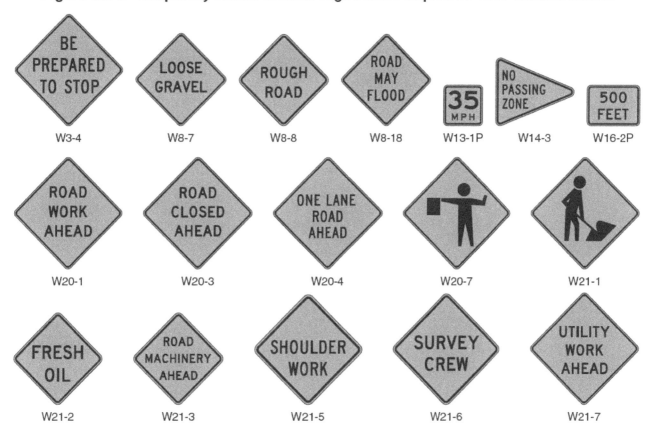

CHAPTER 5H. TRAFFIC CONTROL FOR SCHOOL AREAS

Section 5H.01 Introduction

Support:
01 The provisions for school traffic control devices are contained in Part 7 of this Manual.

Standard:

02 **The sizes of school signs and plaques on low-volume roads shall be in accordance with Section 7B.01 and Table 7B-1.**

PART 6
TEMPORARY TRAFFIC CONTROL

CHAPTER 6A. GENERAL

Section 6A.01 General

Support:

01   Whenever the acronym "TTC" is used in Part 6, it refers to "temporary traffic control."

Standard:

02   **The needs and control of all road users (motorists, bicyclists, and pedestrians within the highway, or on private roads open to public travel (see definition in Section 1A.13), including persons with disabilities in accordance with the Americans with Disabilities Act of 1990 (ADA), Title II, Paragraph 35.130) through a TTC zone shall be an essential part of highway construction, utility work, maintenance operations, and the management of traffic incidents.**

Support:

03   When the normal function of the roadway, or a private road open to public travel, is suspended, TTC planning provides for continuity of the movement of motor vehicle, bicycle, and pedestrian traffic (including accessible passage); transit operations; and access (and accessibility) to property and utilities.

04   The primary function of TTC is to provide for the reasonably safe and effective movement of road users through or around TTC zones while reasonably protecting road users, workers, responders to traffic incidents, and equipment.

05   Of equal importance to the public traveling through the TTC zone is the safety of workers performing the many varied tasks within the work space. TTC zones present constantly changing conditions that are unexpected by the road user. This creates an even higher degree of vulnerability for the workers and incident management responders on or near the roadway (see Section 6D.03). At the same time, the TTC zone provides for the efficient completion of whatever activity interrupted the normal use of the roadway.

06   Consideration for road user safety, worker and responder safety, and the efficiency of road user flow is an integral element of every TTC zone, from planning through completion. A concurrent objective of the TTC is the efficient construction and maintenance of the highway and the efficient resolution of traffic incidents.

07   No one set of TTC devices can satisfy all conditions for a given project or incident. At the same time, defining details that would be adequate to cover all applications is not practical. Instead, Part 6 displays typical applications that depict common applications of TTC devices. The TTC selected for each situation depends on type of highway, road user conditions, duration of operation, physical constraints, and the nearness of the work space or incident management activity to road users.

08   Improved road user performance might be realized through a well-prepared public relations effort that covers the nature of the work, the time and duration of its execution, the anticipated effects upon road users, and possible alternate routes and modes of travel. Such programs have been found to result in a significant reduction in the number of road users traveling through the TTC zone, which reduces the possible number of conflicts.

09   Operational improvements might be realized by using intelligent transportation systems (ITS) in work zones. The use in work zones of ITS technology, such as portable camera systems, highway advisory radio, variable speed limits, ramp metering, traveler information, merge guidance, and queue detection information, is aimed at increasing safety for both workers and road users and helping to ensure a more efficient traffic flow. The use in work zones of ITS technologies has been found to be effective in providing traffic monitoring and management, data collection, and traveler information.

Standard:

10   **TTC plans and devices shall be the responsibility of the authority of a public body or official having jurisdiction for guiding road users. There shall be adequate statutory authority for the implementation and enforcement of needed road user regulations, parking controls, speed zoning, and the management of traffic incidents. Such statutes shall provide sufficient flexibility in the application of TTC to meet the needs of changing conditions in the TTC zone.**

Support:

11   Temporary facilities, including pedestrian routes around worksites, are also covered by the accessibility requirements of the Americans with Disabilities Act of 1990 (ADA) (Public Law 101-336, 104 Stat. 327, July 26, 1990. 42 U.S.C. 12101-12213 (as amended)).

Guidance:

12 *The TTC plan should start in the planning phase and continue through the design, construction, and restoration phases. The TTC plans and devices should follow the principles set forth in Part 6. The management of traffic incidents should follow the principles set forth in Chapter 6I.*

Option:

13 TTC plans may deviate from the typical applications described in Chapter 6H to allow for conditions and requirements of a particular site or jurisdiction.

Support:

14 The provisions of Part 6 apply to both rural and urban areas. A rural highway is normally characterized by lower volumes, higher speeds, fewer turning conflicts, and less conflict with pedestrians. An urban street is typically characterized by relatively low speeds, wide ranges of road user volumes, narrower roadway lanes, frequent intersections and driveways, significant pedestrian activity, and more businesses and houses.

15 The determination as to whether a particular facility at a particular time of day can be considered to be a high-volume roadway or can be considered to be a low-volume roadway is made by the public agency or official having jurisdiction.

CHAPTER 6B. FUNDAMENTAL PRINCIPLES

Section 6B.01 Fundamental Principles of Temporary Traffic Control

Support:
01   Construction, maintenance, utility, and incident zones can all benefit from TTC to compensate for the unexpected or unusual situations faced by road users. When planning for TTC in these zones, it can be assumed that it is appropriate for road users to exercise caution. Even though road users are assumed to be using caution, special care is still needed in applying TTC techniques.

02   Special plans preparation and coordination with transit, other highway agencies, law enforcement and other emergency units, utilities, schools, and railroad companies might be needed to reduce unexpected and unusual road user operation situations.

03   During TTC activities, commercial vehicles might need to follow a different route from passenger vehicles because of bridge, weight, clearance, or geometric restrictions. Also, vehicles carrying hazardous materials might need to follow a different route from other vehicles. The Hazardous Materials and National Network signs are included in Sections 2B.62 and 2B.63, respectively.

04   Experience has shown that following the fundamental principles of Part 6 will assist road users and help protect workers in the vicinity of TTC zones.

Guidance:

05   *Road user and worker safety and accessibility in TTC zones should be an integral and high-priority element of every project from planning through design and construction. Similarly, maintenance and utility work should be planned and conducted with the safety and accessibility of all motorists, bicyclists, pedestrians (including those with disabilities), and workers being considered at all times. If the TTC zone includes a grade crossing, early coordination with the railroad company or light rail transit agency should take place.*

Support:

06   Formulating specific plans for TTC at traffic incidents is difficult because of the variety of situations that can arise.

Guidance:

07   *The following are the seven fundamental principles of TTC:*

1. *General plans or guidelines should be developed to provide safety for motorists, bicyclists, pedestrians, workers, enforcement/emergency officials, and equipment, with the following factors being considered:*
   A. *The basic safety principles governing the design of permanent roadways and roadsides should also govern the design of TTC zones. The goal should be to route road users through such zones using roadway geometrics, roadside features, and TTC devices as nearly as possible comparable to those for normal highway situations.*
   B. *A TTC plan, in detail appropriate to the complexity of the work project or incident, should be prepared and understood by all responsible parties before the site is occupied. Any changes in the TTC plan should be approved by an official who is knowledgeable (for example, trained and/or certified) in proper TTC practices.*

2. *Road user movement should be inhibited as little as practical, based on the following considerations:*
   A. *TTC at work and incident sites should be designed on the assumption that drivers will only reduce their speeds if they clearly perceive a need to do so (see Section 6C.01).*
   B. *Frequent and abrupt changes in geometrics such as lane narrowing, dropped lanes, or main roadway transitions that require rapid maneuvers, should be avoided.*
   C. *Work should be scheduled in a manner that minimizes the need for lane closures or alternate routes, while still getting the work completed quickly and the lanes or roadway open to traffic as soon as possible.*
   D. *Attempts should be made to reduce the volume of traffic using the roadway or freeway to match the restricted capacity conditions. Road users should be encouraged to use alternative routes. For high-volume roadways and freeways, the closure of selected entrance ramps or other access points and the use of signed diversion routes should be evaluated.*
   E. *Bicyclists and pedestrians, including those with disabilities, should be provided with access and reasonably safe passage through the TTC zone.*
   F. *If work operations permit, lane closures on high-volume streets and highways should be scheduled during off-peak hours. Night work should be considered if the work can be accomplished with a series of short-term operations.*
   G. *Early coordination with officials having jurisdiction over the affected cross streets and providing emergency services should occur if significant impacts to roadway operations are anticipated.*

3. *Motorists, bicyclists, and pedestrians should be guided in a clear and positive manner while approaching and traversing TTC zones and incident sites. The following principles should be applied:*

A. Adequate warning, delineation, and channelization should be provided to assist in guiding road users in advance of and through the TTC zone or incident site by using proper pavement marking, signing, or other devices that are effective under varying conditions. Providing information that is in usable formats by pedestrians with visual disabilities should also be considered.
   B. TTC devices inconsistent with intended travel paths through TTC zones should be removed or covered. However, in intermediate-term stationary, short-term, and mobile operations, where visible permanent devices are inconsistent with intended travel paths, devices that highlight or emphasize the appropriate path should be used. Providing traffic control devices that are accessible to and usable by pedestrians with disabilities should be considered.
   C. Flagging procedures, when used, should provide positive guidance to road users traversing the TTC zone.
4. To provide acceptable levels of operations, routine day and night inspections of TTC elements should be performed as follows:
   A. Individuals who are knowledgeable (for example, trained and/or certified) in the principles of proper TTC should be assigned responsibility for safety in TTC zones. The most important duty of these individuals should be to check that all TTC devices of the project are consistent with the TTC plan and are effective for motorists, bicyclists, pedestrians, and workers.
   B. As the work progresses, temporary traffic controls and/or working conditions should be modified, if appropriate, in order to provide mobility and positive guidance to the road user and to provide worker safety. The individual responsible for TTC should have the authority to halt work until applicable or remedial safety measures are taken.
   C. TTC zones should be carefully monitored under varying conditions of road user volumes, light, and weather to check that applicable TTC devices are effective, clearly visible, clean, and in compliance with the TTC plan.
   D. When warranted, an engineering study should be made (in cooperation with law enforcement officials) of reported crashes occurring within the TTC zone. Crash records in TTC zones should be monitored to identify the need for changes in the TTC zone.
5. Attention should be given to the maintenance of roadside safety during the life of the TTC zone by applying the following principles:
   A. To accommodate run-off-the-road incidents, disabled vehicles, or emergency situations, unencumbered roadside recovery areas or clear zones should be provided where practical.
   B. Channelization of road users should be accomplished by the use of pavement markings, signing, and crashworthy, detectable channelizing devices.
   C. Work equipment, workers' private vehicles, materials, and debris should be stored in such a manner to reduce the probability of being impacted by run-off-the-road vehicles.
6. Each person whose actions affect TTC zone safety, from the upper-level management through the field workers, should receive training appropriate to the job decisions each individual is required to make. Only those individuals who are trained in proper TTC practices and have a basic understanding of the principles (established by applicable standards and guidelines, including those of this Manual) should supervise the selection, placement, and maintenance of TTC devices used for TTC zones and for incident management.
7. Good public relations should be maintained by applying the following principles:
   A. The needs of all road users should be assessed such that appropriate advance notice is given and clearly defined alternative paths are provided.
   B. The cooperation of the various news media should be sought in publicizing the existence of and reasons for TTC zones because news releases can assist in keeping the road users well informed.
   C. The needs of abutting property owners, residents, and businesses should be assessed and appropriate accommodations made.
   D. The needs of emergency service providers (law enforcement, fire, and medical) should be assessed and appropriate coordination and accommodations made.
   E. The needs of railroads and transit should be assessed and appropriate coordination and accommodations made.
   F. The needs of operators of commercial vehicles such as buses and large trucks should be assessed and appropriate accommodations made.

Standard:

Before any new detour or temporary route is opened to traffic, all necessary signs shall be in place.

All TTC devices shall be removed as soon as practical when they are no longer needed. When work is suspended for short periods of time, TTC devices that are no longer appropriate shall be removed or covered.

CHAPTER 6C. TEMPORARY TRAFFIC CONTROL ELEMENTS

Section 6C.01 Temporary Traffic Control Plans

Support:

01 A TTC plan describes TTC measures to be used for facilitating road users through a work zone or an incident area. TTC plans play a vital role in providing continuity of effective road user flow when a work zone, incident, or other event temporarily disrupts normal road user flow. Important auxiliary provisions that cannot conveniently be specified on project plans can easily be incorporated into Special Provisions within the TTC plan.

02 TTC plans range in scope from being very detailed to simply referencing typical drawings contained in this Manual, standard approved highway agency drawings and manuals, or specific drawings contained in the contract documents. The degree of detail in the TTC plan depends entirely on the nature and complexity of the situation.

Guidance:

03 *TTC plans should be prepared by persons knowledgeable (for example, trained and/or certified) about the fundamental principles of TTC and work activities to be performed. The design, selection, and placement of TTC devices for a TTC plan should be based on engineering judgment.*

04 *Coordination should be made between adjacent or overlapping projects to check that duplicate signing is not used and to check compatibility of traffic control between adjacent or overlapping projects.*

05 *Traffic control planning should be completed for all highway construction, utility work, maintenance operations, and incident management including minor maintenance and utility projects prior to occupying the TTC zone. Planning for all road users should be included in the process.*

06 *Provisions for effective continuity of accessible circulation paths for pedestrians should be incorporated into the TTC process. Where existing pedestrian routes are blocked or detoured, information should be provided about alternative routes that are usable by pedestrians with disabilities, particularly those who have visual disabilities. Access to temporary bus stops, travel across intersections with accessible pedestrian signals (see Section 4E.09), and other routing issues should be considered where temporary pedestrian routes are channelized. Barriers and channelizing devices that are detectable by people with visual disabilities should be provided.*

Option:

07 Provisions may be incorporated into the project bid documents that enable contractors to develop an alternate TTC plan.

08 Modifications of TTC plans may be necessary because of changed conditions or a determination of better methods of safely and efficiently handling road users.

Guidance:

09 *This alternate or modified plan should have the approval of the responsible highway agency prior to implementation.*

10 *Provisions for effective continuity of transit service should be incorporated into the TTC planning process because often public transit buses cannot efficiently be detoured in the same manner as other vehicles (particularly for short-term maintenance projects). Where applicable, the TTC plan should provide for features such as accessible temporary bus stops, pull-outs, and satisfactory waiting areas for transit patrons, including persons with disabilities, if applicable (see Section 8A.08 for additional light rail transit issues to consider for TTC).*

11 *Provisions for effective continuity of railroad service and acceptable access to abutting property owners and businesses should also be incorporated into the TTC planning process.*

12 *Reduced speed limits should be used only in the specific portion of the TTC zone where conditions or restrictive features are present. However, frequent changes in the speed limit should be avoided. A TTC plan should be designed so that vehicles can travel through the TTC zone with a speed limit reduction of no more than 10 mph.*

13 *A reduction of more than 10 mph in the speed limit should be used only when required by restrictive features in the TTC zone. Where restrictive features justify a speed reduction of more than 10 mph, additional driver notification should be provided. The speed limit should be stepped down in advance of the location requiring the lowest speed, and additional TTC warning devices should be used.*

14 *Reduced speed zoning (lowering the regulatory speed limit) should be avoided as much as practical because drivers will reduce their speeds only if they clearly perceive a need to do so.*

Support:

15 Research has demonstrated that large reductions in the speed limit, such as a 30 mph reduction, increase speed variance and the potential for crashes. Smaller reductions in the speed limit of up to 10 mph cause smaller changes in speed variance and lessen the potential for increased crashes. A reduction in the regulatory speed limit of only up to 10 mph from the normal speed limit has been shown to be more effective.

Section 6C.02 Temporary Traffic Control Zones

Support:

01 A TTC zone is an area of a highway where road user conditions are changed because of a work zone, an incident zone, or a planned special event through the use of TTC devices, uniformed law enforcement officers, or other authorized personnel.

02 A work zone is an area of a highway with construction, maintenance, or utility work activities. A work zone is typically marked by signs, channelizing devices, barriers, pavement markings, and/or work vehicles. It extends from the first warning sign or high-intensity rotating, flashing, oscillating, or strobe lights on a vehicle to the END ROAD WORK sign or the last TTC device.

03 An incident zone is an area of a highway where temporary traffic controls are imposed by authorized officials in response to a traffic incident (see Section 6I.01). It extends from the first warning device (such as a sign, light, or cone) to the last TTC device or to a point where road users return to the original lane alignment and are clear of the incident.

04 A planned special event often creates the need to establish altered traffic patterns to handle the increased traffic volumes generated by the event. The size of the TTC zone associated with a planned special event can be small, such as closing a street for a festival, or can extend throughout a municipality for larger events. The duration of the TTC zone is determined by the duration of the planned special event.

Section 6C.03 Components of Temporary Traffic Control Zones

Support:

01 Most TTC zones are divided into four areas: the advance warning area, the transition area, the activity area, and the termination area. Figure 6C-1 illustrates these four areas. These four areas are described in Sections 6C.04 through 6C.07.

Section 6C.04 Advance Warning Area

Support:

01 The advance warning area is the section of highway where road users are informed about the upcoming work zone or incident area.

Option:

02 The advance warning area may vary from a single sign or high-intensity rotating, flashing, oscillating, or strobe lights on a vehicle to a series of signs in advance of the TTC zone activity area.

Guidance:

03 *Typical distances for placement of advance warning signs on freeways and expressways should be longer because drivers are conditioned to uninterrupted flow. Therefore, the advance warning sign placement should extend on these facilities as far as 1/2 mile or more.*

04 *On urban streets, the effective placement of the first warning sign in feet should range from 4 to 8 times the speed limit in mph, with the high end of the range being used when speeds are relatively high. When a single advance warning sign is used (in cases such as low-speed residential streets), the advance warning area can be as short as 100 feet. When two or more advance warning signs are used on higher-speed streets, such as major arterials, the advance warning area should extend a greater distance (see Table 6C-1).*

05 *Since rural highways are normally characterized by higher speeds, the effective placement of the first warning sign in feet should be substantially longer—from 8 to 12 times the speed limit in mph. Since two or more advance warning signs are normally used for these conditions, the advance warning area should extend 1,500 feet or more for open highway conditions (see Table 6C-1).*

06 *The distances contained in Table 6C-1 are approximate, are intended for guidance purposes only, and should be applied with engineering judgment. These distances should be adjusted for field conditions, if necessary, by increasing or decreasing the recommended distances.*

Figure 6C-1. Component Parts of a Temporary Traffic Control Zone

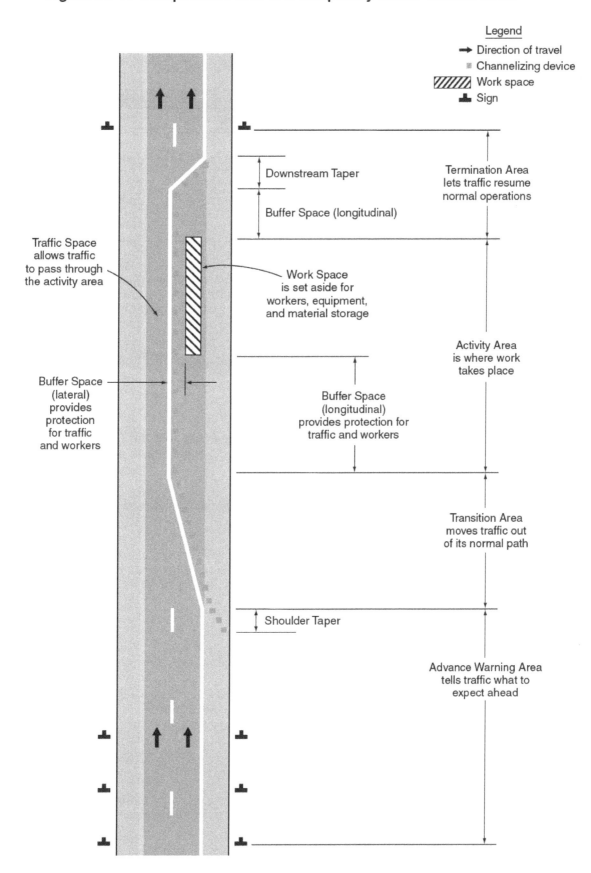

Table 6C-1. Recommended Advance Warning Sign Minimum Spacing

Road Type	Distance Between Signs**		
	A	B	C
Urban (low speed)*	100 feet	100 feet	100 feet
Urban (high speed)*	350 feet	350 feet	350 feet
Rural	500 feet	500 feet	500 feet
Expressway / Freeway	1,000 feet	1,500 feet	2,640 feet

* Speed category to be determined by the highway agency
** The column headings A, B, and C are the dimensions shown in Figures 6H-1 through 6H-46. The A dimension is the distance from the transition or point of restriction to the first sign. The B dimension is the distance between the first and second signs. The C dimension is the distance between the second and third signs. (The "first sign" is the sign in a three-sign series that is closest to the TTC zone. The "third sign" is the sign that is furthest upstream from the TTC zone.)

Support:

07 The need to provide additional reaction time for a condition is one example of justification for increasing the sign spacing. Conversely, decreasing the sign spacing might be justified in order to place a sign immediately downstream of an intersection or major driveway such that traffic turning onto the roadway in the direction of the TTC zone will be warned of the upcoming condition.

Option:

08 Advance warning may be eliminated when the activity area is sufficiently removed from the road users' path so that it does not interfere with the normal flow.

Section 6C.05 Transition Area

Support:

01 The transition area is that section of highway where road users are redirected out of their normal path. Transition areas usually involve strategic use of tapers, which because of their importance are discussed separately in detail.

Standard:

02 **When redirection of the road users' normal path is required, they shall be directed from the normal path to a new path.**

Option:

03 Because it is impractical in mobile operations to redirect the road user's normal path with stationary channelization, more dominant vehicle-mounted traffic control devices, such as arrow boards, portable changeable message signs, and high-intensity rotating, flashing, oscillating, or strobe lights, may be used instead of channelizing devices to establish a transition area.

Section 6C.06 Activity Area

Support:

01 The activity area is the section of the highway where the work activity takes place. It is comprised of the work space, the traffic space, and the buffer space.

02 The work space is that portion of the highway closed to road users and set aside for workers, equipment, and material, and a shadow vehicle if one is used upstream. Work spaces are usually delineated for road users by channelizing devices or, to exclude vehicles and pedestrians, by temporary barriers.

Option:

03 The work space may be stationary or may move as work progresses.

Guidance:

04 *Since there might be several work spaces (some even separated by several miles) within the project limits, each work space should be adequately signed to inform road users and reduce confusion.*

Support:

05 The traffic space is the portion of the highway in which road users are routed through the activity area.

06 The buffer space is a lateral and/or longitudinal area that separates road user flow from the work space or an unsafe area, and might provide some recovery space for an errant vehicle.

Guidance:

07 *Neither work activity nor storage of equipment, vehicles, or material should occur within a buffer space.*

Option:

08 Buffer spaces may be positioned either longitudinally or laterally with respect to the direction of road user flow. The activity area may contain one or more lateral or longitudinal buffer spaces.

09 A longitudinal buffer space may be placed in advance of a work space.

10 The longitudinal buffer space may also be used to separate opposing road user flows that use portions of the same traffic lane, as shown in Figure 6C-2.

11 If a longitudinal buffer space is used, the values shown in Table 6C-2 may be used to determine the length of the longitudinal buffer space.

Support:

12 Typically, the buffer space is formed as a traffic island and defined by channelizing devices.

13 When a shadow vehicle, arrow board, or changeable message sign is placed in a closed lane in advance of a work space, only the area upstream of the vehicle, arrow board, or changeable message sign constitutes the buffer space.

Option:

14 The lateral buffer space may be used to separate the traffic space from the work space, as shown in Figures 6C-1 and 6C-2, or such areas as excavations or pavement-edge drop-offs. A lateral buffer space also may be used between two travel lanes, especially those carrying opposing flows.

Guidance:

15 *The width of a lateral buffer space should be determined by engineering judgment.*

Option:

Table 6C-2. Stopping Sight Distance as a Function of Speed

Speed*	Distance
20 mph	115 feet
25 mph	155 feet
30 mph	200 feet
35 mph	250 feet
40 mph	305 feet
45 mph	360 feet
50 mph	425 feet
55 mph	495 feet
60 mph	570 feet
65 mph	645 feet
70 mph	730 feet
75 mph	820 feet

* Posted speed, off-peak 85th-percentile speed prior to work starting, or the anticipated operating speed

16 When work occurs on a high-volume, highly congested facility, a vehicle storage or staging space may be provided for incident response and emergency vehicles (for example, tow trucks and fire apparatus) so that these vehicles can respond quickly to road user incidents.

Section 6C.07 Termination Area

Support:

01 The termination area is the section of the highway where road users are returned to their normal driving path. The termination area extends from the downstream end of the work area to the last TTC device such as END ROAD WORK signs, if posted.

Option:

02 An END ROAD WORK sign, a Speed Limit sign, or other signs may be used to inform road users that they can resume normal operations.

03 A longitudinal buffer space may be used between the work space and the beginning of the downstream taper.

Section 6C.08 Tapers

Option:

01 Tapers may be used in both the transition and termination areas. Whenever tapers are to be used in close proximity to an interchange ramp, crossroads, curves, or other influencing factors, the length of the tapers may be adjusted.

Support:

02 Tapers are created by using a series of channelizing devices and/or pavement markings to move traffic out of or into the normal path. Types of tapers are shown in Figure 6C-2.

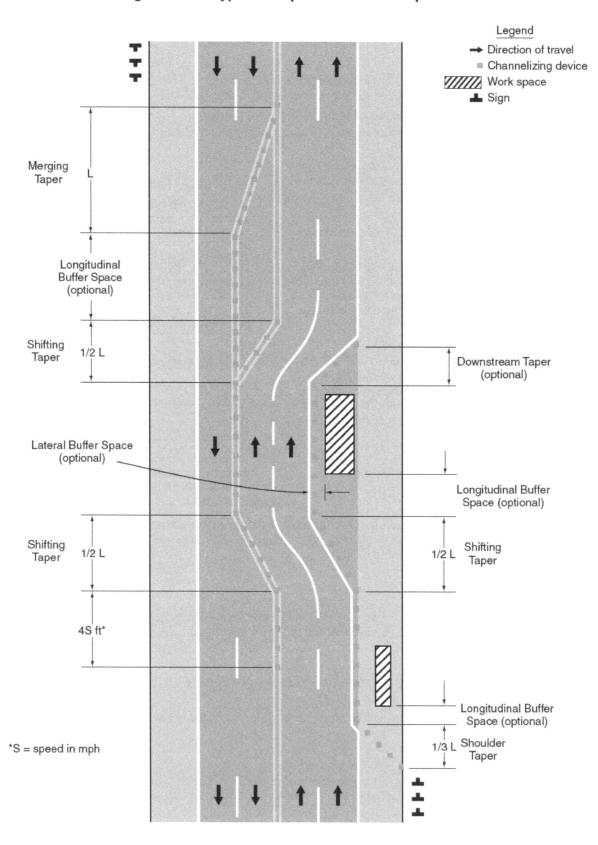

Figure 6C-2. Types of Tapers and Buffer Spaces

03 Longer tapers are not necessarily better than shorter tapers (particularly in urban areas with characteristics such as short block lengths or driveways) because extended tapers tend to encourage sluggish operation and to encourage drivers to delay lane changes unnecessarily. The test concerning adequate lengths of tapers involves observation of driver performance after TTC plans are put into effect.

Guidance:

04 *The appropriate taper length (L) should be determined using the criteria shown in Tables 6C-3 and 6C-4.*

05 *The maximum distance in feet between devices in a taper should not exceed 1.0 times the speed limit in mph.*

Support:

06 A merging taper requires the longest distance because drivers are required to merge into common road space.

Guidance:

07 *A merging taper should be long enough to enable merging drivers to have adequate advance warning and sufficient length to adjust their speeds and merge into an adjacent lane before the downstream end of the transition.*

Support:

08 A shifting taper is used when a lateral shift is needed. When more space is available, a longer than minimum taper distance can be beneficial. Changes in alignment can also be accomplished by using horizontal curves designed for normal highway speeds.

Guidance:

09 *A shifting taper should have a length of approximately 1/2 L (see Tables 6C-3 and 6C-4).*

Table 6C-3. Taper Length Criteria for Temporary Traffic Control Zones

Type of Taper	Taper Length
Merging Taper	at least L
Shifting Taper	at least 0.5 L
Shoulder Taper	at least 0.33 L
One-Lane, Two-Way Traffic Taper	50 feet minimum, 100 feet maximum
Downstream Taper	50 feet minimum, 100 feet maximum

Note: Use Table 6C-4 to calculate L

Table 6C-4. Formulas for Determining Taper Length

Speed (S)	Taper Length (L) in feet
40 mph or less	$L = \dfrac{WS^2}{60}$
45 mph or more	$L = WS$

Where: L = taper length in feet
W = width of offset in feet
S = posted speed limit, or off-peak 85th-percentile speed prior to work starting, or the anticipated operating speed in mph

Support:

10 A shoulder taper might be beneficial on a high-speed roadway where shoulders are part of the activity area and are closed, or when improved shoulders might be mistaken as a driving lane. In these instances, the same type, but abbreviated, closure procedures used on a normal portion of the roadway can be used.

Guidance:

11 *If used, shoulder tapers should have a length of approximately 1/3 L (see Tables 6C-3 and 6C-4). If a shoulder is used as a travel lane, either through practice or during a TTC activity, a normal merging or shifting taper should be used.*

Support:

12 A downstream taper might be useful in termination areas to provide a visual cue to the driver that access is available back into the original lane or path that was closed.

Guidance:

13 *If used, a downstream taper should have a minimum length of 50 feet and a maximum length of 100 feet with devices placed at a spacing of approximately 20 feet.*

Support:

14 The one-lane, two-way taper is used in advance of an activity area that occupies part of a two-way roadway in such a way that a portion of the road is used alternately by traffic in each direction.

Guidance:

15 *Traffic should be controlled by a flagger or temporary traffic control signal (if sight distance is limited), or a STOP or YIELD sign. A short taper having a minimum length of 50 feet and a maximum length of 100 feet with channelizing devices at approximately 20-foot spacing should be used to guide traffic into the one-lane section, and a downstream taper should be used to guide traffic back into their original lane.*

Support:

16 An example of a one-lane, two-way traffic taper is shown in Figure 6C-3.

Section 6C.09 Detours and Diversions

Support:

01 A detour is a temporary rerouting of road users onto an existing highway in order to avoid a TTC zone.

Guidance:

02 Detours should be clearly signed over their entire length so that road users can easily use existing highways to return to the original highway.

Support:

03 A diversion is a temporary rerouting of road users onto a temporary highway or alignment placed around the work area.

Section 6C.10 One-Lane, Two-Way Traffic Control

Standard:

01 **Except as provided in Paragraph 5, when traffic in both directions must use a single lane for a limited distance, movements from each end shall be coordinated.**

Guidance:

02 *Provisions should be made for alternate one-way movement through the constricted section via methods such as flagger control, a flag transfer, a pilot car, traffic control signals, or stop or yield control.*

03 *Control points at each end should be chosen to permit easy passing of opposing lanes of vehicles.*

04 *If traffic on the affected one-lane roadway is not visible from one end to the other, then flagging procedures, a pilot car with a flagger used as described in Section 6C.13, or a traffic control signal should be used to control opposing traffic flows.*

Option:

05 If the work space on a low-volume street or road is short and road users from both directions are able to see the traffic approaching from the opposite direction through and beyond the worksite, the movement of traffic through a one-lane, two-way constriction may be self-regulating.

Section 6C.11 Flagger Method of One-Lane, Two-Way Traffic Control

Guidance:

01 *Except as provided in Paragraph 2, traffic should be controlled by a flagger at each end of a constricted section of roadway. One of the flaggers should be designated as the coordinator. To provide coordination of the control of the traffic, the flaggers should be able to communicate with each other orally, electronically, or with manual signals. These manual signals should not be mistaken for flagging signals.*

Option:

02 When a one-lane, two-way TTC zone is short enough to allow a flagger to see from one end of the zone to the other, traffic may be controlled by either a single flagger or by a flagger at each end of the section.

Guidance:

03 *When a single flagger is used, the flagger should be stationed on the shoulder opposite the constriction or work space, or in a position where good visibility and traffic control can be maintained at all times. When good visibility and traffic control cannot be maintained by one flagger station, traffic should be controlled by a flagger at each end of the section.*

Section 6C.12 Flag Transfer Method of One-Lane, Two-Way Traffic Control

Support:

01 The driver of the last vehicle proceeding into the one-lane section is given a red flag (or other token) and instructed to deliver it to the flagger at the other end. The opposite flagger, upon receipt of the flag, then knows that traffic can be permitted to move in the other direction. A variation of this method is to replace the use of a flag with an official pilot car that follows the last road user vehicle proceeding through the section.

Guidance:

02 *The flag transfer method should be employed only where the one-way traffic is confined to a relatively short length of a road, usually no more than 1 mile in length.*

Figure 6C-3. Example of a One-Lane, Two-Way Traffic Taper

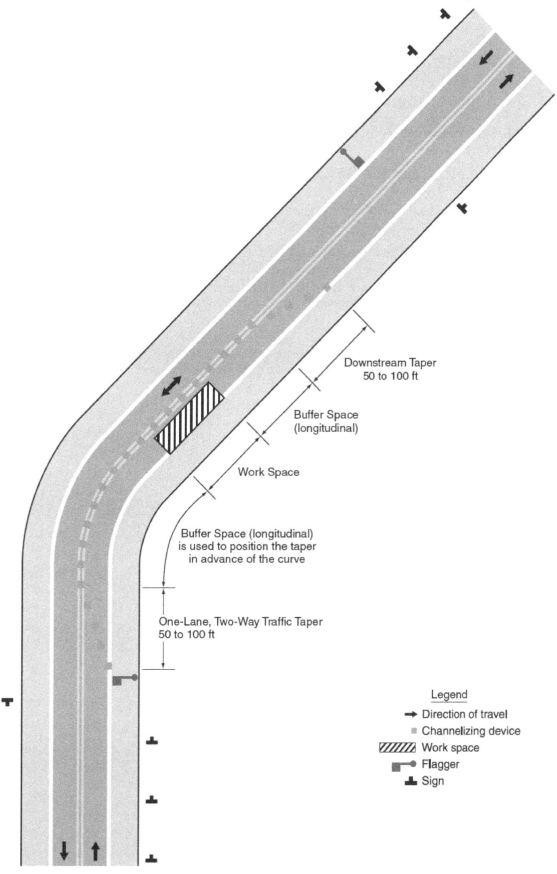

Section 6C.13 Pilot Car Method of One-Lane, Two-Way Traffic Control

Option:

01 A pilot car may be used to guide a queue of vehicles through the TTC zone or detour.

Guidance:

02 *The pilot car should have the name of the contractor or contracting authority prominently displayed.*

Standard:

03 **The PILOT CAR FOLLOW ME (G20-4) sign (see Section 6F.58) shall be mounted on the rear of the pilot vehicle.**

04 **A flagger shall be stationed on the approach to the activity area to control vehicular traffic until the pilot vehicle is available.**

Section 6C.14 Temporary Traffic Control Signal Method of One-Lane, Two-Way Traffic Control

Option:

01 Traffic control signals may be used to control vehicular traffic movements in one-lane, two-way TTC zones (see Figure 6H-12 and Chapter 4H).

Section 6C.15 Stop or Yield Control Method of One-Lane, Two-Way Traffic Control

Option:

01 STOP or YIELD signs may be used to control traffic on low-volume roads at a one-lane, two-way TTC zone when drivers are able to see the other end of the one-lane, two-way operation and have sufficient visibility of approaching vehicles.

Guidance:

02 *If the STOP or YIELD sign is installed for only one direction, then the STOP or YIELD sign should face road users who are driving on the side of the roadway that is closed for the work activity area.*

CHAPTER 6D. PEDESTRIAN AND WORKER SAFETY

Section 6D.01 Pedestrian Considerations

Support:

01 A wide range of pedestrians might be affected by TTC zones, including the young, elderly, and people with disabilities such as hearing, visual, or mobility. These pedestrians need a clearly delineated and usable travel path. Considerations for pedestrians with disabilities are addressed in Section 6D.02.

Standard:

02 **The various TTC provisions for pedestrian and worker safety set forth in Part 6 shall be applied by knowledgeable (for example, trained and/or certified) persons after appropriate evaluation and engineering judgment.**

03 **Advance notification of sidewalk closures shall be provided by the maintaining agency.**

04 **If the TTC zone affects the movement of pedestrians, adequate pedestrian access and walkways shall be provided. If the TTC zone affects an accessible and detectable pedestrian facility, the accessibility and detectability shall be maintained along the alternate pedestrian route.**

Option:

05 If establishing or maintaining an alternate pedestrian route is not feasible during the project, an alternate means of providing for pedestrians may be used, such as adding free bus service around the project or assigning someone the responsibility to assist pedestrians with disabilities through the project limits.

Support:

06 It must be recognized that pedestrians are reluctant to retrace their steps to a prior intersection for a crossing or to add distance or out-of-the-way travel to a destination.

Guidance:

07 *The following three items should be considered when planning for pedestrians in TTC zones:*
   A. Pedestrians should not be led into conflicts with vehicles, equipment, and operations.
   B. Pedestrians should not be led into conflicts with vehicles moving through or around the worksite.
   C. Pedestrians should be provided with a convenient and accessible path that replicates as nearly as practical the most desirable characteristics of the existing sidewalk(s) or footpath(s).

08 *A pedestrian route should not be severed and/or moved for non-construction activities such as parking for vehicles and equipment.*

09 *Consideration should be made to separate pedestrian movements from both worksite activity and vehicular traffic. Unless an acceptable route that does not involve crossing the roadway can be provided, pedestrians should be appropriately directed with advance signing that encourages them to cross to the opposite side of the roadway. In urban and suburban areas with high vehicular traffic volumes, these signs should be placed at intersections (rather than midblock locations) so that pedestrians are not confronted with midblock worksites that will induce them to attempt skirting the worksite or making a midblock crossing.*

Support:

10 Figures 6H-28 and 6H-29 show typical TTC device usage and techniques for pedestrian movement through work zones.

Guidance:

11 *To accommodate the needs of pedestrians, including those with disabilities, the following considerations should be addressed when temporary pedestrian pathways in TTC zones are designed or modified:*
   A. Provisions for continuity of accessible paths for pedestrians should be incorporated into the TTC plan.
   B. Access to transit stops should be maintained.
   C. A smooth, continuous hard surface should be provided throughout the entire length of the temporary pedestrian facility. There should be no curbs or abrupt changes in grade or terrain that could cause tripping or be a barrier to wheelchair use. The geometry and alignment of the facility should meet the applicable requirements of the "Americans with Disabilities Act Accessibility Guidelines for Buildings and Facilities (ADAAG)" (see Section 1A.11).
   D. The width of the existing pedestrian facility should be provided for the temporary facility if practical. Traffic control devices and other construction materials and features should not intrude into the usable width of the sidewalk, temporary pathway, or other pedestrian facility. When it is not possible to maintain a minimum width of 60 inches throughout the entire length of the pedestrian pathway, a 60 x 60-inch passing space should be provided at least every 200 feet to allow individuals in wheelchairs to pass.

E. Blocked routes, alternate crossings, and sign and signal information should be communicated to pedestrians with visual disabilities by providing devices such as audible information devices, accessible pedestrian signals, or barriers and channelizing devices that are detectable to the pedestrians traveling with the aid of a long cane or who have low vision. Where pedestrian traffic is detoured to a TTC signal, engineering judgment should be used to determine if pedestrian signals or accessible pedestrian signals should be considered for crossings along an alternate route.

F. When channelization is used to delineate a pedestrian pathway, a continuous detectable edging should be provided throughout the length of the facility such that pedestrians using a long cane can follow it. These detectable edgings should comply with the provisions of Section 6F.74.

G. Signs and other devices mounted lower than 7 feet above the temporary pedestrian pathway should not project more than 4 inches into accessible pedestrian facilities.

Option:

Whenever it is feasible, closing off the worksite from pedestrian intrusion may be preferable to channelizing pedestrian traffic along the site with TTC devices.

Guidance:

Fencing should not create sight distance restrictions for road users. Fences should not be constructed of materials that would be hazardous if impacted by vehicles. Wooden railing, fencing, and similar systems placed immediately adjacent to motor vehicle traffic should not be used as substitutes for crashworthy temporary traffic barriers.

Ballast for TTC devices should be kept to the minimum amount needed and should be mounted low to prevent penetration of the vehicle windshield.

Movement by work vehicles and equipment across designated pedestrian paths should be minimized and, when necessary, should be controlled by flaggers or TTC. Staging or stopping of work vehicles or equipment along the side of pedestrian paths should be avoided, since it encourages movement of workers, equipment, and materials across the pedestrian path.

Access to the work space by workers and equipment across pedestrian walkways should be minimized because the access often creates unacceptable changes in grade, and rough or muddy terrain, and pedestrians will tend to avoid these areas by attempting non-intersection crossings where no curb ramps are available.

Option:

A canopied walkway may be used to protect pedestrians from falling debris, and to provide a covered passage for pedestrians.

Guidance:

Covered walkways should be sturdily constructed and adequately lighted for nighttime use.

When pedestrian and vehicle paths are rerouted to a closer proximity to each other, consideration should be given to separating them by a temporary traffic barrier.

If a temporary traffic barrier is used to shield pedestrians, it should be designed to accommodate site conditions.

Support:

Depending on the possible vehicular speed and angle of impact, temporary traffic barriers might deflect upon impact by an errant vehicle. Guidance for locating and designing temporary traffic barriers can be found in Chapter 9 of AASHTO's "Roadside Design Guide" (see Section 1A.11).

Standard:

Short intermittent segments of temporary traffic barrier shall not be used because they nullify the containment and redirective capabilities of the temporary traffic barrier, increase the potential for serious injury both to vehicle occupants and pedestrians, and encourage the presence of blunt, leading ends. All upstream leading ends that are present shall be appropriately flared or protected with properly installed and maintained crashworthy cushions. Adjacent temporary traffic barrier segments shall be properly connected in order to provide the overall strength required for the temporary traffic barrier to perform properly.

Normal vertical curbing shall not be used as a substitute for temporary traffic barriers when temporary traffic barriers are needed.

Option:

Temporary traffic barriers or longitudinal channelizing devices may be used to discourage pedestrians from unauthorized movements into the work space. They may also be used to inhibit conflicts with vehicular traffic by minimizing the possibility of midblock crossings.

Support:

25 A major concern for pedestrians is urban and suburban building construction encroaching onto the contiguous sidewalks, which forces pedestrians off the curb into direct conflict with moving vehicles.

Guidance:

26 *If a significant potential exists for vehicle incursions into the pedestrian path, pedestrians should be rerouted or temporary traffic barriers should be installed.*

Support:

27 TTC devices, jersey barriers, and wood or chain link fencing with a continuous detectable edging can satisfactorily delineate a pedestrian path.

Guidance:

28 *Tape, rope, or plastic chain strung between devices are not detectable, do not comply with the design standards in the "Americans with Disabilities Act Accessibility Guidelines for Buildings and Facilities (ADAAG)" (see Section 1A.11), and should not be used as a control for pedestrian movements.*

29 *In general, pedestrian routes should be preserved in urban and commercial suburban areas. Alternative routing should be discouraged.*

30 *The highway agency in charge of the TTC zone should regularly inspect the activity area so that effective pedestrian TTC is maintained.*

Section 6D.02 Accessibility Considerations

Support:

01 Additional information on the design and construction of accessible temporary facilities is found in publications listed in Section 1A.11 (see Publications 12, 38, 39, and 42).

Guidance:

02 *The extent of pedestrian needs should be determined through engineering judgment or by the individual responsible for each TTC zone situation. Adequate provisions should be made for pedestrians with disabilities.*

Standard:

03 **When existing pedestrian facilities are disrupted, closed, or relocated in a TTC zone, the temporary facilities shall be detectable and include accessibility features consistent with the features present in the existing pedestrian facility. Where pedestrians with visual disabilities normally use the closed sidewalk, a barrier that is detectable by a person with a visual disability traveling with the aid of a long cane shall be placed across the full width of the closed sidewalk.**

Support:

04 Maintaining a detectable, channelized pedestrian route is much more useful to pedestrians who have visual disabilities than closing a walkway and providing audible directions to an alternate route involving additional crossings and a return to the original route. Braille is not useful in conveying such information because it is difficult to find. Audible instructions might be provided, but the extra distance and additional street crossings might add complexity to a trip.

Guidance:

05 *Because printed signs and surface delineation are not usable by pedestrians with visual disabilities, blocked routes, alternate crossings, and sign and signal information should be communicated to pedestrians with visual disabilities by providing audible information devices, accessible pedestrian signals, and barriers and channelizing devices that are detectable to pedestrians traveling with the aid of a long cane or who have low vision.*

Support:

06 The most desirable way to provide information to pedestrians with visual disabilities that is equivalent to visual signing for notification of sidewalk closures is a speech message provided by an audible information device. Devices that provide speech messages in response to passive pedestrian actuation are the most desirable. Other devices that continuously emit a message, or that emit a message in response to use of a pushbutton, are also acceptable. signing information can also be transmitted to personal receivers, but currently such receivers are not likely to be carried or used by pedestrians with visual disabilities in TTC zones. Audible information devices might not be needed if detectable channelizing devices make an alternate route of travel evident to pedestrians with visual disabilities.

Guidance:

07 *If a pushbutton is used to provide equivalent TTC information to pedestrians with visual disabilities, the pushbutton should be equipped with a locator tone to notify pedestrians with visual disabilities that a special accommodation is available, and to help them locate the pushbutton.*

Section 6D.03 Worker Safety Considerations

Support:

01 Equally as important as the safety of road users traveling through the TTC zone is the safety of workers. TTC zones present temporary and constantly changing conditions that are unexpected by the road user. This creates an even higher degree of vulnerability for workers on or near the roadway.

02 Maintaining TTC zones with road user flow inhibited as little as possible, and using TTC devices that get the road user's attention and provide positive direction are of particular importance. Likewise, equipment and vehicles moving within the activity area create a risk to workers on foot. When possible, the separation of moving equipment and construction vehicles from workers on foot provides the operator of these vehicles with a greater separation clearance and improved sight lines to minimize exposure to the hazards of moving vehicles and equipment.

Guidance:

03 *The following are the key elements of worker safety and TTC management that should be considered to improve worker safety:*

  A. *Training—all workers should be trained on how to work next to motor vehicle traffic in a way that minimizes their vulnerability. Workers having specific TTC responsibilities should be trained in TTC techniques, device usage, and placement.*
  B. *Temporary Traffic Barriers—temporary traffic barriers should be placed along the work space depending on factors such as lateral clearance of workers from adjacent traffic, speed of traffic, duration and type of operations, time of day, and volume of traffic.*
  C. *Speed Reduction—reducing the speed of vehicular traffic, mainly through regulatory speed zoning, funneling, lane reduction, or the use of uniformed law enforcement officers or flaggers, should be considered.*
  D. *Activity Area—planning the internal work activity area to minimize backing-up maneuvers of construction vehicles should be considered to minimize the exposure to risk.*
  E. *Worker Safety Planning—a trained person designated by the employer should conduct a basic hazard assessment for the worksite and job classifications required in the activity area. This safety professional should determine whether engineering, administrative, or personal protection measures should be implemented. This plan should be in accordance with the Occupational Safety and Health Act of 1970, as amended, "General Duty Clause" Section 5(a)(1) - Public Law 91-596, 84 Stat. 1590, December 29, 1970, as amended, and with the requirement to assess worker risk exposures for each job site and job classification, as per 29 CFR 1926.20 (b)(2) of "Occupational Safety and Health Administration Regulations, General Safety and Health Provisions" (see Section 1A.11).*

Standard:

04 **All workers, including emergency responders, within the right-of-way who are exposed either to traffic (vehicles using the highway for purposes of travel) or to work vehicles and construction equipment within the TTC zone shall wear high-visibility safety apparel that meets the Performance Class 2 or 3 requirements of the ANSI/ISEA 107–2004 publication entitled "American National Standard for High-Visibility Safety Apparel and Headwear" (see Section 1A.11), or equivalent revisions, and labeled as meeting the ANSI 107-2004 standard performance for Class 2 or 3 risk exposure, except as provided in Paragraph 5. A person designated by the employer to be responsible for worker safety shall make the selection of the appropriate class of garment.**

Option:

05 Emergency and incident responders and law enforcement personnel within the TTC zone may wear high-visibility safety apparel that meets the performance requirements of the ANSI/ISEA 207-2006 publication entitled "American National Standard for High-Visibility Public Safety Vests" (see Section 1A.11), or equivalent revisions, and labeled as ANSI 207-2006, in lieu of ANSI/ISEA 107-2004 apparel.

Standard:

06 **When uniformed law enforcement personnel are used to direct traffic, to investigate crashes, or to handle lane closures, obstructed roadways, and disasters, high-visibility safety apparel as described in this Section shall be worn by the law enforcement personnel.**

07 **Except as provided in Paragraph 8, firefighters or other emergency responders working within the right-of-way shall wear high-visibility safety apparel as described in this Section.**

Option:

08 Firefighters or other emergency responders working within the right-of-way and engaged in emergency operations that directly expose them to flame, fire, heat, and/or hazardous materials may wear retroreflective turn-out gear that is specified and regulated by other organizations, such as the National Fire Protection Association.

09 The following are additional elements of TTC management that may be considered to improve worker safety:
  A. Shadow Vehicle—in the case of mobile and constantly moving operations, such as pothole patching and striping operations, a shadow vehicle, equipped with appropriate lights and warning signs, may be used to protect the workers from impacts by errant vehicles. The shadow vehicle may be equipped with a rear-mounted impact attenuator.
  B. Road Closure—if alternate routes are available to handle road users, the road may be closed temporarily. This may also facilitate project completion and thus further reduce worker vulnerability.
  C. Law Enforcement Use—in highly vulnerable work situations, particularly those of relatively short duration, law enforcement units may be stationed to heighten the awareness of passing vehicular traffic and to improve safety through the TTC zone.
  D. Lighting—for nighttime work, the TTC zone and approaches may be lighted.
  E. Special Devices—these include rumble strips, changeable message signs, hazard identification beacons, flags, and warning lights. Intrusion warning devices may be used to alert workers to the approach of errant vehicles.

Support:

10 Judicious use of the special devices described in Item E in Paragraph 9 might be helpful for certain difficult TTC situations, but misuse or overuse of special devices or techniques might lessen their effectiveness.

CHAPTER 6E. FLAGGER CONTROL

Section 6E.01 Qualifications for Flaggers

Guidance:

01 Because flaggers are responsible for public safety and make the greatest number of contacts with the public of all highway workers, they should be trained in safe traffic control practices and public contact techniques. Flaggers should be able to satisfactorily demonstrate the following abilities:

- A. Ability to receive and communicate specific instructions clearly, firmly, and courteously;
- B. Ability to move and maneuver quickly in order to avoid danger from errant vehicles;
- C. Ability to control signaling devices (such as paddles and flags) in order to provide clear and positive guidance to drivers approaching a TTC zone in frequently changing situations;
- D. Ability to understand and apply safe traffic control practices, sometimes in stressful or emergency situations; and
- E. Ability to recognize dangerous traffic situations and warn workers in sufficient time to avoid injury.

Section 6E.02 High-Visibility Safety Apparel

Standard:

01 **For daytime and nighttime activity, flaggers shall wear high-visibility safety apparel that meets the Performance Class 2 or 3 requirements of the ANSI/ISEA 107–2004 publication entitled "American National Standard for High-Visibility Apparel and Headwear" (see Section 1A.11) and labeled as meeting the ANSI 107-2004 standard performance for Class 2 or 3 risk exposure. The apparel background (outer) material color shall be fluorescent orange-red, fluorescent yellow-green, or a combination of the two as defined in the ANSI standard. The retroreflective material shall be orange, yellow, white, silver, yellow-green, or a fluorescent version of these colors, and shall be visible at a minimum distance of 1,000 feet. The retroreflective safety apparel shall be designed to clearly identify the wearer as a person.**

Guidance:

02 *For nighttime activity, high-visibility safety apparel that meets the Performance Class 3 requirements of the ANSI/ISEA 107–2004 publication entitled "American National Standard for High-Visibility Apparel and Headwear" (see Section 1A.11) and labeled as meeting the ANSI 107-2004 standard performance for Class 3 risk exposure should be considered for flagger wear.*

Standard:

03 **When uniformed law enforcement officers are used to direct traffic within a TTC zone, they shall wear high-visibility safety apparel as described in this Section.**

Option:

04 In lieu of ANSI/ISEA 107-2004 apparel, law enforcement personnel within the TTC zone may wear high-visibility safety apparel that meets the performance requirements of the ANSI/ISEA 207-2006 publication entitled "American National Standard for High-Visibility Public Safety Vests" (see Section 1A.11) and labeled as ANSI 207-2006.

Section 6E.03 Hand-Signaling Devices

Guidance:

01 *The STOP/SLOW paddle should be the primary and preferred hand-signaling device because the STOP/SLOW paddle gives road users more positive guidance than red flags. Use of flags should be limited to emergency situations.*

Standard:

02 **The STOP/SLOW paddle shall have an octagonal shape on a rigid handle. STOP/SLOW paddles shall be at least 18 inches wide with letters at least 6 inches high. The STOP (R1-1) face shall have white letters and a white border on a red background. The SLOW (W20-8) face shall have black letters and a black border on an orange background. When used at night, the STOP/SLOW paddle shall be retroreflectorized.**

Guidance:

03 *The STOP/SLOW paddle should be fabricated from light semi-rigid material.*

Support:

04 The optimum method of displaying a STOP or SLOW message is to place the STOP/SLOW paddle on a rigid staff that is tall enough that when the end of the staff is resting on the ground, the message is high enough to be seen by approaching or stopped traffic.

Option:

05 The STOP/SLOW paddle may be modified to improve conspicuity by incorporating either white or red flashing lights on the STOP face, and either white or yellow flashing lights on the SLOW face. The flashing lights may be arranged in any of the following patterns:
   A. Two white or red lights, one centered vertically above and one centered vertically below the STOP legend; and/or two white or yellow lights, one centered vertically above and one centered vertically below the SLOW legend;
   B. Two white or red lights, one centered horizontally on each side of the STOP legend; and/or two white or yellow lights, one centered horizontally on each side of the SLOW legend;
   C. One white or red light centered below the STOP legend; and/or one white or yellow light centered below the SLOW legend;
   D. A series of eight or more small white or red lights no larger than 1/4 inch in diameter along the outer edge of the paddle, arranged in an octagonal pattern at the eight corners of the border of the STOP face; and/or a series of eight or more small white or yellow lights no larger than 1/4 inch in diameter along the outer edge of the paddle, arranged in a diamond pattern along the border of the SLOW face; or
   E. A series of white lights forming the shapes of the letters in the legend.

Standard:

06 **If flashing lights are used on the STOP face of the paddle, their colors shall be all white or all red. If flashing lights are used on the SLOW face of the paddle, their colors shall be all white or all yellow.**

07 **If more than eight flashing lights are used, the lights shall be arranged such that they clearly convey the octagonal shape of the STOP face of the paddle and/or the diamond shape of the SLOW face of the paddle.**

08 **If flashing lights are used on the STOP/SLOW paddle, the flash rate shall be at least 50, but not more than 60, flashes per minute.**

09 **Flags, when used, shall be red or fluorescent orange/red in color, shall be a minimum of 24 inches square, and shall be securely fastened to a staff that is approximately 36 inches in length.**

Guidance:

10 *The free edge of a flag should be weighted so the flag will hang vertically, even in heavy winds.*

Standard:

11 **When used at nighttime, flags shall be retroreflectorized red.**

Option:

12 When flagging in an emergency situation at night in a non-illuminated flagger station, a flagger may use a flashlight with a red glow cone to supplement the STOP/SLOW paddle or flag.

Standard:

13 **When a flashlight is used for flagging in an emergency situation at night in a non-illuminated flagger station, the flagger shall hold the flashlight in the left hand, shall hold the paddle or flag in the right hand as shown in Figure 6E-3, and shall use the flashlight in the following manner to control approaching road users:**
   A. **To inform road users to stop, the flagger shall hold the flashlight with the left arm extended and pointed down toward the ground, and then shall slowly wave the flashlight in front of the body in a slow arc from left to right such that the arc reaches no farther than 45 degrees from vertical.**
   B. **To inform road users to proceed, the flagger shall point the flashlight at the vehicle's bumper, slowly aim the flashlight toward the open lane, then hold the flashlight in that position. The flagger shall not wave the flashlight.**
   C. **To alert or slow traffic, the flagger shall point the flashlight toward oncoming traffic and quickly wave the flashlight in a figure eight motion.**

Section 6E.04 Automated Flagger Assistance Devices

Support:

01 Automated Flagger Assistance Devices (AFADs) enable a flagger(s) to be positioned out of the lane of traffic and are used to control road users through temporary traffic control zones. These devices are designed to be remotely operated either by a single flagger at one end of the TTC zone or at a central location, or by separate flaggers near each device's location.

02 There are two types of AFADs:
- A. An AFAD (see Section 6E.05) that uses a remotely controlled STOP/SLOW sign on either a trailer or a movable cart system to alternately control right-of-way.
- B. An AFAD (see Section 6E.06) that uses remotely controlled red and yellow lenses and a gate arm to alternately control right-of-way.

03 AFADs might be appropriate for short-term and intermediate-term activities (see Section 6G.02). Typical applications include TTC activities such as, but not limited to:
- A. Bridge maintenance;
- B. Haul road crossings; and
- C. Pavement patching.

Standard:

04 **AFADs shall only be used in situations where there is only one lane of approaching traffic in the direction to be controlled.**

05 **When used at night, the AFAD location shall be illuminated in accordance with Section 6E.08.**

Guidance:

06 *AFADs should not be used for long-term stationary work (see Section 6G.02).*

Standard:

07 **Because AFADs are not traffic control signals, they shall not be used as a substitute for or a replacement for a continuously operating temporary traffic control signal as described in Section 6F.84.**

08 **AFADs shall meet the crashworthy performance criteria contained in Section 6F.01.**

Guidance:

09 *If used, AFADs should be located in advance of one-lane, two-way tapers and downstream from the point where approaching traffic is to stop in response to the device.*

Standard:

10 **If used, AFADs shall be placed so that all of the signs and other items controlling traffic movement are readily visible to the driver of the initial approaching vehicle with advance warning signs alerting other approaching traffic to be prepared to stop.**

11 **If used, an AFAD shall be operated only by a flagger (see Section 6E.01) who has been trained on the operation of the AFAD. The flagger(s) operating the AFAD(s) shall not leave the AFAD(s) unattended at any time while the AFAD(s) is being used.**

12 **The use of AFADs shall conform to one of the following methods:**
- **A. An AFAD at each end of the TTC zone (Method 1), or**
- **B. An AFAD at one end of the TTC zone and a flagger at the opposite end (Method 2).**

13 **Except as provided in Paragraph 14, two flaggers shall be used when using either Method 1 or Method 2.**

Option:

14 A single flagger may simultaneously operate two AFADs (Method 1) or may operate a single AFAD on one end of the TTC zone while being the flagger at the opposite end of the TTC zone (Method 2) if both of the following conditions are present:
- A. The flagger has an unobstructed view of the AFAD(s), and
- B. The flagger has an unobstructed view of approaching traffic in both directions.

Guidance:

15 *When an AFAD is used, the advance warning signing should include a ROAD WORK AHEAD (W20-1) sign, a ONE LANE ROAD (W20-4) sign, and a BE PREPARED TO STOP (W3-4) sign.*

Standard:

16 **When the AFAD is not in use, the signs associated with the AFAD, both at the AFAD location and in advance, shall be removed or covered.**

Guidance:

17 *A State or local agency that elects to use AFADs should adopt a policy, based on engineering judgment, governing AFAD applications. The policy should also consider more detailed and/or more restrictive requirements for AFAD use, such as the following:*
- *A. Conditions applicable for the use of Method 1 and Method 2 AFAD operation,*
- *B. Volume criteria,*
- *C. Maximum distance between AFADs,*

D. Conflicting lenses/indications monitoring requirements,
E. Fail safe procedures,
F. Additional signing and pavement markings,
G. Application consistency,
H. Larger signs or lenses to increase visibility, and
I. Use of backplates.

Section 6E.05 STOP/SLOW Automated Flagger Assistance Devices

Standard:

01 A STOP/SLOW Automated Flagger Assistance Device (AFAD) (see Section 6E.04) shall include a STOP/SLOW sign that alternately displays the STOP (R1-1) face and the SLOW (W20-8) face of a STOP/SLOW paddle (see Figure 6E-1).

02 The AFAD's STOP/SLOW sign shall have an octagonal shape, shall be fabricated of rigid material, and shall be mounted with the bottom of the sign a minimum of 6 feet above the pavement on an appropriate support. The size of the STOP/SLOW sign shall be at least 24 x 24 inches with letters at least 8 inches high. The background of the STOP face shall be red with white letters and border. The background of the SLOW face shall be diamond shaped and orange with black letters and border. Both faces of the STOP/SLOW sign shall be retroreflectorized.

03 The AFAD's STOP/SLOW sign shall have a means to positively lock, engage, or otherwise maintain the sign assembly in a stable condition when set in the STOP or SLOW position.

04 The AFAD's STOP/SLOW sign shall be supplemented with active conspicuity devices by incorporating either:
   A. White or red flashing lights within the STOP face and white or yellow flashing lights within the SLOW face meeting the provisions contained in Section 6E.03; or
   B. A Stop Beacon (see Section 4L.05) mounted a maximum of 24 inches above the STOP face and a Warning Beacon (see Section 4L.03) mounted a maximum of 24 inches above, below, or to the side of the SLOW face. The Stop Beacon shall not be flashed or illuminated when the SLOW face is displayed, and the Warning Beacon shall not be flashed or illuminated when the STOP face is displayed. Except for the mounting locations, the beacons shall comply with the provisions of Chapter 4L.

Option:

05 Type B warning light(s) (see Section 6F.83) may be used in lieu of the Warning Beacon during the display of the SLOW face of the AFAD's STOP/SLOW sign.

Standard:

06 If Type B warning lights are used in lieu of a Warning Beacon, they shall flash continuously when the SLOW face is displayed and shall not be flashed or illuminated when the STOP face is displayed.

Option:

07 The faces of the AFAD's STOP/SLOW sign may include louvers to improve the stability of the device in windy or other adverse environmental conditions.

Standard:

08 If louvers are used, the louvers shall be designed such that the full sign face is visible to approaching traffic at a distance of 50 feet or greater.

Guidance:

09 *The STOP/SLOW AFAD should include a gate arm that descends to a down position across the approach lane of traffic when the STOP face is displayed and then ascends to an upright position when the SLOW face is displayed.*

Option:

10 In lieu of a stationary STOP/SLOW sign with a separate gate arm, the STOP/SLOW sign may be attached to a mast arm that physically blocks the approach lane of traffic when the STOP face is displayed and then moves to a position that does not block the approach lane when the SLOW face is displayed.

Standard:

11 Gate arms, if used, shall be fully retroreflectorized on both sides, and shall have vertical alternating red and white stripes at 16-inch intervals measured horizontally as shown in Figure 8C-1. When the arm is in the down position blocking the approach lane:
   A. The minimum vertical aspect of the arm and sheeting shall be 2 inches; and
   B. The end of the arm shall reach at least to the center of the lane being controlled.

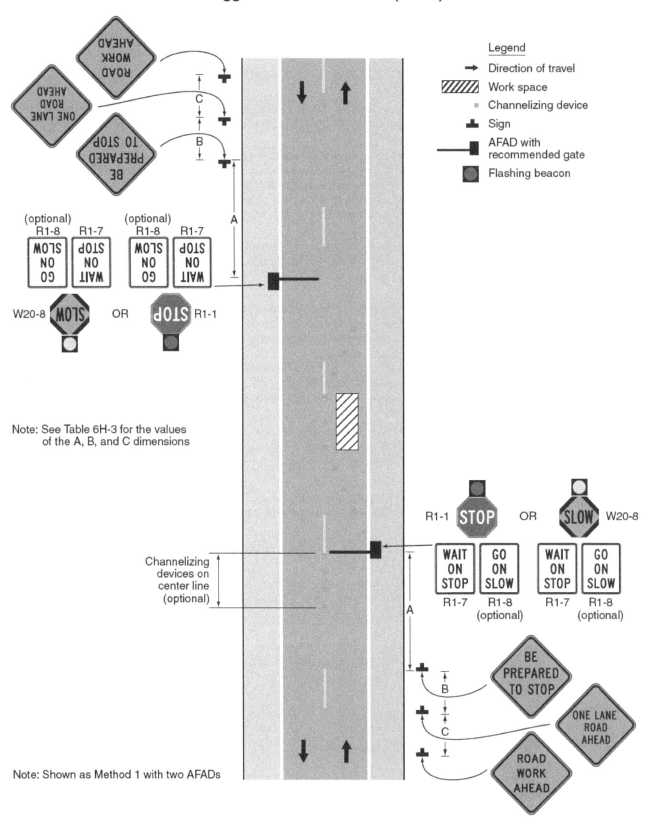

Figure 6E-1. Example of the Use of a STOP/SLOW Automated Flagger Assistance Device (AFAD)

¹² **A WAIT ON STOP (R1-7) sign (see Figure 6E-1) shall be displayed to road users approaching the AFAD.**

Option:

¹³ A GO ON SLOW (R1-8) sign (see Figure 6E-1) may also be displayed to road users approaching the AFAD.

Standard:

¹⁴ **The GO ON SLOW sign, if used, and the WAIT ON STOP sign shall be positioned on the same support structure as the AFAD or immediately adjacent to the AFAD such that they are in the same direct line of view of approaching traffic as the sign faces of the AFAD. Both signs shall have black legends and borders on white backgrounds. Each of these signs shall be rectangular in shape and each shall be at least 24 x 30 inches in size with letters at least 6 inches high.**

¹⁵ **To inform road users to stop, the AFAD shall display the STOP face and the red or white lights, if used, within the STOP face shall flash or the Stop Beacon shall flash. To inform road users to proceed, the AFAD shall display the SLOW face and the yellow or white lights, if used, within the SLOW face shall flash or the Warning Beacon or the Type B warning lights shall flash.**

¹⁶ **If STOP/SLOW AFADs are used to control traffic in a one-lane, two-way TTC zone, safeguards shall be incorporated to prevent the flagger(s) from simultaneously displaying the SLOW face at each end of the TTC zone. Additionally, the flagger(s) shall not display the AFAD's SLOW face until all oncoming vehicles have cleared the one-lane portion of the TTC zone.**

Section 6E.06 Red/Yellow Lens Automated Flagger Assistance Devices

Standard:

⁰¹ **A Red/Yellow Lens Automated Flagger Assistance Device (AFAD) (see Section 6E.04) shall alternately display a steadily illuminated CIRCULAR RED lens and a flashing CIRCULAR YELLOW lens to control traffic without the need for a flagger in the immediate vicinity of the AFAD or on the roadway (see Figure 6E-2).**

⁰² **Red/Yellow Lens AFADs shall have at least one set of CIRCULAR RED and CIRCULAR YELLOW lenses that are 12 inches in diameter. Unless otherwise provided in this Section, the lenses and their arrangement, CIRCULAR RED on top and CIRCULAR YELLOW below, shall comply with the applicable provisions for traffic signal indications in Part 4. If the set of lenses is post-mounted, the bottom of the housing (including brackets) shall be at least 7 feet above the pavement. If the set of lenses is located over any portion of the highway that can be used by motor vehicles, the bottom of the housing (including brackets) shall be at least 15 feet above the pavement.**

Option:

⁰³ Additional sets of CIRCULAR RED and CIRCULAR YELLOW lenses, located over the roadway or on the left-hand side of the approach and operated in unison with the primary set, may be used to improve visibility and/or conspicuity of the AFAD.

Standard:

⁰⁴ **A Red/Yellow Lens AFAD shall include a gate arm that descends to a down position across the approach lane of traffic when the steady CIRCULAR RED lens is illuminated and then ascends to an upright position when the flashing CIRCULAR YELLOW lens is illuminated. The gate arm shall be fully retroreflectorized on both sides, and shall have vertical alternating red and white stripes at 16-inch intervals measured horizontally as shown in Figure 8C-1. When the arm is in the down position blocking the approach lane:**

    A. The minimum vertical aspect of the arm and sheeting shall be 2 inches; and

    B. The end of the arm shall reach at least to the center of the lane being controlled.

⁰⁵ **A Stop Here On Red (R10-6 or R10-6a) sign (see Section 2B.53) shall be installed on the right-hand side of the approach at the point at which drivers are expected to stop when the steady CIRCULAR RED lens is illuminated (see Figure 6E-2).**

⁰⁶ **To inform road users to stop, the AFAD shall display a steadily illuminated CIRCULAR RED lens and the gate arm shall be in the down position. To inform road users to proceed, the AFAD shall display a flashing CIRCULAR YELLOW lens and the gate arm shall be in the upright position.**

⁰⁷ **If Red/Yellow Lens AFADs are used to control traffic in a one-lane, two-way TTC zone, safeguards shall be incorporated to prevent the flagger(s) from actuating a simultaneous display of a flashing CIRCULAR YELLOW lens at each end of the TTC zone. Additionally, the flagger shall not actuate the AFAD's display of the flashing CIRCULAR YELLOW lens until all oncoming vehicles have cleared the one-lane portion of the TTC zone.**

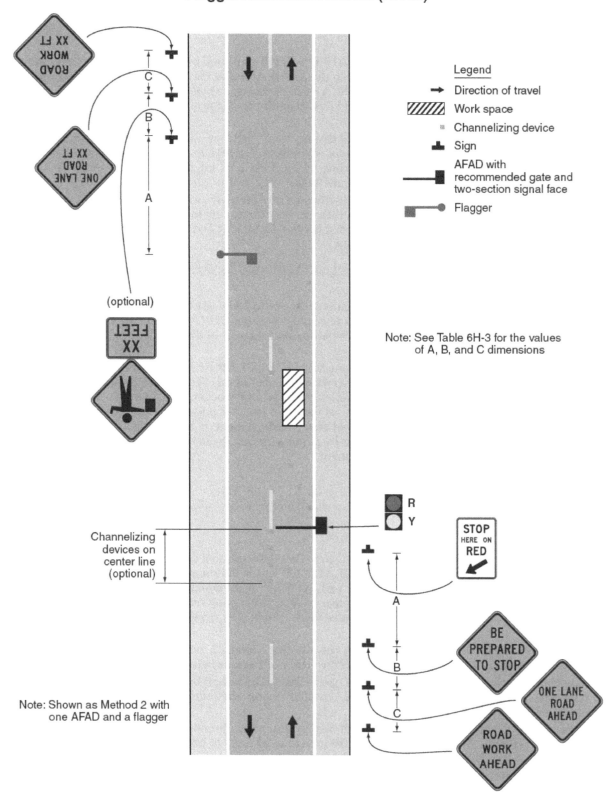

Figure 6E-2. Example of the Use of a Red/Yellow Lens Automated Flagger Assistance Device (AFAD)

08 **A change interval shall be provided as the transition between the display of the flashing CIRCULAR YELLOW indication and the display of the steady CIRCULAR RED indication. During the change interval, the CIRCULAR YELLOW lens shall be steadily illuminated. The gate arm shall remain in the upright position during the display of the steadily illuminated CIRCULAR YELLOW change interval.**

09 **A change interval shall not be provided between the display of the steady CIRCULAR RED indication and the display of the flashing CIRCULAR YELLOW indication.**

Guidance:

10 *The steadily illuminated CIRCULAR YELLOW change interval should have a duration of at least 5 seconds, unless a different duration, within the range of durations recommended by Section 4D.26, is justified by engineering judgment.*

Section 6E.07 Flagger Procedures

Support:

01 The use of paddles and flags by flaggers is illustrated in Figure 6E-3.

Standard:

02 **Flaggers shall use a STOP/SLOW paddle, a flag, or an Automated Flagger Assistance Device (AFAD) to control road users approaching a TTC zone. The use of hand movements alone without a paddle, flag, or AFAD to control road users shall be prohibited except for law enforcement personnel or emergency responders at incident scenes as described in Section 6I.01.**

03 **The following methods of signaling with paddles shall be used:**
   A. To stop road users, the flagger shall face road users and aim the STOP paddle face toward road users in a stationary position with the arm extended horizontally away from the body. The free arm shall be held with the palm of the hand above shoulder level toward approaching traffic.
   B. To direct stopped road users to proceed, the flagger shall face road users with the SLOW paddle face aimed toward road users in a stationary position with the arm extended horizontally away from the body. The flagger shall motion with the free hand for road users to proceed.
   C. To alert or slow traffic, the flagger shall face road users with the SLOW paddle face aimed toward road users in a stationary position with the arm extended horizontally away from the body.

Option:

04 To further alert or slow traffic, the flagger holding the SLOW paddle face toward road users may motion up and down with the free hand, palm down.

Standard:

05 **The following methods of signaling with a flag shall be used:**
   A. To stop road users, the flagger shall face road users and extend the flag staff horizontally across the road users' lane in a stationary position so that the full area of the flag is visibly hanging below the staff. The free arm shall be held with the palm of the hand above shoulder level toward approaching traffic.
   B. To direct stopped road users to proceed, the flagger shall face road users with the flag and arm lowered from the view of the road users, and shall motion with the free hand for road users to proceed. Flags shall not be used to signal road users to proceed.
   C. To alert or slow traffic, the flagger shall face road users and slowly wave the flag in a sweeping motion of the extended arm from shoulder level to straight down without raising the arm above a horizontal position. The flagger shall keep the free hand down.

Guidance:

06 *The flagger should stand either on the shoulder adjacent to the road user being controlled or in the closed lane prior to stopping road users. A flagger should only stand in the lane being used by moving road users after road users have stopped. The flagger should be clearly visible to the first approaching road user at all times. The flagger also should be visible to other road users. The flagger should be stationed sufficiently in advance of the workers to warn them (for example, with audible warning devices such as horns or whistles) of approaching danger by out-of-control vehicles. The flagger should stand alone, away from other workers, work vehicles, or equipment.*

Option:

07 At spot lane closures where adequate sight distance is available for the reasonably safe handling of traffic, the use of one flagger may be sufficient.

Figure 6E-3. Use of Hand-Signaling Devices by Flaggers

PREFERRED METHOD STOP/SLOW Paddle	EMERGENCY SITUATIONS ONLY Red Flag
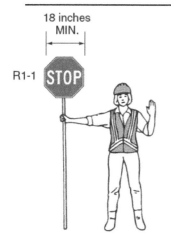	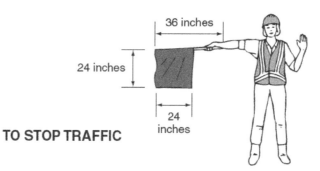
TO STOP TRAFFIC	

 **TO LET TRAFFIC PROCEED**	

TO ALERT AND SLOW TRAFFIC	

Guidance:

08 *When a single flagger is used, the flagger should be stationed on the shoulder opposite the spot lane closure or work space, or in a position where good visibility and traffic control can be maintained at all times.*

Section 6E.08 Flagger Stations

Standard:

01 **Flagger stations shall be located such that approaching road users will have sufficient distance to stop at an intended stopping point.**

Option:

02 The distances shown in Table 6E-1, which provides information regarding the stopping sight distance as a function of speed, may be used for the location of a flagger station. These distances may be increased for downgrades and other conditions that affect stopping distance.

Guidance:

03 *Flagger stations should be located such that an errant vehicle has additional space to stop without entering the work space. The flagger should identify an escape route that can be used to avoid being struck by an errant vehicle.*

Standard:

04 **Except in emergency situations, flagger stations shall be preceded by an advance warning sign or signs. Except in emergency situations, flagger stations shall be illuminated at night.**

Table 6E-1. Stopping Sight Distance as a Function of Speed

Speed*	Distance
20 mph	115 feet
25 mph	155 feet
30 mph	200 feet
35 mph	250 feet
40 mph	305 feet
45 mph	360 feet
50 mph	425 feet
55 mph	495 feet
60 mph	570 feet
65 mph	645 feet
70 mph	730 feet
75 mph	820 feet

* Posted speed, off-peak 85th-percentile speed prior to work starting, or the anticipated operating speed

CHAPTER 6F. TEMPORARY TRAFFIC CONTROL ZONE DEVICES

Section 6F.01 Types of Devices

Guidance:

01　*The design and application of TTC devices used in TTC zones should consider the needs of all road users (motorists, bicyclists, and pedestrians), including those with disabilities.*

Support:

02　FHWA policy requires that all roadside appurtenances such as traffic barriers, barrier terminals and crash cushions, bridge railings, sign and light pole supports, and work zone hardware used on the National Highway System meet the crashworthy performance criteria contained in the National Cooperative Highway Research Program (NCHRP) Report 350, "Recommended Procedures for the Safety Performance Evaluation of Highway Features." The FHWA website at "http://safety.fhwa.dot.gov/programs/roadside_hardware.htm" identifies all such hardware and includes copies of FHWA acceptance letters for each of them. In the case of proprietary items, links are provided to manufacturers' websites as a source of detailed information on specific devices. The website also contains an "Ask the Experts" section where questions on roadside design issues can be addressed.

03　Various Sections of the MUTCD require certain traffic control devices, their supports, and/or related appurtenances to be crashworthy. Such MUTCD crashworthiness provisions apply to all streets, highways, and private roads open to public travel. Also, State Departments of Transportation and local agencies might have expanded the NCHRP Report 350 crashworthy criteria to apply to certain other roadside appurtenances.

04　Crashworthiness and crash testing information on devices described in Part 6 are found in AASHTO's "Roadside Design Guide" (see Section 1A.11).

05　As defined in Section 1A.13, "crashworthy" is a characteristic of a roadside appurtenance that has been successfully crash tested in accordance with a national standard such as the NCHRP Report 350, "Recommended Procedures for the Safety Performance Evaluation of Highway Features."

Standard:

06　**Traffic control devices shall be defined as all signs, signals, markings, and other devices used to regulate, warn, or guide road users, placed on, over, or adjacent to a street, highway, private roads open to public travel (see definition in Section 1A.13), pedestrian facility, or bikeway by authority of a public body or official having jurisdiction.**

07　**All traffic control devices used for construction, maintenance, utility, or incident management operations on a street, highway, or private road open to public travel (see definition in Section 1A.13) shall comply with the applicable provisions of this Manual.**

Section 6F.02 General Characteristics of Signs

Support:

01　TTC zone signs convey both general and specific messages by means of words, symbols, and/or arrows and have the same three categories as all road user signs: regulatory, warning, and guide.

Standard:

02　**The colors for regulatory signs shall follow the Standards for regulatory signs in Table 2A-5 and Chapter 2B. Warning signs in TTC zones shall have a black legend and border on an orange background, except for the Grade Crossing Advance Warning (W10-1) sign which shall have a black legend and border on a yellow background, and except for signs that are required or recommended in Parts 2 or 7 to have fluorescent yellow-green backgrounds. Colors for guide signs shall follow the Standards in Table 2A-5 and Chapter 2D, except for guide signs as otherwise provided in Section 6F.55.**

Option:

03　Where the color orange is required, the fluorescent orange color may also be used.

Support:

04　The fluorescent version of orange provides higher conspicuity than standard orange, especially during twilight.

Option:

05　Existing warning signs that are still applicable may remain in place.

06　In order to maintain the systematic use of yellow or fluorescent yellow-green backgrounds for pedestrian, bicycle, and school warning signs in a jurisdiction, the yellow or fluorescent yellow-green background for pedestrian, bicycle, and school warning signs may be used in TTC zones.

07　Standard orange flags or flashing warning lights may be used in conjunction with signs.

Standard:

08 **When standard orange flags or flashing warning lights are used in conjunction with signs, they shall not block the sign face.**

09 **Except as provided in Section 2A.11, the sizes for TTC signs and plaques shall be as shown in Table 6F-1. The sizes in the minimum column shall only be used on local streets or roadways where the 85th-percentile speed or posted speed limit is less than 35 mph.**

Option:

10 The dimensions of signs and plaques shown in Table 6F-1 may be increased wherever necessary for greater legibility or emphasis.

Standard:

11 **Deviations from standard sizes as prescribed in this Manual shall be in 6-inch increments.**

Support:

12 Sign design details are contained in the "Standard Highway Signs and Markings" book (see Section 1A.11).

13 Section 2A.06 contains additional information regarding the design of signs, including an Option allowing the development of special word message signs if a standard word message or symbol sign is not available to convey the necessary regulatory, warning, or guidance information.

Standard:

14 **All signs used at night shall be either retroreflective with a material that has a smooth, sealed outer surface or illuminated to show the same shape and similar color both day and night.**

15 **The requirement for sign illumination shall not be considered to be satisfied by street, highway, or strobe lighting.**

Option:

16 Sign illumination may be either internal or external.

17 Signs may be made of rigid or flexible material.

Section 6F.03 Sign Placement

Guidance:

01 *Signs should be located on the right-hand side of the roadway unless otherwise provided in this Manual.*

Option:

02 Where special emphasis is needed, signs may be placed on both the left-hand and right-hand sides of the roadway. Signs mounted on portable supports may be placed within the roadway itself. Signs may also be mounted on or above barricades.

Support:

03 The provisions of this Section regarding mounting height apply unless otherwise provided for a particular sign elsewhere in this Manual.

Standard:

04 **The minimum height, measured vertically from the bottom of the sign to the elevation of the near edge of the pavement, of signs installed at the side of the road in rural areas shall be 5 feet (see Figure 6F-1).**

05 **The minimum height, measured vertically from the bottom of the sign to the top of the curb, or in the absence of curb, measured vertically from the bottom of the sign to the elevation of the near edge of the traveled way, of signs installed at the side of the road in business, commercial, or residential areas where parking or pedestrian movements are likely to occur, or where the view of the sign might be obstructed, shall be 7 feet (see Figure 6F-1).**

06 **The minimum height, measured vertically from the bottom of the sign to the sidewalk, of signs installed above sidewalks shall be 7 feet.**

Option:

07 The height to the bottom of a secondary sign mounted below another sign may be 1 foot less than the height provided in Paragraphs 4 through 6.

Guidance:

08 *Neither portable nor permanent sign supports should be located on sidewalks, bicycle facilities, or areas designated for pedestrian or bicycle traffic. If the bottom of a secondary sign that is mounted below another sign is mounted lower than 7 feet above a pedestrian sidewalk or pathway (see Section 6D.02), the secondary sign should not project more than 4 inches into the pedestrian facility.*

Table 6F-1. Temporary Traffic Control Zone Sign and Plaque Sizes (Sheet 1 of 3)

Sign or Plaque	Sign Designation	Section	Conventional Road	Freeway or Expressway	Minimum
Stop	R1-1	6F.06	30 x 30*	—	—
Stop (on Stop/Slow Paddle)	R1-1	6E.03	18 x 18	—	—
Yield	R1-2	6F.06	36 x 36 x 36*	—	30 x 30 x 30
To Oncoming Traffic (plaque)	R1-2aP	6F.06	36 x 30	48 x 36	24 x 18
Wait on Stop	R1-7	6E.05	24 x 30	24 x 30	—
Go on Slow	R1-8	6E.05	24 x 30	24 x 30	—
Speed Limit	R2-1	6F.12	24 x 30*	36 x 48	—
Fines Higher (plaque)	R2-6P	6F.12	24 x 18	36 x 24	—
Fines Double (plaque)	R2-6aP	6F.12	24 x 18	36 x 24	—
$XX Fine (plaque)	R2-6bP	6F.12	24 x 18	36 x 24	—
Begin Higher Fines Zone	R2-10	6F.12	24 x 30	36 x 48	—
End Higher Fines Zone	R2-11	6F.12	24 x 30	36 x 48	—
End Work Zone Speed Limit	R2-12	6F.12	24 x 36	36 x 54	—
Movement Prohibition	R3-1,2,3,4,18,27	6F.06	24 x 24*	36 x 36	—
Mandatory Movement (1 lane)	R3-5	6F.06	30 x 36	—	—
Optional Movement (1 lane)	R3-6	6F.06	30 x 36	—	—
Mandatory Movement (text)	R3-7	6F.06	30 x 30*	—	—
Advance Intersection Lane Control	R3-8	6F.06	Varies x 30	—	—
Do Not Pass	R4-1	6F.06	24 x 30	36 x 48	—
Pass With Care	R4-2	6F.06	24 x 30	36 x 48	—
Keep Right	R4-7	6F.06	24 x 30	36 x 48	—
Narrow Keep Right	R4-7c	6F.06	18 x 30	—	—
Stay in Lane	R4-9	6F.11	24 x 30	36 x 48	—
Do Not Enter	R5-1	6F.06	30 x 30*	36 x 36	—
Wrong Way	R5-1a	6F.06	36 x 24*	42 x 30	—
One Way	R6-1	6F.06	36 x 12*	54 x 18	—
One Way	R6-2	6F.06	24 x 30*	36 x 48	—
No Parking (symbol)	R8-3	6F.06	24 x 24	36 x 36	—
Pedestrian Crosswalk	R9-8	6F.13	36 x 18	—	—
Sidewalk Closed	R9-9	6F.14	24 x 12	—	—
Sidewalk Closed, Use Other Side	R9-10	6F.14	24 x 12	—	—
Sidewalk Closed Ahead, Cross Here	R9-11	6F.14	24 x 18	—	—
Sidewalk Closed, Cross Here	R9-11a	6F.14	24 x 12	—	—
Road Closed	R11-2	6F.08	48 x 30	—	—
Road Closed - Local Traffic Only	R11-3a,3b,4	6F.09	60 x 30	—	—
Weight Limit	R12-1,2	6F.10	24 x 30	36 x 48	—
Weight Limit (with symbols)	R12-5	6F.10	24 x 36	36 x 48	—
Turn and Curve Signs	W1-1,2,3,4	6F.16	36 x 36	48 x 48	30 x 30
Reverse Curve (2 or more lanes)	W1-4b,4c	6F.48	36 x 36	48 x 48	30 x 30
One-Direction Large Arrow	W1-6	6F.16	48 x 24	60 x 30	—
Chevron	W1-8	6F.16	18 x 24	30 x 36	—
Stop Ahead	W3-1	6F.16	36 x 36	48 x 48	30 x 30
Yield Ahead	W3-2	6F.16	36 x 36	48 x 48	30 x 30
Signal Ahead	W3-3	6F.16	36 x 36	48 x 48	30 x 30
Be Prepared to Stop	W3-4	6F.16	36 x 36	48 x 48	30 x 30
Reduced Speed Limit Ahead	W3-5	6F.16	36 x 36	48 x 48	30 x 30

Table 6F-1. Temporary Traffic Control Zone Sign and Plaque Sizes (Sheet 2 of 3)

Sign or Plaque	Sign Designation	Section	Conventional Road	Freeway or Expressway	Minimum
XX MPH Speed Zone Ahead	W3-5a	6F.16	36 x 36	48 x 48	30 x 30
Merging Traffic	W4-1,5	6F.16	36 x 36	48 x 48	36 x 36
Lane Ends	W4-2	6F.24	36 x 36	48 x 48	30 x 30
Added Lane	W4-3,6	6F.16	36 x 36	48 x 48	30 x 30
No Merge Area (plaque)	W4-5P	6F.16	18 x 24	24 x 30	—
Road Narrows	W5-1	6F.16	36 x 36	48 x 48	30 x 30
Narrow Bridge	W5-2	6F.16	36 x 36	48 x 48	30 x 30
One Lane Bridge	W5-3	6F.16	36 x 36	48 x 48	30 x 30
Ramp Narrows	W5-4	6F.26	36 x 36	48 x 48	30 x 30
Divided Highway	W6-1	6F.16	36 x 36	48 x 48	30 x 30
Divided Highway Ends	W6-2	6F.16	36 x 36	48 x 48	30 x 30
Two-Way Traffic	W6-3	6F.32	36 x 36	48 x 48	30 x 30
Two-Way Traffic	W6-4	6F.76	12 x 18	12 x 18	—
Hill (symbol)	W7-1	6F.16	36 x 36	48 x 48	30 x 30
Next XX Miles (plaque)	W7-3aP	6F.53	24 x 18	36 x 30	—
Bump	W8-1	6F.16	36 x 36	48 x 48	30 x 30
Dip	W8-2	6F.16	36 x 36	48 x 48	30 x 30
Pavement Ends	W8-3	6F.16	36 x 36	48 x 48	30 x 30
Soft Shoulder	W8-4	6F.44	36 x 36	48 x 48	30 x 30
Slippery When Wet	W8-5	6F.16	36 x 36	48 x 48	30 x 30
Truck Crossing	W8-6	6F.36	36 x 36	48 x 48	30 x 30
Loose Gravel	W8-7	6F.16	36 x 36	48 x 48	30 x 30
Rough Road	W8-8	6F.16	36 x 36	48 x 48	30 x 30
Low Shoulder	W8-9	6F.44	36 x 36	48 x 48	30 x 30
Uneven Lanes	W8-11	6F.45	36 x 36	48 x 48	30 x 30
No Center Line	W8-12	6F.47	36 x 36	48 x 48	30 x 30
Fallen Rocks	W8-14	6F.16	36 x 36	48 x 48	30 x 30
Grooved Pavement	W8-15	6F.16	36 x 36	48 x 48	30 x 30
Motorcycle (plaque)	W8-15P	6F.54	24 x 18	30 x 24	—
Shoulder Drop Off (symbol)	W8-17	6F.44	36 x 36	48 x 48	30 x 30
Shoulder Drop-Off (plaque)	W8-17P	6F.44	24 x 18	30 x 24	—
Road May Flood	W8-18	6F.16	36 x 36	48 x 48	24 x 24
No Shoulder	W8-23	6F.16	36 x 36	48 x 48	30 x 30
Steel Plate Ahead	W8-24	6F.46	36 x 36	48 x 48	30 x 30
Shoulder Ends	W8-25	6F.16	36 x 36	48 x 48	30 x 30
Lane Ends	W9-1,2	6F.16	36 x 36	48 x 48	30 x 30
Center Lane Closed Ahead	W9-3	6F.23	36 x 36	48 x 48	30 x 30
Grade Crossing Advance Warning	W10-1	6F.16	36 dia.	—	—
Truck	W11-10	6F.36	36 x 36	48 x 48	30 x 30
Double Arrow	W12-1	6F.16	30 x 30	—	—
Low Clearance	W12-2	6F.16	36 x 36	48 x 48	30 x 30
Advisory Speed (plaque)	W13-1P	6F.52	24 x 24	30 x 30	18 x 18
On Ramp (plaque)	W13-4P	6F.25	36 x 36	36 x 36	—
No Passing Zone (pennant)	W14-3	6F.16	48 x 48 x 36	64 x 64 x 48	40 x 40 x 30
XX Feet (plaque)	W16-2P	6F.16	24 x 18	30 x 24	—
Road Work (with distance)	W20-1	6F.18	36 x 36	48 x 48	30 x 30

Table 6F-1. Temporary Traffic Control Zone Sign and Plaque Sizes (Sheet 3 of 3)

Sign or Plaque	Sign Designation	Section	Conventional Road	Freeway or Expressway	Minimum
Detour (with distance)	W20-2	6F.19	36 x 36	48 x 48	30 x 30
Road (Street) Closed (with distance)	W20-3	6F.20	36 x 36	48 x 48	30 x 30
One Lane Road (with distance)	W20-4	6F.21	36 x 36	48 x 48	30 x 30
Lane(s) Closed (with distance)	W20-5,5a	6F.22	36 x 36	48 x 48	30 x 30
Flagger (symbol)	W20-7	6F.31	36 x 36	48 x 48	30 x 30
Flagger	W20-7a	6F.31	36 x 36	48 x 48	30 x 30
Slow (on Stop/Slow Paddle)	W20-8	6E.03	18 x 18	—	—
Workers	W21-1,1a	6F.33	36 x 36	48 x 48	30 x 30
Fresh Oil (Tar)	W21-2	6F.34	36 x 36	48 x 48	30 x 30
Road Machinery Ahead	W21-3	6F.35	36 x 36	48 x 48	30 x 30
Slow Moving Vehicle	W21-4	6G.06	36 x 18	—	—
Shoulder Work	W21-5	6F.37	36 x 36	48 x 48	30 x 30
Shoulder Closed	W21-5a	6F.37	36 x 36	48 x 48	30 x 30
Shoulder Closed (with distance)	W21-5b	6F.37	36 x 36	48 x 48	30 x 30
Survey Crew	W21-6	6F.38	36 x 36	48 x 48	30 x 30
Utility Work Ahead	W21-7	6F.39	36 x 36	48 x 48	30 x 30
Mowing Ahead	W21-8	6G.06	36 x 36	48 x 48	30 x 30
Blasting Zone Ahead	W22-1	6F.41	36 x 36	48 x 48	30 x 30
Turn Off 2-Way Radio and Cell Phone	W22-2	6F.42	42 x 36	42 x 36	—
End Blasting Zone	W22-3	6F.43	42 x 36	42 x 36	36 x 30
Slow Traffic Ahead	W23-1	6F.27	48 x 24	48 x 24	—
New Traffic Pattern Ahead	W23-2	6F.30	36 x 36	48 x 48	30 x 30
Double Reverse Curve (1 lane)	W24-1	6F.49	36 x 36	48 x 48	30 x 30
Double Reverse Curve (2 lanes)	W24-1a	6F.49	36 x 36	48 x 48	30 x 30
Double Reverse Curve (3 lanes)	W24-1b	6F.49	36 x 36	48 x 48	30 x 30
All Lanes	W24-1cP	6F.49	24 x 24	30 x 30	—
Road Work Next XX Miles	G20-1	6F.56	36 x 18	48 x 24	—
End Road Work	G20-2	6F.57	36 x 18	48 x 24	—
Pilot Car Follow Me	G20-4	6F.58	36 x 18	—	—
Work Zone (plaque)	G20-5aP	6F.12	24 x 18	36 x 24	—
Exit Open	E5-2	6F.28	48 x 36	48 x 36	—
Exit Closed	E5-2a	6F.28	48 x 36	48 x 36	—
Exit Only	E5-3	6F.29	48 x 36	48 x 36	—
Detour	M4-8	6F.59	24 x 12	30 x 15	—
End Detour	M4-8a	6F.59	24 x 18	24 x 18	—
End	M4-8b	6F.59	24 x 12	24 x 12	—
Detour	M4-9	6F.59	30 x 24	48 x 36	—
Bike/Pedestrian Detour	M4-9a	6F.59	30 x 24	—	—
Pedestrian Detour	M4-9b	6F.59	30 x 24	—	—
Bike Detour	M4-9c	6F.59	30 x 24	—	—
Detour	M4-10	6F.59	48 x 18	—	—

* See Table 2B-1 for minimum size required for signs facing traffic on multi-lane conventional roads

Notes: 1. Larger signs may be used wherever necessary for greater legibility or emphasis
2. Dimensions are shown in inches and are shown as width x height

Figure 6F-1. Height and Lateral Location of Signs—Typical Installations

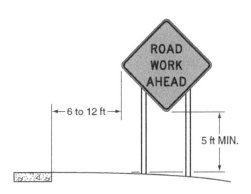

A - RURAL AREA

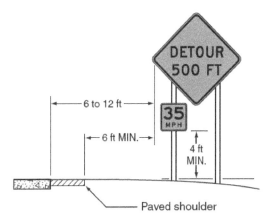

B - RURAL AREA WITH ADVISORY SPEED PLAQUE

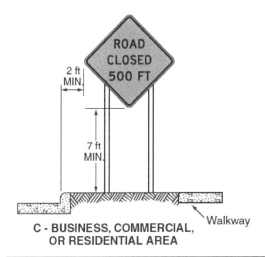

C - BUSINESS, COMMERCIAL, OR RESIDENTIAL AREA

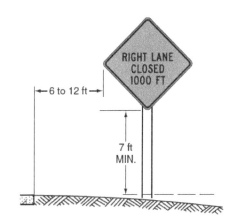

D - BUSINESS, COMMERCIAL, OR RESIDENTIAL AREA (WITHOUT CURB)

Standard:

09 **Where it has been determined that the accommodation of pedestrians with disabilities is necessary, signs shall be mounted and placed in accordance with Section 4.4 of the "Americans with Disabilities Act Accessibility Guidelines for Buildings and Facilities (ADAAG)" (see Section 1A.11).**

10 **Signs mounted on barricades and barricade/sign combinations shall be crashworthy.**

Guidance:

11 *Except as provided in Paragraph 12, signs mounted on portable sign supports that do not meet the minimum mounting heights provided in Paragraphs 4 through 6 should not be used for a duration of more than 3 days.*

Option:

12 The R9-8 through R9-11a series, R11 series, W1-6 through W1-8 series, M4-10, E5-1, or other similar type signs (see Figures 6F-3, 6F-4, and 6F-5) may be used on portable sign supports that do not meet the minimum mounting heights provided in Paragraphs 4 through 6 for longer than 3 days.

Support:

13 Methods of mounting signs other than on posts are illustrated in Figure 6F-2.

Guidance:

14 *Signs mounted on Type 3 Barricades should not cover more than 50 percent of the top two rails or 33 percent of the total area of the three rails.*

Standard:

15 **Sign supports shall be crashworthy. Where large signs having an area exceeding 50 square feet are installed on multiple breakaway posts, the clearance from the ground to the bottom of the sign shall be at least 7 feet.**

Figure 6F-2. Methods of Mounting Signs Other Than on Posts

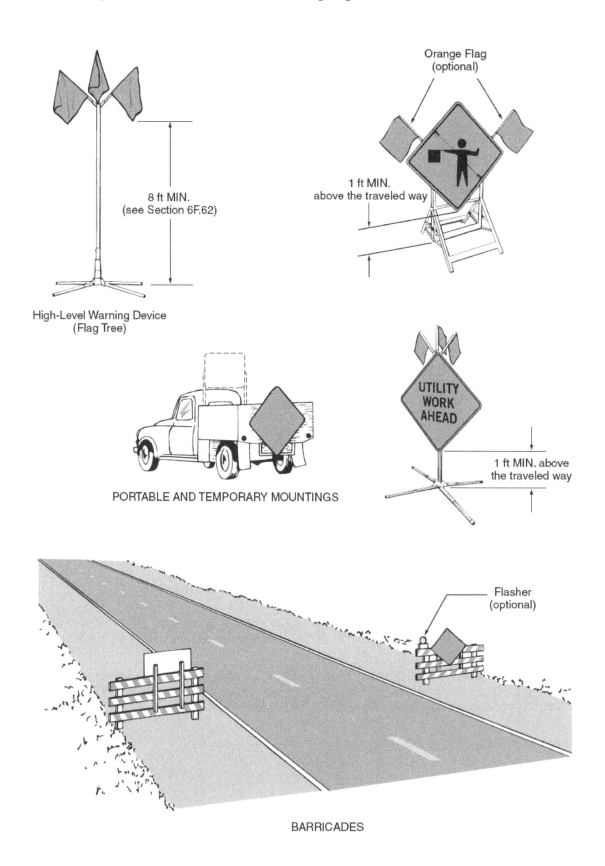

16 **The bottom of a sign mounted on a barricade, or other portable support, shall be at least 1 foot above the traveled way.**

Option:

17 For mobile operations, a sign may be mounted on a work vehicle, a shadow vehicle, or a trailer stationed in advance of the TTC zone or moving along with it.

Support:

18 If alterations are made to specific traffic control device supports that have been successfully crash tested in accordance with NCHRP Report 350, the altered supports might not be considered to be crashworthy.

Section 6F.04 Sign Maintenance

Guidance:

01 *Signs should be properly maintained for cleanliness, visibility, and correct positioning.*

02 *Signs that have lost significant legibility should be promptly replaced.*

Support:

03 Section 2A.08 contains information regarding the retroreflectivity of signs, including the signs that are used in TTC zones.

Section 6F.05 Regulatory Sign Authority

Support:

01 Regulatory signs such as those shown in Figure 6F-3 inform road users of traffic laws or regulations and indicate the applicability of legal requirements that would not otherwise be apparent.

Standard:

02 **Regulatory signs shall be authorized by the public agency or official having jurisdiction and shall conform with Chapter 2B.**

Section 6F.06 Regulatory Sign Design

Standard:

01 **TTC regulatory signs shall comply with the Standards for regulatory signs presented in Part 2 and in the FHWA's "Standard Highway Signs and Markings" book (see Section 1A.11).**

Support:

02 Regulatory signs are generally rectangular with a black legend and border on a white background. Exceptions include the STOP, YIELD, DO NOT ENTER, WRONG WAY, and ONE WAY signs.

Option:

03 The ONE WAY sign may be either a horizontal or vertical rectangular sign.

Section 6F.07 Regulatory Sign Applications

Standard:

01 **If a TTC zone requires regulatory measures different from those existing, the existing permanent regulatory devices shall be removed or covered and superseded by the appropriate temporary regulatory signs. This change shall be made in compliance with applicable ordinances or statutes of the jurisdiction.**

Section 6F.08 ROAD (STREET) CLOSED Sign (R11-2)

Guidance:

01 *The ROAD (STREET) CLOSED (R11-2) sign (see Figure 6F-3) should be used when the roadway is closed to all road users except contractors' equipment or officially authorized vehicles. The R11-2 sign should be accompanied by appropriate warning and detour signing.*

Option:

02 The words BRIDGE OUT (or BRIDGE CLOSED) may be substituted for ROAD (STREET) CLOSED where applicable.

Guidance:

03 *The ROAD (STREET) CLOSED sign should be installed at or near the center of the roadway on or above a Type 3 Barricade that closes the roadway (see Section 6F.68).*

Figure 6F-3. Regulatory Signs and Plaques in Temporary Traffic Control Zones
(Sheet 1 of 2)

Figure 6F-3. Regulatory Signs and Plaques in Temporary Traffic Control Zones
(Sheet 2 of 2)

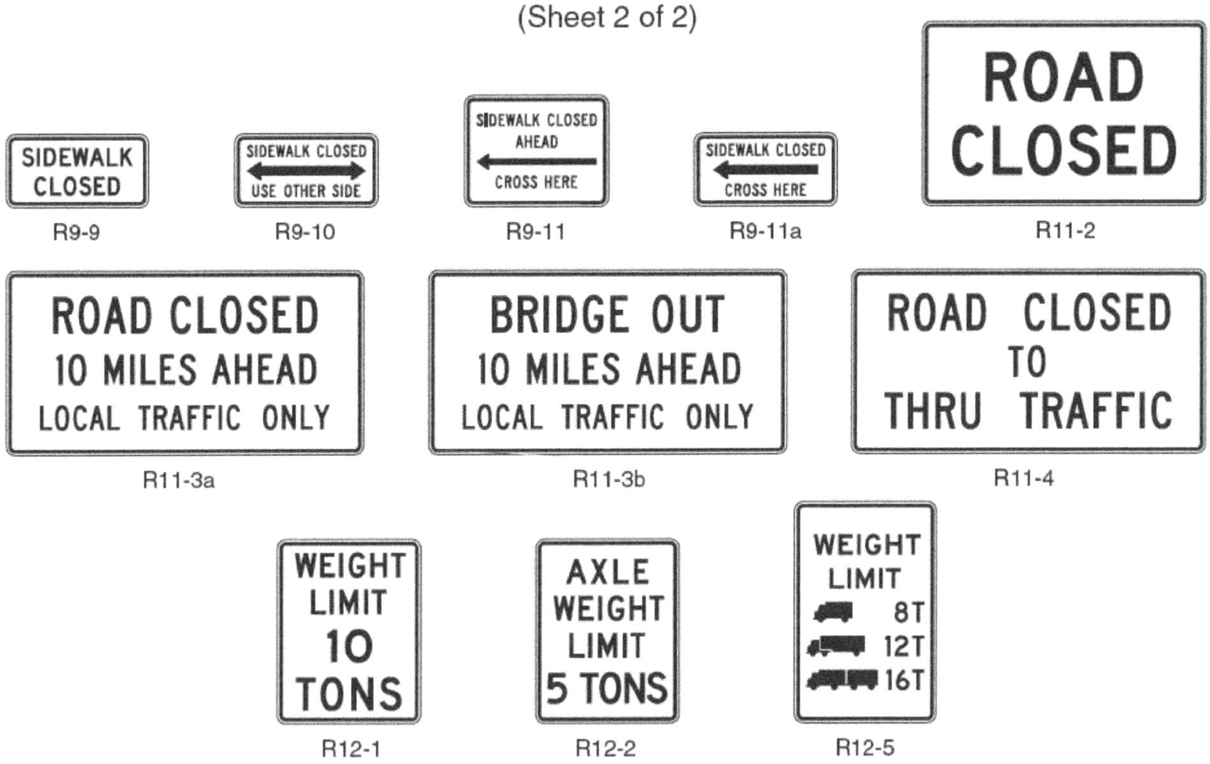

SIDEWALK CLOSED	R9-9
SIDEWALK CLOSED / USE OTHER SIDE (↔)	R9-10
SIDEWALK CLOSED AHEAD / CROSS HERE (←)	R9-11
SIDEWALK CLOSED / CROSS HERE (←)	R9-11a
ROAD CLOSED	R11-2
ROAD CLOSED 10 MILES AHEAD / LOCAL TRAFFIC ONLY	R11-3a
BRIDGE OUT 10 MILES AHEAD / LOCAL TRAFFIC ONLY	R11-3b
ROAD CLOSED TO THRU TRAFFIC	R11-4
WEIGHT LIMIT 10 TONS	R12-1
AXLE WEIGHT LIMIT 5 TONS	R12-2
WEIGHT LIMIT 8T / 12T / 16T	R12-5

Standard:

04 The ROAD (STREET) CLOSED sign shall not be used where road user flow is maintained through the TTC zone with a reduced number of lanes on the existing roadway or where the actual closure is some distance beyond the sign.

Section 6F.09 Local Traffic Only Signs (R11-3a, R11-4)

Guidance:

01 *The Local Traffic Only signs (see Figure 6F-3) should be used where road user flow detours to avoid a closure some distance beyond the sign, but where local road users can use the roadway to the point of closure. These signs should be accompanied by appropriate warning and detour signing.*

02 *In rural applications, the Local Traffic Only sign should have the legend ROAD CLOSED XX MILES AHEAD, LOCAL TRAFFIC ONLY (R11-3a).*

Option:

03 In urban areas, the legend ROAD (STREET) CLOSED TO THRU TRAFFIC (R11-4) or ROAD CLOSED, LOCAL TRAFFIC ONLY may be used.

04 In urban areas, a word message that includes the name of an intersecting street name or well-known destination may be substituted for the words XX MILES AHEAD on the R11-3a sign where applicable.

05 The words BRIDGE OUT (or BRIDGE CLOSED) may be substituted for the words ROAD (STREET) CLOSED on the R11-3a or R11-4 sign where applicable.

Section 6F.10 Weight Limit Signs (R12-1, R12-2, R12-5)

Standard:

01 A Weight Limit sign (see Figure 6F-3), which shows the gross weight or axle weight that is permitted on the roadway or bridge, shall be consistent with State or local regulations and shall not be installed without the approval of the authority having jurisdiction over the highway.

02 When weight restrictions are imposed because of the activity in a TTC zone, a marked detour shall be provided for vehicles weighing more than the posted limit.

Section 6F.11 STAY IN LANE Sign (R4-9)

Option:

01   A STAY IN LANE (R4-9) sign (see Figure 6F-3) may be used where a multi-lane shift has been incorporated as part of the TTC on a highway to direct road users around road work that occupies part of the roadway on a multi-lane highway.

Section 6F.12 Work Zone and Higher Fines Signs and Plaques

Option:

01   A WORK ZONE (G20-5aP) plaque (see Figure 6F-3) may be mounted above a Speed Limit sign to emphasize that a reduced speed limit is in effect within a TTC zone. An END WORK ZONE SPEED LIMIT (R2-12) sign (see Figure 6F-3) may be installed at the downstream end of the reduced speed limit zone.

Guidance:

02   *A BEGIN HIGHER FINES ZONE (R2-10) sign (see Figure 6F-3) should be installed at the upstream end of a work zone where increased fines are imposed for traffic violations, and an END HIGHER FINES ZONE (R2-11) sign (see Figure 6F-3) should be installed at the downstream end of the work zone.*

Option:

03   Alternate legends such as BEGIN (or END) DOUBLE FINES ZONE may also be used for the R2-10 and R2-11 signs.

04   A FINES HIGHER, FINES DOUBLE, or $XX FINE plaque (see Section 2B.17 and Figure 6F-3) may be mounted below the Speed Limit sign if increased fines are imposed for traffic violations within the TTC zone.

05   Individual signs and plaques for work zone speed limits and higher fines may be combined into a single sign or may be displayed as an assembly of signs and plaques.

Section 6F.13 PEDESTRIAN CROSSWALK Sign (R9-8)

Option:

01   The PEDESTRIAN CROSSWALK (R9-8) sign (see Figure 6F-3) may be used to indicate where a temporary crosswalk has been established.

Standard:

02   **If a temporary crosswalk is established, it shall be accessible to pedestrians with disabilities in accordance with Section 6D.02.**

Section 6F.14 SIDEWALK CLOSED Signs (R9-9, R9-10, R9-11, R9-11a)

Guidance:

01   *SIDEWALK CLOSED signs (see Figure 6F-3) should be used where pedestrian flow is restricted. Bicycle/Pedestrian Detour (M4-9a) signs or Pedestrian Detour (M4-9b) signs should be used where pedestrian flow is rerouted (see Section 6F.59).*

02   *The SIDEWALK CLOSED (R9-9) sign should be installed at the beginning of the closed sidewalk, at the intersections preceding the closed sidewalk, and elsewhere along the closed sidewalk as needed.*

03   *The SIDEWALK CLOSED, (ARROW) USE OTHER SIDE (R9-10) sign should be installed at the beginning of the restricted sidewalk when a parallel sidewalk exists on the other side of the roadway.*

04   *The SIDEWALK CLOSED AHEAD, (ARROW) CROSS HERE (R9-11) sign should be used to indicate to pedestrians that sidewalks beyond the sign are closed and to direct them to open crosswalks, sidewalks, or other travel paths.*

05   *The SIDEWALK CLOSED, (ARROW) CROSS HERE (R9-11a) sign should be installed just beyond the point to which pedestrians are being redirected.*

Support:

06   These signs are typically mounted on a detectable barricade to encourage compliance and to communicate with pedestrians that the sidewalk is closed. Printed signs are not useful to many pedestrians with visual disabilities. A barrier or barricade detectable by a person with a visual disability is sufficient to indicate that a sidewalk is closed. If the barrier is continuous with detectable channelizing devices for an alternate route, accessible signing might not be necessary. An audible information device is needed when the detectable barricade or barrier for an alternate channelized route is not continuous.

Section 6F.15 Special Regulatory Signs

Option:

01 Special regulatory signs may be used based on engineering judgment consistent with regulatory requirements.

Guidance:

02 *Special regulatory signs should comply with the general requirements of color, shape, and alphabet size and series. The sign message should be brief, legible, and clear.*

Section 6F.16 Warning Sign Function, Design, and Application

Support:

01 TTC zone warning signs (see Figure 6F-4) notify road users of specific situations or conditions on or adjacent to a roadway that might not otherwise be apparent.

Standard:

02 **TTC warning signs shall comply with the Standards for warning signs presented in Part 2 and in FHWA's "Standard Highway Signs and Markings" book (see Section 1A.11). Except as provided in Paragraph 3, TTC warning signs shall be diamond-shaped with a black legend and border on an orange background, except for the W10-1 sign which shall have a black legend and border on a yellow background, and except for signs that are required or recommended in Parts 2 or 7 to have fluorescent yellow-green backgrounds.**

Option:

03 Warning signs used for TTC incident management situations may have a black legend and border on a fluorescent pink background.

04 Mounting or space considerations may justify a change from the standard diamond shape.

05 In emergencies, available warning signs having yellow backgrounds may be used if signs with orange or fluorescent pink backgrounds are not at hand.

Guidance:

06 *Where roadway or road user conditions require greater emphasis, larger than standard size warning signs should be used, with the symbol or legend enlarged approximately in proportion to the outside dimensions.*

07 *Where any part of the roadway is obstructed or closed by work activities or incidents, advance warning signs should be installed to alert road users well in advance of these obstructions or restrictions.*

08 *Where road users include pedestrians, the provision of supplemental audible information or detectable barriers or barricades should be considered for people with visual disabilities.*

Support:

09 Detectable barriers or barricades communicate very clearly to pedestrians who have visual disabilities that they can no longer proceed in the direction that they are traveling.

Option:

10 Advance warning signs may be used singly or in combination.

11 Where distances are not displayed on warning signs as part of the message, a supplemental plaque with the distance legend may be mounted immediately below the sign on the same support.

Section 6F.17 Position of Advance Warning Signs

Guidance:

01 *Where highway conditions permit, warning signs should be placed in advance of the TTC zone at varying distances depending on roadway type, condition, and posted speed. Table 6C-1 contains information regarding the spacing of advance warning signs. Where a series of two or more advance warning signs is used, the closest sign to the TTC zone should be placed approximately 100 feet for low-speed urban streets to 1,000 feet or more for freeways and expressways.*

02 *Where multiple advance warning signs are needed on the approach to a TTC zone, the ROAD WORK AHEAD (W20-1) sign should be the first advance warning sign encountered by road users.*

Support:

03 Various conditions, such as limited sight distance or obstructions that might require a driver to reduce speed or stop, might require additional advance warning signs.

Option:

04 As an alternative to a specific distance on advance warning signs, the word AHEAD may be used.

Page 588 — 2009 Edition

Figure 6F-4. Warning Signs and Plaques in Temporary Traffic Control Zones
(Sheet 1 of 3)

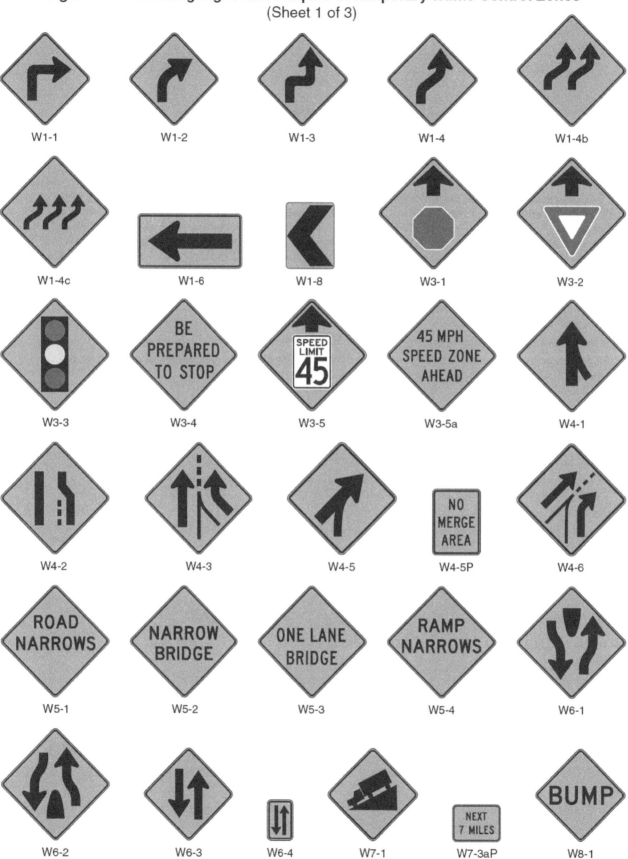

Sect. 6F.17 — December 2009

Figure 6F-4. Warning Signs and Plaques in Temporary Traffic Control Zones
(Sheet 2 of 3)

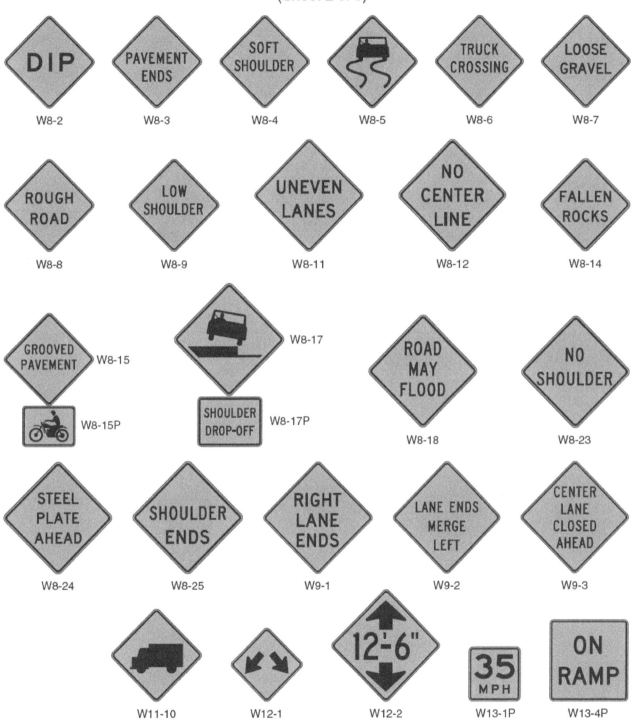

Page 590 — 2009 Edition

Figure 6F-4. Warning Signs and Plaques in Temporary Traffic Control Zones (Sheet 3 of 3)

* An optional STREET WORK word message sign is shown in the "Standard Highway Signs and Markings" book.
** An optional STREET CLOSED word message sign is shown in the "Standard Highway Signs and Markings" book.
*** An optional FLAGGER (W20-7a) word message sign is shown in the "Standard Highway Signs and Markings" book.
**** An optional FRESH TAR word message sign is show in the "Standard Highway Signs and Markings" book.

Support:
05   At TTC zones on lightly-traveled roads, all of the advance warning signs prescribed for major construction might not be needed.

Option:
06   Utility work, maintenance, or minor construction can occur within the TTC zone limits of a major construction project, and additional warning signs may be needed.

Guidance:
07   *Utility, maintenance, and minor construction signing and TTC should be coordinated with appropriate authorities so that road users are not confused or misled by the additional TTC devices.*

Section 6F.18 ROAD (STREET) WORK Sign (W20-1)

Guidance:
01   *The ROAD (STREET) WORK (W20-1) sign (see Figure 6F-4), which serves as a general warning of obstructions or restrictions, should be located in advance of the work space or any detour, on the road where the work is taking place.*

02   *Where traffic can enter a TTC zone from a crossroad or a major (high-volume) driveway, an advance warning sign should be used on the crossroad or major driveway.*

Standard:
03   **The ROAD (STREET) WORK (W20-1) sign shall have the legend ROAD (STREET) WORK, XX FEET, XX MILES, or AHEAD.**

Section 6F.19 DETOUR Sign (W20-2)

Guidance:
01   *The DETOUR (W20-2) sign (see Figure 6F-4) should be used in advance of a road user detour over a different roadway or route.*

Standard:
02   **The DETOUR sign shall have the legend DETOUR, XX FEET, XX MILES, or AHEAD.**

Section 6F.20 ROAD (STREET) CLOSED Sign (W20-3)

Guidance:
01   *The ROAD (STREET) CLOSED (W20-3) sign (see Figure 6F-4) should be used in advance of the point where a highway is closed to all road users, or to all but local road users.*

Standard:
02   **The ROAD (STREET) CLOSED sign shall have the legend ROAD (STREET) CLOSED, XX FEET, XX MILES, or AHEAD.**

Section 6F.21 ONE LANE ROAD Sign (W20-4)

Standard:
01   **The ONE LANE ROAD (W20-4) sign (see Figure 6F-4) shall be used only in advance of that point where motor vehicle traffic in both directions must use a common single lane (see Section 6C.10). It shall have the legend ONE LANE ROAD, XX FEET, XX MILES, or AHEAD.**

Section 6F.22 Lane(s) Closed Signs (W20-5, W20-5a)

Standard:
01   **The Lane(s) Closed sign (see Figure 6F-4) shall be used in advance of that point where one or more through lanes of a multi-lane roadway are closed.**

02   **For a single lane closure, the Lane Closed (W20-5) sign (see Figure 6F-4) shall have the legend RIGHT (LEFT) LANE CLOSED, XX FEET, XX MILES, or AHEAD. Where two adjacent lanes are closed, the W20-5a sign (see Figure 6F-4) shall have the legend XX RIGHT (LEFT) LANES CLOSED, XX FEET, XX MILES, or AHEAD.**

Page 592 — 2009 Edition

Section 6F.23 CENTER LANE CLOSED AHEAD Sign (W9-3)

Guidance:

01 *The CENTER LANE CLOSED AHEAD (W9-3) sign (see Figure 6F-4) should be used in advance of that point where work occupies the center lane(s) and approaching motor vehicle traffic is directed to the right or left of the work zone in the center lane.*

Section 6F.24 Lane Ends Sign (W4-2)

Option:

01 The Lane Ends (W4-2) symbol sign (see Figure 6F-4) may be used to warn drivers of the reduction in the number of lanes for moving motor vehicle traffic in the direction of travel on a multi-lane roadway.

Section 6F.25 ON RAMP Plaque (W13-4P)

Guidance:

01 *When work is being done on a ramp, but the ramp remains open, the ON RAMP (W13-4P) plaque (see Figure 6F-4) should be used to supplement the advance ROAD WORK sign.*

Section 6F.26 RAMP NARROWS Sign (W5-4)

Guidance:

01 *The RAMP NARROWS (W5-4) sign (see Figure 6F-4) should be used in advance of the point where work on a ramp reduces the normal width of the ramp along a part or all of the ramp.*

Section 6F.27 SLOW TRAFFIC AHEAD Sign (W23-1)

Option:

01 The SLOW TRAFFIC AHEAD (W23-1) sign (see Figure 6F-4) may be used on a shadow vehicle, usually mounted on the rear of the most upstream shadow vehicle, along with other appropriate signs for mobile operations to warn of slow moving work vehicles. A ROAD WORK (W20-1) sign may also be used with the SLOW TRAFFIC AHEAD sign.

Section 6F.28 EXIT OPEN and EXIT CLOSED Signs (E5-2, E5-2a)

Option:

01 An EXIT OPEN (E5-2) or EXIT CLOSED (E5-2a) sign (see Figure 6F-5) may be used to supplement other warning signs where work is being conducted in the vicinity of an exit ramp and where the exit maneuver for vehicular traffic using the ramp is different from the normal condition.

Guidance:

02 *When an exit ramp is closed, an EXIT CLOSED sign panel with a black legend and border on an orange background should be placed diagonally across the interchange/intersection guide signs.*

Figure 6F-5. Exit Open and Closed and Detour Signs

Section 6F.29 EXIT ONLY Sign (E5-3)

Option:

01　An EXIT ONLY (E5-3) sign (see Figure 6F-5) may be used to supplement other warning signs where work is being conducted in the vicinity of an exit ramp and where the exit maneuver for vehicular traffic using the ramp is different from the normal condition.

Section 6F.30 NEW TRAFFIC PATTERN AHEAD Sign (W23-2)

Option:

01　A NEW TRAFFIC PATTERN AHEAD (W23-2) sign (see Figure 6F-4) may be used on the approach to an intersection or along a section of roadway to provide advance warning of a change in traffic patterns, such as revised lane usage, roadway geometry, or signal phasing.

Guidance:

02　*To retain its effectiveness, the W23-2 sign should be displayed for up to 2 weeks, and then it should be covered or removed until it is needed again.*

Section 6F.31 Flagger Signs (W20-7, W20-7a)

Guidance:

01　*The Flagger (W20-7) symbol sign (see Figure 6F-4) should be used in advance of any point where a flagger is stationed to control road users.*

Option:

02　A distance legend may be displayed on a supplemental plaque below the Flagger sign. The sign may be used with appropriate legends or in conjunction with other warning signs, such as the BE PREPARED TO STOP (W3-4) sign (see Figure 6F-4).

03　The FLAGGER (W20-7a) word message sign with distance legends may be substituted for the Flagger (W20-7) symbol sign.

Section 6F.32 Two-Way Traffic Sign (W6-3)

Guidance:

01　*When one roadway of a normally divided highway is closed, with two-way vehicular traffic maintained on the other roadway, the Two-Way Traffic (W6-3) sign (see Figure 6F-4) should be used at the beginning of the two-way vehicular traffic section and at intervals to remind road users of opposing vehicular traffic.*

Section 6F.33 Workers Signs (W21-1, W21-1a)

Option:

01　A Workers (W21-1) symbol sign (see Figure 6F-4) may be used to alert road users of workers in or near the roadway.

Guidance:

02　*In the absence of other warning devices, a Workers symbol sign should be used when workers are in the roadway.*

Option:

03　The WORKERS (W21-1a) word message sign may be used as an alternate to the Workers (W21-1) symbol sign.

Section 6F.34 FRESH OIL (TAR) Sign (W21-2)

Guidance:

01　*The FRESH OIL (TAR) (W21-2) sign (see Figure 6F-4) should be used to warn road users of the surface treatment.*

Section 6F.35 ROAD MACHINERY AHEAD Sign (W21-3)

Option:

01　The ROAD MACHINERY AHEAD (W21-3) sign (see Figure 6F-4) may be used to warn of machinery operating in or adjacent to the roadway.

Section 6F.36 Motorized Traffic Signs (W8-6, W11-10)

Option:

01 Motorized Traffic (W8-6, W11-10) signs may be used to alert road users to locations where unexpected travel on the roadway or entries into or departures from the roadway by construction vehicles might occur. The TRUCK CROSSING (W8-6) word message sign may be used as an alternate to the Truck Crossing (W11-10) symbol sign (see Figure 6F-4) where there is an established construction vehicle crossing of the roadway.

Support:

02 These locations might be relatively confined or might occur randomly over a segment of roadway.

Section 6F.37 Shoulder Work Signs (W21-5, W21-5a, W21-5b)

Support:

01 Shoulder Work signs (see Figure 6F-4) warn of maintenance, reconstruction, or utility operations on the highway shoulder where the roadway is unobstructed.

Standard:

02 **The Shoulder Work sign shall have the legend SHOULDER WORK (W21-5), RIGHT (LEFT) SHOULDER CLOSED (W21-5a), or RIGHT (LEFT) SHOULDER CLOSED XX FT or AHEAD (W21-5b).**

Option:

03 The Shoulder Work sign may be used in advance of the point on a non-limited access highway where there is shoulder work. It may be used singly or in combination with a ROAD WORK NEXT XX MILES or ROAD WORK AHEAD sign.

Guidance:

04 *On freeways and expressways, the RIGHT (LEFT) SHOULDER CLOSED XX FT or AHEAD (W21-5b) sign followed by RIGHT (LEFT) SHOULDER CLOSED (W21-5a) sign should be used in advance of the point where the shoulder work occurs and should be preceded by a ROAD WORK AHEAD sign.*

Section 6F.38 SURVEY CREW Sign (W21-6)

Guidance:

01 *The SURVEY CREW (W21-6) sign (see Figure 6F-4) should be used to warn of surveying crews working in or adjacent to the roadway.*

Section 6F.39 UTILITY WORK Sign (W21-7)

Option:

01 The UTILITY WORK (W21-7) sign (see Figure 6F-4) may be used as an alternate to the ROAD (STREET) WORK (W20-1) sign for utility operations on or adjacent to a highway.

Support:

02 Typical examples of where the UTILITY WORK sign is used appear in Figures 6H-4, 6H-6, 6H-10, 6H-15, 6H-18, 6H-21, 6H-22, 6H-26, and 6H-33.

Standard:

03 **The UTILITY WORK sign shall carry the legend UTILITY WORK, XX FEET, XX MILES, or AHEAD.**

Section 6F.40 Signs for Blasting Areas

Support:

01 Radio-Frequency (RF) energy can cause the premature firing of electric detonators (blasting caps) used in TTC zones.

Standard:

02 **Road users shall be warned to turn off mobile radio transmitters and cellular telephones where blasting operations occur. A sequence of signs shall be prominently displayed to direct operators of mobile radio equipment, including cellular telephones, to turn off transmitters in a blasting area. These signs shall be covered or removed when there are no explosives in the area or the area is otherwise secured.**

Section 6F.41 BLASTING ZONE AHEAD Sign (W22-1)

Standard:

01 The BLASTING ZONE AHEAD (W22-1) sign (see Figure 6F-4) shall be used in advance of any TTC zone where explosives are being used. The TURN OFF 2-WAY RADIO AND CELL PHONE and END BLASTING ZONE signs shall be used in sequence with this sign.

Section 6F.42 TURN OFF 2-WAY RADIO AND CELL PHONE Sign (W22-2)

Standard:

01 The TURN OFF 2-WAY RADIO AND CELL PHONE (W22-2) sign (see Figure 6F-4) shall follow the BLASTING ZONE AHEAD sign and shall be placed at least 1,000 feet before the beginning of the blasting zone.

Section 6F.43 END BLASTING ZONE Sign (W22-3)

Standard:

01 The END BLASTING ZONE (W22-3) sign (see Figure 6F-4) shall be placed a minimum of 1,000 feet past the blasting zone.

Option:

02 The END BLASTING ZONE sign may be placed either with or preceding the END ROAD WORK sign.

Section 6F.44 Shoulder Signs and Plaque (W8-4, W8-9, W8-17, and W8-17P)

Option:

01 The SOFT SHOULDER (W8-4) sign (see Figure 6F-4) may be used to warn of a soft shoulder condition.

02 The LOW SHOULDER (W8-9) sign (see Figure 6F-4) may be used to warn of a shoulder condition where there is an elevation difference of 3 inches or less between the shoulder and the travel lane.

Guidance:

03 *The Shoulder Drop Off (W8-17) sign (see Figure 6F-4) should be used when an unprotected shoulder drop-off, adjacent to the travel lane, exceeds 3 inches in depth for a continuous length along the roadway, based on engineering judgment.*

Option:

04 A SHOULDER DROP-OFF (W8-17P) supplemental plaque (see Figure 6F-4) may be mounted below the W8-17 sign.

Section 6F.45 UNEVEN LANES Sign (W8-11)

Guidance:

01 *The UNEVEN LANES (W8-11) sign (see Figure 6F-4) should be used during operations that create a difference in elevation between adjacent lanes that are open to travel.*

Section 6F.46 STEEL PLATE AHEAD Sign (W8-24)

Option:

01 A STEEL PLATE AHEAD (W8-24) sign (see Figure 6F-4) may be used to warn road users that the presence of a temporary steel plate(s) might make the road surface uneven and might create slippery conditions during wet weather.

Section 6F.47 NO CENTER LINE Sign (W8-12)

Guidance:

01 *The NO CENTER LINE (W8-12) sign (see Figure 6F-4) should be used when the work obliterates the center line pavement markings. This sign should be placed at the beginning of the TTC zone and repeated at 2-mile intervals in long TTC zones.*

Support:

02 Section 6F.78 contains information regarding temporary markings.

Section 6F.48 Reverse Curve Signs (W1-4 Series)

Guidance:

01 *In order to give road users advance notice of a lane shift, a Reverse Curve (W1-4, W1-4b, or W1-4c) sign (see Figure 6F-4) should be used when a lane (or lanes) is being shifted to the left or right. If the design speed of the curves is 30 mph or less, a Reverse Turn (W1-3) sign should be used.*

Standard:

02 **If a Reverse Curve (or Turn) sign is used, the direction of the reverse curve (or turn) shall be appropriately illustrated. Except as provided in Paragraph 3, the number of lanes illustrated on the sign shall be the same as the number of through lanes available to road users.**

Option:

03 Where two or more lanes are being shifted, a W1-4 (or W1-3) sign with an ALL LANES (W24-1cP) plaque (see Figure 6F-4) may be used instead of a sign that illustrates the number of lanes.

04 Where more than three lanes are being shifted, the Reverse Curve (or Turn) sign may be rectangular.

Section 6F.49 Double Reverse Curve Signs (W24-1 Series)

Option:

01 The Double Reverse Curve (W24-1, W24-1a, or W24-1b) sign (see Figure 6F-4) may be used where the tangent distance between two reverse curves is less than 600 feet, thus making it difficult for a second Reverse Curve (W1-4 series) sign to be placed between the curves. If the design speed of the curves is 30 mph or less, Double Reverse Turn signs should be used.

Standard:

02 **If a Double Reverse Curve (or Turn) sign is used, the direction of the double reverse curve (or turn) shall be appropriately illustrated. Except as provided in Paragraph 3, the number of lanes illustrated on the sign shall be the same as the number of through lanes available to road users.**

Option:

03 Where two or more lanes are being shifted, a W24-1 (or Double Reverse Turn sign showing one lane) sign with an ALL LANES (W24-1cP) plaque (see Figure 6F-4) may be used instead of a sign that illustrates the number of lanes.

04 Where more than three lanes are being shifted, the Double Reverse Curve (or Turn) sign may be rectangular.

Section 6F.50 Other Warning Signs

Option:

01 Advance warning signs may be used by themselves or with other advance warning signs.

02 Besides the warning signs specifically related to TTC zones, several other warning signs in Part 2 may apply in TTC zones.

Standard:

03 **Except as provided in Section 6F.02, other warning signs that are used in TTC zones shall have black legends and borders on an orange background.**

Section 6F.51 Special Warning Signs

Option:

01 Special warning signs may be used based on engineering judgment.

Guidance:

02 *Special warning signs should comply with the general requirements of color, shape, and alphabet size and series. The sign message should be brief, legible, and clear.*

Section 6F.52 Advisory Speed Plaque (W13-1P)

Option:

01 In combination with a warning sign, an Advisory Speed (W13-1P) plaque (see Figure 6F-4) may be used to indicate a recommended speed through the TTC zone.

Standard:

02 **The Advisory Speed plaque shall not be used in conjunction with any sign other than a warning sign, nor shall it be used alone. When used with orange TTC zone signs, this plaque shall have a black legend and border on an orange background. The sign shall be at least 24 x 24 inches in size when used with a sign that is 36 x 36 inches or larger. Except in emergencies, an Advisory Speed plaque shall not be mounted until the recommended speed is determined by the highway agency.**

Section 6F.53 Supplementary Distance Plaque (W7-3aP)

Option:

01 In combination with a warning sign, a Supplementary Distance (W7-3aP) plaque (see Figure 6F-4) with the legend NEXT XX MILES may be used to indicate the length of highway over which a work activity is being conducted, or over which a condition exists in the TTC zone.

02 In long TTC zones, Supplementary Distance plaques with the legend NEXT XX MILES may be placed in combination with warning signs at regular intervals within the zone to indicate the remaining length of highway over which the TTC work activity or condition exists.

Standard:

03 **The Supplementary Distance plaque with the legend NEXT XX MILES shall not be used in conjunction with any sign other than a warning sign, nor shall it be used alone. When used with orange TTC zone signs, this plaque shall have a black legend and border on an orange background. The sign shall be at least 30 x 24 inches in size when used with a sign that is 36 x 36 inches or larger.**

Guidance:

04 *When used in TTC zones, the Supplementary Distance plaque with the legend NEXT XX MILES should be placed below the initial warning sign designating that, within the approaching zone, a temporary work activity or condition exists.*

Section 6F.54 Motorcycle Plaque (W8-15P)

Option:

01 A Motorcycle (W8-15P) plaque (see Figure 6F-4) may be mounted below a LOOSE GRAVEL (W8-7) sign, a GROOVED PAVEMENT (W8-15) sign, a METAL BRIDGE DECK (W8-16) sign, or a STEEL PLATE AHEAD (W8-24) sign if the warning is intended to be directed primarily to motorcyclists.

Section 6F.55 Guide Signs

Support:

01 Guide signs along highways provide road users with information to help them along their way through the TTC zone. The design of guide signs is presented in Part 2.

Guidance:

02 *The following guide signs should be used in TTC zones as needed:*
  A. Standard route markings, where temporary route changes are necessary,
  B. Directional signs and street name signs, and
  C. Special guide signs relating to the condition or work being done.

Standard:

03 **If additional temporary guide signs are used in TTC zones, they shall have a black legend and border on an orange background.**

Option:

04 Guide signs used in TTC incident management situations may have a black legend and border on a fluorescent pink background.

05 When directional signs and street name signs are used in conjunction with detour routing, these signs may have a black legend and border on an orange background.

06 When permanent directional signs or permanent street name signs are used in conjunction with detour signing, they may have a white legend on a green background.

Section 6F.56 ROAD WORK NEXT XX MILES Sign (G20-1)

Guidance:

01 *The ROAD WORK NEXT XX MILES (G20-1) sign (see Figure 6F-4) should be installed in advance of TTC zones that are more than 2 miles in length.*

Option:

02 The ROAD WORK NEXT XX MILES sign may be mounted on a Type 3 Barricade. The sign may also be used for TTC zones of shorter length.

Standard:

03 **The distance displayed on the ROAD WORK NEXT XX MILES sign shall be stated to the nearest whole mile.**

Section 6F.57 END ROAD WORK Sign (G20-2)

Guidance:

01 *When used, the END ROAD WORK (G20-2) sign (see Figure 6F-4) should be placed near the downstream end of the termination area, as determined by engineering judgment.*

Option:

02 The END ROAD WORK sign may be installed on the back of a warning sign facing the opposite direction of road users or on the back of a Type 3 Barricade.

Section 6F.58 PILOT CAR FOLLOW ME Sign (G20-4)

Standard:

01 **The PILOT CAR FOLLOW ME (G20-4) sign (see Figure 6F-4) shall be mounted in a conspicuous position on the rear of a vehicle used for guiding one-way vehicular traffic through or around a TTC zone (see Section 6C.13).**

Section 6F.59 Detour Signs (M4-8, M4-8a, M4-8b, M4-9, M4-9a, M4-9b, M4-9c, and M4-10)

Standard:

01 **Each detour shall be adequately marked with standard temporary route signs and destination signs.**

Option:

02 Detour signs in TTC incident management situations may have a black legend and border on a fluorescent pink background.

03 The Detour Arrow (M4-10) sign (see Figure 6F-5) may be used where a detour route has been established.

04 The DETOUR (M4-8) sign (see Figure 6F-5) may be mounted at the top of a route sign assembly to mark a temporary route that detours from a highway, bypasses a section closed by a TTC zone, and rejoins the highway beyond the TTC zone.

Guidance:

05 *The Detour Arrow (M4-10) sign should normally be mounted just below the ROAD CLOSED (R11-2, R11-3a, or R11-4) sign. The Detour Arrow sign should include a horizontal arrow pointed to the right or left as required.*

06 *The DETOUR (M4-9) sign (see Figure 6F-5) should be used for unnumbered highways, for emergency situations, for periods of short durations, or where, over relatively short distances, road users are guided along the detour and back to the desired highway without route signs.*

07 *A Street Name sign should be placed above, or the street name should be incorporated into, a DETOUR (M4-9) sign to indicate the name of the street being detoured.*

Option:

08 The END DETOUR (M4-8a) or END (M4-8b) sign (see Figure 6F-5) may be used to indicate that the detour has ended.

Guidance:

09 *When the END DETOUR sign is used on a numbered highway, the sign should be mounted above a route sign after the downstream end of the detour.*

10 *The Pedestrian/Bicycle Detour (M4-9a) sign (see Figure 6F-5) should be used where a pedestrian/bicycle detour route has been established because of the closing of a pedestrian/bicycle facility to through traffic.*

Standard:

11 **If used, the Pedestrian/Bicycle Detour sign shall have an arrow pointing in the appropriate direction.**

Option:

12 The arrow on a Pedestrian/Bicycle Detour sign may be on the sign face or on a supplemental plaque.

13 The Pedestrian Detour (M4-9b) sign or Bicycle Detour (M4-9c) sign (see Figure 6F-5) may be used where a pedestrian or bicycle detour route (not both) has been established because of the closing of the pedestrian or bicycle facility to through traffic.

Section 6F.60 Portable Changeable Message Signs

Support:

01 Portable changeable message signs (PCMS) are TTC devices installed for temporary use with the flexibility to display a variety of messages. In most cases, portable changeable message signs follow the same provisions for design and application as those given for changeable message signs in Chapter 2L. The information in this Section describes situations where the provisions for portable changeable message signs differ from those given in Chapter 2L.

02 Portable changeable message signs are used most frequently on high-density urban freeways, but have applications on all types of highways where highway alignment, road user routing problems, or other pertinent conditions require advance warning and information.

03 Portable changeable message signs have a wide variety of applications in TTC zones including: roadway, lane, or ramp closures; incident management; width restriction information; speed control or reductions; advisories on work scheduling; road user management and diversion; warning of adverse conditions or special events; and other operational control.

04 The primary purpose of portable changeable message signs in TTC zones is to advise the road user of unexpected situations. Portable changeable message signs are particularly useful as they are capable of:

A. Conveying complex messages,
B. Displaying real time information about conditions ahead, and
C. Providing information to assist road users in making decisions prior to the point where actions must be taken.

05 Some typical applications include the following:

A. Where the speed of vehicular traffic is expected to drop substantially;
B. Where significant queuing and delays are expected;
C. Where adverse environmental conditions are present;
D. Where there are changes in alignment or surface conditions;
E. Where advance notice of ramp, lane, or roadway closures is needed;
F. Where crash or incident management is needed; and/or
G. Where changes in the road user pattern occur.

Guidance:

06 *The components of a portable changeable message sign should include: a message sign, control systems, a power source, and mounting and transporting equipment. The front face of the sign should be covered with a protective material.*

Standard:

07 **Portable changeable message signs shall comply with the applicable design and application principles established in Chapter 2A. Portable changeable message signs shall display only traffic operational, regulatory, warning, and guidance information, and shall not be used for advertising messages.**

Support:

08 Section 2L.02 contains information regarding overly simplistic or vague messages that is also applicable to portable changeable message signs.

Standard:

09 **The colors used for legends on portable changeable message signs shall comply with those shown in Table 2A-5.**

Support:

10 Section 2L.04 contains information regarding the luminance, luminance contrast, and contrast orientation that is also applicable to portable changeable message signs.

Guidance:

11 *Portable changeable message signs should be visible from 1/2 mile under both day and night conditions.*

Support:

12 Section 2B.13 contains information regarding the design of portable changeable message signs that are used to display speed limits that change based on operational conditions, or are used to display the speed at which approaching drivers are traveling.

Guidance:

13 *A portable changeable message sign should be limited to three lines of eight characters per line or should consist of a full matrix display.*

14 *Except as provided in Paragraph 15, the letter height used for portable changeable message sign messages should be a minimum of 18 inches.*

Option:

15 For portable changeable message signs mounted on service patrol trucks or other incident response vehicles, a letter height as short as 10 inches may be used. Shorter letter sizes may also be used on a portable changeable message sign used on low speed facilities provided that the message is legible from at least 650 feet.

16 The portable changeable message sign may vary in size.

Guidance:

17 Messages on a portable changeable message sign should consist of no more than two phases, and a phase should consist of no more than three lines of text. Each phase should be capable of being understood by itself, regardless of the order in which it is read. Messages should be centered within each line of legend. If more than one portable changeable message sign is simultaneously legible to road users, then only one of the signs should display a sequential message at any given time.

Support:

18 Road users have difficulties in reading messages displayed in more than two phases on a typical three-line portable changeable message sign.

Standard:

19 **Techniques of message display such as animation, rapid flashing, dissolving, exploding, scrolling, travelling horizontally or vertically across the face of the sign, or other dynamic elements shall not be used.**

Guidance:

20 When a message is divided into two phases, the display time for each phase should be at least 2 seconds, and the sum of the display times for both of the phases should be a maximum of 8 seconds.

21 All messages should be designed with consideration given to the principles provided in this Section and also taking into account the following:

    A. The message should be as brief as possible and should contain three thoughts (with each thought preferably shown on its own line) that convey:
        1. The problem or situation that the road user will encounter ahead,
        2. The location of or distance to the problem or situation, and
        3. The recommended driver action.

    B. If more than two phases are needed to display a message, additional portable changeable message signs should be used. When multiple portable changeable message signs are needed, they should be placed on the same side of the roadway and they should be separated from each other by a distance of at least 1,000 feet on freeways and expressways, and by a distance of at least 500 feet on other types of highways.

Standard:

22 **When the word messages shown in Tables 1A-1 or 1A-2 need to be abbreviated on a portable changeable message sign, the provisions described in Section 1A.15 shall be followed.**

23 **In order to maintain legibility, portable changeable message signs shall automatically adjust their brightness under varying light conditions.**

24 **The control system shall include a display screen upon which messages can be reviewed before being displayed on the message sign. The control system shall be capable of maintaining memory when power is unavailable.**

25 **Portable changeable message signs shall be equipped with a power source and a battery back-up to provide continuous operation when failure of the primary power source occurs.**

26 **The mounting of portable changeable message signs on a trailer, a large truck, or a service patrol truck shall be such that the bottom of the message sign shall be a minimum of 7 feet above the roadway in urban areas and 5 feet above the roadway in rural areas when it is in the operating mode.**

Guidance:

27 Portable changeable message signs should be used as a supplement to and not as a substitute for conventional signs and pavement markings.

28 When portable changeable message signs are used for route diversion, they should be placed far enough in advance of the diversion to allow road users ample opportunity to perform necessary lane changes, to adjust their speed, or to exit the affected highway.

29 Portable changeable message signs should be sited and aligned to provide maximum legibility and to allow time for road users to respond appropriately to the portable changeable message sign message.

30 Portable changeable message signs should be placed off the shoulder of the roadway and behind a traffic barrier, if practical. Where a traffic barrier is not available to shield the portable changeable message sign, it should be placed off the shoulder and outside of the clear zone. If a portable changeable message sign has to be placed on the shoulder of the roadway or within the clear zone, it should be delineated with retroreflective TTC devices.

31 When portable changeable message signs are used in TTC zones, they should display only TTC messages.

Section 6F.61 Arrow Boards
Standard:

01 **An arrow board shall be a sign with a matrix of elements capable of either flashing or sequential displays. This sign shall provide additional warning and directional information to assist in merging and controlling road users through or around a TTC zone.**

Guidance:

02 *An arrow board in the arrow or chevron mode should be used to advise approaching traffic of a lane closure along major multi-lane roadways in situations involving heavy traffic volumes, high speeds, and/or limited sight distances, or at other locations and under other conditions where road users are less likely to expect such lane closures.*

03 *If used, an arrow board should be used in combination with appropriate signs, channelizing devices, or other TTC devices.*

04 *An arrow board should be placed on the shoulder of the roadway or, if practical, farther from the traveled lane. It should be delineated with retroreflective TTC devices. When an arrow board is not being used, it should be removed; if not removed, it should be shielded; or if the previous two options are not feasible, it should be delineated with retroreflective TTC devices.*

Standard:

05 **Arrow boards shall meet the minimum size, legibility distance, number of elements, and other specifications shown in Figure 6F-6.**

Support:

06 Type A arrow boards are appropriate for use on low-speed urban streets. Type B arrow boards are appropriate for intermediate-speed facilities and for maintenance or mobile operations on high-speed roadways. Type C arrow boards are intended to be used on high-speed, high-volume motor vehicle traffic control projects. Type D arrow boards are intended for use on vehicles authorized by the State or local agency.

Standard:

07 **Type A, B, and C arrow boards shall have solid rectangular appearances. A Type D arrow board shall conform to the shape of the arrow.**

08 **All arrow boards shall be finished in non-reflective black. The arrow board shall be mounted on a vehicle, a trailer, or other suitable support.**

Guidance:

09 *The minimum mounting height, measured vertically from the bottom of the board to the roadway below it or to the elevation of the near edge of the roadway, of an arrow board should be 7 feet, except on vehicle-mounted arrow boards, which should be as high as practical.*

10 *A vehicle-mounted arrow board should be provided with remote controls.*

Standard:

11 **Arrow board elements shall be capable of at least a 50 percent dimming from full brilliance. The dimmed mode shall be used for nighttime operation of arrow boards.**

Guidance:

12 *Full brilliance should be used for daytime operation of arrow boards.*

Standard:

13 **The arrow board shall have suitable elements capable of the various operating modes. The color presented by the elements shall be yellow.**

Guidance:

14 *If an arrow board consisting of a bulb matrix is used, the elements should be recess-mounted or equipped with an upper hood of not less than 180 degrees.*

Figure 6F-6. Advance Warning Arrow Board Display Specifications

Operating Mode	Display (Type C arrow board illustrated)
1. At least one of the three following modes shall be provided:	(right arrow shown; left is similar)
Flashing Arrow	Merge Right
Sequential Arrow	Merge Right
Sequential Chevron	Merge Right
2. The following mode shall be provided: Flashing Double Arrow	Merge Right or Left
3. At least one of the following modes shall be provided: Flashing Caution or Alternating Diamond Caution	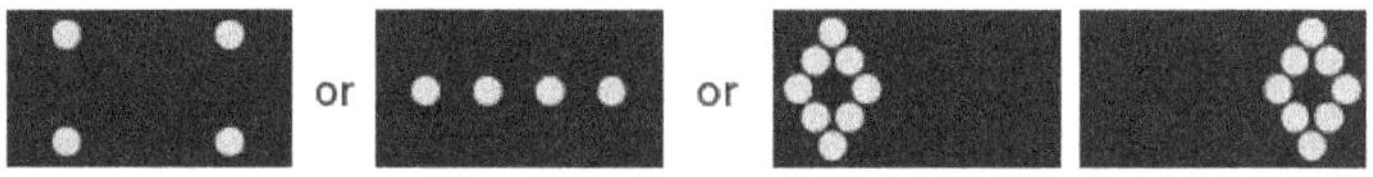 or or Flashing Caution Flashing Caution Alternating Diamond Caution

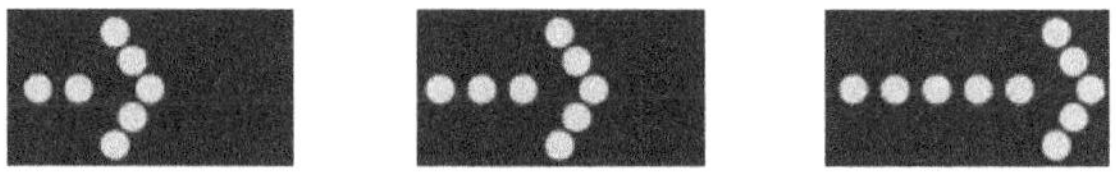

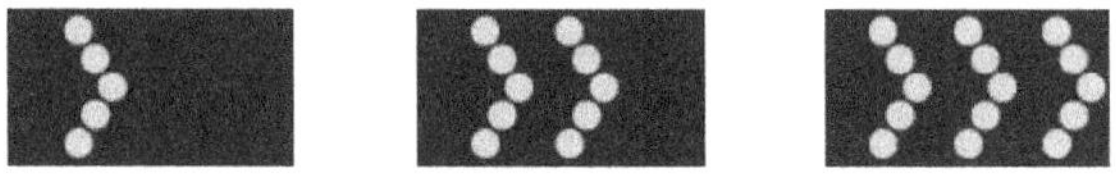

Arrow Board Type	Minimum Size	Minimum Legibility Distance	Minimum Number of Elements
A	48 x 24 inches	1/2 mile	12
B	60 x 30 inches	3/4 mile	13
C	96 x 48 inches	1 mile	15
D	None*	1/2 mile	12

*Length of arrow equals 48 inches, width of arrowhead equals 24 inches

Standard:

15 The minimum element on-time shall be 50 percent for the flashing mode, with equal intervals of 25 percent for each sequential phase. The flashing rate shall be not less than 25 or more than 40 flashes per minute.

16 An arrow board shall have the following three mode selections:
   A. A Flashing Arrow, Sequential Arrow, or Sequential Chevron mode;
   B. A flashing Double Arrow mode; and
   C. A flashing Caution or Alternating Diamond mode.

17 An arrow board in the arrow or chevron mode shall be used only for stationary or moving lane closures on multi-lane roadways.

18 For shoulder work, blocking the shoulder, for roadside work near the shoulder, or for temporarily closing one lane on a two-lane, two-way roadway, an arrow board shall be used only in the caution mode.

Guidance:

19 *For a stationary lane closure, the arrow board should be located on the shoulder at the beginning of the merging taper.*

20 *Where the shoulder is narrow, the arrow board should be located in the closed lane.*

Standard:

21 When arrow boards are used to close multiple lanes, a separate arrow board shall be used for each closed lane.

Guidance:

22 *When arrow boards are used to close multiple lanes, if the first arrow board is placed on the shoulder, the second arrow board should be placed in the first closed lane at the upstream end of the second merging taper (see Figure 6H-37). When the first arrow board is placed in the first closed lane, the second arrow board should be placed in the second closed lane at the downstream end of the second merging taper.*

23 *For mobile operations where a lane is closed, the arrow board should be located to provide adequate separation from the work operation to allow for appropriate reaction by approaching drivers.*

Standard:

24 A vehicle displaying an arrow board shall be equipped with high-intensity rotating, flashing, oscillating, or strobe lights.

25 Arrow boards shall only be used to indicate a lane closure. Arrow boards shall not be used to indicate a lane shift.

Option:

26 A portable changeable message sign may be used to simulate an arrow board display.

Section 6F.62 High-Level Warning Devices (Flag Trees)

Option:

01 A high-level warning device (flag tree) may supplement other TTC devices in TTC zones.

Support:

02 A high-level warning device is designed to be seen over the top of typical passenger cars. A typical high-level warning device is shown in Figure 6F-2.

Standard:

03 A high-level warning device shall consist of a minimum of two flags with or without a Type B high-intensity flashing warning light. The distance from the roadway to the bottom of the lens of the light and to the lowest point of the flag material shall be not less than 8 feet. The flag shall be 16 inches square or larger and shall be orange or fluorescent red-orange in color.

Option:

04 An appropriate warning sign may be mounted below the flags.

Support:

05 High-level warning devices are most commonly used in high-density road user situations to warn road users of short-term operations.

Section 6F.63 Channelizing Devices

Standard:

01 **Designs of various channelizing devices shall be as shown in Figure 6F-7. All channelizing devices shall be crashworthy.**

Support:

02 The function of channelizing devices is to warn road users of conditions created by work activities in or near the roadway and to guide road users. Channelizing devices include cones, tubular markers, vertical panels, drums, barricades, and longitudinal channelizing devices.

03 Channelizing devices provide for smooth and gradual vehicular traffic flow from one lane to another, onto a bypass or detour, or into a narrower traveled way. They are also used to channelize vehicular traffic away from the work space, pavement drop-offs, pedestrian or shared-use paths, or opposing directions of vehicular traffic.

Standard:

04 **Devices used to channelize pedestrians shall be detectable to users of long canes and visible to persons having low vision.**

05 **Where channelizing devices are used to channelize pedestrians, there shall be continuous detectable bottom and top surfaces to be detectable to users of long canes. The bottom of the bottom surface shall be no higher than 2 inches above the ground. The top of the top surface shall be no lower than 32 inches above the ground.**

Option:

06 A gap not exceeding 2 inches between the bottom rail and the ground surface may be used to facilitate drainage.

Guidance:

07 *Where multiple channelizing devices are aligned to form a continuous pedestrian channelizer, connection points should be smooth to optimize long-cane and hand trailing.*

08 *The spacing between cones, tubular markers, vertical panels, drums, and barricades should not exceed a distance in feet equal to 1.0 times the speed limit in mph when used for taper channelization, and a distance in feet equal to 2.0 times the speed limit in mph when used for tangent channelization.*

09 *When channelizing devices have the potential of leading vehicular traffic out of the intended vehicular traffic space as shown in Figure 6H-39, the channelizing devices should be extended a distance in feet of 2.0 times the speed limit in mph beyond the downstream end of the transition area.*

Option:

10 Warning lights (see Section 6F.83) may be added to channelizing devices in areas with frequent fog, snow, or severe roadway curvature, or where visual distractions are present.

Standard:

11 **Warning lights shall flash when placed on channelizing devices used alone or in a cluster to warn of a condition. Except for the sequential flashing warning lights discussed in Paragraphs 12 and 13, warning lights placed on channelizing devices used in a series to channelize road users shall be steady-burn.**

Option:

12 A series of sequential flashing warning lights may be placed on channelizing devices that form a merging taper in order to increase driver detection and recognition of the merging taper.

Standard:

13 **When used, the successive flashing of the sequential warning lights shall occur from the upstream end of the merging taper to the downstream end of the merging taper in order to identify the desired vehicle path. Each warning light in the sequence shall be flashed at a rate of not less than 55 nor more than 75 times per minute.**

14 **The retroreflective material used on channelizing devices shall have a smooth, sealed outer surface that will display a similar color day or night.**

Option:

15 The name and telephone number of the highway agency, contractor, or supplier may be displayed on the non-retroreflective surface of all types of channelizing devices.

Standard:

16 **The letters and numbers of the name and telephone number shall be non-retroreflective and not over 2 inches in height.**

Figure 6F-7. Channelizing Devices

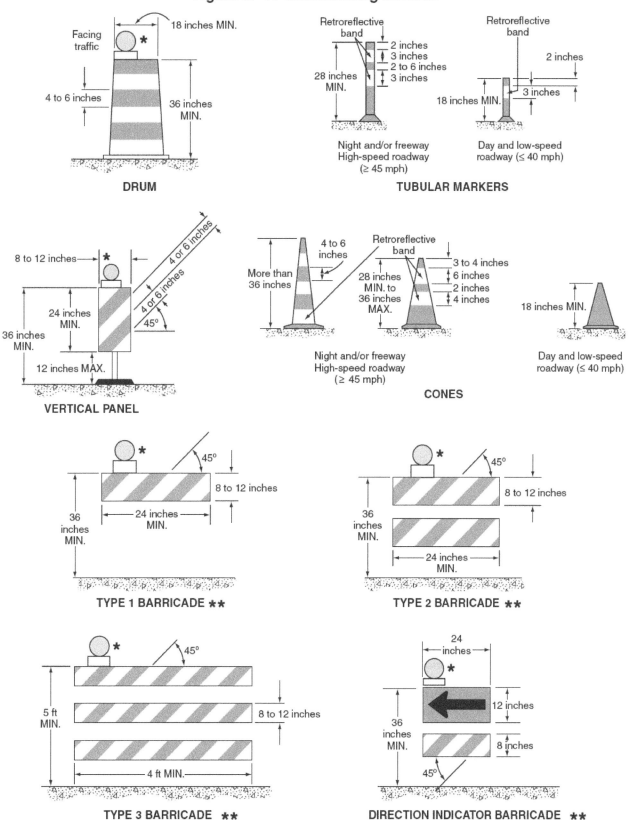

* Warning lights (optional)
** Rail stripe widths shall be 6 inches, except that 4-inch wide stripes may be used if rail lengths are less than 36 inches. The sides of barricades facing traffic shall have retroreflective rail faces.

Guidance:

17 *Particular attention should be given to maintaining the channelizing devices to keep them clean, visible, and properly positioned at all times.*

Standard:

18 **Devices that are damaged or have lost a significant amount of their retroreflectivity and effectiveness shall be replaced.**

Section 6F.64 Cones

Standard:

01 **Cones (see Figure 6F-7) shall be predominantly orange and shall be made of a material that can be struck without causing damage to the impacting vehicle. For daytime and low-speed roadways, cones shall be not less than 18 inches in height. When cones are used on freeways and other high-speed highways or at night on all highways, or when more conspicuous guidance is needed, cones shall be a minimum of 28 inches in height.**

02 **For nighttime use, cones shall be retroreflectorized or equipped with lighting devices for maximum visibility. Retroreflectorization of cones that are 28 to 36 inches in height shall be provided by a 6-inch wide white band located 3 to 4 inches from the top of the cone and an additional 4-inch wide white band located approximately 2 inches below the 6-inch band.**

03 **Retroreflectorization of cones that are more than 36 inches in height shall be provided by horizontal, circumferential, alternating orange and white retroreflective stripes that are 4 to 6 inches wide. Each cone shall have a minimum of two orange and two white stripes with the top stripe being orange. Any non-retroreflective spaces between the orange and white stripes shall not exceed 3 inches in width.**

Option:

04 Traffic cones may be used to channelize road users, divide opposing vehicular traffic lanes, divide lanes when two or more lanes are kept open in the same direction, and delineate short duration maintenance and utility work.

Guidance:

05 *Steps should be taken to minimize the possibility of cones being blown over or displaced by wind or moving vehicular traffic.*

Option:

06 Cones may be doubled up to increase their weight.

Support:

07 Some cones are constructed with bases that can be filled with ballast. Others have specially weighted bases, or weight such as sandbag rings that can be dropped over the cones and onto the base to provide added stability.

Guidance:

08 *Ballast should be kept to the minimum amount needed.*

Section 6F.65 Tubular Markers

Standard:

01 **Tubular markers (see Figure 6F-7) shall be predominantly orange and shall be not less than 18 inches high and 2 inches wide facing road users. They shall be made of a material that can be struck without causing damage to the impacting vehicle.**

02 **Tubular markers shall be a minimum of 28 inches in height when they are used on freeways and other high-speed highways, on all highways during nighttime, or whenever more conspicuous guidance is needed.**

03 **For nighttime use, tubular markers shall be retroreflectorized. Retroreflectorization of tubular markers that have a height of less than 42 inches shall be provided by two 3-inch wide white bands placed a maximum of 2 inches from the top with a maximum of 6 inches between the bands. Retroreflectorization of tubular markers that have a height of 42 inches or more shall be provided by four 4- to 6-inch wide alternating orange and white stripes with the top stripe being orange.**

Guidance:

04 *Tubular markers have less visible area than other devices and should be used only where space restrictions do not allow for the use of other more visible devices.*

05 *Tubular markers should be stabilized by affixing them to the pavement, by using weighted bases, or weights such as sandbag rings that can be dropped over the tubular markers and onto the base to provide added stability. Ballast should be kept to the minimum amount needed.*

Option:

06 Tubular markers may be used effectively to divide opposing lanes of road users, divide vehicular traffic lanes when two or more lanes of moving vehicular traffic are kept open in the same direction, and to delineate the edge of a pavement drop off where space limitations do not allow the use of larger devices.

Standard:

07 **A tubular marker shall be attached to the pavement to display the minimum 2-inch width to the approaching road users.**

Section 6F.66 Vertical Panels

Standard:

01 **Vertical panels (see Figure 6F-7) shall have retroreflective striped material that is 8 to 12 inches in width and at least 24 inches in height. They shall have alternating diagonal orange and white retroreflective stripes sloping downward at an angle of 45 degrees in the direction vehicular traffic is to pass.**

02 **Where the height of the retroreflective material on the vertical panel is 36 inches or more, a stripe width of 6 inches shall be used.**

Option:

03 Where the height of the retroreflective material on the vertical panel is less than 36 inches, a stripe width of 4 inches may be used.

04 Where space is limited, vertical panels may be used to channelize vehicular traffic, divide opposing lanes, or replace barricades.

Section 6F.67 Drums

Standard:

01 **Drums (see Figure 6F-7) used for road user warning or channelization shall be constructed of lightweight, deformable materials. They shall be a minimum of 36 inches in height and have at least an 18-inch minimum width regardless of orientation. Metal drums shall not be used. The markings on drums shall be horizontal, circumferential, alternating orange and white retroreflective stripes 4 to 6 inches wide. Each drum shall have a minimum of two orange and two white stripes with the top stripe being orange. Any non-retroreflectorized spaces between the horizontal orange and white stripes shall not exceed 3 inches wide. Drums shall have closed tops that will not allow collection of construction debris or other debris.**

Support:

02 Drums are highly visible, have good target value, give the appearance of being formidable obstacles and, therefore, command the respect of road users. They are portable enough to be shifted from place to place within a TTC zone in order to accommodate changing conditions, but are generally used in situations where they will remain in place for a prolonged period of time.

Option:

03 Although drums are most commonly used to channelize or delineate road user flow, they may also be used alone or in groups to mark specific locations.

Guidance:

04 *Drums should not be weighted with sand, water, or any material to the extent that would make them hazardous to road users or workers when struck. Drums used in regions susceptible to freezing should have drain holes in the bottom so that water will not accumulate and freeze causing a hazard if struck by a road user.*

Standard:

05 **Ballast shall not be placed on the top of a drum.**

Section 6F.68 Type 1, 2, or 3 Barricades

Support:

01 A barricade is a portable or fixed device having from one to three rails with appropriate markings and is used to control road users by closing, restricting, or delineating all or a portion of the right-of-way.

02 As shown in Figure 6F-7, barricades are classified as Type 1, Type 2, or Type 3.

Standard:

03 **Stripes on barricade rails shall be alternating orange and white retroreflective stripes sloping downward at an angle of 45 degrees in the direction road users are to pass. Except as provided in Paragraph 4, the stripes shall be 6 inches wide.**

Option:

04 When rail lengths are less than 36 inches, 4-inch wide stripes may be used.

Standard:

05 **The minimum length for Type 1 and Type 2 Barricades shall be 24 inches, and the minimum length for Type 3 Barricades shall be 48 inches. Each barricade rail shall be 8 to 12 inches wide. Barricades used on freeways, expressways, and other high-speed roadways shall have a minimum of 270 square inches of retroreflective area facing road users.**

Guidance:

06 *Where barricades extend entirely across a roadway, the stripes should slope downward in the direction toward which road users must turn.*

07 *Where both right and left turns are provided, the barricade stripes should slope downward in both directions from the center of the barricade or barricades.*

08 *Where no turns are intended, the stripes should be positioned to slope downward toward the center of the barricade or barricades.*

09 *Barricade rails should be supported in a manner that will allow them to be seen by the road user, and in a manner that provides a stable support that is not easily blown over or displaced.*

10 *The width of the existing pedestrian facility should be provided for the temporary facility if practical. Traffic control devices and other construction materials and features should not intrude into the usable width of the sidewalk, temporary pathway, or other pedestrian facility. When it is not possible to maintain a minimum width of 60 inches throughout the entire length of the pedestrian pathway, a 60 x 60-inch passing space should be provided at least every 200 feet to allow individuals in wheelchairs to pass.*

11 *Barricade rail supports should not project into pedestrian circulation routes more than 4 inches from the support between 27 and 80 inches from the surface as described in Section 4.4.1 of the "Americans with Disabilities Act Accessibility Guidelines for Buildings and Facilities (ADAAG)" (see Section 1A.11).*

Option:

12 For Type 1 Barricades, the support may include other unstriped horizontal rails necessary to provide stability.

Guidance:

13 *On high-speed expressways or in other situations where barricades may be susceptible to overturning in the wind, ballasting should be used.*

Option:

14 Sandbags may be placed on the lower parts of the frame or the stays of barricades to provide the required ballast.

Support:

15 Type 1 or Type 2 Barricades are intended for use in situations where road user flow is maintained through the TTC zone.

Option:

16 Barricades may be used alone or in groups to mark a specific condition or they may be used in a series for channelizing road users.

17 Type 1 Barricades may be used on conventional roads or urban streets.

Guidance:

18 *Type 2 or Type 3 Barricades should be used on freeways and expressways or other high-speed roadways. Type 3 Barricades should be used to close or partially close a road.*

Option:

19 Type 3 Barricades used at a road closure may be placed completely across a roadway or from curb to curb.

Guidance:

20 *Where provision is made for access of authorized equipment and vehicles, the responsibility for Type 3 Barricades should be assigned to a person who will provide proper closure at the end of each work day.*

Support:

21 When a highway is legally closed but access must still be allowed for local road users, barricades usually are not extended completely across the roadway.

Standard:

22 A sign shall be installed with the appropriate legend concerning permissible use by local road users (see Section 6F.09). Adequate visibility of the barricades from both directions shall be provided.

Option:

23 Signs may be installed on barricades (see Section 6F.03).

Section 6F.69 Direction Indicator Barricades

Standard:

01 **The Direction Indicator Barricade (see Figure 6F-7) shall consist of a One-Direction Large Arrow (W1-6) sign mounted above a diagonal striped, horizontally aligned, retroreflective rail.**

02 **The One-Direction Large Arrow (W1-6) sign shall be black on an orange background. The stripes on the bottom rail shall be alternating orange and white retroreflective stripes sloping downward at an angle of 45 degrees in the direction road users are to pass. The stripes shall be 4 inches wide. The One-Direction Large Arrow (W1-6) sign shall be 24 x 12 inches. The bottom rail shall have a length of 24 inches and a height of 8 inches.**

Option:

03 The Direction Indicator Barricade may be used in tapers, transitions, and other areas where specific directional guidance to drivers is necessary.

Guidance:

04 *If used, Direction Indicator Barricades should be used in series to direct the driver through the transition and into the intended travel lane.*

Section 6F.70 Temporary Traffic Barriers as Channelizing Devices

Support:

01 Temporary traffic barriers are not TTC devices in themselves; however, when placed in a position identical to a line of channelizing devices and marked and/or equipped with appropriate channelization features to provide guidance and warning both day and night, they serve as TTC devices.

Standard:

02 **Temporary traffic barriers serving as TTC devices shall comply with requirements for such devices as set forth throughout Part 6.**

03 **Temporary traffic barriers (see Section 6F.85) shall not be used solely to channelize road users, but also to protect the work space. If used to channelize vehicular traffic, the temporary traffic barrier shall be supplemented with delineation, pavement markings, or channelizing devices for improved daytime and nighttime visibility.**

Guidance:

04 *Temporary traffic barriers should not be used for a merging taper except in low-speed urban areas.*

05 *When it is necessary to use a temporary traffic barrier for a merging taper in low-speed urban areas or for a constricted/restricted TTC zone, the taper length should be designed to optimize road user operations considering the available geometric conditions.*

Standard:

06 **When it is necessary to use a temporary traffic barrier for a merging taper in low-speed urban areas or for a constricted/restricted TTC zone, the taper shall be delineated.**

Guidance:

07 *When used for channelization, temporary traffic barriers should be of a light color for increased visibility.*

Section 6F.71 Longitudinal Channelizing Devices

Support:

01 Longitudinal channelizing devices are lightweight, deformable devices that are highly visible, have good target value, and can be connected together.

Standard:

02 **If used singly as Type 1, 2, or 3 barricades, longitudinal channelizing devices shall comply with the general size, color, stripe pattern, retroreflectivity, and placement characteristics established for the devices described in this Chapter.**

Guidance:

03 *If used to channelize vehicular traffic at night, longitudinal channelizing devices should be supplemented with retroreflective material or delineation for improved nighttime visibility.*

Option:

04 Longitudinal channelizing devices may be used instead of a line of cones, drums, or barricades.

05 Longitudinal channelizing devices may be hollow and filled with water as a ballast.

06 Longitudinal channelizing devices may be used for pedestrian traffic control.

Standard:

07 **If used for pedestrian traffic control, longitudinal channelizing devices shall be interlocked to delineate or channelize flow. The interlocking devices shall not have gaps that allow pedestrians to stray from the channelizing path.**

Guidance:

08 *Longitudinal channelizing devices have not met the crashworthy requirements for temporary traffic barriers and should not be used to shield obstacles or provide positive protection for pedestrians or workers.*

Section 6F.72 Temporary Lane Separators

Option:

01 Temporary lane separators may be used to channelize road users, to divide opposing vehicular traffic lanes, to divide lanes when two or more lanes are open in the same direction, and to provide continuous pedestrian channelization.

Standard:

02 **Temporary lane separators shall be crashworthy. Temporary lane separators shall have a maximum height of 4 inches and a maximum width of 1 foot, and shall have sloping sides in order to facilitate crossover by emergency vehicles.**

Option:

03 Temporary lane separators may be supplemented with any of the approved channelizing devices contained in this Chapter, such as tubular markers, vertical panels, and opposing traffic lane dividers.

Standard:

04 **If appropriate channelizing devices are used to supplement a temporary lane separator, the channelizing devices shall be retroreflectorized to provide nighttime visibility. If channelizing devices are not used, the temporary lane separator shall contain retroreflectorization to enhance its visibility.**

Guidance:

05 *A temporary lane separator should be stabilized by affixing it to the pavement in a manner suitable to its design, while allowing the unit to be shifted from place to place within the TTC zone in order to accommodate changing conditions.*

Standard:

06 **At pedestrian crossing locations, temporary lane separators shall have an opening or be shortened to provide a pathway that is at least 60 inches wide for crossing pedestrians.**

Section 6F.73 Other Channelizing Devices

Option:

01 Channelizing devices other than those described in this Chapter may be used in special situations based on an engineering study.

Guidance:

02 *Other channelizing devices should comply with the general size, color, stripe pattern, retroreflection, and placement characteristics established for the devices described in this Chapter.*

Section 6F.74 Detectable Edging for Pedestrians

Support:

01 Individual channelizing devices, tape or rope used to connect individual devices, other discontinuous barriers and devices, and pavement markings are not detectable by persons with visual disabilities and are incapable of providing detectable path guidance on temporary or realigned sidewalks or other pedestrian facilities.

Guidance:

02 *When it is determined that a facility should be accessible to and detectable by pedestrians with visual disabilities, a continuously detectable edging should be provided throughout the length of the facility such that it can be followed by pedestrians using long canes for guidance. This edging should protrude at least 6 inches above the surface of the sidewalk or pathway, with the bottom of the edging a maximum of 2.5 inches above the surface. This edging should be continuous throughout the length of the facility except for gaps at locations where pedestrians or vehicles will be turning or crossing. This edging should consist of a prefabricated or formed-in-place curbing or other continuous device that is placed along the edge of the sidewalk or walkway. This edging should be firmly attached to the ground or to other devices. Adjacent sections of this edging should be interconnected such that the edging is not displaced by pedestrian or vehicular traffic or work operations, and such that it does not constitute a hazard to pedestrians, workers, or other road users.*

Support:

03 Examples of detectable edging for pedestrians include:
   A. Prefabricated lightweight sections of plastic, metal, or other suitable materials that are interconnected and fixed in place to form a continuous edge.
   B. Prefabricated lightweight sections of plastic, metal, or other suitable materials that are interconnected, fixed in place, and placed at ground level to provide a continuous connection between channelizing devices located at intervals along the edge of the sidewalk or walkway.
   C. Sections of lumber interconnected and fixed in place to form a continuous edge.
   D. Formed-in-place asphalt or concrete curb.
   E. Prefabricated concrete curb sections that are interconnected and fixed in place to form a continuous edge.
   F. Continuous temporary traffic barrier or longitudinal channelizing barricades placed along the edge of the sidewalk or walkway that provides a pedestrian edging at ground level.
   G. Chain link or other fencing equipped with a continuous bottom rail.

Guidance:

04 *Detectable pedestrian edging should be orange, white, or yellow and should match the color of the adjacent channelizing devices or traffic control devices, if any are present.*

Section 6F.75 Temporary Raised Islands

Standard:

01 **Temporary raised islands shall be used only in combination with pavement striping and other suitable channelizing devices.**

Option:

02 A temporary raised island may be used to separate vehicular traffic flows in two-lane, two-way operations on roadways having a vehicular traffic volume range of 4,000 to 15,000 average daily traffic (ADT) and on freeways having a vehicular traffic volume range of 22,000 ADT to 60,000 ADT.

03 Temporary raised islands also may be used in other than two-lane, two-way operations where physical separation of vehicular traffic from the TTC zone is not required.

Guidance:

04 *Temporary raised islands should have the basic dimensions of 4 inches high by at least 12 inches wide and have rounded or chamfered corners.*

05 *The temporary raised islands should not be designed in such a manner that they would cause a motorist to lose control of the vehicle if the vehicle inadvertently strikes the temporary raised island. If struck, pieces of the island should not be dislodged to the extent that they could penetrate the occupant compartment or involve other vehicles.*

Standard:

06 **At pedestrian crossing locations, temporary raised islands shall have an opening or be shortened to provide at least a 60-inch wide pathway for the crossing pedestrian.**

Section 6F.76 Opposing Traffic Lane Divider and Sign (W6-4)

Support:

01 Opposing traffic lane dividers are delineation devices used as center lane dividers to separate opposing vehicular traffic on a two-lane, two-way operation.

Standard:

02 **Opposing traffic lane dividers shall not be placed across pedestrian crossings.**

03 The Opposing Traffic Lane Divider (W6-4) sign (see Figure 6F-4) shall be an upright, retroreflective orange-colored sign placed on a flexible support and sized at least 12 inches wide by 18 inches high.

Section 6F.77 Pavement Markings

Support:

01 Pavement markings are installed or existing markings are maintained or enhanced in TTC zones to provide road users with a clearly defined path for travel through the TTC zone in day, night, and twilight periods under both wet and dry pavement conditions.

Guidance:

02 *The work should be planned and staged to provide for the placement and removal of the pavement markings in a way that minimizes the disruption to traffic flow approaching and through the TTC zone during the placement and removal process.*

Standard:

03 **Existing pavement markings shall be maintained in all long-term stationary (see Section 6G.02) TTC zones in accordance with Chapters 3A and 3B, except as otherwise provided for temporary pavement markings in Section 6F.78. Pavement markings shall match the alignment of the markings in place at both ends of the TTC zone. Pavement markings shall be placed along the entire length of any paved detour or temporary roadway prior to the detour or roadway being opened to road users.**

04 **For long-term stationary operations, pavement markings in the temporary traveled way that are no longer applicable shall be removed or obliterated as soon as practical. Pavement marking obliteration shall remove the non-applicable pavement marking material, and the obliteration method shall minimize pavement scarring. Painting over existing pavement markings with black paint or spraying with asphalt shall not be accepted as a substitute for removal or obliteration.**

Option:

05 Removable, non-reflective, preformed tape that is approximately the same color as the pavement surface may be used where markings need to be covered temporarily.

Section 6F.78 Temporary Markings

Support:

01 Temporary markings are those pavement markings or devices that are placed within TTC zones to provide road users with a clearly defined path of travel through the TTC zone when the permanent markings are either removed or obliterated during the work activities. Temporary markings are typically needed during the reconstruction of a road while it is open to traffic, such as overlays or surface treatments or where lanes are temporarily shifted on pavement that is to remain in place.

Guidance:

02 *Unless justified based on engineering judgment, temporary pavement markings should not remain in place for more than 14 days after the application of the pavement surface treatment or the construction of the final pavement surface on new roadways or over existing pavements.*

03 *The temporary use of edge lines, channelizing lines, lane-reduction transitions, gore markings, and other longitudinal markings, and the various non-longitudinal markings (such as stop lines, railroad crossings, crosswalks, words, symbols, or arrows) should be in accordance with the State's or highway agency's policy.*

Standard:

04 **Warning signs, channelizing devices, and delineation shall be used to indicate required road user paths in TTC zones where it is not possible to provide a clear path by pavement markings.**

05 **Except as otherwise provided in this Section, all temporary pavement markings for no-passing zones shall comply with the requirements of Chapters 3A and 3B. All temporary broken-line pavement markings shall use the same cycle length as permanent markings and shall have line segments that are at least 2 feet long.**

Guidance:

06 *All pavement markings and devices used to delineate road user paths should be reviewed during daytime and nighttime periods.*

Option:

07 Half-cycle lengths with a minimum of 2-foot stripes may be used on roadways with severe curvature (see Section 3A.06) for broken line center lines in passing zones and for lane lines.

08 For temporary situations of 14 days or less, for a two- or three-lane road, no-passing zones may be identified by using DO NOT PASS (R4-1), PASS WITH CARE (R4-2), and NO PASSING ZONE (W14-3) signs (see Sections 2B.28, 2B.29, and 2C.45) rather than pavement markings. Also, DO NOT PASS, PASS WITH CARE, and NO PASSING ZONE signs may be used instead of pavement markings on roads with low volumes for longer periods in accordance with the State's or highway agency's policy.

Guidance:

09 *If used, the DO NOT PASS, PASS WITH CARE, and NO PASSING ZONE signs should be placed in accordance with Sections 2B.28, 2B.29, and 2C.45.*

10 *If used, the NO CENTER LINE sign should be placed in accordance with Section 6F.47.*

Section 6F.79 Temporary Raised Pavement Markers

Option:

01 Retroreflective or internally illuminated raised pavement markers, or non-retroreflective raised pavement markers supplemented by retroreflective or internally illuminated markers, may be substituted for markings of other types in TTC zones.

Standard:

02 **If used, the color and pattern of the raised pavement markers shall simulate the color and pattern of the markings for which they substitute.**

03 **If temporary raised pavement markers are used to substitute for broken line segments, a group of at least three retroreflective markers shall be equally spaced at no greater than N/8 (see Section 3B.14). The value of N for a broken or dotted line shall equal the length of one line segment plus one gap.**

04 **If temporary raised pavement markers are used to substitute for solid lines, the markers shall be equally spaced at no greater than N/4, with retroreflective or internally illuminated units at a spacing no greater than N/2. The value of N referenced for solid lines shall equal the N for the broken or dotted lines that might be adjacent to or might extend the solid lines (see Section 3B.11).**

Option:

05 Temporary raised pavement markers may be used to substitute for broken line segments by using at least two retroreflective markers placed at each end of a segment of 2 to 5 feet in length, using the same cycle length as permanent markings.

Guidance:

06 *Temporary raised pavement markers used on 2- to 5-foot segments to substitute for broken line segments should not be in place for more than 14 days unless justified by engineering judgment.*

07 *Raised pavement markers should be considered for use along surfaced detours or temporary roadways, and other changed or new travel-lane alignments.*

Option:

08 Retroreflective or internally illuminated raised pavement markers, or non-retroreflective raised pavement markers supplemented by retroreflective or internally illuminated markers, may also be used in TTC zones to supplement markings as prescribed in Chapters 3A and 3B.

Section 6F.80 Delineators

Standard:

01 **When used, delineators shall combine with or supplement other TTC devices. They shall be mounted on crashworthy supports so that the reflecting unit is approximately 4 feet above the near roadway edge. The standard color for delineators used along both sides of two-way streets and highways and the right-hand side of one-way roadways shall be white. Delineators used along the left-hand side of one-way roadways shall be yellow.**

Guidance:

02 *Spacing along roadway curves should be as set forth in Section 3F.04 and should be such that several delineators are constantly visible to the driver.*

Option:

03 Delineators may be used in TTC zones to indicate the alignment of the roadway and to outline the required vehicle path through the TTC zone.

Section 6F.81 Lighting Devices

Guidance:

01 *Lighting devices should be provided in TTC zones based on engineering judgment.*

02 *When used to supplement channelization, the maximum spacing for warning lights should be identical to the channelizing device spacing requirements.*

Option:

03 Lighting devices may be used to supplement retroreflectorized signs, barriers, and channelizing devices.

04 During normal daytime maintenance operations, the functions of flashing warning beacons may be provided by high-intensity rotating, flashing, oscillating, or strobe lights on a maintenance vehicle.

Standard:

05 **Although vehicle hazard warning lights are permitted to be used to supplement high-intensity rotating, flashing, oscillating, or strobe lights, they shall not be used instead of high-intensity rotating, flashing, oscillating, or strobe lights.**

Section 6F.82 Floodlights

Support:

01 Utility, maintenance, or construction activities on highways are frequently conducted during nighttime periods when vehicular traffic volumes are lower. Large construction projects are sometimes operated on a double-shift basis requiring night work (see Section 6G.19).

Guidance:

02 *When nighttime work is being performed, floodlights should be used to illuminate the work area, equipment crossings, and other areas.*

Standard:

03 **Except in emergency situations, flagger stations shall be illuminated at night.**

04 **Floodlighting shall not produce a disabling glare condition for approaching road users, flaggers, or workers.**

Guidance:

05 *The adequacy of the floodlight placement and elimination of potential glare should be determined by driving through and observing the floodlighted area from each direction on all approaching roadways after the initial floodlight setup, at night, and periodically.*

Support:

06 Desired illumination levels vary depending upon the nature of the task involved. An average horizontal luminance of 5 foot candles can be adequate for general activities. Tasks requiring high levels of precision and extreme care can require an average horizontal luminance of 20 foot candles.

Section 6F.83 Warning Lights

Support:

01 Type A, Type B, Type C, and Type D 360-degree warning lights are portable, powered, yellow, lens-directed, enclosed lights.

Standard:

02 **Warning lights shall be in accordance with the current ITE "Purchase Specification for Flashing and Steady-Burn Warning Lights" (see Section 1A.11).**

03 **When warning lights are used, they shall be mounted on signs or channelizing devices in a manner that, if hit by an errant vehicle, they will not be likely to penetrate the windshield.**

Guidance:

04 *The maximum spacing for warning lights should be identical to the channelizing device spacing requirements.*

Support:

05 The light weight and portability of warning lights are advantages that make these devices useful as supplements to the retroreflectorization on signs and channelizing devices. The flashing lights are effective in attracting road users' attention.

Option:

06 Warning lights may be used in either a steady-burn or flashing mode.

Standard:

07 **Except for the sequential flashing warning lights that are described in Paragraphs 8 and 9, flashing warning lights shall not be used for delineation, as a series of flashers fails to identify the desired vehicle path.**

Option:

08 A series of sequential flashing warning lights may be placed on channelizing devices that form a merging taper in order to increase driver detection and recognition of the merging taper.

Standard:

09 **If a series of sequential flashing warning lights is used, the successive flashing of the lights shall occur from the upstream end of the merging taper to the downstream end of the merging taper in order to identify the desired vehicle path. Each flashing warning light in the sequence shall be flashed at a rate of not less than 55 or more than 75 times per minute.**

10 **Type A Low-Intensity Flashing warning lights, Type C Steady-Burn warning lights, and Type D 360-degree Steady-Burn warning lights shall be maintained so as to be capable of being visible on a clear night from a distance of 3,000 feet. Type B High-Intensity Flashing warning lights shall be maintained so as to be capable of being visible on a sunny day when viewed without the sun directly on or behind the device from a distance of 1,000 feet.**

11 **Warning lights shall have a minimum mounting height of 30 inches to the bottom of the lens.**

Support:

12 Type A Low-Intensity Flashing warning lights are used to warn road users during nighttime hours that they are approaching or proceeding in a potentially hazardous area.

Option:

13 Type A warning lights may be mounted on channelizing devices.

Support:

14 Type B High-Intensity Flashing warning lights are used to warn road users during both daylight and nighttime hours that they are approaching a potentially hazardous area.

Option:

15 Type B warning lights are designed to operate 24 hours per day and may be mounted on advance warning signs or on independent supports.

16 Type C Steady-Burn warning lights and Type D 360-degree Steady-Burn warning lights may be used during nighttime hours to delineate the edge of the traveled way.

Guidance:

17 *When used to delineate a curve, Type C and Type D 360-degree warning lights should only be used on devices on the outside of the curve, and not on the inside of the curve.*

Section 6F.84 Temporary Traffic Control Signals

Standard:

01 **Temporary traffic control signals (see Section 4D.32) used to control road user movements through TTC zones and in other TTC situations shall comply with the applicable provisions of Part 4.**

Support:

02 Temporary traffic control signals are typically used in TTC zones such as temporary haul road crossings; temporary one-way operations along a one-lane, two-way highway; temporary one-way operations on bridges, reversible lanes, and intersections.

Standard:

03 **A temporary traffic control signal that is used to control traffic through a one-lane, two-way section of roadway shall comply with the provisions of Section 4H.02.**

Guidance:

04 *Where pedestrian traffic is detoured to a temporary traffic control signal, engineering judgment should be used to determine if pedestrian signals or accessible pedestrian signals (see Section 4E.09) are needed for crossing along an alternate route.*

05 *When temporary traffic control signals are used, conflict monitors typical of traditional traffic control signal operations should be used.*

Option:

06 Temporary traffic control signals may be portable or temporarily mounted on fixed supports.

Guidance:

07 *Temporary traffic control signals should only be used in situations where temporary traffic control signals are preferable to other means of traffic control, such as changing the work staging or work zone size to eliminate one-way vehicular traffic movements, using flaggers to control one-way or crossing movements, using STOP or YIELD signs, and using warning devices alone.*

Support:

08 Factors related to the design and application of temporary traffic control signals include the following:
   A. Safety and road user needs;
   B. Work staging and operations;
   C. The feasibility of using other TTC strategies (for example, flaggers, providing space for two lanes, or detouring road users, including bicyclists and pedestrians);
   D. Sight distance restrictions;
   E. Human factors considerations (for example, lack of driver familiarity with temporary traffic control signals);
   F. Road-user volumes including roadway and intersection capacity;
   G. Affected side streets and driveways;
   H. Vehicle speeds;
   I. The placement of other TTC devices;
   J. Parking;
   K. Turning restrictions;
   L. Pedestrians;
   M. The nature of adjacent land uses (such as residential or commercial);
   N. Legal authority;
   O. Signal phasing and timing requirements;
   P. Full-time or part-time operation;
   Q. Actuated, fixed-time, or manual operation;
   R. Power failures or other emergencies;
   S. Inspection and maintenance needs;
   T. Need for detailed placement, timing, and operation records; and
   U. Operation by contractors or by others.

09 Although temporary traffic control signals can be mounted on trailers or lightweight portable supports, fixed supports offer superior resistance to displacement or damage by severe weather, vehicle impact, and vandalism.

Guidance:

10 *Other TTC devices should be used to supplement temporary traffic control signals, including warning and regulatory signs, pavement markings, and channelizing devices.*

11 *Temporary traffic control signals not in use should be covered or removed.*

12 *If a temporary traffic control signal is located within 1/2 mile of an adjacent traffic control signal, consideration should be given to interconnected operation.*

Standard:

13 **Temporary traffic control signals shall not be located within 200 feet of a grade crossing unless the temporary traffic control signal is provided with preemption in accordance with Section 4D.27, or unless a uniformed officer or flagger is provided at the crossing to prevent vehicles from stopping within the crossing.**

Section 6F.85 Temporary Traffic Barriers

Support:

01 Temporary traffic barriers, including shifting portable or movable barriers, are devices designed to help prevent penetration by vehicles while minimizing injuries to vehicle occupants, and to protect workers, bicyclists, and pedestrians.

02 The four primary functions of temporary traffic barriers are:
   A. To keep vehicular traffic from entering work areas, such as excavations or material storage sites;
   B. To separate workers, bicyclists, and pedestrians from motor vehicle traffic;
   C. To separate opposing directions of vehicular traffic; and
   D. To separate vehicular traffic, bicyclists, and pedestrians from the work area such as false work for bridges and other exposed objects.

Option:

03 Temporary traffic barriers may be used to separate two-way vehicular traffic.

Guidance:

04 *Because the protective requirements of a TTC situation have priority in determining the need for temporary traffic barriers, their use should be based on an engineering study.*

Standard:

05 **Temporary traffic barriers shall be supplemented with standard delineation, pavement markings, or channelizing devices for improved daytime and nighttime visibility if they are used to channelize vehicular traffic. The delineation color shall match the applicable pavement marking color.**

06 **Temporary traffic barriers, including their end treatments, shall be crashworthy. In order to mitigate the effect of striking the upstream end of a temporary traffic barrier, the end shall be installed in accordance with AASHTO's "Roadside Design Guide" (see Section 1A.11) by flaring until the end is outside the acceptable clear zone or by providing crashworthy end treatments.**

Option:

07 Warning lights or steady-burn lamps may be mounted on temporary traffic barrier installations.

Support:

08 Movable barriers are capable of being repositioned laterally using a transfer vehicle that travels along the barrier. Movable barriers enable short-term closures to be installed and removed on long-term projects. Providing a barrier-protected work space for short-term closures and providing unbalanced flow to accommodate changes in the direction of peak-period traffic flows are two of the advantages of using movable barriers.

09 Figure 6H-45 shows a temporary reversible lane using movable barriers. The notable feature of the movable barrier is that in both Phase A and Phase B, the lanes used by opposing traffic are separated by a barrier.

10 Figure 6H-34 shows an exterior lane closure using a temporary traffic barrier. Notes 7 though 9 address the option of using a movable barrier. By using a movable barrier, the barrier can be positioned to close the lane during the off-peak periods and can be relocated to open the lane during peak periods to accommodate peak traffic flows. With one pass of the transfer vehicle, the barrier can be moved out of the lane and onto the shoulder. Furthermore, if so desired, with a second pass of the transfer vehicle, the barrier could be moved to the roadside beyond the shoulder.

11 More specific information on the use of temporary traffic barriers is contained in Chapters 8 and 9 of AASHTO's "Roadside Design Guide" (see Section 1A.11).

Section 6F.86 Crash Cushions

Support:

01 Crash cushions are systems that mitigate the effects of errant vehicles that strike obstacles, either by smoothly decelerating the vehicle to a stop when hit head-on, or by redirecting the errant vehicle. The two types of crash cushions that are used in TTC zones are stationary crash cushions and truck-mounted attenuators. Crash cushions in TTC zones help protect the drivers from the exposed ends of barriers, fixed objects, shadow vehicles, and other obstacles. Specific information on the use of crash cushions can be found in AASHTO's "Roadside Design Guide" (see Section 1A.11).

Standard:

02 **Crash cushions shall be crashworthy. They shall also be designed for each application to stop or redirect errant vehicles under prescribed conditions. Crash cushions shall be periodically inspected to verify that they have not been hit or damaged. Damaged crash cushions shall be promptly repaired or replaced to maintain their crashworthiness.**

Support:

03 Stationary crash cushions are used in the same manner as permanent highway installations to protect drivers from the exposed ends of barriers, fixed objects, and other obstacles.

Standard:

04 **Stationary crash cushions shall be designed for the specific application intended.**

05 **Truck-mounted attenuators shall be energy-absorbing devices attached to the rear of shadow trailers or trucks. If used, the shadow vehicle with the attenuator shall be located in advance of the work area, workers, or equipment to reduce the severity of rear-end crashes from errant vehicles.**

Support:

06 Trucks or trailers are often used as shadow vehicles to protect workers or work equipment from errant vehicles. These shadow vehicles are normally equipped with flashing arrows, changeable message signs, and/or high-intensity rotating, flashing, oscillating, or strobe lights located properly in advance of the workers and/or equipment that they are protecting. However, these shadow vehicles might themselves cause injuries to occupants of the errant vehicles if they are not equipped with truck-mounted attenuators.

Guidance:

07 *The shadow truck should be positioned a sufficient distance in advance of the workers or equipment being protected so that there will be sufficient distance, but not so much so that errant vehicles will travel around the shadow truck and strike the protected workers and/or equipment.*

Support:

08 Chapter 9 of AASHTO's "Roadside Design Guide" (see Section 1A.11) contains additional information regarding the use of shadow vehicles.

Guidance:

09 *If used, the truck-mounted attenuator should be used in accordance with the manufacturer's specifications.*

Section 6F.87 Rumble Strips

Support:

01 Transverse rumble strips consist of intermittent, narrow, transverse areas of rough-textured or slightly raised or depressed road surface that extend across the travel lanes to alert drivers to unusual vehicular traffic conditions. Through noise and vibration they attract the driver's attention to such features as unexpected changes in alignment and to conditions requiring a stop.

02 Longitudinal rumble strips consist of a series of rough-textured or slightly raised or depressed road surfaces located along the shoulder to alert road users that they are leaving the travel lanes.

Standard:

03 **If it is desirable to use a color other than the color of the pavement for a longitudinal rumble strip, the color of the rumble strip shall be the same color as the longitudinal line the rumble strip supplements.**

04 **If the color of a transverse rumble strip used within a travel lane is not the color of the pavement, the color of the rumble strip shall be white, black, or orange.**

Option:

05 Intervals between transverse rumble strips may be reduced as the distance to the approached conditions is diminished in order to convey an impression that a closure speed is too fast and/or that an action is imminent. A sign warning drivers of the onset of rumble strips may be placed in advance of any transverse rumble strip installation.

Guidance:

06 *Transverse rumble strips should be placed transverse to vehicular traffic movement. They should not adversely affect overall pavement skid resistance under wet or dry conditions.*

07 *In urban areas, even though a closer spacing might be warranted, transverse rumble strips should be designed in a manner that does not promote unnecessary braking or erratic steering maneuvers by road users.*

08 *Transverse rumble strips should not be placed on sharp horizontal or vertical curves.*

09 *Rumble strips should not be placed through pedestrian crossings or on bicycle routes.*

10 *Transverse rumble strips should not be placed on roadways used by bicyclists unless a minimum clear path of 4 feet is provided at each edge of the roadway or on each paved shoulder as described in AASHTO's "Guide to the Development of Bicycle Facilities" (see Section 1A.11).*

11 *Longitudinal rumble strips should not be placed on the shoulder of a roadway that is used by bicyclists unless a minimum clear path of 4 feet is also provided on the shoulder.*

Section 6F.88 Screens

Support:

01 Screens are used to block the road users' view of activities that can be distracting. Screens might improve safety and motor vehicle traffic flow where volumes approach the roadway capacity because they discourage gawking and reduce headlight glare from oncoming motor vehicle traffic.

Guidance:

02 *Screens should not be mounted where they could adversely restrict road user visibility and sight distance and adversely affect the reasonably safe operation of vehicles.*

Option:

03 Screens may be mounted on the top of temporary traffic barriers that separate two-way motor vehicle traffic.

Guidance:

04 *Design of screens should be in accordance with Chapter 9 of AASHTO's "Roadside Design Guide" (see Section 1A.11).*

CHAPTER 6G. TYPE OF TEMPORARY TRAFFIC CONTROL ZONE ACTIVITIES

Section 6G.01 Typical Applications

Support:

01 Each TTC zone is different. Many variables, such as location of work, highway type, geometrics, vertical and horizontal alignment, intersections, interchanges, road user volumes, road vehicle mix (buses, trucks, and cars), and road user speeds affect the needs of each zone. The goal of TTC in work zones is safety with minimum disruption to road users. The key factor in promoting TTC zone safety is proper judgment.

02 Typical applications (TAs) of TTC zones are organized according to duration, location, type of work, and highway type. Table 6H-1 is an index of these typical applications. These typical applications include the use of various TTC methods, but do not include a layout for every conceivable work situation.

03 Well-designed TTC plans for planned special events will likely be developed from a combination of treatments from several of the typical applications.

Guidance:

04 *For any planned special event that will have an impact on the traffic on any street or highway, a TTC plan should be developed in conjunction with and be approved by the agency or agencies that have jurisdiction over the affected roadways.*

05 *Typical applications should be altered, when necessary, to fit the conditions of a particular TTC zone.*

Option:

06 Other devices may be added to supplement the devices shown in the typical applications, while others may be deleted. The sign spacings and taper lengths may be increased to provide additional time or space for driver response.

Support:

07 Decisions regarding the selection of the most appropriate typical application to use as a guide for a specific TTC zone require an understanding of each situation. Although there are many ways of categorizing TTC zone applications, the four factors mentioned earlier (work duration, work location, work type, and highway type) are used to characterize the typical applications illustrated in Chapter 6H.

Section 6G.02 Work Duration

Support:

01 Work duration is a major factor in determining the number and types of devices used in TTC zones. The duration of a TTC zone is defined relative to the length of time a work operation occupies a spot location.

Standard:

02 **The five categories of work duration and their time at a location shall be:**
   A. Long-term stationary is work that occupies a location more than 3 days.
   B. Intermediate-term stationary is work that occupies a location more than one daylight period up to 3 days, or nighttime work lasting more than 1 hour.
   C. Short-term stationary is daytime work that occupies a location for more than 1 hour within a single daylight period.
   D. Short duration is work that occupies a location up to 1 hour.
   E. Mobile is work that moves intermittently or continuously.

Support:

03 At long-term stationary TTC zones, there is ample time to install and realize benefits from the full range of TTC procedures and devices that are available for use. Generally, larger channelizing devices, temporary roadways, and temporary traffic barriers are used.

Standard:

04 **Since long-term operations extend into nighttime, retroreflective and/or illuminated devices shall be used in long-term stationary TTC zones.**

Guidance:

05 *Inappropriate markings in long-term stationary TTC zones should be removed and replaced with temporary markings.*

Support:

06 In intermediate-term stationary TTC zones, it might not be feasible or practical to use procedures or devices that would be desirable for long-term stationary TTC zones, such as altered pavement markings, temporary traffic barriers, and temporary roadways. The increased time to place and remove these devices in some cases could significantly lengthen the project, thus increasing exposure time.

Standard:

07 **Since intermediate-term operations extend into nighttime, retroreflective and/or illuminated devices shall be used in intermediate-term stationary TTC zones.**

Support:

08 Most maintenance and utility operations are short-term stationary work.

09 As compared to stationary operations, mobile and short-duration operations are activities that might involve different treatments. Devices having greater mobility might be necessary such as signs mounted on trucks. Devices that are larger, more imposing, or more visible can be used effectively and economically. The mobility of the TTC zone is important.

Guidance:

10 *Safety in short-duration or mobile operations should not be compromised by using fewer devices simply because the operation will frequently change its location.*

Option:

11 Appropriately colored or marked vehicles with high-intensity rotating, flashing, oscillating, or strobe lights may be used in place of signs and channelizing devices for short-duration or mobile operations. These vehicles may be augmented with signs or arrow boards.

Support:

12 During short-duration work, it often takes longer to set up and remove the TTC zone than to perform the work. Workers face hazards in setting up and taking down the TTC zone. Also, since the work time is short, delays affecting road users are significantly increased when additional devices are installed and removed.

Option:

13 Considering these factors, simplified control procedures may be warranted for short-duration work. A reduction in the number of devices may be offset by the use of other more dominant devices such as high-intensity rotating, flashing, oscillating, or strobe lights on work vehicles.

Support:

14 Mobile operations often involve frequent short stops for activities such as litter cleanup, pothole patching, or utility operations, and are similar to short-duration operations.

Guidance:

15 *Warning signs and high-intensity rotating, flashing, oscillating, or strobe lights should be used on the vehicles that are participating in the mobile work.*

Option:

16 Flags and/or channelizing devices may additionally be used and moved periodically to keep them near the mobile work area.

17 Flaggers may be used for mobile operations that often involve frequent short stops.

Support:

18 Mobile operations also include work activities where workers and equipment move along the road without stopping, usually at slow speeds. The advance warning area moves with the work area.

Guidance:

19 *When mobile operations are being performed, a shadow vehicle equipped with an arrow board or a sign should follow the work vehicle, especially when vehicular traffic speeds or volumes are high. Where feasible, warning signs should be placed along the roadway and moved periodically as work progresses.*

20 *Under high-volume conditions, consideration should be given to scheduling mobile operations work during off-peak hours.*

21 *If there are mobile operations on a high-speed travel lane of a multi-lane divided highway, arrow boards should be used.*

Standard:

22 **Mobile operations shall have appropriate devices on the equipment (that is, high-intensity rotating, flashing, oscillating, or strobe lights, signs, or special lighting), or shall use a separate vehicle with appropriate warning devices.**

Option:

23 For mobile operations that move at speeds of less than 3 mph, mobile signs or stationary signing that is periodically retrieved and repositioned in the advance warning area may be used.

Section 6G.03 Location of Work

Support:

01 Chapter 6D and Sections 6F.74 and 6G.05 contain additional information regarding the steps to follow when pedestrian or bicycle facilities are affected by the worksite.

02 The choice of TTC needed for a TTC zone depends upon where the work is located. As a general rule, the closer the work is to road users (including bicyclists and pedestrians), the greater the number of TTC devices that are needed. Procedures are described later in this Chapter for establishing TTC zones in the following locations:

- A. Outside the shoulder,
- B. On the shoulder with no encroachment,
- C. On the shoulder with minor encroachment,
- D. Within the median, and
- E. Within the traveled way.

Standard:

03 **When the work space is within the traveled way, except for short-duration and mobile operations, advance warning shall provide a general message that work is taking place and shall supply information about highway conditions. TTC devices shall indicate how vehicular traffic can move through the TTC zone.**

Section 6G.04 Modifications To Fulfill Special Needs

Support:

01 The typical applications in Chapter 6H illustrate commonly encountered situations in which TTC devices are employed.

Option:

02 Other devices may be added to supplement the devices provided in the typical applications, and device spacing may be adjusted to provide additional reaction time. When conditions are less complex than those depicted in the typical applications, fewer devices may be needed.

Guidance:

03 *When conditions are more complex, typical applications should be modified by giving particular attention to the provisions set forth in Chapter 6B and by incorporating appropriate devices and practices from the following list:*

- *A. Additional devices:*
    1. *Signs*
    2. *Arrow boards*
    3. *More channelizing devices at closer spacing (see Section 6F.74 for information regarding detectable edging for pedestrians)*
    4. *Temporary raised pavement markers*
    5. *High-level warning devices*
    6. *Portable changeable message signs*
    7. *Temporary traffic control signals (including pedestrian signals and accessible pedestrian signals)*
    8. *Temporary traffic barriers*
    9. *Crash cushions*
    10. *Screens*
    11. *Rumble strips*
    12. *More delineation*

B. Upgrading of devices:
   1. A full complement of standard pavement markings
   2. Brighter and/or wider pavement markings
   3. Larger and/or brighter signs
   4. Channelizing devices with greater conspicuity
   5. Temporary traffic barriers in place of channelizing devices
C. Improved geometrics at detours or crossovers
D. Increased distances:
   1. Longer advance warning area
   2. Longer tapers
E. Lighting:
   1. Temporary roadway lighting
   2. Steady-burn lights used with channelizing devices
   3. Flashing lights for isolated hazards
   4. Illuminated signs
   5. Floodlights
F. Pedestrian routes and temporary facilities
G. Bicycle diversions and temporary facilities

Section 6G.05 Work Affecting Pedestrian and Bicycle Facilities

Support:

01 It is not uncommon, particularly in urban areas, that road work and the associated TTC will affect existing pedestrian or bicycle facilities. It is essential that the needs of all road users, including pedestrians with disabilities, are considered in TTC zones.

02 In addition to specific provisions identified in Sections 6G.06 through 6G.14, there are a number of provisions that might be applicable for all of the types of activities identified in this Chapter.

Guidance:

03 *Where pedestrian or bicycle usage is high, the typical applications should be modified by giving particular attention to the provisions set forth in Chapter 6D, this Chapter, Section 6F.74, and in other Sections of Part 6 related to accessibility and detectability provisions in TTC zones.*

04 *Pedestrians should be separated from the worksite by appropriate devices that maintain the accessibility and detectability for pedestrians with disabilities.*

05 *Bicyclists and pedestrians should not be exposed to unprotected excavations, open utility access, overhanging equipment, or other such conditions.*

06 *Except for short duration and mobile operations, when a highway shoulder is occupied, a SHOULDER WORK (W21-5) sign should be placed in advance of the activity area. When work is performed on a paved shoulder 8 feet or more in width, channelizing devices should be placed on a taper having a length that conforms to the requirements of a shoulder taper. Signs should be placed such that they do not narrow any existing pedestrian passages to less than 48 inches.*

07 *Pedestrian detours should be avoided since pedestrians rarely observe them and the cost of providing accessibility and detectability might outweigh the cost of maintaining a continuous route. Whenever possible, work should be done in a manner that does not create a need to detour pedestrians from existing routes or crossings.*

Standard:

08 **Where pedestrian routes are closed, alternate pedestrian routes shall be provided.**

09 **When existing pedestrian facilities are disrupted, closed, or relocated in a TTC zone, the temporary facilities shall be detectable and shall include accessibility features consistent with the features present in the existing pedestrian facility.**

Section 6G.06 Work Outside of the Shoulder

Support:

01 When work is being performed off the roadway (beyond the shoulders, but within the right-of-way), little or no TTC might be needed. TTC generally is not needed where work is confined to an area 15 feet or more from the edge of the traveled way. However, TTC is appropriate where distracting situations exist, such as vehicles parked on the shoulder, vehicles accessing the worksite via the highway, and equipment traveling on or crossing the roadway to perform the work operations (for example, mowing). For work beyond the shoulder, see Figure 6H-1.

Guidance:

02 Where the situations described in Paragraph 1 exist, a single warning sign, such as ROAD WORK AHEAD (W20-1), should be used. If the equipment travels on the roadway, the equipment should be equipped with appropriate flags, high-intensity rotating, flashing, oscillating, or strobe lights, and/or a SLOW MOVING VEHICLE (W21-4) sign.

Option:

03 If work vehicles are on the shoulder, a SHOULDER WORK (W21-5) sign may be used. For mowing operations, the sign MOWING AHEAD (W21-8) may be used.

04 Where the activity is spread out over a distance of more than 2 miles, the SHOULDER WORK (W21-5) sign may be repeated every 1 mile.

05 A supplementary plaque with the message NEXT XX MILES (W7-3aP) may be used.

Guidance:

06 A general warning sign like ROAD MACHINERY AHEAD (W21-3) should be used if workers and equipment must occasionally move onto the shoulder.

Section 6G.07 Work on the Shoulder with No Encroachment

Support:

01 The provisions of this Section apply to short-term through long-term stationary operations.

Standard:

02 **When paved shoulders having a width of 8 feet or more are closed, at least one advance warning sign shall be used. In addition, channelizing devices shall be used to close the shoulder in advance to delineate the beginning of the work space and direct motor vehicle traffic to remain within the traveled way.**

Guidance:

03 When paved shoulders having a width of 8 feet or more are closed on freeways and expressways, road users should be warned about potential disabled vehicles that cannot get off the traveled way. An initial general warning sign, such as ROAD WORK AHEAD (W20-1), should be used, followed by a RIGHT or LEFT SHOULDER CLOSED (W21-5a) sign. Where the downstream end of the shoulder closure extends beyond the distance that can be perceived by road users, a supplementary plaque bearing the message NEXT XX FEET (W16-4P) or MILES (W7-3aP) should be placed below the SHOULDER CLOSED (W21-5a) sign. On multi-lane, divided highways, signs advising of shoulder work or the condition of the shoulder should be placed only on the side of the affected shoulder.

04 When an improved shoulder is closed on a high-speed roadway, it should be treated as a closure of a portion of the road system because road users expect to be able to use it in emergencies. Road users should be given ample advance warning that shoulders are closed for use as refuge areas throughout a specified length of the approaching TTC zone. The sign(s) should read SHOULDER CLOSED (W21-5a) with distances indicated. The work space on the shoulder should be closed off by a taper or channelizing devices with a length of 1/3 L using the formulas in Tables 6C-3 and 6C-4.

05 When the shoulder is not occupied but work has adversely affected its condition, the LOW SHOULDER (W8-9) or SOFT SHOULDER (W8-4) sign should be used, as appropriate.

06 Where the condition extends over a distance in excess of 1 mile, the sign should be repeated at 1-mile intervals.

Option:

07 In addition, a supplementary plaque bearing the message NEXT XX MILES (W7-3aP) may be used. Temporary traffic barriers may be needed to inhibit encroachment of errant vehicles into the work space and to protect workers.

Standard:

08 **When used for shoulder work, arrow boards shall operate only in the caution mode.**

Support:

09 A typical application for stationary work operations on shoulders is shown in Figure 6H-3. Short duration or mobile work on shoulders is shown in Figure 6H-4. Work on freeway shoulders is shown in Figure 6H-5.

Section 6G.08 Work on the Shoulder with Minor Encroachment

Support:

01 Chapter 6D and Sections 6F.74 and 6G.05 contain additional information regarding the steps to follow when pedestrian or bicycle facilities are affected by the worksite.

Guidance:

02 *When work takes up part of a lane, vehicular traffic volumes, vehicle mix (buses, trucks, cars, and bicycles), speed, and capacity should be analyzed to determine whether the affected lane should be closed. Unless the lane encroachment permits a remaining lane width of 10 feet, the lane should be closed.*

03 *Truck off-tracking should be considered when determining whether the minimum lane width of 10 feet is adequate.*

Option:

04 A lane width of 9 feet may be used for short-term stationary work on low-volume, low-speed roadways when vehicular traffic does not include longer and wider heavy commercial vehicles.

Support:

05 Figure 6H-6 illustrates a method for handling vehicular traffic where the stationary or short duration work space encroaches slightly into the traveled way.

Section 6G.09 Work Within the Median

Support:

01 Chapter 6D and Sections 6F.74 and 6G.05 contain additional information regarding the steps to follow when pedestrian or bicycle facilities are affected by the worksite.

Guidance:

02 *If work in the median of a divided highway is within 15 feet from the edge of the traveled way for either direction of travel, TTC should be used through the use of advance warning signs and channelizing devices.*

Section 6G.10 Work Within the Traveled Way of a Two-Lane Highway

Support:

01 Chapter 6D and Sections 6F.74 and 6G.05 contain additional information regarding the steps to follow when pedestrian or bicycle facilities are affected by the worksite.

02 Detour signs are used to direct road users onto another roadway. At diversions, road users are directed onto a temporary roadway or alignment placed within or adjacent to the right-of-way. Typical applications for detouring or diverting road users on two-lane highways are shown in Figures 6H-7, 6H-8, and 6H-9. Figure 6H-7 illustrates the controls around an area where a section of roadway has been closed and a diversion has been constructed. Channelizing devices and pavement markings are used to indicate the transition to the temporary roadway.

Guidance:

03 *When a detour is long, Detour (M4-8, M4-9) signs should be installed to remind and reassure road users periodically that they are still successfully following the detour.*

04 *When an entire roadway is closed, as illustrated in Figure 6H-8, a detour should be provided and road users should be warned in advance of the closure, which in this example is a closure 10 miles from the intersection. If local road users are allowed to use the roadway up to the closure, the ROAD CLOSED AHEAD, LOCAL TRAFFIC ONLY (R11-3a) sign should be used. The portion of the road open to local road users should have adequate signing, marking, and delineation.*

05 *Detours should be signed so that road users will be able to traverse the entire detour route and back to the original roadway as shown in Figure 6H-9.*

Support:

06 Techniques for controlling vehicular traffic under one-lane, two-way conditions are described in Section 6C.10.

Option:

07 Flaggers may be used as shown in Figure 6H-10.

08 STOP/YIELD sign control may be used on roads with low traffic volumes as shown in Figure 6H-11.

09 A temporary traffic control signal may be used as shown in Figure 6H-12.

Section 6G.11 Work Within the Traveled Way of an Urban Street

Support:

01　Chapter 6D and Sections 6F.74 and 6G.05 contain additional information regarding the steps to follow when pedestrian or bicycle facilities are affected by the worksite.

02　In urban TTC zones, decisions are needed on how to control vehicular traffic, such as how many lanes are required, whether any turns need to be prohibited at intersections, and how to maintain access to business, industrial, and residential areas.

03　Pedestrian traffic needs separate attention. Chapter 6D contains information regarding pedestrian movements near TTC zones.

Standard:

04　**If the TTC zone affects the movement of bicyclists, adequate access to the roadway or shared-use paths shall be provided (see Part 9).**

05　**Where transit stops are affected or relocated because of work activity, both pedestrian and vehicular access to the affected or relocated transit stops shall be provided.**

Guidance:

06　*If a designated bicycle route is closed because of the work being done, a signed alternate route should be provided. Bicyclists should not be directed onto the path used by pedestrians.*

07　*Worksites within the intersection should be protected against inadvertent pedestrian incursion by providing detectable channelizing devices.*

Support:

08　Utility work takes place both within and outside the roadway to construct and maintain services such as power, gas, light, water, or telecommunications. Operations often involve intersections, since that is where many of the network junctions occur. The work force is usually small, only a few vehicles are involved, and the number and types of TTC devices placed in the TTC zone is usually minimal.

Standard:

09　**All TTC devices shall be retroreflective or illuminated if utility work is performed during nighttime hours.**

Guidance:

10　*As discussed under short-duration projects, however, the reduced number of devices in utility work zones should be offset by the use of high-visibility devices, such as high-intensity rotating, flashing, oscillating, or strobe lights on work vehicles or high-level warning devices.*

Support:

11　Figures 6H-6, 6H-10, 6H-15, 6H-18, 6H-21, 6H-22, 6H-23, 6H-26, and 6H-33 are examples of typical applications for utility operations. Other typical applications might apply as well.

Section 6G.12 Work Within the Traveled Way of a Multi-Lane, Non-Access Controlled Highway

Support:

01　Chapter 6D and Sections 6F.74 and 6G.05 contain additional information regarding the steps to follow when pedestrian or bicycle facilities are affected by the worksite.

02　Work on multi-lane (two or more lanes of moving motor vehicle traffic in one direction) highways is divided into right-lane closures, left-lane closures, interior-lane closures, multiple-lane closures, and closures on five-lane roadways.

Standard:

03　**When a lane is closed on a multi-lane road for other than a mobile operation, a transition area containing a merging taper shall be used.**

Guidance:

04　*When justified by an engineering study, temporary traffic barriers (see Section 6F.70) should be used to prevent incursions of errant vehicles into hazardous areas or work space.*

Support:

05　Figure 6H-34 illustrates a lane closure in which temporary traffic barriers are used.

Option:

06　When the right lane is closed, TTC similar to that shown in Figure 6H-33 may be used for undivided or divided four-lane roads.

Guidance:

07 *If morning and evening peak hour vehicular traffic volumes in the two directions are uneven and the greater volume is on the side where the work is being done in the right-hand lane, consideration should be given to closing the inside lane for opposing vehicular traffic and making the lane available to the side with heavier vehicular traffic, as shown in Figure 6H-31.*

08 *If the larger vehicular traffic volume changes to the opposite direction at a different time of the day, the TTC should be changed to allow two lanes for opposing vehicular traffic by moving the devices from the opposing lane to the center line. When it is necessary to create a temporary center line that is not consistent with the pavement markings, channelizing devices should be used and closely spaced.*

Option:

09 When closing a left lane on a multi-lane undivided road, as vehicular traffic flow permits, the two interior lanes may be closed, as shown in Figure 6H-30, to provide drivers and workers additional lateral clearance and to provide access to the work space.

Standard:

10 **When only the left lane is closed on undivided roads, channelizing devices shall be placed along the center line as well as along the adjacent lane.**

Guidance:

11 *When an interior lane is closed, an adjacent lane should also be considered for closure to provide additional space for vehicles and materials and to facilitate the movement of equipment within the work space.*

12 *When multiple lanes in one direction are closed, a capacity analysis should be made to determine the number of lanes needed to accommodate motor vehicle traffic needs. Vehicular traffic should be moved over one lane at a time. As shown in Figure 6H-37, the tapers should be separated by a distance of 2L, with L being determined by the formulas in Tables 6C-3 and 6C-4.*

Option:

13 If operating speeds are 40 mph or less and the space approaching the work area does not permit moving traffic over one lane at a time, a single continuous taper may be used.

Standard:

14 **When a directional roadway is closed, inapplicable WRONG WAY signs and markings, and other existing traffic control devices at intersections within the temporary two-lane, two-way operations section shall be covered, removed, or obliterated.**

Option:

15 When half the road is closed on an undivided highway, both directions of vehicular traffic may be accommodated as shown in Figure 6H-32. When both interior lanes are closed, temporary traffic controls may be used as provided in Figure 6H-30. When a roadway must be closed on a divided highway, a median crossover may be used (see Section 6G.16).

Support:

16 TTC for lane closures on five-lane roads is similar to other multi-lane undivided roads. Figure 6H-32 can be adapted for use on five-lane roads. Figure 6H-35 can be used on a five-lane road for short duration and mobile operations.

Section 6G.13 Work Within the Traveled Way at an Intersection

Support:

01 Chapter 6D and Sections 6F.74 and 6G.05 contain additional information regarding the steps to follow when pedestrian or bicycle facilities are affected by the worksite.

02 The typical applications for intersections are classified according to the location of the work space with respect to the intersection area (as defined by the extension of the curb or edge lines). The three classifications are near side, far side, and in-the-intersection. Work spaces often extend into more than one portion of the intersection. For example, work in one quadrant often creates a near-side work space on one street and a far-side work space on the cross street. In such instances, an appropriate TTC plan is obtained by combining features shown in two or more of the intersection and pedestrian typical applications.

03 TTC zones in the vicinity of intersections might block movements and interfere with normal road user flows. Such conflicts frequently occur at more complex signalized intersections having such features as traffic signal heads over particular lanes, lanes allocated to specific movements, multiple signal phases, signal detectors for actuated control, and accessible pedestrian signals and detectors.

Guidance:

04 *The effect of the work upon signal operation should be considered, and temporary corrective actions should be taken, if necessary, such as revising signal phasing and/or timing to provide adequate capacity, maintaining or adjusting signal detectors, and relocating signal heads to provide adequate visibility as described in Part 4.*

Standard:

05 **When work will occur near an intersection where operational, capacity, or pedestrian accessibility problems are anticipated, the highway agency having jurisdiction shall be contacted.**

Guidance:

06 *For work at an intersection, advance warning signs, devices, and markings should be used on all cross streets, as appropriate. The typical applications depict urban intersections on arterial streets. Where the posted speed limit, the off-peak 85th-percentile speed prior to the work starting, or the anticipated speed exceeds 40 mph, additional warning signs should be used in the advance warning area.*

07 *Pedestrian crossings near TTC sites should be separated from the worksite by appropriate barriers that maintain the accessibility and detectability for pedestrians with disabilities.*

Support:

08 Near-side work spaces, as depicted in Figure 6H-21, are simply handled as a midblock lane closure. A problem that might occur with near-side lane closure is a reduction in capacity, which during certain hours of operation could result in congestion and backups.

Option:

09 When near-side work spaces are used, an exclusive turn lane may be used for through vehicular traffic.

10 Where space is restricted in advance of near-side work spaces, as with short block spacings, two warning signs may be used in the advance warning area, and a third action-type warning or a regulatory sign (such as Keep Left) may be placed within the transition area.

Support:

11 Far-side work spaces, as depicted in Figures 6H-22 through 6H-25, involve additional treatment because road users typically enter the activity area by straight-through and left- or right-turning movements.

Guidance:

12 *When a lane through an intersection must be closed on the far side, it should also be closed on the near-side approach to preclude merging movements within the intersection.*

Option:

13 If there are a significant number of vehicles turning from a near-side lane that is closed on the far side, the near-side lane may be converted to an exclusive turn lane.

Support:

14 Figures 6H-26 and 6H-27 provide guidance on applicable procedures for work performed within the intersection.

Option:

15 If the work is within the intersection, any of the following strategies may be used:
   A. A small work space so that road users can move around it, as shown in Figure 6H-26;
   B. Flaggers or uniformed law enforcement officers to direct road users, as shown in Figure 6H-27;
   C. Work in stages so the work space is kept to a minimum; and
   D. Road closures or upstream diversions to reduce road user volumes.

Guidance:

16 *Depending on road user conditions, a flagger(s) and/or a uniformed law enforcement officer(s) should be used to control road users.*

Section 6G.14 Work Within the Traveled Way of a Freeway or Expressway

Support:

01 Problems of TTC might occur under the special conditions encountered where vehicular traffic must be moved through or around TTC zones on high-speed, high-volume roadways. Although the general principles outlined in the previous Sections of this Manual are applicable to all types of highways, high-speed, access-controlled highways need special attention in order to accommodate vehicular traffic while also protecting road users and workers. The road user volumes, road vehicle mix (buses, trucks, cars, and bicycles, if permitted), and speed of vehicles on these facilities require that careful TTC procedures be implemented, for example, to induce critical merging maneuvers well in advance of work spaces and in a manner that creates minimum turbulence and

delay in the vehicular traffic stream. These situations often require more conspicuous devices than specified for normal rural highway or urban street use. However, the same important basic considerations of uniformity and standardization of general principles apply for all roadways.

02 Work under high-speed, high-volume vehicular traffic on a controlled access highway is complicated by the roadway design and operational features. The presence of a median that establishes separate roadways for directional vehicular traffic flow might prohibit the closing of one of the roadways or the diverting of vehicular traffic to the other roadway. Lack of access to and from adjacent roadways prohibits rerouting of vehicular traffic away from the work space in many cases. Other conditions exist where work must be limited to night hours, thereby necessitating increased use of warning lights, illumination of work spaces, and advance warning systems.

03 TTC for a typical lane closure on a divided highway is shown in Figure 6H-33. Temporary traffic controls for short duration and mobile operations on freeways are shown in Figure 6H-35. A typical application for shifting vehicular traffic lanes around a work space is shown in Figure 6H-36. TTC for multiple and interior lane closures on a freeway is shown in Figures 6H-37 and 6H-38.

Guidance:

04 *The method for closing an interior lane when the open lanes have the capacity to carry vehicular traffic should be as shown in Figure 6H-37. When the capacity of the other lanes is needed, the method shown in Figure 6H-38 should be used.*

Section 6G.15 Two-Lane, Two-Way Traffic on One Roadway of a Normally Divided Highway

Support:

01 Two-lane, two-way operation on one roadway of a normally divided highway is a typical procedure that requires special consideration in the planning, design, and work phases, because unique operational problems (for example, increasing the risk of head-on crashes) can arise with the two-lane, two-way operation.

Standard:

02 **When two-lane, two-way traffic control must be maintained on one roadway of a normally divided highway, opposing vehicular traffic shall be separated with either temporary traffic barriers (concrete safety-shape or approved alternate), channelizing devices, or a temporary raised island throughout the length of the two-way operation. The use of markings and complementary signing, by themselves, shall not be used.**

Support:

03 Figure 6H-39 shows the procedure for two-lane, two-way operation. Treatments for entrance and exit ramps within the two-way roadway segment of this type of work are shown in Figures 6H-40 and 6H-41.

Section 6G.16 Crossovers

Guidance:

01 *The following are considered good guiding principles for the design of crossovers:*

   A. *Tapers for lane drops should be separated from the crossovers, as shown in Figure 6H-39.*
   B. *Crossovers should be designed for speeds no lower than 10 mph below the posted speed, the off-peak 85th-percentile speed prior to the work starting, or the anticipated operating speed of the roadway, unless unusual site conditions require that a lower design speed be used.*
   C. *A good array of channelizing devices, delineators, and full-length, properly placed pavement markings should be used to provide drivers with a clearly defined travel path.*
   D. *The design of the crossover should accommodate all vehicular traffic, including trucks and buses.*

Support:

02 Temporary traffic barriers and the excessive use of TTC devices cannot compensate for poor geometric and roadway cross-section design of crossovers.

Section 6G.17 Interchanges

Guidance:

01 *Access to interchange ramps on limited-access highways should be maintained even if the work space is in the lane adjacent to the ramps. Access to exit ramps should be clearly marked and delineated with channelizing devices. For long-term projects, conflicting pavement markings should be removed and new ones placed. Early coordination with officials having jurisdiction over the affected cross streets and providing emergency services should occur before ramp closings.*

Option:

02   If access is not possible, ramps may be closed by using signs and Type 3 Barricades. As the work space changes, the access area may be changed, as shown in Figure 6H-42. A TTC zone in the exit ramp may be handled as shown in Figure 6H-43.

03   When a work space interferes with an entrance ramp, a lane may need to be closed on the freeway (see Figure 6H-44). A TTC zone in the entrance ramp may require shifting ramp vehicular traffic (see Figure 6H-44).

Section 6G.18 Work in the Vicinity of a Grade Crossing
Standard:

01   **When grade crossings exist either within or in the vicinity of a TTC zone, lane restrictions, flagging, or other operations shall not create conditions where vehicles can be queued across the tracks. If the queuing of vehicles across the tracks cannot be avoided, a uniformed law enforcement officer or flagger shall be provided at the crossing to prevent vehicles from stopping on the tracks, even if automatic warning devices are in place.**

Support:

02   Figure 6H-46 shows work in the vicinity of a grade crossing.

03   Section 8A.08 contains additional information regarding temporary traffic control zones in the vicinity of grade crossings.

Guidance:

04   *Early coordination with the railroad company or light rail transit agency should occur before work starts.*

Section 6G.19 Temporary Traffic Control During Nighttime Hours
Support:

01   Chapter 6D and Sections 6F.74 and 6G.05 contain additional information regarding the steps to follow when pedestrian or bicycle facilities are affected by the worksite.

02   Conducting highway construction and maintenance activities during night hours could provide an advantage when traditional daytime traffic control strategies cannot achieve an acceptable balance between worker and public safety, traffic and community impact, and constructability. The two basic advantages of working at night are reduced traffic congestion and less involvement with business activities. However, the two basic conditions that must normally be met for night work to offer any advantage are reduced traffic volumes and easy set up and removal of the traffic control patterns on a nightly basis.

03   Shifting work activities to night hours, when traffic volumes are lower and normal business is less active, might offer an advantage in some cases, as long as the necessary work can be completed and the worksite restored to essentially normal operating conditions to carry the higher traffic volume during non-construction hours.

04   Although working at night might offer advantages, it also includes safety issues. Reduced visibility inherent in night work impacts the performance of both drivers and workers. Because traffic volumes are lower and congestion is minimized, speeds are often higher at night necessitating greater visibility at a time when visibility is reduced. Finally, the incidence of impaired (alcohol or drugs), fatigued, or drowsy drivers might be higher at night.

05   Working at night also involves other factors, including construction productivity and quality, social impacts, economics, and environmental issues. A decision to perform construction or maintenance activities at night normally involves some consideration of the advantages to be gained compared to the safety and other issues that might be impacted.

Guidance:

06   *Considering the safety issues inherent to night work, consideration should be given to enhancing traffic controls (see Section 6G.04) to provide added visibility and driver guidance, and increased protection for workers.*

07   *In addition to the enhancements listed in Section 6G.04, consideration should be given to providing additional lights and retroreflective markings to workers, work vehicles, and equipment.*

Option:

08   Where reduced traffic volumes at night make it feasible, the entire roadway may be closed by detouring traffic to alternate facilities, thus removing the traffic risk from the activity area.

Guidance:

09   *Consideration should be given to stationing uniformed law enforcement officers and lighted patrol cars at night work locations where there is a concern that high speeds or impaired drivers might result in undue risks for workers or other drivers.*

Standard:

Except in emergencies, temporary lighting shall be provided at all flagger stations.

Support:

Desired illumination levels vary depending upon the nature of the task involved. An average horizontal luminance of 5 foot candles can be adequate for general activities. An average horizontal luminance of 10 foot candles can be adequate for activities around equipment. Tasks requiring high levels of precision and extreme care can require an average horizontal luminance of 20 foot candles.

CHAPTER 6H. TYPICAL APPLICATIONS

Section 6H.01 Typical Applications

Support:

01 Chapter 6G contains discussions of typical TTC activities. This Chapter presents typical applications for a variety of situations commonly encountered. While not every situation is addressed, the information illustrated can generally be adapted to a broad range of conditions. In many instances, an appropriate TTC plan is achieved by combining features from various typical applications. For example, work at an intersection might present a near-side work zone for one street and a far-side work zone for the other street. These treatments are found in two different typical applications, while a third typical application shows how to handle pedestrian crosswalk closures. For convenience in using the typical application diagrams, Tables 6C-1 and 6C-4 are reproduced in this Chapter as Tables 6H-3 and 6H-4, respectively.

02 Procedures for establishing TTC zones vary with such conditions as road configuration, location of the work, work activity, duration of work, road user volumes, road vehicle mix (buses, trucks, cars, motorcycles, and bicycles), and road user speeds.

03 In general, the procedures illustrated represent minimum solutions for the situations depicted. Except for the notes (which are clearly classified using headings as being Standard, Guidance, Option, or Support), the information presented in the typical applications can generally be regarded as Guidance.

Option:

04 Other devices may be added to supplement the devices and device spacing may be adjusted to provide additional reaction time or delineation. Fewer devices may be used based on field conditions.

Support:

05 Figures and tables found throughout Part 6 provide information for the development of TTC plans. Also, Table 6H-3 is used for the determination of sign spacing and other dimensions for various area and roadway types.

06 Table 6H-1 is an index of the 46 typical applications. Typical applications are shown on the right-hand page with notes on the facing page to the left. The legend for the symbols used in the typical applications is provided in Table 6H-2. In many of the typical applications, sign spacings and other dimensions are indicated by letters using the criteria provided in Table 6H-3. The formulas for determining taper lengths are provided in Table 6H-4.

07 Most of the typical applications show TTC devices for only one direction.

Table 6H-1. Index to Typical Applications

Typical Application Description	Typical Application Number
Work Outside of the Shoulder (see Section 6G.06)	
Work Beyond the Shoulder	TA-1
Blasting Zone	TA-2
Work on the Shoulder (see Sections 6G.07 and 6G.08)	
Work on the Shoulders	TA-3
Short Duration or Mobile Operation on a Shoulder	TA-4
Shoulder Closure on a Freeway	TA-5
Shoulder Work with Minor Encroachment	TA-6
Work Within the Traveled Way of a Two-Lane Highway (see Section 6G.10)	
Road Closed with a Diversion	TA-7
Roads Closed with an Off-Site Detour	TA-8
Overlapping Routes with a Detour	TA-9
Lane Closure on a Two-Lane Road Using Flaggers	TA-10
Lane Closure on a Two-Lane Road with Low Traffic Volumes	TA-11
Lane Closure on a Two-Lane Road Using Traffic Control Signals	TA-12
Temporary Road Closure	TA-13
Haul Road Crossing	TA-14
Work in the Center of a Road with Low Traffic Volumes	TA-15
Surveying Along the Center Line of a Road with Low Traffic Volumes	TA-16
Mobile Operations on a Two-Lane Road	TA-17
Work Within the Traveled Way of an Urban Street (see Section 6G.11)	
Lane Closure on a Minor Street	TA-18
Detour for One Travel Direction	TA-19
Detour for a Closed Street	TA-20
Work Within the Traveled Way at an Intersection and on Sidewalks (see Section 6G.13)	
Lane Closure on the Near Side of an Intersection	TA-21
Right-Hand Lane Closure on the Far Side of an Intersection	TA-22
Left-Hand Lane Closure on the Far Side of an Intersection	TA-23
Half Road Closure on the Far Side of an Intersection	TA-24
Multiple Lane Closures at an Intersection	TA-25
Closure in the Center of an Intersection	TA-26
Closure at the Side of an Intersection	TA-27
Sidewalk Detour or Diversion	TA-28
Crosswalk Closures and Pedestrian Detours	TA-29
Work Within the Traveled Way of a Multi-Lane, Non-Access Controlled Highway (see Section 6G.12)	
Interior Lane Closure on a Multi-Lane Street	TA-30
Lane Closure on a Street with Uneven Directional Volumes	TA-31
Half Road Closure on a Multi-Lane, High-Speed Highway	TA-32
Stationary Lane Closure on a Divided Highway	TA-33
Lane Closure with a Temporary Traffic Barrier	TA-34
Mobile Operation on a Multi-Lane Road	TA-35
Work Within the Traveled Way of a Freeway or Expressway (see Section 6G.14)	
Lane Shift on a Freeway	TA-36
Double Lane Closure on a Freeway	TA-37
Interior Lane Closure on a Freeway	TA-38
Median Crossover on a Freeway	TA-39
Median Crossover for an Entrance Ramp	TA-40
Median Crossover for an Exit Ramp	TA-41
Work in the Vicinity of an Exit Ramp	TA-42
Partial Exit Ramp Closure	TA-43
Work in the Vicinity of an Entrance Ramp	TA-44
Temporary Reversible Lane Using Movable Barriers	TA-45
Work in the Vicinity of a Grade Crossing (see Section 6G.18)	
Work in the Vicinity of a Grade Crossing	TA-46

Table 6H-2. Meaning of Symbols on Typical Application Diagrams

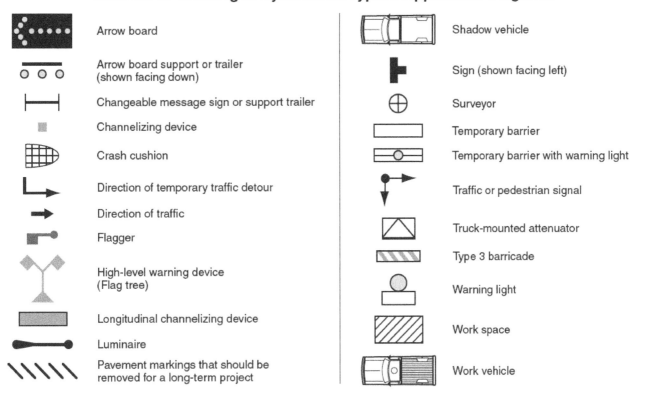

Symbol	Meaning	Symbol	Meaning
	Arrow board		Shadow vehicle
	Arrow board support or trailer (shown facing down)		Sign (shown facing left)
	Changeable message sign or support trailer		Surveyor
	Channelizing device		Temporary barrier
	Crash cushion		Temporary barrier with warning light
	Direction of temporary traffic detour		Traffic or pedestrian signal
	Direction of traffic		Truck-mounted attenuator
	Flagger		Type 3 barricade
	High-level warning device (Flag tree)		Warning light
	Longitudinal channelizing device		Work space
	Luminaire		Work vehicle
	Pavement markings that should be removed for a long-term project		

Table 6H-3. Meaning of Letter Codes on Typical Application Diagrams

Road Type	Distance Between Signs**		
	A	B	C
Urban (low speed)*	100 feet	100 feet	100 feet
Urban (high speed)*	350 feet	350 feet	350 feet
Rural	500 feet	500 feet	500 feet
Expressway / Freeway	1,000 feet	1,500 feet	2,640 feet

* Speed category to be determined by highway agency

** The column headings A, B, and C are the dimensions shown in Figures 6H-1 through 6H-46. The A dimension is the distance from the transition or point of restriction to the first sign. The B dimension is the distance between the first and second signs. The C dimension is the distance between the second and third signs. (The "first sign" is the sign in a three-sign series that is closest to the TTC zone. The "third sign" is the sign that is furthest upstream from the TTC zone.)

Table 6H-4. Formulas for Determining Taper Length

Speed (S)	Taper Length (L) in feet
40 mph or less	$L = \dfrac{WS^2}{60}$
45 mph or more	$L = WS$

Where: L = taper length in feet
W = width of offset in feet
S = posted speed limit, or off-peak 85th-percentile speed prior to work starting, or the anticipated operating speed in mph

Notes for Figure 6H-1—Typical Application 1
Work Beyond the Shoulder

Guidance:
1. *If the work space is in the median of a divided highway, an advance warning sign should also be placed on the left side of the directional roadway.*

Option:
2. The ROAD WORK AHEAD sign may be replaced with other appropriate signs such as the SHOULDER WORK sign. The SHOULDER WORK sign may be used for work adjacent to the shoulder.
3. The ROAD WORK AHEAD sign may be omitted where the work space is behind a barrier, more than 24 inches behind the curb, or 15 feet or more from the edge of any roadway.
4. For short-term, short duration or mobile operation, all signs and channelizing devices may be eliminated if a vehicle with activated high-intensity rotating, flashing, oscillating, or strobe lights is used.
5. Vehicle hazard warning signals may be used to supplement high-intensity rotating, flashing, oscillating, or strobe lights.

Standard:
6. **Vehicle hazard warning signals shall not be used instead of the vehicle's high-intensity rotating, flashing, oscillating, or strobe lights.**

Figure 6H-1. Work Beyond the Shoulder (TA-1)

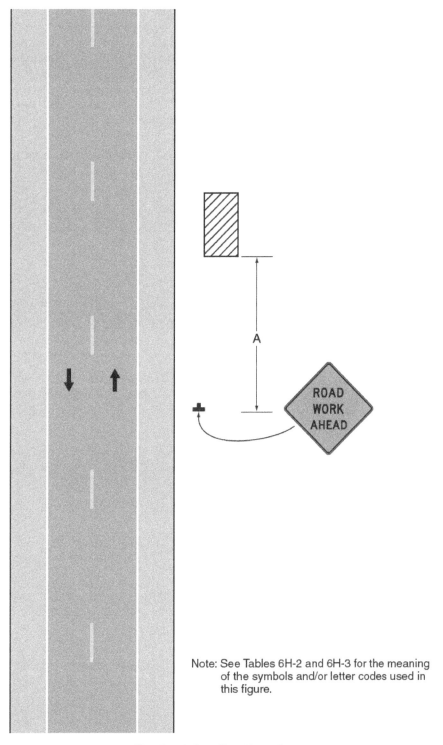

Note: See Tables 6H-2 and 6H-3 for the meaning of the symbols and/or letter codes used in this figure.

Typical Application 1

Notes for Figure 6H-2—Typical Application 2
Blasting Zone

Standard:
1. Whenever blasting caps are used within 1,000 feet of a roadway, the signing shown shall be used.
2. The signs shall be covered or removed when there are no explosives in the area or the area is otherwise secure.
3. Whenever a side road intersects the roadway between the BLASTING ZONE AHEAD sign and the END BLASTING ZONE sign, or a side road is within 1,000 feet of any blasting cap, similar signing, as on the mainline, shall be installed on the side road.
4. Prior to blasting, the blaster in charge shall determine whether road users in the blasting zone will be endangered by the blasting operation. If there is danger, road users shall not be permitted to pass through the blasting zone during blasting operations.

Guidance:
5. *On a divided highway, the signs should be mounted on both sides of the directional roadways.*

Figure 6H-2. Blasting Zone (TA-2)

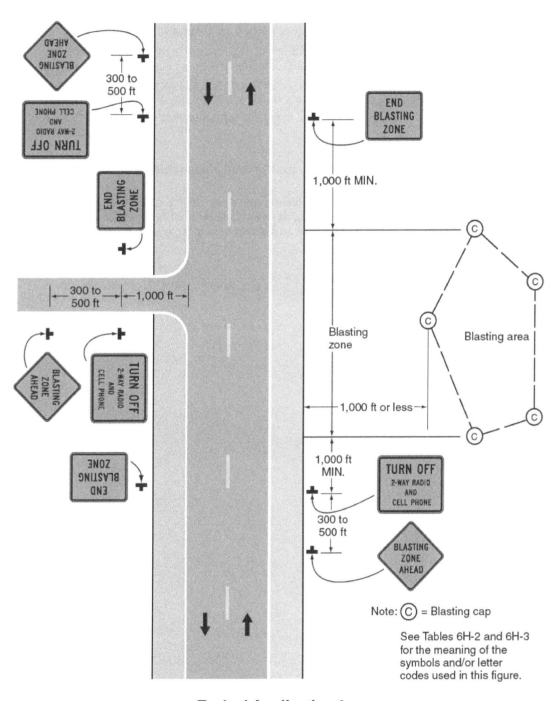

Typical Application 2

Notes for Figure 6H-3—Typical Application 3
Work on the Shoulders

Guidance:
1. *A SHOULDER WORK sign should be placed on the left side of the roadway for a divided or one-way street only if the left shoulder is affected.*

Option:
2. The Workers symbol signs may be used instead of SHOULDER WORK signs.
3. The SHOULDER WORK AHEAD sign on an intersecting roadway may be omitted where drivers emerging from that roadway will encounter another advance warning sign prior to this activity area.
4. For short duration operations of 60 minutes or less, all signs and channelizing devices may be eliminated if a vehicle with activated high-intensity rotating, flashing, oscillating, or strobe lights is used.
5. Vehicle hazard warning signals may be used to supplement high-intensity rotating, flashing, oscillating, or strobe lights.

Standard:
6. **Vehicle hazard warning signals shall not be used instead of the vehicle's high-intensity rotating, flashing, oscillating, or strobe lights.**
7. **When paved shoulders having a width of 8 feet or more are closed, at least one advance warning sign shall be used. In addition, channelizing devices shall be used to close the shoulder in advance to delineate the beginning of the work space and direct vehicular traffic to remain within the traveled way.**

Figure 6H-3. Work on the Shoulders (TA-3)

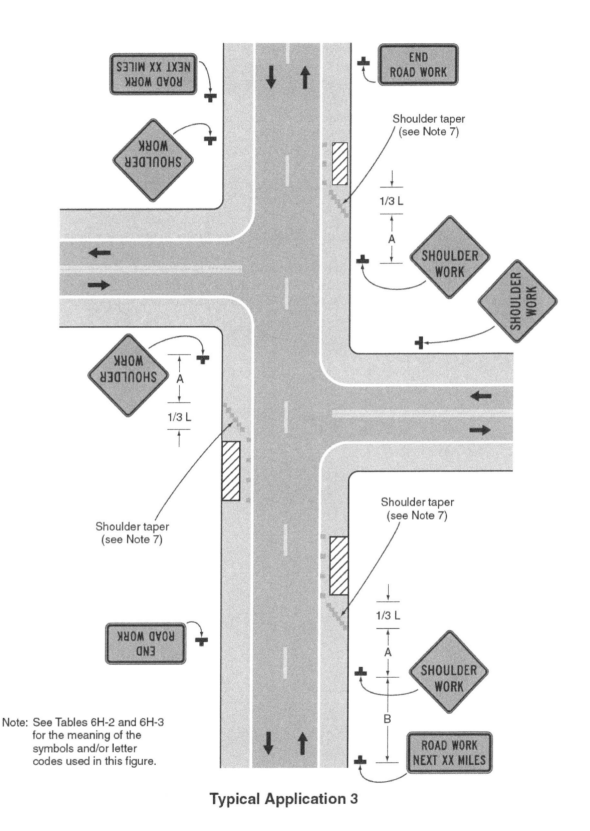

Note: See Tables 6H-2 and 6H-3 for the meaning of the symbols and/or letter codes used in this figure.

Typical Application 3

Notes for Figure 6H-4—Typical Application 4
Short Duration or Mobile Operation on a Shoulder

Guidance:
1. *In those situations where multiple work locations within a limited distance make it practical to place stationary signs, the distance between the advance warning sign and the work should not exceed 5 miles.*
2. *In those situations where the distance between the advance signs and the work is 2 miles to 5 miles, a Supplemental Distance plaque should be used with the ROAD WORK AHEAD sign.*

Option:
3. The ROAD WORK NEXT XX MILES sign may be used instead of the ROAD WORK AHEAD sign if the work locations occur over a distance of more than 2 miles.
4. Stationary warning signs may be omitted for short duration or mobile operations if the work vehicle displays high-intensity rotating, flashing, oscillating, or strobe lights.
5. Vehicle hazard warning signals may be used to supplement high-intensity rotating, flashing, oscillating, or strobe lights.

Standard:
6. **Vehicle hazard warning signals shall not be used instead of the vehicle's high-intensity rotating, flashing, oscillating, or strobe lights.**
7. If an arrow board is used for an operation on the shoulder, the caution mode shall be used.
8. Vehicle-mounted signs shall be mounted in a manner such that they are not obscured by equipment or supplies. Sign legends on vehicle-mounted signs shall be covered or turned from view when work is not in progress.

Figure 6H-4. Short-Duration or Mobile Operation on a Shoulder (TA-4)

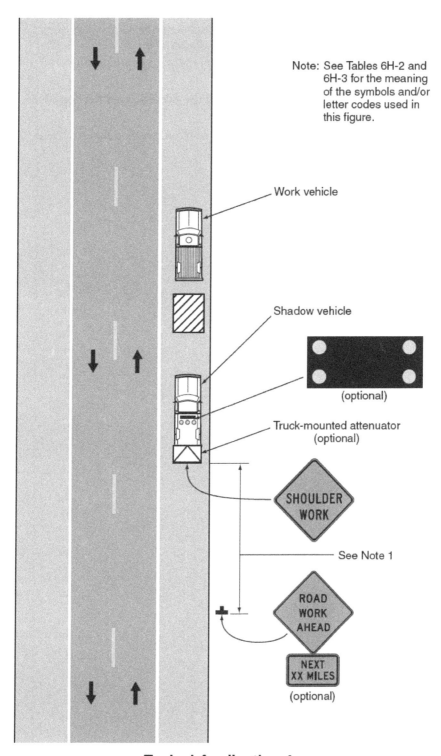

Typical Application 4

Notes for Figure 6H-5—Typical Application 5
Shoulder Closure on a Freeway

Guidance:
1. *SHOULDER CLOSED signs should be used on limited-access highways where there is no opportunity for disabled vehicles to pull off the roadway.*
2. *If drivers cannot see a pull-off area beyond the closed shoulder, information regarding the length of the shoulder closure should be provided in feet or miles, as appropriate.*
3. *The use of a temporary traffic barrier should be based on engineering judgment.*

Standard:

4. Temporary traffic barriers, if used, shall comply with the provisions of Section 6F.85.

Option:
5. The barrier shown in this typical application is an example of one method that may be used to close a shoulder of a long-term project.
6. The warning lights shown on the barrier may be used.

Figure 6H-5. Shoulder Closure on a Freeway (TA-5)

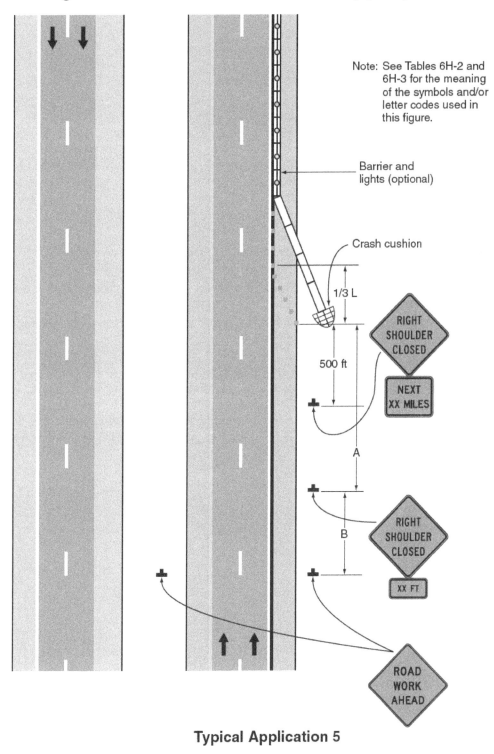

Typical Application 5

Notes for Figure 6H-6—Typical Application 6
Shoulder Work with Minor Encroachment

Guidance:
1. All lanes should be a minimum of 10 feet in width as measured to the near face of the channelizing devices.
2. The treatment shown should be used on a minor road having low speeds. For higher-speed traffic conditions, a lane closure should be used.

Option:
3. For short-term use on low-volume, low-speed roadways with vehicular traffic that does not include longer and wider heavy commercial vehicles, a minimum lane width of 9 feet may be used.
4. Where the opposite shoulder is suitable for carrying vehicular traffic and of adequate width, lanes may be shifted by use of closely-spaced channelizing devices, provided that the minimum lane width of 10 feet is maintained.
5. Additional advance warning may be appropriate, such as a ROAD NARROWS sign.
6. Temporary traffic barriers may be used along the work space.
7. The shadow vehicle may be omitted if a taper and channelizing devices are used.
8. A truck-mounted attenuator may be used on the shadow vehicle.
9. For short-duration work, the taper and channelizing devices may be omitted if a shadow vehicle with activated high-intensity rotating, flashing, oscillating, or strobe lights is used.
10. Vehicle hazard warning signals may be used to supplement high-intensity rotating, flashing, oscillating, or strobe lights.

Standard:
11. **Vehicle-mounted signs shall be mounted in a manner such that they are not obscured by equipment or supplies. Sign legends on vehicle-mounted signs shall be covered or turned from view when work is not in progress.**
12. **Shadow and work vehicles shall display high-intensity rotating, flashing, oscillating, or strobe lights.**
13. **Vehicle hazard warning signals shall not be used instead of the vehicle's high-intensity rotating, flashing, oscillating, or strobe lights.**

Figure 6H-6. Shoulder Work with Minor Encroachment (TA-6)

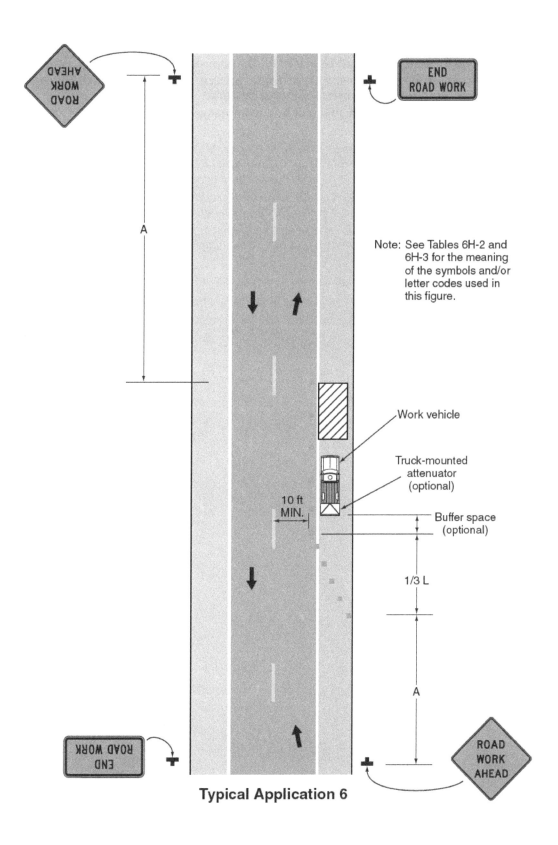

Typical Application 6

Notes for Figure 6H-7—Typical Application 7
Road Closure with a Diversion

Support:
1. Signs and object markers are shown for one direction of travel only.

Standard:
2. **Devices similar to those depicted shall be placed for the opposite direction of travel.**
3. **Pavement markings no longer applicable to the traffic pattern of the roadway shall be removed or obliterated before any new traffic patterns are open to traffic.**
4. **Temporary barriers and end treatments shall be crashworthy.**

Guidance:
5. *If the tangent distance along the temporary diversion is more than 600 feet, a Reverse Curve sign, left first, should be used instead of the Double Reverse Curve sign, and a second Reverse Curve sign, right first, should be placed in advance of the second reverse curve back to the original alignment.*
6. *When the tangent section of the diversion is more than 600 feet, and the diversion has sharp curves with recommended speeds of 30 mph or less, Reverse Turn signs should be used.*
7. *Where the temporary pavement and old pavement are different colors, the temporary pavement should start on the tangent of the existing pavement and end on the tangent of the existing pavement.*

Option:
8. Flashing warning lights and/or flags may be used to call attention to the warning signs.
9. On sharp curves, large arrow signs may be used in addition to other advance warning signs.
10. Delineators or channelizing devices may be used along the diversion.

Figure 6H-7. Road Closure with a Diversion (TA-7)

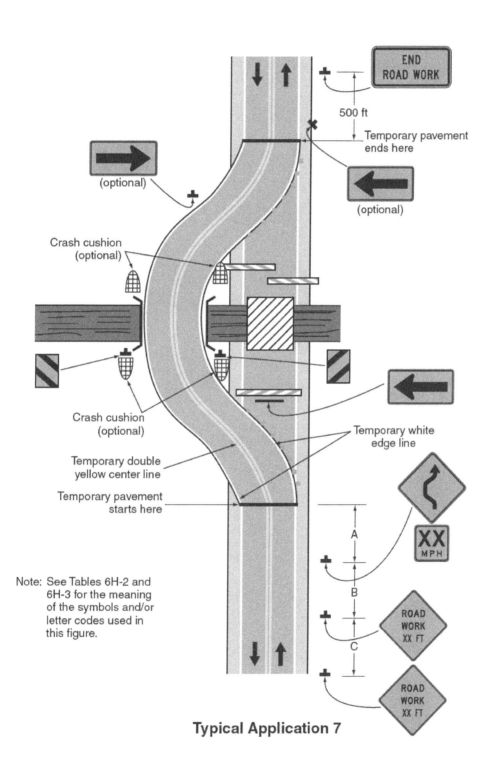

Typical Application 7

Notes for Figure 6H-8—Typical Application 8
Road Closure with an Off-Site Detour

Guidance:
1. *Regulatory traffic control devices should be modified as needed for the duration of the detour.*

Option:
2. If the road is opened for some distance beyond the intersection and/or there are significant origin/destination points beyond the intersection, the ROAD CLOSED and DETOUR signs on Type 3 Barricades may be located at the edge of the traveled way.
3. A Route Sign Directional assembly may be placed on the far left corner of the intersection to augment or replace the one shown on the near right corner.
4. Flashing warning lights and/or flags may be used to call attention to the advance warning signs.
5. Cardinal direction plaques may be used with route signs.

Figure 6H-8. Road Closure with an Off-Site Detour (TA-8)

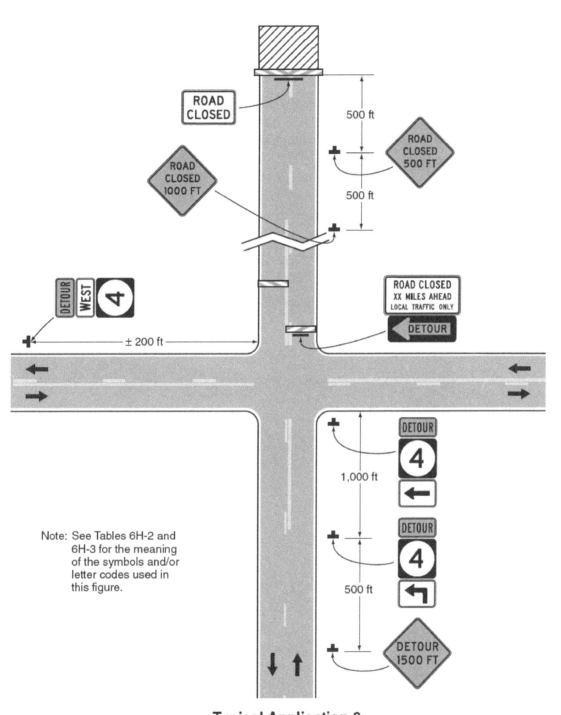

Typical Application 8

Notes for Figure 6H-9—Typical Application 9
Overlapping Routes with a Detour

Support:
1. TTC devices are shown for one direction of travel only.

Standard:
2. Devices similar to those depicted shall be placed for the opposite direction of travel.

Guidance:
3. *STOP or YIELD signs displayed to side roads should be installed as needed along the temporary route.*

Option:
4. Flashing warning lights and/or flags may be used to call attention to the advance warning signs.
5. Flashing warning lights may be used on the Type 3 Barricades.
6. Cardinal direction plaques may be used with route signs.

Figure 6H-9. Overlapping Routes with a Detour (TA-9)

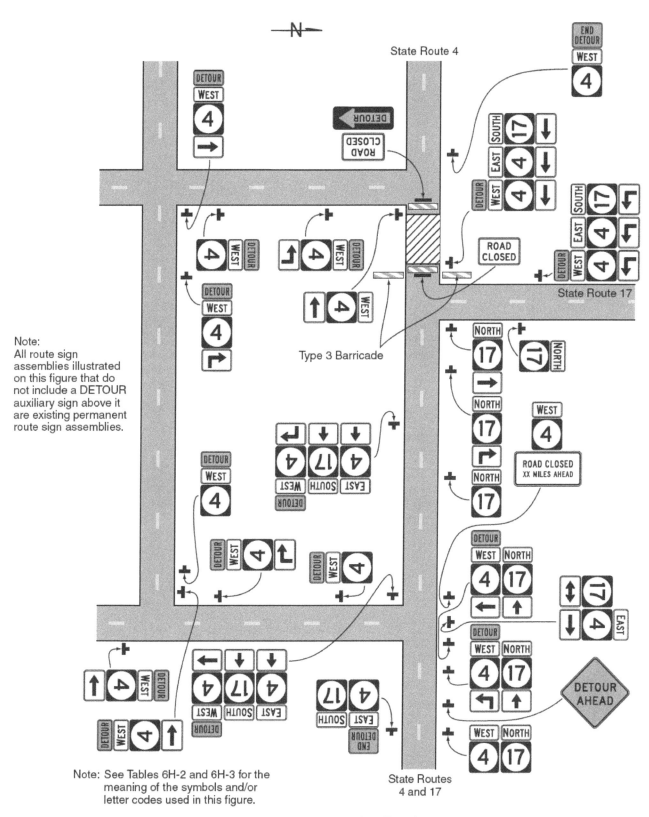

Typical Application 9

Notes for Figure 6H-10—Typical Application 10
Lane Closure on a Two-Lane Road Using Flaggers

Option:

1. For low-volume situations with short work zones on straight roadways where the flagger is visible to road users approaching from both directions, a single flagger, positioned to be visible to road users approaching from both directions, may be used (see Chapter 6E).
2. The ROAD WORK AHEAD and the END ROAD WORK signs may be omitted for short-duration operations.
3. Flashing warning lights and/or flags may be used to call attention to the advance warning signs. A BE PREPARED TO STOP sign may be added to the sign series.

Guidance:

4. *The buffer space should be extended so that the two-way traffic taper is placed before a horizontal (or crest vertical) curve to provide adequate sight distance for the flagger and a queue of stopped vehicles.*

Standard:

5. **At night, flagger stations shall be illuminated, except in emergencies.**

Guidance:

6. *When used, the BE PREPARED TO STOP sign should be located between the Flagger sign and the ONE LANE ROAD sign.*
7. *When a grade crossing exists within or upstream of the transition area and it is anticipated that queues resulting from the lane closure might extend through the grade crossing, the TTC zone should be extended so that the transition area precedes the grade crossing.*
8. *When a grade crossing equipped with active warning devices exists within the activity area, provisions should be made for keeping flaggers informed as to the activation status of these warning devices.*
9. *When a grade crossing exists within the activity area, drivers operating on the left-hand side of the normal center line should be provided with comparable warning devices as for drivers operating on the right-hand side of the normal center line.*
10. *Early coordination with the railroad company or light rail transit agency should occur before work starts.*

Option:

11. A flagger or a uniformed law enforcement officer may be used at the grade crossing to minimize the probability that vehicles are stopped within 15 feet of the grade crossing, measured from both sides of the outside rails.

Figure 6H-10. Lane Closure on a Two-Lane Road Using Flaggers (TA-10)

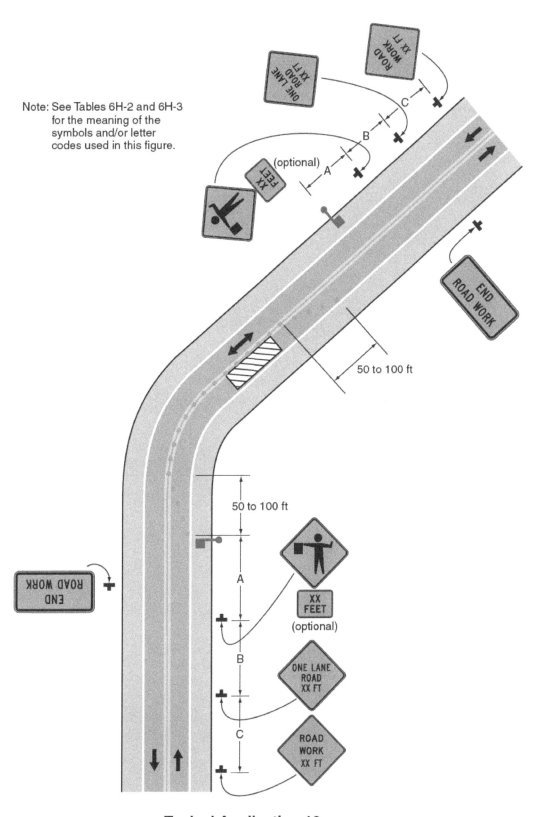

Typical Application 10

Notes for Figure 6H-11—Typical Application 11
Lane Closure on a Two-Lane Road with Low Traffic Volumes

Option:

1. This TTC zone application may be used as an alternate to the TTC application shown in Figure 6H-10 (using flaggers) when the following conditions exist:
    a. Vehicular traffic volume is such that sufficient gaps exist for vehicular traffic that must yield.
    b. Road users from both directions are able to see approaching vehicular traffic through and beyond the worksite and have sufficient visibility of approaching vehicles.
2. The Type B flashing warning lights may be placed on the ROAD WORK AHEAD and the ONE LANE ROAD AHEAD signs whenever a night lane closure is necessary.

Figure 6H-11. Lane Closure on a Two-Lane Road with Low Traffic Volumes (TA-11)

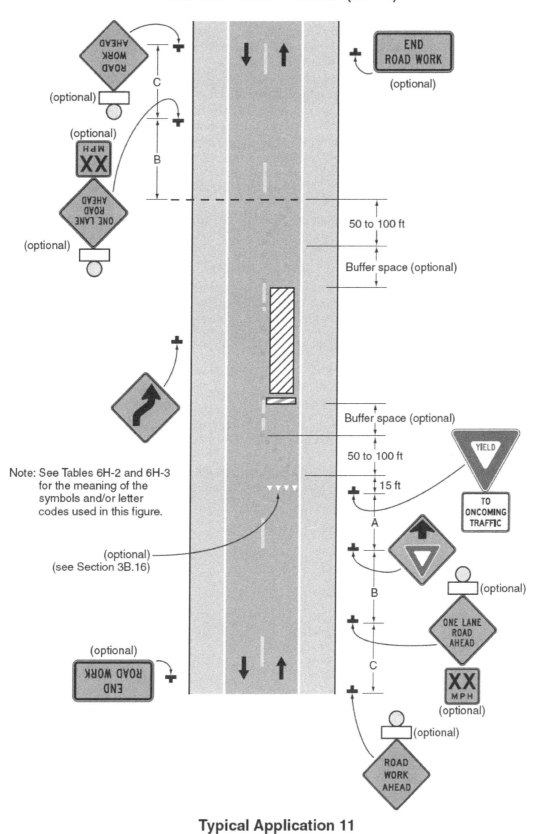

Typical Application 11

Notes for Figure 6H-12—Typical Application 12
Lane Closure on a Two-Lane Road Using Traffic Control Signals

Standard:
1. Temporary traffic control signals shall be installed and operated in accordance with the provisions of Part 4. Temporary traffic control signals shall meet the physical display and operational requirements of conventional traffic control signals.
2. Temporary traffic control signal timing shall be established by authorized officials. Durations of red clearance intervals shall be adequate to clear the one-lane section of conflicting vehicles.
3. When the temporary traffic control signal is changed to the flashing mode, either manually or automatically, red signal indications shall be flashed to both approaches.
4. Stop lines shall be installed with temporary traffic control signals for intermediate and long-term closures. Existing conflicting pavement markings and raised pavement marker reflectors between the activity area and the stop line shall be removed. After the temporary traffic control signal is removed, the stop lines and other temporary pavement markings shall be removed and the permanent pavement markings restored.
5. Safeguards shall be incorporated to avoid the possibility of conflicting signal indications at each end of the TTC zone.

Guidance:
6. *Where no-passing lines are not already in place, they should be added.*
7. *Adjustments in the location of the advance warning signs should be made as needed to accommodate the horizontal or vertical alignment of the roadway, recognizing that the distances shown for sign spacings are minimums. Adjustments in the height of the signal heads should be made as needed to conform to the vertical alignment.*

Option:
8. Flashing warning lights shown on the ROAD WORK AHEAD and the ONE LANE ROAD AHEAD signs may be used.
9. Removable pavement markings may be used.

Support:
10. Temporary traffic control signals are preferable to flaggers for long-term projects and other activities that would require flagging at night.
11. The maximum length of activity area for one-way operation under temporary traffic control signal control is determined by the capacity required to handle the peak demand.

Figure 6H-12. Lane Closure on a Two-Lane Road Using Traffic Control Signals (TA-12)

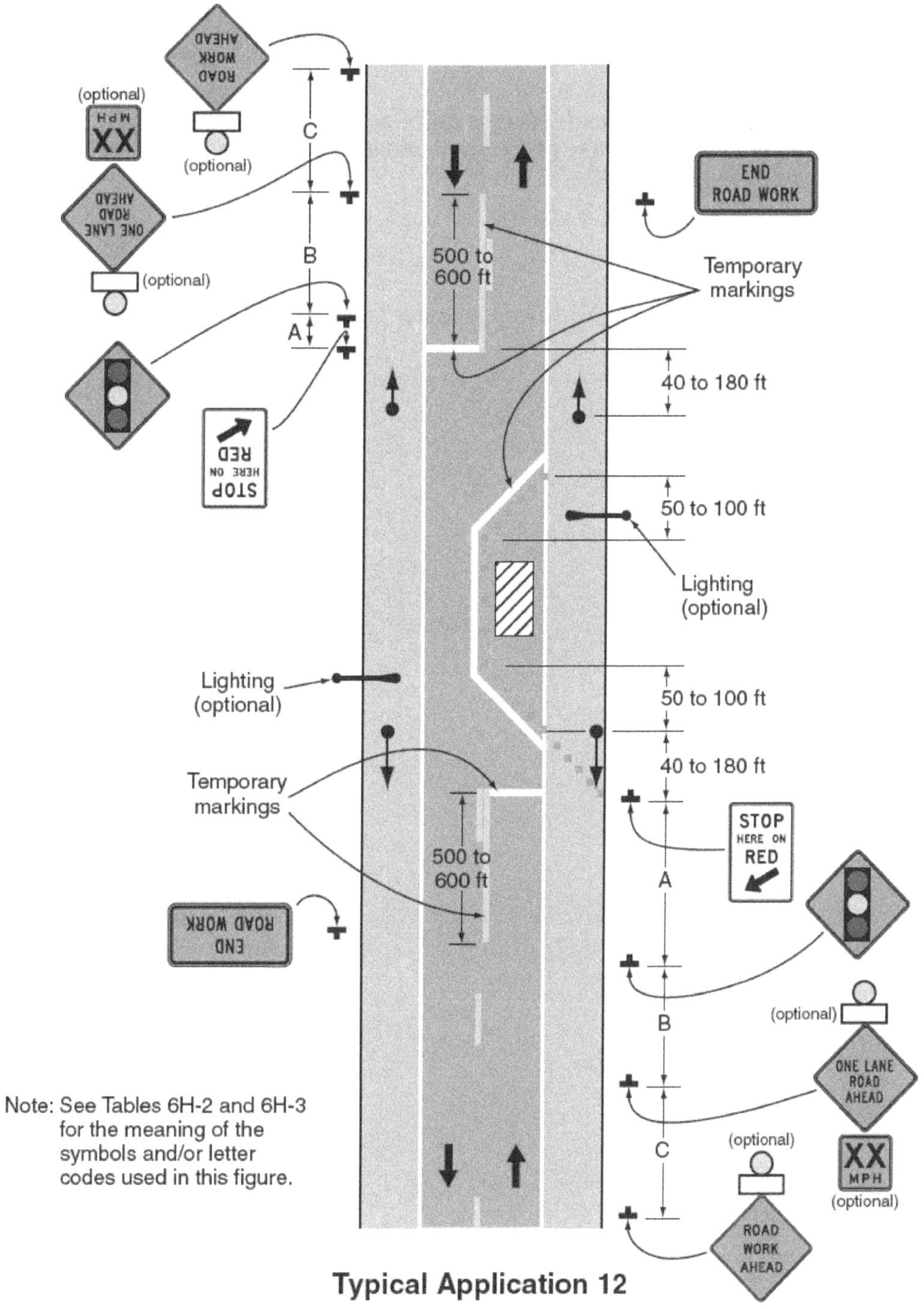

Typical Application 12

Note: See Tables 6H-2 and 6H-3 for the meaning of the symbols and/or letter codes used in this figure.

Notes for Figure 6H-13—Typical Application 13
Temporary Road Closure

Support:
1. Conditions represented are a planned closure not exceeding 20 minutes during the daytime.

Standard:

2. A flagger or uniformed law enforcement officer shall be used for this application. The flagger, if used for this application, shall follow the procedures provided in Sections 6E.07 and 6E.08.

Guidance:

3. The uniformed law enforcement officer, if used for this application, should follow the procedures provided in Sections 6E.07 and 6E.08.

Option:
4. A BE PREPARED TO STOP sign may be added to the sign series.

Guidance:

5. When used, the BE PREPARED TO STOP sign should be located before the Flagger symbol sign.

Figure 6H-13. Temporary Road Closure (TA-13)

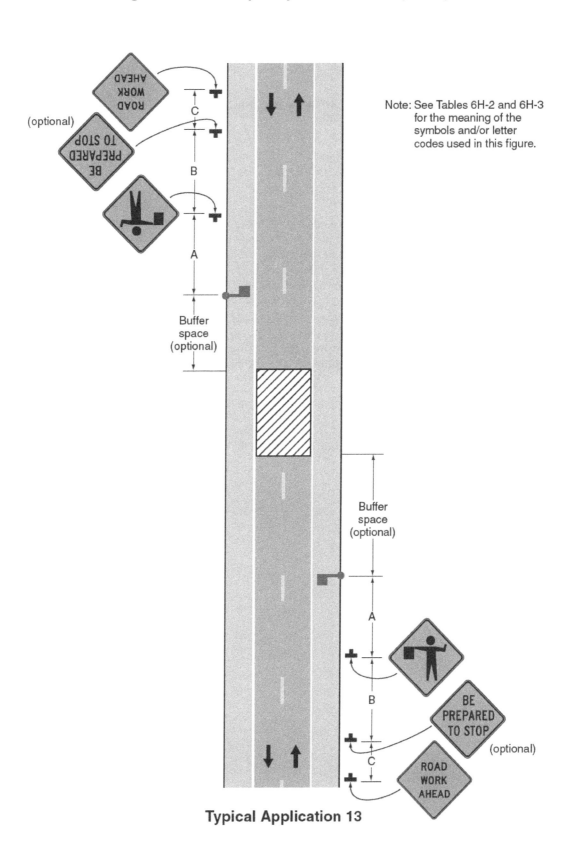

Typical Application 13

Notes for Figure 6H-14—Typical Application 14
Haul Road Crossing

Guidance:
1. Floodlights should be used to illuminate haul road crossings where existing light is inadequate.
2. Where no-passing lines are not already in place, they should be added.

Standard:
3. The traffic control method selected shall be used in both directions.

<u>Flagging Method</u>
4. When a road used exclusively as a haul road is not in use, the haul road shall be closed with Type 3 Barricades and the Flagger symbol signs covered.
5. The flagger shall follow the procedures provided in Sections 6E.07 and 6E.08.
6. At night, flagger stations shall be illuminated, except in emergencies.

<u>Signalized Method</u>
7. When a road used exclusively as a haul road is not in use, the haul road shall be closed with Type 3 Barricades. The signals shall either flash yellow on the main road or be covered, and the Signal Ahead and STOP HERE ON RED signs shall be covered or hidden from view.
8. The temporary traffic control signals shall control both the highway and the haul road and shall meet the physical display and operational requirements of conventional traffic control signals as described in Part 4. Traffic control signal timing shall be established by authorized officials.
9. Stop lines shall be used on existing highway with temporary traffic control signals.
10. Existing conflicting pavements markings between the stop lines shall be removed. After the temporary traffic control signal is removed, the stop lines and other temporary pavement markings shall be removed and the permanent pavement markings restored.

Figure 6H-14. Haul Road Crossing (TA-14)

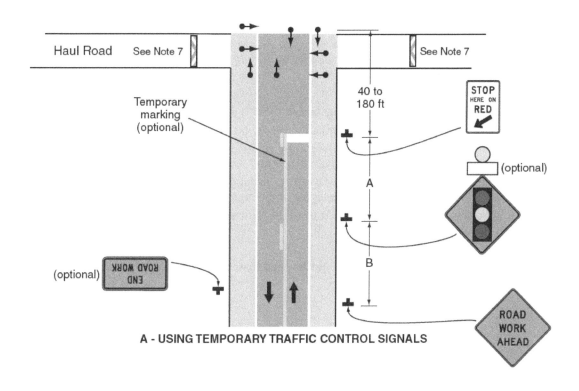

A - USING TEMPORARY TRAFFIC CONTROL SIGNALS

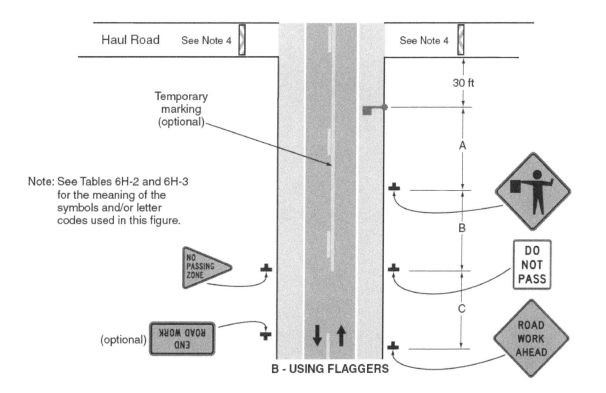

B - USING FLAGGERS

Typical Application 14

Notes for Figure 6H-15—Typical Application 15
Work in the Center of a Road with Low Traffic Volumes

Guidance:
1. *The lanes on either side of the center work space should have a minimum width of 10 feet as measured from the near edge of the channelizing devices to the edge of the pavement or the outside edge of the paved shoulder.*

Option:
2. Flashing warning lights and/or flags may be used to call attention to the advance warning signs.
3. If the closure continues overnight, warning lights may be used on the channelizing devices.
4. A lane width of 9 feet may be used for short-term stationary work on low-volume, low-speed roadways when motor vehicle traffic does not include longer and wider heavy commercial vehicles.
5. A work vehicle displaying high-intensity rotating, flashing, oscillating, or strobe lights may be used instead of the channelizing devices forming the tapers or the high-level warning devices.
6. Vehicle hazard warning signals may be used to supplement high-intensity rotating, flashing, oscillating, or strobe lights.

Standard:
7. **Vehicle hazard warning signals shall not be used instead of the vehicle's high-intensity rotating, flashing, oscillating, or strobe lights.**

Figure 6H-15. Work in the Center of a Road with Low Traffic Volumes (TA-15)

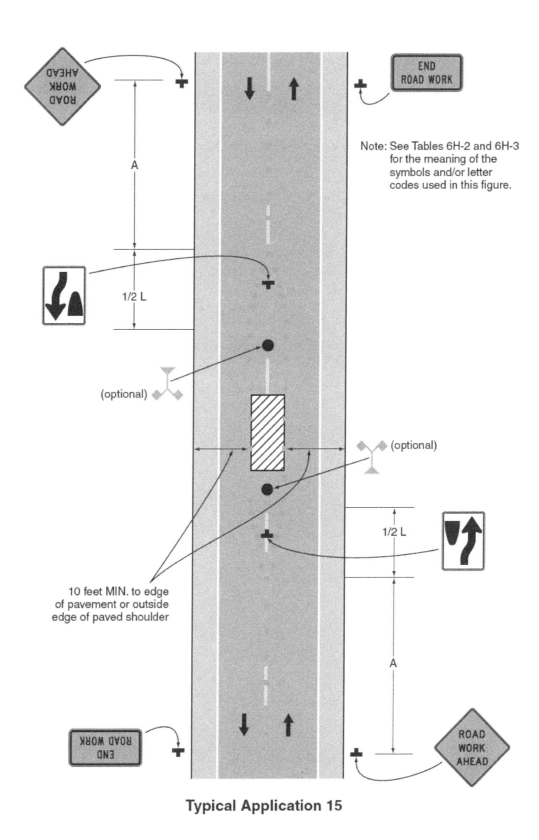

Typical Application 15

Notes for Figure 6H-16—Typical Application 16
Surveying Along the Center Line of a Road with Low Traffic Volumes

Guidance:
1. *The lanes on either side of the center work space should have a minimum width of 10 feet as measured from the near edge of the channelizing devices to the edge of the pavement or the outside edge of the paved shoulder.*
2. *Cones should be placed 6 to 12 inches on either side of the center line.*
3. *A flagger should be used to warn workers who cannot watch road users.*

Standard:

4. **For surveying on the center line of a high-volume road, one lane shall be closed using the information illustrated in Figure 6H-10.**

Option:
5. A high-level warning device may be used to protect a surveying device, such as a target on a tripod.
6. Cones may be omitted for a cross-section survey.
7. ROAD WORK AHEAD signs may be used in place of the SURVEY CREW AHEAD signs.
8. Flags may be used to call attention to the advance warning signs.
9. If the work is along the shoulder, the flagger may be omitted.
10. For a survey along the edge of the road or along the shoulder, cones may be placed along the edge line.
11. A BE PREPARED TO STOP sign may be added to the sign series.

Guidance:
12. *When used, the BE PREPARED TO STOP sign should be located before the Flagger symbol sign.*

Figure 6H-16. Surveying Along the Center Line of a Road with Low Traffic Volumes (TA-16)

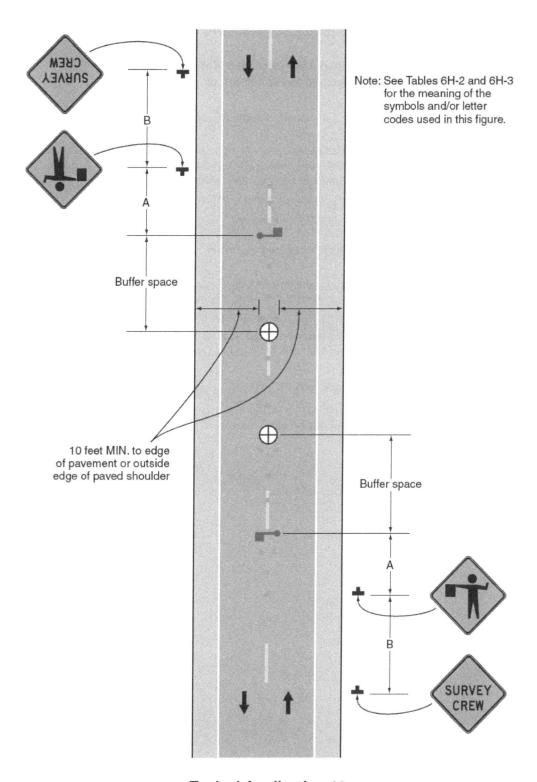

Typical Application 16

Notes for Figure 6H-17—Typical Application 17
Mobile Operations on a Two-Lane Road

Standard:
1. **Vehicle-mounted signs shall be mounted in a manner such that they are not obscured by equipment or supplies. Sign legends on vehicle-mounted signs shall be covered or turned from view when work is not in progress.**
2. **Shadow and work vehicles shall display high-intensity rotating, flashing, oscillating, or strobe lights.**
3. **If an arrow board is used, it shall be used in the caution mode.**

Guidance:

4. *Where practical and when needed, the work and shadow vehicles should pull over periodically to allow vehicular traffic to pass.*
5. *Whenever adequate stopping sight distance exists to the rear, the shadow vehicle should maintain the minimum distance from the work vehicle and proceed at the same speed. The shadow vehicle should slow down in advance of vertical or horizontal curves that restrict sight distance.*
6. *The shadow vehicles should also be equipped with two high-intensity flashing lights mounted on the rear, adjacent to the sign.*

Option:

7. The distance between the work and shadow vehicles may vary according to terrain, paint drying time, and other factors.
8. Additional shadow vehicles to warn and reduce the speed of oncoming or opposing vehicular traffic may be used. Law enforcement vehicles may be used for this purpose.
9. A truck-mounted attenuator may be used on the shadow vehicle or on the work vehicle.
10. If the work and shadow vehicles cannot pull over to allow vehicular traffic to pass frequently, a DO NOT PASS sign may be placed on the rear of the vehicle blocking the lane.

Support:

11. Shadow vehicles are used to warn motor vehicle traffic of the operation ahead.

Standard:

12. **Vehicle hazard warning signals shall not be used instead of the vehicle's high-intensity rotating, flashing, oscillating, or strobe lights.**

Figure 6H-17. Mobile Operations on a Two-Lane Road (TA-17)

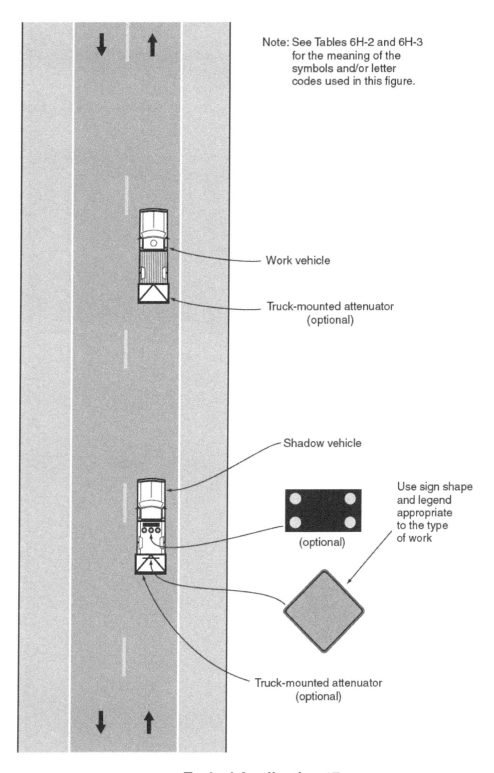

Typical Application 17

Notes for Figure 6H-18—Typical Application 18
Lane Closure on a Minor Street

Standard:

   1. **This TTC shall be used only for low-speed facilities having low traffic volumes.**

Option:

   2. Where the work space is short, where road users can see the roadway beyond, and where volume is low, vehicular traffic may be self-regulating.

Standard:

   3. **Where vehicular traffic cannot effectively self-regulate, one or two flaggers shall be used as illustrated in Figure 6H-10.**

Option:

   4. Flashing warning lights and/or flags may be used to call attention to the advance warning signs.

   5. A truck-mounted attenuator may be used on the work vehicle and the shadow vehicle.

Figure 6H-18. Lane Closure on a Minor Street (TA-18)

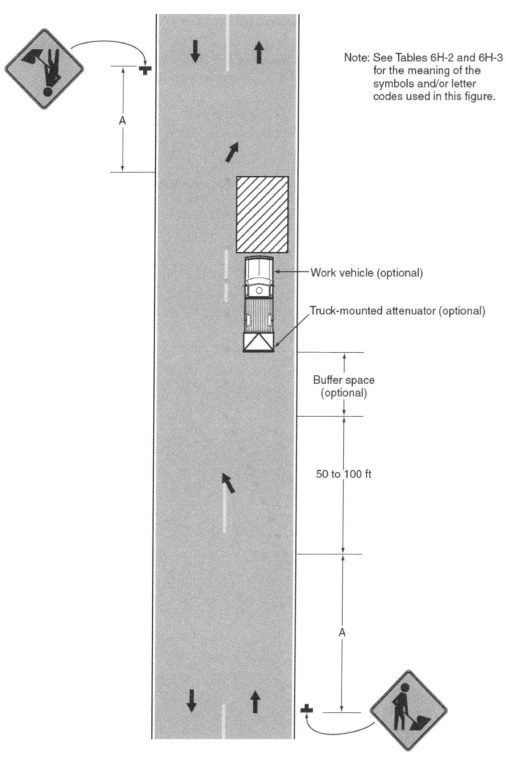

Typical Application 18

Notes for Figure 6H-19—Typical Application 19
Detour for One Travel Direction

Guidance:
1. *This plan should be used for streets without posted route numbers.*
2. *On multi-lane streets, Detour signs with an Advance Turn Arrow should be used in advance of a turn.*

Option:
3. The STREET CLOSED legend may be used in place of ROAD CLOSED.
4. Additional DO NOT ENTER signs may be used at intersections with intervening streets.
5. Warning lights may be used on Type 3 Barricades.
6. Detour signs may be located on the far side of intersections.
7. A Street Name sign may be mounted with the Detour sign. The Street Name sign may be either white on green or black on orange.

Standard:
8. When used, the Street Name sign shall be placed above the Detour sign.

Figure 6H-19. Detour for One Travel Direction (TA-19)

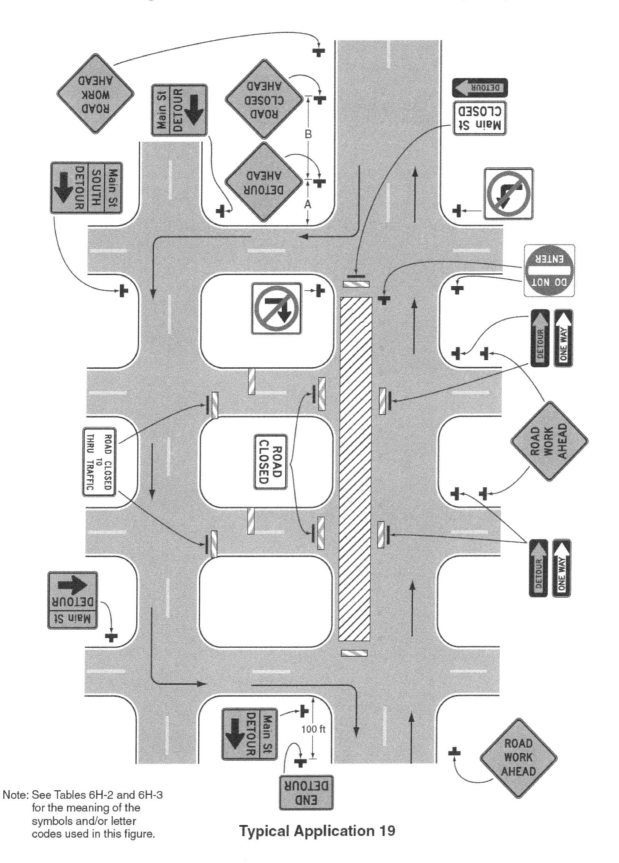

Note: See Tables 6H-2 and 6H-3 for the meaning of the symbols and/or letter codes used in this figure.

Typical Application 19

Notes for Figure 6H-20—Typical Application 20
Detour for a Closed Street

Guidance:
1. *This plan should be used for streets without posted route numbers.*
2. *On multi-lane streets, Detour signs with an Advance Turn Arrow should be used in advance of a turn.*

Option:
3. Flashing warning lights and/or flags may be used to call attention to the advance warning signs.
4. Flashing warning lights may be used on Type 3 Barricades.
5. Detour signs may be located on the far side of intersections. A Detour sign with an advance arrow may be used in advance of a turn.
6. A Street Name sign may be mounted with the Detour sign. The Street Name sign may be either white on green or black on orange.

Standard:
7. When used, the Street Name sign shall be placed above the Detour sign.

Support:
8. See Figure 6H-9 for the information for detouring a numbered highway.

Figure 6H-20. Detour for a Closed Street (TA-20)

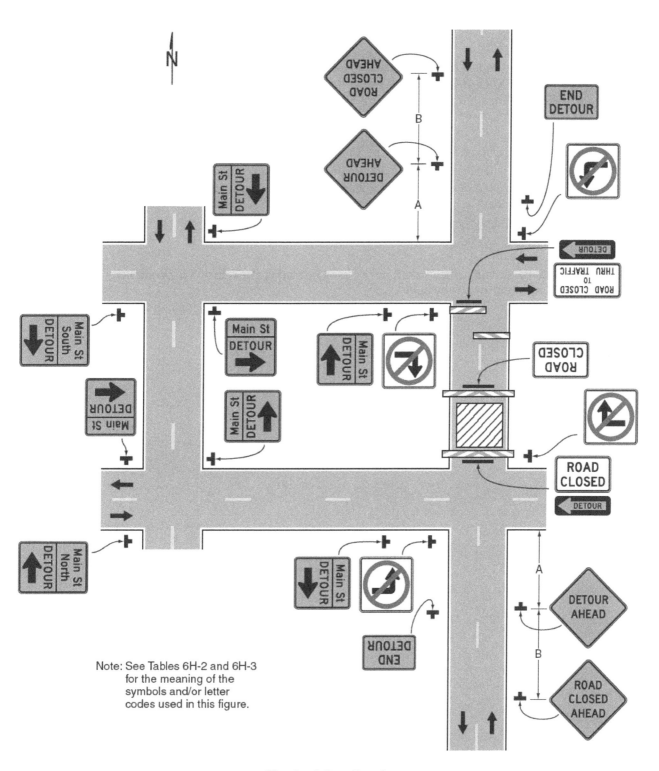

Note: See Tables 6H-2 and 6H-3 for the meaning of the symbols and/or letter codes used in this figure.

Typical Application 20

Notes for Figure 6H-21—Typical Application 21
Lane Closure on the Near Side of an Intersection

Standard:
1. **The merging taper shall direct vehicular traffic into either the right-hand or left-hand lane, but not both.**

Guidance:
2. *In this typical application, a left taper should be used so that right-turn movements will not impede through motor vehicle traffic. However, the reverse should be true for left-turn movements.*
3. *If the work space extends across a crosswalk, the crosswalk should be closed using the information and devices shown in Figure 6H-29.*

Option:
4. Flashing warning lights and/or flags may be used to call attention to the advance warning signs.
5. A shadow vehicle with a truck-mounted attenuator may be used.
6. A work vehicle with high-intensity rotating, flashing, oscillating, or strobe lights may be used with the high-level warning device.
7. Vehicle hazard warning signals may be used to supplement high-intensity rotating, flashing, oscillating, or strobe lights.

Standard:
8. **Vehicle hazard warning signals shall not be used instead of the vehicle's high-intensity rotating, flashing, oscillating, or strobe lights.**

Figure 6H-21. Lane Closure on the Near Side of an Intersection (TA-21)

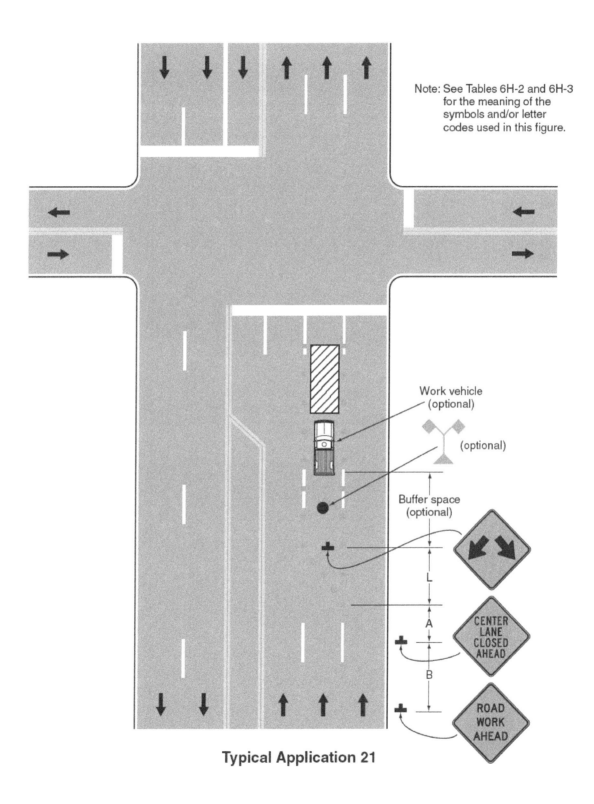

Typical Application 21

Notes for Figure 6H-22—Typical Application 22
Right-Hand Lane Closure on the Far Side of an Intersection

Guidance:
1. *If the work space extends across a crosswalk, the crosswalk should be closed using the information and devices shown in Figure 6H-29.*

Option:
2. The normal procedure is to close on the near side of the intersection any lane that is not carried through the intersection. However, when this results in the closure of a right-hand lane having significant right turning movements, then the right-hand lane may be restricted to right turns only, as shown. This procedure increases the through capacity by eliminating right turns from the open through lane.
3. For intersection approaches reduced to a single lane, left-turning movements may be prohibited to maintain capacity for through vehicular traffic.
4. Flashing warning lights and/or flags may be used to call attention to the advance warning signs.
5. Where the turning radius is large, it may be possible to create a right-turn island using channelizing devices or pavement markings.

Figure 6H-22. Right-Hand Lane Closure on the Far Side of an Intersection (TA-22)

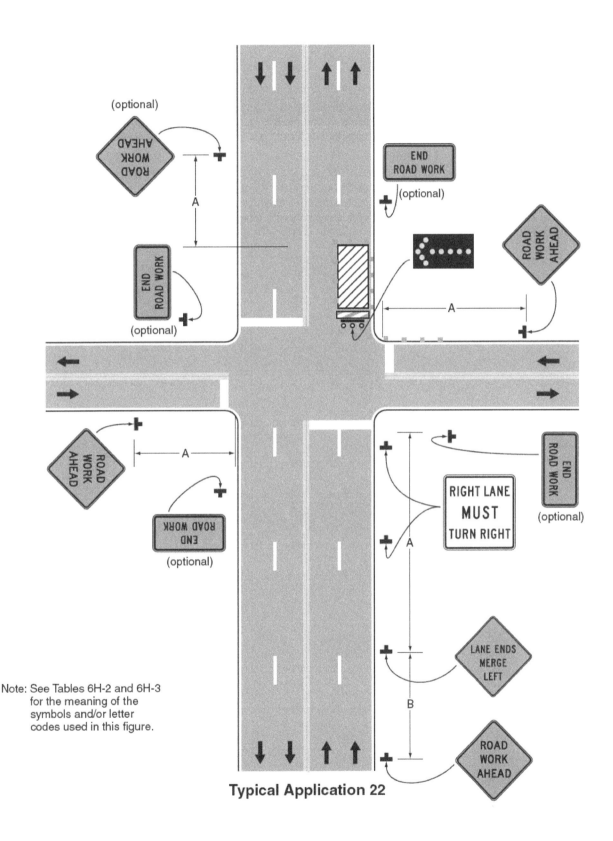

Note: See Tables 6H-2 and 6H-3 for the meaning of the symbols and/or letter codes used in this figure.

Typical Application 22

Notes for Figure 6H-23—Typical Application 23
Left-Hand Lane Closure on the Far Side of an Intersection

Guidance:

1. *If the work space extends across a crosswalk, the crosswalk should be closed using the information and devices shown in Figure 6H-29.*

Option:

2. Flashing warning lights and/or flags may be used to call attention to the advance warning signs.
3. The normal procedure is to close on the near side of the intersection any lane that is not carried through the intersection. However, when this results in the closure of a left lane having significant left-turning movements, then the left lane may be reopened as a turn bay for left turns only, as shown.

Support:

4. By first closing off the left lane and then reopening it as a turn bay, the left-turn bay allows storage of turning vehicles so that the movement of through traffic is not impeded. A left-turn bay that is long enough to accommodate all turning vehicles during a traffic signal cycle will provide the maximum benefit for through traffic. Also, an island is created with channelizing devices that allows the LEFT LANE MUST TURN LEFT sign to be repeated on the left adjacent to the lane that it controls.

Figure 6H-23. Left-Hand Lane Closure on the Far Side of an Intersection (TA-23)

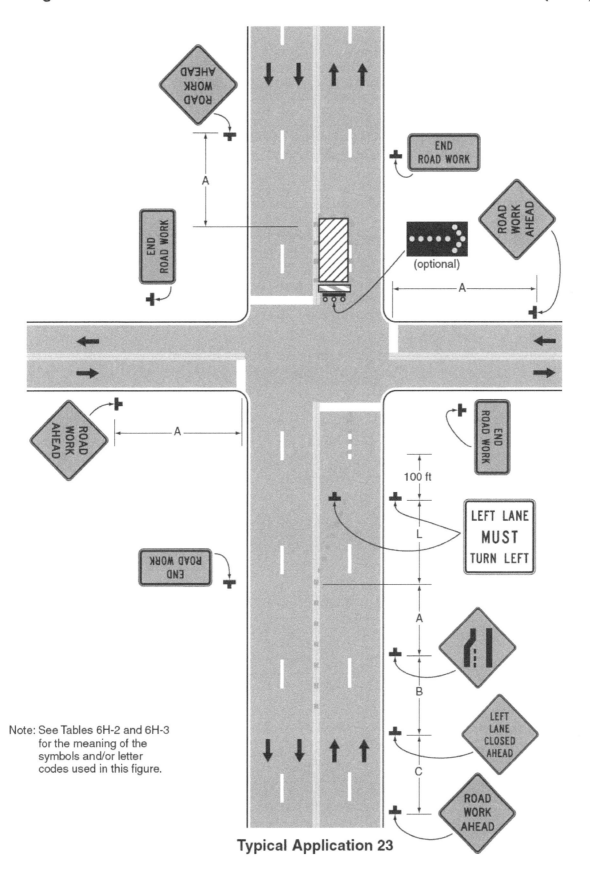

Typical Application 23

Note: See Tables 6H-2 and 6H-3 for the meaning of the symbols and/or letter codes used in this figure.

Notes for Figure 6H-24—Typical Application 24
Half Road Closure on the Far Side of an Intersection

Guidance:
1. If the work space extends across a crosswalk, the crosswalk should be closed using the information and devices shown in Figure 6H-29.
2. When turn prohibitions are implemented, two turn prohibition signs should be used, one on the near side and, space permitting, one on the far side of the intersection.

Option:
3. A buffer space may be used between opposing directions of vehicular traffic as shown in this application.
4. The normal procedure is to close on the near side of the intersection any lane that is not carried through the intersection. However, if there is a significant right-turning movement, then the right-hand lane may be restricted to right turns only, as shown.
5. Where the turning radius is large, a right-turn island using channelizing devices or pavement markings may be used.
6. There may be insufficient space to place the back-to-back Keep Right sign and No Left Turn symbol signs at the end of the row of channelizing devices separating opposing vehicular traffic flows. In this situation, the No Left Turn symbol sign may be placed on the right and the Keep Right sign may be omitted.
7. For intersection approaches reduced to a single lane, left-turning movements may be prohibited to maintain capacity for through vehicular traffic.
8. Flashing warning lights and/or flags may be used to call attention to advance warning signs.
9. Temporary pavement markings may be used to delineate the travel path through the intersection.

Support:
10. Keeping the right-hand lane open increases the through capacity by eliminating right turns from the open through lane.
11. A temporary turn island reinforces the nature of the temporary exclusive right-turn lane and enables a second RIGHT LANE MUST TURN RIGHT sign to be placed in the island.

Figure 6H-24. Half Road Closure on the Far Side of an Intersection (TA-24)

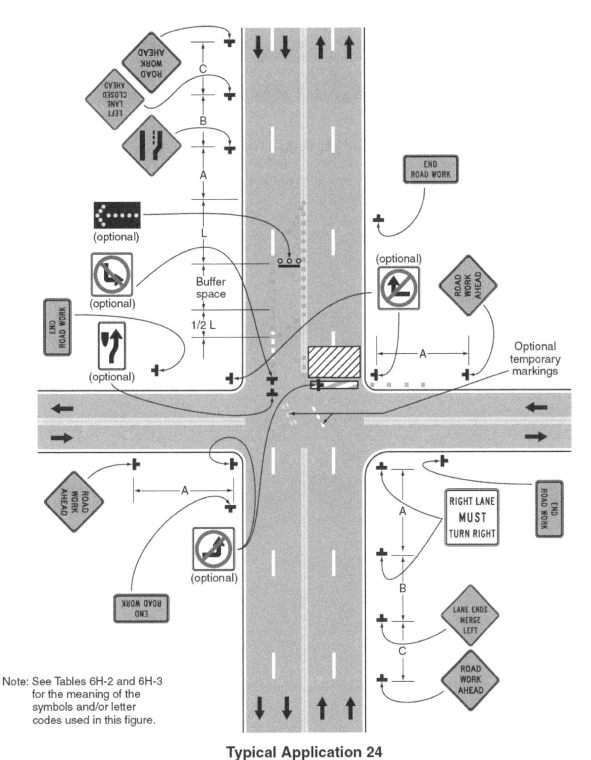

Note: See Tables 6H-2 and 6H-3 for the meaning of the symbols and/or letter codes used in this figure.

Typical Application 24

Notes for Figure 6H-25—Typical Application 25
Multiple Lane Closures at an Intersection

Guidance:
1. *If the work space extends across a crosswalk, the crosswalk should be closed using the information and devices shown in Figure 6H-29.*
2. *If the left through lane is closed on the near-side approach, the LEFT LANE MUST TURN LEFT sign should be placed in the median to discourage through vehicular traffic from entering the left-turn bay.*

Support:
3. The normal procedure is to close on the near side of the intersection any lane that is not carried through the intersection.

Option:
4. If the left-turning movement that normally uses the closed turn bay is small and/or the gaps in opposing vehicular traffic are frequent, left turns may be permitted on that approach.
5. Flashing warning lights and/or flags may be used to call attention to the advance warning signs.

Figure 6H-25. Multiple Lane Closures at an Intersection (TA-25)

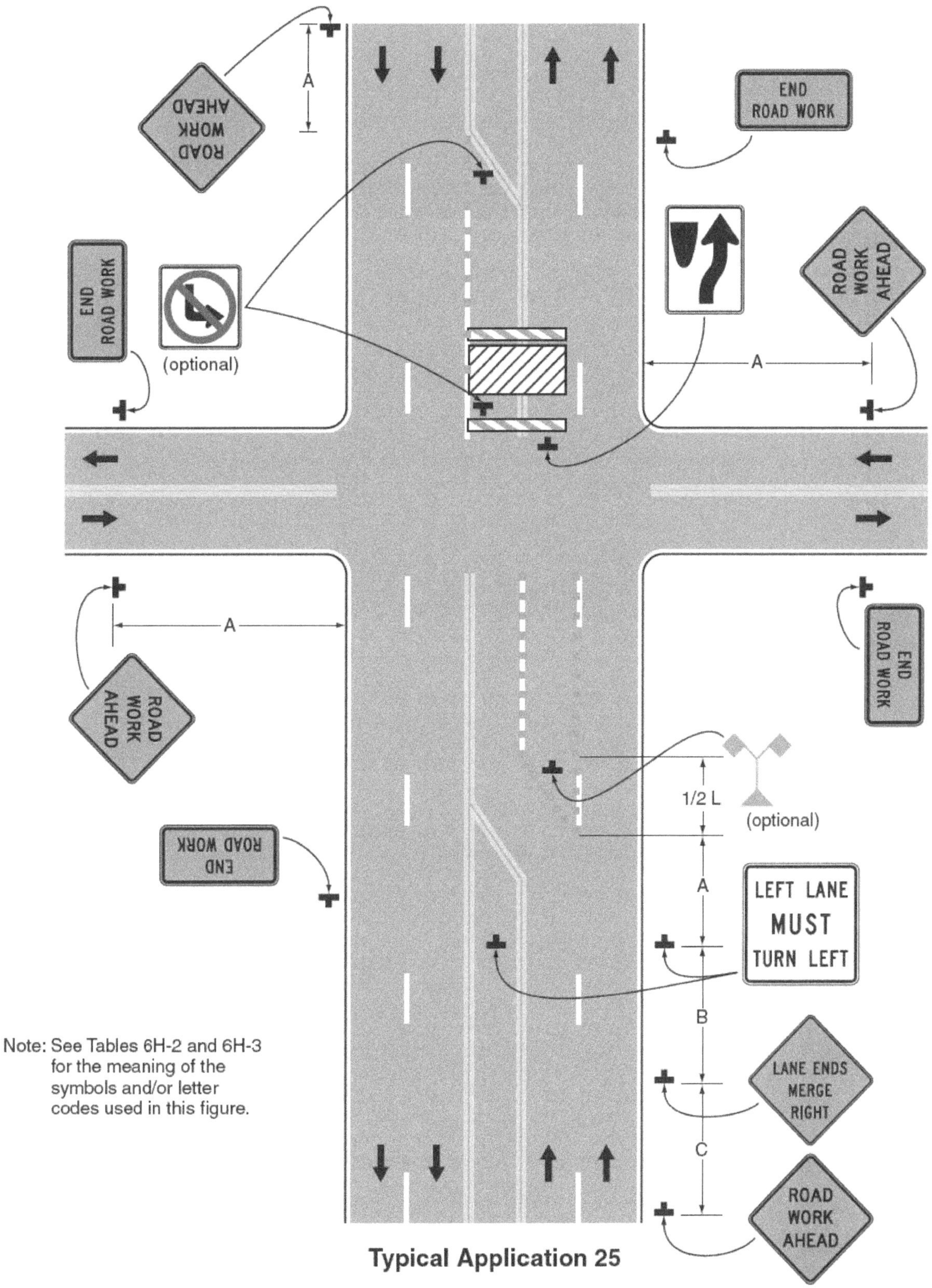

Typical Application 25

Note: See Tables 6H-2 and 6H-3 for the meaning of the symbols and/or letter codes used in this figure.

Notes for Figure 6H-26—Typical Application 26
Closure in the Center of an Intersection

Guidance:
1. *All lanes should be a minimum of 10 feet in width as measured to the near face of the channelizing devices.*

Option:
2. A high-level warning device may be placed in the work space, if there is sufficient room.
3. For short-term use on low-volume, low-speed roadways with vehicular traffic that does not include longer and wider heavy commercial vehicles, a minimum lane width of 9 feet may be used.
4. Flashing warning lights and/or flags may be used to call attention to advance warning signs.
5. Unless the streets are wide, it may be physically impossible to turn left, especially for large vehicles. Left turns may be prohibited as required by geometric conditions.
6. For short-duration work operations, the channelizing devices may be eliminated if a vehicle displaying high-intensity rotating, flashing, oscillating, or strobe lights is positioned in the work space.
7. Vehicle hazard warning signals may be used to supplement high-intensity rotating, flashing, oscillating, or strobe lights.

Standard:
8. **Vehicle hazard warning signals shall not be used instead of the vehicle's high-intensity rotating, flashing, oscillating, or strobe lights.**

Figure 6H-26. Closure in the Center of an Intersection (TA-26)

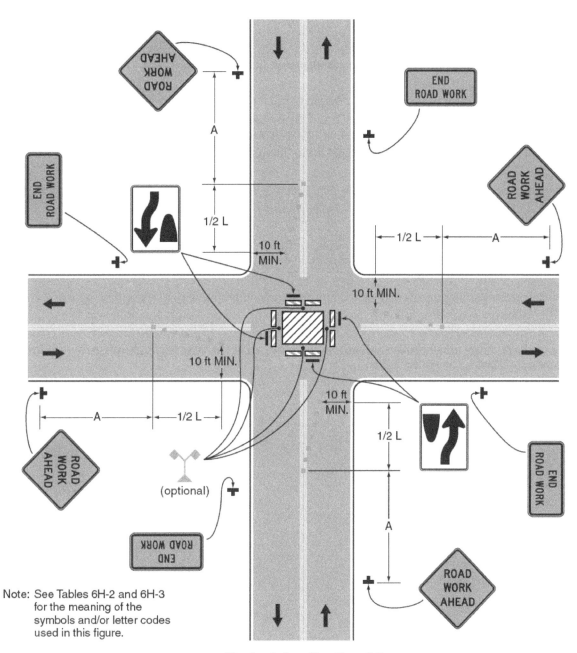

Note: See Tables 6H-2 and 6H-3 for the meaning of the symbols and/or letter codes used in this figure.

Typical Application 26

Notes for Figure 6H-27—Typical Application 27
Closure at the Side of an Intersection

Guidance:
1. *The situation depicted can be simplified by closing one or more of the intersection approaches. If this cannot be done, and/or when capacity is a problem, through vehicular traffic should be directed to other roads or streets.*
2. *Depending on road user conditions, flagger(s) or uniformed law enforcement officer(s) should be used to direct road users within the intersection.*

Standard:
   3. At night, flagger stations shall be illuminated, except in emergencies.

Option:
   4. Flashing warning lights and/or flags may be used to call attention to the advance warning signs.
   5. For short-duration work operations, the channelizing devices may be eliminated if a vehicle displaying high-intensity rotating, flashing, oscillating, or strobe lights is positioned in the work space.
   6. A BE PREPARED TO STOP sign may be added to the sign series.

Guidance:
7. *When used, the BE PREPARED TO STOP sign should be located before the Flagger symbol sign.*
8. *ONE LANE ROAD AHEAD signs should also be used to provide adequate advance warning.*

Support:
   9. Turns can be prohibited as required by vehicular traffic conditions. Unless the streets are wide, it might be physically impossible to make certain turns, especially for large vehicles.

Option:
   10. Vehicle hazard warning signals may be used to supplement high-intensity rotating, flashing, oscillating, or strobe lights.

Standard:
   11. Vehicle hazard warning signals shall not be used instead of the vehicle's high-intensity rotating, flashing, oscillating, or strobe lights.

Figure 6H-27. Closure at the Side of an Intersection (TA-27)

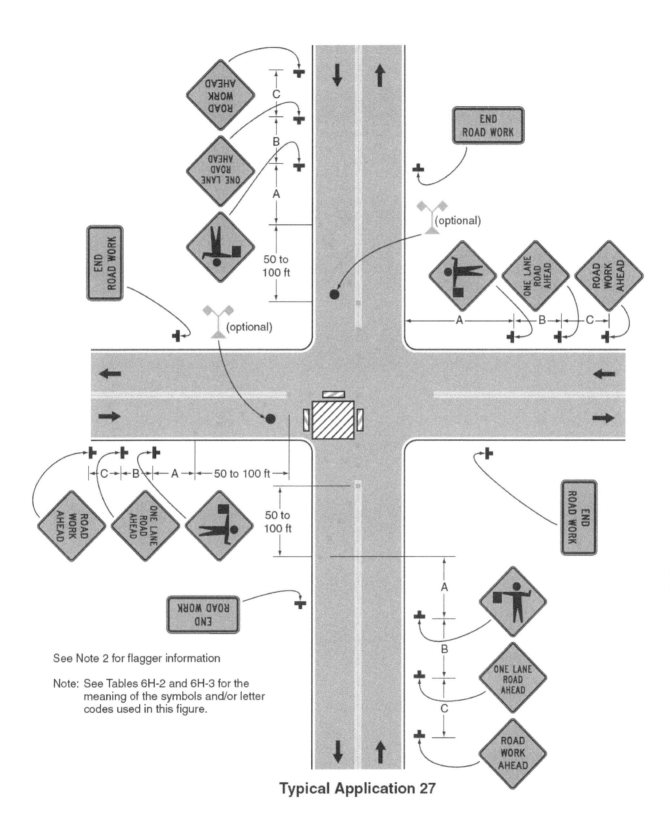

See Note 2 for flagger information

Note: See Tables 6H-2 and 6H-3 for the meaning of the symbols and/or letter codes used in this figure.

Typical Application 27

Notes for Figure 6H-28—Typical Application 28
Sidewalk Detour or Diversion

Standard:
1. **When crosswalks or other pedestrian facilities are closed or relocated, temporary facilities shall be detectable and shall include accessibility features consistent with the features present in the existing pedestrian facility.**

Guidance:
2. *Where high speeds are anticipated, a temporary traffic barrier and, if necessary, a crash cushion should be used to separate the temporary sidewalks from vehicular traffic.*
3. *Audible information devices should be considered where midblock closings and changed crosswalk areas cause inadequate communication to be provided to pedestrians who have visual disabilities.*

Option:
4. Street lighting may be considered.
5. Only the TTC devices related to pedestrians are shown. Other devices, such as lane closure signing or ROAD NARROWS signs, may be used to control vehicular traffic.
6. For nighttime closures, Type A Flashing warning lights may be used on barricades that support signs and close sidewalks.
7. Type C Steady-Burn or Type D 360-degree Steady-Burn warning lights may be used on channelizing devices separating the temporary sidewalks from vehicular traffic flow.
8. Signs, such as KEEP RIGHT (LEFT), may be placed along a temporary sidewalk to guide or direct pedestrians.

Figure 6H-28. Sidewalk Detour or Diversion (TA-28)

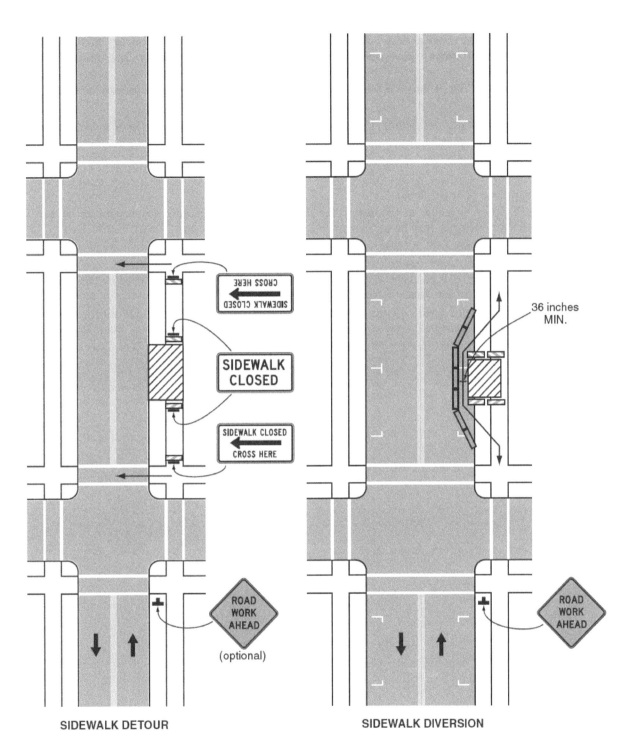

Typical Application 28

Note: See Tables 6H-2 and 6H-3 for the meaning of the symbols and/or letter codes used in this figure.

Notes for Figure 6H-29—Typical Application 29
Crosswalk Closures and Pedestrian Detours

Standard:
1. **When crosswalks or other pedestrian facilities are closed or relocated, temporary facilities shall be detectable and shall include accessibility features consistent with the features present in the existing pedestrian facility.**
2. **Curb parking shall be prohibited for at least 50 feet in advance of the midblock crosswalk.**

Guidance:
3. *Audible information devices should be considered where midblock closings and changed crosswalk areas cause inadequate communication to be provided to pedestrians who have visual disabilities.*
4. *Pedestrian traffic signal displays controlling closed crosswalks should be covered or deactivated.*

Option:
5. Street lighting may be considered.
6. Only the TTC devices related to pedestrians are shown. Other devices, such as lane closure signing or ROAD NARROWS signs, may be used to control vehicular traffic.
7. For nighttime closures, Type A Flashing warning lights may be used on barricades supporting signs and closing sidewalks.
8. Type C Steady-Burn or Type D 360-degree Steady-Burn warning lights may be used on channelizing devices separating the work space from vehicular traffic.
9. In order to maintain the systematic use of the fluorescent yellow-green background for pedestrian, bicycle, and school warning signs in a jurisdiction, the fluorescent yellow-green background for pedestrian, bicycle, and school warning signs may be used in TTC zones.

Figure 6H-29. Crosswalk Closures and Pedestrian Detours (TA-29)

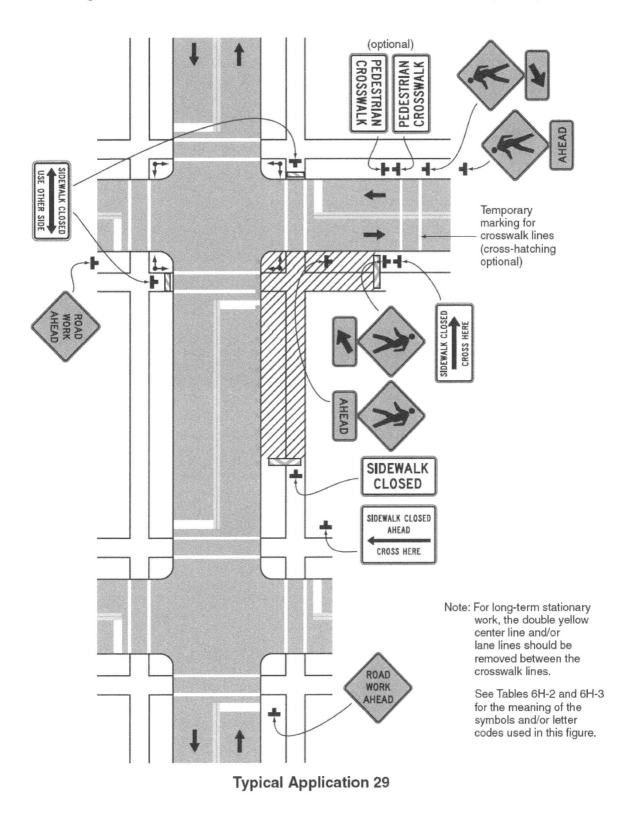

Typical Application 29

Notes for Figure 6H-30—Typical Application 30
Interior Lane Closure on a Multi-Lane Street

Guidance:
1. *This information applies to low-speed, low-volume urban streets. Where speed or volume is higher, additional signing such as LEFT LANE CLOSED XX FT should be used between the signs shown.*

Option:
2. The closure of the adjacent interior lane in the opposing direction may not be necessary, depending upon the activity being performed and the work space needed for the operation.
3. Shadow vehicles with a truck-mounted attenuator may be used.

Figure 6H-30. Interior Lane Closure on a Multi-Lane Street (TA-30)

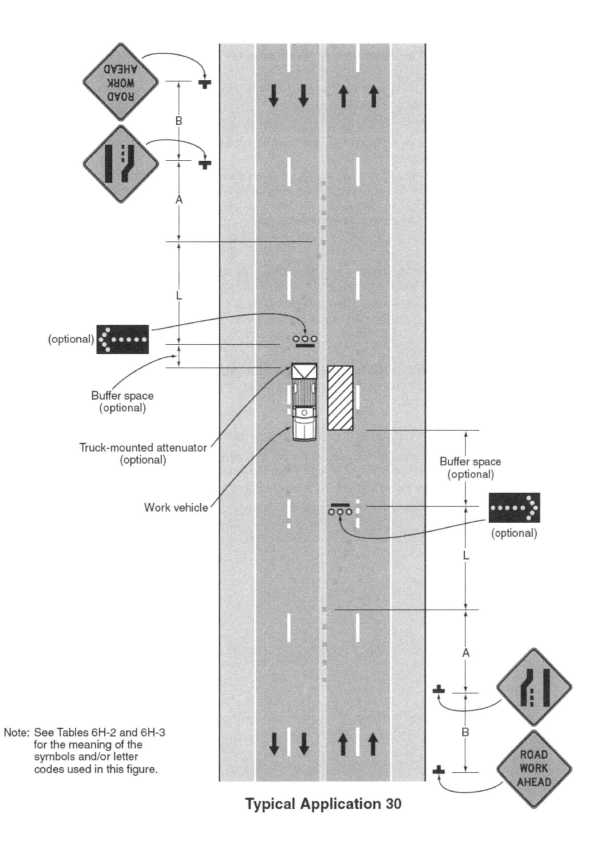

Typical Application 30

Note: See Tables 6H-2 and 6H-3 for the meaning of the symbols and/or letter codes used in this figure.

Notes for Figure 6H-31—Typical Application 31
Lane Closure on a Street with Uneven Directional Volumes

Standard:
1. **The illustrated information shall be used only when the vehicular traffic volume indicates that two lanes of vehicular traffic shall be maintained in the direction of travel for which one lane is closed.**

Option:
2. The procedure may be used during a peak period of vehicular traffic and then changed to provide two lanes in the other direction for the other peak.

Guidance:
3. *For high speeds, a LEFT LANE CLOSED XX FT sign should be added for vehicular traffic approaching the lane closure, as shown in Figure 6H-32.*
4. *Conflicting pavement markings should be removed for long-term projects. For short-term and intermediate-term projects where this is not practical, the channelizing devices in the area where the pavement markings conflict should be placed at a maximum spacing of 1/2 S feet where S is the speed in mph. Temporary markings should be installed where needed.*
5. *If the lane shift has curves with recommended speeds of 30 mph or less, Reverse Turn signs should be used.*
6. *Where the shifted section is long, a Reverse Curve sign should be used to show the initial shift and a second sign should be used to show the return to the normal alignment.*
7. *If the tangent distance along the temporary diversion is less than 600 feet, the Double Reverse Curve sign should be used at the location of the first Two Lane Reverse Curve sign. The second Two Lane Reverse Curve sign should be omitted.*

Standard:
8. **The number of lanes illustrated on the Reverse Curve or Double Reverse Curve signs shall be the same as the number of through lanes available to road users, and the direction of the reverse curves shall be appropriately illustrated.**

Option:
9. A longitudinal buffer space may be used in the activity area to separate opposing vehicular traffic.
10. Where two or more lanes are being shifted, a W1-4 (or W1-3) sign with an ALL LANES (W24-1cP) plaque (see Figure 6F-4) may be used instead of a sign that illustrates the number of lanes.
11. Where more than three lanes are being shifted, the Reverse Curve (or Turn) sign may be rectangular.
12. A work vehicle or a shadow vehicle may be equipped with a truck-mounted attenuator.

Figure 6H-31. Lane Closures on a Street with Uneven Directional Volumes (TA-31)

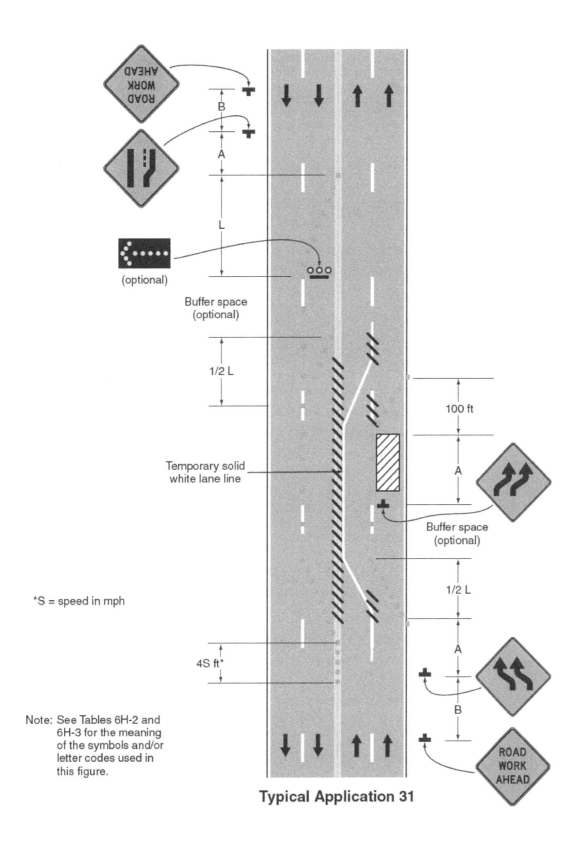

Typical Application 31

Notes for Figure 6H-32—Typical Application 32
Half Road Closure on a Multi-Lane, High-Speed Highway

Standard:
1. Pavement markings no longer applicable shall be removed or obliterated as soon as practical. Except for intermediate-term and short-term situations, temporary markings shall be provided to clearly delineate the temporary travel path. For short-term and intermediate-term situations where it is not feasible to remove and restore pavement markings, channelization shall be made dominant by using a very close device spacing.

Guidance:
2. When paved shoulders having a width of 8 feet or more are closed, channelizing devices should be used to close the shoulder in advance of the merging taper to direct vehicular traffic to remain within the traveled way.
3. Where channelizing devices are used instead of pavement markings, the maximum spacing should be 1/2 S feet where S is the speed in mph.
4. If the tangent distance along the temporary diversion is less than 600 feet, a Double Reverse Curve sign should be used instead of the first Reverse Curve sign, and the second Reverse Curve sign should be omitted.

Option:
5. Warning lights may be used to supplement channelizing devices at night.
6. A truck-mounted attenuator may be used on the work vehicle and/or the shadow vehicle.

Figure 6H-32. Half Road Closure on a Multi-Lane, High-Speed Highway (TA-32)

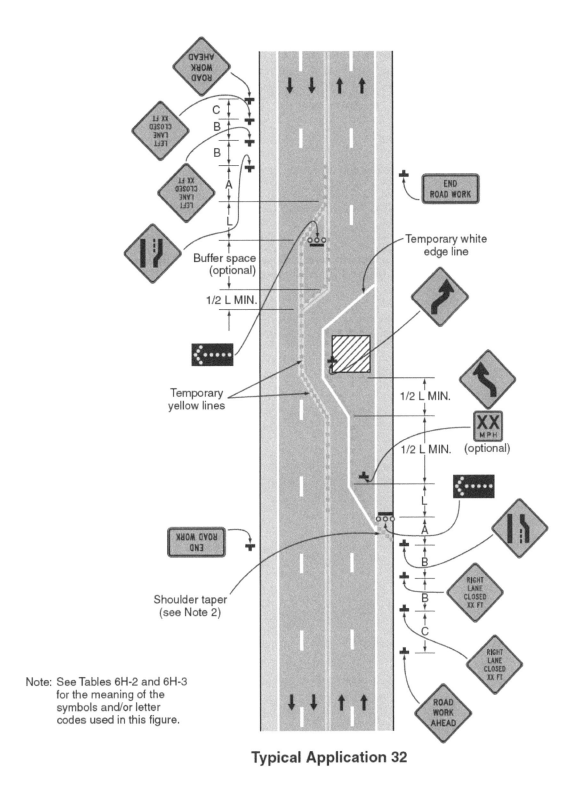

Typical Application 32

Notes for Figure 6H-33—Typical Application 33
Stationary Lane Closure on a Divided Highway

Standard:
1. This information also shall be used when work is being performed in the lane adjacent to the median on a divided highway. In this case, the LEFT LANE CLOSED signs and the corresponding Lane Ends signs shall be substituted.
2. When a side road intersects the highway within the TTC zone, additional TTC devices shall be placed as needed.

Guidance:
3. *When paved shoulders having a width of 8 feet or more are closed, channelizing devices should be used to close the shoulder in advance of the merging taper to direct vehicular traffic to remain within the traveled way.*

Option:
4. A truck-mounted attenuator may be used on the work vehicle and/or shadow vehicle.

Support:
5. Where conditions permit, restricting all vehicles, equipment, workers, and their activities to one side of the roadway might be advantageous.

Standard:
6. An arrow board shall be used when a freeway lane is closed. When more than one freeway lane is closed, a separate arrow board shall be used for each closed lane.

Figure 6H-33. Stationary Lane Closure on a Divided Highway (TA-33)

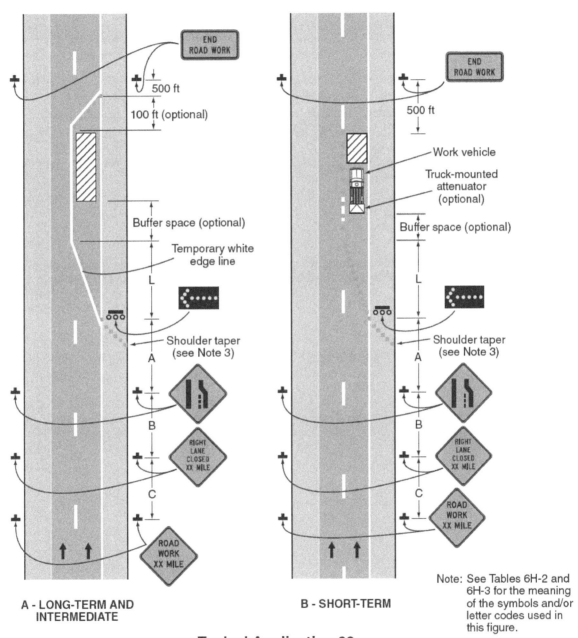

Typical Application 33

Notes for Figure 6H-34—Typical Application 34
Lane Closure with a Temporary Traffic Barrier

Standard:
1. This information also shall be used when work is being performed in the lane adjacent to the median on a divided highway. In this case, the LEFT LANE CLOSED signs and the corresponding Lane Ends signs shall be substituted.

Guidance:
2. *For long-term lane closures on facilities with permanent edge lines, a temporary edge line should be installed from the upstream end of the merging taper to the downstream end of the downstream taper, and conflicting pavement markings should be removed.*
3. *The use of a barrier should be based on engineering judgment.*

Standard:
4. Temporary traffic barriers, if used, shall comply with the provisions of Section 6F.85.
5. The barrier shall not be placed along the merging taper. The lane shall first be closed using channelizing devices and pavement markings.

Option:
6. Type C Steady-Burn warning lights may be placed on channelizing devices and the barrier parallel to the edge of pavement for nighttime lane closures.
7. The barrier shown in this typical application is an example of one method that may be used to close a lane for a long-term project. If the work activity permits, a movable barrier may be used and relocated to the shoulder during non-work periods or peak-period vehicular traffic conditions, as appropriate.

Standard:
8. If a movable barrier is used, the temporary white edge line shown in the typical application shall not be used. During the period when the right-hand lane is opened, the sign legends and the channelization shall be changed to indicate that only the shoulder is closed, as illustrated in Figure 6H-5. The arrow board, if used, shall be placed at the downstream end of the shoulder taper and shall display the caution mode.

Guidance:
9. *If a movable barrier is used, the shift should be performed in the following manner. When closing the lane, the lane should be initially closed with channelizing devices placed along a merging taper using the same information employed for a stationary lane closure. The lane closure should then be extended with the movable-barrier transfer vehicle moving with vehicular traffic. When opening the lane, the movable-barrier transfer vehicle should travel against vehicular traffic from the termination area to the transition area. The merging taper should then be removed using the same information employed for a stationary lane closure.*

Figure 6H-34. Lane Closure with a Temporary Traffic Barrier (TA-34)

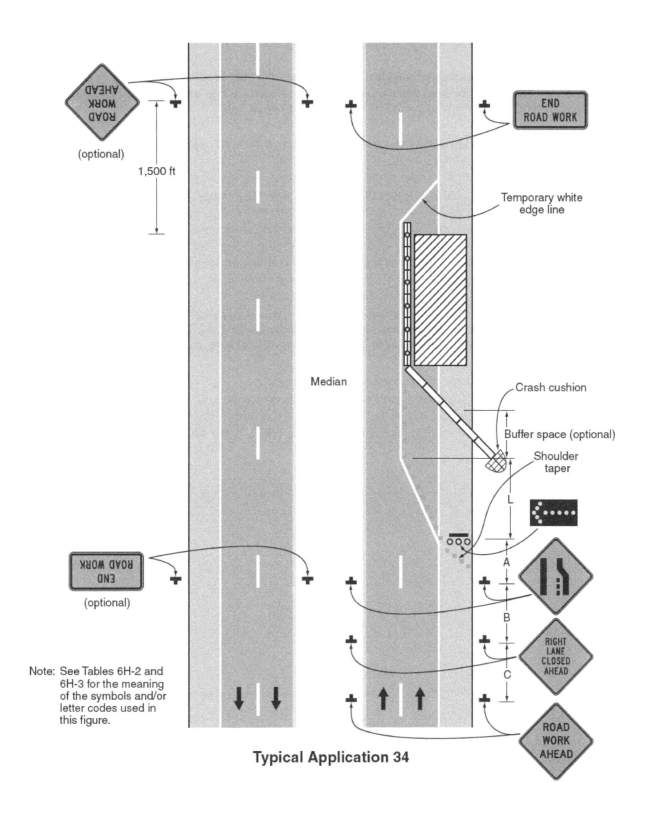

Typical Application 34

Notes for Figure 6H-35—Typical Application 35
Mobile Operation on a Multi-Lane Road

Standard:
1. **Arrow boards shall, as a minimum, be Type B, with a size of 60 x 30 inches.**
2. **Vehicle-mounted signs shall be mounted in a manner such that they are not obscured by equipment or supplies. Sign legends on vehicle-mounted signs shall be covered or turned from view when work is not in progress.**
3. **Shadow and work vehicles shall display high-intensity rotating, flashing, oscillating, or strobe lights.**
4. **An arrow board shall be used when a freeway lane is closed. When more than one freeway lane is closed, a separate arrow board shall be used for each closed lane.**

Guidance:
5. *Vehicles used for these operations should be made highly visible with appropriate equipment, such as flags, signs, or arrow boards.*
6. *Shadow Vehicle 1 should be equipped with an arrow board and truck-mounted attenuator.*
7. *Shadow Vehicle 2 should be equipped with an arrow board. An appropriate lane closure sign should be placed on Shadow Vehicle 2 so as not to obscure the arrow board.*
8. *Shadow Vehicle 2 should travel at a varying distance from the work operation so as to provide adequate sight distance for vehicular traffic approaching from the rear.*
9. *The spacing between the work vehicles and the shadow vehicles, and between each shadow vehicle should be minimized to deter road users from driving in between.*
10. *Work should normally be accomplished during off-peak hours.*
11. *When the work vehicle occupies an interior lane (a lane other than the far right or far left) of a directional roadway having a right-hand shoulder 10 feet or more in width, Shadow Vehicle 2 should drive the right-hand shoulder with a sign indicating that work is taking place in the interior lane.*

Option:
12. A truck-mounted attenuator may be used on Shadow Vehicle 2.
13. On high-speed roadways, a third shadow vehicle (not shown) may be used with Shadow Vehicle 1 in the closed lane, Shadow Vehicle 2 straddling the edge line, and Shadow Vehicle 3 on the shoulder.
14. Where adequate shoulder width is not available, Shadow Vehicle 3 may also straddle the edge line.

Figure 6H-35. Mobile Operation on a Multi-Lane Road (TA-35)

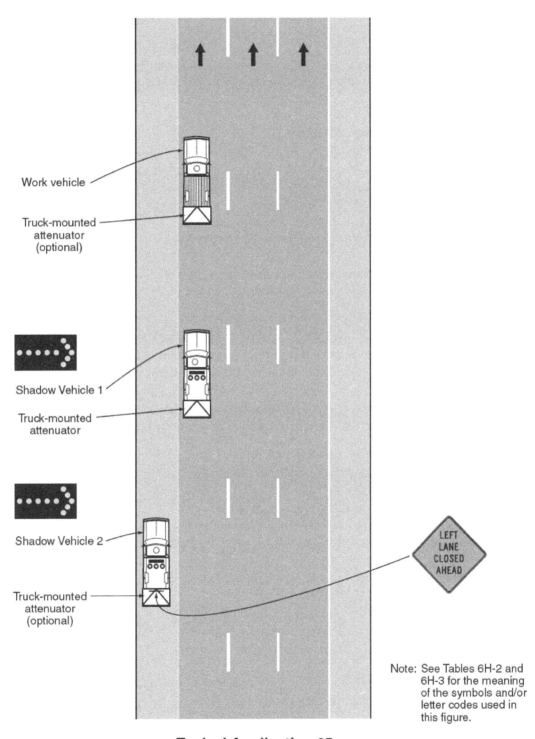

Typical Application 35

Notes for Figure 6H-36—Typical Application 36
Lane Shift on a Freeway

Guidance:
1. *The lane shift should be used when the work space extends into either the right-hand or left-hand lane of a divided highway and it is not practical, for capacity reasons, to reduce the number of available lanes.*

Support:
2. When a lane shift is accomplished by using (1) geometry that meets the design speed at which the permanent highway was designed, (2) full normal cross-section (full lane width and full shoulders), and (3) complete pavement markings, then only the initial general work-zone warning sign is required.

Guidance:
3. *When the conditions in Note 2 are not met, the information shown in the typical application should be employed and all the following notes apply.*

Standard:
4. **Temporary traffic barriers, if used, shall comply with the provisions of Section 6F.85.**
5. **The barrier shall not be placed along the shifting taper. The lane shall first be shifted using channelizing devices and pavement markings.**

Guidance:
6. *A warning sign should be used to show the changed alignment.*

Standard:
7. **The number of lanes illustrated on the Reverse Curve signs shall be the same as the number of through lanes available to road users, and the direction of the reverse curves shall be appropriately illustrated.**

Option:
8. Where two or more lanes are being shifted, a W1-4 (or W1-3) sign with an ALL LANES (W24-1cP) plaque (see Figure 6F-4) may be used instead of a sign that illustrates the number of lanes.
9. Where more than three lanes are being shifted, the Reverse Curve (or Turn) sign may be rectangular.

Guidance:
10. *Where the shifted section is longer than 600 feet, one set of Reverse Curve signs should be used to show the initial shift and a second set should be used to show the return to the normal alignment. If the tangent distance along the temporary diversion is less than 600 feet, a Double Reverse Curve sign should be used instead of the first Reverse Curve sign, and the second Reverse Curve sign should be omitted.*
11. *If a STAY IN LANE sign is used, then solid white lane lines should be used.*

Standard:
12. **The minimum width of the shoulder lane shall be 10 feet.**
13. **For long-term stationary work, existing conflicting pavement markings shall be removed and temporary markings shall be installed before traffic patterns are changed.**

Option:
14. For short-term stationary work, lanes may be delineated by channelizing devices or removable pavement markings instead of temporary markings.

Guidance:
15. *If the shoulder cannot adequately accommodate trucks, trucks should be directed to use the travel lanes.*
16. *The use of a barrier should be based on engineering judgment.*

Option:
17. Type C Steady-Burn warning lights may be placed on channelizing devices and the barrier parallel to the edge of the pavement for nighttime lane closures.

Figure 6H-36. Lane Shift on a Freeway (TA-36)

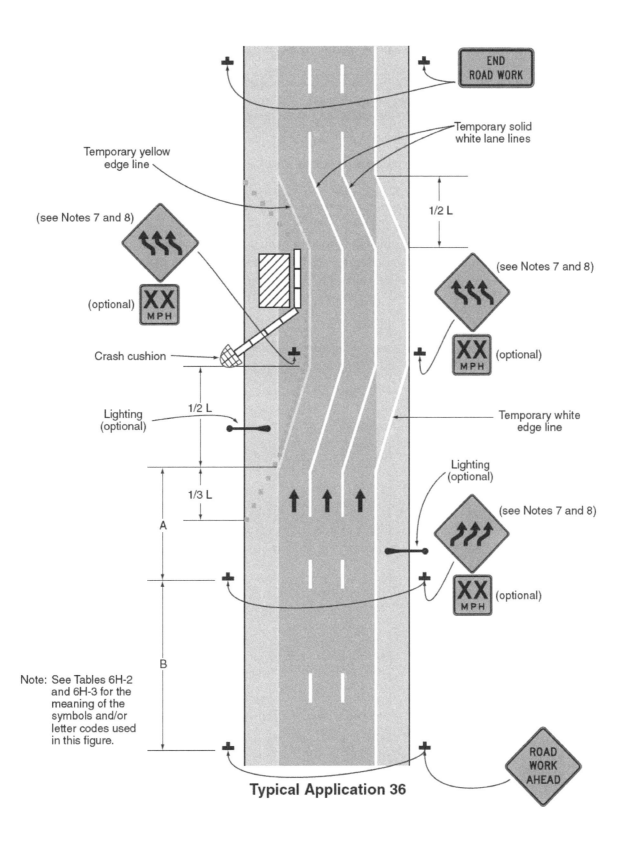

Typical Application 36

Notes for Figure 6H-37—Typical Application 37
Double Lane Closure on a Freeway

Standard:
1. **An arrow board shall be used when a freeway lane is closed. When more than one freeway lane is closed, a separate arrow board shall be used for each closed lane.**

Guidance:
2. *Ordinarily, the preferred position for the second arrow board is in the closed exterior lane at the upstream end of the second merging taper. However, the second arrow board should be placed in the closed interior lane at the downstream end of the second merging taper in the following situations:*
    a. *When a shadow vehicle is used in the interior closed lane, and the second arrow board is mounted on the shadow vehicle;*
    b. *If alignment or other conditions create any confusion as to which lane is closed by the second arrow board; and*
    c. *When the first arrow board is placed in the closed exterior lane at the downstream end of the first merging taper (the alternative position when the shoulder is narrow).*

Option:
3. Flashing warning lights and/or flags may be used to call attention to the initial warning signs.
4. A truck-mounted attenuator may be used on the shadow vehicle.
5. If a paved shoulder having a minimum width of 10 feet and sufficient strength is available, the left and adjacent interior lanes may be closed and vehicular traffic carried around the work space on the right-hand lane and a right-hand shoulder.

Guidance:
6. *When a shoulder lane is used that cannot adequately accommodate trucks, trucks should be directed to use the normal travel lanes.*

Figure 6H-37. Double Lane Closure on a Freeway (TA-37)

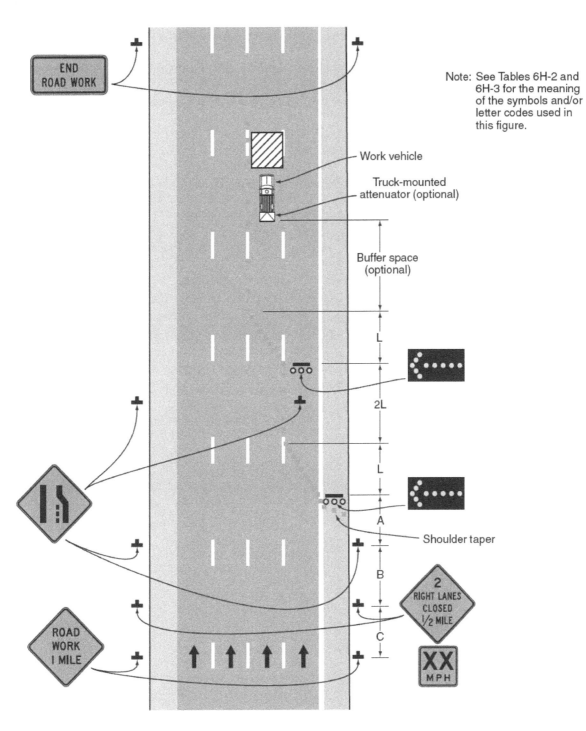

Typical Application 37

Notes for Figure 6H-38—Typical Application 38
Interior Lane Closure on a Freeway

Standard:
1. An arrow board shall be used when a freeway lane is closed. When more than one freeway lane is closed, a separate arrow board shall be used for each closed lane.
2. If temporary traffic barriers are installed, they shall comply with the provisions and requirements in Section 6F.85.
3. The barrier shall not be placed along the shifting taper. The lane shall first be shifted using channelizing devices and pavement markings.
4. For long-term stationary work, existing conflicting pavement markings shall be removed and temporary markings shall be installed before traffic patterns are changed.

Guidance:
5. *For a long-term closure, a barrier should be used to provide additional safety to the operation in the closed interior lane. A buffer space should be used at the upstream end of the closed interior lane.*
6. *The first arrow board displaying an arrow pointing to the right should be on the left-hand shoulder at the beginning of the taper. The arrow board displaying a double arrow should be centered in the closed interior lane and placed at the downstream end of the shifting taper.*
7. *If the two arrow boards create confusion, the 2L distance between the end of the merging taper and beginning of the shift taper should be extended so that road users can focus on one arrow board at a time.*
8. *The placement of signs should not obstruct or obscure arrow boards.*
9. *For long-term use, the dashed lane lines should be made solid white in the two-lane section.*

Option:
10. As an alternative to initially closing the left-hand lane, as shown in the typical application, the right-hand lane may be closed in advance of the interior lane closure with appropriate channelization and signs.
11. A short, single row of channelizing devices in advance of the vehicular traffic split to restrict vehicular traffic to their respective lanes may be added.
12. DO NOT PASS signs may be used.
13. If a paved shoulder having a minimum width of 10 feet and sufficient strength is available, the left-hand and center lanes may be closed and motor vehicle traffic carried around the work space on the right-hand lane and a right-hand shoulder.

Guidance:
14. *When a shoulder lane is used that cannot adequately accommodate trucks, trucks should be directed to use the normal travel lanes.*

Figure 6H-38. Interior Lane Closure on a Freeway (TA-38)

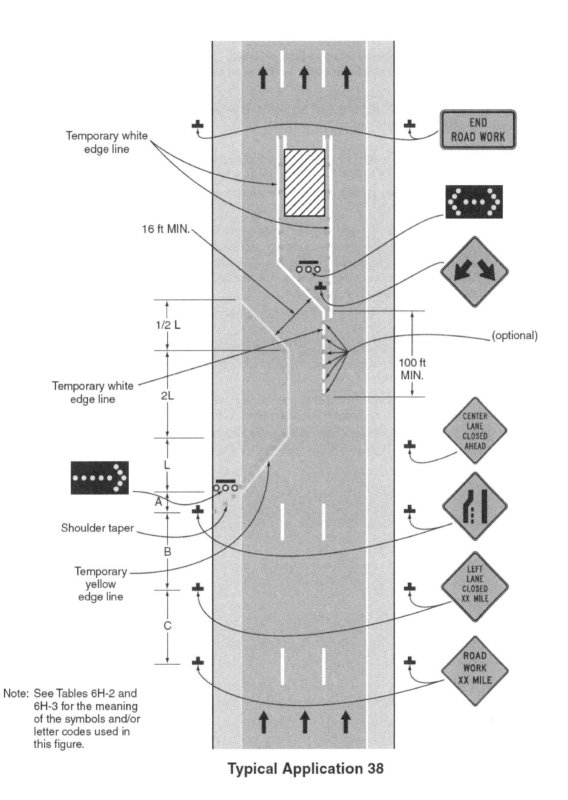

Typical Application 38

Note: See Tables 6H-2 and 6H-3 for the meaning of the symbols and/or letter codes used in this figure.

Notes for Figure 6H-39—Typical Application 39
Median Crossover on a Freeway

Standard:
1. **Channelizing devices or temporary traffic barriers shall be used to separate opposing vehicular traffic.**
2. **An arrow board shall be used when a freeway lane is closed. When more than one freeway lane is closed, a separate arrow board shall be used for each closed lane.**

Guidance:

3. *For long-term work on high-speed, high-volume highways, consideration should be given to using a temporary traffic barrier to separate opposing vehicular traffic.*

Option:

4. When a temporary traffic barrier is used to separate opposing vehicular traffic, the Two-Way Traffic, Do Not Pass, KEEP RIGHT, and DO NOT ENTER signs may be eliminated.
5. The alignment of the crossover may be designed as a reverse curve.

Guidance:

6. *When the crossover follows a curved alignment, the design criteria contained in the AASHTO "Policy on the Geometric Design of Highways and Streets" (see Section 1A.11) should be used.*
7. *When channelizing devices have the potential of leading vehicular traffic out of the intended traffic space, the channelizing devices should be extended a distance in feet of 2.0 times the speed limit in mph beyond the downstream end of the transition area as depicted.*
8. *Where channelizing devices are used, the Two-Way Traffic signs should be repeated every 1 mile.*

Option:

9. NEXT XX MILES Supplemental Distance plaques may be used with the Two-Way Traffic signs, where XX is the distance to the downstream end of the two-way section.

Support:

10. When the distance is sufficiently short that road users entering the section can see the downstream end of the section, they are less likely to forget that there is opposing vehicular traffic.
11. The sign legends for the four pairs of signs approaching the lane closure for the non-crossover direction of travel are not shown. They are similar to the series shown for the crossover direction, except that the left lane is closed.

Figure 6H-39. Median Crossover on a Freeway (TA-39)

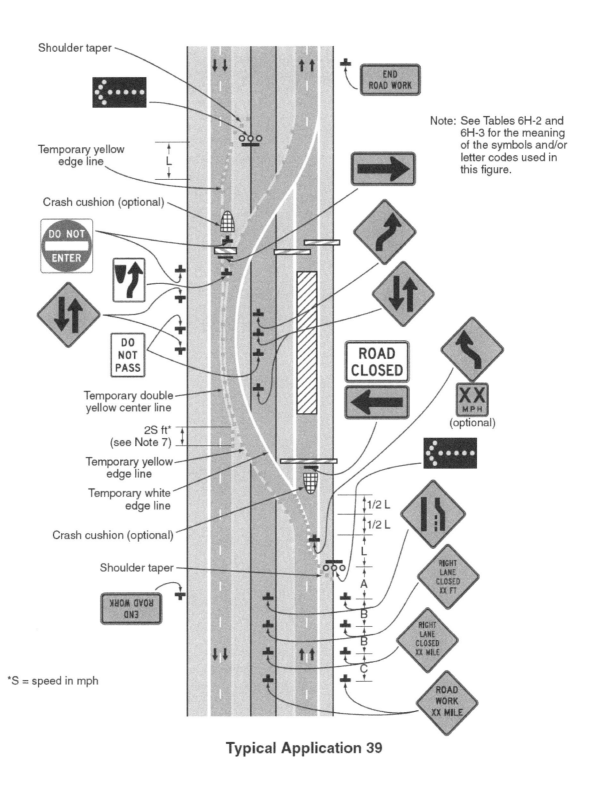

Typical Application 39

Notes for Figure 6H-40—Typical Application 40
Median Crossover for an Entrance Ramp

Guidance:
1. The typical application illustrated should be used for carrying an entrance ramp across a closed directional roadway of a divided highway.
2. A temporary acceleration lane should be used to facilitate merging.
3. When used, the YIELD or STOP sign should be located far enough forward to provide adequate sight distance of oncoming mainline vehicular traffic to select an acceptable gap, but should not be located so far forward that motorists will be encouraged to stop in the path of the mainline traffic. If needed, yield or stop lines should be installed across the ramp to indicate the point at which road users should yield or stop. Also, a longer acceleration lane should be provided beyond the sign to reduce the gap size needed.

Option:
4. If vehicular traffic conditions allow, the ramp may be closed.
5. A broken edge line may be carried across the temporary entrance ramp to assist in defining the through vehicular traffic lane.
6. When a temporary traffic barrier is used to separate opposing vehicular traffic, the Two-Way Traffic signs and the DO NOT ENTER signs may be eliminated.

Figure 6H-40. Median Crossover for an Entrance Ramp (TA-40)

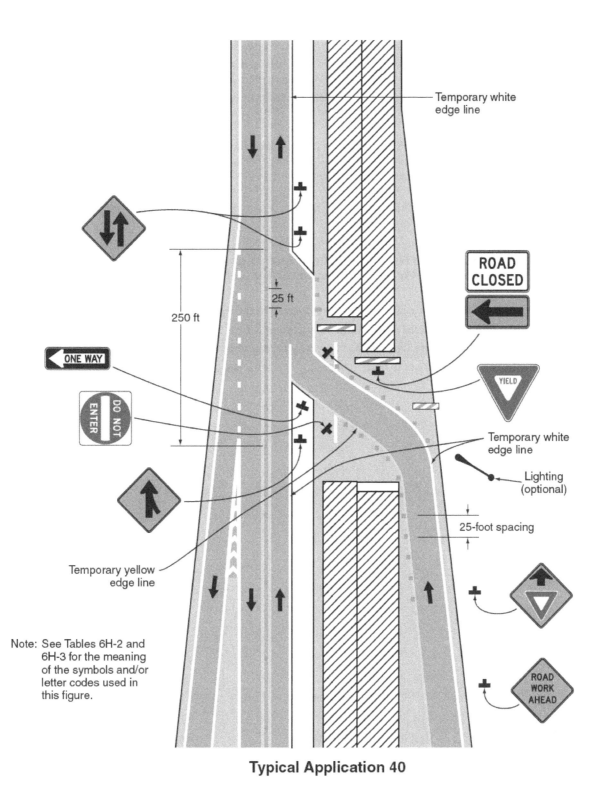

Typical Application 40

Note: See Tables 6H-2 and 6H-3 for the meaning of the symbols and/or letter codes used in this figure.

Notes for Figure 6H-41—Typical Application 41
Median Crossover for an Exit Ramp

Guidance:
1. This typical application should be used for carrying an exit ramp across a closed directional roadway of a divided highway. The design criteria contained in the AASHTO "Policy on the Geometric Design of Highways and Streets" (see Section 1A.11) should be used for determining the curved alignment.
2. The guide signs should indicate that the ramp is open, and where the temporary ramp is located. Conversely, if the ramp is closed, guide signs should indicate that the ramp is closed.
3. When the exit is closed, a black on orange EXIT CLOSED sign panel should be placed diagonally across the interchange/intersection guide signs and channelizing devices should be placed to physically close the ramp.
4. In the situation (not shown) where channelizing devices are placed along the mainline roadway, the devices' spacing should be reduced in the vicinity of the off ramp to emphasize the opening at the ramp itself. Channelizing devices and/or temporary pavement markings should be placed on both sides of the temporary ramp where it crosses the median and the closed roadway.
5. Advance guide signs providing information related to the temporary exit should be relocated or duplicated adjacent to the temporary roadway.

Standard:
6. **A temporary EXIT sign shall be located in the temporary gore. For better visibility, it shall be mounted a minimum of 7 feet from the pavement surface to the bottom of the sign.**

Option:
7. Guide signs referring to the exit may need to be relocated to the median.
8. The temporary EXIT sign placed in the temporary gore may be either black on orange or white on green.
9. In some instances, a temporary deceleration lane may be useful in facilitating the exiting maneuver.
10. When a temporary traffic barrier is used to separate opposing vehicular traffic, the Two-Way Traffic signs may be omitted.

Figure 6H-41. Median Crossover for an Exit Ramp (TA-41)

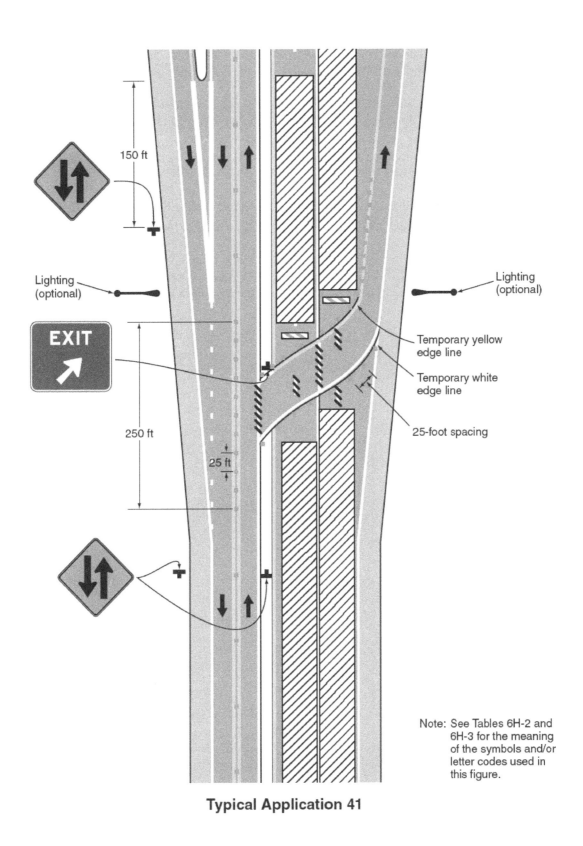

Typical Application 41

Notes for Figure 6H-42—Typical Application 42
Work in the Vicinity of an Exit Ramp

Guidance:
1. The guide signs should indicate that the ramp is open, and where the temporary ramp is located. However, if the ramp is closed, guide signs should indicate that the ramp is closed.
2. When the exit ramp is closed, a black on orange EXIT CLOSED sign panel should be placed diagonally across the interchange/intersection guide signs.
3. The design criteria contained in the AASHTO "Policy on the Geometric Design of Highways and Streets" (see Section 1A.11) should be used for determining the alignment.

Standard:
   4. **A temporary EXIT sign shall be located in the temporary gore. For better visibility, it shall be mounted a minimum of 7 feet from the pavement surface to the bottom of the sign.**

Option:
   5. The temporary EXIT sign placed in the temporary gore may be either black on orange or white on green.
   6. An alternative procedure that may be used is to channelize exiting vehicular traffic onto the right-hand shoulder and close the lane as necessary.

Standard:
   7. **An arrow board shall be used when a freeway lane is closed. When more than one freeway lane is closed, a separate arrow board shall be used for each closed lane.**

Figure 6H-42. Work in the Vicinity of an Exit Ramp (TA-42)

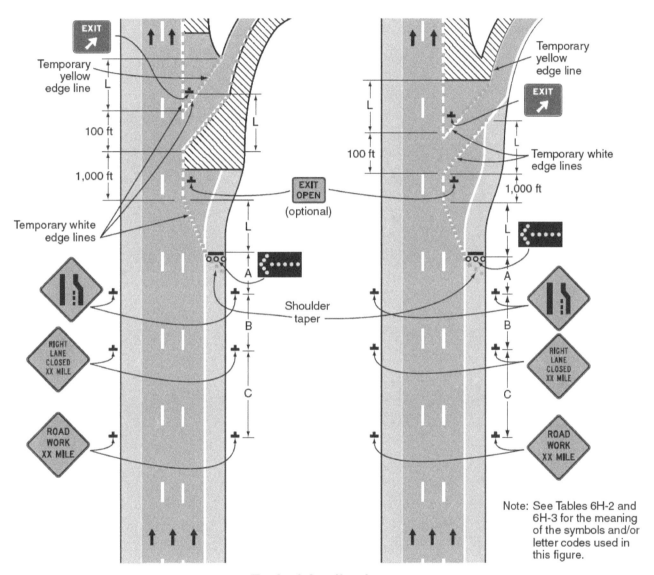

Typical Application 42

Notes for Figure 6H-43—Typical Application 43
Partial Exit Ramp Closure

Guidance:
1. *Truck off-tracking should be considered when determining whether the minimum lane width of 10 feet is adequate (see Section 6G.08).*

Figure 6H-43. Partial Exit Ramp Closure (TA-43)

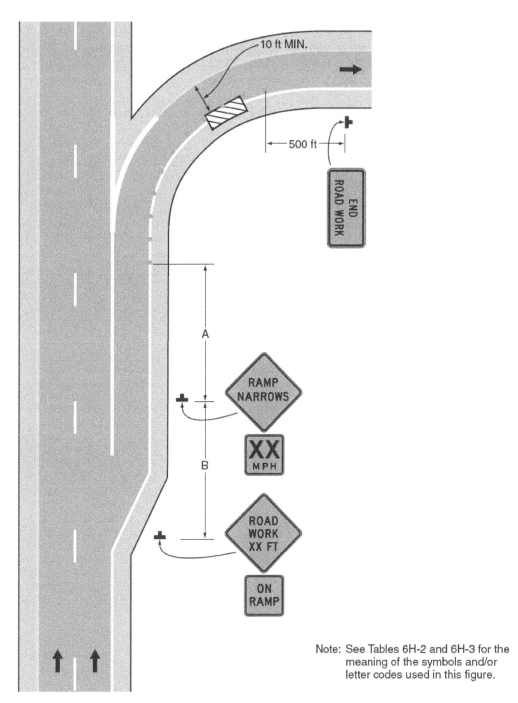

Typical Application 43

Note: See Tables 6H-2 and 6H-3 for the meaning of the symbols and/or letter codes used in this figure.

Notes for Figure 6H-44—Typical Application 44
Work in the Vicinity of an Entrance Ramp

Guidance:
1. *An acceleration lane of sufficient length should be provided whenever possible as shown on the left diagram.*

Standard:

2. **For the information shown on the diagram on the right-hand side of the typical application, where inadequate acceleration distance exists for the temporary entrance, the YIELD sign shall be replaced with STOP signs (one on each side of the approach).**

Guidance:

3. *When used, the YIELD or STOP sign should be located so that ramp vehicular traffic has adequate sight distance of oncoming mainline vehicular traffic to select an acceptable gap in the mainline vehicular traffic flow, but should not be located so far forward that motorists will be encouraged to stop in the path of the mainline traffic. Also, a longer acceleration lane should be provided beyond the sign to reduce the gap size needed. If insufficient gaps are available, consideration should be given to closing the ramp.*

4. *Where STOP signs are used, a temporary stop line should be placed across the ramp at the desired stop location.*

5. *The mainline merging taper with the arrow board at its starting point should be located sufficiently in advance so that the arrow board is not confusing to drivers on the entrance ramp, and so that the mainline merging vehicular traffic from the lane closure has the opportunity to stabilize before encountering the vehicular traffic merging from the ramp.*

6. *If the ramp curves sharply to the right, warning signs with advisory speeds located in advance of the entrance terminal should be placed in pairs (one on each side of the ramp).*

Option:

7. A Stop Beacon (see Section 4L.05) or a Type B high-intensity warning flasher with a red lens may be placed above the STOP sign.

8. Where the acceleration distance is significantly reduced, a supplemental plaque may be placed below the Yield Ahead sign reading NO MERGE AREA.

Standard:

9. **An arrow board shall be used when a freeway lane is closed. When more than one freeway lane is closed, a separate arrow board shall be used for each closed lane.**

Figure 6H-44. Work in the Vicinity of an Entrance Ramp (TA-44)

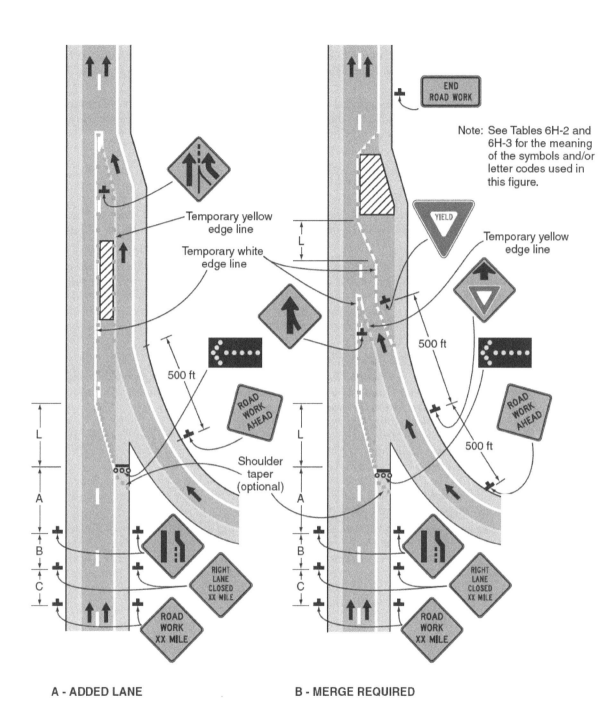

Typical Application 44

Notes for Figure 6H-45—Typical Application 45
Temporary Reversible Lane Using Movable Barriers

Support:
1. This application addresses one of several uses for movable barriers (see Section 6F.85) in highway work zones. In this example, one side of a 6-lane divided highway is closed to perform the work operation, and vehicular traffic is carried in both directions on the remaining 3-lane roadway by means of a median crossover.

   To accommodate unbalanced peak-period vehicular traffic volumes, the direction of travel in the center lane is switched to the direction having the greater volume, with the transfer typically being made twice daily. Thus, there are four vehicular traffic phases described as follows:
   a. Phase A—two travel lanes northbound and one lane southbound;
   b. Transition A to B—one travel lane in each direction;
   c. Phase B—one travel lane northbound and two lanes southbound; and
   d. Transition B to A—one travel lane in each direction.

   The typical application on the left illustrates the placement of devices during Phase A. The typical application on the right shows conditions during the transition (Transition A to B) from Phase A to Phase B.

Guidance:
2. *For the reversible-lane situation depicted, the ends of the movable barrier should terminate in a protected area or a crash cushion should be provided. During Phase A, the transfer vehicle should be parked behind the downstream end of the movable barrier for southbound traffic as shown in the typical application on the left. During Phase B, the transfer vehicle should be parked behind between the downstream ends of the movable barriers at the north end of the TTC zone as shown in the typical application on the right.*

   The transition shift from Phase A to B should be as follows:
   a. *Change the signs in the northbound advance warning area and transition area from a LEFT LANE CLOSED AHEAD to a 2 LEFT LANES CLOSED AHEAD. Change the mode of the second northbound arrow board from Caution to Right Arrow.*
   b. *Place channelizing devices to close the northbound center lane.*
   c. *Move the transfer vehicle from south to north to shift the movable barrier from the west side to the east side of the reversible lane.*
   d. *Remove the channelizing devices closing the southbound center lane.*
   e. *Change the signs in the southbound transition area and advance warning area from a 2 LEFT LANES CLOSED AHEAD to a LEFT LANE CLOSED AHEAD. Change the mode of the second southbound arrow board from Right Arrow to Caution.*

3. *Where the lane to be opened and closed is an exterior lane (adjacent to the edge of the traveled way or the work space), the lane closure should begin by closing the lane with channelizing devices placed along a merging taper using the same information employed for a stationary lane closure. The lane closure should then be extended with the movable-barrier transfer vehicle moving with vehicular traffic. When opening the lane, the transfer vehicle should travel against vehicular traffic. The merging taper should be removed in a method similar to a stationary lane closure.*

Option:
4. The procedure may be used during a peak period of vehicular traffic and then changed to provide two lanes in the other direction for the other peak.
5. A longitudinal buffer space may be used in the activity area to separate opposing vehicular traffic.
6. A work vehicle or a shadow vehicle may be equipped with a truck-mounted attenuator.

Standard:
7. **An arrow board shall be used when a freeway lane is closed. When more than one freeway lane is closed, a separate arrow board shall be used for each closed lane.**

Figure 6H-45. Temporary Reversible Lane Using Movable Barriers (TA-45)

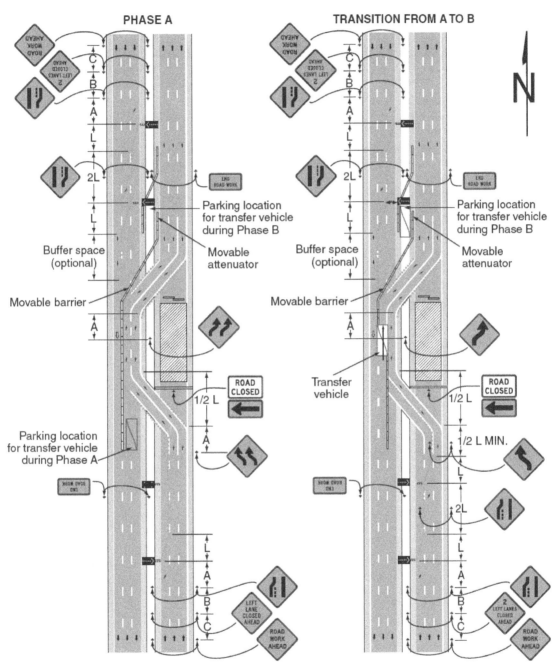

Typical Application 45

Note: See Tables 6H-2 and 6H-3 for the meaning of the symbols and/or letter codes used in this figure. Although leader lines point to the signs on the right-hand side of the roadway, most of these signs should be installed on both sides of the roadway.

Notes for Figure 6H-46—Typical Application 46
Work in the Vicinity of a Grade Crossing

Guidance:
1. When grade crossings exist either within or in the vicinity of roadway work activities, extra care should be taken to minimize the probability of conditions being created, by lane restrictions, flagging, or other operations, where vehicles might be stopped within the grade crossing, considered as being 15 feet on either side of the closest and farthest rail.

Standard:
2. **If the queuing of vehicles across active rail tracks cannot be avoided, a uniformed law enforcement officer or flagger shall be provided at the grade crossing to prevent vehicles from stopping within the grade crossing (as described in Note 1), even if automatic warning devices are in place.**

Guidance:
3. Early coordination with the railroad company or light rail transit agency should occur before work starts.
4. In the example depicted, the buffer space of the activity area should be extended upstream of the grade crossing (as shown) so that a queue created by the flagging operation will not extend across the grade crossing.
5. The DO NOT STOP ON TRACKS sign should be used on all approaches to a grade crossing within the limits of a TTC zone.

Option:
6. Flashing warning lights and/or flags may be used to call attention to the advance warning signs.
7. A BE PREPARED TO STOP sign may be added to the sign series.

Guidance:
8. When used, the BE PREPARED TO STOP sign should be located before the Flagger symbol sign.

Standard:
9. **At night, flagger stations shall be illuminated, except in emergencies.**

Figure 6H-46. Work in the Vicinity of a Grade Crossing (TA-46)

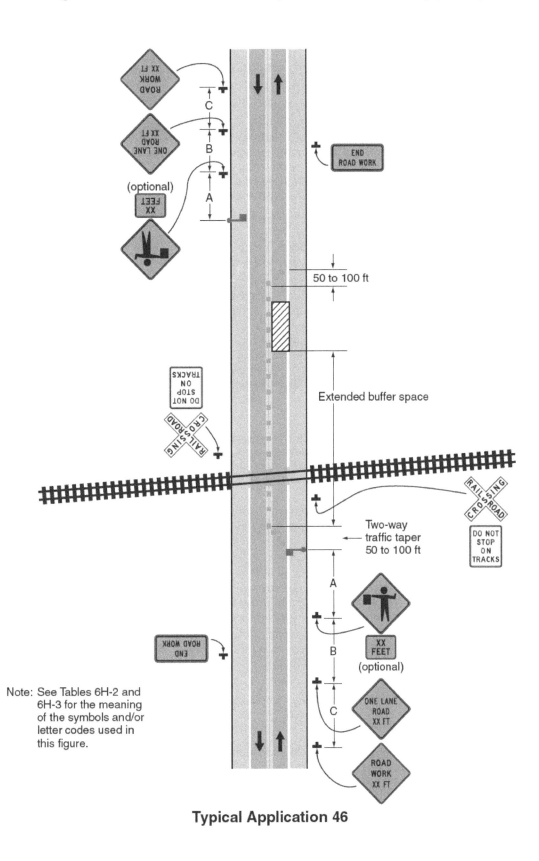

Typical Application 46

CHAPTER 6I. CONTROL OF TRAFFIC THROUGH TRAFFIC INCIDENT MANAGEMENT AREAS

Section 6I.01 General

Support:

01 The National Incident Management System (NIMS) requires the use of the Incident Command System (ICS) at traffic incident management scenes.

02 A traffic incident is an emergency road user occurrence, a natural disaster, or other unplanned event that affects or impedes the normal flow of traffic.

03 A traffic incident management area is an area of a highway where temporary traffic controls are installed, as authorized by a public authority or the official having jurisdiction of the roadway, in response to a road user incident, natural disaster, hazardous material spill, or other unplanned incident. It is a type of TTC zone and extends from the first warning device (such as a sign, light, or cone) to the last TTC device or to a point where vehicles return to the original lane alignment and are clear of the incident.

04 Traffic incidents can be divided into three general classes of duration, each of which has unique traffic control characteristics and needs. These classes are:
   A. Major—expected duration of more than 2 hours,
   B. Intermediate—expected duration of 30 minutes to 2 hours, and
   C. Minor—expected duration under 30 minutes.

05 The primary functions of TTC at a traffic incident management area are to inform road users of the incident and to provide guidance information on the path to follow through the incident area. Alerting road users and establishing a well defined path to guide road users through the incident area will serve to protect the incident responders and those involved in working at the incident scene and will aid in moving road users expeditiously past or around the traffic incident, will reduce the likelihood of secondary traffic crashes, and will preclude unnecessary use of the surrounding local road system. Examples include a stalled vehicle blocking a lane, a traffic crash blocking the traveled way, a hazardous material spill along a highway, and natural disasters such as floods and severe storm damage.

Guidance:

06 *In order to reduce response time for traffic incidents, highway agencies, appropriate public safety agencies (law enforcement, fire and rescue, emergency communications, emergency medical, and other emergency management), and private sector responders (towing and recovery and hazardous materials contractors) should mutually plan for occurrences of traffic incidents along the major and heavily traveled highway and street system.*

07 *On-scene responder organizations should train their personnel in TTC practices for accomplishing their tasks in and near traffic and in the requirements for traffic incident management contained in this Manual. On-scene responders should take measures to move the incident off the traveled roadway or to provide for appropriate warning. All on-scene responders and news media personnel should constantly be aware of their visibility to oncoming traffic and wear high-visibility apparel.*

08 *Emergency vehicles should be safe-positioned (see definition in Section 1A.13) such that traffic flow through the incident scene is optimized. All emergency vehicles that subsequently arrive should be positioned in a manner that does not interfere with the established temporary traffic flow.*

09 *Responders arriving at a traffic incident should estimate the magnitude of the traffic incident, the expected time duration of the traffic incident, and the expected vehicle queue length, and then should set up the appropriate temporary traffic controls for these estimates.*

Option:

10 Warning and guide signs used for TTC traffic incident management situations may have a black legend and border on a fluorescent pink background (see Figure 6I-1).

Support:

11 While some traffic incidents might be anticipated and planned for, emergencies and disasters might pose more severe and unpredictable problems. The ability to quickly install proper temporary traffic controls might greatly reduce the effects of an incident, such as secondary crashes or excessive traffic delays. An essential part of fire, rescue, spill clean-up, highway agency, and enforcement activities is the proper control of road users through the traffic incident management area in order to protect responders, victims, and other personnel at the site. These operations might need corroborating legislative authority for the implementation and enforcement of appropriate road user regulations, parking controls, and speed zoning. It is desirable for these statutes to provide sufficient flexibility in the authority for, and implementation of, TTC to respond to the needs of changing conditions found in traffic incident management areas.

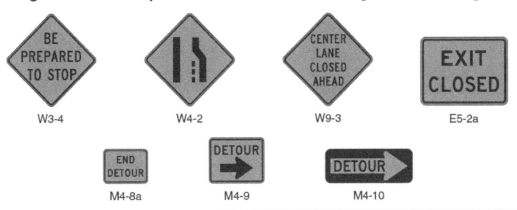

Figure 6I-1. Examples of Traffic Incident Management Area Signs

Option:

For traffic incidents, particularly those of an emergency nature, TTC devices on hand may be used for the initial response as long as they do not themselves create unnecessary additional hazards.

Section 6I.02 Major Traffic Incidents

Support:

Major traffic incidents are typically traffic incidents involving hazardous materials, fatal traffic crashes involving numerous vehicles, and other natural or man-made disasters. These traffic incidents typically involve closing all or part of a roadway facility for a period exceeding 2 hours.

Guidance:

If the traffic incident is anticipated to last more than 24 hours, applicable procedures and devices set forth in other Chapters of Part 6 should be used.

Support:

A road closure can be caused by a traffic incident such as a road user crash that blocks the traveled way. Road users are usually diverted through lane shifts or detoured around the traffic incident and back to the original roadway. A combination of traffic engineering and enforcement preparations is needed to determine the detour route, and to install, maintain or operate, and then to remove the necessary traffic control devices when the detour is terminated. Large trucks are a significant concern in such a detour, especially when detouring them from a controlled-access roadway onto local or arterial streets.

During traffic incidents, large trucks might need to follow a route separate from that of automobiles because of bridge, weight, clearance, or geometric restrictions. Also, vehicles carrying hazardous material might need to follow a different route from other vehicles.

Some traffic incidents such as hazardous material spills might require closure of an entire highway. Through road users must have adequate guidance around the traffic incident. Maintaining good public relations is desirable. The cooperation of the news media in publicizing the existence of, and reasons for, traffic incident management areas and their TTC can be of great assistance in keeping road users and the general public well informed.

The establishment, maintenance, and prompt removal of lane diversions can be effectively managed by interagency planning that includes representatives of highway and public safety agencies.

Guidance:

All traffic control devices needed to set up the TTC at a traffic incident should be available so that they can be readily deployed for all major traffic incidents. The TTC should include the proper traffic diversions, tapered lane closures, and upstream warning devices to alert traffic approaching the queue and to encourage early diversion to an appropriate alternative route.

Attention should be paid to the upstream end of the traffic queue such that warning is given to road users approaching the back of the queue.

If manual traffic control is needed, it should be provided by qualified flaggers or uniformed law enforcement officers.

Option:

10 If flaggers are used to provide traffic control for an incident management situation, the flaggers may use appropriate traffic control devices that are readily available or that can be brought to the traffic incident scene on short notice.

Guidance:

11 *When light sticks or flares are used to establish the initial traffic control at incident scenes, channelizing devices (see Section 6F.63) should be installed as soon thereafter as practical.*

Option:

12 The light sticks or flares may remain in place if they are being used to supplement the channelizing devices.

Guidance:

13 *The light sticks, flares, and channelizing devices should be removed after the incident is terminated.*

Section 6I.03 Intermediate Traffic Incidents

Support:

01 Intermediate traffic incidents typically affect travel lanes for a time period of 30 minutes to 2 hours, and usually require traffic control on the scene to divert road users past the blockage. Full roadway closures might be needed for short periods during traffic incident clearance to allow traffic incident responders to accomplish their tasks.

02 The establishment, maintenance, and prompt removal of lane diversions can be effectively managed by interagency planning that includes representatives of highway and public safety agencies.

Guidance:

03 *All traffic control devices needed to set up the TTC at a traffic incident should be available so that they can be readily deployed for intermediate traffic incidents. The TTC should include the proper traffic diversions, tapered lane closures, and upstream warning devices to alert traffic approaching the queue and to encourage early diversion to an appropriate alternative route.*

04 *Attention should be paid to the upstream end of the traffic queue such that warning is given to road users approaching the back of the queue.*

05 *If manual traffic control is needed, it should be provided by qualified flaggers or uniformed law enforcement officers.*

Option:

06 If flaggers are used to provide traffic control for an incident management situation, the flaggers may use appropriate traffic control devices that are readily available or that can be brought to the traffic incident scene on short notice.

Guidance:

07 *When light sticks or flares are used to establish the initial traffic control at incident scenes, channelizing devices (see Section 6F.63) should be installed as soon thereafter as practical.*

Option:

08 The light sticks or flares may remain in place if they are being used to supplement the channelizing devices.

Guidance:

09 *The light sticks, flares, and channelizing devices should be removed after the incident is terminated.*

Section 6I.04 Minor Traffic Incidents

Support:

01 Minor traffic incidents are typically disabled vehicles and minor crashes that result in lane closures of less than 30 minutes. On-scene responders are typically law enforcement and towing companies, and occasionally highway agency service patrol vehicles.

02 Diversion of traffic into other lanes is often not needed or is needed only briefly. It is not generally possible or practical to set up a lane closure with traffic control devices for a minor traffic incident. Traffic control is the responsibility of on-scene responders.

Guidance:

03 *When a minor traffic incident blocks a travel lane, it should be removed from that lane to the shoulder as quickly as possible.*

Section 6I.05 Use of Emergency-Vehicle Lighting

Support:

01 The use of emergency-vehicle lighting (such as high-intensity rotating, flashing, oscillating, or strobe lights) is essential, especially in the initial stages of a traffic incident, for the safety of emergency responders and persons involved in the traffic incident, as well as road users approaching the traffic incident. Emergency-vehicle lighting, however, provides warning only and provides no effective traffic control. The use of too many lights at an incident scene can be distracting and can create confusion for approaching road users, especially at night. Road users approaching the traffic incident from the opposite direction on a divided facility are often distracted by emergency-vehicle lighting and slow their vehicles to look at the traffic incident posing a hazard to themselves and others traveling in their direction.

02 The use of emergency-vehicle lighting can be reduced if good traffic control has been established at a traffic incident scene. This is especially true for major traffic incidents that might involve a number of emergency vehicles. If good traffic control is established through placement of advanced warning signs and traffic control devices to divert or detour traffic, then public safety agencies can perform their tasks on scene with minimal emergency-vehicle lighting.

Guidance:

03 *Public safety agencies should examine their policies on the use of emergency-vehicle lighting, especially after a traffic incident scene is secured, with the intent of reducing the use of this lighting as much as possible while not endangering those at the scene. Special consideration should be given to reducing or extinguishing forward facing emergency-vehicle lighting, especially on divided roadways, to reduce distractions to oncoming road users.*

04 *Because the glare from floodlights or vehicle headlights can impair the nighttime vision of approaching road users, any floodlights or vehicle headlights that are not needed for illumination, or to provide notice to other road users of an incident response vehicle being in an unexpected location, should be turned off at night.*

(This page left intentionally blank)

PART 7
TRAFFIC CONTROL FOR SCHOOL AREAS

CHAPTER 7A. GENERAL

Section 7A.01 Need for Standards
Support:
01 Regardless of the school location, the best way to achieve effective traffic control is through the uniform application of realistic policies, practices, and standards developed through engineering judgment or studies.
02 Pedestrian safety depends upon public understanding of accepted methods for efficient traffic control. This principle is especially important in the control of pedestrians, bicycles, and other vehicles in the vicinity of schools. Neither pedestrians on their way to or from school nor other road users can be expected to move safely in school areas unless they understand both the need for traffic controls and how these controls function for their benefit.
03 Procedures and devices that are not uniform might cause confusion among pedestrians and other road users, prompt wrong decisions, and contribute to crashes. To achieve uniformity of traffic control in school areas, comparable traffic situations need to be treated in a consistent manner. Each traffic control device and control method described in Part 7 fulfills a specific function related to specific traffic conditions.
04 A uniform approach to school area traffic controls assures the use of similar controls for similar situations, which promotes appropriate and uniform behavior on the part of motorists, pedestrians, and bicyclists.
05 A school traffic control plan permits the orderly review of school area traffic control needs, and the coordination of school/pedestrian safety education and engineering measures. Engineering measures alone do not always result in the intended change in student and road user behavior.
Guidance:
06 *A school route plan for each school serving elementary to high school students should be prepared in order to develop uniformity in the use of school area traffic controls and to serve as the basis for a school traffic control plan for each school.*
07 *The school route plan, developed in a systematic manner by the school, law enforcement, and traffic officials responsible for school pedestrian safety, should consist of a map (see Figure 7A-1) showing streets, the school, existing traffic controls, established school walk routes, and established school crossings.*
08 *The type(s) of school area traffic control devices used, either warning or regulatory, should be related to the volume and speed of vehicular traffic, street width, and the number and age of the students using the crossing.*
09 *School area traffic control devices should be included in a school traffic control plan.*
Support:
10 Reduced speed limit signs for school areas and crossings are included in this Manual solely for the purpose of standardizing signing for these zones and not as an endorsement of mandatory reduced speed zones.
11 "School" and "school zone" are defined in Section 1A.13.

Section 7A.02 School Routes and Established School Crossings
Support:
01 To establish a safer route to and from school for schoolchildren, the application of planning criterion for school walk routes might make it necessary for children to walk an indirect route to an established school crossing located where there is existing traffic control and to avoid the use of a direct crossing where there is no existing traffic control.
Guidance:
02 *School walk routes should be planned to take advantage of existing traffic controls.*
03 *The following factors should be considered when determining the feasibility of requiring children to walk a longer distance to a crossing with existing traffic control:*
 A. *The availability of adequate sidewalks or other pedestrian walkways to and from the location with existing control,*
 B. *The number of students using the crossing,*
 C. *The age levels of the students using the crossing, and*
 D. *The total extra walking distance.*

Section 7A.03 School Crossing Control Criteria
Support:
01 The frequency of gaps in the traffic stream that are sufficient for student crossing is different at each crossing location. When the delay between the occurrences of adequate gaps becomes excessive, students might become impatient and endanger themselves by attempting to cross the street during an inadequate gap. In these instances, the creation of sufficient gaps needs to be considered to accommodate the crossing demand.
02 A recommended method for determining the frequency and adequacy of gaps in the traffic stream is given in the "Traffic Control Devices Handbook" (see Section 1A.11).

Figure 7A-1. Example of School Route Plan Map

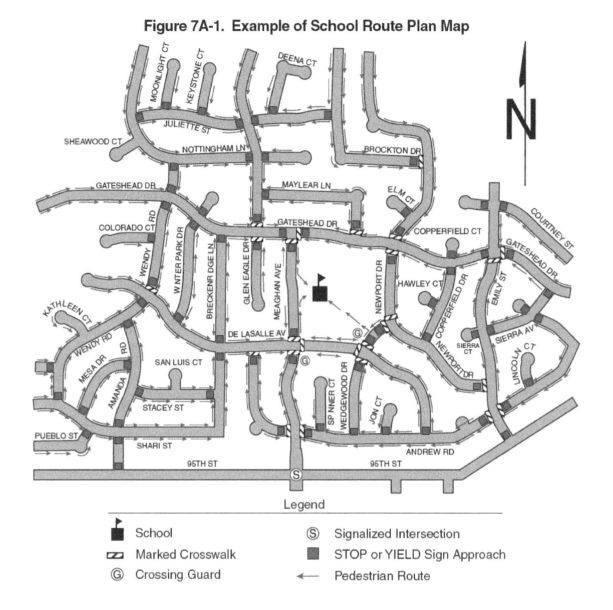

Section 7A.04 Scope

Standard:

01 **Part 7 sets forth basic principles and prescribes standards that shall be followed in the design, application, installation, and maintenance of all traffic control devices (including signs, signals, and markings) and other controls (including adult crossing guards) required for the special pedestrian conditions in school areas.**

Support:

02 Sections 1A.01 and 1A.08 contain information regarding unauthorized devices and messages. Sections 1A.02 and 1A.07 contain information regarding the application of standards. Section 1A.05 contains information regarding the maintenance of traffic control devices. Section 1A.08 contains information regarding placement authority for traffic control devices. Section 1A.09 contains information regarding engineering studies and the assistance that is available to jurisdictions that do not have engineers on their staffs who are trained and/or experienced in traffic control devices.

03 Provisions contained in Chapter 2A and Section 2B.06 are applicable in school areas.

04 Part 3 contains provisions regarding pavement markings that are applicable in school areas.

05 Part 4 contains provisions regarding highway traffic signals that are applicable in school areas. The School Crossing signal warrant is described in Section 4C.06.

CHAPTER 7B. SIGNS

Section 7B.01 Size of School Signs

Standard:

01 Except as provided in Section 2A.11, the sizes of signs and plaques to be used on conventional roadways in school areas shall be as shown in Table 7B-1.

02 The sizes in the Conventional Road column shall be used unless engineering judgment determines that a minimum or oversized sign size would be more appropriate.

03 The sizes in the Minimum column shall be used only where traffic volumes are low and speeds are 30 mph or lower, as determined by engineering judgment.

04 The sizes in the Oversized column shall be used on expressways.

Guidance:

05 *The sizes in the Oversized column should be used on roadways that have four or more lanes with posted speed limits of 40 mph or higher.*

Option:

06 The sizes in the Oversized column may also be used at other locations that require increased emphasis, improved recognition, or increased legibility.

07 Signs and plaques larger than those shown in Table 7B-1 may be used (see Section 2A.11).

Table 7B-1. School Area Sign and Plaque Sizes

Sign	Sign Designation	Section	Conventional Road	Minimum	Oversized
School	S1-1	7B.08	36 x 36	30 x 30	48 x 48
School Bus Stop Ahead	S3-1	7B.13	36 x 36	30 x 30	48 x 48
School Bus Turn Ahead	S3-2	7B.14	36 x 36	30 x 30	48 x 48
Reduced School Speed Limit Ahead	S4-5, S4-5a	7B.16	36 x 36	30 x 30	48 x 48
School Speed Limit XX When Flashing	S5-1	7B.15	24 x 48	—	36 x 72
End School Zone	S5-2	7B.09	24 x 30	—	36 x 48
End School Speed Limit	S5-3	7B.15	24 x 30	—	36 x 48
In-Street Ped Crossing	R1-6, R1-6a, R1-6b, R1-6c	7B.11, 7B.12	12 x 36	—	—
Speed Limit (School Use)	R2-1	7B.15	24 x 30	—	36 x 48
Begin Higher Fines Zone	R2-10	7B.10	24 x 30	—	36 x 48
End Higher Fines Zone	R2-11	7B.10	24 x 30	—	36 x 48

Plaque	Sign Designation	Section	Conventional Road	Minimum	Oversized
X:XX to X:XX AM / X:XX to X:XX PM	S4-1P	7B.15	24 x 10	—	36 x 18
When Children Are Present	S4-2P	7B.15	24 x 10	—	36 x 18
School	S4-3P	7B.09, 7B.15	24 x 8	—	36 x 12
When Flashing	S4-4P	7B.15	24 x 10	—	36 x 18
Mon-Fri	S4-6P	7B.15	24 x 10	—	36 x 18
All Year	S4-7P	7B.09	24 x 12	—	30 x 18
Fines Higher	R2-6P	7B.10	24 x 18	—	36 x 24
XX Feet	W16-2P	7B.08	24 x 18	—	30 x 24
XX Ft	W16-2aP	7B.08	24 x 12	—	30 x 18
Turn Arrow	W16-5P	7B.08, 7B.09, 7B.11	24 x 12	—	30 x 18
Advance Turn Arrow	W16-6P	7B.08, 7B.09, 7B.11	24 x 12	—	30 x 18
Diagonal Arrow	W16-7P	7B.12	24 x 12	—	30 x 18
Diagonal Arrow (optional size)	W16-7P	7B.12	21 x 15	—	—
Ahead	W16-9P	7B.11	24 x 12	—	30 x 18

Note: 1. Larger sizes may be used when appropriate
2. Dimensions are shown in inches and are shown as width x height
3. Minimum sign sizes for multi-lane conventional roads shall be as shown in the Conventional Road column

Section 7B.02 Illumination and Reflectorization
Standard:
01   **The signs used for school area traffic control shall be retroreflectorized or illuminated.**

Section 7B.03 Position of Signs
Support:
01   Sections 2A.16 and 2A.17 contain provisions regarding the placements and locations of signs.
02   Section 2A.19 contains provisions regarding the lateral offsets of signs.

Option:
03   In-roadway signs for school traffic control areas may be used consistent with the requirements of Sections 2B.12, 7B.08, and 7B.12.

Section 7B.04 Height of Signs
Support:
01   Section 2A.18 contains provisions regarding the mounting height of signs.

Section 7B.05 Installation of Signs
Support:
01   Section 2A.16 contains provisions regarding the installation of signs.

Section 7B.06 Lettering
Support:
01   The "Standard Highway Signs and Markings" book (see Section 1A.11) contains information regarding sign lettering.

Section 7B.07 Sign Color for School Warning Signs
Standard:
01   **School warning signs, including the "SCHOOL" portion of the School Speed Limit (S5-1) sign and including any supplemental plaques used in association with these warning signs, shall have a fluorescent yellow-green background with a black legend and border unless otherwise provided in this Manual for a specific sign.**

Section 7B.08 School Sign (S1-1) and Plaques
Support:
01   Many state and local jurisdictions find it beneficial to advise road users that they are approaching a school that is adjacent to a highway, where additional care is needed, even though no school crossing is involved and the speed limit remains unchanged. Additionally, some jurisdictions designate school zones that have a unique legal standing in that fines for speeding or other traffic violations within designated school zones are increased or special enforcement techniques such as photo radar systems are used. It is important and sometimes legally necessary to mark the beginning and end points of these designated school zones so that the road user is given proper notice.

02   The School (S1-1) sign (see Figure 7B-1) has the following four applications:
   A. School Area – the S1-1 sign can be used to warn road users that they are approaching a school area that might include school buildings or grounds, a school crossing, or school related activity adjacent to the highway.
   B. School Zone – the S1-1 sign can be used to identify the location of the beginning of a designated school zone (see Section 7B.09).
   C. School Advance Crossing – if combined with an AHEAD (W16-9P) plaque or an XX FEET (W16-2P or W16-2aP) plaque to comprise the School Advance Crossing assembly, the S1-1 sign can be used to warn road users that they are approaching a crossing where schoolchildren cross the roadway (see Section 7B.11).
   D. School Crossing – if combined with a diagonal downward pointing arrow (W16-7P) plaque to comprise the School Crossing assembly, the S1-1 sign can be used to warn approaching road users of the location of a crossing where schoolchildren cross the roadway (see Section 7B.12).

Option:
03   If a school area is located on a cross street in close proximity to the intersection, a School (S1-1) sign with a supplemental arrow (W16-5P or W16-6P) plaque may be installed on each approach of the street or highway to warn road users making a turn onto the cross street that they will encounter a school area soon after making the turn.

Figure 7B-1. School Area Signs

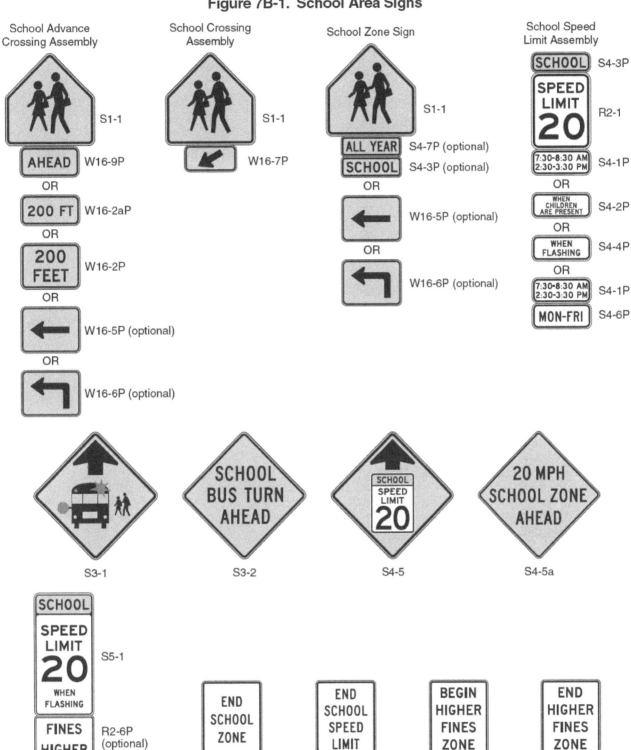

Section 7B.09 School Zone Sign (S1-1) and Plaques (S4-3P, S4-7P) and END SCHOOL ZONE Sign (S5-2)

Standard:

01 **If a school zone has been designated under State or local statute, a School (S1-1) sign (see Figure 7B-1) shall be installed to identify the beginning point(s) of the designated school zone (see Figure 7B-2).**

Option:

02 A School Zone (S1-1) sign may be supplemented with a SCHOOL (S4-3P) plaque (see Figure 7B-1).

03 A School Zone (S1-1) sign may be supplemented with an ALL YEAR (S4-7P) plaque (see Figure 7B-1) if the school operates on a 12-month schedule.

04 The downstream end of a designated school zone may be identified with an END SCHOOL ZONE (S5-2) sign (see Figures 7B-1 and 7B-2).

05 If a school zone is located on a cross street in close proximity to the intersection, a School Zone (S1-1) sign with a supplemental arrow (W16-5P or W16-6P) plaque may be installed on each approach of the street or highway to warn road users making a turn onto the cross street that they will encounter a school zone soon after making the turn.

Section 7B.10 Higher Fines Zone Signs (R2-10, R2-11) and Plaques

Standard:

01 **Where increased fines are imposed for traffic violations within a designated school zone, a BEGIN HIGHER FINES ZONE (R2-10) sign (see Figure 7B-1) or a FINES HIGHER (R2-6P), FINES DOUBLE (R2-6aP), or $XX FINE (R2-6bP) plaque (see Figure 2B-3) shall be installed as a supplement to the School Zone (S1-1) sign to identify the beginning point of the higher fines zone (see Figures 7B-2 and 7B-3).**

Option:

02 Where appropriate, one of the following plaques may be mounted below the sign that identifies the beginning point of the higher fines zone:

   A. An S4-1P plaque (see Figure 7B-1) specifying the times that the higher fines are in effect,
   B. A WHEN CHILDREN ARE PRESENT (S4-2P) plaque (see Figure 7B-1), or
   C. A WHEN FLASHING (S4-4P) plaque (see Figure 7B-1) if used in conjunction with a yellow flashing beacon.

Standard:

03 **Where a BEGIN HIGHER FINES ZONE (R2-10) sign or a FINES HIGHER (R2-6P) plaque supplementing a School Zone (S1-1) sign is posted to notify road users of increased fines for traffic violations, an END HIGHER FINES ZONE (R2-11) sign (see Figure 7B-1) or an END SCHOOL ZONE (S5-2) sign shall be installed at the downstream end of the zone to notify road users of the termination of the increased fines zone (see Figures 7B-2 and 7B-3).**

Section 7B.11 School Advance Crossing Assembly

Standard:

01 **The School Advance Crossing assembly (see Figure 7B-1) shall consist of a School (S1-1) sign supplemented with an AHEAD (W16-9P) plaque or an XX FEET (W16-2P or W16-2aP) plaque.**

02 **Except as provided in Paragraph 3, a School Advance Crossing assembly shall be used in advance (see Table 2C-4 for advance placement guidelines) of the first School Crossing assembly (see Section 7B.12) that is encountered in each direction as traffic approaches a school crosswalk (see Figure 7B-4).**

Option:

03 The School Advance Crossing assembly may be omitted (see Figure 7B-5) where a School Zone (S1-1) sign (see Section 7B.09) is installed to identify the beginning of a school zone in advance of the School Crossing assembly.

04 If a school crosswalk is located on a cross street in close proximity to an intersection, a School Advance Crossing assembly with a supplemental arrow (W16-5P or W16-6P) plaque may be installed on each approach of the street or highway to warn road users making a turn onto the cross street that they will encounter a school crosswalk soon after making the turn.

05 A 12-inch reduced size in-street School (S1-1) sign (see Figure 7B-6), installed in compliance with the mounting height and special mounting support requirements for In-Street Pedestrian Crossing (R1-6 or R1-6a) signs (see Section 2B.12), may be used in advance of a school crossing to supplement the post-mounted school warning signs. A 12 x 6-inch reduced size AHEAD (W16-9P) plaque may be mounted below the reduced size in-street School (S1-1) sign.

Figure 7B-2. Example of Signing for a Higher Fines School Zone without a School Crossing

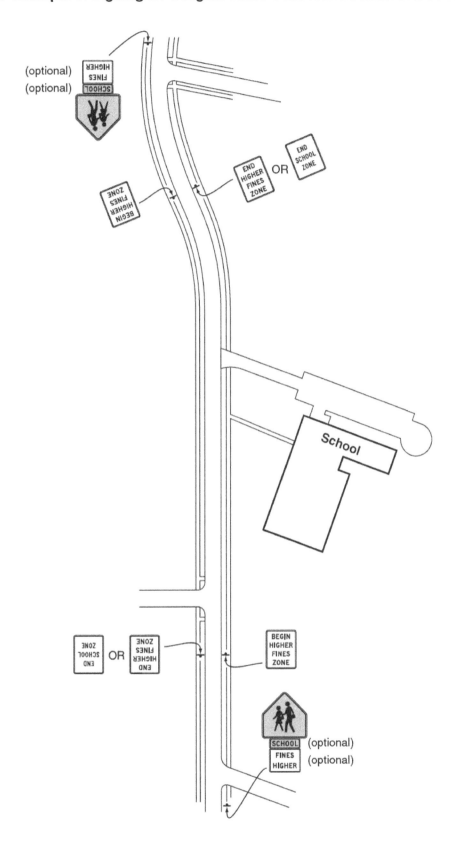

Figure 7B-3. Example of Signing for a Higher Fines School Zone with a School Speed Limit

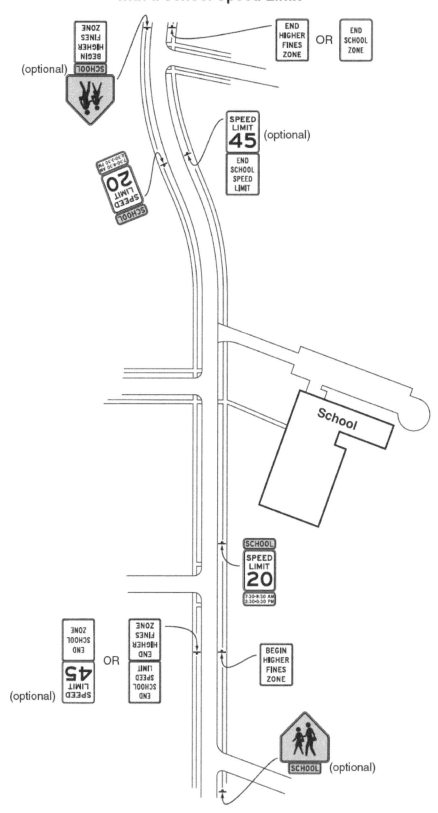

Figure 7B-4. Example of Signing for a School Crossing Outside of a School Zone

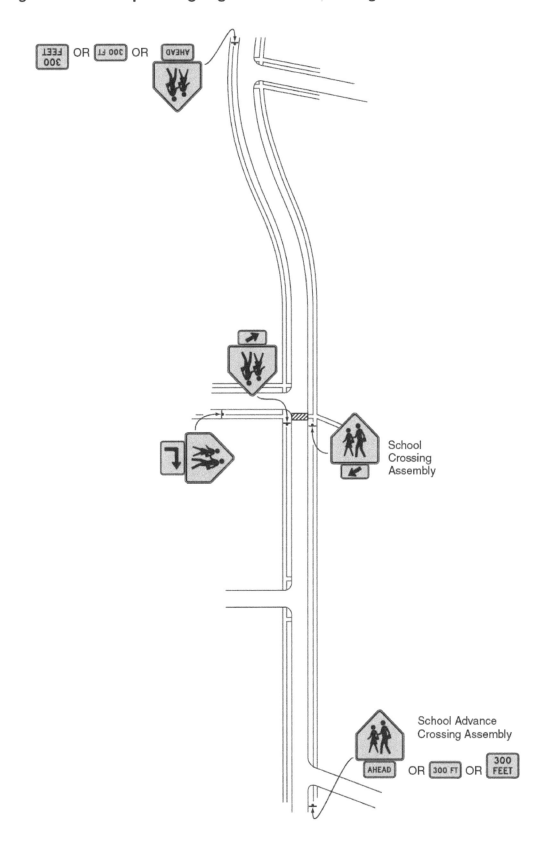

Figure 7B-5. Example of Signing for a School Zone with a School Speed Limit and a School Crossing

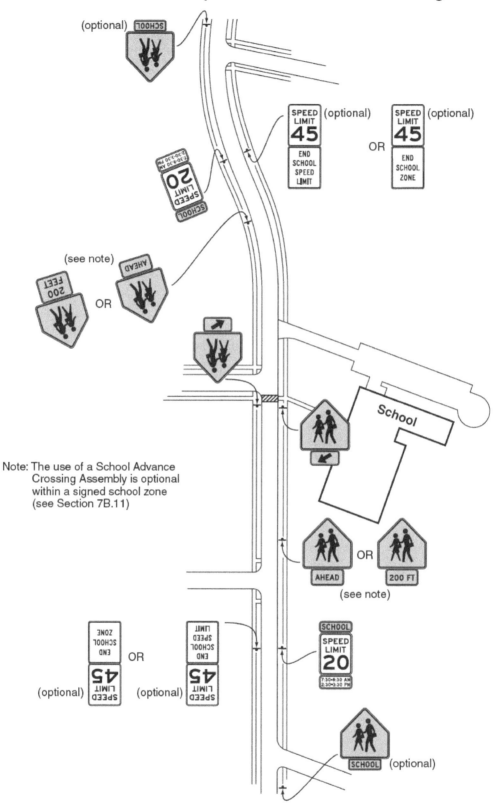

Figure 7B-6. In-Street Signs in School Areas

A - In advance of the school crossing

B - At the school crossing

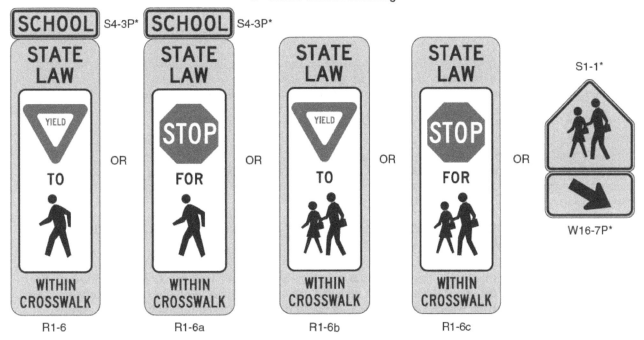

Notes:
1. The use of the STATE LAW legend is optional on the R1-6 series signs (see Section 7B.12).
2. The use of the SCHOOL plaque above the R1-6 and R1-6a signs is optional.

Section 7B.12 School Crossing Assembly

Standard:

01   If used, the School Crossing assembly (see Figure 7B-1) shall be installed at the school crossing (see Figures 7B-4 and 7B-5), or as close to it as possible, and shall consist of a School (S1-1) sign supplemented with a diagonal downward pointing arrow (W16-7P) plaque to show the location of the crossing.

02   The School Crossing assembly shall not be used at crossings other than those adjacent to schools and those on established school pedestrian routes.

03   The School Crossing assembly shall not be installed on approaches controlled by a STOP or YIELD sign.

Option:

04   The In-Street Pedestrian Crossing (R1-6 or R1-6a) sign (see Section 2B.12 and Figure 7B-6) or the In-Street Schoolchildren Crossing (R1-6b or R1-6c) sign (see Figure 7B-6) may be used at unsignalized school crossings. If used at a school crossing, a 12 x 4-inch SCHOOL (S4-3P) plaque (see Figure 7B-6) may be mounted above the sign. The STATE LAW legend on the R1-6 series signs may be omitted.

05   The Overhead Pedestrian Crossing (R1-9 or R1-9a) sign (see Section 2B.12 and Figure 2B-2) may be modified to replace the standard pedestrian symbol with the standard schoolchildren symbol and may be used at unsignalized school crossings. The STATE LAW legend on the R1-9 series signs may be omitted.

06   A 12-inch reduced size in-street School (S1-1) sign (see Figure 7B-6) may be used at an unsignalized school crossing instead of the In-Street Pedestrian Crossing (R1-6 or R1-6a) or the In-Street Schoolchildren Crossing (R1-6b or R1-6c) sign. A 12 x 6-inch reduced size diagonal downward pointing arrow (W16-7P) plaque may be mounted below the reduced size in-street School (S1-1) sign.

Standard:

07   **If an In-Street Pedestrian Crossing sign, an In-Street Schoolchildren Crossing sign, or a reduced size in-street School (S1-1) sign is placed in the roadway, the sign support shall comply with the mounting height and special mounting support requirements for In-Street Pedestrian Crossing (R1-6 or R1-6a) signs (see Section 2B.12).**

08   **The In-Street Pedestrian Crossing sign, the In-Street Schoolchildren Crossing sign, the Overhead Pedestrian Crossing sign, and the reduced size in-street School (S1-1) sign shall not be used at signalized locations.**

Section 7B.13 School Bus Stop Ahead Sign (S3-1)

Guidance:

01   *The School Bus Stop Ahead (S3-1) sign (see Figure 7B-1) should be installed in advance of locations where a school bus, when stopped to pick up or discharge passengers, is not visible to road users for an adequate distance and where there is no opportunity to relocate the school bus stop to provide adequate sight distance.*

Section 7B.14 SCHOOL BUS TURN AHEAD Sign (S3-2)

Option:

01   The SCHOOL BUS TURN AHEAD (S3-2) sign (see Figure 7B-1) may be installed in advance of locations where a school bus turns around on a roadway at a location not visible to approaching road users for a distance as determined by the "0" column under Condition B of Table 2C-4, and where there is no opportunity to relocate the school bus turn around to provide the distance provided in Table 2C-4.

Section 7B.15 School Speed Limit Assembly (S4-1P, S4-2P, S4-3P, S4-4P, S4-6P, S5-1) and END SCHOOL SPEED LIMIT Sign (S5-3)

Standard:

01   **A School Speed Limit assembly (see Figure 7B-1) or a School Speed Limit (S5-1) sign (see Figure 7B-1) shall be used to indicate the speed limit where a reduced school speed limit zone has been established based upon an engineering study or where a reduced school speed limit is specified for such areas by statute. The School Speed Limit assembly or School Speed Limit sign shall be placed at or as near as practical to the point where the reduced school speed limit zone begins (see Figures 7B-3 and 7B-5).**

02   **If a reduced school speed limit zone has been established, a School (S1-1) sign shall be installed in advance (see Table 2C-4 for advance placement guidelines) of the first School Speed Limit sign assembly or S5-1 sign that is encountered in each direction as traffic approaches the reduced school speed limit zone (see Figures 7B-3 and 7B-5).**

03   **Where increased fines are imposed for traffic violations within a reduced school speed limit zone, a FINES HIGHER (R2-6P), FINES DOUBLE (R2-6aP), or $XX FINE (R2-6bP) plaque (see Figure 2B-3) shall be installed as a supplement to the reduced school speed limit sign to notify road users.**

04   **Except as provided in Paragraph 5, the downstream end of an authorized and posted reduced school speed limit zone shall be identified with an END SCHOOL SPEED LIMIT (S5-3) sign (see Figures 7B-1 and 7B-5).**

Option:

05   If a reduced school speed limit zone ends at the same point as a higher fines zone, an END SCHOOL ZONE (S5-2) sign may be used instead of a combination of an END HIGHER FINES ZONE (R2-11) sign and an END SCHOOL SPEED LIMIT (S5-3) sign.

06   A standard Speed Limit sign showing the speed limit for the section of highway that is downstream from the authorized and posted reduced school speed limit zone may be mounted on the same post above the END SCHOOL SPEED LIMIT (S5-3) sign or the END SCHOOL ZONE (S5-2) sign.

Guidance:

07   *The beginning point of a reduced school speed limit zone should be at least 200 feet in advance of the school grounds, a school crossing, or other school related activities; however, this 200-foot distance should be increased if the reduced school speed limit is 30 mph or higher.*

Standard:

08 **The School Speed Limit assembly shall be either a fixed-message sign assembly or a changeable message sign.**

09 **The fixed-message School Speed Limit assembly shall consist of a top plaque (S4-3P) with the legend SCHOOL, a Speed Limit (R2-1) sign, and a bottom plaque (S4-1P, S4-2P, S4-4P, or S4-6P) indicating the specific periods of the day and/or days of the week that the special school speed limit is in effect (see Figure 7B-1).**

Option:

10 Changeable message signs (see Chapter 2L and Section 6F.60) may be used to inform drivers of the school speed limit. If the sign is internally illuminated, it may have a white legend on a black background. Changeable message signs with flashing beacons may be used for situations where greater emphasis of the special school speed limit is needed.

Guidance:

11 *Even though it might not always be practical because of special features to make changeable message signs conform in all respects to the standards in this Manual for fixed-message signs, during the periods that the school speed limit is in effect, their basic shape, message, legend layout, and colors should comply with the standards for fixed-message signs.*

12 *A confirmation light or device to indicate that the speed limit message is in operation should be considered for inclusion on the back of the changeable message sign.*

Standard:

13 **Fluorescent yellow-green pixels shall be used when the "SCHOOL" message is displayed on a changeable message sign for a school speed limit.**

Option:

14 Changeable message signs may use blank-out messages or other methods in order to display the school speed limit only during the periods it applies.

15 Changeable message signs that display the speed of approaching drivers (see Section 2B.13) may be used in a school speed limit zone.

16 A Speed Limit Sign Beacon (see Section 4L.04) also may be used, with a WHEN FLASHING legend, to identify the periods that the school speed limit is in effect.

Section 7B.16 Reduced School Speed Limit Ahead Sign (S4-5, S4-5a)

Guidance:

01 *A Reduced School Speed Limit Ahead (S4-5, S4-5a) sign (see Figure 7B-1) should be used to inform road users of a reduced speed zone where the speed limit is being reduced by more than 10 mph, or where engineering judgment indicates that advance notice would be appropriate.*

Standard:

02 **If used, the Reduced School Speed Limit Ahead sign shall be followed by a School Speed Limit sign or a School Speed Limit assembly.**

03 **The speed limit displayed on the Reduced School Speed Limit Ahead sign shall be identical to the speed limit displayed on the subsequent School Speed Limit sign or School Speed Limit assembly.**

Section 7B.17 Parking and Stopping Signs (R7 and R8 Series)

Option:

01 Parking and stopping regulatory signs may be used to prevent parked or waiting vehicles from blocking pedestrians' views, and drivers' views of pedestrians, and to control vehicles as a part of the school traffic plan.

Support:

02 Parking signs and other signs governing the stopping and standing of vehicles in school areas cover a wide variety of regulations. Typical examples of regulations are as follows:

    A. No Parking X:XX AM to X:XX PM School Days Only,
    B. No Stopping X:XX AM to X:XX PM School Days Only,
    C. XX Min Loading X:XX AM to X:XX PM School Days Only, and
    D. No Standing X:XX AM to X:XX PM School Days Only.

03 Sections 2B.46, 2B.47, and 2B.48 contain information regarding the signing of parking regulations in school zone areas.

CHAPTER 7C. MARKINGS

Section 7C.01 Functions and Limitations

Support:

01 Markings have definite and important functions in a proper scheme of school area traffic control. In some cases, they are used to supplement the regulations or warnings provided by other devices, such as traffic signs or signals. In other instances, they are used alone and produce results that cannot be obtained by the use of any other device. In such cases they serve as an effective means of conveying certain regulations, guidance, and warnings that could not otherwise be made clearly understandable.

02 Pavement markings have some potential limitations. They might be obscured by snow, might not be clearly visible when wet, and might not be durable when subjected to heavy traffic. In spite of these potential limitations, they have the advantage, under favorable conditions, of conveying warnings or information to the road user without diverting attention from the road.

Section 7C.02 Crosswalk Markings

Guidance:

01 *Crosswalks should be marked at all intersections on established routes to a school where there is substantial conflict between motorists, bicyclists, and student movements; where students are encouraged to cross between intersections; where students would not otherwise recognize the proper place to cross; or where motorists or bicyclists might not expect students to cross (see Figure 7A-1).*

02 *Crosswalk lines should not be used indiscriminately. An engineering study considering the factors described in Section 3B.18 should be performed before a marked crosswalk is installed at a location away from a traffic control signal or an approach controlled by a STOP or YIELD sign.*

03 *Because non-intersection school crossings are generally unexpected by the road user, warning signs (see Sections 7B.11 and 7B.12) should be installed for all marked school crosswalks at non-intersection locations. Adequate visibility of students by approaching motorists and of approaching motorists by students should be provided by parking prohibitions or other appropriate measures.*

Support:

04 Section 3B.18 contains provisions regarding the placement and design of crosswalks, and Section 3B.16 contains provisions regarding the placement and design of the stop lines and yield lines that are associated with them. Provisions regarding the curb markings that can be used to establish parking regulations on the approaches to crosswalks are contained in Section 3B.23.

Section 7C.03 Pavement Word, Symbol, and Arrow Markings

Option:

01 If used, the SCHOOL word marking may extend to the width of two approach lanes (see Figure 7C-1).

Guidance:

02 *If the two-lane SCHOOL word marking is used, the letters should be 10 feet or more in height.*

Support:

03 Section 3B.20 contains provisions regarding other word, symbol, and arrow pavement markings that can be used to guide, warn, or regulate traffic.

Figure 7C-1. Two-Lane Pavement Marking of "SCHOOL"

CHAPTER 7D. CROSSING SUPERVISION

Section 7D.01 Types of Crossing Supervision
Support:
01   There are three types of school crossing supervision:
   A. Adult control of pedestrians and vehicles by adult crossing guards,
   B. Adult control of pedestrians and vehicles by uniformed law enforcement officers, and
   C. Student and/or parent control of only pedestrians with student and/or parent patrols.

02   Information regarding the organization, administration, and operation of a school safety patrol program is contained in the "AAA School Safety Patrol Operations Manual" (see Section 1A.11).

Section 7D.02 Adult Crossing Guards
Option:
01   Adult crossing guards may be used to provide gaps in traffic at school crossings where an engineering study has shown that adequate gaps need to be created (see Section 7A.03), and where authorized by law.

Section 7D.03 Qualifications of Adult Crossing Guards
Support:
01   High standards for selection of adult crossing guards are essential because they are responsible for the safety of and the efficient crossing of the street by schoolchildren within and in the immediate vicinity of school crosswalks.

Guidance:
02   *Adult crossing guards should possess the following minimum qualifications:*
   A. *Average intelligence;*
   B. *Good physical condition, including sight, hearing, and ability to move and maneuver quickly in order to avoid danger from errant vehicles;*
   C. *Ability to control a STOP paddle effectively to provide approaching road users with a clear, fully direct view of the paddle's STOP message during the entire crossing movement;*
   D. *Ability to communicate specific instructions clearly, firmly, and courteously;*
   E. *Ability to recognize potentially dangerous traffic situations and warn and manage students in sufficient time to avoid injury.*
   F. *Mental alertness;*
   G. *Neat appearance;*
   H. *Good character;*
   I. *Dependability; and*
   J. *An overall sense of responsibility for the safety of students.*

Section 7D.04 Uniform of Adult Crossing Guards
Standard:
01   **Law enforcement officers performing school crossing supervision and adult crossing guards shall wear high-visibility retroreflective safety apparel labeled as ANSI 107-2004 standard performance for Class 2 as described in Section 6E.02.**

Section 7D.05 Operating Procedures for Adult Crossing Guards
Standard:
01   **Adult crossing guards shall not direct traffic in the usual law enforcement regulatory sense. In the control of traffic, they shall pick opportune times to create a sufficient gap in the traffic flow. At these times, they shall stand in the roadway to indicate that pedestrians are about to use or are using the crosswalk, and that all vehicular traffic must stop.**

02   **Adult crossing guards shall use a STOP paddle. The STOP paddle shall be the primary hand-signaling device.**

03   **The STOP (R1-1) paddle shall be an octagonal shape. The background of the STOP face shall be red with at least 6-inch series upper-case white letters and border. The paddle shall be at least 18 inches in size and have the word message STOP on both sides. The paddle shall be retroreflectorized or illuminated when used during hours of darkness.**

Option:

04 The STOP paddle may be modified to improve conspicuity by incorporating white or red flashing lights on both sides of the paddle. Among the types of flashing lights that may be used are individual LEDs or groups of LEDs.

05 The white or red flashing lights or LEDs may be arranged in any of the following patterns:
   A. Two white or red lights centered vertically above and below the STOP legend,
   B. Two white or red lights centered horizontally on each side of the STOP legend,
   C. One white or red light centered below the STOP legend,
   D. A series of eight or more small white or red lights having a diameter of 1/4 inch or less along the outer edge of the paddle, arranged in an octagonal pattern at the eight corners of the STOP paddle (more than eight lights may be used only if the arrangement of the lights is such that it clearly conveys the octagonal shape of the STOP paddle), or
   E. A series of white lights forming the shapes of the letters in the legend.

Standard:

06 **If flashing lights are used on the STOP paddle, the flash rate shall be at least 50, but no more than 60, flash periods per minute.**

PART 8
TRAFFIC CONTROL FOR RAILROAD AND LIGHT RAIL TRANSIT GRADE CROSSINGS

CHAPTER 8A. GENERAL

Section 8A.01 Introduction
Support:
01 Whenever the acronym "LRT" is used in Part 8, it refers to "light rail transit."

02 Part 8 describes the traffic control devices that are used at highway-rail and highway-LRT grade crossings. Unless otherwise provided in the text or on a figure or table, the provisions of Part 8 are applicable to both highway-rail and highway-LRT grade crossings. When the phrase "grade crossing" is used by itself without the prefix "highway-rail" or "highway-LRT," it refers to both highway-rail and highway-LRT grade crossings.

03 Traffic control for grade crossings includes all signs, signals, markings, other warning devices, and their supports along highways approaching and at grade crossings. The function of this traffic control is to promote safety and provide effective operation of rail and/or LRT and highway traffic at grade crossings.

04 For purposes of design, installation, operation, and maintenance of traffic control devices at grade crossings, it is recognized that the crossing of the highway and rail tracks is situated on a right-of-way available for the joint use of both highway traffic and railroad or LRT traffic.

05 The highway agency or authority with jurisdiction and the regulatory agency with statutory authority, if applicable, jointly determine the need and selection of devices at a grade crossing.

06 In Part 8, the combination of devices selected or installed at a specific grade crossing is referred to as a "traffic control system."

Standard:
07 **The traffic control devices, systems, and practices described in this Manual shall be used at all grade crossings open to public travel, consistent with Federal, State, and local laws and regulations.**

Support:
08 Part 8 also describes the traffic control devices that are used in locations where light rail LRT vehicles are operating along streets and highways in mixed traffic with automotive vehicles.

09 LRT is a mode of metropolitan transportation that employs LRT vehicles (commonly known as light rail vehicles, streetcars, or trolleys) that operate on rails in streets in mixed traffic, and LRT traffic that operates in semi-exclusive rights-of-way, or in exclusive rights-of-way. Grade crossings with LRT can occur at intersections or at midblock locations, including public and private driveways.

10 An initial educational campaign along with an ongoing program to continue to educate new drivers is beneficial when introducing LRT operations to an area and, hence, new traffic control devices.

11 LRT alignments can be grouped into one of the following three types:
   A. Exclusive: An LRT right-of-way that is grade-separated or protected by a fence or traffic barrier. Motor vehicles, pedestrians, and bicycles are prohibited within the right-of-way. Subways and aerial structures are included within this group. This type of alignment does not have grade crossings and is not further addressed in Part 8.
   B. Semi-exclusive: An LRT alignment that is in a separate right-of-way or along a street or railroad right-of-way where motor vehicles, pedestrians, and bicycles have limited access and cross at designated locations only.
   C. Mixed-use: An alignment where LRT operates in mixed traffic with all types of road users. This includes streets, transit malls, and pedestrian malls where the right-of-way is shared.

Standard:
12 **Where LRT and railroads use the same tracks or adjacent tracks, the traffic control devices, systems, and practices for highway-rail grade crossings shall be used.**

Support:
13 To promote an understanding of common terminology between highway and railroad and LRT signaling issues, definitions and acronyms pertaining to Part 8 are provided in Sections 1A.13 and 1A.14.

Section 8A.02 Use of Standard Devices, Systems, and Practices at Highway-Rail Grade Crossings
Support:
01 Because of the large number of significant variables to be considered, no single standard system of traffic control devices is universally applicable for all highway-rail grade crossings.

Guidance:

01 *The appropriate traffic control system to be used at a highway-rail grade crossing should be determined by an engineering study involving both the highway agency and the railroad company.*

Option:

03 The engineering study may include the Highway-Rail Intersection (HRI) components of the National Intelligent Transportation Systems (ITS) architecture, which is a USDOT accepted method for linking the highway, vehicles, and traffic management systems with rail operations and wayside equipment.

Support:

04 More detail on Highway-Rail Intersection components is available from the USDOT's Federal Railroad Administration, 1200 New Jersey Avenue, SE, Washington, DC 20590, or www.fra.dot.gov.

Standard:

05 **Traffic control devices, systems, and practices shall be consistent with the design and application of the Standards contained in this Manual.**

06 **Before any new highway-rail grade crossing traffic control system is installed or before modifications are made to an existing system, approval shall be obtained from the highway agency with the jurisdictional and/or statutory authority, and from the railroad company.**

Guidance:

07 *To stimulate effective responses from road users, these devices, systems, and practices should use the five basic considerations employed generally for traffic control devices and described fully in Section 1A.02: design, placement, operation, maintenance, and uniformity.*

Support:

08 Many other details of highway-rail grade crossing traffic control systems that are not set forth in Part 8 are contained in the publications listed in Section 1A.11, including the "2000 AREMA Communications & Signals Manual" published by the American Railway Engineering & Maintenance-of-Way Association (AREMA) and the 2006 edition of "Preemption of Traffic Signals Near Railroad Crossings" published by the Institute of Transportation Engineers (ITE).

Section 8A.03 Use of Standard Devices, Systems, and Practices at Highway-LRT Grade Crossings

Support:

01 The combination of devices selected or installed at a specific highway-LRT grade crossing is referred to as a Light Rail Transit Traffic Control System.

02 Because of the large number of significant variables to be considered, no single standard system of traffic control devices is universally applicable for all highway-LRT grade crossings.

03 For the safety and integrity of operations by highway and LRT users, the highway agency with jurisdiction, the regulatory agency with statutory authority, if applicable, and the LRT authority jointly determine the need and selection of traffic control devices and the assignment of priority to LRT at a highway-LRT grade crossing.

04 The normal rules of the road and traffic control priority identified in the "Uniform Vehicle Code" govern the order assigned to the movement of vehicles at an intersection unless the local agency determines that it is appropriate to assign a higher priority to LRT. Examples of different types of LRT priority control include separate traffic control signal phases for LRT movements, restriction of movement of roadway vehicles in favor of LRT operations, and preemption of highway traffic signal control to accommodate LRT movements.

Guidance:

05 *The appropriate traffic control system to be used at a highway-LRT grade crossing should be determined by an engineering study conducted by the LRT or highway agency in cooperation with other appropriate State and local organizations.*

Standard:

06 **Traffic control devices, systems, and practices shall be consistent with the design and application of the Standards contained in this Manual.**

07 **The traffic control devices, systems, and practices described in this Manual shall be used at all highway-LRT grade crossings.**

08 **Before any new highway-LRT grade crossing traffic control system is installed or before modifications are made to an existing system, approval shall be obtained from the highway agency with the jurisdictional and/or statutory authority, and from the LRT agency.**

Guidance:

09 *To stimulate effective responses from road users, these devices, systems, and practices should use the five basic considerations employed generally for traffic control devices and described fully in Section 1A.02: design, placement, operation, maintenance, and uniformity.*

Support:

10 Many other details of highway-LRT grade crossing traffic control systems that are not set forth in Part 8 are contained in the publications listed in Section 1A.11.

Standard:

11 **Highway-LRT grade crossings in semi-exclusive alignments shall be equipped with a combination of automatic gates and flashing-light signals, or flashing-light signals only, or traffic control signals, unless an engineering study indicates that the use of Crossbuck Assemblies, STOP signs, or YIELD signs alone would be adequate.**

Option:

12 Highway-LRT grade crossings in mixed-use alignments may be equipped with traffic control signals unless an engineering study indicates that the use of Crossbuck Assemblies, STOP signs, or YIELD signs alone would be adequate.

Support:

13 Sections 8B.03 and 8B.04 contain provisions regarding the use and placement of Crossbuck signs and Crossbuck Assemblies. Section 8B.05 describes the appropriate conditions for the use of STOP or YIELD signs alone at a highway-LRT grade crossing. Sections 8C.10 and 8C.11 contain provisions regarding the use of traffic control signals at highway-LRT grade crossings.

Section 8A.04 Uniform Provisions

Standard:

01 **All signs used in grade crossing traffic control systems shall be retroreflectorized or illuminated as described in Section 2A.07 to show the same shape and similar color to an approaching road user during both day and night.**

02 **No sign or signal shall be located in the center of an undivided highway, unless it is crashworthy (breakaway, yielding, or shielded with a longitudinal barrier or crash cushion) or unless it is placed on a raised island.**

Guidance:

03 *Any signs or signals placed on a raised island in the center of an undivided highway should be installed with a clearance of at least 2 feet from the outer edge of the raised island to the nearest edge of the sign or signal, except as permitted in Section 2A.19.*

04 *Where the distance between tracks, measured along the highway between the inside rails, exceeds 100 feet, additional signs or other appropriate traffic control devices should be used to inform approaching road users of the long distance to cross the tracks.*

Section 8A.05 Grade Crossing Elimination

Guidance:

01 *Because grade crossings are a potential source of crashes and congestion, agencies should conduct engineering studies to determine the cost and benefits of eliminating these crossings.*

Standard:

02 **When a grade crossing is eliminated, the traffic control devices for the crossing shall be removed.**

03 **If the existing traffic control devices at a multiple-track grade crossing become improperly placed or inaccurate because of the removal of some of the tracks, the existing devices shall be relocated and/or modified.**

Guidance:

04 *Any grade crossing that cannot be justified should be eliminated.*

05 *Where a roadway is removed from a grade crossing, the roadway approaches in the railroad or LRT right-of-way should also be removed and appropriate signs and object markers should be placed at the roadway end in accordance with Section 2C.66.*

06 *Where a railroad or LRT is eliminated at a grade crossing, the tracks should be removed or covered.*

Option:

07 Based on engineering judgment, the TRACKS OUT OF SERVICE (R8-9) sign (see Figure 8B-1) may be temporarily installed until the tracks are removed or covered. The length of time before the tracks will be removed or covered may be considered in making the decision as to whether to install the sign.

Section 8A.06 Illumination at Grade Crossings

Support:

01 Illumination is sometimes installed at or adjacent to a grade crossing in order to provide better nighttime visibility of trains or LRT equipment and the grade crossing (for example, where a substantial amount of railroad or LRT operations are conducted at night, where grade crossings are blocked for extended periods of time, or where crash history indicates that road users experience difficulty in seeing trains or LRT equipment or traffic control devices during hours of darkness).

02 Recommended types and locations of luminaires for illuminating grade crossings are contained in the American National Standards Institute's (ANSI) "Practice for Roadway Lighting RP-8," which is available from the Illuminating Engineering Society (see Section 1A.11).

Section 8A.07 Quiet Zone Treatments at Highway-Rail Grade Crossings

Support:

01 49 CFR Part 222 (Use of Locomotive Horns at Highway-Rail Grade Crossings; Final Rule) prescribes Quiet Zone requirements and treatments.

Standard:

02 **Any traffic control device and its application where used as part of a Quiet Zone shall comply with all applicable provisions of the MUTCD.**

Section 8A.08 Temporary Traffic Control Zones

Support:

01 Temporary traffic control planning provides for continuity of operations (such as movement of traffic, pedestrians and bicycles, transit operations, and access to property/utilities) when the normal function of a roadway at a grade crossing is suspended because of temporary traffic control operations.

Standard:

02 **Traffic controls for temporary traffic control zones that include grade crossings shall be as outlined in Part 6.**

03 **When a grade crossing exists either within or in the vicinity of a temporary traffic control zone, lane restrictions, flagging (see Chapter 6E), or other operations shall not be performed in a manner that would cause highway vehicles to stop on the railroad or LRT tracks, unless a flagger or uniformed law enforcement officer is provided at the grade crossing to minimize the possibility of highway vehicles stopping on the tracks, even if automatic warning devices are in place.**

Guidance:

04 *Public and private agencies, including emergency services, businesses, and railroad or LRT companies, should meet to plan appropriate traffic detours and the necessary signing, marking, and flagging requirements for operations during temporary traffic control zone activities. Consideration should be given to the length of time that the grade crossing is to be closed, the type of rail or LRT and highway traffic affected, the time of day, and the materials and techniques of repair.*

05 *The agencies responsible for the operation of the LRT and highway should be contacted when the initial planning begins for any temporary traffic control zone that might directly or indirectly influence the flow of traffic on mixed-use facilities where LRT and road users operate.*

06 *Temporary traffic control operations should minimize the inconvenience, delay, and crash potential to affected traffic. Prior notice should be given to affected public or private agencies, emergency services, businesses, railroad or LRT companies, and road users before the free movement of road users or rail traffic is infringed upon or blocked.*

07 *Temporary traffic control zone activities should not be permitted to extensively prolong the closing of the grade crossing.*

08 *The width, grade, alignment, and riding quality of the highway surface at a grade crossing should, at a minimum, be restored to correspond with the quality of the approaches to the grade crossing.*

Support:

09 Section 6G.18 contains additional information regarding temporary traffic control zones in the vicinity of grade crossings, and Figure 6H-46 shows an example of a typical situation that might be encountered.

CHAPTER 8B. SIGNS AND MARKINGS

Section 8B.01 Purpose

Support:
01 Passive traffic control systems, consisting of signs and pavement markings only, identify and direct attention to the location of a grade crossing and advise road users to slow down or stop at the grade crossing as necessary in order to yield to any rail traffic occupying, or approaching and in proximity to, the grade crossing.

02 Signs and markings regulate, warn, and guide the road users so that they, as well as LRT vehicle operators on mixed-use alignments, can take appropriate action when approaching a grade crossing.

Standard:
03 **The design and location of signs shall comply with the provisions of Part 2. The design and location of pavement markings shall comply with the provisions of Part 3.**

Section 8B.02 Sizes of Grade Crossing Signs

Standard:
01 **The sizes of grade crossing signs shall be as shown in Table 8B-1.**

Option:
02 Signs larger than those shown in Table 8B-1 may be used (see Section 2A.11).

Section 8B.03 Grade Crossing (Crossbuck) Sign (R15-1) and Number of Tracks Plaque (R15-2P) at Active and Passive Grade Crossings

Standard:
01 **The Grade Crossing (R15-1) sign (see Figure 8B-1), commonly identified as the Crossbuck sign, shall be retroreflectorized white with the words RAILROAD CROSSING in black lettering, mounted as shown in Figure 8B-2.**

Support:
02 In most States, the Crossbuck sign requires road users to yield the right-of-way to rail traffic at a grade crossing.

Standard:
03 **As a minimum, one Crossbuck sign shall be used on each highway approach to every highway-rail grade crossing, alone or in combination with other traffic control devices.**

Option:
04 A Crossbuck sign may be used on a highway approach to a highway-LRT grade crossing on a semi-exclusive or mixed-use alignment, alone or in combination with other traffic control devices.

Standard:
05 **If automatic gates are not present and if there are two or more tracks at a grade crossing, the number of tracks shall be indicated on a supplemental Number of Tracks (R15-2P) plaque (see Figure 8B-1) of inverted T shape mounted below the Crossbuck sign in the manner shown in Figure 8B-2.**

06 **On each approach to a highway-rail grade crossing and, if used, on each approach to a highway-LRT grade crossing, the Crossbuck sign shall be installed on the right-hand side of the highway on each approach to the grade crossing. Where restricted sight distance or unfavorable highway geometry exists on an approach to a grade crossing, an additional Crossbuck sign shall be installed on the left-hand side of the highway, possibly placed back-to-back with the Crossbuck sign for the opposite approach, or otherwise located so that two Crossbuck signs are displayed for that approach.**

07 **A strip of retroreflective white material not less than 2 inches in width shall be used on the back of each blade of each Crossbuck sign for the length of each blade, at all grade crossings where Crossbuck signs have been installed, except those where Crossbuck signs have been installed back-to-back.**

Guidance:
08 *Crossbuck signs should be located with respect to the highway pavement or shoulder in accordance with the criteria in Chapter 2A and Figures 2A-2 and 2A-3, and should be located with respect to the nearest track in accordance with Figure 8C-2.*

09 *The minimum lateral offset for the nearest edge of the Crossbuck sign should be 6 feet from the edge of the shoulder or 12 feet from the edge of the traveled way in rural areas (whichever is greater), and 2 feet from the face of the curb in urban areas.*

Table 8B-1. Grade Crossing Sign and Plaque Minimum Sizes

Sign or Plaque	Sign Designation	Section	Conventional Road		Expressway	Minimum	Oversized
			Single Lane	Multi-Lane			
Stop	R1-1	8B.04, 8B.05	30 x 30	36 x 36	36 x 36	—	48 x 48
Yield	R1-2	8B.04, 8B.05	36 x 36 x 36	48 x 48 x 48	48 x 48 x 48	30 x 30 x 30	—
No Right Turn Across Tracks	R3-1a	8B.08	24 x 30	30 x 36	—	—	—
No Left Turn Across Tracks	R3-2a	8B.08	24 x 30	30 x 36	—	—	—
Do Not Stop on Tracks	R8-8	8B.09	24 x 30	24 x 30	36 x 48	—	36 x 48
Tracks Out of Service	R8-9	8B.10	24 x 24	24 x 24	36 x 36	—	36 x 36
Stop Here When Flashing	R8-10	8B.11	24 x 36	24 x 36	—	—	36 x 48
Stop Here When Flashing	R8-10a	8B.11	24 x 30	24 x 30	—	—	36 x 42
Stop Here on Red	R10-6	8B.12	24 x 36	24 x 36	—	—	36 x 48
Stop Here on Red	R10-6a	8B.12	24 x 30	24 x 30	—	—	36 x 42
Grade Crossing (Crossbuck)	R15-1	8B.03	48 x 9	48 x 9	—	—	—
Number of Tracks (plaque)	R15-2P	8B.03	27 x 18	27 x 18	—	—	—
Exempt (plaque)	R15-3P	8B.07	24 x 12	24 x 12	—	—	—
Light Rail Only Right Lane	R15-4a	8B.13	24 x 30	24 x 30	—	—	—
Light Rail Only Left Lane	R15-4b	8B.13	24 x 30	24 x 30	—	—	—
Light Rail Only Center Lane	R15-4c	8B.13	24 x 30	24 x 30	—	—	—
Light Rail Do Not Pass	R15-5	8B.14	24 x 30	24 x 30	—	—	—
Do Not Pass Stopped Train	R15-5a	8B.14	24 x 30	24 x 30	—	—	—
No Motor Vehicles On Tracks Symbol	R15-6	8B.15	24 x 24	24 x 24	—	—	—
Do Not Drive On Tracks	R15-6a	8B.15	24 x 30	24 x 30	—	—	—
Light Rail Divided Highway Symbol	R15-7	8B.16	24 x 24	24 x 24	—	—	—
Light Rail Divided Highway Symbol (T-Intersection)	R15-7a	8B.16	24 x 24	24 x 24	—	—	—
Look	R15-8	8B.17	36 x 18	36 x 18	—	—	—
Grade Crossing Advance Warning	W10-1	8B.06	36 Dia.	36 Dia.	48 Dia.	—	48 Dia.
Exempt (plaque)	W10-1aP	8B.07	24 x 12	24 x 12	—	—	—
Grade Crossing and Intersection Advance Warning	W10-2,3,4	8B.06	36 x 36	36 x 36	48 x 48	—	48 x 48
Low Ground Clearance	W10-5	8B.23	36 x 36	36 x 36	48 x 48	—	48 x 48
Low Ground Clearance (plaque)	W10-5P	8B.23	30 x 24	30 x 24	—	—	—
Light Rail Activated Blank-Out Symbol	W10-7	8B.19	24 x 24	24 x 24	—	—	—
Trains May Exceed 80 MPH	W10-8	8B.20	36 x 36	36 x 36	48 x 48	—	48 x 48
No Train Horn	W10-9	8B.21	36 x 36	36 x 36	48 x 48	—	48 x 48
No Train Horn (plaque)	W10-9P	8B.21	30 x 24	30 x 24	—	—	—
Storage Space Symbol	W10-11	8B.24	36 x 36	36 x 36	48 x 48	—	48 x 48
Storage Space XX Feet Between Tracks & Highway	W10-11a	8B.24	30 x 36	30 x 36	—	—	—
Storage Space XX Feet Between Highway & Tracks Behind You	W10-11b	8B.24	30 x 36	30 x 36	—	—	—
Skewed Crossing	W10-12	8B.25	36 x 36	36 x 36	48 x 48	—	48 x 48
No Gates or Lights (plaque)	W10-13P	8B.22	30 x 24	30 x 24	—	—	—
Next Crossing (plaque)	W10-14P	8B.23	30 x 24	30 x 24	—	—	—
Use Next Crossing (plaque)	W10-14aP	8B.23	30 x 24	30 x 24	—	—	—
Rough Crossing (plaque)	W10-15P	8B.23	30 x 24	30 x 24	—	—	36 x 30

Notes:
1. Larger signs may be used when appropriate
2. Dimensions in inches are shown as width x height
3. Table 9B-1 shows the minimum sizes that may be used for grade crossing signs and plaques that face shared-use paths and pedestrian facilities

Figure 8B-1. Regulatory Signs and Plaques for Grade Crossings

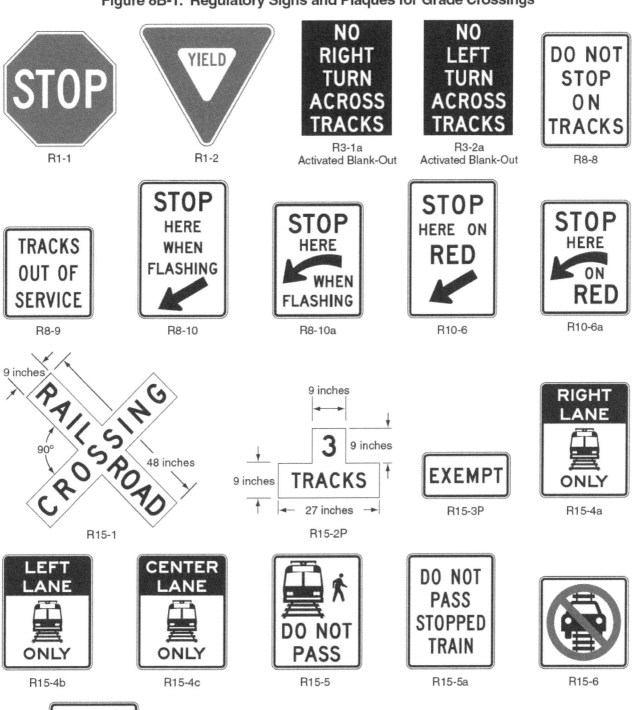

Figure 8B-2. Crossbuck Assembly with a YIELD or STOP Sign on the Crossbuck Sign Support

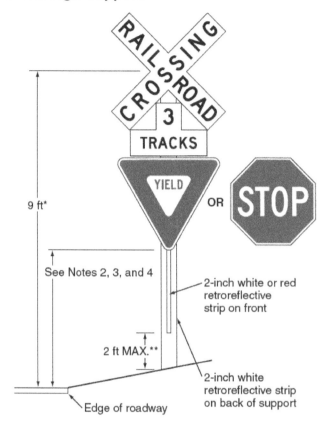

*Height may be varied as required by local conditions and may be increased to accommodate signs mounted below the Crossbuck sign

**Measured to the ground level at the base of the support

Notes:
1. YIELD or STOP signs are used only at passive crossings. A STOP sign is used only if an engineering study determines that it is appropriate for that particular approach.
2. Mounting height shall be at least 4 feet for installations of YIELD or STOP signs on existing Crossbuck sign supports.
3. Mounting height shall be at least 7 feet for new installations in areas with pedestrian movements or parking.

---

10 *Where unusual conditions make variations in location and lateral offset appropriate, engineering judgment should be used to provide the best practical combination of view and safety clearances.*

Section 8B.04 Crossbuck Assemblies with YIELD or STOP Signs at Passive Grade Crossings

Standard:

01 A grade crossing Crossbuck Assembly shall consist of a Crossbuck (R15-1) sign, and a Number of Tracks (R15-2P) plaque if two or more tracks are present, that complies with the provisions of Section 8B.03, and either a YIELD (R1-2) or STOP (R1-1) sign installed on the same support, except as provided in Paragraph 8. If used at a passive grade crossing, a YIELD or STOP sign shall be installed in compliance with the provisions of Part 2, Section 2B.10, and Figures 8B-2 and 8B-3.

02 At all public highway-rail grade crossings that are not equipped with the active traffic control systems that are described in Chapter 8C, except crossings where road users are directed by an authorized person on the ground to not enter the crossing at all times that an approaching train is about to occupy the crossing, a Crossbuck Assembly shall be installed on the right-hand side of the highway on each approach to the highway-rail grade crossing.

03 If a Crossbuck sign is used on a highway approach to a public highway-LRT grade crossing that is not equipped with the active traffic control systems that are described in Chapter 8C, a Crossbuck Assembly shall be installed on the right-hand side of the highway on each approach to the highway-LRT grade crossing.

Figure 8B-3. Crossbuck Assembly with a YIELD or STOP Sign on a Separate Sign Support (Sheet 1 of 2)

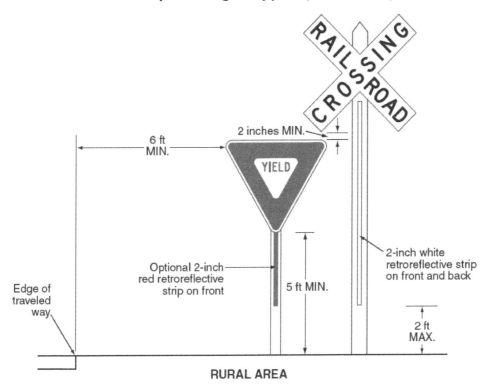

RURAL AREA

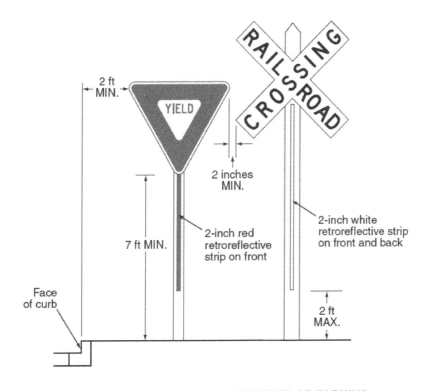

AREA WITH PEDESTRIAN MOVEMENTS OR PARKING

Notes:
1. YIELD signs are used only at passive crossings.
2. Place the face of the signs in the same plane and place the YIELD sign closest to the traveled way. Provide a 2-inch minimum separation between the edge of the Crossbuck sign and the edge of the YIELD sign.

Figure 8B-3. Crossbuck Assembly with a YIELD or STOP Sign on a Separate Sign Support (Sheet 2 of 2)

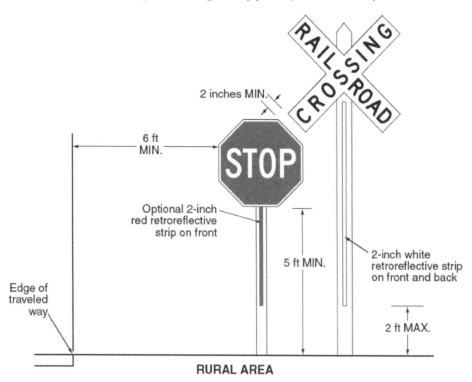

RURAL AREA

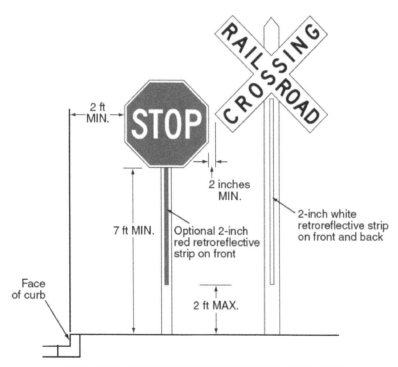

AREA WITH PEDESTRIAN MOVEMENTS OR PARKING

Notes:
1. STOP signs are used only at passive crossings and only if an engineering study determines that it is appropriate for that particular approach.
2. Place the face of the signs in the same plane and place the STOP sign closest to the traveled way. Provide a 2-inch minimum separation between the edge of the Crossbuck sign and the edge of the STOP sign.

04　Where restricted sight distance or unfavorable highway geometry exists on an approach to a grade crossing that has a Crossbuck Assembly, or where there is a one-way multi-lane approach, an additional Crossbuck Assembly shall be installed on the left-hand side of the highway.

05　A YIELD sign shall be the default traffic control device for Crossbuck Assemblies on all highway approaches to passive grade crossings unless an engineering study performed by the regulatory agency or highway authority having jurisdiction over the roadway approach determines that a STOP sign is appropriate.

Guidance:

06　*The use of STOP signs at passive grade crossings should be limited to unusual conditions where requiring all highway vehicles to make a full stop is deemed essential by an engineering study. Among the factors that should be considered in the engineering study are the line of sight to approaching rail traffic (giving due consideration to seasonal crops or vegetation beyond both the highway and railroad or LRT rights-of-ways), the number of tracks, the speeds of trains or LRT equipment and highway vehicles, and the crash history at the grade crossing.*

Support:

07　Sections 8A.02 and 8A.03 contain information regarding the responsibilities of the highway agency and the railroad company or LRT agency regarding the selection, design, and operation of traffic control devices placed at grade crossings.

Option:

08　If a YIELD or STOP sign is installed for a Crossbuck Assembly at a grade crossing, it may be installed on the same support as the Crossbuck sign or it may be installed on a separate support at a point where the highway vehicle is to stop, or as near to that point as practical, but in either case, the YIELD or STOP sign is considered to be a part of the Crossbuck Assembly.

Standard:

09　**If a YIELD or STOP sign is installed on an existing Crossbuck sign support, the minimum height, measured vertically from the bottom of the YIELD or STOP sign to the top of the curb, or in the absence of curb, measured vertically from the bottom of the YIELD or STOP sign to the elevation of the near edge of the traveled way, shall be 4 feet (see Figure 8B-2).**

10　**If a Crossbuck Assembly is installed on a new sign support (see Figure 8B-2) or if the YIELD or STOP sign is installed on a separate support (see Figure 8B-3), the minimum height, measured vertically from the bottom of the YIELD or STOP sign to the top of the curb, or in the absence of curb, measured vertically from the bottom of the YIELD or STOP sign to the elevation of the near edge of the traveled way, shall be 7 feet if the Crossbuck Assembly is installed in an area where parking or pedestrian movements are likely to occur.**

Guidance:

11　*If a YIELD or STOP sign is installed for a Crossbuck Assembly at a grade crossing on a separate support than the Crossbuck sign (see Figure 8B-3), the YIELD or STOP sign should be placed at a point where the highway vehicle is to stop, or as near to that point as practical, but no closer than 15 feet measured perpendicular from the nearest rail.*

Support:

12　The meaning of a Crossbuck Assembly that includes a YIELD sign is that a road user approaching the grade crossing needs to be prepared to decelerate, and when necessary, yield the right-of-way to any rail traffic that might be occupying the crossing or might be approaching and in such close proximity to the crossing that it would be unsafe for the road user to cross.

13　Certain commercial motor vehicles and school buses are required to stop at all grade crossings in accordance with 49 CFR 392.10 even if a YIELD sign (or just a Crossbuck sign) is posted.

14　The meaning of a Crossbuck Assembly that includes a STOP sign is that a road user approaching the grade crossing must come to a full and complete stop not less than 15 feet short of the nearest rail, and remain stopped while the road user determines if there is rail traffic either occupying the crossing or approaching and in such close proximity to the crossing that the road user must yield the right-of-way to rail traffic. The road user is permitted to proceed when it is safe to cross.

Standard:

15　**A vertical strip of retroreflective white material, not less than 2 inches in width, shall be used on each Crossbuck support at passive grade crossings for the full length of the back of the support from the Crossbuck sign or Number of Tracks plaque to within 2 feet above the ground, except as provided in Paragraph 16.**

Option:

16 The vertical strip of retroreflective material may be omitted from the back sides of Crossbuck sign supports installed on one-way streets.

17 If a YIELD or STOP sign is installed on the same support as the Crossbuck sign, a vertical strip of red (see Section 2A.21) or white retroreflective material that is at least 2 inches wide may be used on the front of the support from the YIELD or STOP sign to within 2 feet above the ground.

Standard:

18 **If a Crossbuck sign support at a passive grade crossing does not include a YIELD or STOP sign (either because the YIELD or STOP sign is placed on a separate support or because a YIELD or STOP sign is not present on the approach), a vertical strip of retroreflective white material, not less than 2 inches in width, shall be used for the full length of the front of the support from the Crossbuck sign or Number of Tracks plaque to within 2 feet above the ground.**

19 **At all grade crossings where YIELD or STOP signs are installed, Yield Ahead (W3-2) or Stop Ahead (W3-1) signs shall also be installed if the criteria for their installation in Section 2C.36 is met.**

Support:

20 Section 8B.28 contains provisions regarding the use of stop lines or yield lines at grade crossings.

Section 8B.05 Use of STOP (R1-1) or YIELD (R1-2) Signs without Crossbuck Signs at Highway-LRT Grade Crossings

Standard:

01 **For all highway-LRT grade crossings where only STOP (R1-1) or YIELD (R1-2) signs are installed, the placement shall comply with the requirements of Section 2B.10. Stop Ahead (W3-1) or Yield Ahead (W3-2) Advance Warning signs (see Figure 2C-6) shall also be installed if the criteria for their installation given in Section 2C.36 is met.**

Guidance:

02 *The use of only STOP or YIELD signs for road users at highway-LRT grade crossings should be limited to those crossings where the need and feasibility is established by an engineering study. Such crossings should have all of the following characteristics:*

   A. The crossing roadways should be secondary in character (such as a minor street with one lane in each direction, an alley, or a driveway) with low traffic volumes and low speed limits. The specific thresholds of traffic volumes and speed limits should be determined by the local agencies.
   B. LRT speeds do not exceed 25 mph.
   C. The line of sight for an approaching LRT operator is adequate from a sufficient distance such that the operator can sound an audible signal and bring the LRT equipment to a stop before arriving at the crossing.
   D. The road user has sufficient sight distance at the stop line to permit the vehicle to cross the tracks before the arrival of the LRT equipment.
   E. If at an intersection of two roadways, the intersection does not meet the warrants for a traffic control signal as provided in Chapter 4C.
   F. The LRT tracks are located such that highway vehicles are not likely to stop on the tracks while waiting to enter a cross street or highway.

Section 8B.06 Grade Crossing Advance Warning Signs (W10 Series)

Standard:

01 **A Highway-Rail Grade Crossing Advance Warning (W10-1) sign (see Figure 8B-4) shall be used on each highway in advance of every highway-rail grade crossing, and every highway-LRT grade crossing in semi-exclusive alignments, except in the following circumstances:**

   A. On an approach to a grade crossing from a T-intersection with a parallel highway if the distance from the edge of the track to the edge of the parallel roadway is less than 100 feet and W10-3 signs are used on both approaches of the parallel highway;
   B. On low-volume, low-speed highways crossing minor spurs or other tracks that are infrequently used and road users are directed by an authorized person on the ground to not enter the crossing at all times that approaching rail traffic is about to occupy the crossing;
   C. In business or commercial areas where active grade crossing traffic control devices are in use; or
   D. Where physical conditions do not permit even a partially effective display of the sign.

02 **The placement of the Grade Crossing Advance Warning sign shall be in accordance with Section 2C.05 and Table 2C-4.**

Figure 8B-4. Warning Signs and Plaques for Grade Crossings

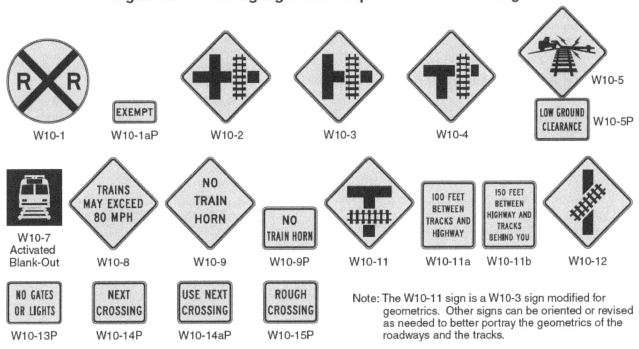

03 A Yield Ahead (W3-2) or Stop Ahead (W3-1) Advance Warning sign (see Figure 2C-6) shall also be installed if the criteria for their installation given in Section 2C.36 is met. If a Yield Ahead or Stop Ahead sign is installed on the approach to the crossing, the W10-1 sign shall be installed upstream from the Yield Ahead or Stop Ahead sign. The Yield Ahead or Stop Ahead sign shall be located in accordance with Table 2C-4. The minimum distance between the signs shall be in accordance with Section 2C.05 and Table 2C-4.

Option:

04 On divided highways and one-way streets, an additional W10-1 sign may be installed on the left-hand side of the roadway.

Standard:

05 **If the distance between the tracks and a parallel highway, from the edge of the tracks to the edge of the parallel roadway, is less than 100 feet, W10-2, W10-3, or W10-4 signs (see Figure 8B-4) shall be installed on each approach of the parallel highway to warn road users making a turn that they will encounter a grade crossing soon after making a turn, and a W10-1 sign for the approach to the tracks shall not be required to be between the tracks and the parallel highway.**

06 **If the W10-2, W10-3, or W10-4 signs are used, sign placement in accordance with the guidelines for Intersection Warning signs in Table 2C-4 using the speed of through traffic shall be measured from the highway intersection.**

Guidance:

07 *If the distance between the tracks and the parallel highway, from the edge of the tracks to the edge of the parallel roadway, is 100 feet or more, a W10-1 sign should be installed in advance of the grade crossing, and the W10-2, W10-3, or W10-4 signs should not be used on the parallel highway.*

Section 8B.07 EXEMPT Highway-Rail Grade Crossing Plaques (R15-3P, W10-1aP)

Option:

01 When authorized by law or regulation, a supplemental EXEMPT (R15-3P) plaque (see Figure 8B-1) with a white background may be used below the Crossbuck sign or Number of Tracks plaque, if present, at the grade crossing, and a supplemental EXEMPT (W10-1aP) plaque (see Figure 8B-4) with a yellow background may be used below the Grade Crossing Advance Warning (W10 series) sign.

02 Where neither the Crossbuck sign nor the advance warning signs exist for a particular highway-LRT grade crossing, an EXEMPT (R15-3P) plaque with a white background may be placed on its own post on the near right-hand side of the approach to the crossing.

Support:

03 These supplemental plaques inform drivers of highway vehicles carrying passengers for hire, school buses carrying students, or highway vehicles carrying hazardous materials that a stop is not required at certain designated grade crossings, except when rail traffic is approaching or occupying the grade crossing, or the driver's view is blocked.

Section 8B.08 Turn Restrictions During Preemption

Guidance:

01 *At a signalized intersection that is located within 200 feet of a highway-rail grade crossing, measured from the edge of the track to the edge of the roadway, where the intersection traffic control signals are preempted by the approach of a train, all existing turning movements toward the highway-rail grade crossing should be prohibited during the signal preemption sequences.*

Option:

02 A blank-out or changeable message sign and/or appropriate highway traffic signal indication or other similar type sign may be used to prohibit turning movements toward the highway-rail grade crossing during preemption. The R3-1a and R3-2a signs shown in Figure 8B-1 may be used for this purpose.

Support:

03 LRT operations can include the use of activated blank-out sign technology for turn prohibition signs. The signs are typically used on roads paralleling a semi-exclusive or mixed-use LRT alignment where road users might turn across the LRT tracks. A blank-out sign displays its message only when activated. When not activated, the sign face is blank.

Guidance:

04 *An LRT-activated blank-out turn prohibition (R3-1a or R3-2a) sign should be used where an intersection adjacent to a highway-LRT crossing is controlled by STOP signs, or is controlled by traffic control signals with permissive turn movements for road users crossing the tracks.*

Option:

05 An LRT-activated blank-out turn prohibition (R3-1a or R3-2a) sign may be used for turning movements that cross the tracks.

06 As an alternative to LRT-activated blank-out turn prohibition signs at intersections with traffic control signals, exclusive traffic control signal phases such that all movements that cross the tracks have a steady red indication may be used in combination with No Turn on Red (R10-11, R10-11a, or R10-11b) signs (see Section 2B.53).

Standard:

07 **Turn prohibition signs that are associated with preemption shall be visible or activated only when the grade crossing restriction is in effect.**

Section 8B.09 DO NOT STOP ON TRACKS Sign (R8-8)

Guidance:

01 A DO NOT STOP ON TRACKS (R8-8) sign (see Figure 8B-1) should be installed whenever an engineering study determines that the potential for highway vehicles stopping on the tracks at a grade crossing is significant. Placement of the R8-8 sign should be determined as part of the engineering study. The sign, if used, should be located on the right-hand side of the highway on either the near or far side of the grade crossing, depending upon which position provides better visibility to approaching drivers.

02 If a STOP or YIELD sign is installed at a location, including at a circular intersection, that is downstream from the grade crossing such that highway vehicle queues are likely to extend beyond the tracks, a DO NOT STOP ON TRACKS sign (R8-8) should be used.

Option:

03 DO NOT STOP ON TRACKS signs may be placed on both sides of the track.

04 On divided highways and one-way streets, a second DO NOT STOP ON TRACKS sign may be placed on the near or far left-hand side of the highway at the grade crossing to further improve visibility of the sign.

Section 8B.10 TRACKS OUT OF SERVICE Sign (R8-9)

Option:

01 The TRACKS OUT OF SERVICE (R8-9) sign (see Figure 8B-1) may be used at a grade crossing instead of a Crossbuck (R15-1) sign and a Number of Tracks (R15-2P) plaque or instead of a Crossbuck Assembly when

railroad or LRT tracks have been temporarily or permanently abandoned, but only until such time that the tracks are removed or covered.

Standard:

01 **When tracks are out of service, traffic control devices and gate arms shall be removed and the signal heads shall be removed or hooded or turned from view to clearly indicate that they are not in operation.**

03 **The R8-9 sign shall be removed when the tracks have been removed or covered or when the grade crossing is returned to service.**

Section 8B.11 STOP HERE WHEN FLASHING Signs (R8-10, R8-10a)

Option:

01 The STOP HERE WHEN FLASHING (R8-10, R8-10a) sign (see Figure 8B-1) may be used at a grade crossing to inform drivers of the location of the stop line or the point at which to stop when the flashing-light signals (see Section 8C.02) are activated.

Section 8B.12 STOP HERE ON RED Signs (R10-6, R10-6a)

Support:

01 The STOP HERE ON RED (R10-6, R10-6a) sign (see Figure 8B-1) defines and facilitates observance of stop lines at traffic control signals.

Option:

02 A STOP HERE ON RED sign may be used at locations where highway vehicles frequently violate the stop line or where it is not obvious to road users where to stop.

Guidance:

03 *If possible, stop lines should be placed at a point where the highway vehicle driver has adequate sight distance along the track.*

Section 8B.13 Light Rail Transit Only Lane Signs (R15-4 Series)

Support:

01 The Light Rail Transit Only Lane (R15-4 series) signs (see Figure 8B-1) are used for multi-lane operations, where road users might need additional guidance on lane use and/or restrictions.

Option:

02 Light Rail Transit Only Lane signs may be used on a roadway lane limited to only LRT use to indicate the restricted use of a lane in semi-exclusive and mixed alignments.

Guidance:

03 *If used, the R15-4a, R15-4b, and R15-4c signs should be installed on posts adjacent to the roadway containing the LRT tracks or overhead above the LRT only lane.*

Option:

04 If the trackway is paved, preferential lane markings (see Chapter 3D) may be installed but only in combination with Light Rail Transit Only Lane signs.

Support:

05 The trackway is the continuous way designated for LRT, including the entire dynamic envelope. Section 8B.29 contains more information regarding the dynamic envelope.

Section 8B.14 Do Not Pass Light Rail Transit Signs (R15-5, R15-5a)

Support:

01 A Do Not Pass Light Rail Transit (R15-5) sign (see Figure 8B-1) is used to indicate that motor vehicles are not allowed to pass LRT vehicles that are loading or unloading passengers where there is no raised platform or physical separation from the lanes upon which other motor vehicles are operating.

Option:

02 The R15-5 sign may be used in mixed-use alignments and may be mounted overhead where there are multiple lanes.

03 Instead of the R15-5 symbol sign, a regulatory sign with the word message DO NOT PASS STOPPED TRAIN (R15-5a) may be used (see Figure 8B-1).

Guidance:

04 *If used, the R15-5 sign should be located immediately before the LRT boarding area.*

Section 8B.15 No Motor Vehicles On Tracks Signs (R15-6, R15-6a)

Support:
01 The No Motor Vehicles On Tracks (R15-6) sign (see Figure 8B-1) is used where there are adjacent traffic lanes separated from the LRT lane by a curb or pavement markings.

Guidance:
02 *The DO NOT ENTER (R5-1) sign should be used where a road user could wrongly enter an LRT only street.*

Option:
03 A No Motor Vehicles On Tracks sign may be used to deter motor vehicles from driving on the trackway. It may be installed on a 3-foot flexible post between double tracks, on a post alongside the tracks, or overhead.

04 Instead of the R15-6 symbol sign, a regulatory sign with the word message DO NOT DRIVE ON TRACKS (R15-6a) may be used (see Figure 8B-1).

05 A reduced size of 12 x 12 inches may be used if the R15-6 sign is installed between double tracks.

Standard:
06 **The smallest size for the R15-6 sign shall be 12 x 12 inches.**

Section 8B.16 Divided Highway with Light Rail Transit Crossing Signs (R15-7 Series)

Option:
01 The Divided Highway with Light Rail Transit Crossing (R15-7) sign (see Figure 8B-1) may be used as a supplemental sign on the approach legs of a roadway that intersects with a divided highway where LRT equipment operates in the median. The sign may be placed beneath a STOP sign or mounted separately.

Guidance
02 *The number of tracks displayed on the R15-7 sign should be the same as the actual number of tracks.*

Standard:
03 **When the Divided Highway With Light Rail Transit Crossing sign is used at a four-legged intersection, the R15-7 sign shall be used. When used at a T-intersection, the R15-7a sign shall be used.**

Section 8B.17 LOOK Sign (R15-8)

Option:
01 At grade crossings, the LOOK (R15-8) sign (see Figure 8B-1) may be mounted as a supplemental plaque on the Crossbuck support, or on a separate post in the immediate vicinity of the grade crossing on the railroad or LRT right-of-way.

Guidance:
02 *A LOOK sign should not be mounted as a supplemental plaque on a Crossbuck Assembly that has a YIELD or STOP sign mounted on the same support as the Crossbuck.*

Section 8B.18 Emergency Notification Sign (I-13)

Guidance:
01 *Emergency Notification (I-13) signs (see Figure 8B-5) should be installed at all highway-rail grade crossings, and at all highway-LRT grade crossings on semi-exclusive alignments, to provide information to road users so that they can notify the railroad company or LRT agency about emergencies or malfunctioning traffic control devices.*

Standard:
02 **When Emergency Notification signs are used at a highway-rail grade crossing, they shall, at a minimum, include the USDOT grade crossing inventory number and the emergency contact telephone number.**

03 **When Emergency Notification signs are used at a highway-LRT grade crossing, they shall, at a minimum, include a unique crossing identifier and the emergency contact telephone number.**

04 **Emergency Notification Signs shall have a white legend and border on a blue background.**

05 **The Emergency Notification signs shall be positioned so as to not obstruct any traffic control devices or limit the view of rail traffic approaching the grade crossing.**

Figure 8B-5. Example of an Emergency Notification Sign

Guidance:

06 *Emergency Notification signs should be retroreflective.*

07 *Emergency Notification signs should be oriented so as to face highway vehicles stopped on or at the grade crossing or on the traveled way near the grade crossing.*

08 *At station crossings, Emergency Notification signs or information should be posted in a conspicuous location.*

09 *Emergency Notification signs mounted on Crossbuck Assemblies or signal masts should only be large enough to provide the necessary contact information. Use of larger signs that might obstruct the view of rail traffic or other highway vehicles should be avoided.*

Section 8B.19 Light Rail Transit Approaching-Activated Blank-Out Warning Sign (W10-7)

Support:

01 The Light Rail Transit Approaching-Activated Blank-Out (W10-7) warning sign (see Figure 8B-4) supplements the traffic control devices to warn road users crossing the tracks of approaching LRT equipment.

Option:

02 A Light Rail Transit Approaching-Activated Blank-Out warning sign may be used at signalized intersections near highway-LRT grade crossings or at crossings controlled by STOP signs or automatic gates.

Section 8B.20 TRAINS MAY EXCEED 80 MPH Sign (W10-8)

Guidance:

01 *Where trains are permitted to travel at speeds exceeding 80 mph, a TRAINS MAY EXCEED 80 MPH (W10-8) sign (see Figure 8B-4) should be installed facing road users approaching the highway-rail grade crossing.*

02 *If used, the TRAINS MAY EXCEED 80 MPH signs should be installed between the Grade Crossing Advance Warning (W10 series) sign (see Figure 8B-4) and the highway-rail grade crossing on all approaches to the highway-rail grade crossing. The locations should be determined based on specific site conditions.*

Section 8B.21 NO TRAIN HORN Sign or Plaque (W10-9, W10-9P)

Standard:

01 **Either a NO TRAIN HORN (W10-9) sign (see Figure 8B-4) or a NO TRAIN HORN (W10-9P) plaque shall be installed in each direction at each highway-rail grade crossing where a quiet zone has been established in compliance with 49 CFR Part 222. If a W10-9P plaque is used, it shall supplement and be mounted directly below the Grade Crossing Advance Warning (W10 series) sign (see Figure 8B-4).**

Section 8B.22 NO GATES OR LIGHTS Plaque (W10-13P)

Option:

01 The NO GATES OR LIGHTS (W10-13P) sign plaque (see Figure 8B-4) may be mounted below the Grade Crossing Advance Warning (W10 series) sign at grade crossings that are not equipped with automated signals.

Section 8B.23 Low Ground Clearance Grade Crossing Sign (W10-5)

Guidance:

01 *If the highway profile conditions are sufficiently abrupt to create a hang-up situation for long wheelbase vehicles or for trailers with low ground clearance, the Low Ground Clearance Grade Crossing (W10-5) sign (see Figure 8B-4) should be installed in advance of the grade crossing.*

Standard:

02 **Because this symbol might not be readily recognizable by the public, the Low Ground Clearance Grade Crossing (W10-5) warning sign shall be accompanied by an educational plaque, LOW GROUND CLEARANCE. The LOW GROUND CLEARANCE educational plaque shall remain in place for at least 3 years after the initial installation of the W10-5 sign (see Section 2A.12).**

Guidance:

03 *Auxiliary plaques such as AHEAD, NEXT CROSSING, or USE NEXT CROSSING (with appropriate arrows), or a supplemental distance plaque should be placed below the W10-5 sign at the nearest intersecting highway where a vehicle can detour or at a point on the highway wide enough to permit a U-turn.*

04 *If engineering judgment of roadway geometric and operating conditions confirms that highway vehicle speeds across the tracks should be below the posted speed limit, a W13-1P advisory speed plaque should be posted.*

Option:

05 If the grade crossing is rough, word message signs such as BUMP, DIP, or ROUGH CROSSING may be installed. A W13-1P advisory speed plaque may be installed below the word message sign in advance of rough crossings.

Support:

06 Information on ground clearance requirements at grade crossings is available in the "American Railway Engineering and Maintenance-of-Way Association's Engineering Manual," or the American Association of State Highway and Transportation Officials' "Policy on Geometric Design of Highways and Streets" (see Section 1A.11).

Section 8B.24 Storage Space Signs (W10-11, W10-11a, W10-11b)

Guidance:

01 *A Storage Space (W10-11) sign supplemented by a word message storage distance (W10-11a) sign (see Figure 8B-4) should be used where there is a highway intersection in close proximity to the grade crossing and an engineering study determines that adequate space is not available to store a design vehicle(s) between the highway intersection and the train or LRT equipment dynamic envelope.*

02 *The Storage Space (W10-11 and W10-11a) signs should be mounted in advance of the grade crossing at an appropriate location to advise drivers of the space available for highway vehicle storage between the highway intersection and the grade crossing.*

Option:

03 A Storage Space (W10-11b) sign (see Figure 8B-4) may be mounted beyond the grade crossing at the highway intersection under the STOP or YIELD sign or just prior to the signalized intersection to remind drivers of the storage space between the tracks and the highway intersection.

Section 8B.25 Skewed Crossing Sign (W10-12)

Option:

01 The Skewed Crossing (W10-12) sign (see Figure 8B-4) may be used at a skewed grade crossing to warn road users that the tracks are not perpendicular to the highway.

Guidance:

02 *If the Skewed Crossing sign is used, the symbol should show the direction of the crossing (near left to far right as shown in Figure 8B-4, or the mirror image if the track goes from far left to near right). If the Skewed Crossing sign is used where the angle of the crossing is significantly different than 45 degrees, the symbol should show the approximate angle of the crossing.*

Standard:

03 **The Skewed Crossing sign shall not be used as a replacement for the required Advance Warning (W10-1) sign. If used, the Skewed Crossing sign shall supplement the W10-1 sign and shall be mounted on a separate post.**

Section 8B.26 Light Rail Transit Station Sign (I-12)

Option:

01 The Light Rail Transit Station (I-12) sign (see Figure 2H-1) may be used to direct road users to an LRT station or boarding location. It may be supplemented by the name of the transit system and by arrows as provided in Section 2D.08.

Section 8B.27 Pavement Markings

Standard:

01 **All grade crossing pavement markings shall be retroreflectorized white. All other markings shall be in accordance with Part 3.**

02 **On paved roadways, pavement markings in advance of a grade crossing shall consist of an X, the letters RR, a no-passing zone marking (on two-lane, two-way highways with center line markings in compliance with Section 3B.01), and certain transverse lines as shown in Figures 8B-6 and 8B-7.**

03 **Identical markings shall be placed in each approach lane on all paved approaches to grade crossings where signals or automatic gates are located, and at all other grade crossings where the posted or statutory highway speed is 40 mph or greater.**

04 **Pavement markings shall not be required at grade crossings where the posted or statutory highway speed is less than 40 mph if an engineering study indicates that other installed devices provide suitable**

Figure 8B-6. Example of Placement of Warning Signs and Pavement Markings at Grade Crossings

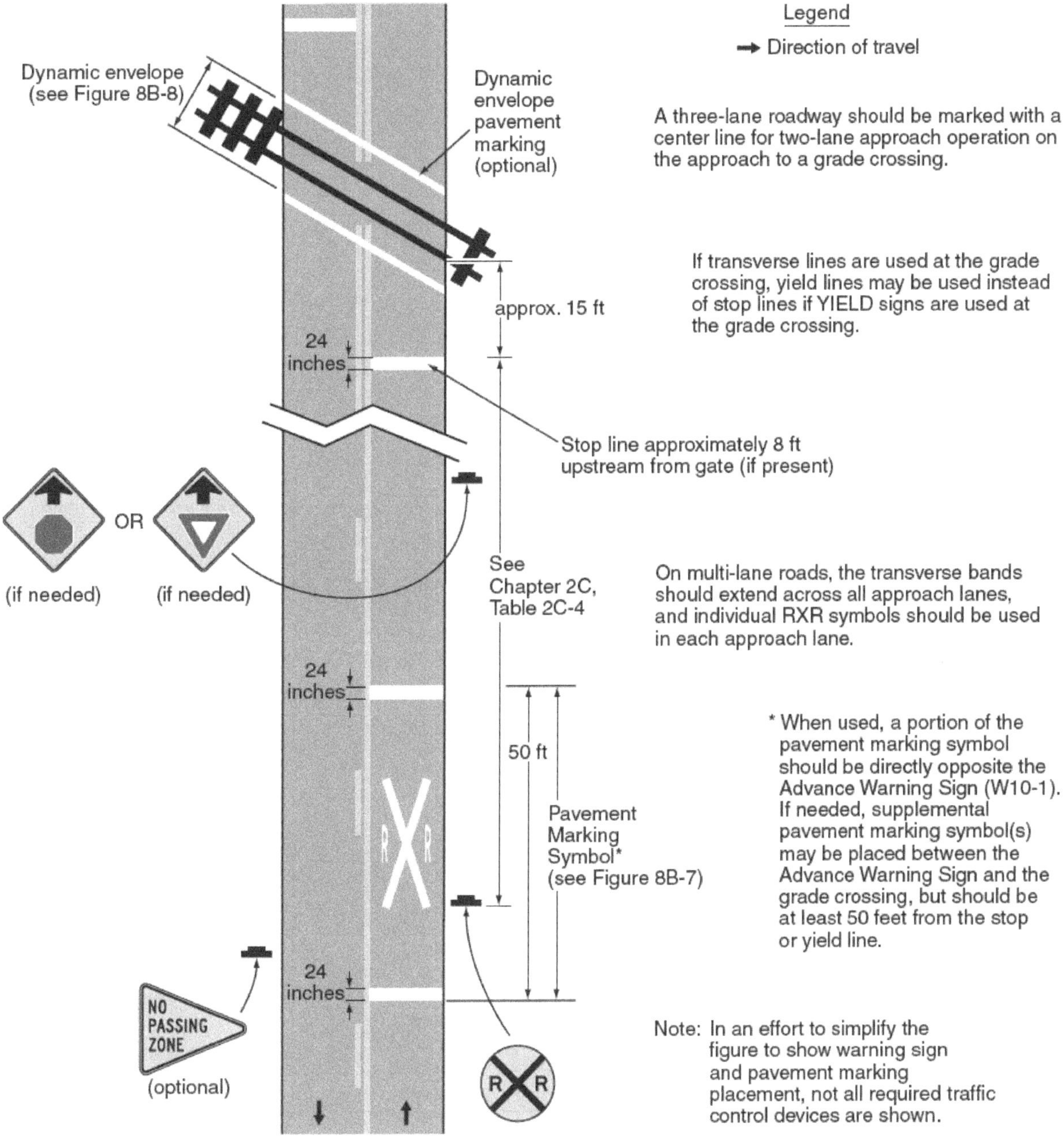

Legend
→ Direction of travel

Dynamic envelope (see Figure 8B-8)

Dynamic envelope pavement marking (optional)

A three-lane roadway should be marked with a center line for two-lane approach operation on the approach to a grade crossing.

approx. 15 ft

24 inches

If transverse lines are used at the grade crossing, yield lines may be used instead of stop lines if YIELD signs are used at the grade crossing.

Stop line approximately 8 ft upstream from gate (if present)

OR

(if needed)   (if needed)

See Chapter 2C, Table 2C-4

On multi-lane roads, the transverse bands should extend across all approach lanes, and individual RXR symbols should be used in each approach lane.

24 inches

50 ft

Pavement Marking Symbol* (see Figure 8B-7)

* When used, a portion of the pavement marking symbol should be directly opposite the Advance Warning Sign (W10-1). If needed, supplemental pavement marking symbol(s) may be placed between the Advance Warning Sign and the grade crossing, but should be at least 50 feet from the stop or yield line.

24 inches

NO PASSING ZONE
(optional)

Note: In an effort to simplify the figure to show warning sign and pavement marking placement, not all required traffic control devices are shown.

---

warning and control. **Pavement markings shall not be required at grade crossings in urban areas if an engineering study indicates that other installed devices provide suitable warning and control.**

Guidance:

05 *When pavement markings are used, a portion of the X symbol should be directly opposite the Grade Crossing Advance Warning sign. The X symbol and letters should be elongated to allow for the low angle at which they will be viewed.*

Option:

06 When justified by engineering judgment, supplemental pavement marking symbol(s) may be placed between the Grade Crossing Advance Warning sign and the grade crossing.

Figure 8B-7. Grade Crossing Pavement Markings

A - Grade crossing pavement marking symbol

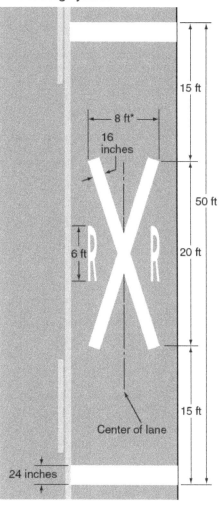

*Width may vary according to lane width

Note: Refer to Figure 8B-6 for placement

B - Grade crossing alternative (narrow) pavement marking symbol

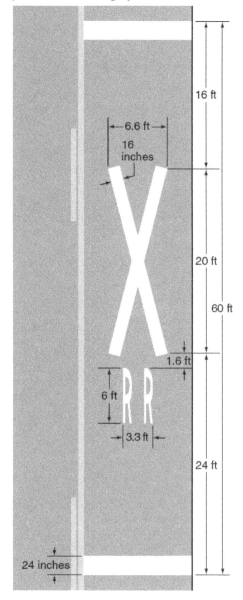

Section 8B.28 Stop and Yield Lines

Standard:

01　**On paved roadways at grade crossings that are equipped with active control devices such as flashing-light signals, gates, or traffic control signals, a stop line (see Section 3B.16) shall be installed to indicate the point behind which highway vehicles are or might be required to stop.**

Guidance:

02　*On paved roadway approaches to passive grade crossings where a STOP sign is installed in conjunction with the Crossbuck sign, a stop line should be installed to indicate the point behind which highway vehicles are required to stop or as near to that point as practical.*

03　*If a stop line is used, it should be a transverse line at a right angle to the traveled way and should be placed approximately 8 feet in advance of the gate (if present), but no closer than 15 feet in advance of the nearest rail.*

Option:
04 On paved roadway approaches to passive grade crossings where a YIELD sign is installed in conjunction with the Crossbuck sign, a yield line (see Section 3B.16) or a stop line may be installed to indicate the point behind which highway vehicles are required to yield or stop or as near to that point as practical.
Guidance:
05 *If a yield line is used, it should be a transverse line (see Figure 3B-16) at a right angle to the traveled way and should be placed no closer than 15 feet in advance of the nearest rail (see Figure 8B-7).*

Section 8B.29 Dynamic Envelope Markings

Support:
01 The dynamic envelope (see Figures 8B-8 and 8B-9) markings indicate the clearance required for the train or LRT equipment overhang resulting from any combination of loading, lateral motion, or suspension failure.

Option:
02 Dynamic envelope markings may be installed at all grade crossings, unless a Four-Quadrant Gate system (see Section 8C.06) is used.

Standard:
03 **If used, pavement markings for indicating the dynamic envelope shall comply with the provisions of Part 3 and shall be a 4-inch normal solid white line or contrasting pavement color and/or contrasting pavement texture.**

Guidance:
04 *If pavement markings are used to convey the dynamic envelope, they should be placed completely outside of the dynamic envelope. If used, dynamic envelope pavement markings should be placed on the highway 6 feet*

Figure 8B-8. Example of Dynamic Envelope Pavement Markings at Grade Crossings

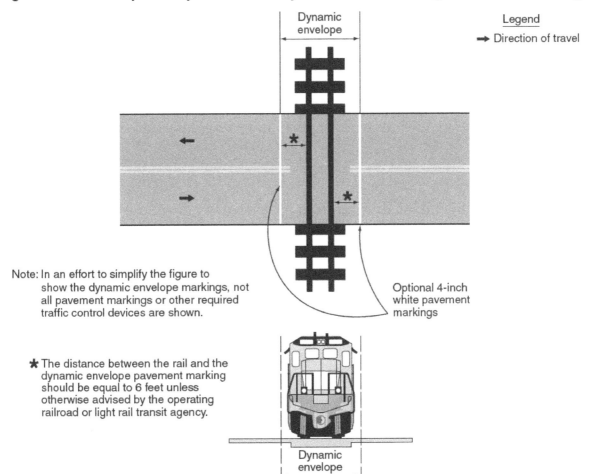

Figure 8B-9. Examples of Light Rail Transit Vehicle Dynamic Envelope Markings for Mixed-Use Alignments

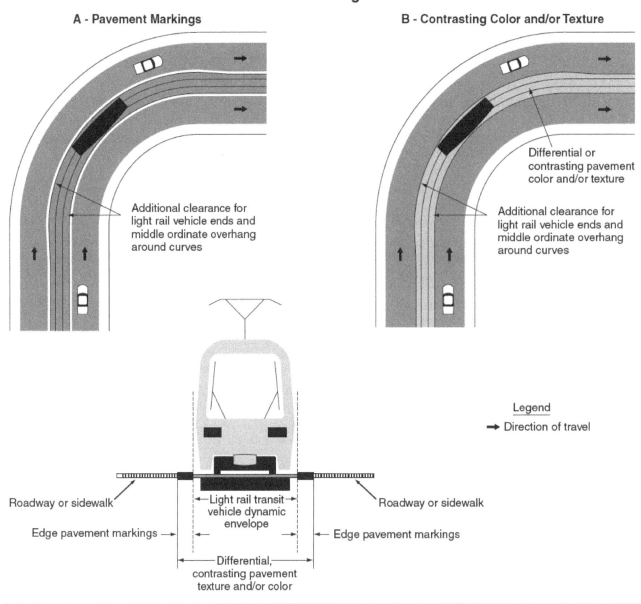

from and parallel to the nearest rail unless the operating railroad company or LRT agency advises otherwise. The pavement markings should extend across the roadway as shown in Figure 8B-8. The dynamic envelope pavement markings should not be placed perpendicular to the roadway at skewed grade crossings.

Option:

05   In semi-exclusive LRT alignments, the dynamic envelope markings may be along the LRT trackway between intersections where the trackway is immediately adjacent to travel lanes and no physical barrier is present.

06   In mixed-use LRT alignments, the dynamic envelope markings may be continuous between intersections (see Figure 8B-9).

07   In mixed-use LRT alignments, pavement markings for adjacent travel or parking lanes may be used instead of dynamic envelope markings if the lines are outside the dynamic envelope.

CHAPTER 8C. FLASHING-LIGHT SIGNALS, GATES, AND TRAFFIC CONTROL SIGNALS

Section 8C.01 Introduction

Support:

01 Active traffic control systems inform road users of the approach or presence of rail traffic at grade crossings. These systems include four-quadrant gate systems, automatic gates, flashing-light signals, traffic control signals, actuated blank-out and variable message signs, and other active traffic control devices.

02 A composite drawing (see Figure 8C-1) shows a post-mounted flashing-light signal (two light units mounted in a horizontal line), a flashing-light signal mounted on an overhead structure, and an automatic gate assembly.

Option:

03 Post-mounted and overhead flashing-light signals may be used separately or in combination with each other as determined by an engineering study. Also, flashing-light signals may be used without automatic gate assemblies, as determined by an engineering study.

Standard:

04 **The meaning of flashing-light signals and gates shall be as stated in the "Uniform Vehicle Code" (see Sections 11-701 and 11-703 of the UVC), which is available from the National Committee on Uniform Traffic Laws and Ordinances (see Page i for the address).**

05 **Location and clearance dimensions for flashing-light signals and gates shall be as shown in Figure 8C-1.**

06 **When there is a curb, a horizontal offset of at least 2 feet shall be provided from the face of the vertical curb to the closest part of the signal or gate arm in its upright position. When a cantilevered-arm flashing-light signal is used, the vertical clearance shall be at least 17 feet above the crown of the highway to the lowest point of the signal unit.**

07 **Where there is a shoulder, but no curb, a horizontal offset of at least 2 feet from the edge of a paved or surfaced shoulder shall be provided, with an offset of at least 6 feet from the edge of the traveled way.**

08 **Where there is no curb or shoulder, the minimum horizontal offset shall be 6 feet from the edge of the traveled way.**

Guidance:

09 *Equipment housings (controller cabinets) should have a lateral offset of at least 30 feet from the edge of the highway, and where railroad or LRT property and conditions allow, at least 25 feet from the nearest rail.*

10 *If a pedestrian route is provided, sufficient clearance from supports, posts, and gate mechanisms should be maintained for pedestrian travel.*

11 *When determined by an engineering study, a lateral escape route to the right of the highway in advance of the grade crossing traffic control devices should be kept free of guardrail or other ground obstructions. Where guardrail is not deemed necessary or appropriate, barriers should not be used for protecting signal supports.*

12 *The same lateral offset and roadside safety features should apply to flashing-light signal and automatic gate locations on both the right-hand and left-hand sides of the roadway.*

Option:

13 In industrial or other areas involving only low-speed highway traffic or where signals are vulnerable to damage by turning truck traffic, guardrail may be installed to provide protection for the signal assembly.

Guidance:

14 *Where both traffic control signals and flashing-light signals (with or without automatic gates) are in operation at the same highway-LRT grade crossing, the operation of the devices should be coordinated to avoid any display of conflicting signal indications.*

Support:

15 LRT typically operates through grade crossings in semi-exclusive and mixed-use alignments at speeds between 10 and 65 mph.

16 When LRT speed is cited in this Part, it refers to the maximum speed at which LRT equipment is permitted to traverse a particular grade crossing.

Section 8C.02 Flashing-Light Signals

Support:

01 Section 8C.03 contains additional information regarding flashing-light signals at highway-LRT grade crossings in semi-exclusive and mixed-use alignments.

Figure 8C-1. Composite Drawing of Active Traffic Control Devices for Grade Crossings Showing Clearances

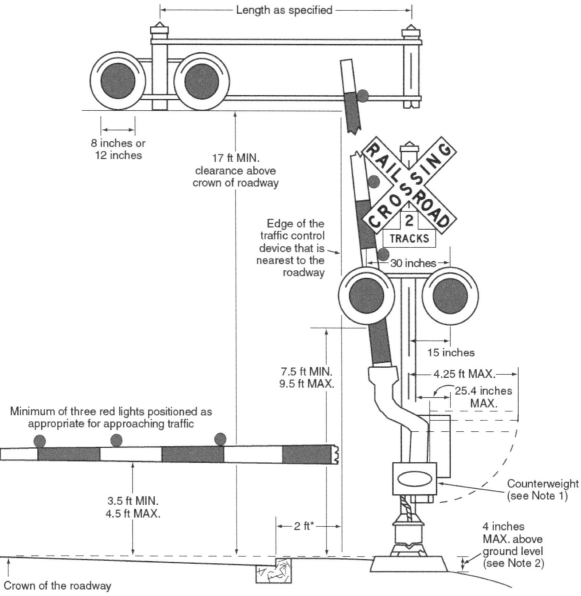

*For locating this reference line on an approach that does not have a curb, see Section 8C.01.

Notes:
1. Where gates are located in the median, additional median width may be required to provide the minimum clearance for the counterweight supports.
2. The top of the signal foundation should be no more than 4 inches above the surface of the ground and should be at the same elevation as the crown of the roadway. Where site conditions would not allow this to be achieved, the shoulder side slope should be re-graded or the height of the signal post should be adjusted to meet the 17-foot vertical clearance requirement.

Standard:

02 **If used, the flashing-light signal assembly (shown in Figure 8C-1) on the side of the highway shall include a standard Crossbuck (R15-1) sign, and where there is more than one track, a supplemental Number of Tracks (R15-2P) plaque, all of which indicate to motorists, bicyclists, and pedestrians the location of a grade crossing.**

Option:

03 At highway-rail grade crossings, bells or other audible warning devices may be included in the assembly and may be operated in conjunction with the flashing lights to provide additional warning for pedestrians, bicyclists, and/or other non-motorized road users.

Standard:

04 **When indicating the approach or presence of rail traffic, the flashing-light signal shall display toward approaching highway traffic two red lights mounted in a horizontal line flashing alternately.**

05 **If used, flashing-light signals shall be placed to the right of approaching highway traffic on all highway approaches to a grade crossing. They shall be located laterally with respect to the highway in compliance with Figure 8C-1 except where such location would adversely affect signal visibility.**

06 **If used at a grade crossing with highway traffic in both directions, back-to-back pairs of lights shall be placed on each side of the tracks. On multi-lane one-way streets and divided highways, flashing-light signals shall be placed on the approach side of the grade crossing on both sides of the roadway or shall be placed above the highway.**

07 **Each red signal unit in the flashing-light signal shall flash alternately. The number of flashes per minute for each lamp shall be 35 minimum and 65 maximum. Each lamp shall be illuminated approximately the same length of time. Total time of illumination of each pair of lamps shall be the entire operating time. Flashing-light units shall use either 8-inch or 12-inch nominal diameter lenses.**

Guidance:

08 *In choosing between the 8-inch or 12-inch nominal diameter lenses for use in grade crossing flashing-light signals, consideration should be given to the principles stated in Section 4D.07.*

Standard:

09 **Grade crossing flashing-light signals shall operate at a low voltage using storage batteries either as a primary or stand-by source of electrical energy. Provision shall be made to provide a source of energy for charging batteries.**

Option:

10 Additional pairs of flashing-light units may be mounted on the same supporting post and directed toward vehicular traffic approaching the grade crossing from other than the principal highway route, such as where there are approaching routes on highways closely adjacent to and parallel to the track(s).

Standard:

11 **References to lenses in this Section shall not be used to limit flashing-light signal optical units to incandescent lamps within optical assemblies that include lenses.**

Support:

12 Research has resulted in flashing-light signal optical units that are not lenses, such as, but not limited to, light emitting diode (LED) flashing-light signal modules.

Option:

13 Flashing-light signals may be installed on overhead structures or cantilevered supports as shown in Figure 8C-1 where needed for additional emphasis, or for better visibility to approaching traffic, particularly on multi-lane approaches or highways with profile restrictions.

14 If it is determined by an engineering study that one set of flashing lights on the cantilever arm is not sufficiently visible to road users, one or more additional sets of flashing lights may be mounted on the supporting post and/or on the cantilever arm.

Standard:

15 **Breakaway or frangible bases shall not be used for overhead structures or cantilevered supports.**

16 **Except as otherwise provided in Paragraphs 13 through 15, flashing-light signals mounted overhead shall comply with the applicable provisions of this Section.**

Section 8C.03 Flashing-Light Signals at Highway-LRT Grade Crossings

Support:

01 Section 8C.02 contains additional provisions regarding the design and operation of flashing-light signals, including those installed at highway-LRT grade crossings.

Standard:

02 **Highway-LRT grade crossings in semi-exclusive alignments shall be equipped with flashing-light signals where LRT speeds exceed 35 mph. Flashing-light signals shall be clearly visible to motorists, pedestrians, and bicyclists.**

03 **If flashing-light signals are in operation at a highway-LRT crossing that is used by pedestrians, bicyclists, and/or other non-motorized road users, an audible device such as a bell shall also be provided and shall be operated in conjunction with the flashing-light signals.**

Guidance:

04 *Where the crossing is at a location other than an intersection and LRT speeds exceed 25 mph, flashing-light signals should be installed.*

Option:

05 Traffic control signals may be used instead of flashing-light signals at highway-LRT grade crossings within highway-highway intersections where LRT speeds do not exceed 35 mph. Traffic control signals or flashing-light signals may be used where the crossing is at a location other than an intersection, where LRT speeds do not exceed 25 mph, and when the roadway is a low-volume street where prevailing speeds do not exceed 25 mph.

Section 8C.04 Automatic Gates

Support:

01 An automatic gate is a traffic control device used in conjunction with flashing-light signals.

Standard:

02 **The automatic gate (see Figure 8C-1) shall consist of a drive mechanism and a fully retroreflectorized red- and white-striped gate arm with lights. When in the down position, the gate arm shall extend across the approaching lanes of highway traffic.**

03 **In the normal sequence of operation, unless constant warning time detection or other advanced system requires otherwise, the flashing-light signals and the lights on the gate arm (in its normal upright position) shall be activated immediately upon detection of approaching rail traffic. The gate arm shall start its downward motion not less than 3 seconds after the flashing-light signals start to operate, shall reach its horizontal position at least 5 seconds before the arrival of the rail traffic, and shall remain in the down position as long as the rail traffic occupies the grade crossing.**

04 **When the rail traffic clears the grade crossing, and if no other rail traffic is detected, the gate arm shall ascend to its upright position, following which the flashing-light signals and the lights on the gate arm shall cease operation.**

05 **Gate arms shall be fully retroreflectorized on both sides and shall have vertical stripes alternately red and white at 16-inch intervals measured horizontally.**

Support:

06 It is acceptable to replace a damaged gate with a gate having vertical stripes even if the other existing gates at the same grade crossing have diagonal stripes; however, it is also acceptable to replace a damaged gate with a gate having diagonal stripes if the other existing gates at the same grade crossing have diagonal stripes in order to maintain consistency per the provisions of Paragraph 24 of the Introduction.

Standard:

07 **Gate arms shall have at least three red lights as provided in Figure 8C-1.**

08 **When activated, the gate arm light nearest the tip shall be illuminated continuously and the other lights shall flash alternately in unison with the flashing-light signals.**

09 **The entrance gate arm mechanism shall be designed to fail safe in the down position.**

Guidance:

10 *The gate arm should ascend to its upright position in 12 seconds or less.*

11 *In its normal upright position, when no rail traffic is approaching or occupying the grade crossing, the gate arm should be either vertical or nearly so (see Figure 8C-1).*

12 *In the design of individual installations, consideration should be given to timing the operation of the gate arm to accommodate large and/or slow-moving highway vehicles.*

¹³ *The gates should cover the approaching highway to block all highway vehicles from being driven around the gate without crossing the center line.*

Option:

¹⁴ The effectiveness of gates may be enhanced by the use of channelizing devices or raised median islands to discourage driving around lowered automatic gates.

¹⁵ Where gates are located in the median, additional median width may be required to provide the minimum clearance for the counterweight supports.

¹⁶ Automatic gates may be supplemented by cantilevered flashing-light signals (see Figure 8C-1) where there is a need for additional emphasis or better visibility.

Section 8C.05 Use of Automatic Gates at LRT Grade Crossings

Guidance:

⁰¹ *Highway-LRT grade crossings in semi-exclusive alignments should be equipped with automatic gates and flashing-light signals (see Sections 8C.02 and 8C.03) where LRT speeds exceed 35 mph.*

Option:

⁰² Where a highway-LRT grade crossing is at a location other than an intersection, where LRT speeds exceed 25 mph, automatic gates and flashing-light signals may be installed.

⁰³ Traffic control signals may be used instead of automatic gates at highway-LRT grade crossings within highway-highway intersections where LRT speeds do not exceed 35 mph. Traffic control signals or flashing-light signals without automatic gates may be used where the crossing is at a location other than an intersection and where LRT speeds do not exceed 25 mph and the roadway is a low-volume street where prevailing speeds do not exceed 25 mph.

Section 8C.06 Four-Quadrant Gate Systems

Option:

⁰¹ Four-Quadrant Gate systems may be installed to improve safety at grade crossings based on an engineering study when less restrictive measures, such as automatic gates and median islands, are not effective.

Standard:

⁰² **A Four-Quadrant Gate system shall consist of entrance and exit gates that control and block road users on all lanes entering and exiting the grade crossing.**

⁰³ **The Four-Quadrant Gate system shall use a series of drive mechanisms and fully retroreflectorized red- and white-striped gate arms with lights, and when in the down position the gate arms extend individually across the entrance and exit lanes of the roadway as shown in Figure 8C-2. Standards contained in Sections 8C.01 through 8C.03 for flashing-light signals shall be followed for signal specifications, location, and clearance distances.**

⁰⁴ **In the normal sequence of operation, unless constant warning time detection or other advanced system requires otherwise, the flashing-light signals and the lights on the gate arms (in their normal upright positions) shall be activated immediately upon the detection of approaching rail traffic. The gate arms for the entrance lanes of traffic shall start their downward motion not less than 3 seconds after the flashing-light signals start to operate and shall reach their horizontal position at least 5 seconds before the arrival of the rail traffic. Exit gate arm activation and downward motion shall be based on detection or timing requirements established by an engineering study of the individual site. The gate arms shall remain in the down position as long as the rail traffic occupies the grade crossing.**

⁰⁵ **When the rail traffic clears the grade crossing, and if no other rail traffic is detected, the gate arms shall ascend to their upright positions, following which the flashing-light signals and the lights on the gate arms shall cease operation.**

⁰⁶ **Gate arm design, colors, and lighting requirements shall be in accordance with the Standards contained in Section 8C.04.**

⁰⁷ **Except as provided in Paragraph 19, the exit gate arm mechanism shall be designed to fail-safe in the up position.**

⁰⁸ **At locations where gate arms are offset a sufficient distance for highway vehicles to drive between the entrance and exit gate arms, median islands (see Figure 8C-2) shall be installed in accordance with the needs established by an engineering study.**

Guidance:

⁰⁹ *The gate arm should ascend to its upright position in 12 seconds or less.*

Figure 8C-2. Example of Location Plan for Flashing-Light Signals and Four-Quadrant Gates

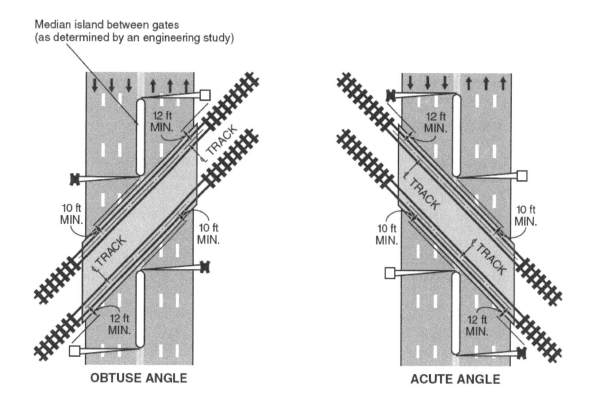

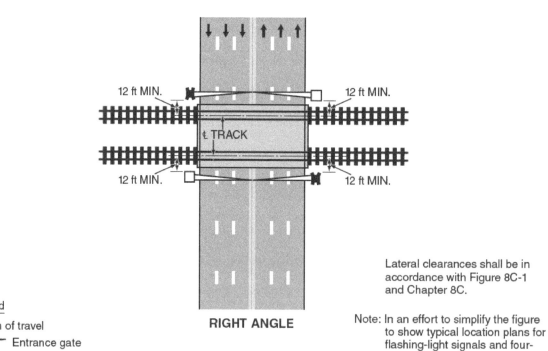

Lateral clearances shall be in accordance with Figure 8C-1 and Chapter 8C.

Note: In an effort to simplify the figure to show typical location plans for flashing-light signals and four-quadrant gates, not all traffic control devices are shown on this figure.

10. *Four-Quadrant Gate systems should only be used in locations with constant warning time detection.*
11. *The operating mode of the exit gates should be determined based upon an engineering study, with input from the affected railroad company or LRT agency.*
12. *If the Timed Exit Gate Operating Mode is used, the engineering study, with input from the affected railroad company or LRT agency, should also determine the Exit Gate Clearance Time (see definition in Section 1A.13).*
13. *If the Dynamic Exit Gate Operating Mode is used, highway vehicle intrusion detection devices that are part of a system that incorporates processing logic to detect the presence of highway vehicles within the minimum track clearance distance should be installed to control exit gate operation.*
14. *Regardless of which exit gate operating mode is used, the Exit Gate Clearance Time should be considered when determining additional time requirements for the Minimum Warning Time.*
15. *If a Four-Quadrant Gate system is used at a location that is adjacent to an intersection that could cause highway vehicles to queue within the minimum track clearance distance, the Dynamic Exit Gate Operating Mode should be used unless an engineering study indicates otherwise.*
16. *If a Four-Quadrant Gate system is interconnected with a highway traffic signal, backup or standby power should be considered for the highway traffic signal. Also, circuitry should be installed to prevent the highway traffic signal from leaving the track clearance green interval until all of the gates are lowered.*
17. *At locations where sufficient space is available, exit gates should be positioned downstream from the track a distance that provides a safety zone long enough to accommodate at least one design vehicle between the exit gate and the nearest rail.*
18. *Four-Quadrant Gate systems should include remote health (status) monitoring capable of automatically notifying railroad or LRT signal maintenance personnel when anomalies have occurred within the system.*

Option:

19. Exit gate arms may fail in the down position if the grade crossing is equipped with remote health (status) monitoring.
20. Four-Quadrant Gate installations may include median islands between opposing lanes on an approach to a grade crossing.

Guidance:

21. *Where sufficient space is available, median islands should be at least 60 feet in length.*

Section 8C.07 Wayside Horn Systems

Option:

01. A wayside horn system (see definition in Section 1A.13) may be installed in compliance with 49 CFR Part 222 to provide audible warning directed toward the road users at a highway-rail or highway-LRT grade crossing or at a pathway grade crossing.

Standard:

02. **Wayside horn systems used at grade crossings where the locomotive horn is not sounded shall be equipped and shall operate in compliance with the requirements of Appendix E to 49 CFR Part 222.**

Guidance:

03. *The same lateral clearance and roadside safety features should apply to wayside horn systems as described in the Standards contained in Section 8C.01. Wayside horn systems, when mounted on a separate pole assembly, should be installed no closer than 15 feet from the center of the nearest track and should be positioned to not obstruct the motorists' line of sight of the flashing-light signals.*

Section 8C.08 Rail Traffic Detection

Standard:

01. **The devices employed in active traffic control systems shall be actuated by some form of rail traffic detection.**
02. **Rail traffic detection circuits, insofar as practical, shall be designed on the fail-safe principle.**
03. **Flashing-light signals shall operate for at least 20 seconds before the arrival of any rail traffic, except as provided in Paragraph 4.**

Option:

04. On tracks where all rail traffic operates at less than 20 mph and where road users are directed by an authorized person on the ground to not enter the crossing at all times that approaching rail traffic is about to occupy the crossing, a shorter signal operating time for the flashing-light signals may be used.

05 Additional warning time may be provided when determined by an engineering study.

Guidance:

06 *Where the speeds of different rail traffic on a given track vary considerably under normal operation, special devices or circuits should be installed to provide reasonably uniform notice in advance of all rail traffic movements over the grade crossing. Special control features should be used to eliminate the effects of station stops and switching operations within approach control circuits to prevent excessive activation of the traffic control devices while rail traffic is stopped on or switching upon the approach track control circuits.*

Section 8C.09 Traffic Control Signals at or Near Highway-Rail Grade Crossings

Option:

01 Traffic control signals may be used instead of flashing-light signals to control road users at industrial highway-rail grade crossings and other places where train movements are very slow, such as in switching operations.

Standard:

02 **The appropriate provisions of Part 4 relating to traffic control signal design, installation, and operation shall be applicable where traffic control signals are used to control road users instead of flashing-light signals at highway-rail grade crossings.**

03 **Traffic control signals shall not be used instead of flashing-light signals to control road users at a mainline highway-rail grade crossing.**

Guidance:

04 *If a highway-rail grade crossing is equipped with a flashing-light signal system and is located within 200 feet of an intersection or midblock location controlled by a traffic control signal, the traffic control signal should be provided with preemption in accordance with Section 4D.27.*

05 *Coordination with the flashing-light signal system, queue detection, or other alternatives should be considered for traffic control signals located farther than 200 feet from the highway-rail grade crossing. Factors to be considered should include traffic volumes, highway vehicle mix, highway vehicle and train approach speeds, frequency of trains, and queue lengths.*

06 *The highway agency or authority with jurisdiction and the regulatory agency with statutory authority, if applicable, should jointly determine the preemption operation and the timing of traffic control signals interconnected with highway-rail grade crossings adjacent to signalized highway intersections.*

Support:

07 Section 4D.27 includes a recommendation that traffic control signals that are adjacent to highway-rail grade crossings and that are coordinated with the flashing-light signals or that include railroad preemption features be provided with a back-up power supply.

Standard:

08 **Information regarding the type of preemption and any related timing parameters shall be provided to the railroad company so that they can design the appropriate train detection circuitry.**

09 **If preemption is provided, the normal sequence of traffic control signal indications shall be preempted upon the approach of trains to avoid entrapment of highway vehicles on the highway-rail grade crossing.**

10 **This preemption feature shall have an electrical circuit of the closed-circuit principle, or a supervised communication circuit between the control circuits of the highway-rail grade crossing warning system and the traffic control signal controller. The traffic control signal controller preemptor shall be activated via the supervised communication circuit or the electrical circuit that is normally energized by the control circuits of the highway-rail grade crossing warning system. The approach of a train to a highway-rail grade crossing shall de-energize the electrical circuit or activate the supervised communication circuit, which in turn shall activate the traffic control signal controller preemptor. This shall establish and maintain the preemption condition during the time the highway-rail grade crossing warning system is activated, except that when crossing gates exist, the preemption condition shall be maintained until the crossing gates are energized to start their upward movement. When multiple or successive preemptions occur, train activation shall receive first priority.**

Guidance:

11 *If a highway-rail grade crossing is located within 50 feet (or within 75 feet for a highway that is regularly used by multi-unit highway vehicles) of an intersection controlled by a traffic control signal, the use of pre-signals to control traffic approaching the grade crossing should be considered.*

Standard:

12 **If used, the pre-signals shall display a steady red signal indication during the track clearance portion of a signal preemption sequence to prohibit additional highway vehicles from crossing the railroad track.**

Guidance:

13 *Consideration should be given to using visibility-limited signal faces (see definition in Section 1A.13) at the intersection for the downstream signal faces that control the approach that is equipped with pre-signals.*

Option:

14 The pre-signal phase sequencing may be timed with an offset from the downstream signalized intersection such that the railroad track area and the area between the railroad track and the downstream signalized intersection is generally kept clear of stopped highway vehicles.

Standard:

15 **If a pre-signal is installed at an interconnected highway-rail grade crossing near a signalized intersection, a STOP HERE ON RED (R10-6) sign shall be installed near the pre-signal or at the stop line if used. If there is a nearby signalized intersection with insufficient clear storage distance for a design vehicle, or the highway-rail grade crossing does not have gates, a No Turn on Red (R10-11, R10-11a, or R10-11b) sign (see Section 2B.53) shall be installed for the approach that crosses the railroad track, if applicable.**

Option:

16 At locations where a highway-rail grade crossing is located more than 50 feet (or more than 75 feet for a highway regularly used by multi-unit highway vehicles) from an intersection controlled by a traffic control signal, a pre-signal may be used if an engineering study determines a need.

17 If highway traffic signals must be located within close proximity to the flashing-light signal system, the highway traffic signals may be mounted on the same overhead structure as the flashing-light signals.

Support:

18 Section 4C.10 describes the Intersection Near a Grade Crossing signal warrant that is intended for use at a location where the proximity to the intersection of a grade crossing on an intersection approach controlled by a STOP or YIELD sign is the principal reason to consider installing a traffic control signal.

19 Section 4D.27 describes additional considerations regarding preemption of traffic control signals at or near highway-rail grade crossings.

Section 8C.10 Traffic Control Signals at or Near Highway-LRT Grade Crossings

Support:

01 There are two types of traffic control signals for controlling vehicular and LRT movements at interfaces of the two modes. The first is the standard traffic control signal described in Part 4, which is the focus of this Section. The other type of signal is referred to as an LRT signal and is discussed in Section 8C.11.

Standard:

02 **The provisions of Part 4 and Section 8C.09 relating to traffic control signal design, installation, and operation, including interconnection with nearby automatic gates or flashing-light signals, shall be applicable as appropriate where traffic control signals are used at highway-LRT grade crossings.**

03 **If traffic control signals are in operation at a crossing that is used by pedestrians, bicyclists, and/or other non-motorized road users, an audible device such as a bell shall also be provided and shall be operated in conjunction with the traffic control signals.**

Guidance:

04 *When a highway-LRT grade crossing equipped with a flashing-light signal system is located within 200 feet of an intersection or midblock location controlled by a traffic control signal, the traffic control signal should be provided with preemption in accordance with Section 4D.27.*

05 *Coordination with the flashing-light signal system should be considered for traffic control signals located more than 200 feet from the crossing. Factors to be considered should include traffic volumes, highway vehicle mix, highway vehicle and LRT approach speeds, frequency of LRT traffic, and queue lengths.*

06 *If the highway traffic signal has emergency-vehicle preemption capability, it should be coordinated with LRT operation.*

07 *Where LRT operates in a wide median, highway vehicles crossing the tracks and being controlled by both near and far side traffic signal faces should receive a protected left-turn green phase from the far side signal face to clear highway vehicles from the crossing when LRT equipment is approaching the crossing.*

Option:

⁰⁸ Green indications may be provided during LRT phases for highway vehicle, pedestrian, and bicycle movements that do not conflict with LRT movements.

⁰⁹ Traffic control signals may be installed in addition to four-quadrant gate systems and automatic gates at a highway-LRT crossing if the crossing occurs within a highway-highway intersection and if the traffic control signals meet the warrants described in Chapter 4C.

¹⁰ At a location other than an intersection, when LRT speeds are less than 25 mph, traffic control signals alone may be used to control road users at highway-LRT grade crossings only when justified by an engineering study.

¹¹ Typical circumstances may include:
A. Geometric conditions preclude the installation of highway-LRT grade crossing warning devices.
B. LRT vehicles share the same roadway with road users.
C. Traffic control signals already exist.

Support:

¹² Section 4D.27 contains information regarding traffic control signals at or near highway-LRT grade crossings that are not equipped with highway-LRT grade crossing warning devices.

¹³ Section 4C.10 describes the Intersection Near a Grade Crossing signal warrant that is intended for use at a location where the proximity to the intersection of a grade crossing on an intersection approach controlled by a STOP or YIELD sign is the principal reason to consider installing a traffic control signal.

Guidance:

¹⁴ *When a highway-LRT grade crossing exists within a signalized intersection, consideration should be given to providing separate turn signal faces (see definition in Section 1A.13) for the movements crossing the tracks.*

Standard:

¹⁵ **Separate turn signal faces that are provided for turn movements toward the crossing shall display a steady red indication during the approach and/or passage of LRT traffic.**

Guidance:

¹⁶ *When a signalized intersection that is located within 200 feet of a highway-LRT grade crossing is preempted, all existing turning movements toward the highway-LRT grade crossing should be prohibited.*

Support:

¹⁷ Section 8B.08 contains information regarding the prohibition of turning movements toward the crossing during preemption.

¹⁸ Part 4 contains information regarding signal phasing and timing requirements.

Section 8C.11 Use of Traffic Control Signals for Control of LRT Vehicles at Grade Crossings

Guidance:

⁰¹ *LRT movements in semi-exclusive alignments at non-gated grade crossings that are equipped with traffic control signals should be controlled by special LRT signal indications.*

⁰² *LRT traffic control signals that are used to control LRT movements only should display the signal indications illustrated in Figure 8C-3.*

Support:

⁰³ Section 4D.27 contains information about the use of the signal indications shown in Figure 8C-3 for the control of exclusive bus movements at "queue jumper lanes" and for the control of exclusive bus rapid transit movements on semi-exclusive or mixed-use alignments.

Option:

⁰⁴ Standard traffic control signals may be used instead of LRT traffic control signals to control the movement of LRT vehicles (see Section 8C.10).

Standard:

⁰⁵ **If a separate set of standard traffic control signal indications (red, yellow, and green circular and arrow indications) is used to control LRT movements, the indications shall be positioned so they are not visible to motorists, pedestrians, and bicyclists (see Section 4D.12).**

⁰⁶ **If the LRT crossing control is separate from the intersection control, the two shall be interconnected. The LRT signal phase shall not be terminated until after the LRT vehicle has cleared the crossing.**

Option:

LRT signals may be used at grade crossings and at intersections in mixed-use alignments in conjunction with standard traffic control signals where special LRT signal phases are used to accommodate turning LRT vehicles or where additional LRT clearance time is desirable.

Guidance:

LRT signal faces should be separated vertically or horizontally from the nearest highway traffic signal face for the same approach by at least 3 feet.

Figure 8C-3. Light Rail Transit Signals

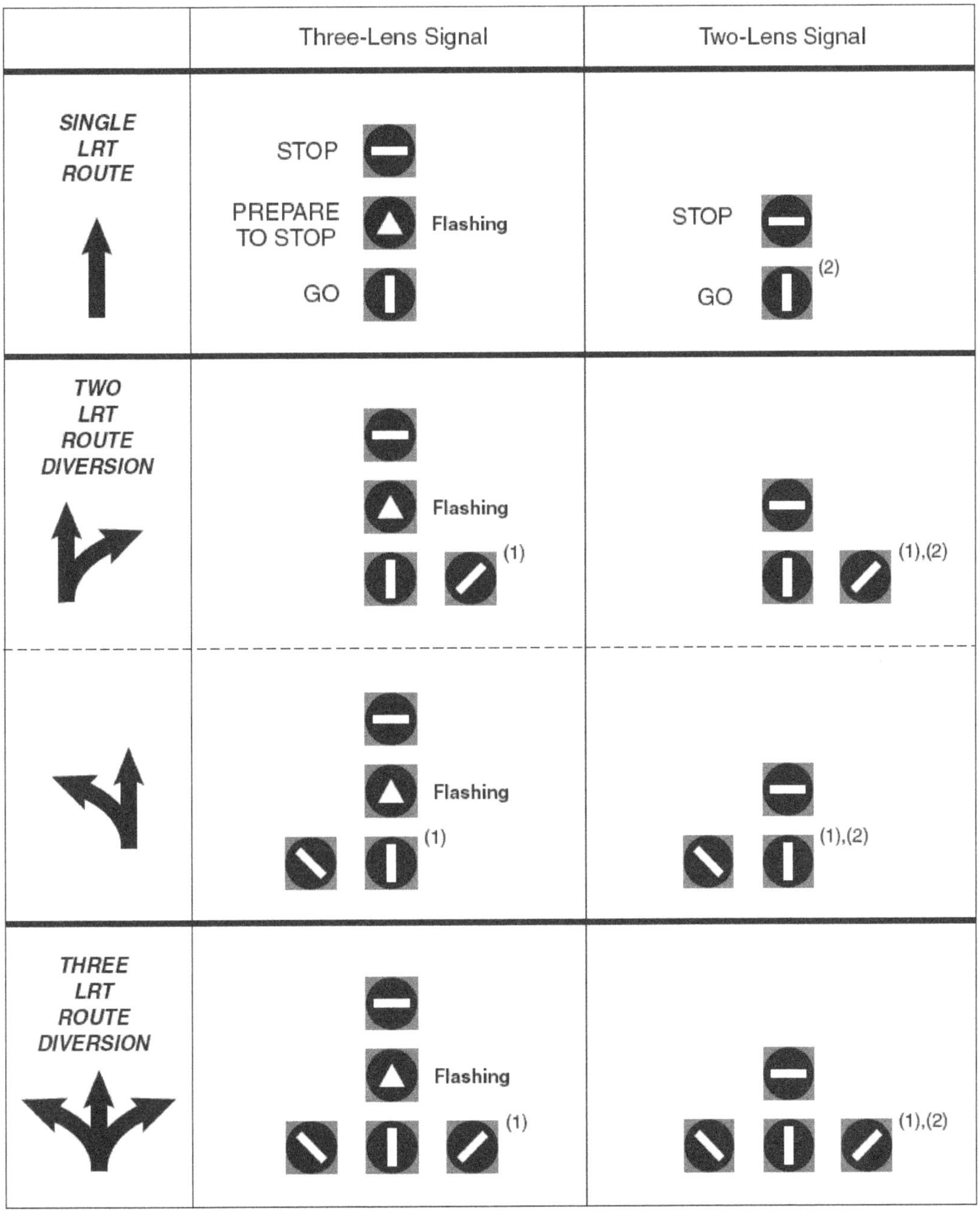

	Three-Lens Signal	Two-Lens Signal
SINGLE LRT ROUTE (↑)	STOP ⊖ PREPARE TO STOP △ Flashing GO ▯	STOP ⊖ GO ▯ (2)
TWO LRT ROUTE DIVERSION	⊖ △ Flashing ▯ ⊘ (1)	⊖ ▯ ⊘ (1),(2)
	⊖ △ Flashing ⊘ ▯ (1)	⊖ ⊘ ▯ (1),(2)
THREE LRT ROUTE DIVERSION	⊖ △ Flashing ⊘ ▯ ⊘ (1)	⊖ ⊘ ▯ ⊘ (1),(2)

Notes:
All aspects (or signal indications) are white.
(1) Could be in single housing.
(2) "Go" lens may be used in flashing mode to indicate "prepare to stop".

Section 8C.12 Grade Crossings Within or In Close Proximity to Circular Intersections

Support:

01 At circular intersections, such as roundabouts and traffic circles, that include or are within close proximity to a grade crossing, a queue of vehicular traffic could cause highway vehicles to stop on the grade crossing.

Standard:

02 **Where circular intersections include or are within 200 feet of a grade crossing, an engineering study shall be made to determine if queuing could impact the grade crossing. If traffic queues impact the grade crossing, provisions shall be made to clear highway traffic from the grade crossing prior to the arrival of rail traffic.**

Support:

03 Among the actions that can be taken to keep the grade crossing clear of traffic or to clear traffic from the grade crossing prior to the arrival of rail traffic are the following:

    A. Elimination of the circular intersection,
    B. Geometric design revisions,
    C. Grade crossing regulatory and warning devices,
    D. Highway traffic signals,
    E. Traffic metering devices,
    F. Activated signs, or
    G. A combination of these or other actions.

Section 8C.13 Pedestrian and Bicycle Signals and Crossings at LRT Grade Crossings

Guidance:

01 *Where LRT tracks are immediately adjacent to other tracks or a road, pedestrian signalization should be designed to avoid having pedestrians wait between sets of tracks or between the tracks and the road. If adequate space exists for a pedestrian refuge and is justified based on engineering judgment, additional pedestrian signal heads, signing, and detectors should be installed (see Section 4E.08).*

Standard:

02 **When used at LRT crossings, pedestrian signal heads shall comply with the provisions of Section 4E.04.**

Guidance:

03 *Flashing-light signals (see Figure 8C-4) with a Crossbuck (R15-1) sign and an audible device should be installed at pedestrian and bicycle crossings where an engineering study has determined that the sight distance is not sufficient for pedestrians and bicyclists to complete their crossing prior to the arrival of the LRT traffic at the crossing, or where LRT speeds exceed 35 mph.*

04 *If an engineering study shows that flashing-light signals with a Crossbuck sign and an audible device would not provide sufficient notice of an approaching LRT traffic, the LOOK (R15-8) sign (see Figure 8C-4) and/or pedestrian gates should be considered (see Figures 8C-5 through 8C-7).*

Support:

05 A pedestrian gate is similar to an automatic gate except the gate arm is shorter.

06 The swing gate alerts pedestrians to the LRT tracks that are to be crossed. Swing gates are designed to open away from the tracks, requiring users to pull the gate open to cross, but permitting a quick exit from the trackway, and to automatically close.

Option:

07 Swing gates may be installed across pedestrian and bicycle walkways (see Figure 8C-8).

08 Pedestrian barriers at offset crossings may be used at pedestrian and bicycle crossings as passive devices that force users to face approaching LRT before entering the trackway (see Figures 8C-9 and 8C-10).

Figure 8C-4. Example of Flashing-Light Signal Assembly for Pedestrian Crossings

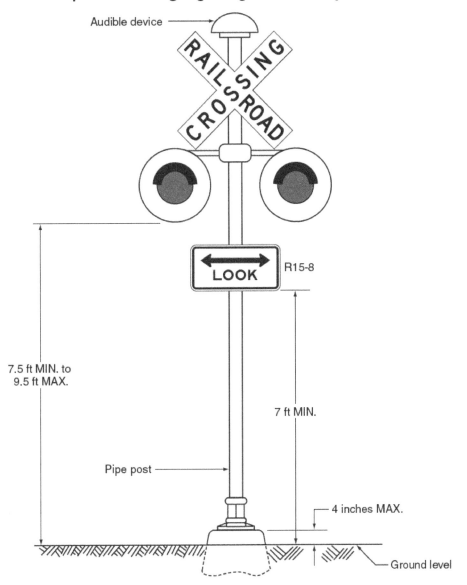

Figure 8C-5. Example of a Shared Pedestrian/Roadway Gate

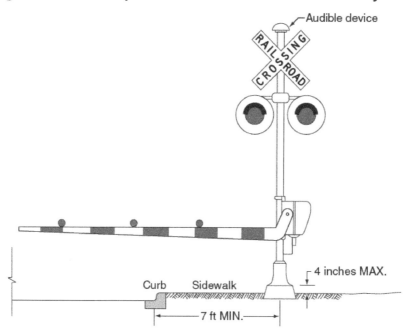

Figure 8C-6. Example of a Separate Pedestrian Gate

Note: The provision of a separate pedestrian gate is optional based upon site-specific conditions. If a separate pedestrian gate is provided, the need for a separate Crossbuck sign, audible device, and flashing-light signals should be determined based upon site-specific conditions such as the proximity of the sidewalk or shared-use path to the roadway grade crossing devices.

* For locating this reference line on an approach that does not have a curb, see Section 8C.01.

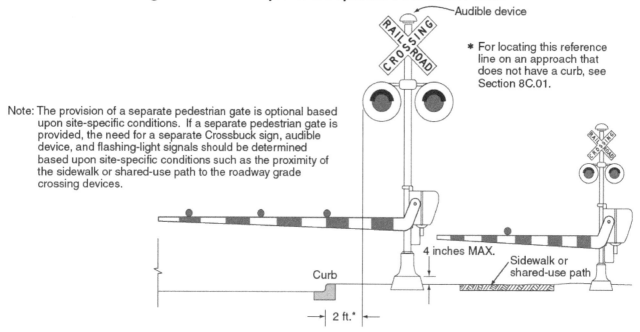

Figure 8C-7. Examples of Placement of Pedestrian Gates

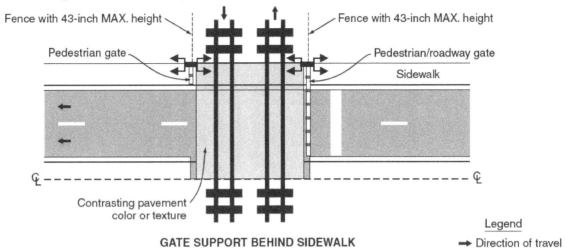

GATE SUPPORT BEHIND SIDEWALK

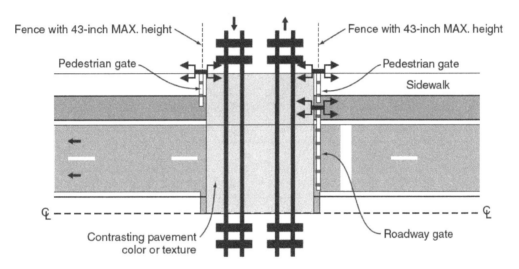

GATE SUPPORT BETWEEN SIDEWALK AND ROADWAY

Legend
→ Direction of travel

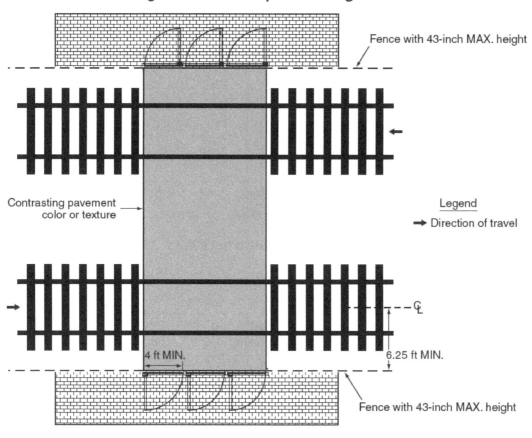

Figure 8C-8. Example of Swing Gates

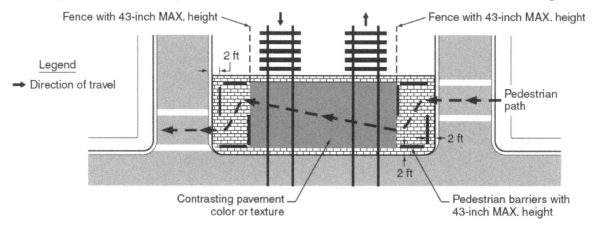

Figure 8C-9. Example of Pedestrian Barriers at an Offset Grade Crossing

Figure 8C-10. Examples of Pedestrian Barrier Installation at an Offset Non-Intersection Grade Crossing

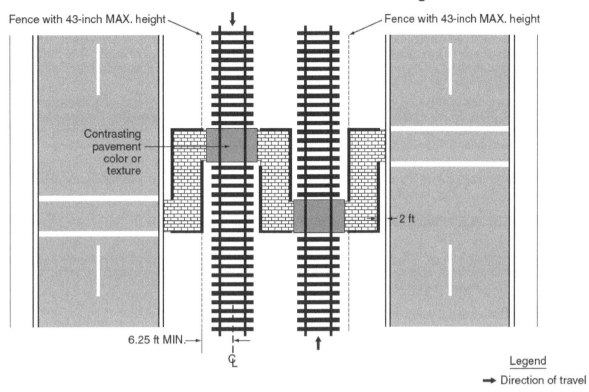
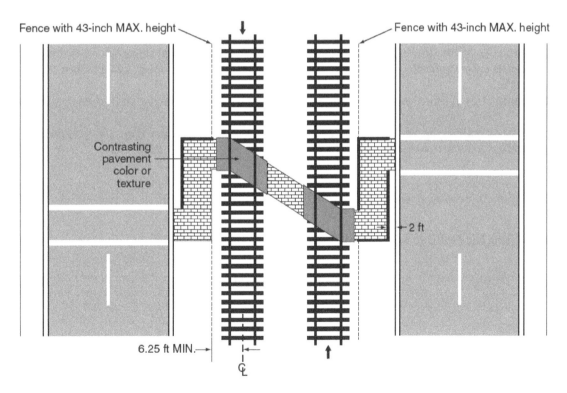

CHAPTER 8D. PATHWAY GRADE CROSSINGS

Section 8D.01 Purpose

Support:

01 Traffic control for pathway grade crossings includes all signs, signals, markings, other warning devices, and their supports at pathway grade crossings and along pathway approaches to grade crossings. The function of this traffic control is to promote safety and provide effective operation of both rail and pathway traffic at pathway grade crossings.

02 Except as specifically provided in this Chapter, sidewalks are considered to be part of a highway-rail or highway-LRT grade crossing rather than a pathway grade crossing, and are covered by the provisions of Chapters 8B and 8C rather than by the provisions of this Chapter. However, many of the treatments outlined in this Chapter are applicable to sidewalks adjacent to highway-rail or highway-LRT grade crossings, including detectable warnings, swing gates, and automatic gates.

03 Crosswalks at intersections where pedestrians cross LRT tracks in mixed-use alignments are covered by the provisions of Section 3B.18 rather than by the provisions of this Chapter.

Section 8D.02 Use of Standard Devices, Systems, and Practices

Guidance:

01 *The public agency with jurisdiction over the pathway and the regulatory agency with statutory authority, if applicable, should jointly determine the need and selection of devices at a pathway grade crossing, including the appropriate traffic control system to be used.*

Section 8D.03 Pathway Grade Crossing Signs and Markings

Standard:

01 **Pathway grade crossing signs shall be standard in shape, legend, and color.**

02 **Traffic control devices mounted adjacent to pathways at a height of less than 8 feet measured vertically from the bottom edge of the device to the elevation of the near edge of the pathway surface shall have a minimum lateral offset of 2 feet from the near edge of the device to the near edge of the pathway (see Figure 9B-1).**

03 **The minimum mounting height for post-mounted signs on pathways shall be 4 feet, measured vertically from the bottom edge of the sign to the elevation of the near edge of the pathway surface (see Figure 9B-1).**

04 **Pathway grade crossing traffic control devices shall be located a minimum of 12 feet from the center of the nearest track.**

05 **The minimum sizes of pathway grade crossing signs shall be as shown in the shared-use path column in Table 9B-1.**

06 **When overhead traffic control devices are used on pathways, the clearance from the bottom edge of the device to the pathway surface directly under the sign or device shall be at least 8 feet.**

Guidance:

07 *If pathway users include those who travel faster than pedestrians, such as bicyclists or skaters, the use of warning signs and pavement markings in advance of the pathway grade crossing (see Figure 8D-1) should be considered.*

Section 8D.04 Stop Lines, Edge Lines, and Detectable Warnings

Guidance:

01 *If used at pathway grade crossings, the pathway stop line should be a transverse line at the point where a pathway user is to stop. The pathway stop line should be placed at least 2 feet further from the nearest rail than the gate, counterweight, or flashing-light signals (if any of these are present) is placed, and at least 12 feet from the nearest rail.*

Option:

02 Edge lines (see Section 3B.06) may be used on approach to and across the tracks at a pathway grade crossing, a sidewalk at a highway-rail or highway-LRT grade crossing, or a station crossing to delineate the designated pathway user route.

Support:

03 Edge line delineation can be beneficial where the distance across the tracks is long, commonly because of a skewed grade crossing or because of multiple tracks, or where the pathway surface is immediately adjacent to a traveled way.

Figure 8D-1. Example of Signing and Markings for a Pathway Grade Crossing

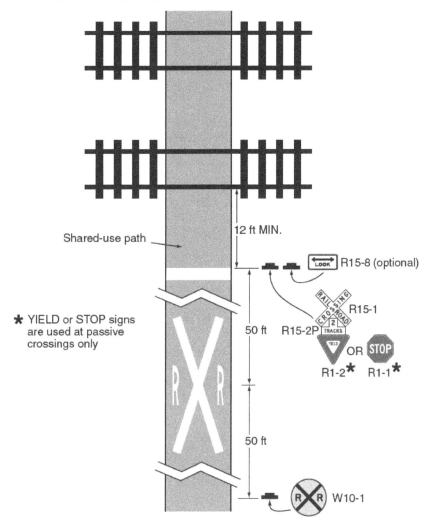

04 Detectable warning surfaces (see Section 3B.18) that contrast visually with adjacent walking surfaces, either light-on-dark or dark-on-light, can be used to warn pedestrians about the locations of the tracks at a grade crossing. The "Americans with Disabilities Act Accessibility Guidelines for Buildings and Facilities (ADAAG)" (see Section 1A.11) contains specifications for design and placement of detectable warning surfaces.

Section 8D.05 Passive Devices for Pathway Grade Crossings
Standard:

01 **Except as provided in Paragraph 2, where active traffic control devices are not used, a Crossbuck Assembly shall be installed on each approach to a pathway grade crossing.**

Option:

02 The Crossbuck Assembly may be omitted at station crossings and on the approaches to a pathway grade crossing that is located within 25 feet of the traveled way at a highway-rail or highway-LRT grade crossing.

Guidance:

03 *The pathway user's ability to detect the presence of approaching rail traffic should be considered in determining the type and placement of traffic control devices or design features (such as fencing or swing gates).*

04 *Nighttime visibility should be considered if design features (such as fencing or swing gates) are used to channelize pathway users.*

05 *If automatic gates and swing gates are used, the pathway should be channelized to direct users to the entrance to and exit from the pathway grade crossing.*

Standard:

06 **If used, swing gates shall be designed to open away from the track(s) so that pathway users can quickly push the gate open when moving away from the track(s). If used, swing gates shall be designed to automatically return to the closed position after each use.**

Option:

07 When used in conjunction with automatic gates at pathway grade crossings, swing gates may be equipped with a latching device that permits the gate to be opened only from the track side of the gate.

Support:

08 The "Americans with Disabilities Act Accessibility Guidelines for Buildings and Facilities (ADAAG)" (see Section 1A.11) contains information regarding spring hinges and door and gate opening forces for swing gates.

Section 8D.06 Active Traffic Control Systems for Pathway Grade Crossings

Standard:

01 **If used at a pathway grade crossing, an active traffic control system shall include flashing-light signals for each direction of the pathway. A bell or other audible warning device shall also be provided.**

Option:

02 Separate active traffic control devices may be omitted at a pathway grade crossing that is located within 25 feet of the traveled way of a highway-rail or highway-LRT grade crossing that is equipped with an active traffic control system.

Standard:

03 **If used at pathway grade crossings, alternately flashing red lights shall be aligned horizontally and the light units shall have a diameter of at least 4 inches. The minimum mounting height of the flashing red lights shall be 4 feet, measured vertically from the bottom edge of the lights to the elevation of the near edge of the pathway surface.**

Option:

04 Traffic control devices may be installed between the tracks at multiple track crossings at stations.

Standard:

05 **The mounting height for flashing lights that are installed between the tracks at multiple track crossings at stations shall be a minimum of 1 foot, measured vertically from the bottom edge of the lights to the elevation of the near edge of the pathway surface.**

Option:

06 Automatic gates may be used at pathway grade crossings.

Guidance:

07 *If used at a pathway grade crossing, the height of the automatic gate arm when in the down position should be a minimum of 2.5 feet and a maximum of 4 feet above the sidewalk.*

08 *If used, the gate configuration, which might include a combination of automatic gates and swing gates, should provide for full width coverage of the pathway on both approaches to the track.*

Standard:

09 **Where a sidewalk is located between the edge of a roadway and the support for a gate arm that extends across the sidewalk and into the roadway, the location, placement, and height prescribed for vehicular gates shall be used (see Section 8C.04).**

Guidance:

10 *If a separate automatic gate is used for a sidewalk, the height of the gate arm when in the down position should be a minimum of 2.5 feet and a maximum of 4 feet above the sidewalk.*

11 *If a separate automatic gate is used for a sidewalk at a highway-rail or highway-LRT grade crossing, instead of a supplemental or auxiliary gate arm installed as a part of the same mechanism as the vehicular gate, a separate mechanism should be provided for the sidewalk gate to prevent a pedestrian from raising the vehicular gate.*

PART 9
TRAFFIC CONTROL FOR BICYCLE FACILITIES

CHAPTER 9A. GENERAL

Section 9A.01 Requirements for Bicyclist Traffic Control Devices
Support:
01 General information and definitions concerning traffic control devices are found in Part 1.

Section 9A.02 Scope
Support:
01 Part 9 covers signs, pavement markings, and highway traffic signals specifically related to bicycle operation on both roadways and shared-use paths.
Guidance:
02 *Parts 1, 2, 3, and 4 should be reviewed for general provisions, signs, pavement markings, and signals.*
Standard:
03 **The absence of a marked bicycle lane or any of the other traffic control devices discussed in this Chapter on a particular roadway shall not be construed to mean that bicyclists are not permitted to travel on that roadway.**

Section 9A.03 Definitions Relating to Bicycles
Support:
01 Definitions and acronyms pertaining to Part 9 are provided in Sections 1A.13 and 1A.14.

Section 9A.04 Maintenance
Guidance:
01 *All signs, signals, and markings, including those on bicycle facilities, should be properly maintained to command respect from both the motorist and the bicyclist. When installing signs and markings on bicycle facilities, an agency should be designated to maintain these devices.*

Section 9A.05 Relation to Other Documents
Support:
01 "The Uniform Vehicle Code and Model Traffic Ordinance" published by the National Committee on Uniform Traffic Laws and Ordinances (see Section 1A.11) has provisions for bicycles and is the basis for the traffic control devices included in this Manual.
02 Informational documents used during the development of the signing and marking recommendations in Part 9 include the following:
   A. "Guide for Development of Bicycle Facilities," which is available from the American Association of State Highway and Transportation Officials (see Page i for the address); and
   B. State and local government design guides.
03 Other publications that relate to the application of traffic control devices in general are listed in Section 1A.11.

Section 9A.06 Placement Authority
Support:
01 Section 1A.08 contains information regarding placement authority for traffic control devices.

Section 9A.07 Meaning of Standard, Guidance, Option, and Support
Support:
01 The introduction to this Manual contains information regarding the meaning of the headings Standard, Guidance, Option, and Support, and the use of the words "shall," "should," and "may."

Section 9A.08 Colors
Support:
01 Section 1A.12 contains information regarding the color codes.

CHAPTER 9B. SIGNS

Section 9B.01 Application and Placement of Signs
Standard:

01 **Bicycle signs shall be standard in shape, legend, and color.**

02 **All signs shall be retroreflectorized for use on bikeways, including shared-use paths and bicycle lane facilities.**

03 **Where signs serve both bicyclists and other road users, vertical mounting height and lateral placement shall be as provided in Part 2.**

04 **Where used on a shared-use path, no portion of a sign or its support shall be placed less than 2 feet laterally from the near edge of the path, or less than 8 feet vertically over the entire width of the shared-use path (see Figure 9B-1).**

05 **Mounting height for post-mounted signs on shared-use paths shall be a minimum of 4 feet, measured vertically from the bottom of the sign to the elevation of the near edge of the path surface (see Figure 9B-1).**

Guidance:

06 *Signs for the exclusive use of bicyclists should be located so that other road users are not confused by them.*

07 *The clearance for overhead signs on shared-use paths should be adjusted when appropriate to accommodate path users requiring more clearance, such as equestrians, or typical maintenance or emergency vehicles.*

Section 9B.02 Design of Bicycle Signs
Standard:

01 **If the sign or plaque applies to motorists and bicyclists, then the size shall be as shown for conventional roads in Tables 2B-1, 2C-2, or 2D-1.**

02 **The minimum sign and plaque sizes for shared-use paths shall be those shown in Table 9B-1, and shall be used only for signs and plaques installed specifically for bicycle traffic applications. The minimum sign and plaque sizes for bicycle facilities shall not be used for signs or plaques that are placed in a location that would have any application to other vehicles.**

Option:

03 Larger size signs and plaques may be used on bicycle facilities when appropriate (see Section 2A.11).

Figure 9B-1. Sign Placement on Shared-Use Paths

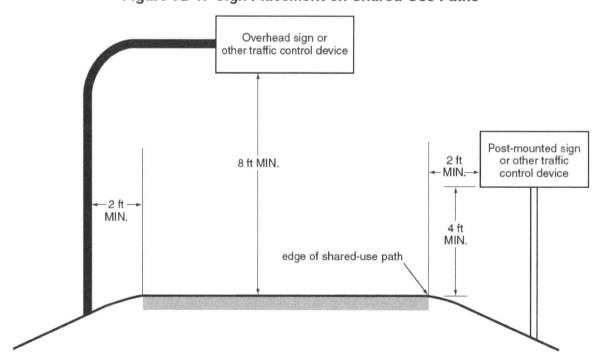

Table 9B-1. Bicycle Facility Sign and Plaque Minimum Sizes (Sheet 1 of 2)

Sign or Plaque	Sign Designation	Section	Shared-Use Path	Roadway
Stop	R1-1	2B.05, 9B.03	18 x 18	30 x 30
Yield	R1-2	2B.08, 9B.03	18 x 18 x 18	30 x 30 x 30
Bike Lane	R3-17	9B.04	—	24 x 18
Bike Lane (plaques)	R3-17aP, R3-17bP	9B.04	—	24 x 8
Movement Restriction	R4-1,2,3,7,16	2B.28,29,30,32; 9B.14	12 x 18	18 x 24
Begin Right Turn Lane Yield to Bikes	R4-4	9B.05	—	36 x 30
Bicycles May Use Full Lane	R4-11	9B.06	—	30 x 30
Bicycle Wrong Way	R5-1b	9B.07	12 x 18	12 x 18
No Motor Vehicles	R5-3	9B.08	24 x 24	24 x 24
No Bicycles	R5-6	9B.09	18 x 18	24 x 24
No Parking Bike Lane	R7-9,9a	9B.10	—	12 x 18
No Pedestrians	R9-3	9B.09	18 x 18	18 x 18
Ride With Traffic (plaque)	R9-3cP	9B.07	12 x 12	12 x 12
Bicycle Regulatory	R9-5,6	9B.11	12 x 18	12 x 18
Shared-Use Path Restriction	R9-7	9B.12	12 x 18	—
No Skaters	R9-13	9B.09	18 x 18	18 x 18
No Equestrians	R9-14	9B.09	18 x 18	18 x 18
Push Button for Green Light	R10-4	9B.11	9 x 12	9 x 12
To Request Green Wait on Symbol	R10-22	9B.13	12 x 18	12 x 18
Bike Push Button for Green Light	R10-24	9B.11	9 x 15	9 x 15
Push Button to Turn On Warning Lights	R10-25	9B.11	9 x 12	9 x 12
Bike Push Button for Green Light (arrow)	R10-26	9B.11	9 x 15	9 x 15
Grade Crossing (Crossbuck)	R15-1	8B.03, 9B.14	24 x 4.5	48 x 9
Number of Tracks (plaque)	R15-2P	8B.03, 9B.14	13.5 x 9	27 x 18
Look	R15-8	8B.17, 9B.14	18 x 9	36 x 18
Turn and Curve Warning	W1-1,2,3,4,5	2C.04, 9B.15	18 x 18	24 x 24
Arrow Warning	W1-6,7	2C.12, 2C.47, 9B.15	24 x 12	36 x 18
Intersection Warning	W2-1,2,3,4,5	2C.46, 9B.16	18 x 18	24 x 24
Stop, Yield, Signal Ahead	W3-1,2,3	2C.36, 9B.19	18 x 18	30 x 30
Narrow Bridge	W5-2	2C.20, 9B.19	18 x 18	30 x 30
Path Narrows	W5-4a	9B.19	18 x 18	—
Hill	W7-5	9B.19	18 x 18	30 x 30
Bump or Dip	W8-1,2	2C.28, 9B.17	18 x 18	24 x 24
Pavement Ends	W8-3	2C.30, 9B.17	18 x 18	30 x 30
Bicycle Surface Condition	W8-10	9B.17	18 x 18	30 x 30
Slippery When Wet (plaque)	W8-10P	9B.17	12 x 9	12 x 9
Grade Crossing Advance Warning	W10-1	8B.06, 9B.19	24 Dia.	36 Dia.
No Train Horn (plaque)	W10-9P	8B.21, 9B.19	18 x 12	30 x 24
Skewed Crossing	W10-12	8B.25, 9B.19	18 x 18	36 x 36
Bicycle Warning	W11-1	9B.18	18 x 18	24 x 24
Pedestrian Crossing	W11-2	2C.50, 9B.19	18 x 18	24 x 24
Combination Bike and Ped Crossing	W11-15	9B.18	18 x 18	30 x 30
Trail Crossing (plaque)	W11-15P	9B.18	18 x 12	24 x 18
Low Clearance	W12-2	2C.27, 9B.19	18 x 18	30 x 30
Playground	W15-1	2C.51, 9B.19	18 x 18	24 x 24
Share the Road (plaque)	W16-1P	2C.60, 9B.19	—	18 x 24

Table 9B-1. Bicycle Facility Sign and Plaque Minimum Sizes (Sheet 2 of 2)

Sign or Plaque	Sign Designation	Section	Shared-Use Path	Roadway
XX Feet (plaque)	W16-2P	2C.55, 9B.18	18 x 12	24 x 18
XX Ft (plaque)	W16-2aP	2C.55, 9B.18	18 x 9	24 x 12
Diagonal Arrow (plaque)	W16-7P	9B.18	—	24 x 12
Ahead (plaque)	W16-9P	9B.18	—	24 x 12
Destination (1 line)	D1-1, D1-1a	2D.37, 9B.20	varies x 6	varies x 18
Bicycle Destination (1 line)	D1-1b, D1-1c	9B.20	varies x 6	varies x 6
Destination (2 lines)	D1-2, D1-2a	2D.37, 9B.20	varies x 12	varies x 30
Bicycle Destination (2 lines)	D1-2b, D1-2c	9B.20	varies x 12	varies x 12
Destination (3 lines)	D1-3, D1-3a	2D.37, 9B.20	varies x 18	varies x 42
Bicycle Destination (3 lines)	D1-3b, D1-3c	9B.20	varies x 18	varies x 18
Street Name	D3-1	2D.43, 9B.20	varies x 6	varies x 8
Bicycle Parking Area	D4-3	9B.23	12 x 18	12 x 18
Reference Location (1-digit)	D10-1	2H.02, 9B.24	6 x 12	10 x 18
Intermediate Reference Location (1-digit)	D10-1a	2H.02, 9B.24	6 x 18	10 x 27
Reference Location (2-digit)	D10-2	2H.02, 9B.24	6 x 18	10 x 27
Intermediate Reference Location (2-digit)	D10-2a	2H.02, 9B.24	6 x 24	10 x 36
Reference Location (3-digit)	D10-3	2H.02, 9B.24	6 x 24	10 x 36
Intermediate Reference Location (3-digit)	D10-3a	2H.02, 9B.24	6 x 30	10 x 48
Bike Route	D11-1, D11-1c	9B.20	24 x 18	24 x 18
Bicycles Permitted	D11-1a	9B.25	18 x 18	—
Bike Route (plaque)	D11-1bP	9B.25	18 x 6	—
Pedestrians Permitted	D11-2	9B.25	18 x 18	—
Skaters Permitted	D11-3	9B.25	18 x 18	—
Equestrians Permitted	D11-4	9B.25	18 x 18	—
Bicycle Route	M1-8, M1-8a	9B.21	12 x 18	18 x 24
U.S. Bicycle Route	M1-9	9B.21	12 x 18	18 x 24
Bicycle Route Auxiliary Signs	M2-1; M3-1,2,3,4; M4-1,1a,2,3,5,6,7,7a,8,14	9B.22	12 x 6	12 x 6
Bicycle Route Arrow Signs	M5-1,2; M6-1,2,3,4,5,6,7	9B.22	12 x 9	12 x 9
Type 3 Object Markers	OM3-L,C,R	2C.63, 9B.26	6 x 18	12 x 36

Notes: 1. Larger signs may be used when appropriate
2. Dimensions are shown in inches and are shown as width x height

Guidance:

04 *Except for size, the design of signs and plaques for bicycle facilities should be identical to that provided in this Manual for signs and plaques for streets and highways.*

Support:

05 Uniformity in design of bicycle signs and plaques includes shape, color, symbols, arrows, wording, lettering, and illumination or retroreflectorization.

Section 9B.03 STOP and YIELD Signs (R1-1, R1-2)

Standard:

01 **STOP (R1-1) signs (see Figure 9B-2) shall be installed on shared-use paths at points where bicyclists are required to stop.**

02 **YIELD (R1-2) signs (see Figure 9B-2) shall be installed on shared-use paths at points where bicyclists have an adequate view of conflicting traffic as they approach the sign, and where bicyclists are required to yield the right-of-way to that conflicting traffic.**

Figure 9B-2. Regulatory Signs and Plaques for Bicycle Facilities

Option:

03   A 30 x 30-inch STOP sign or a 36 x 36 x 36-inch YIELD sign may be used on shared-use paths for added emphasis.

Guidance:

04   *Where conditions require path users, but not roadway users, to stop or yield, the STOP or YIELD sign should be placed or shielded so that it is not readily visible to road users.*

05   *When placement of STOP or YIELD signs is considered, priority at a shared-use path/roadway intersection should be assigned with consideration of the following:*
   A. *Relative speeds of shared-use path and roadway users,*
   B. *Relative volumes of shared-use path and roadway traffic, and*
   C. *Relative importance of shared-use path and roadway.*

06   *Speed should not be the sole factor used to determine priority, as it is sometimes appropriate to give priority to a high-volume shared-use path crossing a low-volume street, or to a regional shared-use path crossing a minor collector street.*

07   *When priority is assigned, the least restrictive control that is appropriate should be placed on the lower priority approaches. STOP signs should not be used where YIELD signs would be acceptable.*

Section 9B.04 Bike Lane Signs and Plaques (R3-17, R3-17aP, R3-17bP)

Standard:

01   **The BIKE LANE (R3-17) sign and the R3-17aP and R3-17bP plaques (see Figure 9B-2) shall be used only in conjunction with marked bicycle lanes as described in Section 9C.04.**

Guidance:

02   *If used, Bike Lane signs and plaques should be used in advance of the upstream end of the bicycle lane, at the downstream end of the bicycle lane, and at periodic intervals along the bicycle lane as determined by engineering judgment based on prevailing speed of bicycle and other traffic, block length, distances from adjacent intersections, and other considerations.*

Section 9B.05 BEGIN RIGHT TURN LANE YIELD TO BIKES Sign (R4-4)

Option:

01   Where motor vehicles entering an exclusive right-turn lane must weave across bicycle traffic in bicycle lanes, the BEGIN RIGHT TURN LANE YIELD TO BIKES (R4-4) sign (see Figure 9B-2) may be used to inform both the motorist and the bicyclist of this weaving maneuver (see Figures 9C-1, 9C-4, and 9C-5).

Guidance:

02   *The R4-4 sign should not be used when bicyclists need to move left because of a right-turn lane drop situation.*

Section 9B.06 Bicycles May Use Full Lane Sign (R4-11)

Option:

01   The Bicycles May Use Full Lane (R4-11) sign (see Figure 9B-2) may be used on roadways where no bicycle lanes or adjacent shoulders usable by bicyclists are present and where travel lanes are too narrow for bicyclists and motor vehicles to operate side by side.

02   The Bicycles May Use Full Lane sign may be used in locations where it is important to inform road users that bicyclists might occupy the travel lane.

03   Section 9C.07 describes a Shared Lane Marking that may be used in addition to or instead of the Bicycles May Use Full Lane sign to inform road users that bicyclists might occupy the travel lane.

Support:

04   The Uniform Vehicle Code (UVC) defines a "substandard width lane" as a "lane that is too narrow for a bicycle and a vehicle to travel safely side by side within the same lane."

Section 9B.07 Bicycle WRONG WAY Sign and RIDE WITH TRAFFIC Plaque (R5-1b, R9-3cP)

Option:

01   The Bicycle WRONG WAY (R5-1b) sign and RIDE WITH TRAFFIC (R9-3cP) plaque (see Figure 9B-2) may be placed facing wrong-way bicycle traffic, such as on the left side of a roadway.

02   This sign and plaque may be mounted back-to-back with other signs to minimize visibility to other traffic.

Guidance:

03 *The RIDE WITH TRAFFIC plaque should be used only in conjunction with the Bicycle WRONG WAY sign, and should be mounted directly below the Bicycle WRONG WAY sign.*

Section 9B.08 NO MOTOR VEHICLES Sign (R5-3)

Option:

01 The NO MOTOR VEHICLES (R5-3) sign (see Figure 9B-2) may be installed at the entrance to a shared-use path.

Section 9B.09 Selective Exclusion Signs

Option:

01 Selective Exclusion signs (see Figure 9B-2) may be installed at the entrance to a roadway or facility to notify road or facility users that designated types of traffic are excluded from using the roadway or facility.

Standard:

02 **If used, Selective Exclusion signs shall clearly indicate the type of traffic that is excluded.**

Support:

03 Typical exclusion messages include:
   A. No Bicycles (R5-6),
   B. No Pedestrians (R9-3),
   C. No Skaters (R9-13), and
   D. No Equestrians (R9-14).

Option:

04 Where bicyclists, pedestrians, and motor-driven cycles are all prohibited, it may be more desirable to use the R5-10a word message sign that is described in Section 2B.39.

Section 9B.10 No Parking Bike Lane Signs (R7-9, R7-9a)

Standard:

01 **If the installation of signs is necessary to restrict parking, standing, or stopping in a bicycle lane, appropriate signs as described in Sections 2B.46 through 2B.48, or the No Parking Bike Lane (R7-9 or R7-9a) signs (see Figure 9B-2) shall be installed.**

Section 9B.11 Bicycle Regulatory Signs (R9-5, R9-6, R10-4, R10-24, R10-25, and R10-26)

Option:

01 The R9-5 sign (see Figure 9B-2) may be used where the crossing of a street by bicyclists is controlled by pedestrian signal indications.

02 Where it is not intended for bicyclists to be controlled by pedestrian signal indications, the R10-4, R10-24, or R10-26 sign (see Figure 9B-2 and Section 2B.52) may be used.

Guidance:

03 *If used, the R9-5, R10-4, R10-24, or R10-26 signs should be installed near the edge of the sidewalk in the vicinity of where bicyclists will be crossing the street.*

Option:

04 If bicyclists are crossing a roadway where In-Roadway Warning Lights (see Section 4N.02) or other warning lights or beacons have been provided, the R10-25 sign (see Figure 9B-2) may be used.

05 The R9-6 sign (see Figure 9B-2) may be used where a bicyclist is required to cross or share a facility used by pedestrians and is required to yield to the pedestrians.

Section 9B.12 Shared-Use Path Restriction Sign (R9-7)

Option:

01 The Shared-Use Path Restriction (R9-7) sign (see Figure 9B-2) may be installed to supplement a solid white pavement marking line (see Section 9C.03) on facilities that are to be shared by pedestrians and bicyclists in order to provide a separate designated pavement area for each mode of travel. The symbols may be switched as appropriate.

Guidance:

02 *If two-way operation is permitted on the facility for pedestrians and/or bicyclists, the designated pavement area that is provided for each two-way mode of travel should be wide enough to accommodate both directions of travel for that mode.*

Section 9B.13 Bicycle Signal Actuation Sign (R10-22)

Option:

01 The Bicycle Signal Actuation (R10-22) sign (see Figure 9B-2) may be installed at signalized intersections where markings are used to indicate the location where a bicyclist is to be positioned to actuate the signal (see Section 9C.05).

Guidance:

02 *If the Bicycle Signal Actuation sign is installed, it should be placed at the roadside adjacent to the marking to emphasize the connection between the marking and the sign.*

Section 9B.14 Other Regulatory Signs

Option:

01 Other regulatory signs described in Chapter 2B may be installed on bicycle facilities as appropriate.

Section 9B.15 Turn or Curve Warning Signs (W1 Series)

Guidance:

01 *To warn bicyclists of unexpected changes in shared-use path direction, appropriate turn or curve (W1-1 through W1-7) signs (see Figure 9B-3) should be used.*

02 *The W1-1 through W1-5 signs should be installed at least 50 feet in advance of the beginning of the change of alignment.*

Section 9B.16 Intersection Warning Signs (W2 Series)

Option:

01 Intersection Warning (W2-1 through W2-5) signs (see Figure 9B-3) may be used on a roadway, street, or shared-use path in advance of an intersection to indicate the presence of an intersection and the possibility of turning or entering traffic.

Guidance:

02 *When engineering judgment determines that the visibility of the intersection is limited on the shared-use path approach, Intersection Warning signs should be used.*

03 *Intersection Warning signs should not be used where the shared-use path approach to the intersection is controlled by a STOP sign, a YIELD sign, or a traffic control signal.*

Section 9B.17 Bicycle Surface Condition Warning Sign (W8-10)

Option:

01 The Bicycle Surface Condition Warning (W8-10) sign (see Figure 9B-3) may be installed where roadway or shared-use path conditions could cause a bicyclist to lose control of the bicycle.

02 Signs warning of other conditions that might be of concern to bicyclists, including BUMP (W8-1), DIP (W8-2), PAVEMENT ENDS (W8-3), and any other word message that describes conditions that are of concern to bicyclists, may also be used.

03 A supplemental plaque may be used to clarify the specific type of surface condition.

Section 9B.18 Bicycle Warning and Combined Bicycle/Pedestrian Signs (W11-1 and W11-15)

Support:

01 The Bicycle Warning (W11-1) sign (see Figure 9B-3) alerts the road user to unexpected entries into the roadway by bicyclists, and other crossing activities that might cause conflicts. These conflicts might be relatively confined, or might occur randomly over a segment of roadway.

Option:

02 The combined Bicycle/Pedestrian (W11-15) sign (see Figure 9B-3) may be used where both bicyclists and pedestrians might be crossing the roadway, such as at an intersection with a shared-use path. A TRAIL X-ING (W11-15P) supplemental plaque (see Figure 9B-3) may be mounted below the W11-15 sign.

03 A supplemental plaque with the legend AHEAD or XX FEET may be used with the Bicycle Warning or combined Bicycle/Pedestrian sign.

Guidance:

04 *If used in advance of a specific crossing point, the Bicycle Warning or combined Bicycle/Pedestrian sign should be placed at a distance in advance of the crossing location that conforms with the guidance given in Table 2C-4.*

Figure 9B-3. Warning Signs and Plaques and Object Markers for Bicycle Facilities

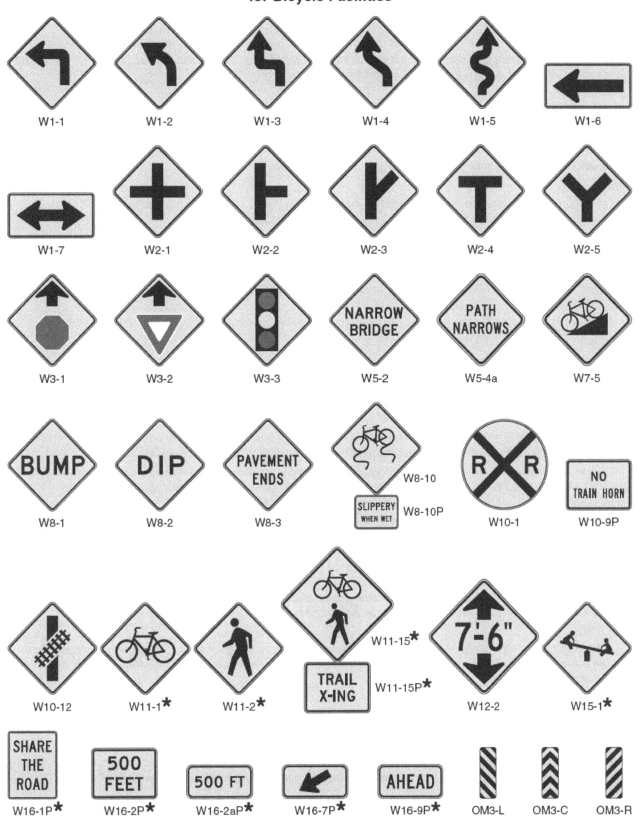

* A fluorescent yellow-green background color may be used for this sign or plaque. The background color of the plaque should match the color of the warning sign that it supplements.

Standard:

⁰⁵ **Bicycle Warning and combined Bicycle/Pedestrian signs, when used at the location of the crossing, shall be supplemented with a diagonal downward pointing arrow (W16-7P) plaque (see Figure 9B-3) to show the location of the crossing.**

Option:

⁰⁶ A fluorescent yellow-green background color with a black legend and border may be used for Bicycle Warning and combined Bicycle/Pedestrian signs and supplemental plaques.

Guidance:

⁰⁷ *When the fluorescent yellow-green background color is used, a systematic approach featuring one background color within a zone or area should be used. The mixing of standard yellow and fluorescent yellow-green backgrounds within a zone or area should be avoided.*

Section 9B.19 Other Bicycle Warning Signs

Option:

⁰¹ Other bicycle warning signs (see Figure 9B-3) such as PATH NARROWS (W5-4a) and Hill (W7-5) may be installed on shared-use paths to warn bicyclists of conditions not readily apparent.

⁰² In situations where there is a need to warn motorists to watch for bicyclists traveling along the highway, the SHARE THE ROAD (W16-1P) plaque (see Figure 9B-3) may be used in conjunction with the W11-1 sign.

Guidance:

⁰³ *If used, other advance bicycle warning signs should be installed at least 50 feet in advance of the beginning of the condition.*

⁰⁴ *Where temporary traffic control zones are present on bikeways, appropriate signs from Part 6 should be used.*

Option:

⁰⁵ Other warning signs described in Chapter 2C may be installed on bicycle facilities as appropriate.

Section 9B.20 Bicycle Guide Signs (D1-1b, D1-1c, D1-2b, D1-2c, D1-3b, D1-3c, D11-1, D11-1c)

Option:

⁰¹ Bike Route Guide (D11-1) signs (see Figure 9B-4) may be provided along designated bicycle routes to inform bicyclists of bicycle route direction changes and to confirm route direction, distance, and destination.

⁰² If used, Bike Route Guide signs may be repeated at regular intervals so that bicyclists entering from side streets will have an opportunity to know that they are on a bicycle route. Similar guide signing may be used for shared roadways with intermediate signs placed for bicyclist guidance.

⁰³ Alternative Bike Route Guide (D11-1c) signs may be used to provide information on route direction, destination, and/or route name in place of the "BIKE ROUTE" wording on the D11-1 sign (see Figures 9B-4 and 9B-6).

⁰⁴ Destination (D1-1, D1-1a) signs, Street Name (D3) signs, or Bicycle Destination (D1-1b, D1-1c, D1-2b, D1-2c, D1-3b, D1-3c) signs (see Figure 9B-4) may be installed to provide direction, destination, and distance information as needed for bicycle travel. If several destinations are to be shown at a single location, they may be placed on a single sign with an arrow (and the distance, if desired) for each name. If more than one destination lies in the same direction, a single arrow may be used for the destinations.

Guidance:

⁰⁵ *Adequate separation should be made between any destination or group of destinations in one direction and those in other directions by suitable design of the arrow, spacing of lines of legend, heavy lines entirely across the sign, or separate signs.*

Standard:

⁰⁶ **An arrow pointing to the right, if used, shall be at the extreme right-hand side of the sign. An arrow pointing left or up, if used, shall be at the extreme left-hand side of the sign. The distance numerals, if used, shall be placed to the right of the destination names.**

⁰⁷ **On Bicycle Destination signs, a bicycle symbol shall be placed next to each destination or group of destinations. If an arrow is at the extreme left, the bicycle symbol shall be placed to the right of the respective arrow.**

Guidance:

⁰⁸ *Unless a sloping arrow will convey a clearer indication of the direction to be followed, the directional arrows should be horizontal or vertical.*

Figure 9B-4. Guide Signs and Plaques for Bicycle Facilities (Sheet 1 of 2)

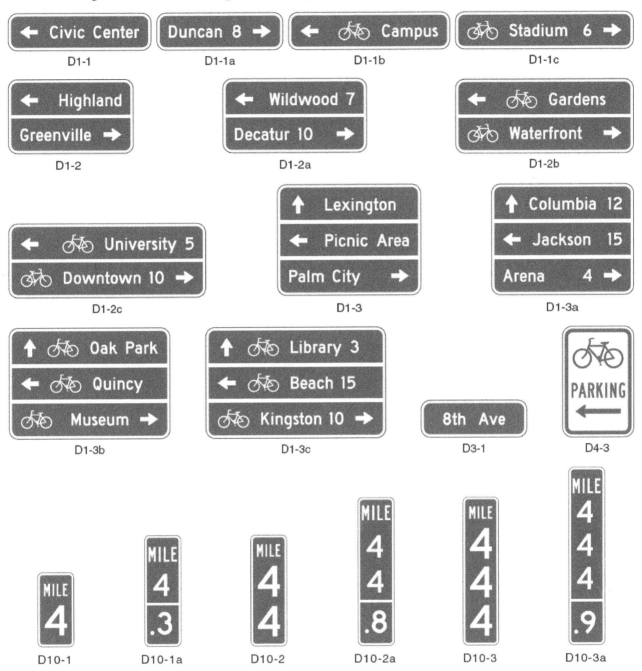

09 *The bicycle symbol should be to the left of the destination legend.*

10 *If several individual name signs are assembled into a group, all signs in the assembly should have the same horizontal width.*

11 *Because of their smaller size, Bicycle Destination signs should not be used as a substitute for vehicular destination signs when the message is also intended to be seen by motorists.*

Support:

12 Figure 9B-5 shows an example of the signing for the beginning and end of a designated bicycle route on a shared-use path. Figure 9B-6 shows an example of signing for an on-roadway bicycle route. Figure 9B-7 shows examples of signing and markings for a shared-use path crossing.

Figure 9B-4. Guide Signs and Plaques for Bicycle Facilities (Sheet 2 of 2)

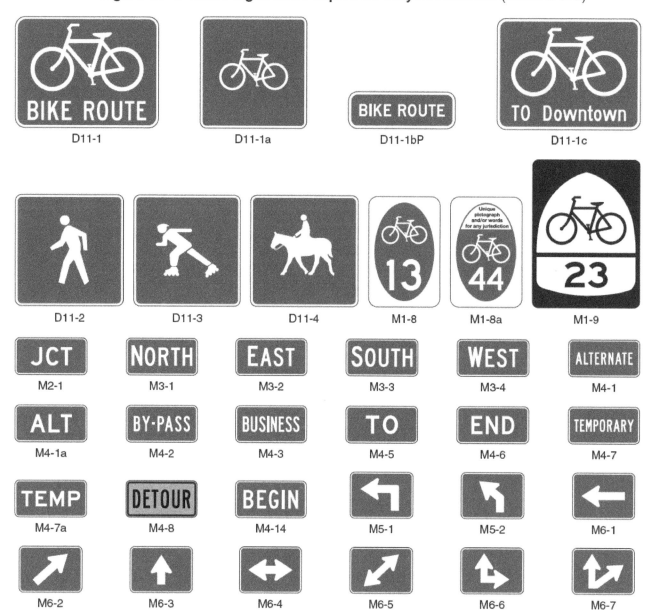

Section 9B.21 Bicycle Route Signs (M1-8, M1-8a, M1-9)

Option:

01 To establish a unique identification (route designation) for a State or local bicycle route, the Bicycle Route (M1-8, M1-8a) sign (see Figure 9B-4) may be used.

Standard:

02 **The Bicycle Route (M1-8) sign shall contain a route designation and shall have a green background with a retroreflectorized white legend and border. The Bicycle Route (M1-8a) sign shall contain the same information as the M1-8 sign and in addition shall include a pictograph or words that are associated with the route or with the agency that has jurisdiction over the route.**

Guidance:

03 *Bicycle routes, which might be a combination of various types of bikeways, should establish a continuous routing.*

Figure 9B-5. Example of Signing for the Beginning and End of a Designated Bicycle Route on a Shared-Use Path

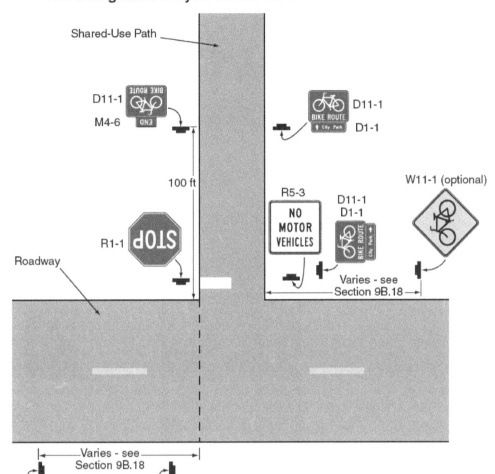

04   *Where a designated bicycle route extends through two or more States, a coordinated submittal by the affected States for an assignment of a U.S. Bicycle Route number designation should be sent to the American Association of State Highway and Transportation Officials (see Page i for the address).*

Standard:

05   **The U.S. Bicycle Route (M1-9) sign (see Figure 9B-4) shall contain the route designation as assigned by AASHTO and shall have a black legend and border with a retroreflectorized white background.**

Guidance:

06   *If used, the Bicycle Route or U.S. Bicycle Route signs should be placed at intervals frequent enough to keep bicyclists informed of changes in route direction and to remind motorists of the presence of bicyclists.*

Option:

07   Bicycle Route or U.S. Bicycle Route signs may be installed on shared roadways or on shared-use paths to provide guidance for bicyclists.

08   The Bicycle Route Guide (D11-1) sign (see Figure 9B-4) may be installed where no unique designation of routes is desired.

Figure 9B-6. Example of Bicycle Guide Signing

Section 9B.22 Bicycle Route Sign Auxiliary Plaques

Option:

01 Auxiliary plaques may be used in conjunction with Bike Route Guide signs, Bicycle Route signs, or U.S. Bicycle Route signs as needed.

Guidance:

02 *If used, Junction (M2-1), Cardinal Direction (M3 series), and Alternative Route (M4 series) auxiliary plaques (see Figure 9B-4) should be mounted above the appropriate Bike Route Guide signs, Bicycle Route signs, or U.S. Bicycle Route signs.*

03 *If used, Advance Turn Arrow (M5 series) and Directional Arrow (M6 series) auxiliary plaques (see Figure 9B-4) should be mounted below the appropriate Bike Route Guide sign, Bicycle Route sign, or U.S. Bicycle Route sign.*

04 *Except for the M4-8 plaque, all route sign auxiliary plaques should match the color combination of the route sign that they supplement.*

Figure 9B-7. Examples of Signing and Markings for a Shared-Use Path Crossing

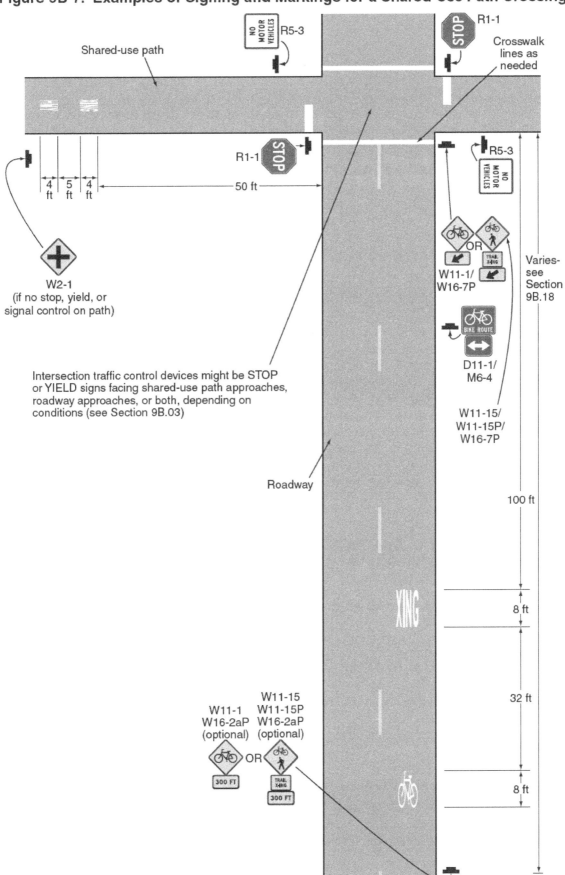

05 *Route sign auxiliary plaques carrying word legends that are used on bicycle routes should have a minimum size of 12 x 6 inches. Route sign auxiliary plaques carrying arrow symbols that are used on bicycle routes should have a minimum size of 12 x 9 inches.*

Option:

06 With route signs of larger sizes, auxiliary plaques may be suitably enlarged, but not such that they exceed the width of the route sign.

07 A route sign and any auxiliary plaques used with it may be combined on a single sign.

08 Destination (D1-1b and D1-1c) signs (see Figure 9B-4) may be mounted below Bike Route Guide signs, Bicycle Route signs, or U.S. Bicycle Route signs to furnish additional information, such as directional changes in the route, or intermittent distance and destination information.

Section 9B.23 Bicycle Parking Area Sign (D4-3)

Option:

01 *The Bicycle Parking Area (D4-3) sign (see Figure 9B-4) may be installed where it is desirable to show the direction to a designated bicycle parking area. The arrow may be reversed as appropriate.*

Standard:

02 **The legend and border of the Bicycle Parking Area sign shall be green on a retroreflectorized white background.**

Section 9B.24 Reference Location Signs (D10-1 through D10-3) and Intermediate Reference Location Signs (D10-1a through D10-3a)

Support:

01 There are two types of reference location signs:
   A. Reference Location (D10-1, 2, and 3) signs show an integer distance point along a shared-use path; and
   B. Intermediate Reference Location (D10-1a, 2a, and 3a) signs also show a decimal between integer distance points along a shared-use path.

Option:

02 Reference Location (D10-1 to D10-3) signs (see Figure 9B-4) may be installed along any section of a shared-use path to assist users in estimating their progress, to provide a means for identifying the location of emergency incidents and crashes, and to aid in maintenance and servicing.

03 To augment the reference location sign system, Intermediate Reference Location (D10-1a to D10-3a) signs (see Figure 9B-4), which show the tenth of a mile with a decimal point, may be installed at one tenth of a mile intervals, or at some other regular spacing.

Standard:

04 **If Intermediate Reference Location (D10-1a to D10-3a) signs are used to augment the reference location sign system, the reference location sign at the integer mile point shall display a decimal point and a zero numeral.**

05 **If placed on shared-use paths, reference location signs shall contain 4.5-inch white numerals on a green background that is at least 6 inches wide with a white border. The signs shall contain the word MILE in 2.25-inch white letters.**

06 **Reference location signs shall have a minimum mounting height of 2 feet, measured vertically from the bottom of the sign to the elevation of the near edge of the shared-use path, and shall not be governed by the mounting height requirements prescribed in Section 9B.01.**

Option:

07 Reference location signs may be installed on one side of the shared-use path only and may be installed back-to-back.

08 If a reference location sign cannot be installed in the correct location, it may be moved in either direction as much as 50 feet.

Guidance:

09 *If a reference location sign cannot be placed within 50 feet of the correct location, it should be omitted.*

10 *Zero distance should begin at the south and west terminus points of shared-use paths.*

Support:

11 Section 2H.05 contains additional information regarding reference location signs.

Section 9B.25 Mode-Specific Guide Signs for Shared-Use Paths (D11-1a, D11-2, D11-3, D11-4)

Option:

01 Where separate pathways are provided for different types of users, Mode-Specific Guide (D11-1a, D11-3, D11-4) signs (see Figure 9B-4) may be used to guide different types of users to the traveled way that is intended for their respective modes.

02 Mode-Specific Guide signs may be installed at the entrance to shared-use paths where the signed mode(s) are permitted or encouraged, and periodically along these facilities as needed.

03 The Bicycles Permitted (D11-1a) sign, when combined with the BIKE ROUTE supplemental plaque (D11-1bP), may be substituted for the D11-1 Bicycle Route Guide sign on paths and shared roadways.

04 When some, but not all, non-motorized user types are encouraged or permitted on a shared-use path, Mode-Specific Guide signs may be placed in combination with each other, and in combination with signs (see Section 9B.09) that prohibit travel by particular modes.

Support:

05 Figure 9B-8 shows an example of signing where separate pathways are provided for different non-motorized user types.

Section 9B.26 Object Markers

Option:

01 Fixed objects adjacent to shared-use paths may be marked with Type 1, Type 2, or Type 3 object markers (see Figure 9B-3) such as those described in Section 2C.63. If the object marker is not intended to also be seen by motorists, a smaller version of the Type 3 object marker may be used (see Table 9B-1).

Standard:

02 **Obstructions in the traveled way of a shared-use path shall be marked with retroreflectorized material or appropriate object markers.**

03 **All object markers shall be retroreflective.**

04 **On Type 3 object markers, the alternating black and retroreflective yellow stripes shall be sloped down at an angle of 45 degrees toward the side on which traffic is to pass the obstruction.**

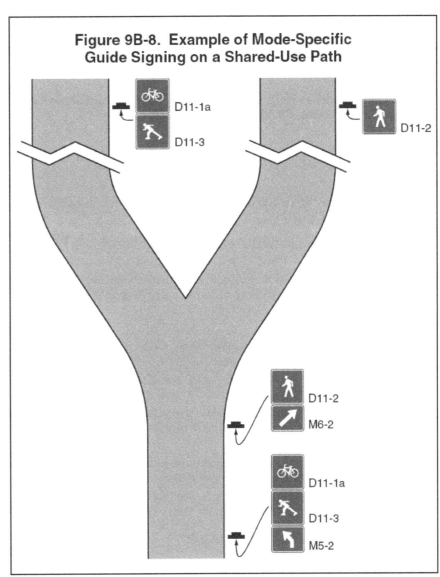

Figure 9B-8. Example of Mode-Specific Guide Signing on a Shared-Use Path

CHAPTER 9C. MARKINGS

Section 9C.01 Functions of Markings

Support:

01 Markings indicate the separation of the lanes for road users, assist the bicyclist by indicating assigned travel paths, indicate correct position for traffic control signal actuation, and provide advance information for turning and crossing maneuvers.

Section 9C.02 General Principles

Guidance:

01 Bikeway design guides (see Section 9A.05) should be used when designing markings for bicycle facilities.

Standard:

02 **Markings used on bikeways shall be retroreflectorized.**

Guidance:

03 Pavement marking word messages, symbols, and/or arrows should be used in bikeways where appropriate. Consideration should be given to selecting pavement marking materials that will minimize loss of traction for bicycles under wet conditions.

Standard:

04 **The colors, width of lines, patterns of lines, symbols, and arrows used for marking bicycle facilities shall be as defined in Sections 3A.05, 3A.06, and 3B.20.**

Support:

05 Figures 9B-7 and 9C-1 through 9C-9 show examples of the application of lines, word messages, symbols, and arrows on designated bikeways.

Option:

06 A dotted line may be used to define a specific path for a bicyclist crossing an intersection (see Figure 9C-1) as described in Sections 3A.06 and 3B.08.

Section 9C.03 Marking Patterns and Colors on Shared-Use Paths

Option:

01 Where shared-use paths are of sufficient width to designate two minimum width lanes, a solid yellow line may be used to separate the two directions of travel where passing is not permitted, and a broken yellow line may be used where passing is permitted (see Figure 9C-2).

Guidance:

02 Broken lines used on shared-use paths should have the usual 1-to-3 segment-to-gap ratio. A nominal 3-foot segment with a 9-foot gap should be used.

03 If conditions make it desirable to separate two directions of travel on shared-use paths at particular locations, a solid yellow line should be used to indicate no passing and no traveling to the left of the line.

04 Markings as shown in Figure 9C-2 should be used at the location of obstructions in the center of the path, including vertical elements intended to physically prevent unauthorized motor vehicles from entering the path.

Option:

05 A solid white line may be used on shared-use paths to separate different types of users. The R9-7 sign (see Section 9B.12) may be used to supplement the solid white line.

06 Smaller size letters and symbols may be used on shared-use paths. Where arrows are needed on shared-use paths, half-size layouts of the arrows may be used (see Section 3B.20).

Section 9C.04 Markings For Bicycle Lanes

Support:

01 Pavement markings designate that portion of the roadway for preferential use by bicyclists. Markings inform all road users of the restricted nature of the bicycle lane.

Standard:

02 **Longitudinal pavement markings shall be used to define bicycle lanes.**

Guidance:

03 If used, bicycle lane word, symbol, and/or arrow markings (see Figure 9C-3) should be placed at the beginning of a bicycle lane and at periodic intervals along the bicycle lane based on engineering judgment.

Figure 9C-1. Example of Intersection Pavement Markings—Designated Bicycle Lane with Left-Turn Area, Heavy Turn Volumes, Parking, One-Way Traffic, or Divided Highway

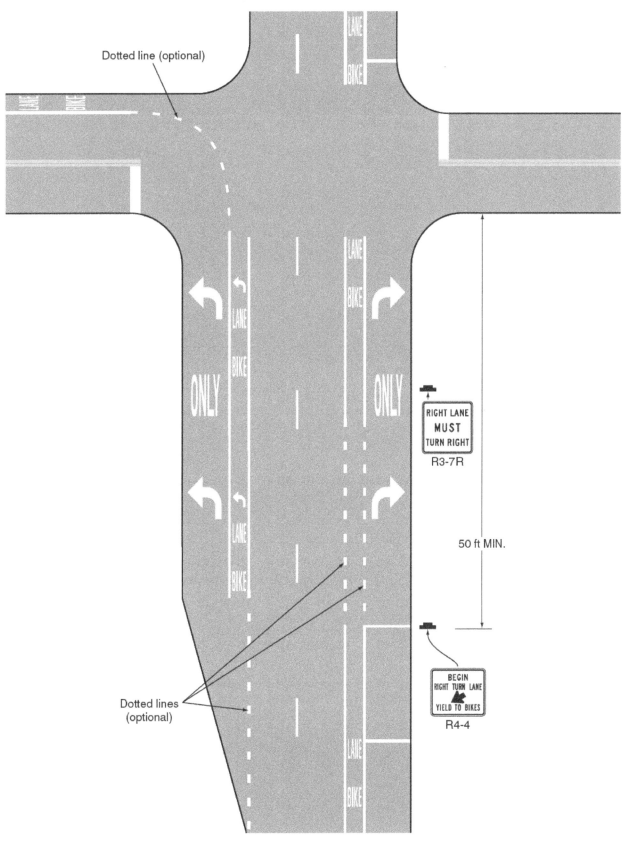

Figure 9C-2. Examples of Center Line Markings for Shared-Use Paths

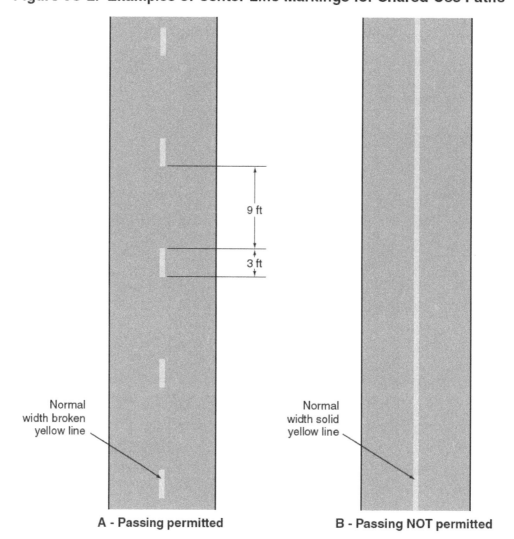

A - Passing permitted

B - Passing NOT permitted

Standard:

04 **If the bicycle lane symbol marking is used in conjunction with word or arrow messages, it shall precede them.**

Option:

05 If the word, symbol, and/or arrow pavement markings shown in Figure 9C-3 are used, Bike Lane signs (see Section 9B.04) may also be used, but to avoid overuse of the signs not necessarily adjacent to every set of pavement markings.

Standard:

06 **A through bicycle lane shall not be positioned to the right of a right turn only lane or to the left of a left turn only lane.**

Support:

07 A bicyclist continuing straight through an intersection from the right of a right-turn lane or from the left of a left-turn lane would be inconsistent with normal traffic behavior and would violate the expectations of right- or left-turning motorists.

Guidance:

08 *When the right through lane is dropped to become a right turn only lane, the bicycle lane markings should stop at least 100 feet before the beginning of the right-turn lane. Through bicycle lane markings should resume to the left of the right turn only lane.*

Figure 9C-3. Word, Symbol, and Arrow Pavement Markings for Bicycle Lanes

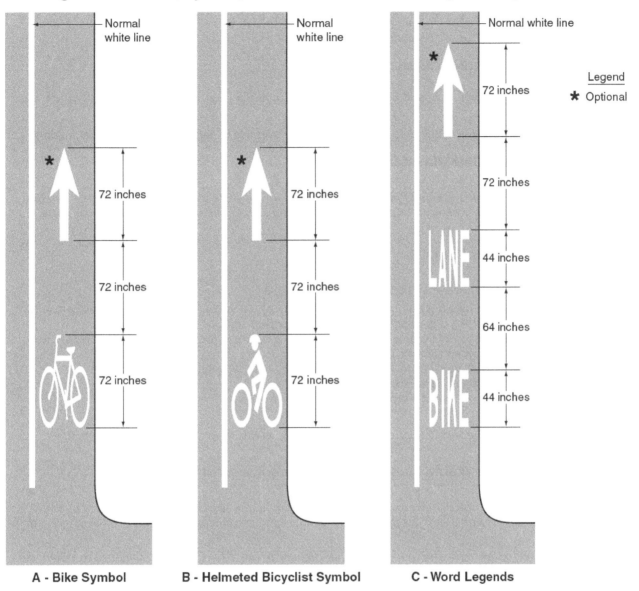

09 *An optional through-right turn lane next to a right turn only lane should not be used where there is a through bicycle lane. If a capacity analysis indicates the need for an optional through-right turn lane, the bicycle lane should be discontinued at the intersection approach.*

10 *Posts or raised pavement markers should not be used to separate bicycle lanes from adjacent travel lanes.*

Support:

11 Using raised devices creates a collision potential for bicyclists by placing fixed objects immediately adjacent to the travel path of the bicyclist. In addition, raised devices can prevent vehicles turning right from merging with the bicycle lane, which is the preferred method for making the right turn. Raised devices used to define a bicycle lane can also cause problems in cleaning and maintaining the bicycle lane.

Standard:

12 **Bicycle lanes shall not be provided on the circular roadway of a roundabout.**

Guidance:

13 *Bicycle lane markings should stop at least 100 feet before the crosswalk, or if no crosswalk is provided, at least 100 feet before the yield line, or if no yield line is provided, then at least 100 feet before the edge of the circulatory roadway.*

Support:
14 Examples of bicycle lane markings at right-turn lanes are shown in Figures 9C-1, 9C-4, and 9C-5. Examples of pavement markings for bicycle lanes on a two-way street are shown in Figure 9C-6. Pavement word message, symbol, and arrow markings for bicycle lanes are shown in Figure 9C-3.

Section 9C.05 Bicycle Detector Symbol
Option:
01 A symbol (see Figure 9C-7) may be placed on the pavement indicating the optimum position for a bicyclist to actuate the signal.
02 An R10-22 sign (see Section 9B.13 and Figure 9B-2) may be installed to supplement the pavement marking.

Section 9C.06 Pavement Markings for Obstructions
Guidance:
01 *In roadway situations where it is not practical to eliminate a drain grate or other roadway obstruction that is inappropriate for bicycle travel, white markings applied as shown in Figure 9C-8 should be used to guide bicyclists around the condition.*

Section 9C.07 Shared Lane Marking
Option:
01 The Shared Lane Marking shown in Figure 9C-9 may be used to:
   A. Assist bicyclists with lateral positioning in a shared lane with on-street parallel parking in order to reduce the chance of a bicyclist's impacting the open door of a parked vehicle,
   B. Assist bicyclists with lateral positioning in lanes that are too narrow for a motor vehicle and a bicycle to travel side by side within the same traffic lane,
   C. Alert road users of the lateral location bicyclists are likely to occupy within the traveled way,
   D. Encourage safe passing of bicyclists by motorists, and
   E. Reduce the incidence of wrong-way bicycling.

Guidance:
02 *The Shared Lane Marking should not be placed on roadways that have a speed limit above 35 mph.*

Standard:
03 **Shared Lane Markings shall not be used on shoulders or in designated bicycle lanes.**

Guidance:
04 *If used in a shared lane with on-street parallel parking, Shared Lane Markings should be placed so that the centers of the markings are at least 11 feet from the face of the curb, or from the edge of the pavement where there is no curb.*
05 *If used on a street without on-street parking that has an outside travel lane that is less than 14 feet wide, the centers of the Shared Lane Markings should be at least 4 feet from the face of the curb, or from the edge of the pavement where there is no curb.*
06 *If used, the Shared Lane Marking should be placed immediately after an intersection and spaced at intervals not greater than 250 feet thereafter.*

Option:
07 Section 9B.06 describes a Bicycles May Use Full Lane sign that may be used in addition to or instead of the Shared Lane Marking to inform road users that bicyclists might occupy the travel lane.

Figure 9C-4. Example of Bicycle Lane Treatment at a Right Turn Only Lane

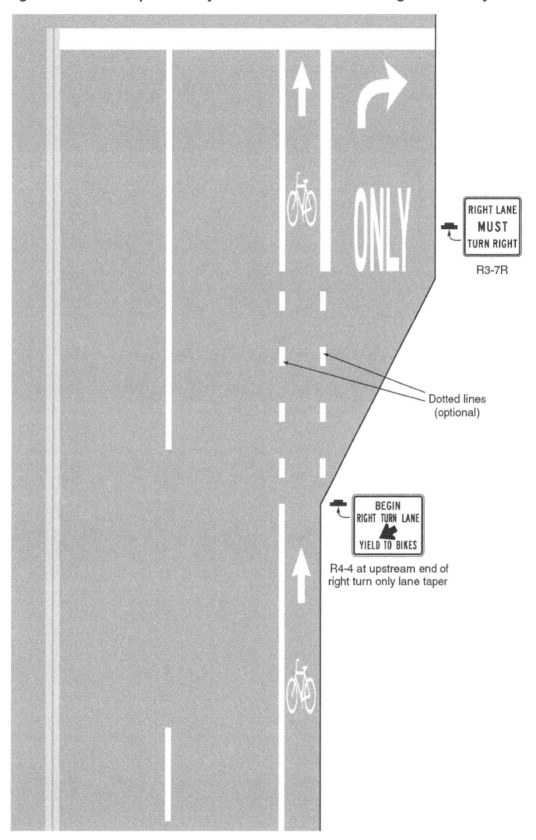

Figure 9C-5. Example of Bicycle Lane Treatment at Parking Lane into a Right Turn Only Lane

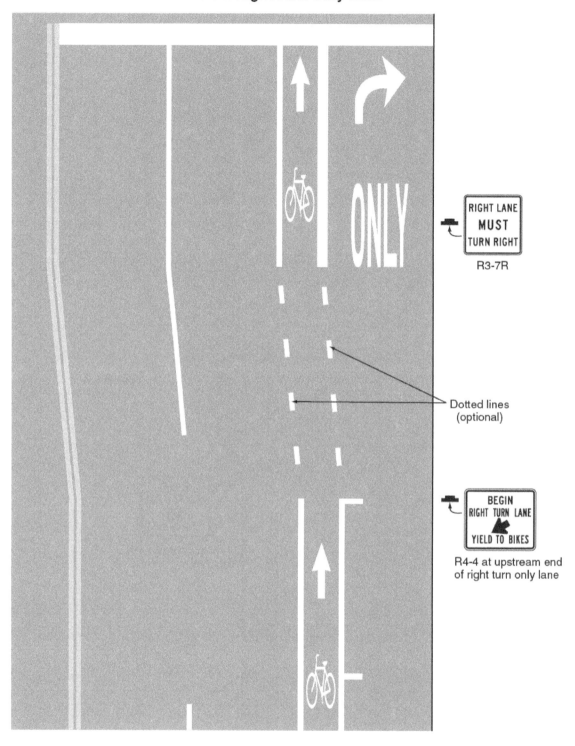

Figure 9C-6. Example of Pavement Markings for Bicycle Lanes on a Two-Way Street

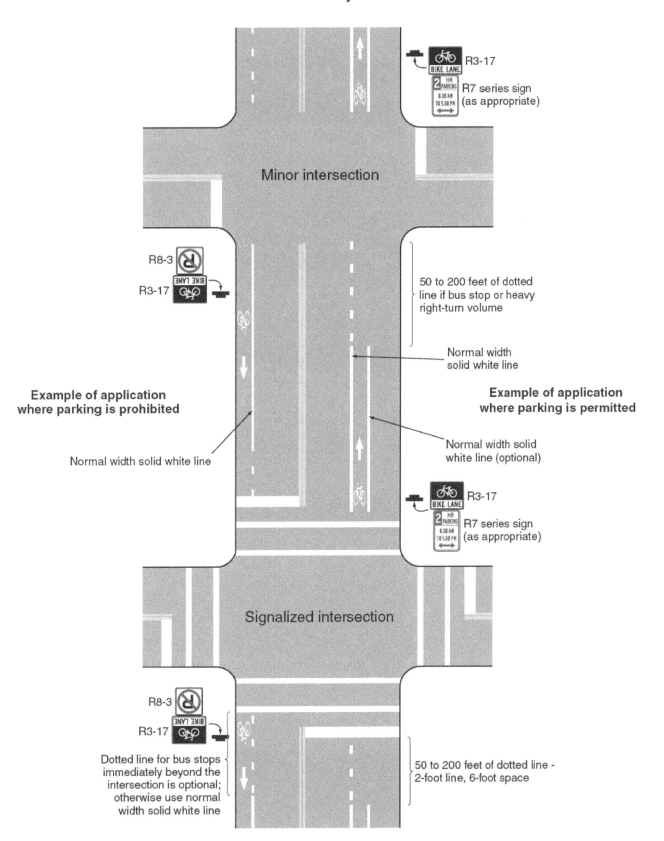

Figure 9C-7. Bicycle Detector Pavement Marking

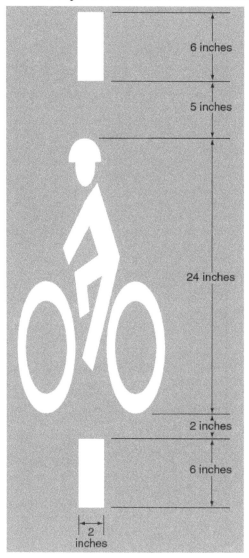

Figure 9C-8. Examples of Obstruction Pavement Markings

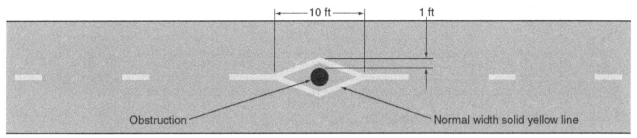

A - Obstruction within the path

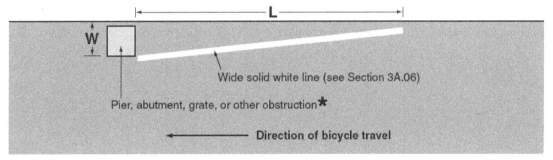

B - Obstruction at edge of path or roadway

L = WS, where W is the offset in feet and S is bicycle approach speed in mph

*Provide an additional foot of offset for a raised obstruction and use the formula L = (W+1) S for the taper length

Figure 9C-9. Shared Lane Marking

CHAPTER 9D. SIGNALS

Section 9D.01 Application

Support:
01 Part 4 contains information regarding signal warrants and other requirements relating to signal installations.

Option:
02 For purposes of signal warrant evaluation, bicyclists may be counted as either vehicles or pedestrians.

Section 9D.02 Signal Operations for Bicycles

Standard:
01 **At installations where visibility-limited signal faces are used, signal faces shall be adjusted so bicyclists for whom the indications are intended can see the signal indications. If the visibility-limited signal faces cannot be aimed to serve the bicyclist, then separate signal faces shall be provided for the bicyclist.**
02 **On bikeways, signal timing and actuation shall be reviewed and adjusted to consider the needs of bicyclists.**